Power System Protection in Smart Grid Environment

Power System Protection in Smart Grid Environment

Ramesh Bansal
ECE Department
University of Sharjah, UAE

CRC Press
Taylor & Francis Group
Boca Raton London New York

CRC Press is an imprint of the
Taylor & Francis Group, an **informa** business

CRC Press
Taylor & Francis Group
52 Vanderbilt Avenue,
New York, NY 10017

Printed on acid-free paper
Printed by CPI Group (UK) Ltd, Croydon CR0 4YY

International Standard Book Number-13: 978-1-1380-3241-5 (Hardback)

Library of Congress Cataloging-in-Publication Data

Names: Bansal, Ramesh, author.
Title: Power system protection in smart grid environment / Ramesh Bansal.
Description: Boca Raton : Taylor & Francis, a CRC title, part of the Taylor & Francis imprint, a member of the Taylor & Francis Group, the academic division of T&F Informa, plc, 2019. | Includes bibliographical references and index.
Identifiers: LCCN 2018034375 | ISBN 9781138032415 (hardback : alk. paper) | ISBN 9780429401756 (ebook)
Subjects: LCSH: Smart power grids. | Electric power systems--Protection.
Classification: LCC TK3105 .B36 2019 | DDC 621.31/042--dc23
LC record available at https://lccn.loc.gov/2018034375

Visit the Taylor & Francis Web site at
http://www.taylorandfrancis.com

and the CRC Press Web site at
http://www.crcpress.com

Contents

Preface

With the integration of renewable energy sources into conventional power system, power flow has become bidirectional. These changes mean significant changes in the protection schemes to protect the power system networks. Significant developments in information and communication power system are leading towards smart grid.

Power System Protection in Smart Grid Environment contains 18 chapters arranged in 4 sections. It gives a comprehensive analysis of power systems protection in smart grid environments. Section I includes 5 chapters covering faults analysis and power system protection devices, i.e., fuses, circuit breakers, instrument transformers and protective relays. Section II contains 4 chapters and presents protection of feeders, busbars, long transmission lines and FACTS devices. The 5 chapters of Section III cover protection of transformers, generators, and induction motors, and substation automation and control and overvoltage and earthing protection. Section IV, in 4 chapters, covers power quality and equipment protection, reliability of distribution system, wide area and system protection, and renewable DG protection.

Power System Protection in Smart Grid Environment is a power system protection textbook that is suitable at both the under- and postgraduate levels. It provides a comprehensive reference material for power system professionals and engineers. The book is supported by a large number of solved examples and tutorial problems. A brief description of the contents follows.

Chapter 1 introduces the smart grid system with an emphasis on the smart grid architecture, grid technology and distributed energy resources, etc. It also provides some insights into the application of protective devices in the smart grid operation. Network reduction and bus impedance matrix methods for the calculation of different types of faults are presented in Chapter 2, which also includes the application of the concept of symmetrical components for the analysis of unbalanced faults and detailed worked examples on various methods of fault analysis. Chapter 3 presents different types of fuses along with the material used and factors affecting the performance of a fuse and types of circuit breakers and testing methods.

Chapter 4 deals with the various aspects of instrument transformers (current and potential transformers) in detail. Basic connections and complex vector diagrams and design aspects are discussed. Several solved examples are presented in order to give a better insight into instrument transformers. Chapter 5 provides an overview of the protective relaying system using overcurrent and differential protection functions. It includes information about the different types of overcurrent protection relays, characteristic curve analysis on the time-overcurrent protection functions and grading principles. The balanced current and voltage differential protection functions, percentage restraint differential principles and use of case studies on time-overcurrent protection, relay-to-relay coordination and transformer differential protection functions are analyzed using DigSilent simulation models. In addition, solved and unsolved examples on protective relaying system are provided.

Chapter 6 explains in detail the protection of a medium voltage network using overcurrent protection that includes network layout, different protective devices, relay technology, protection operating curves and auto-reclosing. The emphasizes equipment let-through energy exposure, how to determine conductor limits and then the application of this concept to both radial and interconnected (or multisource) networks. Examples are provided to show how to implement the grading methods and to illustrate the advantages and disadvantages of each method. Chapter 7 presents the various arrangements of busbars, for example, single busbar arrangement, ring main arrangement, with advantages and disadvantages of these arrangements. Various faults associated with the busbar are illustrated with solved examples.

Chapter 8 provides an overview of the distance protective relaying system for long, high transmission lines. It provides the basic principle of the distance protection function and analyses the characteristic curve of the impedance, admittance and reactance type relays. Communication-assisted distance protection relaying schemes are analysed for the remote tripping of the circuit breakers connected at both ends of the long transmission line. The case study on distance protection function is analysed using the DigSilent simulation model. The lab-scale test bench setup is implemented to test the performance of the distance protection function of the SEL421 intelligent electronic device (IED). The test bench uses the Omicron CMC356 test set to generate the required current and voltage signals for the considered test case. In addition, solved examples on distance protective relaying system are provided. Reactors, static var compensators (SVCs) and STATCOM protection strategies are presented in Chapter 9.

Chapter 10 discusses various types of protection applied to the transformer, including the simple fuse, overcurrent relays, and the differential schemes. Various types of fault are discussed, and the appropriate protection strategies are outlined along with solved and unsolved examples. Chapter 11 presents generator protection from the electrical, mechanical faults and adverse effects that arise from loss of synchronization, loss of field winding, over- and undervoltage, over- and underfrequency, overheating, overspeed, loss of excitation, earth faults, etc.

Chapter 12 presents the different aspects of induction motor protection. The chapter describes the system approach that can be used in the smart grid environment through the smart metering system or accurate smart sensor devices. The chapter also discusses the importance of the induction motor strategy during principal protection for either standard grid or smart grid environment. Chapter 13 presents substation automation and control of modern power systems. The IEC 61850 standard for achieving the modern nonproprietary automation and control in smart grids is explained. The application of the standardized configuration language in the engineering of IEC 61850-based systems and two redundancy protocols adopted for use in power system automation are also presented. Chapter 14 discusses several causes of overvoltage, and an earthing approach is presented along with case studies and numerical calculations.

Chapter 15 discusses power quality problems and provides analysis on the causes, effects of harmonics and mitigation solutions in various components of the power networks. Definition of terminologies used in power quality analysis, and case studies about methods to solve harmonics are presented.

Chapter 16 presents reliability assessment of the distribution system in the presence of the protective devices. The integration of protective devices into the distribution system can be utilized to minimize the economic impacts of power outages, but this will increase the investment costs of the protective devices. The validity of the concepts and techniques are demonstrated by analyzing the results obtained from different case studies.

Chapter 17 discusses various aspects of the design, development, and application of wide area monitoring, protection and control (WAMPAC) for grid visualization, analytics, and protection/control tools for monitoring and preserving system integrity during normal operating conditions and after a disturbance. Chapter 18 provides a detailed review of the protection issues related to the renewable distributed generation. Islanding protection and anti-islanding methods are described. Protection schemes commonly used in microgrids, wind energy systems and PV systems are presented.

I am very grateful to a number of individuals who have contributed to this book. In particular, I would like to thank all the authors for their contributions and the reviewers for reviewing the book chapters and thus improving the quality of *Power System Protection in Smart Grid Environment.* I sincerely thank Dr. Gagandeep Singh and Mauli Sharma of CRC for their help in the timely publication of the book. I would also like to thank Dr. Arvind, Dr. Patrick, Dr. Senthil, Dr. Jackson, Abhishek, Ranjay, Yuvraj, Nzeba, and Aishwarya for reviewing and proofreading book chapters. I would also like to thank the University of Sharjah and University of Pretoria authorities and staff members for maintaining a cordial atmosphere and providing the facilities for completion of the book. I would like to express my gratitude and sincere regards to my family members who have provided me with great support during my preparation of this book.

Ramesh Bansal

Editor

Ramesh Bansal has more than 25 years of experience in teaching, research, and industry. Currently he is a Professor and Chairman of Department of Electrical and Computer Engineering at University of Sharjah. Previously he was Professor and Group Head (Power) in the Department of Electrical, Electronic and Computer Engineering at University of Pretoria (UP). Prior to his appointment at UP he was employed by the University of Queensland in Australia; the University of the South Pacific in Fiji; the Birla Institute of Technology and Science in Pilani, India; and the Civil Construction Wing of All-India Radio. He has worked with Powerlink, an Australian government-owned corporation responsible for Queensland's high-voltage electricity transmission network.

Professor Bansal has significant industrial experience collaborating with power utilities around the world. These utilities include NTPC (a 40 GW Indian power generation company), Powerlink, and ESKOM. Professor Bansal has published more than 300 journal articles, presented papers at conferences, and has contributed to books and book chapters. He has supervised 18 Ph.D. students and is currently supervising several Ph.D. students. His many research interests are in the areas of renewable energy and conventional power systems, including wind, photovoltaics (PV), hybrid power, distributed generation, grid integration of renewable energy, power systems analysis and power system protection. Professor Bansal is an editor or associate editor of highly regarded journals, including *IET-Renewable Power Generation* (regional editor for Africa), *Electric Power Components and Systems*, and *Technology and Economics of Smart Grids and Sustainable Energy*. He is a fellow and a chartered engineer of the Institution of Engineering and Technology in the United Kingdom, a fellow of Engineers Australia, a fellow of the Institution of Engineers (India) and a senior member of the Institute of Electrical and Electronics Engineers (IEEE).

Contributors

T. Adefarati
Department of Electrical, Electronic and Computer Engineering
University of Pretoria
Pretoria, South Africa

Adeyemi Charles Adewole
Centre for Substation Automation and Energy Management Systems (CSAEMS)
Cape Peninsula University of Technology
Cape Town, South Africa

Ramesh Bansal
Department of Electrical and Computer Engineering
University of Sharjah, UAE

Aishwarya Biju
Department of Electrical, Electronic, and Computer Engineering
University of Pretoria
Pretoria, South Africa

Abhishek Chauhan
Department of Electrical Engineering
Institute of Technology
Dehradun, India

J. J. Justo
Department of Electrical Engineering
University of Dar es Salaam
Dar es Salaam, Tanzania

Senthil Krishnamurthy
Cape Peninsula University of Technology
Center for Substation Automation and Energy Management Systems (CSAEMS)
Department of Electrical, Electronic and Computer Engineering
Bellville, South Africa

Abhishek Kumar
College of Electrical Engineering
Zhejiang University
Hangzhou, China

Patrick T. Manditereza
Department of Electrical, Electronic and Computer Engineering
Central University of Technology
Free State, South Africa

N. T. Mbungu
Department of Electrical, Electronic and Computer Engineering
University of Pretoria
Pretoria, South Africa

Raj Naidoo
Department of Electrical, Electronic and Computer Engineering
University of Pretoria
Pretoria, South Africa

K. A. Nzeba
Department of Electrical, Electronic, and Computer Engineering
University of Pretoria
Pretoria, South Africa

Rishabh Dev Shukla
Electrical Engineering Department
Budge Budge Institute of Technology
Kolkata, India

Arvind R. Singh
Department of Electrical, Electronic and Computer Engineering
University of Pretoria
Pretoria, South Africa

Asheesh Kumar Singh
Department of Electrical Engineering
Allahabad, India

Rajiv Singh
Department of Electrical Engineering
College of Technology
G.B. Pant University of Agriculture & Technology
Dehradun, India

Ranjay Singh
Department of Electrical, Electronic and Computer Engineering
University of Pretoria
Pretoria, South Africa

Martin J. Slabbert
Department of Electrical, Electronic and Computer Engineering
University of Pretoria
Pretoria, South Africa

Padmanabh Thakur
Department of Electrical & Electronics Engineering
Graphic Era University
Dehradun, India

Ramesh K. Tripathi
Electrical Engineering Department
Motilal Nehru National Institute of Technology
Allahabad, India

D. H. Tungadio
Department of Electrical, Electronic and Computer Engineering
University of Pretoria
Pretoria, South Africa

Raynitchka Tzoneva
Centre for Substation Automation and Energy Management Systems (CSAEMS)
Cape Peninsula University of Technology
Cape Town, South Africa

Section I

Faults Analysis and Power System Protection Devices

1

An Overview of Smart Grid in Protection Perspective

T. Adefarati and Ramesh Bansal

CONTENTS

1.1 Introduction

The smart grid is an efficient technology that intelligently monitors the behaviour and actions of the consumers that are connected to the power system with the application of network automation and control as a measure to efficiently deliver sustainable, economic and secure electrical power supplies [1,2]. The major components that constitute the smart grid are consumers, markets, service providers, centralized operation system, generation sources, transmission system and distribution system, as presented in Figure 1.1. A smart grid utilizes state of the art technologies coupled with intelligent monitoring, control and communication facilities as well as emerging self-healing technologies are used to achieve the objectives of a power system. The smart grid can be used to reduce the number of power outages and costs that are associated with power outages as well as to reduce the impacts of terrorism or other sabotage on the grid. The application of intelligent solutions at the transmission and distribution (T&D) sections of the smart grid improves the capacity of the power system to accommodate renewable energy distributed generation technologies and improves the security of the power supply. This platform has the ability to run with various types of generating units and a number of applications at the same time, as shown in Figure 1.2. Smart grid is an interconnection of modern communication and information technologies along with control and automation processes across entire electrical power system that consists of generation, transmission, distribution and consumer. The purpose of smart grid is to make existing power system facilities more robust, dependable and effective by utilizing intelligent technologies, supporting peak load management, guaranteeing sustainability of power supply via renewable energy incorporation, encouraging a pollution-free environment and motivating proficient power generation, transmission, distribution and consumption

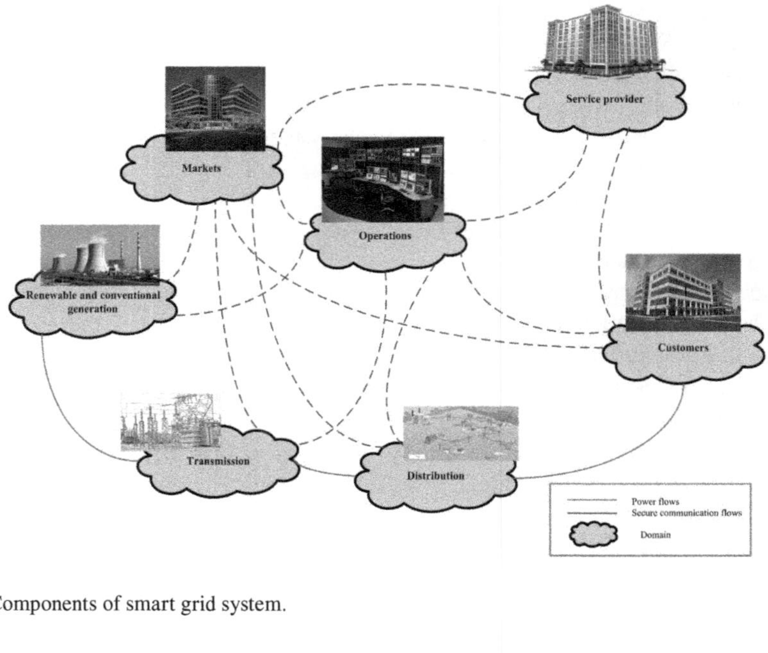

FIGURE 1.1 Components of smart grid system.

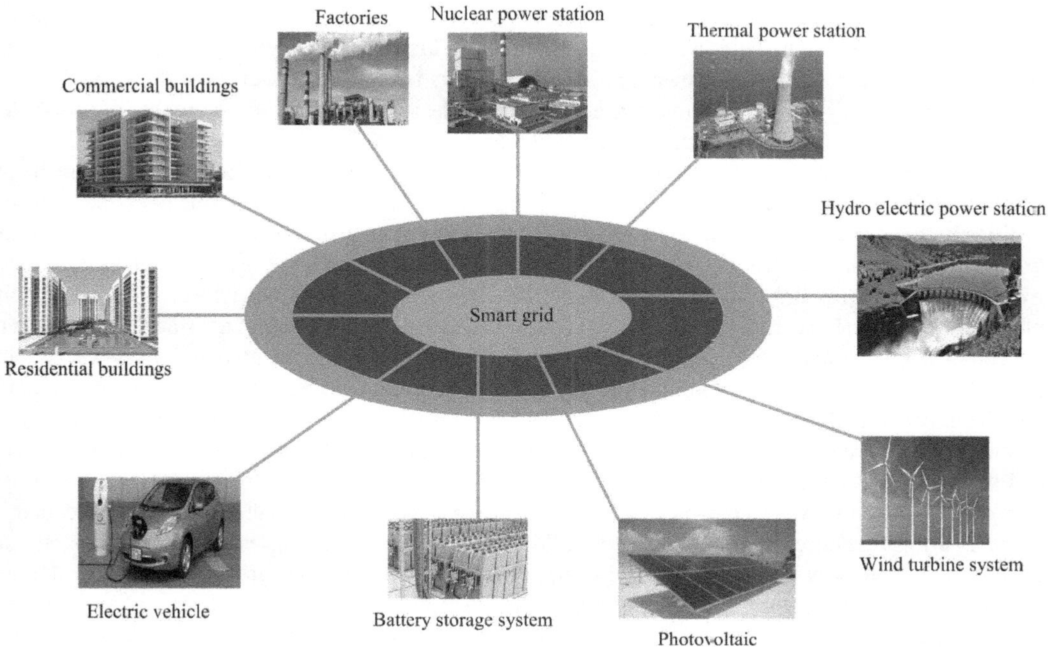

FIGURE 1.2 Smart grid system with various generating units and applications.

of energy across the value chain of electricity. In some cases, active network management is introduced to facilitate the additional number of renewable energy resources into the T&D systems. The power system can be prevented from being loaded beyond its optimal limits and save capacity with the application of system constraints such as power limits, voltage limits, thermal limits, fault level limits, etc.

Apart from this, the smart grid has a number of devices that can be used where the output power of the generating units is dispersed and requires remote measurement and communication facilities to monitor the changeover between grid-connected and independent operation as a measure to ensure a continuous power supply at the load points. The smart grid is responsible for reliable and efficient delivery of electricity from generating units to the consumers. The benefit of this is to prevent a loss of connection between the loads in a geographical location and the bulk power system so that it will not affect a number of customers in such locations. This indicates that the smart grid offers flexible, improved energy delivery, and efficient, reliable and cost-effective operation for both localized and concentrated loads. To harness the benefits of the smart grid system, it is necessary to coordinate all the local energy networks in different geographical locations into a single entity with the application of a smart grid features. This facilitates higher levels of security, quality and availability of power supply; improves economic generation of electricity; improves safety and reduces the impacts of greenhouse gas (GHG) emission from the conventional generating units.

Smart grid integrates control, monitoring and communication proficiencies into the T&D systems to maximize the optimal operation of the power system while reducing the power consumption [3]. The basic concept of a smart grid is to take advantage of new technologies, such as plug-in hybrid electric vehicles, various forms of distributed generation technology, smart metering, power management systems and distribution, feeder, substation and customer automation [4]. Smart grid is referred as an improved electricity supply chain that allows millions of people to get their sources of power from wind energy, nuclear energy, coal, hydro, natural gas, and other green sources. The application of the modern technologies has made the conventional power system smarter and able to operate with internet facilities, so that control and automation systems can be used to monitor and control the generation, distribution and transmission systems. The main reason for replacing the conventional power system with the smart grid is to have autonomous control over the grid and to make it self-healing instead of constructing new power stations, substations and T&D lines. The application of the smart grid system allows the power utilities to utilize the existing power systems efficiently. This reduces congestion of the T&D lines by supplementing the power demand by a number of customers from various sources of distributed energy resources and battery storage systems. The automation of the smart grid permits a number of distributed technologies to be safely and efficiently integrated into the grid [5].

The concept of smart grid depends on the communication and control technologies that are incorporated into the power system to maximize the technical, economic and environmental benefits to the utilities and customers. This can be achieved by extension in the aforementioned technologies to the existing energy infrastructure. The purpose of employing smart features in the grid is to enhance the quality of the power supply and customer satisfaction. The enhanced communication between various sections of the power system is expected to result in an overall increase in the efficiency of the smart grid. Owing to the complex nature of each section of the power system, it is reasonable to integrate intelligence technologies in all parts of the power system though the smart grid that can facilitate efficiently the point-to-point intelligent two-way delivery system from generating sources to the load points through incorporation of renewable energy resources and T&D systems. In this way, smart grid technology improves the efficiency and sustainability of the generating units to meet the power demand based on the specifications of the customers. It also motivates customers to take part in the power management process. The acceptance of the smart grid requires an important effort, allocation of resources and investment to demonstrate the effectiveness of the new infrastructure and integration of renewable energy sources as well as the adoption of smart tools to manage electricity flows, energy efficiency and conservation.

A smart grid is an automated power system that allows the costumers and networks to be observed in real time. It has the ability for dual flow of energy and information from the power generating units to the customers, and vice versa. The automation of the smart grid permits all the customers to have access to an efficient, environmentally friendly and reliable power supply. The smart grid is designed with the purpose of accomplishing many technical and economic benefits with the integration of the communication facilities into the existing conventional power system. This allows unhindered flow of power and information within the system with a high level of coordination. The incorporation of power, control and communication technologies into the traditional power system allows the remote sensing and control of the power system [6]. Moreover, the smart grid consists of customers' automation, distribution system automation, transmission system automation, power market, energy service provider, data communication, control centres, information management and centralized operation system to form a complete system automation, as shown in Figure 1.3. This allows easy monitoring and management of the smart grid and provision of a reliable, affordable and sustainable power supply [7]. The benefits have motivated

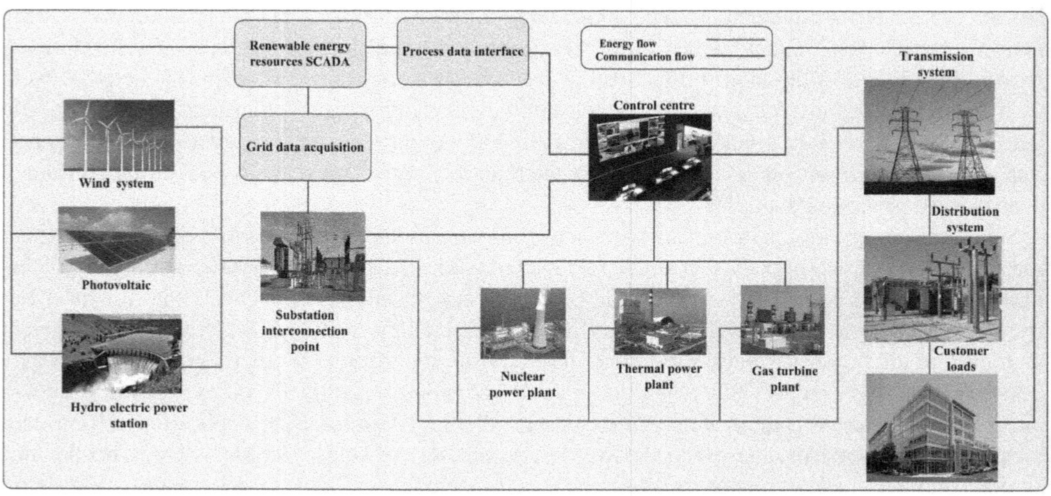

FIGURE 1.3 Smart grid architecture.

many countries to formulate a number of the developmental agenda and vision for expansion of the smart grid on the global note. Under many circumstances, the grid development activities have been focused on the smart grid and peak load management, owing to the fact that the power outages have negative impacts on the optimal operation of a power system. In view of this, smart grid is designed to improve the quality and reliability of the power supply, reduce operation and maintenance costs, increase the economic and environmental benefits and improve power management and utilization [8,9].

Taking into consideration the significance of smart grids, the past and present status of the application, protection and development of the smart grid are reviewed and reported in this chapter. The smart grid has a lot of protection risks due to the fact that the power system was initially designed without considering the bidirectional power flow [10,11]. This chapter introduces the scope and functionalities of smart grid with the application of the protective devices to improve the optimal operation of the power systems. The potential of the protective devices and their effects on the smart grid operation are discussed with the major challenges and strategies to protect the smart grid against electrical and mechanical faults. This chapter also provides detailed discussion of the smart grid system and future plans for expansion of the conventional power systems to incorporate smart grid features into their applications.

1.2 Major Functions of a Smart Grid System

Smart grid is a network that intelligently integrates the consumption patterns of the consumers into the power system as a measure to efficiently deliver sustainable, economic and secure electrical power supplies. The smart grid system uses communication and control facilities to enhance the reliability, security and efficiency of the power supply from the generation system and delivery of power supply to the consumers at the load centres with a growing number of distributed generation and storage resources. Some features such as sensing, embedded processing and digital communication facilities are used in the smart grid as a prerequisite to make the system observable, controllable, automated and fully integrated. The development of the smart grid has caused a bidirectional flow of electricity owing to the incorporation of the distributed energy technologies. The automation of a conventional power system has reduced the risks of overvoltages, congestion, growth of consumption peaks and GHG emissions. These objectives can only be achieved by introduction of a market mechanisms in terms of the peak load management and demand response to the smart grid. Apart from this, the power utilities need to make their systems smarter as a way to manage a set of new constraints that are formulated for optimal operation of the networks. Because the smart grid system allows flexible management of the network, the concept of the smart grid is based on the scheme where technological advances are combined with communication and control technologies that make the operation of the power system more efficient [11]. The major functions of the smart grid are as follows [8,12–14]:

1. The smart grid facilitates the connection and operation of various sizes and technologies of the generating units.
2. It allows consumers to play a prominent role in the optimization of the operating costs of the power system through net metering and feed-in tariff.
3. It furnishes consumers with more important information to make the best decisions about the choice of power supply.
4. It considerably reduces the environmental impact of the conventional generating system.
5. It improves the power system reliability, quality and security of the power supply.
6. The smart grid allows the power system operators to economically and efficiently deliver electricity at the load points.
7. It allows consumers to use electricity as economically as possible.
8. It increases the efficiency of the entire electrical delivery system.
9. It reduces power system delivery process and greenhouse gas emissions.
10. It optimizes the operation of the grid system.

11. It is environmentally friendly and offers a transparent interface with operators.
12. It allows online monitoring of power consumption.
13. It provides an enabling platform for the integration of renewable generation technologies.
14. It is easy to compute the electricity by using time of use and time of day tariff.
15. It saves energy cost by logically controlling loads.
16. It is the most productive and most dependable method for conveying power to consumers by adjusting the load with different power sources.
17. It responds quickly to the problems caused by climate change and component failures as a way to avert blackouts.
18. It provides various advantages for the customers to efficiently and economically use electricity.
19. Capital spending for generation and T&D investments can be deferred.
20. It encourages the participation of consumers in energy management.

1.3 Features of the Smart Grid

Smart grid technologies consist of advanced technologies spanning over the whole power system, starting from the power stations through the T&D systems to different kinds of consumers. Moreover, the smart grid utilizes a variety of technologies and communication facilities for a cordial interaction between grid operators, consumers and the personnel that operate the power stations [15]. The smart grid comprises a number of operational and energy schemes that include smart appliances, sustainable power resources, smart meters and energy efficient sources to detect, analyse, respond and restore network sections [16]. It also supports the grid reliability, security and power quality. The control of the power system generation and distribution is one of the cogent significant features of the smart grid [17]. The automation of the existing power infrastructures with the integration of new technologies has increased the overall features of the smart grid and new synergistic benefits, as shown in Figure 1.4. The features of the smart grid are explained in detail in the following subsections.

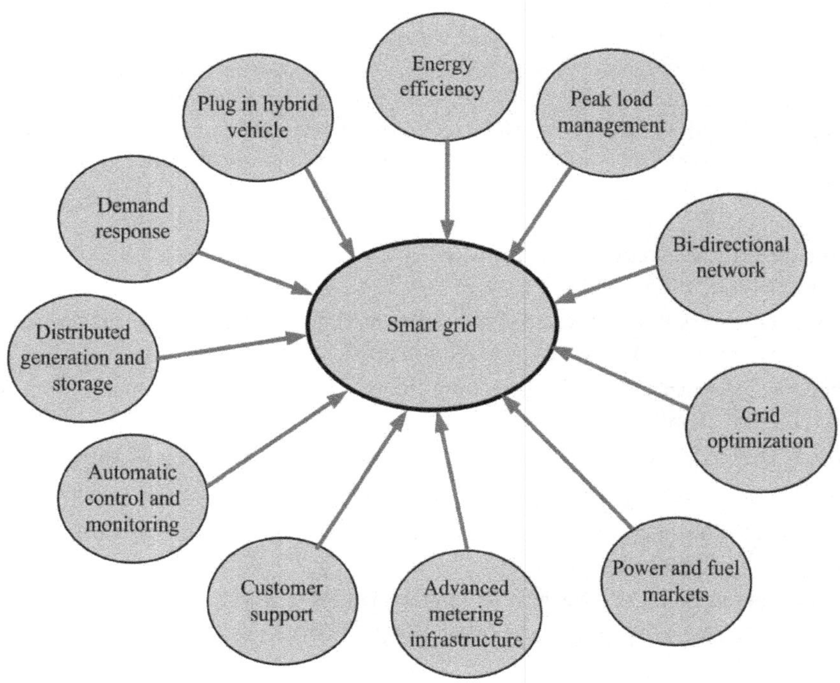

FIGURE 1.4 Smart grid system.

1.3.1 Demand Response

Demand response is a mechanism used by the utilities to motivate consumers to willingly reduce power consumption usage during a specific period of the day, i.e., peak periods when electricity price is high or during emergencies for prevention of a blackout. Smart grid supports interaction of power generation from various generating units and power demand in a real time. Owing to the advances in the level of technology, smart grid has many communication and control facilities to access generating units, transmission systems, substations, distribution systems and customers. The communication configuration of the smart grid allows information to flow from the costumers to the utilities, and vice versa. The power utilities design their power systems to meet the total amount of power demanded by the customers with the application of reserve merging [18]. With the advent of the smart grid, most of the power systems are incorporated with a number of standby generating units for the purpose of responding to a sudden change in the power consumption at the peak load periods. The demand response can be used for a variety of loads such as commercial, residential loads and industrial loads [18].

The demand response also provides an opportunity for the consumers to play an important role in the operation of the smart grid, through reduction of their power consumption during peak periods in response to the financial incentives introduced by the utilities [19]. At times, demand response can be used by the power system operators as a tool or platform to balance the power generated from different sources and power demand. Demand response programs are being used by some power system planners and operators, as the dynamic platforms to balance power supply and load demand [20]. This measure reduces the cost of energy and in turn causes a considerable reduction in the price of electricity. The power utilities use demand response as a valuable option whose proficiencies and effects are increased by modernization of the grid. Owing to this, communication and control devices are incorporated into the smart grid to detect peak load periods in the power system and in turn trigger the automatic switch to isolate non-essential loads during the aforementioned periods. This reduces the possibility of congestion of the network and the resulting power failure. The application of the smart meters makes it easy for various customers to alter their consumption patterns during peak periods based on the information that is available on the consumers' power consumption, cost of electricity and the offer of time-based rates Based on the capacity of the generating units, demand response is utilized to shift energy consumption from peak to off-peak at times of high production and low demand. Demand response also assists the power utilities to save a lot of money with reduction of peak load demand and the capacity to defer construction of new distribution, transmission and generation systems, especially those reserved for the peak load periods [21].

1.3.2 Load Balancing

The power consumption of individual consumers connected to the grid is time variant based on factors such as environmental and weather conditions, the type of consumers, pattern of production, behaviour of consumers and economic structure. The smart grid system is designed to quickly respond to a sudden change in the power demand due to the availability of many standby generating units. Apart from this, it is deliberately scheduled to project the number of standby units that are needed for the peak load periods and how to utilize load management to meet the power requirements of the consumers at all times [18]. A smart grid system allows a mathematical algorithm to predict the load demand at any time and sends warning signals to individual customers to shut down non-essential load during the peak periods [22]. At times, different tariff regimes are utilized with the application of smart meters to reduce congestion of the grid [23].

Smart grid is a concept that allows matching of power generation from the numerous sources with the power demand. It uses the number of the generating units that can start or stop automatically and operate effectively at a selected load [18]. The generating units are projected to be independent of one another and suitable for base and peak load operations. This indicates that load balancing encourages a stable and reliable power supply. However, any excess power demand that is beyond the capacity of the generating units will result in a frequency deviation and a prolonged power interruption [18]. In view of this, it is mandatory for the operators of the power systems to ensure adequate matching of the power output of the generating units with the load demand on the grid. Otherwise, it will cause damage that will become unbearable for the utilities and customers. The load balance has become a global challenge due to the high demand for uninterrupted power supply and utilization of many complex electronic gadgets. The intermittent nature of the renewable energy resources and change in the operating variables of the conventional generating units have compelled the utilities to increase their reserve margin [18], and to ensure that sudden and unanticipated power demands are met without affecting the operation of the smart grid.

1.3.3 Reliability

The smart grid uses a variety of technologies, communication, monitoring and signal processing facilities that ensure a rapid fault detection and self-healing without the involvement of the personnel that operate the power system [24]. Smart grid technologies are used by the utilities to speed up power restoration procedures as a result of the faults that originated from a severe thunderstorm, component failure, a tree falling on lines, equipment failure and lightning [8,25]. This reduces the number of people affected by power outages and the economic damage associated with the power outages. This indicates that the application of the smart grid can improve the overall service reliability of a power system. Apart from making the supply of electricity more reliable, the smart grid also reduces the vulnerability of the power system to terrorist sabotage and natural disaster [18]. The economic effect of improving the grid reliability with the application of the smart grid features can be estimated with a lot of key performance indicators that have been explained in Chapter 16 of this book.

1.3.4 Peak Curtailment

The application of smart meters and communication facilities can be deliberately used by the utilities to inform numerous customers that connect to their networks how to utilize electricity economically. This allows peak curtailment when the tariff of electricity is high during the peak demand periods and encourages a high power demand during the off-peak periods [26]. It motivates consumers to consume less electricity during peak load demand periods by communicating directly to them about the high price premium of consuming a large quantity of electricity at peak periods [18]. The objective of smart grid is accomplished by reducing the power consumption of customers, thus preventing the power system from being overloaded. Apart from this, it also notifies the customers when to charge their electric vehicles at the charging stations and adjusts the temperature setting of air conditioning based on the tariff regimes introduced by the utilities [22]. The application of smart grid also reduces the spanning reserve that utilities use as a standby unit with the application of central control, power management services and the free market mechanism.

1.3.5 Flexibility in Network Topology

The smart grid is designed for two-way flow of energy and communication between the power producers and consumers. If a local consumer generates more than it is consuming, the excess power will be sent to the grid [16,18]. The power utilities face multiple challenges as a result of the proliferation of distributed generation technology. One of the challenges is the safety and reliability issue that arises from the reverse power flow [27]. A smart grid is purposely designed to control the possibility of bidirectional power flow, which allows utilization of multiple energy sources such as photovoltaic systems, wind systems, fuel cells, electric vehicles, battery storage systems, pumped hydroelectric power, etc. The power system flexibility can be enhanced by balancing supply and demand by integrating distributed energy resources [26].

1.3.6 Efficiency

Smart grids provide a number of opportunities for consumers and utilities to save energy and improve the efficiency of energy with the application of demand side and power management system. The highly intelligent smart grid, with the integration of the smart meter systems, allows consumers to have a better understanding of energy usage and to save a lot of money due to energy saving opportunities [18].

The consumer demand response and a better understanding of the status of the electrical grid by the utilities can improve grid efficiency and reduce T&D line congestion during peak periods [28]. This improves utilization of the generating units and grid as well as reduction of energy bills. Moreover, a smart grid can enhance transportation efficiency by facilitating charging of plug-in electric vehicles. This is achieved by paying more attention to energy efficiency and net demand reductions.

1.3.7 Market Enabling

The smart grid permits a logical exchange of information between the utilities and the consumers based on the price of electricity and their capacities to pay. It also allows the utilities and the consumers to be elastic in their operational transactions. A good market platform created by the utilities and consumers allows the consumers to be more strategic based on the tariff regimes at the peak and off-peak periods [18]. This indicates that consumers will only pay for essential loads at the peak load periods, while the nonessential load will be shielded at this particular period. The independent power producers (IPPs) that are flexible in their operations will derive a maximum profit by selling their electricity strategically at high prices during the peak load periods. Inflexible IPPs are liable to get a fluctuating tariff in light of the level of demand and the operational state of the generating units. The general effect is an improvement in energy efficiency and a reduction in energy consumption that is sensitive to the time-varying limitations of the power supply [18].

1.3.8 Sustainability

Smart grid has made a considerable contribution to the grid operation in different parts of the world with the application of renewable and nonrenewable energy sources [29,30]. The flexibility of the smart grid topology allows integration of a number of distributed generation technologies such as wind turbines, photovoltaic cells, electric vehicles, hydro turbines, microturbines, reciprocating engines, gas turbines, steam turbines, nuclear power, battery storage systems, etc. [18]. The share of renewable energy resources is increasing globally, and the smart grid, in collaboration with some policies and regulations, can transform the power system and modify the grid infrastructure to support a sustainable energy development. For this reason, smart grid allows integration of a large number of sustainable energy sources into the grid.

1.3.9 Environmental

The application of renewable energy with smart grid features reduces the impact of global climate change [31]. It also offers a genuine path towards environmental improvement and impact of electric power serving humanity. The high penetration of renewable energy sources into the power system minimizes overall environmental impacts and ozone layer depletion [32]. A significant reduction in the amount of emissions can be achieved with the utilization of demand response, electric vehicles, demand side management and renewable energy distributed resources.

1.3.10 Power Quality Management

Power quality management (PQM) is concerned with the control of the quality of the power supply. Power quality problems are the major causes of equipment downtime, equipment malfunction and damage. Various power quality disturbances and their effects on equipment and electric installation, as well as possible mitigation solutions, are well known today. However, PQM addresses the following issues: voltage flickering, unbalanced phases voltages, power factor, current and voltage instability and harmonic distorted power supply. This facilitates efficient and reliable operation of the power system, reduces losses, improves customer satisfaction and reduces equipment failures. PQM includes voltage/volt ampere reactive VAR control, load balancing, harmonics controller, etc. Implementation of smart grid features in the power system can be used to manage and continuously improve power quality, reducing unexpected downtime and optimizing equipment lifetime and operating conditions in the electrical networks.

1.4 Smart Grid Technologies

Smart grid technologies are centred on the design, development, control and monitoring of remote communication equipment fully integrated with the internally developed network management software. The communication devices in the smart grid consist of a global system for mobile communications, satellite, low-power radio technology, networks, etc. The smart grid is a modern approach for technical and operational management of the national, regional and domestic utility grids as well as systems within the security industry, remote machine monitoring, remote sensor stations and pollution measuring equipment. Smart grid technologies involve the management system, protection system, communication system, energy efficiency scheme, GHG emissions reduction system, network automation, power quality and the security of the power supply [17]. The utilization of the smart grid technologies in the traditional power system increases the capacity of the system as well as the flexibility of the network. In addition, the smart grid architecture allows the integration of state-of-the-art technologies by using communication, monitoring and control facilities. This section reviews the developing smart grid technologies and their effects on the power system operations [33]. The main parts of smart grid technologies used in the power system operations are explained in the following subsections.

1.4.1 Communication System for the Smart Grid

The need for a number of the state-of-the-art technologies in the traditional power system to operate as a self-healing and self-monitoring systems is imminent due the large demand of uninterrupted power supply and sustainability of energy on a global scale [6]. In view of this, integration of communication technologies such as digital control and information technology into the conventional power system has presented many critical constraints to improve the grid's reliability, security and efficiency. Several communication technologies are purposely installed at different sections of the grid to convey information from the customers to the utilities, and vice versa. They enhance the efficiency of data exchange between the electricity users and utilities as much as the architecture of communication system permits numerous signals to be transmitted concurrently. This reduces the operating cost and makes the communication easier by using internet protocol (IP), local area network (LAN), Ehernet, wide area network (WAN), etc., as presented in Figure 1.5.

The purpose of a traditional power grid is to convey electric power from generating units to the load centres [33]. This requirement can easily be achieved by expanding the existing network with the application of a number of digital communication facilities. This improves the functions of the existing power system with smart grid features and emerging technologies. Therefore, the digitization of the traditional power system with the smart grid features has provided a platform for improvement of the power system flexibility, sustainability and resilience. The communication architecture supports online data acquisition for remote sensing and control of a smart grid, as shown in Figure 1.5. The application of communication facilities takes into consideration the control, information and data exchange to improve the availability of power supply, resources utilization and security [17]. The communication framework of the smart grid and how information is being exchanged among the generation domain, transmission domain and customer domain is presented in Figure 1.6. The listed domains consist of control centre, substation automation system, distribution control centre, distribution automation system, distributed energy resources system automation and customer automation system for optimal operation of the smart grid [34]. The communication interfaces between the domains allow information to be shared easily with the regional system operator, energy service provider and power market [34]. This shows that the operations of the energy service provider and market mechanism are being controlled or monitored by the regional system operator [35]. Also, the three domains are regulated by the regional system operator with the aid of information exchanged by using advanced metering infrastructure (AMI), distribution control centre and generation control centre, as illustrated in Figure 1.6.

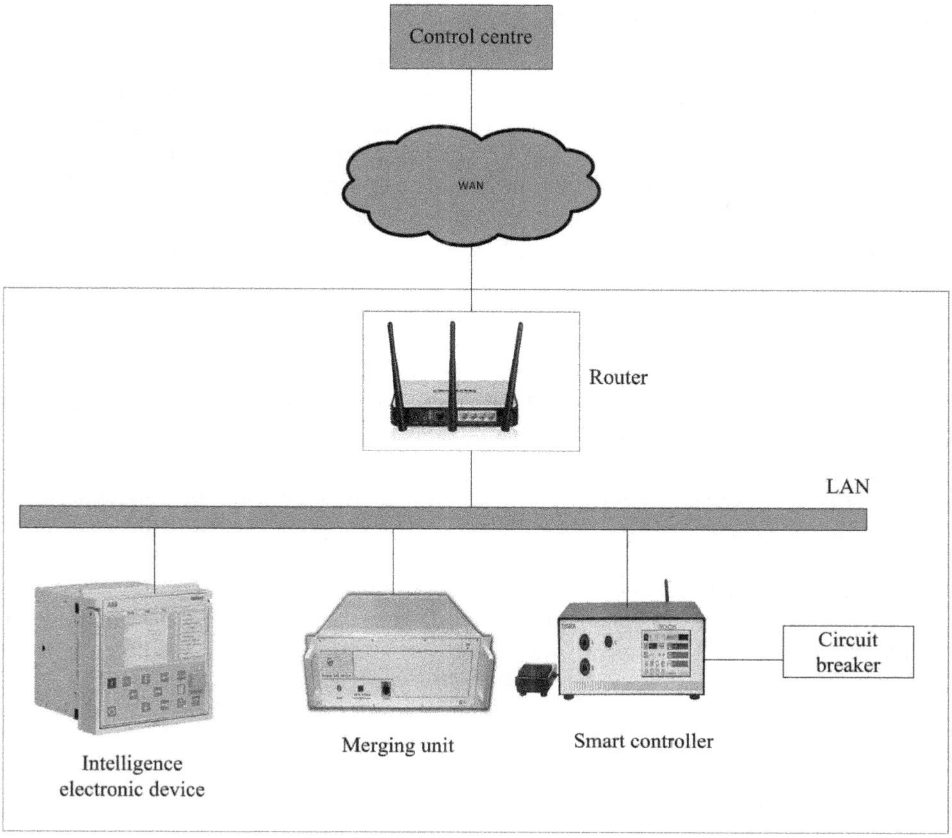

FIGURE 1.5 Communication system of a smart grid with digitized facilities.

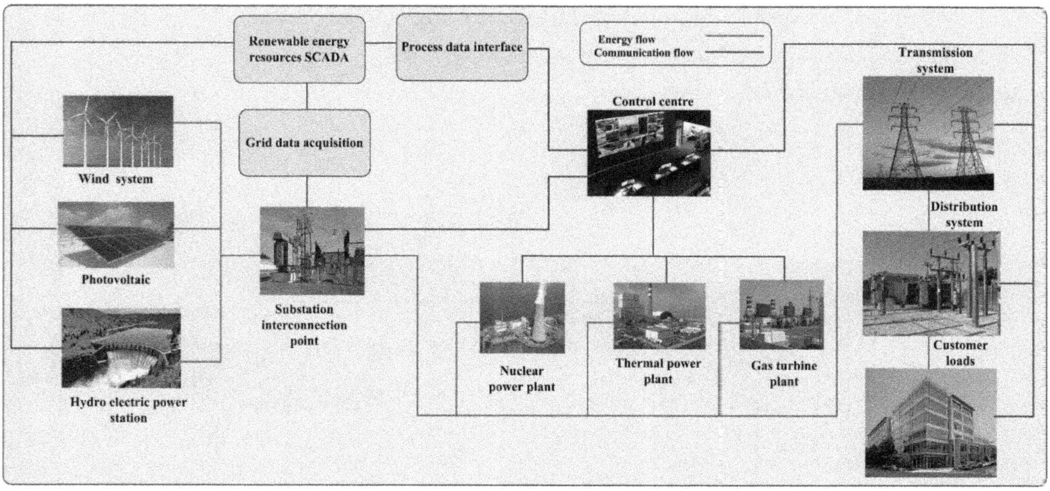

FIGURE 1.6 Communication architecture of smart grid.

1.4.1.1 Supervisory Control and Data Acquisition

A supervisory control and data acquisition (SCADA) system consists of application software, control centres, programmable logic controllers, human machine interface, communication facilities, sensors and remote terminal units used to support grid reliability, security, power quality, power plant effectiveness and data integrity [36,37]. The main purpose of utilizing SCADA in the smart grid include data acquisition and processing, remote sensing and control, data analysis, etc. The SCADA system processes, distributes and displays the data. It also assists operators in analysis of the data and makes a significant decision [16]. The SCADA system has been utilized in the power sector for efficient operations and energy management of a power system, as presented in Figure 1.7. It can be used to monitor a larger area that has a number of generation, transmission and distribution facilities. It permits utilities to quantify the values of voltage and current as well as to control resources in their systems through an automated power management system with quick decision making, effective fault location and identification and power restoration. In the grid, the SCADA system is a common tool for collection of measurements and status data and at the same time sending control commands to switching devices. Based on the data collection, an energy management system (EMS) provides analytical tools for operators to determine the system state and take appropriate actions [33]. A SCADA system can be utilized for collection of consumption data from the smart meters for accurate customer billing. The SCADA system has been extended to the distribution network with the application of demand side management and smart meter to optimize energy demand, shift peak energy and reduce network congestion. Some of the advantages of utilizing SCADA in the smart grid are as follows:

1. Increases reliability.
2. Reduces operating and maintenance costs.
3. Accurate forecasts of power demand.
4. Reduces human influence and chances of error.
5. Assists power system operators to make the best decisions.
6. Troubleshoots faults easily.
7. Automated meter readings.
8. Effective load management of the power system.
9. Quick restoration of the power supply after interruptions.
10. Deferred capital expenditures for upgrading the existing power systems.
11. Monitoring of power system components conditions.
12. Reduces downtime in the power sectors.

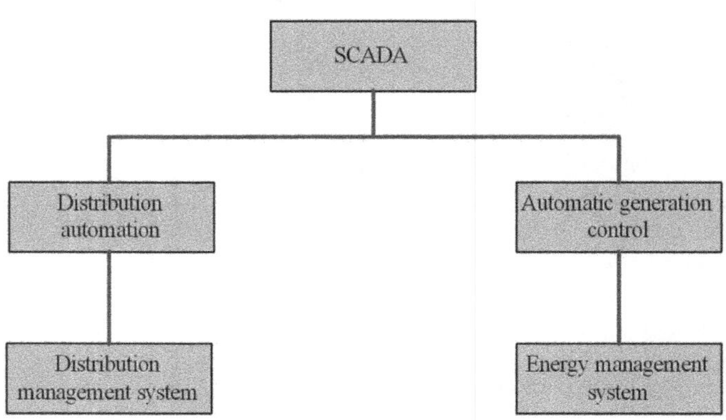

FIGURE 1.7 Applications of SCADA in the power system.

1.5 Sensing and Measurement

The need for an urgent change in the power system delivery has prompted the development and utilization of sensing and measuring technologies in the power system. Owing to the present trend of the smart grid, new communication technologies, sensors and smart meters are used to support more complex measurements and more frequent meter reading [17]. They also expedite direct interactions between the utilities and customers. The fundamental objectives of using sensing and measurement devices in the power system is to assess the level of power system congestion and grid stability, prevent power energy theft and provide control strategies support [18]. The various types of sensing and measurement technologies that have had an impact on grid modernization are available in large quantities today. However, their utilization has been hindered owing to the high costs of production, operation and maintenance. The sensing and measurement technologies consist of smart meters, digital protective relays, consumer portals and agents, real-time pricing tools, major equipment health monitors, wide area monitoring systems, electronic instrument transformers, electromagnetic signature measurement/analysis, phasor measurement units, advanced switches and cables, line sag monitors, meter reading equipment and time-of-use and real-time pricing tools [18]. The power utilities can use sensing and measurement devices in conjunction with the pricing tools to shape the pattern of customer's power usage, which results in cost savings, and both utilities and customers will benefit from this. The platform allows the utilities to control the price signals while the customers decide the particular times and what method to be adopted to modify their power consumption patterns.

The impacts of these technologies further strengthen their implementations in the power system. The application of sensing and measurement technologies can be used to achieve the following [11,23]

1. Empower the electric power market.
2. Allow customers to make reasonable choices.
3. Allow a number of customers to be involved in power generation activities through feed-in tariff and net metering.
4. Improve the savings of capital and operating costs.
5. Improve environmental benefits and power system efficiency.
6. Economic and public benefits from improved safety, reliability and power quality.

1.6 Smart Meter

A smart meter is a system that records the energy consumption of each customer periodically and automatically communicates the information to the utilities for billing and monitoring, as shown in Figure 1.8. The smart meters are designed to record the energy consumption of the consumers on hourly basis and report to the utilities at regular intervals. In addition, a smart meter can be used to digitally send meter readings to the utilities for more accurate energy bills. The two-way communication between the utilities and customers facilitates an effective control for consumers' peak load management. The multiple functions of a smart meter allow it to be used for prevention and detection of power theft; at the same time, it can be utilized to detect where equipment failures have taken place [38]. The non-technical power loss that has become a global challenge when providing reliable electrical service can be overcome with the application of smart meters. Smart meters have been embedded in many facilities, which enables various signals to be obtained from demand response, remote load connection and disconnection, and operation of critical and non-critical loads [33]. The two-way communication systems allow warning signals to be sent from the control centre to consumers and from the consumers to the control centre. The smart meters can be used to report tampering actions to the utilities via the control centres [18]. Also, smart meters can be integrated into the street light system for automatic rather than manual operation and thus save energy and operating costs.

FIGURE 1.8 Two-way communication of smart meter.

1.7 Phasor Measurement Unit

A phasor measurement unit (PMU) is utilized in the smart grid to measure the electrical waves in the grid by means of a time source for synchronization. It is an important component in collecting real-time monitoring information for observability and controllability of smart grid operations. Owing to the importance of the smart grid, the synchrophasor has been produced in large quantities to improve the observability of the power system. A PMU is designed so that it allows synchronized real-time measurements of numerous remote points in the grid [39]. A PMU network is composed of PMUs at substations and generation stations, phasor data concentration (PDC) for collection of information, local communication networks and SCADA at control centres. The network is globally utilized in wide area measurement systems (WAMSs). The consequence of PMU network failure is so serious that it can cause a severe power outage. The PMU applications have been projected to enhance the reliability of the grid. Therefore, the reliability of a PMU network can be quantified using reliability indices [39]. A PMU is thought to be a standout amongst the most essential measuring devices that can be integrated into a protective relay or protection system.

The PMUs and synchrophasors make a significant contribution to the evaluation of transient processes in the power system. The PMUs are utilized to measure the phase and amplitude current and voltage of the grid. The high-precision time synchronization allows comparing measured values from different substations, and conclusions will be drawn based on the system state and dynamic events such as power swing conditions. The PMUs can also be used to determine current and voltage phasors, provide them with highly accurate time stamps and transmit them for analysis together with other measured values, such as frequency and speed of frequency change using the IEEE C37.118 communication

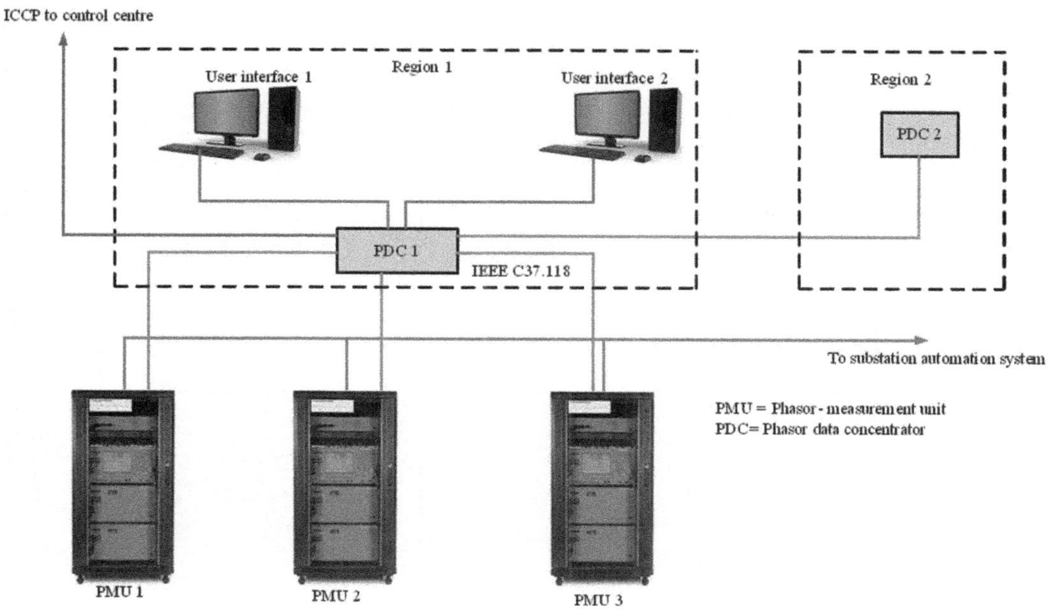

FIGURE 1.9 Applications of phasor measurement unit in a power system.

protocol, which are typically sent to the control centre. The phasor data processor is a WAMS that is used to measure different variables with the application of a synchrophasor from phasor estimation units serving as sensors. It assists with a rapid acknowledgment of the present system circumstance and transparently and instantly shows power swings and transient phenomena. The device supports control centre work force in evaluating critical grid conditions and making an appropriate decision, as shown in Figure 1.9. As every single estimated result is stored, the power system disturbances can easily be analyzed.

1.8 Distributed Energy Resources

The power system sector faces many critical challenges such as high energy demand, high T&D losses, fuel price fluctuations, unreliable power supply and execution of environmental strategies to reduce ozone layer depletion and climatic changes, etc. Consequently, the open electricity market has shifted the traditional pattern by which power is generated, transmitted and distributed to the smart grid. The new paradigm allows consumers to be flexible in purchase of electricity based on the guidelines put in place by different utilities in the deregulated power sector. This has led to the development of electric power production from grid-interactive renewable energy sources, which are inherently volatile, exploration of new possibilities for energy storage, development of electric vehicles (EVs), recognition of consumers and utilities as smart energy decision makers and advancement of energy efficiency in real time, etc.

Distributed energy resources (DERs) such as wind, solar, hydro turbine, reciprocating engines, microturbine, gas turbine, etc., are used in the distribution system. The DERs are owned, operated and controlled by consumers and utilities for home consumption as well as trading in the electrical energy market [12]. Owing to the recent innovation in technology, the energy management system can be utilized to manage a group of DER systems via WAN/LAN in the smart grid system [13]. A smart grid that is embedded with control and communication devices supports greater utilization of renewable energy technology by providing more consistent, real-time control over how that energy is routed within the grid [14]. This technology enables the grid to be more effective in the storage of the intermittent energy from renewable sources that can later be used [40]. The utilization of renewable

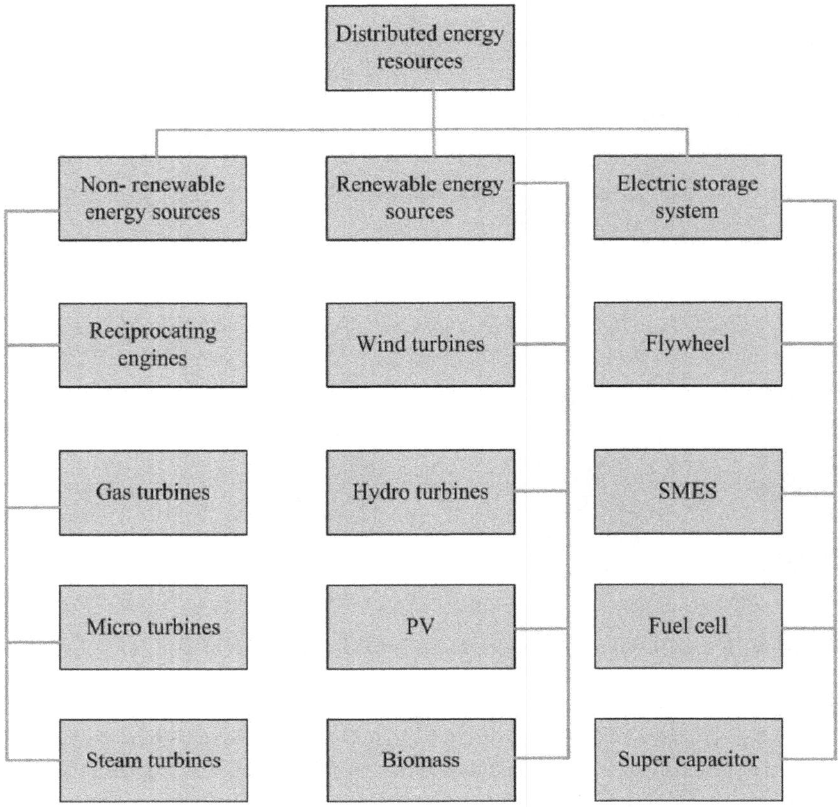

FIGURE 1.10 Energy sources of the smart grid.

energy sources provides not only ancillary generation support to utilities but also benefits consumers. Net metering and feed-in tariff are introduced by the stakeholders in the power sector to enable local consumers to connect their facilities to the grid. This will enable them to become energy providers due to the fact that they have utilized renewable energy resources to improve the sustainability of energy and green energy benefits [41,42]. The renewable energy sources, nonrenewable energy sources and electric storage system shown in Figure 1.10 can be used in the design of the smart grid based on the availability of the energy resources [30,43].

1.8.1 Energy Storage System

The energy storage system (ESS) is utilized as an energy reservoir that injects electricity to maintain the key operating parameters of the grid during a contingency, such as an abrupt loss of power from the renewable energy and conventional sources. It also provides thrust for the use of the surplus renewable energy available during off-peak hours, which can be stored. The ESS is an effective method to store energy and can be utilized on demand. Owing to this, the manufacturers have improved the technologies of the ESS based on the sudden change of energy requirements of consumers. Hence, the ESS provides a number of technological methods as a measure to achieve a reliable power supply and create a more resilient energy infrastructure as well as to improve the cost savings of the utilities and consumers. The different types of energy storage technologies presented in Figure 1.11 are used as a measure to understand the various techniques presently being used globally.

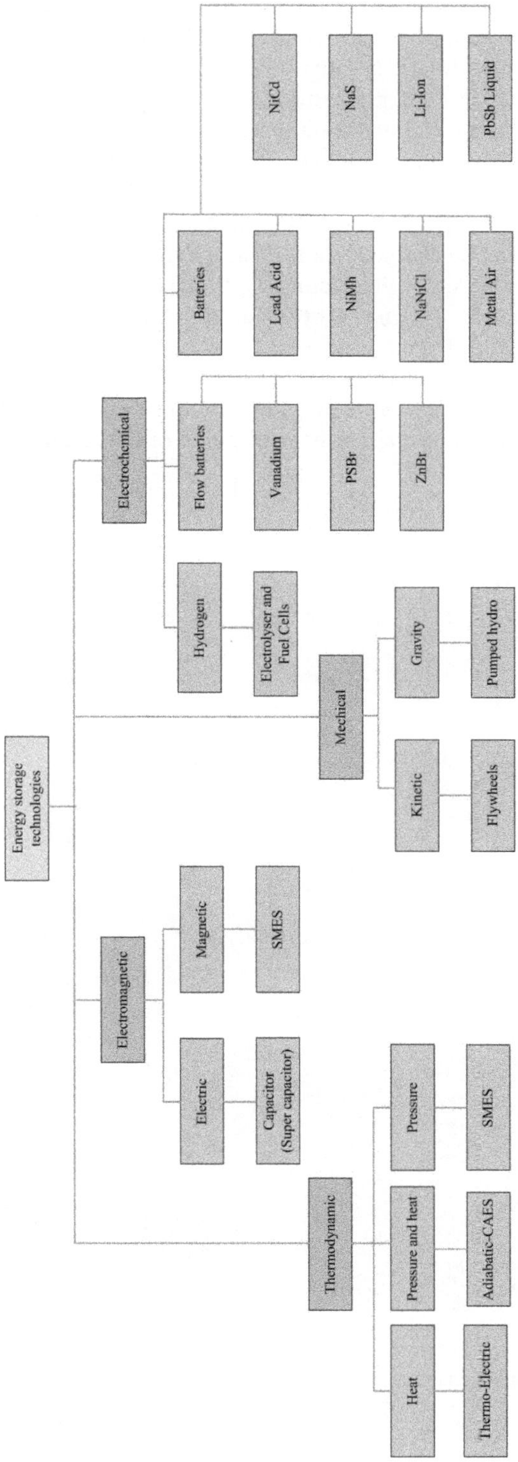

FIGURE 1.11 Energy storage technologies.

1.8.2 Electric Vehicle

The global energy crisis and the environmental impact of the conventional internal combustion engine have motivated adoption of electric vehicles (EVs). The recent commercial production of EVs and development of smart grid concept have increased the roles of EVs in the form of vehicle to grid technology. The integration of EVs into the smart grid allows bidirectional flow of power between the grid and the vehicle. The smart grid offers a number of technical and economic benefits to the utilities and customers: It is environmentally friendly, minimizes GHG emissions, is cost effective, has health benefits, provides power grid regulation, improves safety, spinning reserve, low maintenance, energy security, reduces noise pollution, peak load shaving, diversifies the choice fuel available for transportation, load levelling, is safe to drive, and reactive power compensation. The EV uses a rechargeable battery storage system that can be charged to a full capacity by connecting it to the external electric power source. This can be achieved by connecting the EV to the smart grid with the application of a connector system that is specially designed for this function. Most developed EVs can recover a portion of the energy storage used through regenerative breaking. At the point when regenerative braking is applied, EVs can normally recoup 5%–15% of the energy used to move the vehicles to the vehicle speed that precedes breaking. The photovoltaic panels are installed on vehicle roofs to provide additional power to accessory loads in vehicles. The application of the smart grid for charging of EVs is presented in Figure 1.12.

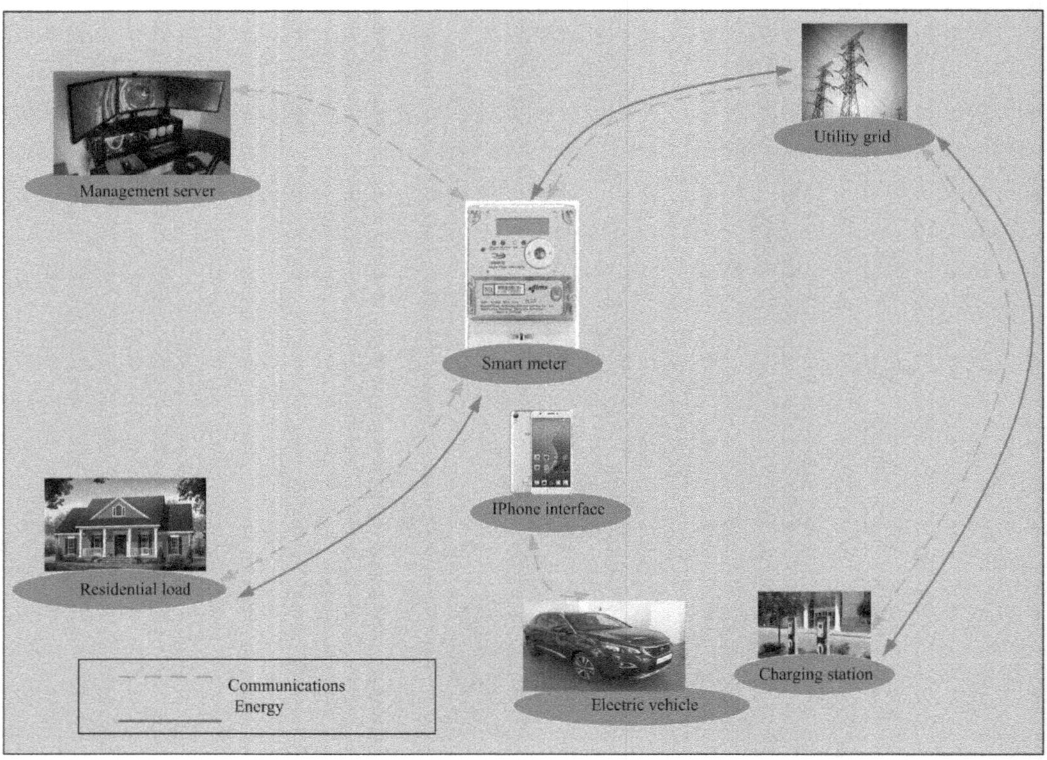

FIGURE 1.12 Electric vehicle in the smart grid.

1.9 Peak Load Management

Peak load management (PLM) is a process of balancing the power supply in the network with the power demand [44]. It can only be accomplished by adjusting the electrical load instead of power output of the generating units. The PLM allows the independent power producers to meet the power demand during the peak periods, which reduces the operating costs associated with peak the operation of the power stations [8]. Moreover, PLM can be used to minimize GHG emissions and make the operation of the generation units efficient due to the fact that the number of generators assigned for peak load operations often produce more GHG emissions and are less efficient than base load generating units. PLM is an initiative through which utilities can reduce the energy consumption of consumers with the application of automated intelligent load management and intelligent peak demand energy storage system, and by using special tariffs to influence consumer behaviour [44]. This plays an important part in decreasing the cost of energy and minimizing the congestion of existing facilities. With the application of PLM, power utilities can minimize their unscheduled power demand and higher cost of energy at peak period. The management of peak power demand can easily be achieved by utilizing a number of smart meters installed at the premises of every customer.

1.9.1 Automated Intelligent Load Management

Substantial peak power reductions can be achieved without any inconvenience to the business or its clients by managing the starting and stopping of certain equipment and appliances, such as heat pumps, geysers, boilers, air conditioners, refrigeration units and pumps.

1.9.2 Intelligent Peak Demand Energy Storage System

Intelligent peak demand energy storage system is purposely designed to continuously monitor the value of energy utilization and discharges from a battery storage system when surges are required. During the off-peak periods, when excess power is available, renewable energy resources are meant to charge the battery storage system. It should straighten the spikes in power utilization and reduce energy consumption to a certain extent. This is the best approach to saving on the electricity bill without switching off the appliances or altering consumers' consumption pattern.

1.10 Smart Grid Automation

Smart grid is a system that provides an interactive platform among the power producers, service providers, consumers, etc. [45]. It requires reliable, fast and cost-effective communication, and control and protective devices to achieve the objectives of utilizing smart grid instead of the conventional power systems. The smart grid devices provide many vital functions for the grid automation [11]. Furthermore, the smart grid applications, such as the integrated voltage and reactive power control, fault detection, isolation and restoration in distribution automation, smart meter and demand response offer increased operational functionality of the power system. The benefits of the smart grid features can be fully realized by automation of the substations, distribution systems and feeders. Financial benefits for the operators and customers include a reduction in the cost of power supply, more efficient utilization of the network, elongation of the life span of the components, economical use of energy and deferrals for upgrading of the network. In addition, smart grid automation with the application of intelligence components allows the peak loads to be controlled and minimized. This reduces purchase of the more expensive peaking power from less efficient power plants, thereby reducing the utilities costs of operation. Smart grid automation allows the utilities to monitor, coordinate and operate their power systems in a real-time mode from remote locations. Smart grid automation allows a quick troubleshooting of faults and an accurate solution to the grid power outages. It exhibits a clear utilization of information technologies in the operation of the grid system. Moreover, it requires innovative technology, automation and control system as a measure to accomplish the following objectives:

1. Guarantees proficient operation of the power system as a measure to maximize the profit of the utilities.
2. Reduces operations and maintenance costs.
3. Maintains and strengthens grid integration.
4. Improves the availability of the power supply.
5. Provides a secured network with control and communication facilities.
6. Provides a good platform for the centralized operator to monitor, control and manage all the components that are associated with the operation of the network.

The components that constitute a smart grid automation include distributed automation, feeder automation, customer automation and substation automation, as shown in Figure 1.13. The implementation of smart grid automation permits the utilities to have total control over the power system. This in turn can be utilized to improve the reliability, efficiency and quality of the power supply.

1.10.1 Distributed Automation

The main reason for the automation of the distribution system is to enhance the efficiency, power quality and reliability of the power system. This has a direct impact on the sensitive loads and the continuity of power supply at the load points. However, the functionality of the embedded smart grid devices and the number of benefits provided by them should be analysed based on their performances. The sudden increase in power demand and the dilapidated nature of the traditional power system can impede the delivery of power at the load points. This has a negative impact on a reliable supply of power. In view of this, power utilities can manage these challenges economically with the application

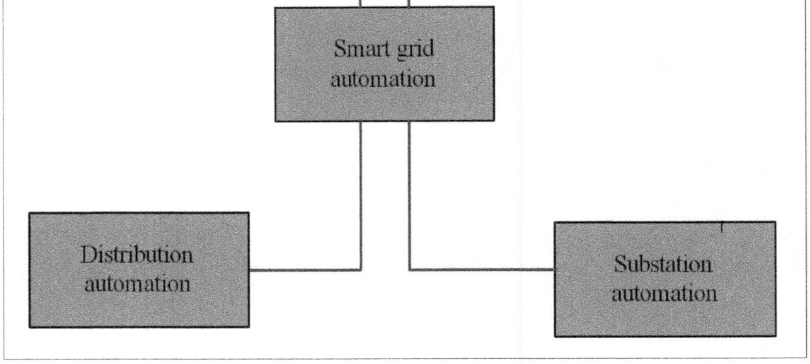

FIGURE 1.13 Components of smart grid automation.

of distributed automation. Distributed automation provides the best possible way of improving the reliability of the system despite increasing complexity of operation and control of the network [46]. This is achieved with the application of the communication devices, switching devices, energy management system, relays, smart meters, and distributed energy technologies. The network interfaces allow remote sensing and control of the distribution network via a centralized control centre. The remote-controlled devices that are normally used in distribution system automation are capacitor banks, feeder switches, voltage regulators, etc. Distributed automation can be used for fault detection, fault segments isolation and power supply recovery as well as to control the Volt-VAR that increases the reliability of the power system.

1.10.1.1 Application of Distribution System Automation

The purpose of the distribution automation is to reduce the number of power outages that result from network faults. Some of the functions of distribution system automation are [8]:

1. Reconfiguration of the network.
2. Fault identification and isolation and power supply restoration.
3. Load balancing.

1.10.2 Substation Automation

Substations are important parts of the power system that step up or step down voltage based on the requirements of the utilities and customers. Substations can be used by the utilities to perform other functions that guarantee safe and reliable delivery of power supply. Substations can be utilized to disconnect parts of the network for maintenance purposes or to supervise power flow within the system for different reasons. Any significant advancement in substation innovation is seen as one of the notable parts of the smart grid development [47]. Compared with other sections of the power system, substations have the largest amount of information that is paramount for the operation of the smart grid [11]. Owing to this, substations require a large number of pieces of equipment to function effectively, i.e., power transformers, isolators, circuit breakers, instrument transformers, bus bars, switches, lightning arrestors, capacitor banks, a network of protective relays, etc. Moreover, substations must be protected from an extensive variety of dangers such as over-voltages, over-frequency, fault current, surges, and component failures. Substations have many intelligent electronic devices embedded within the operational data and nonoperational data that can be utilized to improve the reliability and efficiency of the power system as well as to increase economic benefits to the utilities [48]. The data can also be utilized for predictive maintenance management and as a means to increase the life spans of electrical components. Substation automation is one of the most important parts of the smart grid that needs an urgent deployment of resources to benefit both utilities and customers. Automation of substations includes supervisory control and monitoring of circuit breakers, load tap changers, regulators and substation capacitor banks. The remote data acquisition function is a must for effective use of the supervisor control function. The application of some communication, protection, control, PMU, human machine interface, intelligent electronic device, Ethernet and data processing devices for the substation automation leads to the following benefits:

1. Reduces the engineering cost with the application of protective devices.
2. Improves the compatibility of different protective devices from various manufacturers.
3. Assists the utilities to construct a reliable, intelligent and efficient power system.
4. Automatic operation and reporting significantly reduces the engineering time required to gather, analyse and produce compliance reports.
5. Saves time with automatic collection of fault records and determination of the fault location.
6. Automatic analysis and reporting of power quality criteria violations.

1.10.3 Feeder Automation

The feeder automation includes monitoring, supervision and control of electrical components that are associated with the operation of the smart grid system. This allows the remote operation of components installed at the feeder to reduce outage time based on the fault location and feeder configuration without any human involvement in the operation of the system. Feeder automation is required in the traditional power system to perform the fault detection and location, isolation and restoration activities by employing self-healing and smart grid technology. It also provides remote control and monitoring devices to quantify the ability of the feeders to restore power to the affected customers within a short duration through an alternative feeder. The maintenance team can be contacted through an online application to rectify the affected feeder while restored feeders operate normally. The application of communication and control devices and application software in the smart grid power system minimizes feeder downtime. This is achieved by automatically restoring operation of the serviceable feeder sections while isolating those requiring repair. The feeder automation is designed to enhance the reliability of customer service significantly as well as reduce outage duration and monitoring and management of power demand [46]. The feeder automation reduces the number of labours, operating costs and payback period for the project as well as defers capital investment in upgrading of the T&D systems. The application of the state of the art technologies in the feeders allows the operators to have remote control over all the devices that constitute the system from the control centres. The power system operators can directly monitor feeder details, important information on transformer loads, customer power demands and maintenance schedule information without going into the field. The customer's power consumption can be generated based on the information available on each feeder.

1.10.3.1 Advantages of Feeder Automation

The advantages of feeder automation are as follows:

1. It minimizes outage durations and the associated costs.
2. It minimizes the number, frequency and duration of power outages.
3. With the incorporation of the communication facilities, it can be used to defer capital investment of new stations/facilities.
4. It is flexible and adds value to the existing power system.
5. It allows the best usage of labour hours
6. It reduces monitoring and control costs.
7. It reduces capital and maintenance costs.
8. Improving the quality of distributed energy.
9. It gets an immediate return on investment.
10. It reduces operational costs.

1.10.4 Customer Automation

Customer automation includes a set of technologies, processes, devices and applications that improve the operation of the power system. Customer automation is driven by the sudden demand for global electricity that is expected to increase by 25% by 2030. Moreover, the recent trends in smart grid customer automation have been reflected in the development of communication technologies such as the internet, cellular and wireless phone services, etc. The higher expectations of customers with regards to power availability, increasing costs of energy and access to energy information have prompted a number of utilities to encourage customer automation. Access to energy information can save a number of hours for both utilities and customers. Customer automation can also create numerous supports that are not possible when the processes are guided by human competency. Customer automation technologies assist utilities in reducing operating costs and improving the quality of services. This indicates that customer automation should be seen as an opportunity to qualify and assign someone to the right points where human touch is essential.

1.10.4.1 Benefits of Customer Automation

Implementing customer automation processes in the smart grid has several advantages:

1. It removes or minimizes human involvement in a given area of service.
2. It removes the potential for redundancy and unnecessary efforts.
3. It reduces human error.
4. It reduces the costs of operation.
5. It reduces customer involvement.
6. It provides better services to the customers.
7. It provides a level of responsibility that is difficult to achieve manually.
8. It reduces stress on human resources.

1.10.5 Cyber Security

Smart grid cyber security is vital to protect electrical infrastructure from cyber intrusion. It is utilized globally to address the issues that are related to accidental compromises of the power system facilities owing to human error, component failures and natural disasters as well as deliberate attacks from terrorists and disgruntled workers. The main purpose of the smart grid cyber security is the development of risk management strategy that will allow interoperability of solutions across different components of the smart grid. Smart grid cyber security addresses the cyber security needs of the utilities and consumers by promoting technology transfer of best practices and standards in the area of microgrid power system. Generation, transmission and distribution automation systems collect operational information from dispersed locations on centrally located servers. This information exchange between centrally located servers and field equipment is achieved through various communication media on open standards. Such systems are also connected to the corporate network for sharing energy information to the different business model. The use of open standard and connectivity to public networks has exposed power systems to cyberattacks. Cyber security is an important part of a smart grid.

1.11 Grid Code

A grid code is a technical requirement that power stations, consumers, distribution and transmission systems which are connected to the public power networks must meet as measures to guarantee safe, reliable and economic operation of the power system. The grid code is compiled by the regulating body that is in charge of the power system operation and integrity. At times, a grid code specifies the required parameters of the power system such as power factor limits, control function requirements, reliability benchmarks, voltage regulation, frequency response, power quality, reactive power capability, voltage control functions, protection and fault levels, etc. The electricity network of each country is managed and governed according to grid code prepared by the regulatory body of each country. Hence, the grid code of each country lays the stringent requirements and network performance indicators that must be complied with accordingly.

1.11.1 Grid Code Compliance

Some grid codes that are available for all sectors of the power system are the following:

1. Planning code
2. Connection code
3. Operating code
4. Scheduling and dispatch code

5. Balancing code
6. Data registration code
7. Power generation, transmission and development codes
8. Renewable energy resources connection code
9. Network code
10. Metering code
11. Information exchange code

1.12 Protection System in the Smart Grid

A rapid increase in population, economic development and high demand of a continuous power supply have encouraged many power utilities to incorporate a number of protective devices into their power systems. The integration of the state of the art of the protective devices into the smart grid system can be used to provide a fast response to electrical faults. The recent trend in the integrated circuit technology has given the utilities total control and manageability of the components that constitute their power systems. The purpose of a protection system in the smart grid is to enhance reliability and to improve handling of faults and the utilisation of the power grid with reduced risk and number of outages. Moreover, protective devices are major components of the smart grid that guarantee a reliable disconnection of the faulty sections of the network as a measure to reduce the consequences of power outages. Based on the new technologies, the protective devices have been embedded with the smart grid features that enable them to communicate in real time and utilize information from other sections of the power system for self-healing. This indicates that the faults information from the protective devices is used by the utilities to identify the location, time and number of faults in the grid and resume normal operation after the faults have been cleared manually or remotely. The application of a protection relay and fault analysing system is utilized to collect the fault information and analyse the data obtained from the power system. Different types of protection schemes for the smart grid are presented in Figure 1.14. The application of a protection system in the smart grid has prompted fault detection and self-healing to prevent power outages and voltage fluctuations. This improves the power quality and frequency and reduces the number of power outages [17].

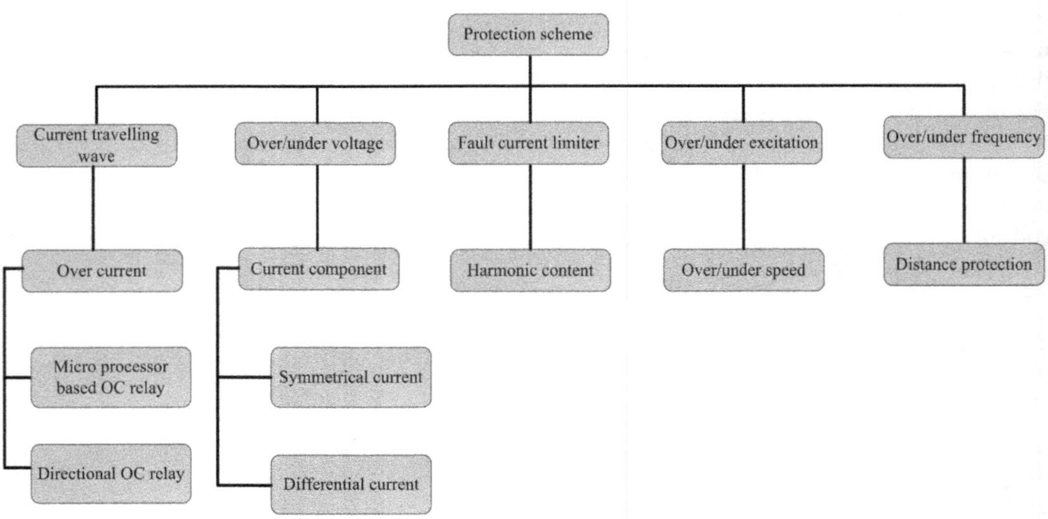

FIGURE 1.14 Different type of protection schemes for smart grid.

1.13 Importance of Protection in the Smart Grid

A smart grid is an intelligent system that utilises automation and control facilities, smart meter, communication technologies, etc., to meet the challenges associated with the delivery of power at the load points [49]. The challenges are due to the number of generating units and the high demand for uninterrupted power demand from a sudden increase in state-of-the-art equipment. The benefits of utilizing a smart grid system in the modern power systems can fully be realised with the application of protective devices that come in different sizes and functions. In this chapter, the requirements and impacts of protective devices in the smart grid are fully discussed. The protection system is designed by the respective manufacturers to ensure that each protective device satisfies the technical specifications of the smart grid. The protective devices should be able to perform the protection tasks based on their selectivity, sensitivity and discrimination characteristics for internal and external faults. The faulty section of a power system can be isolated from the entire network within a short duration with the application of a number of protective devices and their technical coordination. The benefits of a protection system in the smart grid are as follows:

1. Enables bidirectional energy flows and hence improve the capability to interconnect systems.
2. Protects equipment, people and property.
3. Maximizes usage of the network.
4. Separates a faulty section from the entire network so that the power system will perform its function satisfactorily without any severe damage owing to the fault currents.
5. Restores normal operation.
6. Minimizes operating cost of the system.
7. Protects damage due to faults.
8. Improves energy efficiency.
9. Minimizes overall energy consumption.
10. High-speed fault clearance with a correct selectivity.
11. High sensitivity to faults and insensitivity to maximum load currents.
12. Operates effectively under normal operating conditions.
13. Improves service quality and reliability of power supply.

1.14 Challenges of Protective Devices in the Smart Grid

The implementation of the smart grid has become a great challenge due to the coordination of the protective devices that are newly introduced in the grid. The newly introduced protective devices within the smart grid features find it difficult to operate optimally in the traditional power system due to lack of coordination among the protective devices [8]. Many challenges such as bi-directional flow of current, sudden change in the fault current level, high costs of smart grid components and change in the mode of operation of the traditional power system have made the operation of the smart grid difficult [36]. The only way that a smart grid can achieve optimal operation is to introduce a properly coordinated scheme for all the protective devices in the system [8]. The systematic coordination of the protective devices starts with coordination of overcurrent relays, inverse time relays, directional overcurrent relays, differential relays, voltage restraint overcurrent relay, reverse power relays, negative phase sequence relays, frequency relays, power relays, Buchholz relays, overflux relay, etc. These relays are utilized in the smart grid to sense all the abnormal conditions of the power system and close their contacts as measures to complete their respective circuit breakers' circuits. This makes the circuit breakers trip and thus isolate the faulty sections from the rest of the operating segments of the power system. The proposed technique considers the selectivity, sensitivity and proper protection function of the protective devices.

The purpose of the protection system coordination is to ensure proper coordination among the protective devices. In properly coordinated systems, faults can be easily removed within a very short time without affecting the optimal operation of the system. This will guarantee safe and reliable operation of the power system. The elimination of faults can easily be achieved in the traditional power system that operates with a unidirectional power flow. However, the smart grid system with the integration of a number of renewable energy resources has changed the topology of the traditional power system with a bidirectional power flow and the characteristics of the system in terms of the level of faults current. This will have a large effect on the security and reliable operation of the smart grid. In a situation where the existing traditional power system is upgraded to a smart grid, the existing protective devices will be subjected to coordination problems due to the number of distributed generation technologies that have been incorporated into the system. The lack of proper coordination among the protective devices will invariably reduce the safety and reliability of the system. The large integration of the renewable energy resources into the conventional power system will consistently affect the protection coordination owing to the sudden change in the fault levels, multiple current flow paths during faults and false tripping of the protective devices. These conditions decrease the capability of the protection system in the power system [17]. The application of the smart grid features has made the operation of the power system highly reliable and secured [8].

1.15 Tutorial Problems

1. What is the smart grid?
2. Why is it essential to have national grid codes in a country?
3. What are the factors contributing to a smart grid?
4. Why is it necessary for a country to build a smart grid infrastructure?
5. List the components of the smart grid and their functions.
6. What are applications of the smart grid?
7. Compare the smart grid and a traditional power system.
8. Who are the main stakeholders in the smart grid?
9. What are the key drivers of the smart grid?
10. What are the functions of the sensors and self-healing in the smart grid?
11. What are the main challenges of the smart grid?
12. Explain how to manage peak load demand in the smart grid.
13. What are the benefits of demand response in the smart grid?
14. What are the risks that associated with utilizing smart grid technologies?
15. What technologies are used in demand response activities?
16. Define demand response.
17. Explain the effects of demand response on the distributed energy resources and grid reliability.

1.16 Conclusion

The smart grid is an efficient technology that intelligently monitors the behaviour and actions of consumers connected to the system with the application of system automation and control as measures to efficiently deliver sustainable, economic and secure electricity supplies. Smart grid integrates monitoring, analysis, control and communication capabilities into the electrical delivery system to maximize the optimal operation of the power system while reducing energy consumption. Smart grid supports interaction of power generation from various generating units and power demand in real time. Owing to technology advances, a smart grid has many communication and control facilities to access generating units, transmission systems, substations, distribution systems and customers. The smart grid utilizes a variety of technologies and communication facilities for smooth interaction between the grid operators, consumers and power station personnel. The smart grid has a lot of protection risks due to the fact that the power system was initially designed without considering the bi-directional power flow.

REFERENCES

1. "Defining and solving the energy allocation problem with continuous prosumers," https://www.iiia.csic.es/sites/default/files/5499.pdf, accessed June 2018.
2. M. G. Molina, "Energy storage and power electronics technologies: A strong combination to empower the transformation to the smart grid," *Proceedings of the IEEE*, vol. 105, no. 11, pp. 2191–2219, 2017.
3. Q. Sun et al., "Review of smart grid comprehensive assessment systems," *Energy Procedia*, vol. 12, pp. 219–229, 2011.
4. "Smart grid: What is it and why is it important?" https://www.nema.org/Policy/Energy/Smartgrid/Pages/default.aspx, accessed June 2018.
5. "Distributed demand-side optimization in the smart grid," https://tdx.cat/bitstream/handle/10803/145974/TIA1de1.pdf?sequence=1, accessed June 2018.
6. M. Emmanuel and R. Rayudu, "Communication technologies for smart grid applications: A survey," *Journal of Network and Computer Applications*, vol. 74, pp. 133–148, 2016.
7. "Sustainable energy solutions for South African local government," http://www.sustainable.org.za/userfiles/smart%20grids.pdf, accessed June 2018.
8. A. S. Ali, *Smart Grids: Opportunities, Developments, and Trends*, Springer, New York, US, 2013.
9. T. Ayodele, A. Jimoh, J. Munda, and J. Agee, "The influence of wind power on the small signal stability of a power system," in *Proceedings of the International Conference on Renewable Energy and Power Quality (ICREPQ'11)*, 2011.
10. V. C. Gungor, B. Lu, and G. P. Hancke, "Opportunities and challenges of wireless sensor networks in smart grid," *IEEE Transactions on Industrial Electronics*, vol. 57, no. 10, pp. 3557–3564, 2010.
11. K. C. Budka, J. G. Deshpande, and M. Thottan, *Communication Networks for Smart Grids: Making Smart Grid Real*, Springer Science & Business Media, New York, US, 2014.
12. Z. Li and Y. Li, "Understanding and control of power grids," in *Autonomous Systems: Developments and Trends*, Springer, New York, US, 2012, pp. 49–58.
13. X. Wei, Z. Yu-hui, and Z. Jie-lin, "Energy-efficient distribution in smart grid," in *IEEE International Conference on Sustainable Power Generation and Supply, SUPERGEN'09*, pp. 1–6, 2009.
14. X. Fang, S. Misra, G. Xue, and D. Yang, "Smart grid—The new and improved power grid: A survey," *IEEE Communications Surveys & Tutorials*, vol. 14, no. 4, pp. 944–980, 2012.
15. M. E. El-Hawary, "The smart grid—State-of-the-art and future trends," *Electric Power Components and Systems*, vol. 42, no. 3–4, pp. 239–250, 2014.
16. H. Y. Shwe and P. H. J. Chong, "Scalable distributed cloud data storage service for internet of things," in *International IEEE Conferences on Ubiquitous Intelligence & Computing, Advanced and Trusted Computing, Scalable Computing and Communications, Cloud and Big Data Computing, Internet of People, and Smart World Congress (UIC/ATC/ScalCom/CBDCom/IoP/SmartWorld)*, pp. 869–873, 2016.
17. B. M. Buchholz and Z. Styczynski, *Smart Grids: Fundamentals and Technologies in Electricity Networks*, Springer, New York, US, 2016.
18. "Smart grid," https://en.wikipedia.org/wiki?curid=13201685, accessed June 2018.
19. P. Murty, *Operation and Control in Power Systems*, BS Publications, Hyderabad, India, 2011.
20. "Energy regulations and electricity deregulation in Japan," https://www.japanindustrynews.com/2016/04/energy-regulations-electricity-deregulation-japan/, accessed June 2018.
21. "Demand response," https://www.energy.gov/oe/activities/technology-development/grid-modernization-and-smart-grid/demand-response, accessed June 2018.
22. N. A. Sinitsyn, S. Kundu, and S. Backhaus, "Safe protocols for generating power pulses with heterogeneous populations of thermostatically controlled loads," *Energy Conversion and Management*, vol. 67, pp. 297–308, 2013.

23. H. Daki, A. El Hannani, A. Aqqal, A. Haidine, and A. Dahbi, "Big data management in smart grid: Concepts, requirements and implementation," *Journal of Big Data*, vol. 4, no. 1, p. 13, 2017.
24. Y.-F. Huang, S. Werner, J. Huang, N. Kashyap, and V. Gupta, "State estimation in electric power grids: Meeting new challenges presented by the requirements of the future grid," *IEEE Signal Processing Magazine*, vol. 29, no. 5, pp. 33–43, 2012.
25. P. Mukhopadhyay and H. Chawla, "Approach to make smart grid a reality," in *IEEEInternational Conference on Advances in Energy Conversion Technologies* (*ICAECT*), pp. 77–82, 2014.
26. L. T. Berger and K. Iniewski, *Smart Grid: Applications, Communications, and Security*. Wiley, Hoboken, New Jersey, US, 2012.
27. B. Tomoiagă, M. Chindriş, A. Sumper, A. Sudria-Andreu, and R. Villafafila-Robles, "Pareto optimal reconfiguration of power distribution systems using a genetic algorithm based on NSGA-II," *Energies*, vol. 6, no. 3, pp. 1439–1455, 2013.
28. "Application of automated controls for voltage and reactive power management—Initial results," https://www.smartgrid.gov/files/VVO_Report_-_Final.pdf, accessed June 2018.
29. T. Adefarati, N. B. Papy, M. Thopil, and H. Tazvinga, "Non-renewable distributed generation technologies: A review," in *Handbook of Distributed Generation*, Springer, New York, US, 2017, pp. 69–105.
30. T. Adefarati, R. C. Bansal, and J. J. Justo, "Reliability and economic evaluation of a microgrid power system," *Energy Procedia*, vol. 142, pp. 43–48, 2017.
31. T. Adefarati, R. C. Bansal, and J. J. Justo, "Techno-economic analysis of a PV–wind–battery–diesel standalone power system in a remote area," *The Journal of Engineering*, vol. 2017, no. 13, pp. 740–744, 2017.
32. H. Tazvinga, M. Thopil, P. B. Numbi, and T. Adefarati, "Distributed renewable energy technologies," in *Handbook of Distributed Generation*, Springer, Cham, US, 2017, pp. 3–67.
33. C. C. Sun, A. Hahn, and C. C. Liu, "Cyber security of a power grid: State-of-the-art," *International Journal of Electrical Power & Energy Systems*, vol. 99, pp. 45–56, 2018.
34. P. P. Parikh, M. G. Kanabar, and T. S. Sidhu, "Opportunities and challenges of wireless communication technologies for smart grid applications," in *IEEE Power and Energy Society General Meeting*, pp. 1–7, 2010.
35. M. G. Kanabar, I. Voloh, and D. McGinn, "Reviewing smart grid standards for protection, control, and monitoring applications," in *IEEE PES on Innovative Smart Grid Technologies* (*ISGT*), pp. 1–8, 2012.
36. "Enhancing reliability in passive anti-islanding protection schemes for distribution systems with distributed generation," https://ir.lib.uwo.ca/cgi/viewcontent.cgi?referer=&httpsredir=1&=&article=1894&context=etd, accessed June 2018.
37. Y. Yan, Y. Qian, H. Sharif, and D. Tipper, "A survey on smart grid communication infrastructures: Motivations, requirements and challenges," *IEEE Communications Surveys & Tutorials*, vol. 15, no. 1, pp. 5–20, 2013.
38. F. Li et al., "Smart transmission grid: Vision and framework," *IEEE Transactions on Smart Grid*, vol. 1, no. 2, pp. 168–177, 2010.
39. S. A. Ali, N. Aamir, M. S. Suleman, W. N. Hasan, M. Shaikh, and M. A. Memon, "Real time implementation of non-DFT based three phase phasor measurement unit as per IEEE standard C37.118.1," in *IEEE International Conference on Environment and Electrical Engineering and Industrial and Commercial Power Systems Europe* (*EEEIC/I&CPS Europe*), pp. 1–6, 2017.
40. T. Adefarati and R. C. Bansal, "Reliability and economic assessment of a microgrid power system with the integration of renewable energy resources," *Applied Energy*, vol. 206, pp. 911–933, 2017.
41. T. Adefarati and R. C. Bansal, "Integration of renewable distributed generators into the distribution system: A review," *IET Renewable Power Generation*, vol. 10, no. 7, pp. 873–884, 2016.
42. T. Adefarati and R. C. Bansal, "Reliability assessment of distribution system with the integration of renewable distributed generation," *Applied Energy*, vol. 185, pp. 158–171, 2017.
43. T. Adefarati and R. C. Bansal, "The impacts of PV-wind-diesel-electric storage hybrid system on the reliability of a power system," *Energy Procedia*, vol. 105, pp. 616–621, 2017.
44. "Load management," https://en.wikipedia.org/wiki/Load_management, accessed June 2018.

45. L. Ardito, G. Procaccianti, G. Menga, and M. Morisio, "Smart grid technologies in Europe: An overview," *Energies*, vol. 6, no. 1, pp. 251–281, 2013.
46. S. Koozehkanani, S. Salemi, and S. Sadr, "Optimal implementation of feeder automation in medium voltage distribution networks," in *Electrical Power Distribution Networks Conference (EPDC), 2015 20th Conference on*, IEEE, 2015, pp. 16–21.
47. "Intelligent decision support system for energy management in demand response programs and residential and industrial sectors of the smart grid," https://espace.curtin.edu.au/bitstream/handle/20.500.11937/1358/240088_Sianaki%202015.pdf?isAllowed=y&sequence=2, accessed June 2018.
48. B. E. Bilgin and V. Gungor, "Performance evaluations of ZigBee in different smart grid environments," *Computer Networks*, vol. 56, no. 8, pp. 2196–2205, 2012.
49. "Secured communication scheme for SCADA in smart grid environment," http://www.sersc.org/journals/JSE/vol7_no6_2010/2.pdf, accessed June 2018.

2

Fault Analysis

Patrick T. Manditereza

CONTENTS

2.1 Introduction

The purpose of an electrical power system is to generate and supply electrical energy to consumers. The system should be designed and managed to deliver this energy to the utilisation points with both reliability and economy. The electric power system comprises many diverse items of equipment (generation, transmission/distribution, and consumer loads) that are very expensive. The complete power system hence represents a very large capital investment. No matter how well designed is the power system, faults will always occur. Faults carrying high current can cause damage to a plant if they continue for more than a few seconds. The provision of adequate protection to detect and disconnect elements of the power system in the event of fault is therefore an integral part of power system design. The protection needs to be installed to detect fault occurrence and isolate the faulty equipment so that damage to the faulty equipment is limited,

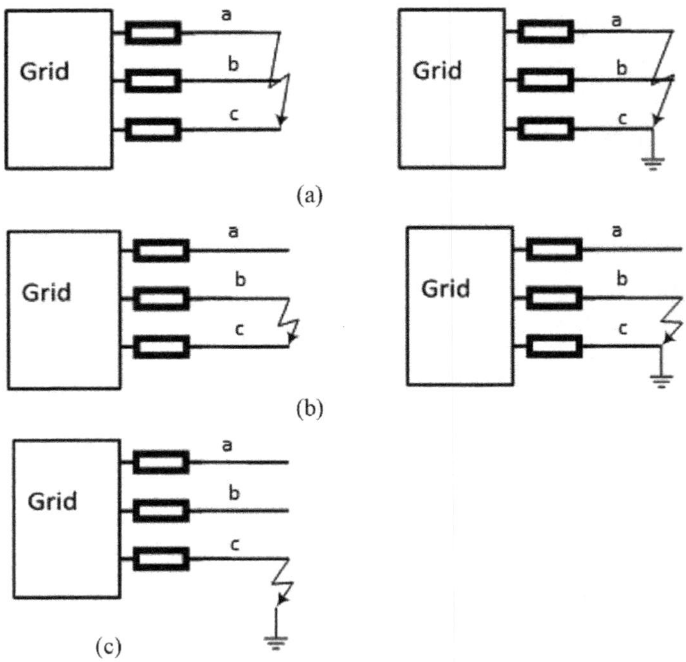

FIGURE 2.1 Types of short circuit faults: (a) three-phase fault, (b) two-phase fault, and (c) one-phase fault.

area shutdown is minimized, disruption of supplies to adjacent nonfaulty equipment is minimized, and injuries to personnel are prevented, hence ensuring a safe and reliable electricity supply system.

Faults occur as a result of the deterioration or failure of the insulation due to one or any of many causes such as lightning strike, ageing or physical damage. The insulation failure may be between two phases, or between one phase and earth, or between multiple phases and earth, creating short circuit fault types as illustrated in Figure 2.1. Statistically, the most common type of fault is the short-circuiting of a single-phase conductor to earth. Faults involving all three phases are referred to as balanced or symmetrical faults. The normal single-phase equivalent circuits are used to calculate the magnitude of the resulting fault currents. Faults not involving all three phases are referred to as unbalanced or unsymmetrical faults. Calculation of these unbalanced faults is facilitated by applying the concept of symmetrical components.

Some abnormal conditions such as overloading, undervoltage, overvoltage, and single-phasing in three phase systems also need to be identified and removed because, if sustained, they may lead to insulation failure and short circuit conditions. The main objective of fault analysis is to determine the magnitude of fault currents that flow when faults of various types occur on the power network. This facilitates specification of the type and current settings of the protection to be used and the ratings, or sizes, of other protective devices, such as fuses, circuit breakers, and current transformers. Network reduction and bus impedance matrix methods for the calculation of different types of faults are discussed in this chapter, including application of the concept of symmetrical components for the analysis of unbalanced faults.

2.2 The Per Unit System—A Review

The per unit system is used to simplify the analysis of power networks [1]. A brief review of the per unit system is given in this section. The per unit system:

- Eliminates the need to transfer circuit quantities from one side of the transformer to the other; i.e., the per unit transformer impedance is the same on either side of the transformer.
- Reduces the use of $\sqrt{3}$ in three-phase system calculations.

- Simplifies computation and makes it possible to quickly check the correctness of the computed values since the parameters and variables fall within a narrow numerical range when expressed in the per unit system.

In the per unit system the actual values of quantities are expressed as fractions of reference quantities, such as rated or full-load values. The per unit (p.u.) value is defined as:

$$\text{per unit (p.u.) value} = \frac{\text{the actual value (in any unit)}}{\text{the base or reference value in the same unit}} \tag{2.1}$$

The p.u. value may also be expressed as a percentage of the reference value. The system parameters of voltage (V), current (I), volt-amperes (VA) and impedance (Z) are all related. Hence, selection of the base values for any two of these parameters determines the base values of the remaining two. Usually base VA and base V are selected to specify the base values.

Using single phase quantities, the base current is given by:

$$I_{base} = \frac{VA_{base}}{V_{base}} \tag{2.2}$$

where $V_{base} = V_{phase}$

Base impedance is given by:

$$Z_{base} = \frac{V_{base}}{I_{base}} \tag{2.3}$$

Three phase systems are solved using the equivalent per-phase circuits: V_{ph}, R, X. However, three-phase (nameplate) data is usually given and may be used directly in the calculations. For example, base current may be expressed as:

$$I_{base} = \frac{baseVA_{1-\varnothing}}{baseV_{1-\varnothing}} = \frac{baseVA_{3-\varnothing}/\sqrt{3}}{baseV_L/\sqrt{3}} = \frac{baseVA_{3\varnothing}}{\sqrt{3} \times baseV_L} \tag{2.4}$$

where V_L is the line voltage.

Similarly it can be shown that:

$$Z_{base} = \frac{(baseV_L)^2}{baseVA_{3\varnothing}} \tag{2.5}$$

Use of the per unit system requires that a single base VA be selected for a particular calculation. However, the p.u. impedance of system components are usually given based on the VA rating of that component and may thus need to be expressed on a base other than the rated. It can be seen from (2.1), (2.3), and (2.5) that the per unit impedance is proportional to the base VA and inversely proportional to the square of the base voltage. Hence, the impedance referred to a new base is given by:

$$\text{p.u.}Z_{new} = \text{p.u.}Z_{given}\left(\frac{baseV_{given}}{baseV_{new}}\right)^2\left(\frac{baseVA_{new}}{baseVA_{given}}\right) \tag{2.6}$$

2.3 Synchronous Machine Reactances

Fault current magnitude is determined by the internal electromotive force (e.m.f.) of the power generating machines in the network and the impedance between the machines and the fault. However, if a short circuit is applied at the terminals of a synchronous generator, or at a point close to the source, the resulting fault current is illustrated by the simplified oscillogram in Figure 2.2.

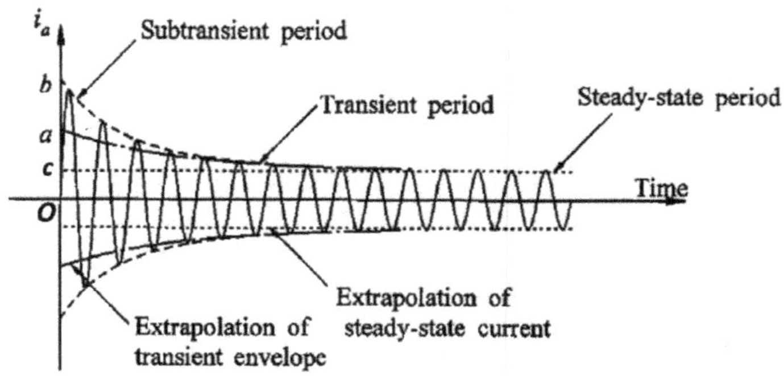

FIGURE 2.2 Fault current at short-circuited terminals of synchronous machine.

It can be seen that the fault current varies considerably with time, implying a varying machine reactance. This variation is due to the effect of armature current on the flux that generates the voltage in the machine [2]. This implies a continuously varying machine reactance. However, three envelopes can be identified in the oscillogram that are separated by two distinct discontinuities, and the existence of three (constant) reactances can be assumed to account for this phenomenon. The machine can thus be represented by one of three distinct reactances, depending on the time frame under consideration. The subtransient reactance (X'') is used to represents the very initial condition when the current is varying very rapidly. A few cycles later, the transient reactance (X') is applicable. The synchronous reactance (X_S) is used under steady state conditions.

From the oscillogram, the reactances can be calculated thus:

Subtransient reactance:

$$X'' = \frac{E}{Ob/\sqrt{2}}, \tag{2.7}$$

where E is the generated e.m.f. and Ob is initial fault current given by the intersection of the subtransient period envelop with the y-axis, and

Transient reactance:

$$X' = \frac{E}{Oa/\sqrt{2}}, \tag{2.8}$$

where Oa is peak transient period current given by the intersection of the transient period envelope with the y-axis, and

Synchronous reactance:

$$X_s = \frac{E}{Oc/\sqrt{2}}, \tag{2.9}$$

where Oc is steady state value of the fault current.

2.4 Effect of Large Motors on Fault Level

Induction and synchronous motors constitute a large proportion of industrial load. The construction of these motors is similar to that of the corresponding generators. It is therefore important to analyze the impact of these motors on the fault current under fault conditions, whether they contribute fault current or not. Looking at the construction and operational behaviour of the synchronous motor, when a short circuit occurs, the motor no longer receives electric energy from source. However, the motor has a field winding that remains energized during the fault. The motor also continues to rotate due to the inertia of the load and the rotor. Hence, the motor continues to generate internal voltage, and the synchronous motor contributes current to the fault.

The induction motor, however, does not have a field winding. Hence, when electric supply is disrupted under fault conditions, the induced field flux collapses immediately. Even though the rotor continues to rotate, no e.m.f. is generated. Hence, given a power network that also consists of synchronous and induction motors, the synchronous motor should be considered as a generator for the purpose of determining the magnitude of fault current. The induction motor can be neglected as its contribution is negligible, only resulting from the residual flux in the rotor poles, which collapses rapidly.

2.5 Network Reduction Technique for Balanced Fault Calculation

The short circuit fault current and fault level is immediately obtained if the impedance from the source of voltage to the point of fault is known [3]. Considering an initially unloaded generator, as shown in Figure 2.3, if a short circuit is applied across the three terminals as illustrated in the figure, the resulting fault current is given by (2.10):

$$I_F = \frac{E}{Z} \tag{2.10}$$

where Z is the source impedance and E is the source voltage.

If source voltage E and full load current I_{FL} are selected as the base values, then:

$$Z_{\text{p.u.}} = \frac{Z.I_{FL}}{E}$$

and

$$Z = \frac{E.Z_{\text{p.u.}}}{I_{FL}} \tag{2.11}$$

Substituting (2.11) into (2.10):

$$I_F = \frac{E.I_{FL}}{E.Z_{\text{p.u.}}} = \frac{I_{FL}}{Z_{\text{p.u.}}} \tag{2.12}$$

The three-phase short circuit (fault) level is then given by:

$$\text{Fault level} = \sqrt{3}V.I_F = \frac{\sqrt{3}V.I_{FL}}{Z_{\text{p.u.}}} = \frac{\text{Base } VA}{Z_{\text{p.u.}}} \tag{2.13}$$

The calculation of fault current ignores the direct-current component, the magnitude of which depends on the instant in the cycle that the short circuit occurs. After fault inception, the dc component decays exponentially with time constant $\tau = L/R$ $(= X/\omega R)$, as shown in Figure 2.4.

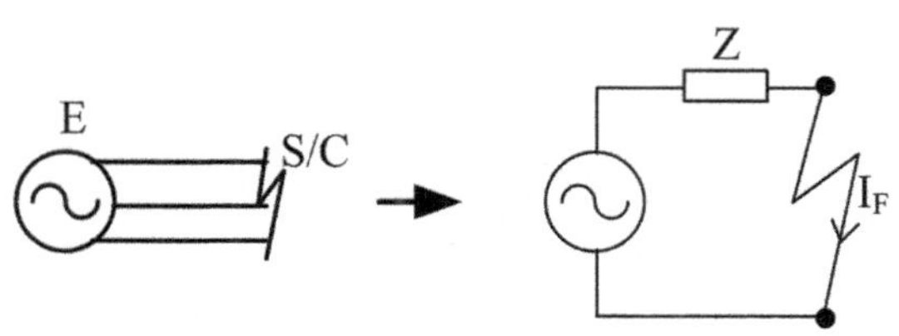

FIGURE 2.3 Short circuit fault current.

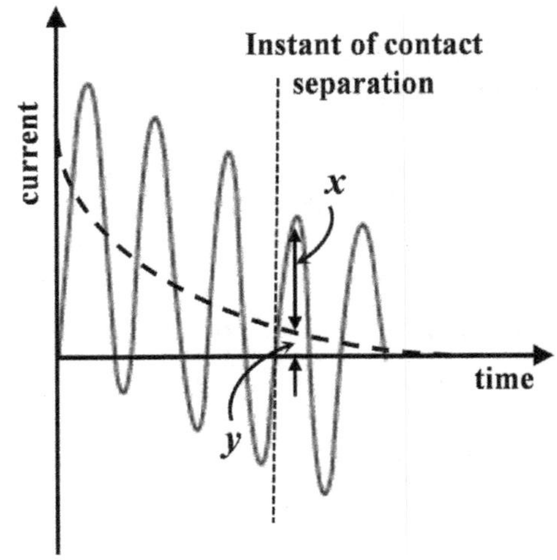

FIGURE 2.4 Oscillogram of fault current including direct-current component.

To include the DC component, the symmetrical root mean square (r.m.s.) value obtained from the fault calculation is multiplied by a certain factor that depends on the operating time of the circuit breaker [4]. The value obtained is the asymmetrical r.m.s. value of the fault current. If the circuit breaker operates in steady state, i.e., after the dc component has decayed to zero, the multiplying factor is 1. For a 2-cycle circuit breaker, for example, the multiplying factor is 1.4.

Example 2.1

A three-phase fault occurs at busbar *E* of the power system shown in Figure 2.5. The per unit impedance of each item of equipment is based on its *VA* rating. Taking a common base of 100 Mega Volt Ampere (MVA), determine the fault MVA and the current in amperes at the point of fault.

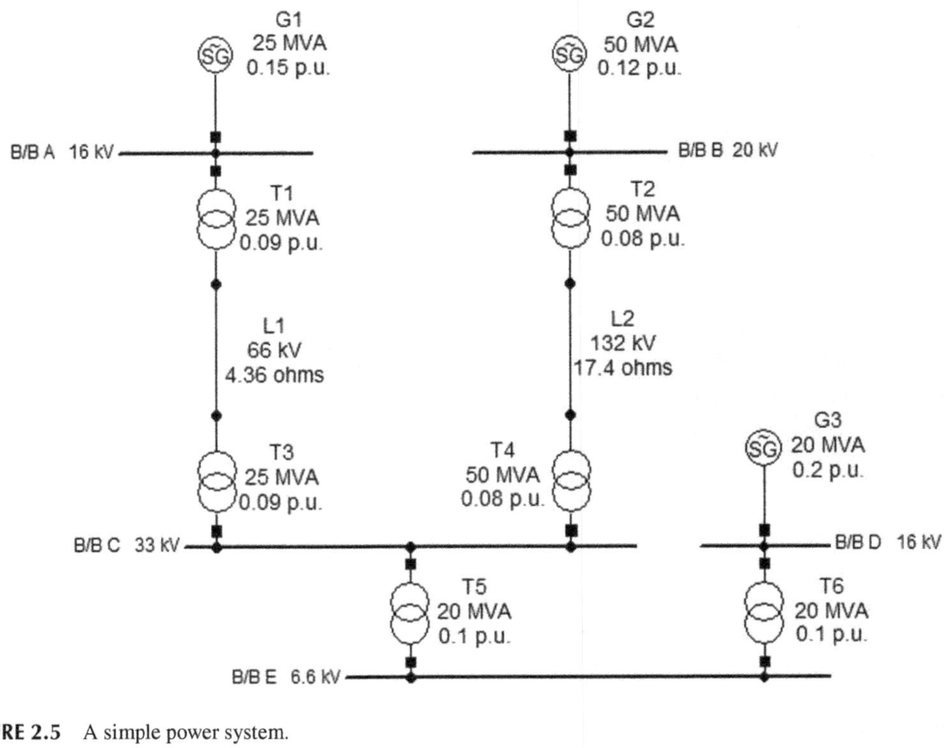

FIGURE 2.5 A simple power system.

Solution

Convert impedances to same base MVA (100 MVA):

$$X_{G1} = j0.6; X_{G2} = j0.24; X_{G3} = 1\ \text{p.u.}$$

$$X_{T1} = j0.36; X_{T2} = j0.16; X_{T3} = j0.36; X_{T4} = j0.16; X_{T5} = j0.5; X_{T6} = j0.5\ \text{p.u.}$$

$$Z_{base}(66\ \text{kV}) = (V_B)^2/S_B = 43.56\ \Omega; X_{L1} = j0.1\ \text{p.u.}$$

$$Z_{base}(132\ \text{kV}) = (V_B)^2/S_B = 174.24\ \Omega; X_{L2} = j0.1\ \text{p.u.}$$

Reactance diagram (Figure 2.6).
Reduce network to single source, single impedance:

$$X_{eq} = j0.85\ \text{p.u.}$$

$$\text{Fault level} = S_B/X_{eq} = 117.65\ \text{MVA}$$

and

$$\text{Fault current} = \frac{\text{Fault level}}{\text{sqrt}(3)*6.6\ \text{kV}} = 10.29\ \text{kA}$$

Example 2.2

Equipment ratings for the three-bus power system shown in Figure 2.7 are as follows:

Generator $G1$: 10 MVA, 3.3 kV, $x = j0.15$ p.u.
Synchronous motor: 6 MVA, 11 kV, $x = j0.20$ p.u.
Induction motor: 5 MVA, 11 kV, $x = j0.20$ p.u.
Transformer $T1$: 15 MVA, 3.3/11 kV, $x = j0.15$ p.u.
Transformer $T2$: 20 MVA, 66/11 kV, $x = j0.15$ p.u.
Line $L1$: $x = j8\ \Omega$
Line $L2$: $x = j4\ \Omega$
Line $L3$: $x = j3\ \Omega$
Line $L4$: $x = j5\ \Omega$
Line $L5$: $x = j2\ \Omega$

Using a 20 MVA base, calculate the fault current (in amperes) for a three-phase short circuit at busbar F.

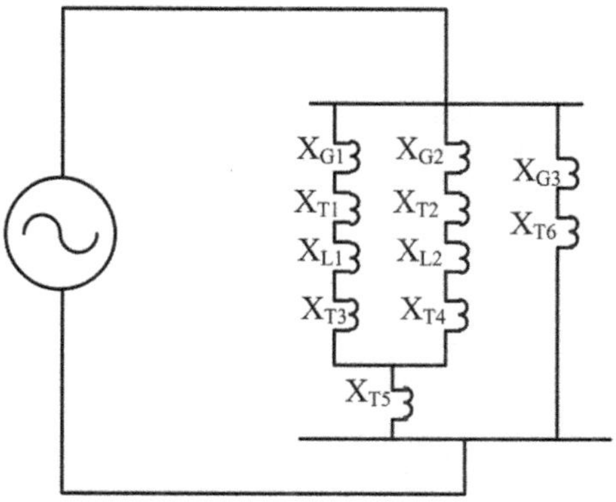

FIGURE 2.6 Reactance diagram for the simple power system.

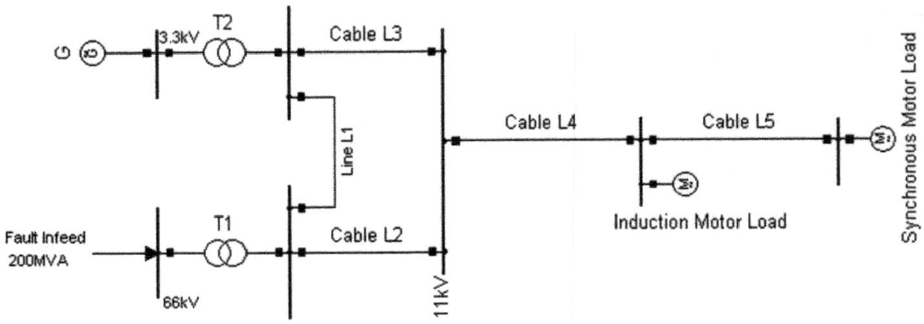

FIGURE 2.7 A simple power network with motor loads.

Solution

Convert impedances to same base MVA (20 MVA):

$$X_G = j0.30 \text{ p.u.}; X_{SM} = j0.67 \text{ p.u.}; X_{T1} = j0.20 \text{ p.u.}$$

$$Z_{base}(11\,\text{kV}) = (V_B)^2 / S_B = 6.05\,\Omega;$$

$$X_{L1} = j1.32 \text{ p.u.}; X_{L2} = j0.66 \text{ p.u.}; X_{L3} = j0.50 \text{ p.u.}$$

$$X_{L4} = j0.83 \text{ p.u.}; X_{L5} = j0.33 \text{p.u.}$$

66 kV bus source impedance, *Xs* = (Base VA)/(Fault infeed) = *j*0.10 p.u.
Reactance diagram (Figure 2.8).
Reduce network to single source, single impedance:

$$X_{eq} = j0.38 \text{ p.u.}$$

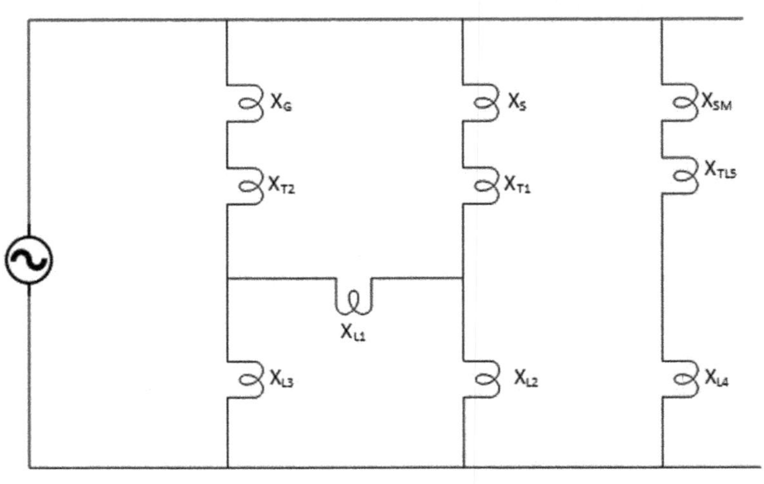

FIGURE 2.8 Reactance diagram for the power network with motor loads.

$$\text{Fault level} = S_B / X_{eq} = 52.63\ \text{MVA}$$

and

$$\text{Fault current} = \frac{\text{Fault level}}{\text{sqrt}\ (3)*11\ \text{kV}} = 2.76\ \text{kA}$$

2.6 Methods of Reducing Fault Levels

It is essential to determine the magnitude of fault currents in order to facilitate specification of the ratings of other protective devices, such as circuit breakers and current transformers. Very high fault current level would require that circuit breakers of high MVA rating be installed, which might become uneconomical. It is therefore economical to reduce the magnitude of the fault levels [5]. This section discusses several methods that can be employed to reduce the magnitude of the fault level.

2.6.1 Current Limiting Reactors

Artificial reactors can be inserted into the power network in order to increase effective reactance from the sources to the point of fault. Various connection arrangements are possible, as illustrated in Figure 2.9. Figure 2.9a shows reactors connected in series with the generators, while in Figure 2.9c series reactors are connected to the transmission lines. The reactors can be connected between the busbars as shown in Figure 2.9c or a tie-bar arrangement can be used as shown in Figure 2.9d.

Two types of reactors can be used—the iron-cored type or the air-cored type. However, the iron-cored type may saturate, leading to reduction of the reactance. The air-cored type is a better option as saturation cannot take place and the reactance is independent of current.

The air-cored type can take two forms—dry type and oil-immersed type. The dry type uses natural or forced ventilation for cooling. It must be magnetically screened from adjacent equipment; otherwise, large eddy currents can be induced, leading to increased power losses. This dry type is limited to low voltage applications. The oil-immersed reactors, as the name implies, use oil as insulating and cooling medium. This type has high thermal capacity, there is no magnetic leakage flux outside the tank, and it is insulated against flashovers. The oil-immersed type is can be applied to all voltage levels.

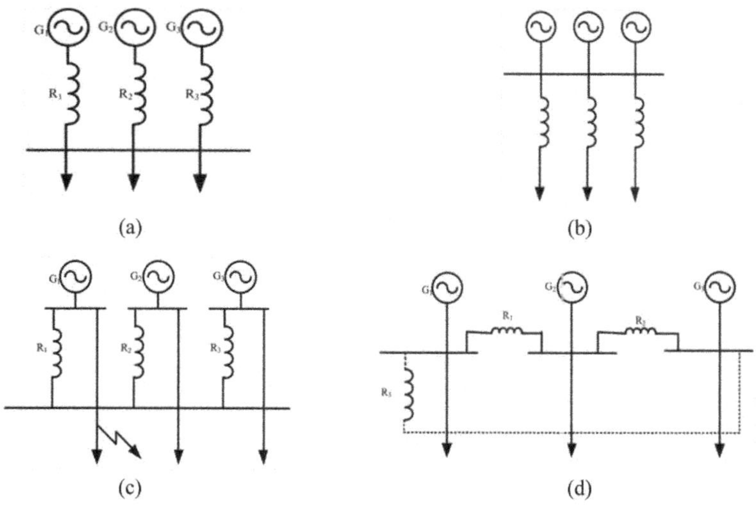

FIGURE 2.9 Various connection arrangements: (a) generator reactors, (b) feeder reactors, (c) reactors between busbars, and (d) tie-bar arrangement.

2.6.2 Reduced Number of Parallel Paths

Power system strengthening and expansion projects may involve building a second line, for example, in order to increase the power transfer in response to increased demand. This may also involve adding a second transformer to a substation. This leads to the paralleling of a line or transformer with a second component, resulting in reduction of the effective reactance from the source, which in turn results in increased fault levels that demand replacement of circuit breakers and other devices with higher rated versions, and this is obviously not economical. This problem can be mitigated by reducing the number of lines or devices that are operated in parallel. As illustrated in Figure 2.10, for example, the busbar is split into two sections that are connected through a circuit breaker called a bus-section (BS) circuit breaker. The substation busbar is sectioned by opening the BS breaker under normal operating conditions. The two transformers (and feeders) are therefore effectively not paralleled, thus increasing the effective impedance to the point of fault.

2.6.3 High Voltage Direct Current Transmission Links

An alternative to meeting increased demand by building additional high voltage AC lines is to add a high voltage direct current (HVDC) transmission line instead [6]. Due to the characteristics and operation of the DC link, no increase in short circuit levels is noted. That is, the HVDC system does not contribute current to the AC short circuit beyond its rated current. In the event of a fault on the DC line, after a momentary transient due to the discharge of the line capacitance, the current is limited by automatic grid control. Also the DC line does not draw excessive current from the AC system.

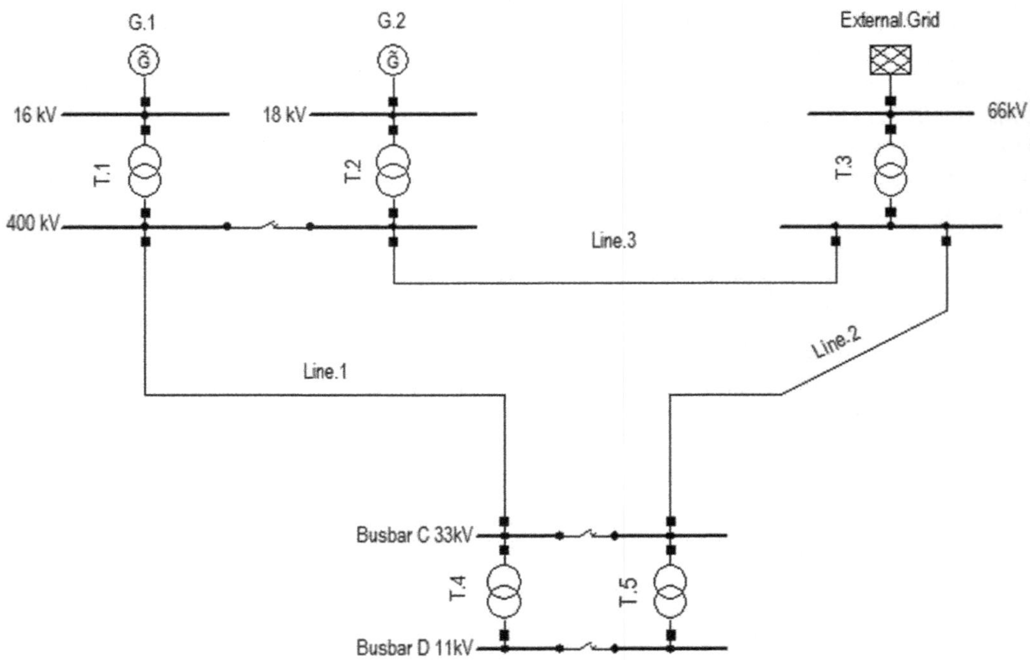

FIGURE 2.10 Section of power network showing paralleled components.

2.6.4 Short Circuit Current Limiter (SCCL) Using FACTS Devices

A flexible AC transmission system (FACTS) is a system composed of power electronic and other static equipment that provides the control of one or more AC transmission system parameters to enhance the controllability and increase the power transfer capability [7]. For example, Figure 2.11 shows the use of a FACTS device to increase the power transfer capability by reducing the effective line reactance by adding a thyristor controlled series compensator (TCSC). In this application, the FACTS device operates with near-zero impedance in steady-state conditions for increased power transfer; in case of a short circuit, it is switched to the limiting-reactor impedance within a few milliseconds by firing the thyristor to short-circuit the capacitor.

Example 2.3

The source end of a power system that includes two 11 kV generators (G1 and G2), one 33/11 kV transformer (T1) and a current limiting reactor (X_1) is shown in Figure 2.12. The MVA rating and reactance (in per unit of the respective rating) of each component is shown in the figure. Find the absolute value of the reactance of X_1 in ohms such that the short circuit level for a three-phase fault on the outgoing line at point F does not exceed 300 MVA. Assume 1.0 p.u. prefault voltage everywhere in the system. Choose 15 MVA as the common system base. The source impedance at T1 is very small and may be neglected.

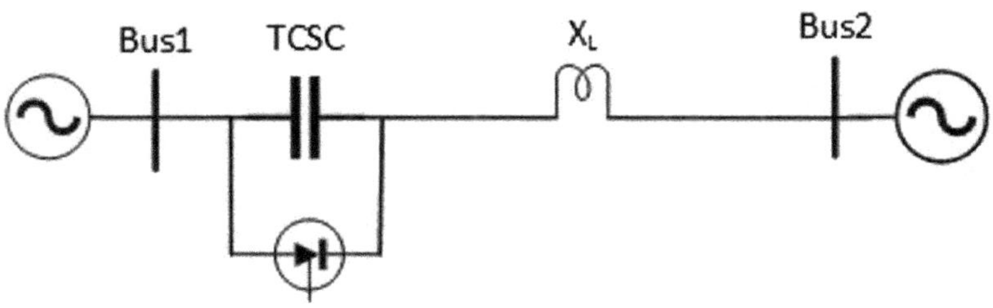

FIGURE 2.11 SCCL with FACTS devices.

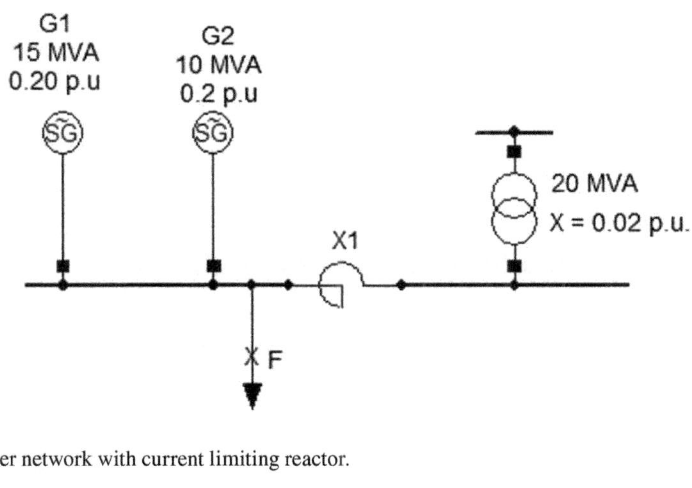

FIGURE 2.12 Power network with current limiting reactor.

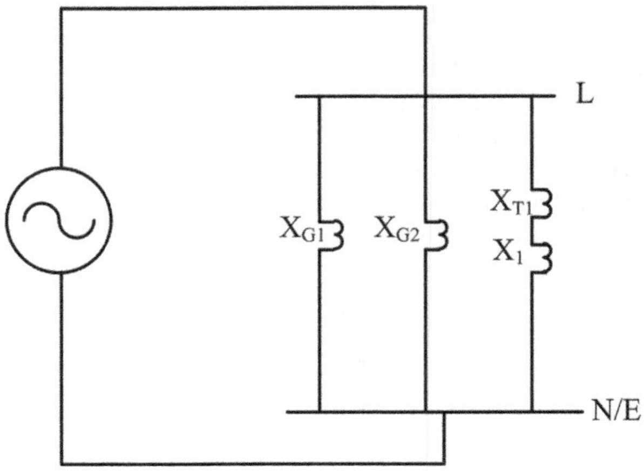

FIGURE 2.13 Reactance diagram for the power network with current limiting reactor.

Solution

Convert impedances to same base MVA (15 MVA):

$$X_{G2} = j0.3 \text{ p.u.}$$

$$X_{T1} = j0.015 \text{ p.u.}$$

Reactance diagram (Figure 2.13).
Suppose $X_1 = jx$.
Reduce the network to obtain,

$$X_{eq} = \frac{j0.12(j0.015 + jx)}{j(0.135 + x)}$$

$$S_{max} = 300 \text{ MVA}$$

Hence, $X_{eq} = (S_B/S_{max}) = 0.05$ p.u.
And hence:

$$x = 0.071 \text{ p.u.}$$

$$Z_{base}\,(11\,\text{kV}) = (V_B)^2/S_B = 8.07\,\Omega$$

Giving $X_1 = 0.570\ \Omega$.

2.7 Bus Impedance Matrix Method of Fault Calculation

Fault calculation using network reduction techniques, as discussed in Section 2.5, becomes cumbersome when the network under consideration becomes large. The fault currents can be easily computed for any N-bus network by using the bus impedance matrix method [6].

2.7.1 Nodal Equations

Consider the one-line diagram consisting of three buses shown in Figure 2.14a. The corresponding reactance diagram is shown in Figure 2.14b. The Norton equivalent circuit is obtained by redrawing the circuit with e.m.f. sources replaced with current sources and impedances replaced by admittances, as shown in Figure 2.14c.

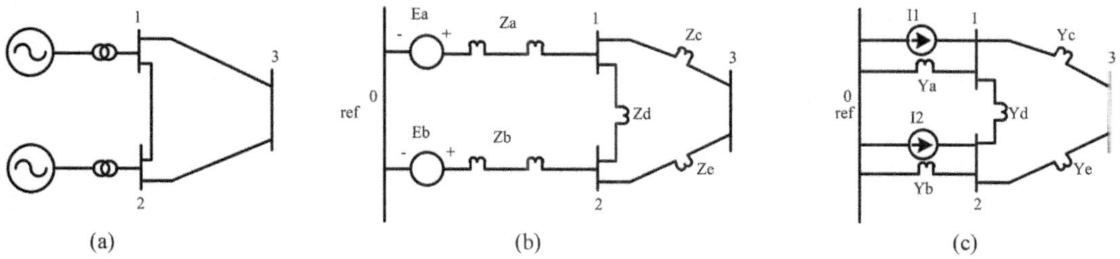

FIGURE 2.14 Simple circuit for the formulation of nodal equations. (a) one-line diagram, (b) reactance diagram, and (c) Norton equivalent circuit.

Applying Kirchhoff's current law (KCL) at node 1 (i.e., equating current into node to currents away from node):

$$I_1 = V_1Y_a + (V_1 - V_2)Y_d + (V_1 - V_3) \tag{2.14}$$

And for node 2:

$$I_2 = V_2Y_b + (V_2 - V_1)Y_d + (V_2 - V_3)Y_e \tag{2.15}$$

And for node 3:

$$0 = (V_3 - V_1)Y_c + (V_3 - V_2)Y_e \tag{2.16}$$

Rearranging (2.11) to (2.13):

$$I_1 = V_1(Y_a + Y_d + Y_c) - V_2Y_d - V_3Y_c \tag{2.17}$$

$$I_2 = -V_1Y_d + V_2(Y_b + Y_d + Y_e) - V_3Y_e \tag{2.18}$$

$$0 = -V_1Y_c - V_2Y_e + V_3(Y_c + Y_e) \tag{2.19}$$

Writing (2.17) to (2.19) in matrix form gives:

$$\begin{bmatrix} I_1 \\ I_2 \\ I_3 \end{bmatrix} = \begin{bmatrix} Y_{11} & Y_{12} & Y_{13} \\ Y_{21} & Y_{22} & Y_{23} \\ Y_{31} & Y_{32} & Y_{33} \end{bmatrix} \begin{bmatrix} V_1 \\ V_2 \\ V_3 \end{bmatrix} \tag{2.20}$$

where $Y_{11} = (Y_a + Y_d + Y_c)$; $Y_{12} = -Y_d$; $Y_{13} = -Y_c$ … , etc.

The Y matrix is designated Y_{bus} and called the bus admittance matrix. The admittances Y_{11}, Y_{22}, Y_{33} (on the diagonal) are called self-admittances. Other (off-diagonal) admittances are called mutual admittances or transfer admittances.

Note that the admittance matrix Y_{bus} may be formulated simply by inspection. By inspecting Equations 2.17 through 2.19, it can be deduced that the current into a node is equal to the sum of several products:

- The voltage at that node times the sum of all admittances that terminate at that node.
- The negative of the voltage at another node times the admittance connected directly between the other node and the node at which the equation is being formulated. This is repeated for all the nodes.

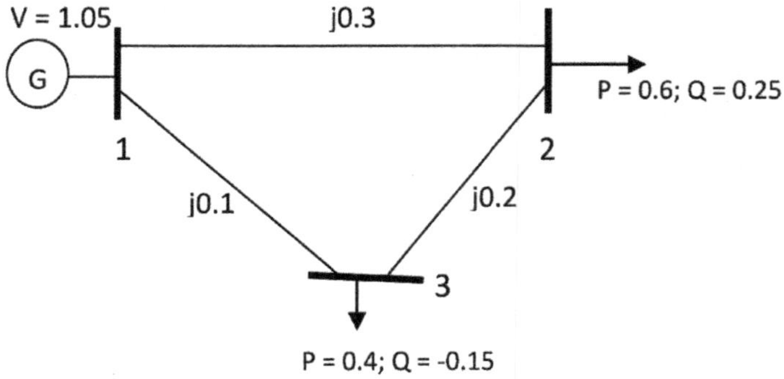

FIGURE 2.15 Three bus power network.

Example 2.4

A three bus power network is presented in Figure 2.15. Data relevant for a load flow analysis on this system is given in p.u., all on the same base. Form the Y_{bus} matrix for this system.

Solution

The line admittances are:

$$Y_{L12} = -j3.33; Y_{L13} = -j10; Y_{L23} = -j5$$

Applying (2.20), obtain:

$$Y_{bus} = j\begin{bmatrix} -10 & 3.33 & 6.67 \\ 3.33 & -7.33 & 4 \\ 6.67 & 4 & -10.67 \end{bmatrix}$$

2.7.2 The Bus Impedance Matrix in Fault Calculation

Consider the system shown in Figure 2.14b under normal conditions. Figure 2.16 shows the same system but with a voltage source V_f connected at bus 3. This voltage is equal to the voltage at that point (bus 3) under normal conditions (before the fault occurs). This generator has no effect on the currents flowing before the fault occurs; i.e., E_a, E_b, and V_f cause normal load currents to flow.

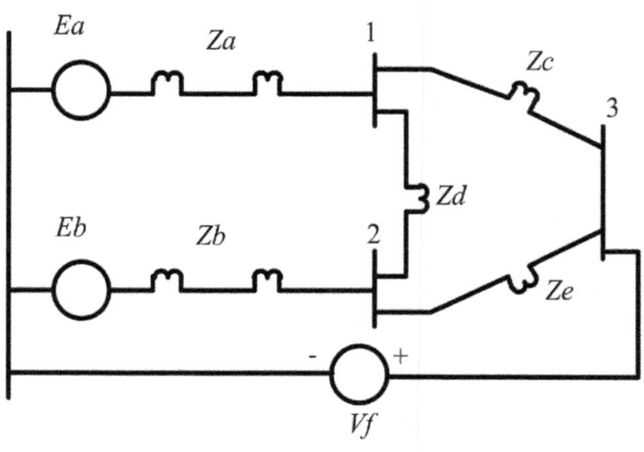

FIGURE 2.16 System in Figure 2.14b but voltage source V_f connected at bus 3.

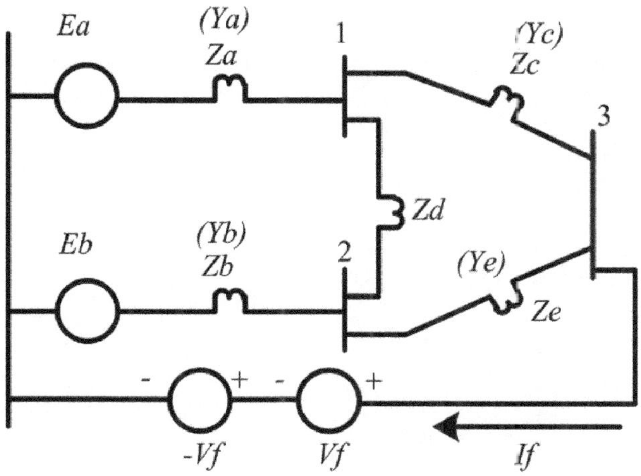

FIGURE 2.17 System in Figure 2.14 with voltage source ($-V_f$) connected in series with V_f.

However, if a second voltage source ($-V_f$) having an e.m.f. of magnitude equal to but in anti-phase with V_f is added in series to that path, as shown in Figure 2.17, a short circuit is created and a branch current I_f flows; i.e., $-V_f$ causes the fault current I_f to flow (with E_a, E_b, and V_f considered short-circuited).

Now rewriting the node equations in (2.20) with $-V_f$ the only source gives:

$$\begin{bmatrix} 0 \\ 0 \\ -I_f \end{bmatrix} = \begin{bmatrix} Y_{11} & Y_{12} & Y_{13} \\ Y_{21} & Y_{22} & Y_{23} \\ Y_{31} & Y_{32} & Y_{33} \end{bmatrix} \begin{bmatrix} \Delta V_1 \\ \Delta V_2 \\ -V_f \end{bmatrix} \tag{2.21}$$

It should be noted that the only current entering or leaving a node from a source is I_f (from node 3). The Δ in the voltages indicates voltage changes at the nodes due to the fault (or due to voltage $-V_f$).

Now, by inverting the bus admittance matrix, the bus impedance matrix can be obtained:

$$[I] = [Y][V] \tag{2.22}$$

$$[Y]^{-1}[I] = [Y]^{-1}[Y][V] = [V] \tag{2.23}$$

That is:

$$\begin{bmatrix} \Delta V_1 \\ \Delta V_2 \\ -V_f \end{bmatrix} = \begin{bmatrix} Z_{11} & Z_{12} & Z_{13} \\ Z_{21} & Z_{22} & Z_{23} \\ Z_{31} & Z_{32} & Z_{33} \end{bmatrix} \begin{bmatrix} 0 \\ 0 \\ -I_f \end{bmatrix} \tag{2.24}$$

From (2.24) it can be seen that:

$$I_f = \frac{V_f}{Z_{33}} \tag{2.25}$$

That is, the fault current at a node in the network is given by the voltage at that point before the fault occurs divided by the self-impedance at that point. The voltage drops at the buses can also be determined; thus:

$$V_1 = -I_f Z_{13} \qquad V_2 = -I_f Z_{23} \qquad V_3 = -I_f Z_{33} \tag{2.25}$$

That is, the voltage drop at a node is given by the product of the fault current and the transfer impedance between that node and the point of fault.

Example 2.5

Calculate the current feeding into a three-phase fault at bus 2 of the system shown in Figure 2.18. The system data is given below:

Generator 1: 100 MVA, $X_1 = 0.2$ p.u.
Generator 2: 100 MVA, $X_1 = 0.15$ p.u.
Transformer 1: 100 MVA, $X_1 = 0.08$ p.u.
Transformer 2: 100 MVA, $X_1 = 0.10$ p.u.
Transmission line: $X_1 = 0.15$ p.u. 100 MVA base

Solution

Reactance diagram (Figure 2.19):

Ybus1 =

$$\begin{matrix} 0-10.2400i & 0+6.6700i \\ 0+6.6700i & 0-10.6700i \end{matrix}$$

$$C^T = \begin{matrix} 0-10.6700i & 0-6.6700i \\ 0-6.6700i & 0-10.2400i \end{matrix}$$

Det (Ybus1) = − 64.7719

Zbus1 = inv (Ybus1) =

$$\begin{matrix} 0+0.1647i & 0+0.1030i \\ 0+0.1030i & 0+0.1581i \end{matrix}$$

Hence:

$$I_f = \frac{V_f}{Z_{22}} = \frac{1}{j0.1581} = -j6.33\,\text{p.u.}$$

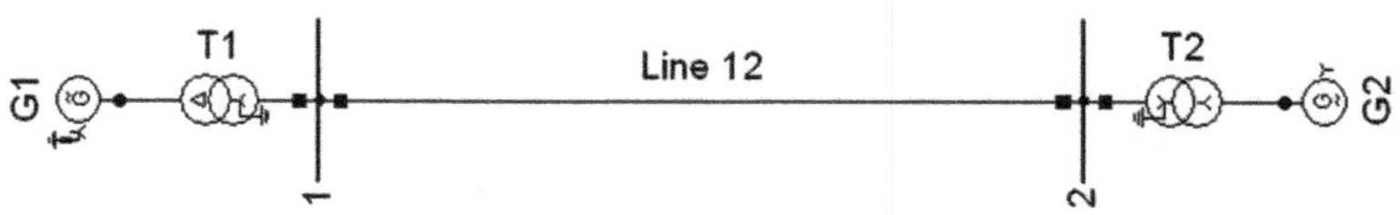

FIGURE 2.18 Two bus power network.

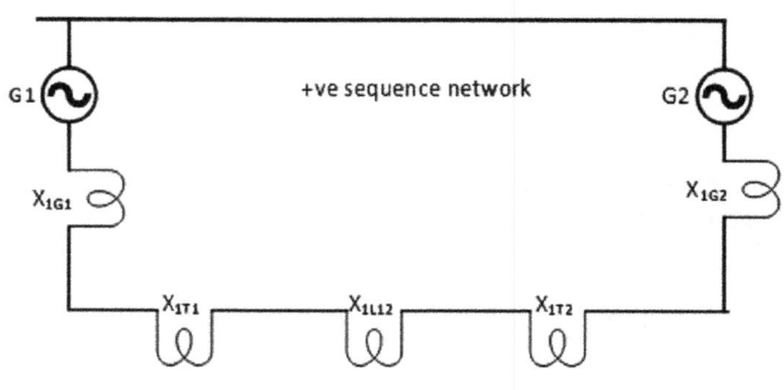

FIGURE 2.19 Reactance diagram for the two bus power network.

2.7.3 Introduction to the *Z*-Building Procedure

In practice, direct matrix inversion, as discussed in Section 2.7.2, is not feasible for typical power systems because of the large number of buses. As an alternative, a computationally attractive method for building or assembling the bus impedance matrix, called *Z building* [8], may be used. One of the Z-building procedures is introduced in this section. This procedure applies some rules that are fundamentally used to calculate the impedance matrix Z_{bus}^{new} of a modification of an originally specified network with bus impedance matrix Z_{bus}. The rules for performing three types of modifications to a specified r-bus network are considered below.

The first modification involves adding a branch with impedance Z_b from a new $(r + 1)$th bus (node) to the reference node. The new bus impedance matrix Z_{bus}^{new} is given by the following $(r+1)\times(r+1)$ matrix:

$$Z_{bus}^{new} = \begin{bmatrix} Z_{bus} & 0 \\ 0^T & Z_b \end{bmatrix} \tag{2.27}$$

The second modification involves adding a branch Z_b from a new $(r + 1)$th node to the ith node. The rule for this modification is as follows: Suppose that the ith column of Z_{bus} is Z_i and the iith element of Z_{bus} is z_{ii}, then:

$$Z_{bus}^{new} = \begin{bmatrix} Z_{bus} & Z_i \\ Z_i^T & Z_b + z_{ii} \end{bmatrix} \tag{2.28}$$

The third modification involves adding a branch Z_b between existing ith and jth nodes. The rule for this modification is as follows: Suppose that the ith and jth columns of Z_{bus} are Z_i and Z_j, and the iith, jjth and ijth elements of Z_{bus} are z_{ii}, z_{jj} and z_{ij}, respectively. Then:

$$Z_{bus}^{new} = Z_{bus} - \gamma\, bb^T \tag{2.29}$$

$$\text{where} \quad b = Z_i - Z_j \quad \text{and} \quad \gamma = (Z_b + z_{ii} + z_{jj} - 2z_{ij})^{-1}$$

Example 2.6

An existing three-bus power network (consisting of busbars 1, 2, and 3), shown in Figure 2.20, has the following bus impedance matrix:

$$Z_{bus} = j\begin{bmatrix} 0.35 & 0.39 & 0.35 \\ 0.39 & 0.54 & 0.39 \\ 0.35 & 0.39 & 0.45 \end{bmatrix}$$

A new generator, *G*3, with reactance *j*0.25 p.u. is to be connected to this network through the generator transformer, *T*3, of reactance *j*0.2 p.u., and transmission line, *L*24, of reactance 0.1 p.u. The reactances are all referred to the same base. Determine the three-phase fault current at the new busbar 4.

Solution

Step 1: Original matrix:

$$Z_{bus} = j\begin{bmatrix} 0.35 & 0.39 & 0.35 \\ 0.39 & 0.54 & 0.39 \\ 0.35 & 0.39 & 0.45 \end{bmatrix}$$

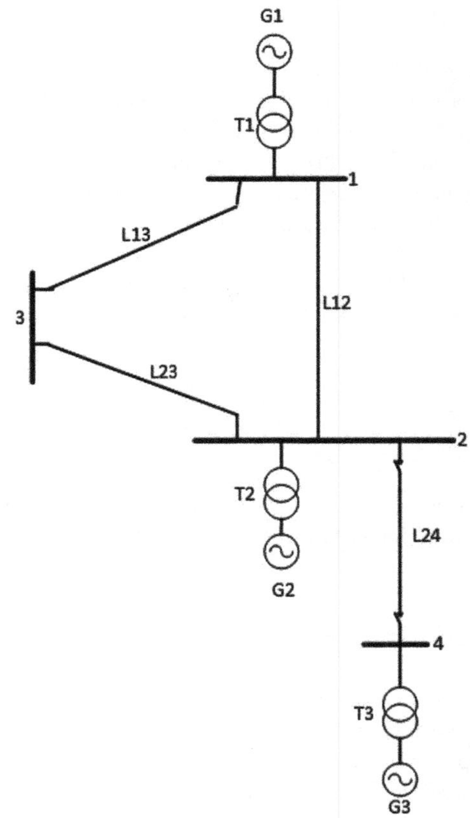

FIGURE 2.20 Four bus power network.

Step 2: Add branch from new node 4 to reference node:

Apply

$$Z_{bus}^{n} = \begin{bmatrix} Z_{bus} & 0 \\ 0^{T} & Z_{b} \end{bmatrix}$$

$$(Z_b = j0.45)$$

$$Z_{bus} = j\begin{bmatrix} 0.35 & 0.39 & 0.35 & 0 \\ 0.39 & 0.54 & 0.39 & 0 \\ 0.35 & 0.39 & 0.45 & 0 \\ 0 & 0 & 0 & 0.45 \end{bmatrix}$$

Step 3: Add branch between nodes 2 and 4:

Apply

$$Z_{bus}^{n} = Z_{bus} - \gamma\, bb^{T}$$

where $b = Z_i - Z_j$ and $\gamma = (Z_b + z_{ii} + z_{jj} - 2z_{ij})^{-1}$

$$(i = 2,\ j = 4,\ Z_b = j0.1)$$

and obtain,

Z2 =

0 + 0.3900*i*
0 + 0.5400*i*
0 + 0.3900*i*
0

Z4 =

0
0
0
0 + 0.4500*i*

b =

0 + 0.3900*i*
0 + 0.5400*i*
0 + 0.3900*i*
0 − 0.4500*i*

Now, $Z_b = 0 + 0.1000i$

$$\gamma = 0 - 0.9174i$$

$b^T = 0 + 0.3900i \quad 0 + 0.5400i \quad 0 + 0.3900i \quad 0 - 0.4500i$

$\gamma bb^T =$

0 + 0.1395*i*	0 + 0.1932*i*	0 + 0.1395*i*	0 − 0.1610*i*
0 + 0.1932*i*	0 + 0.2675*i*	0 + 0.1932*i*	0 − 0.2229*i*
0 + 0.1395*i*	0 + 0.1932*i*	0 + 0.1395*i*	0 − 0.1610*i*
0 − 0.1610*i*	0 − 0.2229*i*	0 − 0.1610*i*	0 + 0.1858*i*

$Z_{bus}^{New} =$

0 + 0.2105*i*	0 + 0.1968*i*	0 + 0.2105*i*	0 + 0.1610*i*
0 + 0.1968*i*	0 + 0.2725*i*	0 + 0.1968*i*	0 + 0.2229*i*
0 + 0.2105*i*	0 + 0.1968*i*	0 + 0.3105*i*	0 + 0.1610*i*
0 + 0.1610*i*	0 + 0.2229*i*	0 + 0.1610*i*	0 + 0.2642*i*

Fault current at Bus 4 $= V_f / Z_{44} = 1/(j0.2642) = -j3.78$ p.u.

2.7.4 The *Z*-Building Procedure

The three rules governing network modifications can be used to build the bus impedance matrix, Z_{bus}, of a network, starting with a simple network and then adding elements until the network is completed. This involves the following steps:

Step 1: Number the nodes of the given network, starting with those nodes at the ends of branches connected to the reference bus.

Step 2: Start with a network composed only of all those branches connected to the reference node. This initial $Z_{bus(0)}$ is diagonal with the impedance values of the branches on the diagonal, consistent with the repeated use of (2.27).

Step 3: Add a new node to the *i*th node of the existing network using (2.28). Continue until all the nodes of the complete network are attached.

Step 4: Add a branch between *i*th and *j*th nodes using (2.29). Continue until all the remaining branches are connected.

Example 2.7

Calculate the current (in per unit) feeding into a three-phase fault at the busbar labeled "East" of the three-bus system shown in Figure 2.21. The system data is given in Table 2.1.

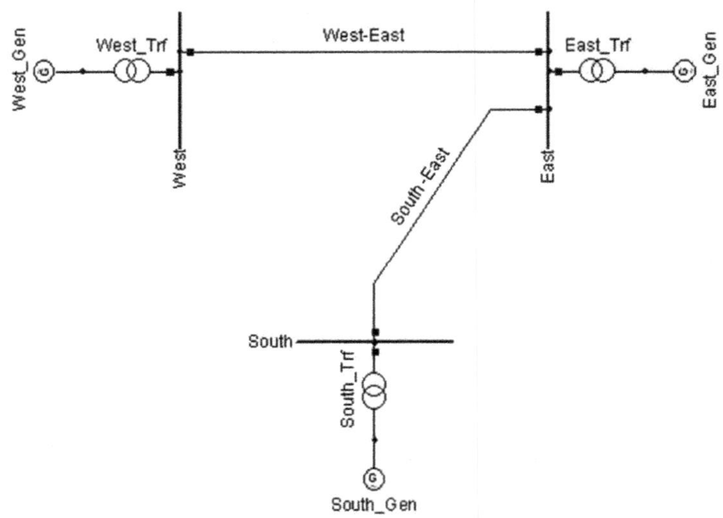

FIGURE 2.21 A simple three bus power network.

TABLE 2.1
Data for the Simple Three Bus Power Net work

Item	Impedances
West_Gen	$X = j0.15$ p.u.
East_Gen	$X = j0.20$ p.u.
South_Gen	$X = j0.20$ p.u.
West_Trf	$X = j0.30$ p.u.
East_Trf	$X = j0.20$ p.u.
South_Trf	$X = j0.25$ p.u.
West-East	$X = j0.10$ p.u.
South-East	$X = j0.15$ p.u.

Solution

Step 1: Number the nodes (Figure 2.22).

Step 2: Write down the bus impedance matrix for the network consisting only of those branches connected directly to the reference bus:

$$Z_{bus(0)} = j\begin{bmatrix} 0.45 & 0 & 0 \\ 0 & 0.40 & 0 \\ 0 & 0 & 0.45 \end{bmatrix}$$

Step 3: Add branch between nodes 1 and 2:

Apply

$$Z_{bus}^{n} = Z_{bus} - \gamma\, bb^{T}$$

$$\text{where} \quad b = Z_i - Z_j \quad \text{and} \quad \gamma = (Z_b + z_{ii} + z_{jj} - 2z_{ij})^{-1}$$

$$(i = 1,\ j = 2,\ Z_b = j0.10)$$

and obtain

$Z1 =$

$0 + 0.4500i$

0

0

$Z2 =$

0

$0 + 0.4000i$

0

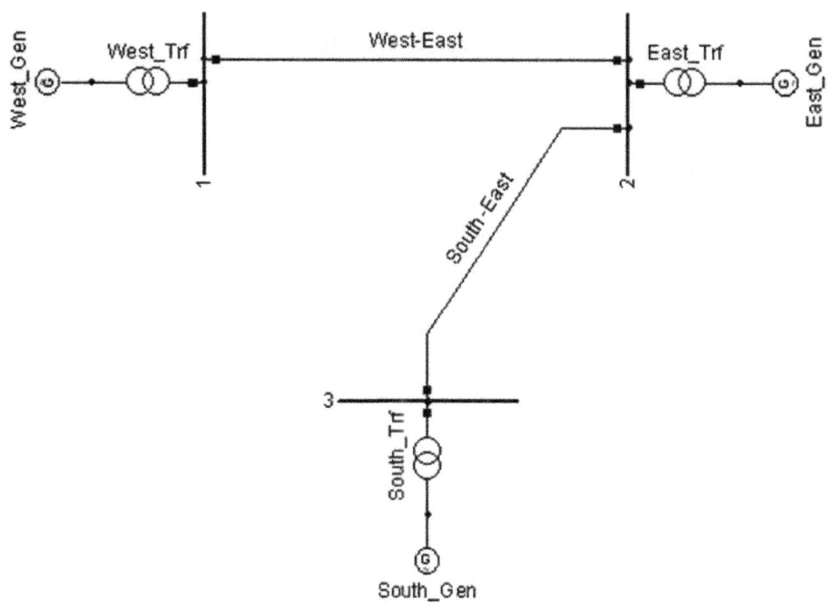

FIGURE 2.22 Numbering of power network busbars.

$$b = \begin{matrix} 0+0.4500i \\ 0-0.4000i \\ 0 \end{matrix}$$

$$Z_b = 0+0.1000i$$

$$\gamma = 0-1.0526i$$

$$b^T = 0+0.4500i \quad 0-0.4000i \quad 0$$

$$\gamma bb^T = \begin{matrix} 0+0.2132i & 0-0.1895i & 0 \\ 0-0.1895i & 0+0.1684i & 0 \\ 0 & 0 & 0 \end{matrix}$$

$$Z_{bus}^{New} = \begin{matrix} 0+0.2368i & 0+0.1895i & 0 \\ 0+0.1895i & 0+0.2316i & 0 \\ 0 & 0 & 0+0.4500i \end{matrix}$$

Step 4: Add branch between nodes 2 and 3:

$$(i = 2,\ j = 3,\ Z_b = j0.15)$$

and obtain

$$Z2 = \begin{matrix} 0+0.1895i \\ 0+0.2316i \\ 0 \end{matrix}$$

$$Z3 = \begin{matrix} 0 \\ 0 \\ 0+0.4500i \end{matrix}$$

$$b = \begin{matrix} 0+0.1895i \\ 0+0.2316i \\ 0-0.4500i \end{matrix}$$

$$Z_b = 0+0.1500i$$

$$\gamma = 0-1.2025i$$

$$b^T = 0+0.1895i \quad 0+0.2316i \quad 0-0.4500i$$

$$\gamma bb^T =$$

$$\begin{matrix} 0+0.0432i & 0+0.0528i & 0-0.1025i \\ 0+0.0528i & 0+0.0645i & 0-0.1253i \\ 0-0.1025i & 0-0.1253i & 0+0.2435i \end{matrix}$$

$$Z_{bus}^{New} =$$

$$\begin{matrix} 0+0.1936i & 0+0.1367i & 0+0.1025i \\ 0+0.1367i & 0+0.1671i & 0+0.1253i \\ 0+0.1025i & 0+0.1253i & 0+0.2065i \end{matrix}$$

$$I_f = \frac{E}{Z_{22}} = -j5.98 \text{ p.u.}$$

2.8 Symmetrical Components

The concept of symmetrical components is used to facilitate the analysis of unbalanced faults [9]. It is a technique that resolves a system of three unbalanced phasors (of a three-phase system) into three balanced systems of phasors. The advantages of this transformation is that the unbalanced three-phase network can be resolved into three decoupled networks, called sequence networks. This powerful mathematical tool greatly simplifies the analysis of unbalanced faults. The concept of symmetrical components is illustrated in Figure 2.23.

An unbalanced fault results in unbalanced phasors appearing on the three phases, as illustrated in Figure 2.23, which shows the phase voltages (V_R, V_Y, and V_B). It can be seen that the three voltages differ in magnitude, and the phase angle has also shifted from 120°. Applying the concept of symmetrical components resolves the three unbalanced phasors into three systems of balanced phasors known as the positive phase sequence system, negative phase sequence system and zero phase sequence system, respectively. Subscript 1 is normally used to indicate a positive phase sequence component, while

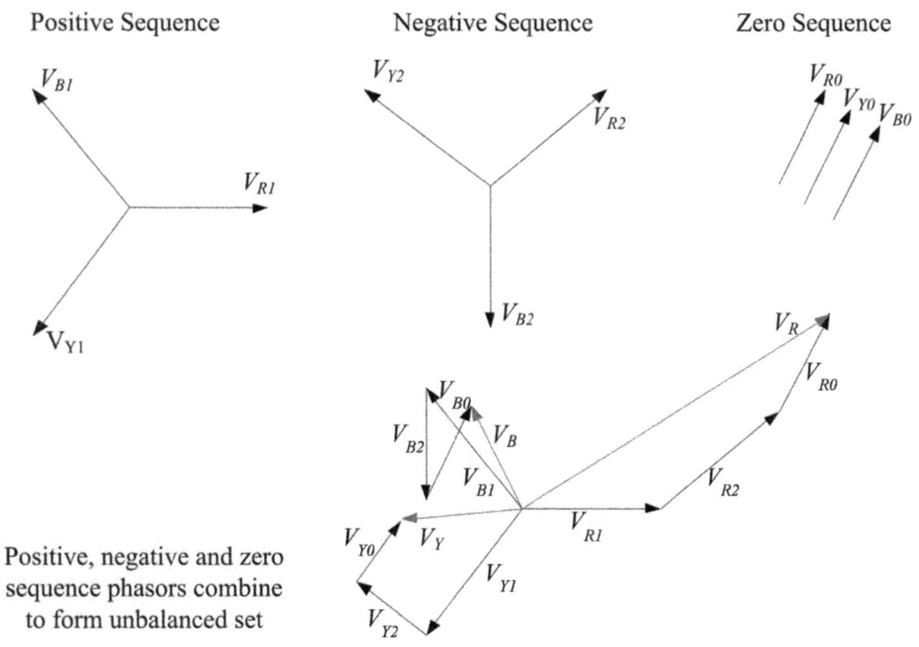

FIGURE 2.23 The concept of symmetrical components.

subscript 2 indicates a negative phase and 0 (zero) indicates a zero phase sequence component. Adding the phase sequence components such as V_{R1}, V_{R2}, and V_{R0}, for example, gives the original unbalanced phasor V_R, as illustrated in Figure 2.23.

Mathematically, the original phasors may be expressed in terms of their components as:

$$V_R = V_{R1} + V_{R2} + V_{RO} \tag{2.30}$$

$$V_Y = V_{Y1} + V_{Y2} + V_{YO} \tag{2.31}$$

$$V_B = V_{B1} + V_{B2} + V_{BO} \tag{2.32}$$

Looking at the positive phase sequence system in Figure 2.14, it can be seen that, if the R-phase sequence component V_{R1} is known, then the other two positive sequence components V_{Y1} and V_{B1} can be found simply by shifting V_{R1} through 120° or 240°, respectively. Mathematically, this can be done by intruding the "a" operator. If a phasor is multiplied by the "a" operator, the phasor is shifted through 120°. Multiplying by "a^2" rotates the phasor through 240°. That is:

$$a := 1\angle 120^\circ = 1.e^{j2\pi/3} = -0.5 + j0.866$$

in the three notations: polar, complex and rectangular. Also:

$$a^2 := 1\angle 240^\circ = -0.5 - j0.866$$

$$a^3 := 1\angle 360^\circ = 1$$

When expressing symmetrical components, it is therefore only necessary to list the R-phase components. Thus:

$$V_R = V_{R1} + V_{R2} + V_{RO} \tag{2.33}$$

$$V_Y = a^2 V_{R1} + aV_{R2} + V_{RO} \tag{2.34}$$

$$V_B = aV_{R1} + a^2 V_{R2} + V_{RO} \tag{2.35}$$

or in matrix form:

$$\begin{bmatrix} V_R \\ V_Y \\ V_B \end{bmatrix} = \begin{bmatrix} 1 & 1 & 1 \\ 1 & a^2 & a \\ 1 & a & a^2 \end{bmatrix} \begin{bmatrix} V_{RO} \\ V_{R1} \\ V_{R2} \end{bmatrix} \tag{2.36}$$

The same concept applies to currents, thus:

$$\begin{bmatrix} I_R \\ I_Y \\ I_B \end{bmatrix} = \begin{bmatrix} 1 & 1 & 1 \\ 1 & a^2 & a \\ 1 & a & a^2 \end{bmatrix} \begin{bmatrix} I_{RO} \\ I_{R1} \\ I_{R2} \end{bmatrix} \tag{2.37}$$

Matrix inversion can be applied to find the symmetrical components from the unbalanced phasors:

$$\begin{bmatrix} I_{RO} \\ I_{R1} \\ I_{R2} \end{bmatrix} = \frac{1}{3} \begin{bmatrix} 1 & 1 & 1 \\ 1 & a & a^2 \\ 1 & a^2 & a \end{bmatrix} \begin{bmatrix} I_R \\ I_Y \\ I_B \end{bmatrix} \tag{2.38}$$

Example 2.8

A three-phase, unbalanced, delta-connected load draws 100 A of line current from a balanced three-phase supply. Using the method of symmetrical components, calculate the magnitude and phase angle of the phase sequence components of the currents in phases *Y* and *B* if the current in phase *R* drops down to zero due to an open-circuit fault. Assume Red Yellow Blue (RYB) phase sequence. Given:

$$\begin{bmatrix} I_{R0} \\ I_{R1} \\ I_{R2} \end{bmatrix} = \frac{1}{3}\begin{bmatrix} 1 & 1 & 1 \\ 1 & a & a^2 \\ 1 & a^2 & a \end{bmatrix}\begin{bmatrix} I_R \\ I_Y \\ I_B \end{bmatrix}$$

Solution

Due to the open circuit on R-phase:

$$I_R = 0, I_Y = -I_C = 100 \text{ A}$$

$$I_{R0} = I_{YO} = I_{BO} = 0 \text{ A}$$

$$I_{R1} = 57.74\angle 90° \text{ A}$$

$$I_{Y1} = 57.74\angle -30° \text{ A}$$

$$I_{B1} = 57.74\angle 210° \text{ A}$$

$$I_{R2} = 57.74\angle -90° \text{ A}$$

$$I_{Y2} = 57.74\angle -210° \text{ A}$$

$$I_{B2} = 57.74\angle 30° \text{ A}$$

2.8.1 Use of Symmetrical Components in Unbalanced Fault Analysis

It can be shown that the symmetrical components of unbalanced currents produce voltage drops of like sequence only. The positive sequence currents produce positive sequence voltage drops; i.e., $V_{R1} = I_{R1}Z$. Similarly, the negative sequence currents produce negative sequence voltage drops ($V_{R2} = I_{R2}Z$) and zero sequence currents produce zero sequence voltage drops ($V_{R0} = I_{R0}Z$).

Since phase sequence currents produce voltage drops of like sequence only, circuit impedances of like sequence may thus be defined. That is, positive sequence impedance may be defined as the impedance of a network to positive sequence currents to produce positive sequence voltage drop; i.e., $V_{R1} = I_{R1}Z_1$. Similarly, negative sequence impedance is the impedance to negative sequence currents (i.e., $V_{R2} = I_{R2}Z_2$) as well as zero sequence impedance (i.e., $V_{R0} = I_{R0}Z_0$).

Thus, it may be considered that positive sequence currents flow in an independent network composed of positive sequence impedances only, called the positive sequence network. The negative sequence and zero sequence networks may be similarly defined. The concept of symmetrical components requires calculating the symmetrical components of the current in each line of the network and then combining them to obtain the actual values. These symmetrical currents flow in a network consisting of impedances of like sequence only. These phase sequence networks are derived from the real network in which the unbalanced fault currents are flowing. The sequence networks carrying the sequence currents I_{R0}, I_{R1}, I_{R2} (for R-phase) are interconnected in some manner to represent the various unbalanced fault conditions.

The objective of fault analysis using the concept of phase sequence components is therefore to decompose the real system into three sequence networks and then couple the networks only at the point of the unbalance (i.e., the fault).

The positive sequence network is the same as the real balanced equivalent network used for three-phase symmetrical fault analysis. The negative sequence network is almost the same as the positive sequence network except that the sources are shorted out. That is, there are no generated negative sequence voltages from the generators.

2.8.2 Representation of Plant in Phase Sequence Networks

This section discusses the representation or modelling of the various power system components in each of the sequence networks.

1. *Synchronous machine*: Real generators produce positive sequence voltages only; therefore only the positive sequence network has a voltage source.
   a. The positive sequence network model of the synchronous machine is the same as the normal equivalent circuit model, as shown in Figure 2.24a.
   b. The negative sequence network model is almost the same as the positive sequence network model except that the voltage source is shorted out since no negative sequence voltages are generated, as shown in Figure 2.24b.
   c. The zero sequence network model of the synchronous machine is derived with the aid of Figure 2.25a, which shows the three windings of the machine with the star point grounded through impedance Z_N. Referring to (2.38), note that:

$$I_{R0} = \frac{1}{3}\left(I_R + I_Y + I_B\right) \tag{2.39}$$

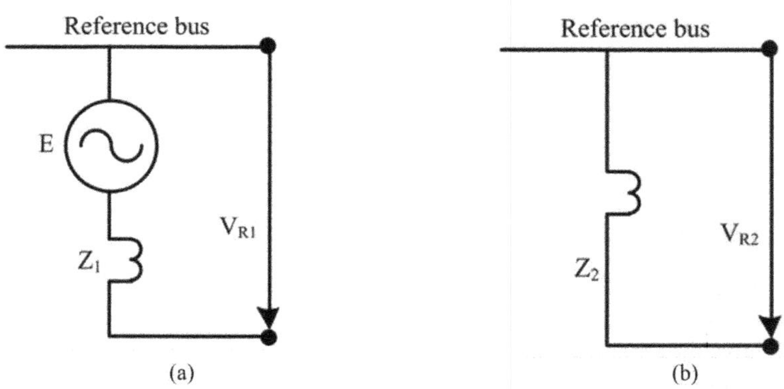

FIGURE 2.24 (a) Positive sequence network model of synchronous machine, (b) negative sequence network model of synchronous machine.

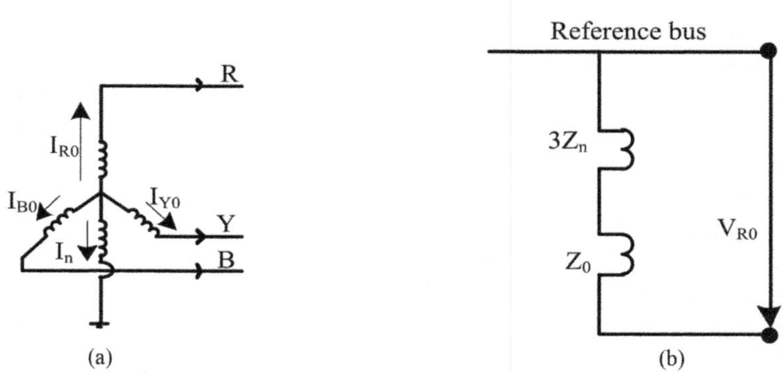

FIGURE 2.25 Zero sequence network model of synchronous machine.

For a three-wire (or balanced) system, $(I_R + I_Y + I_B) = 0$. Hence, no zero sequence currents exist. However, for a four-wire system, $I_{R0} = I_N/3$ or $I_N = 3\,I_{R0}$. That is, the neutral current is three times the zero sequence (phase) current. Therefore, the current flowing through the earthing impedance Z_N is $I_N = 3\,I_{R0}$. Hence, the voltage drop between the terminal of the R-phase winding and earth in Figure 2.25a is given by:

$$\text{Volt drop} = I_N Z_N + I_{R0} Z_0 = 3I_{R0} Z_N + I_{R0} Z_0 = I_{R0}(Z_0 + 3Z_N) \tag{2.40}$$

Z_0 is the zero sequence impedance of the winding. Equation 2.40 is consistent with the circuit in Figure 2.25b, which represents the zero sequence network model of the synchronous machine.

2. *Lines and cables*: The positive and negative sequence impedances of lines and cables are the normal balanced values. The zero sequence impedance depends on the nature of the return path, which is influenced by the line construction type and existence or absence of return earth wire.
3. *Transformers*: The positive and negative sequence impedances are the normal balanced values. The zero sequence impedance depends on the nature of the connection of the windings and how the transformer is grounded. Figure 2.26 shows the zero sequence representations of transformers for various winding arrangements. These representations are based on the principle that zero sequence currents in the windings on one side of transformer must produce the corresponding ampere turns in the other winding. So, if current cannot flow in one winding, then it cannot flow in the other.

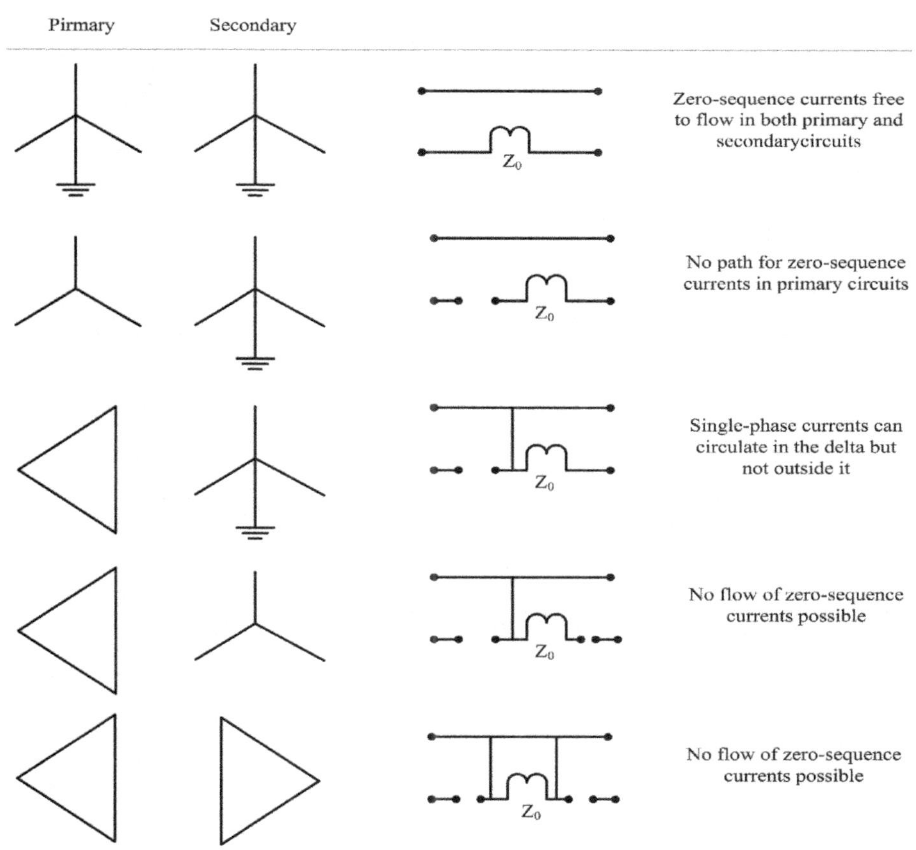

FIGURE 2.26 Zero sequence network models of transformers of various winding configurations.

The inclusion of impedances in the star-point-to-earth connection in a transformer modifies the sequence diagrams. The earthing impedance Z_N in the earth path is represented by impedance $3Z_N$ in the zero-sequence network (connected in the shunt path to ground).

Example 2.9

Draw the sequence network diagrams of the system shown in the single-line diagram of Figure 2.27.

Solution (Figure 2.28).

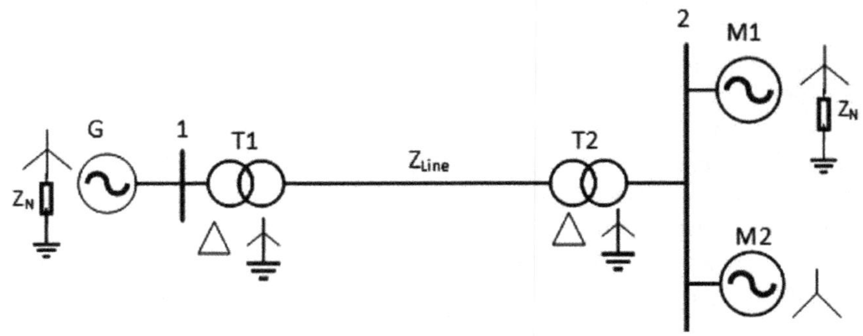

FIGURE 2.27 Single line diagram of two bus power network.

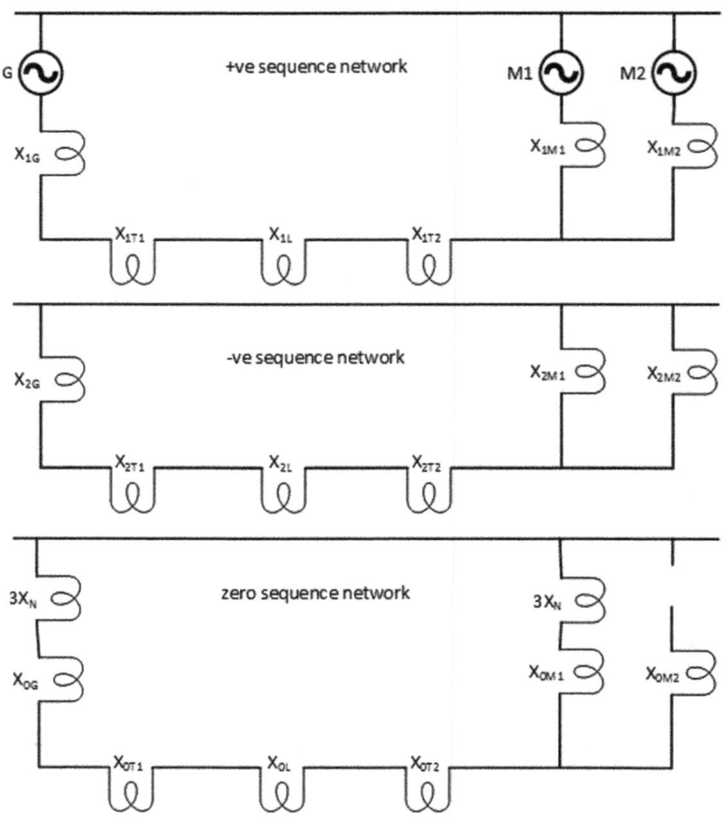

FIGURE 2.28 Phase sequence networks for Example 2.9.

2.9 Unsymmetrical Faults

This section discusses the application of the concept of symmetrical components to the solution of unbalanced faults. Most of the faults that occur on the power system are unbalanced faults, which cause unbalanced currents to flow in the system. These faults may be short circuits, faults to ground through some impedance, or open conductors. However, open circuit faults are not discussed further.

2.9.1 Single-Phase to Earth Fault

Assuming a single-phase to earth fault on the R-phase, as illustrated in Figure 2.29:
$V_R = 0$ at point of fault and $I_Y = I_B = 0$

From (2.38), and if I_1, I_2, and I_0 are the phase sequence components of I_R:

$$I_0 = \frac{1}{3}\left(I_R + I_Y + I_B\right), \tag{2.41}$$

$$I_1 = \frac{1}{3}\left(I_R + aI_Y + a^2 I_B\right), \tag{2.42}$$

$$I_2 = \frac{1}{3}\left(I_R + a^2 I_Y + aI_B\right) \tag{2.43}$$

Hence:

$$I_0 = \frac{I_R}{3} = I_1 = I_2 \tag{2.44}$$

Also:

$$V_R = E - I_1 Z_1 - I_2 Z_2 - I_0 Z_0 = 0$$

Eliminating I_0 and I_2:

$$E - I_1(Z_1 + Z_2 + Z_0) = 0$$

and

$$I_1 = \frac{E}{(Z_1 + Z_2 + Z_0)} \tag{2.45}$$

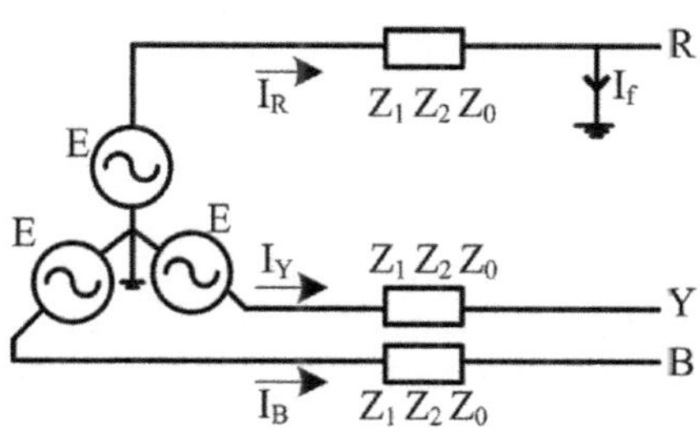

FIGURE 2.29 Single-phase to earth fault on R-phase.

Equation 2.45 is consistent with the equivalent circuit shown in Figure 2.30.

From (2.44) and (2.45), the fault current is given by:

$$I_F = I_R = 3I_1 = \frac{3E}{(Z_1 + Z_2 + Z_0)} \tag{2.46}$$

Example 2.10

Consider the one-line diagram of the simple power system shown in Figure 2.31. System data in per unit (p.u.) on appropriate MVA base are given in Table 2.2. The neutral of each generator is grounded through a current limiting reactor of 0.03 p.u. on 100 MVA base. All transformer neutrals are solidly grounded. The generators are operating at no load with a voltage of 1.05 p.u. Calculate the per unit value of the current flowing in the fault for a 1-phase to earth fault at busbar 1.

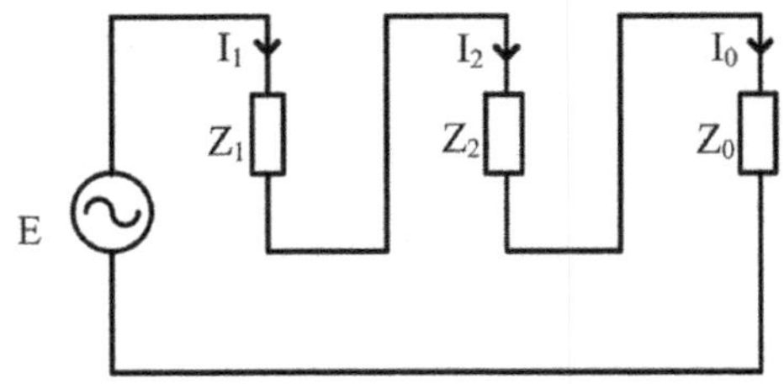

FIGURE 2.30 Equivalent circuit for a single-phase to earth fault.

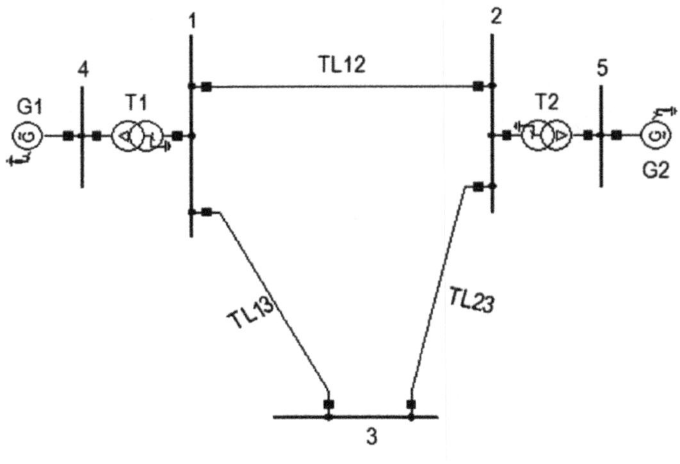

FIGURE 2.31 One-line diagram of a simple power system.

TABLE 2.2

Data for the Power Network

Item	Specification	Sequence Impedance
G1	400 MVA, 16 kV, *Y*-connected	$x_1 = x_2 = j0.15$, $x_0 = j0.05$ p.u.
G2	400 MVA, 16 kV, *Y*-connected	$x_1 = x_2 = j0.15$, $x_0 = 0.05$ p.u.
T1	400 MVA, 16/400 kV, Δ/*Y*	$x_1 = x_2 = x_0 = j0.08$ p.u.
T2	400 MVA, 16/400 kV, Δ/*Y*	$x_1 = x_2 = x_0 = j0.08$ p.u.
TL12	250 MVA, 400 kV	$x_1 = x_2 = j0.05$, $x_0 = j0.15$ p.u.
TL13/TL23	250 MVA, 400 kV	$x_1 = x_2 = j0.025$, $x_0 = j0.075$ p.u.

Solution

Convert quantities to 100 MVA base:

For $G1/G2: X_1 = X_2 = j0.038$ p.u.; $X_0 = j0.013$ p.u.
For $T1/T2: X_1 = X_2 = X_0 = j0.02$ p.u.
For $TL12: X_1 = X_2 = j0.02$ p.u.; $X_0 = j0.06$ p.u.
For $TL13/TL23: X_1 = X_2 = j0.01$ p.u.; $X_0 = j0.03$ p.u.

Phase sequence networks (Figure 2.32):
Reduce the networks to obtain:

$$Z_1 = Z_2 = j0.031 \text{ p.u.}$$

$$Z_0 = j0.014 \text{ p.u.}$$

Hence:

$$I_F = \frac{3.E}{Z_1 + Z_2 + Z_0} = -j41.44 \text{ p.u.}$$

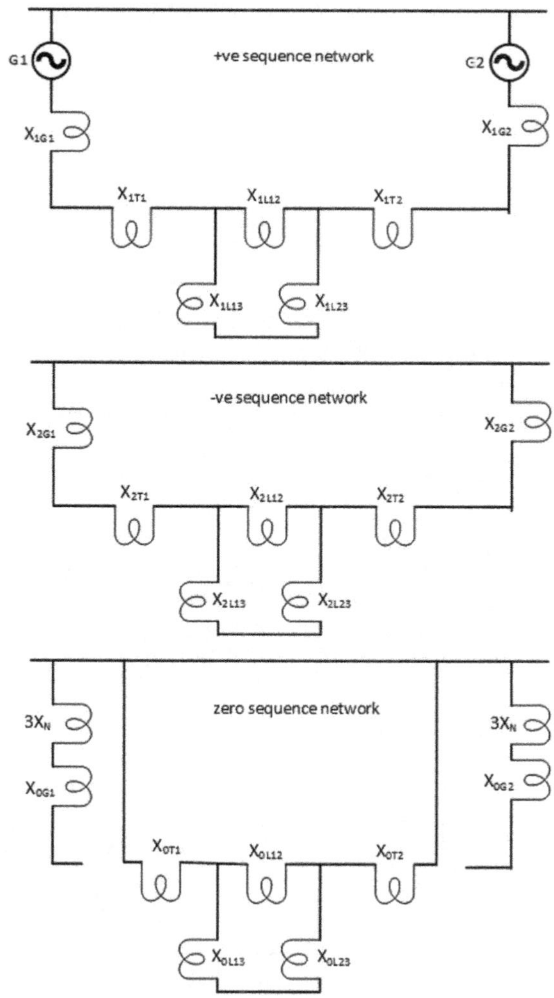

FIGURE 2.32 Phase sequence networks for Example 2.10.

2.9.2 Phase-to-Phase Fault

Assuming a phase-to-phase fault between the Y and B phases, as illustrated in Figure 2.33:

$V_Y = V_B$ at point of fault; $I_Y = -I_B$ and $I_R = 0$

From (2.38), and if I_1, I_2, and I_0 are the phase sequence components of I_R:

$$I_0 = \frac{1}{3}(I_R + I_Y + I_B) = 0, \qquad (2.47)$$

$$I_1 = \frac{1}{3}\left(I_R + aI_Y + a^2 I_B\right) = \frac{1}{3} I_Y \left(a - a^2\right), \qquad (2.48)$$

$$I_2 = \frac{1}{3}\left(I_R + a^2 I_Y + aI_B\right) = \frac{1}{3} I_Y \left(a^2 - a\right) \qquad (2.49)$$

Therefore, $I_1 = -I_2$

and

$V_Y = V_B$; therefore:

$$V_Y = E_Y - I_{Y1}Z_1 - I_{Y2}Z_2 - I_{Y0}Z_0 = a^2E - a^2 I_1 Z_1 - aI_2Z_2 - I_0Z_0 \qquad (2.50)$$

$$V_B = E_B - I_{B1}Z_1 - I_{B2}Z_2 - I_{B0}Z_0 = aE - aI_1Z_1 - a^2I_2Z_2 - I_0Z_0 \qquad (2.51)$$

Equating (2.50) and (2.51):

$$a^2E - a^2I_1Z_1 - aI_2Z_2 - I_0Z_0 = aE - aI_1Z_1 - a^2I_2Z_2 - I_0Z_0$$

From which:

$$I_1 = \frac{E}{(Z_1 + Z_2)} \qquad (2.52)$$

Equation 2.52 is consistent with the equivalent circuit shown in Figure 2.34. Note that the zero sequence network is not required for the calculation of phase-to-phase fault current.

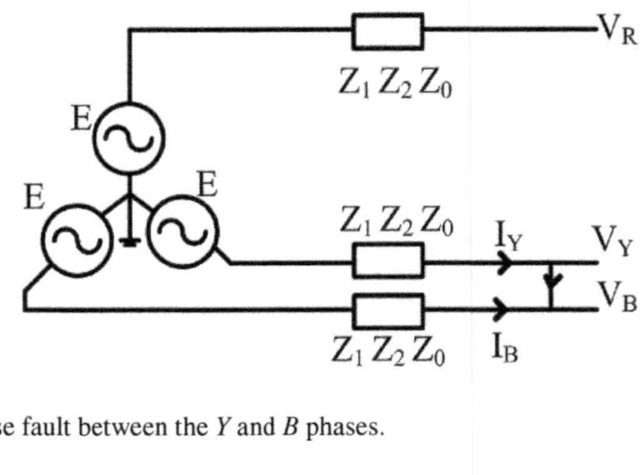

FIGURE 2.33 Phase-to-phase fault between the Y and B phases.

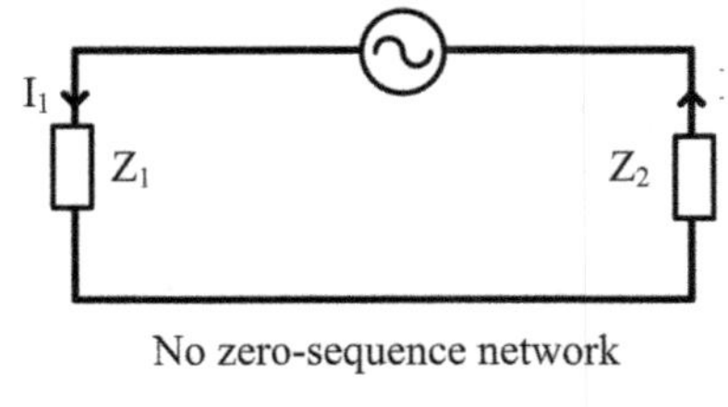

FIGURE 2.34 Equivalent circuit for phase-to-phase fault.

Fault current, $I_f = I_Y = -I_B$
From (2.37),

$$I_Y = I_0 + a^2 I_1 + a I_2 \tag{2.53}$$

Example 2.11

For the system in Figure 2.35, calculate the current (in per unit) fed into a phase-to-phase fault at busbar 3, given the data in Table 2.3.

Sequence networks (Figure 2.36).

Redraw the sequence networks (after delta-star transformation: $Z_Y = Z_\Delta/3$).

Impedances to point of fault:

$$Z_1 = j0.28\ \Omega$$

$$Z_2 = j0.25\ \Omega$$

$$I_1 = \frac{E}{Z_1 + Z_2} = -j1.89\ \text{p.u.} = -I_2$$

$$I_0 = 0$$

Fault current:

$$I_f = I_Y = -I_B = I_0 + a^2 I_1 + a I_2 = -3.28\ \text{p.u.}$$

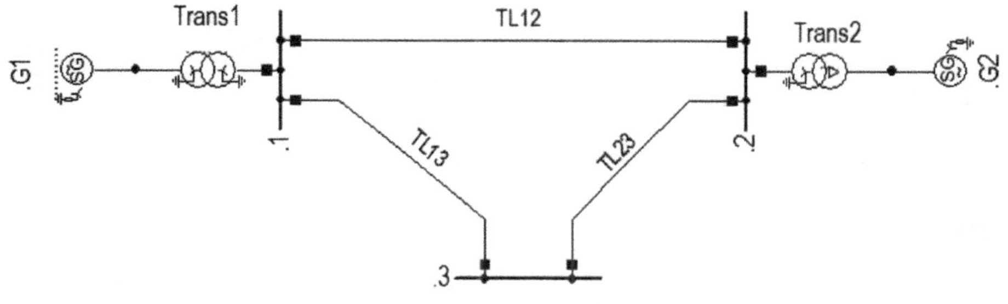

FIGURE 2.35 Three bus power network for Example 2.11.

TABLE 2.3
Network Data

Item	Specification	Sequence Impedance
G1	100 MVA, 25 kV, Y-connected through *j*0.03 p.u. earthing impedance	$x_1 = 0.25$, $x_2 = 0.20$, $x_0 = 0.05$ p.u.
G2	100 MVA, 13.8 kV, Y-connected through *j*0.03 p.u. earthing impedance	$x_1 = 0.25$, $x_2 = 0.20$, $x_0 = 0.05$ p.u.
TR1	100 MVA, 25/230 kV, Y/Y (both solidly grounded)	$x_1 = x_2 = 0.1$, $x_0 = 0.05$ p.u.
TR2	100 MVA, 13.8/230 kV, Y/Δ (*Y*-solidly grounded)	$x_1 = x_2 = x_0 = 0.375$ p.u.
Lines (all)	100 MVA, 230 kV	$x_1 = x_2 = 0.1$, $x_0 = 0.3$ p.u.

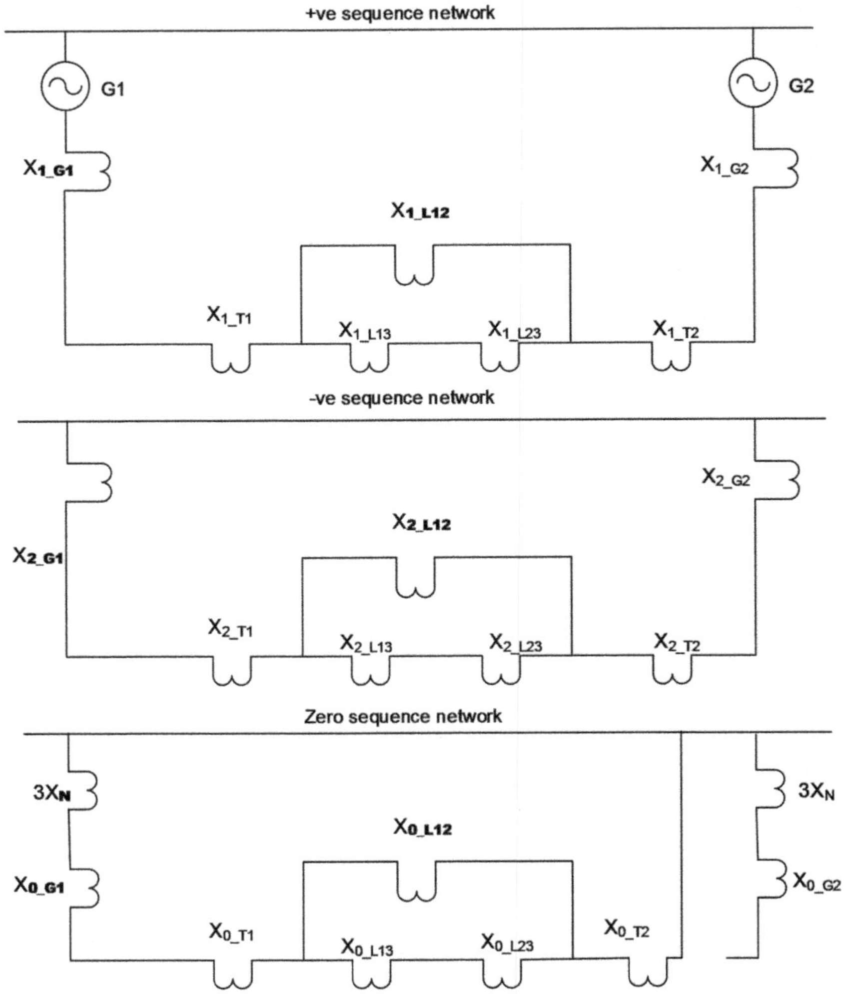

FIGURE 2.36 Phase sequence networks for Example 2.11.

2.9.3 Double-Phase to Earth Fault

Assuming a phase-to-phase to ground fault on the Y and B phases, as illustrated in Figure 2.37. From (2.37), and if I_1, I_2, and I_0 are the phase sequence components of I_R:

$$I_R = \left(I_1 + I_2 + I_0\right) = 0, \tag{2.54}$$

Therefore, from (2.50) and (2.51):

$$V_Y = a^2E - a^2I_1Z_1 - aI_2Z_2 - I_0Z_0 = 0 \tag{2.55}$$

$$V_B = aE - aI_1Z_1 - a^2I_2Z_2 - I_0Z_0 = 0 \tag{2.56}$$

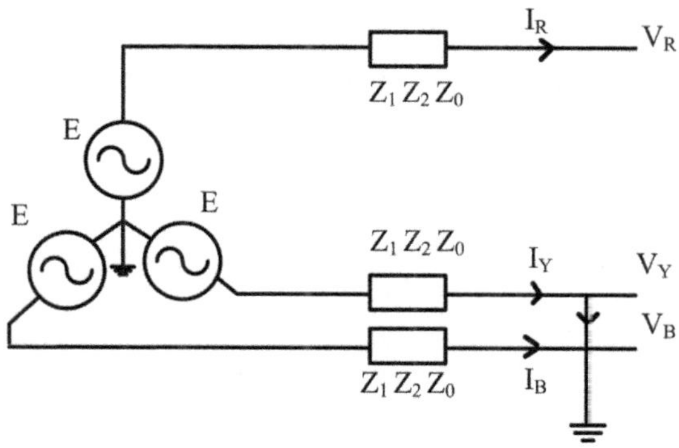

FIGURE 2.37 Phase-to-phase to ground fault on the Y and B phases.

$V_Y = V_B$ at point of fault, hence from (2.55) and (2.56):

$$I_1 = \frac{E}{Z_1 + \left(\dfrac{Z_2 Z_0}{Z_2 + Z_0}\right)} \tag{2.57}$$

$$I_2 = -I_1 \frac{Z_0}{(Z_2 + Z_0)} \tag{2.58}$$

$$I_0 = -I_1 \frac{Z_2}{(Z_2 + Z_0)} \tag{2.59}$$

Equations 2.57 through 2.59 are consistent with the equivalent circuit diagram in Figure 2.38. The fault current is given by:

$$I_f = I_Y + I_B \tag{2.60}$$

From (2.37),

$$I_Y = I_0 + a^2 I_1 + a I_2, \tag{2.61}$$

$$I_B = I_0 + a I_1 + a^2 I_2 \tag{2.62}$$

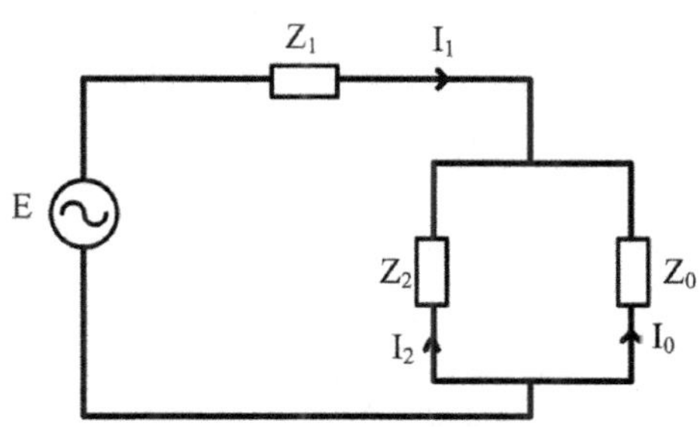

FIGURE 2.38 Equivalent circuit diagram for double-phase to ground fault.

Example 2.12

For the system in Figure 2.39, calculate the current (in p.u.) fed into a double-phase to earth fault at busbar 1. The network data is shown in the figure.

Solution

Convert impedances to 1000 MVA base:

Base impedance (765 kV):

$$Z_B = \frac{V_B^2}{S_B} = 585.2\,\Omega$$

Hence, line impedances: $x_1 = x_2 = 0.07$ p.u., $x_0 = 0.17$ p.u.

$$T3\!: x_1 = x_2 = x_0 = 0.24 \text{ p.u.};$$

$$M3\!: x_1 = x_2 = 0.30 \text{ p.u.}, x_0 = 0.10 \text{ p.u}$$

$$T4\!: x_1 = x_2 = x_0 = 0.15;$$

$$M4\!: x_1 = x_2 = 0.24 \text{ p.u.}, x_0 = 0.09 \text{ p.u.}$$

Sequence networks (Figure 2.40):

$x_1 = x_2$ for all the components in the given system; draw only the +ve sequence, or −ve sequence, networks.

Redraw the sequence networks (after delta-star transformation: $Z_Y = Z_\Delta/3$).

Impedances to point of fault:

$$Z_1 = Z_2 = j0.10\,\Omega$$

$$Z_0 = j0.07\,\Omega$$

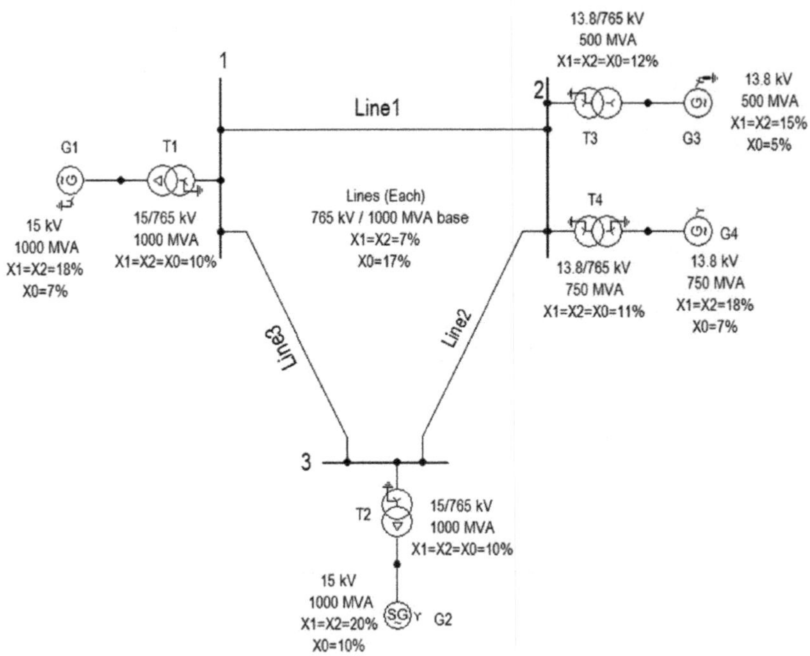

FIGURE 2.39 One-line diagram of power network with network data.

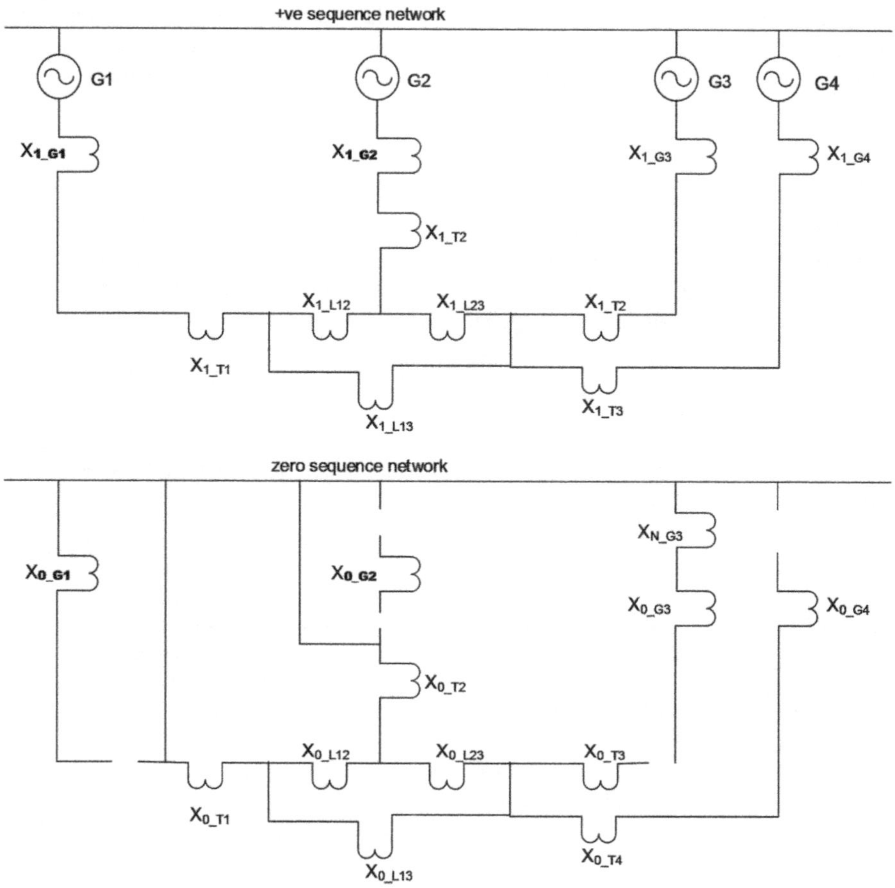

FIGURE 2.40 Phase sequence networks. for Example 2.12.

Phase current components:

$$I_1 = \frac{E}{Z_1 + \left(\frac{Z_2 Z_0}{Z_2 + Z_0}\right)} = -j0.041$$

$$I_2 = -I_1 \frac{Z_0}{Z_2 + Z_0} = j0.017$$

$$I_0 = j0.024$$

$$I_Y = I_0 + a^2 I_1 + a I_2 = -0.051 + j0.036$$

$$I_B = I_0 + a I_1 + a^2 I_2 = 0.051 + j0.036$$

$$I_f = I_Y + I_B = j0.072 \text{ p.u.}$$

2.10 The Bus Impedance Matrix in Unbalanced Fault Calculation

The bus impedance matrix may be used to determine symmetrical fault currents or unsymmetrical fault currents. The problem is to determine the bus impedance matrix elements. For symmetrical faults, these are the normal (positive sequence) impedances. For unsymmetrical faults, the negative and zero sequence impedances also need to be included. With unbalanced faults, the three bus impedance matrices: Z_{bus}(+ve sequence), Z_{bus}(−ve sequence) and Z_{bus}(zero sequence) need to be determined.

Symmetrical fault current is determined by applying (2.25) at the faulty bus. That is,

$$I_f = \frac{V_f}{Z_{33}}$$

where the fault is at busbar 3 of the network, for example.

A single-phase to ground unsymmetrical fault is determined by applying (2.46). Using the bus impedance matrix method, the Z_1, Z_2 and Z_0 impedance values are obtained from the three-phase sequence bus impedance matrices, and the fault current is given by:

$$I_f = \frac{3E}{(Z_{33-1} + Z_{33-2} + Z_{33-0})}, \tag{2.63}$$

for example, where the fault is at bus 3 of a network. The subscripts 1, 2 and 0 refer to the positive, negative or zero sequence impedances.

Example 2.13

Figure 2.41 shows a 765 kV network. The network data is shown in the figure. The positive bus impedance matrix for this network is given by Equation 2.64. Using the Z-building bus impedance matrix algorithm, calculate the current feeding into a single-phase to ground fault at busbar 2 when Line 3 is disconnected (i.e., out of service). Use 1000 MVA base.

$$Z_{bus\_1} = j\begin{bmatrix} 0.1040 & 0.0810 & 0.0829 \\ 0.0810 & 0.1009 & 0.0815 \\ 0.0829 & 0.0815 & 0.1049 \end{bmatrix} \tag{2.64}$$

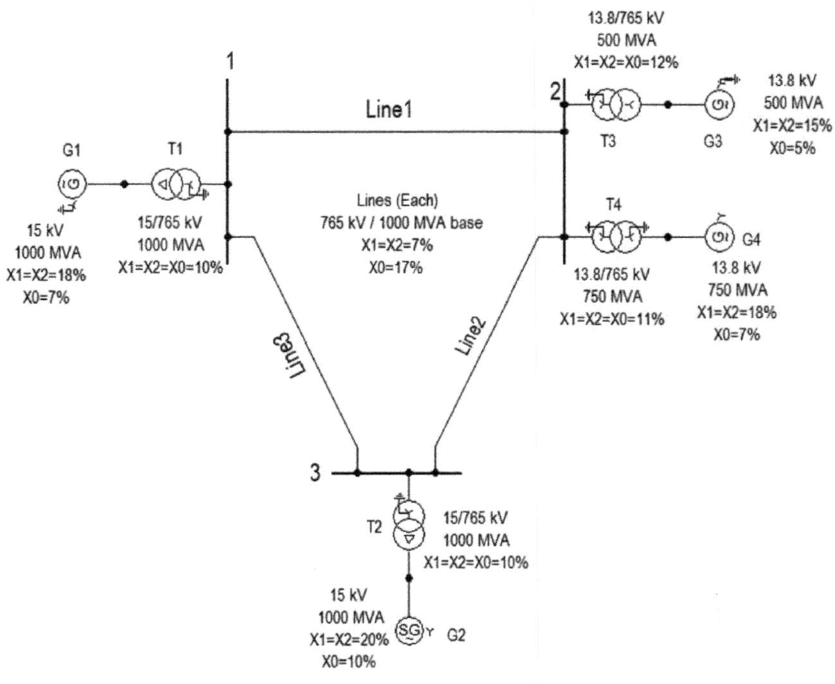

FIGURE 2.41 Three node power network with network data.

Solution

New positive sequence bus impedance matrix: Disconnecting Line 3 is equivalent to adding line of reactance $-j0.07$ p.u. between buses 1 and 3.

Step 1: Add branch between nodes 1 and 3:
Apply

$$Z_{bus}^{n} = Z_{bus} - \gamma bb^{T}$$

$$\text{where} \quad b = Z_i - Z_j \quad \text{and} \quad \gamma = (Z_b + z_{ii} + z_{jj} - 2z_{ij})^{-1}$$

$$(i = 1,\ j = 3,\ Z_b = -\,j0.07)$$

and obtain

$$Z1 = \begin{matrix} 0.0000 + 0.1040i \\ 0.0000 + 0.0810i \\ 0.0000 + 0.0829i \end{matrix}$$

$$Z3 = \begin{matrix} 0.0000 + 0.0829i \\ 0.0000 + 0.0815i \\ 0.0000 + 0.1049i \end{matrix}$$

$$b = \begin{matrix} 0.0000 + 0.0211i \\ 0.0000 - 0.0005i \\ 0.0000 - 0.0220i \end{matrix}$$

$$b^{T} = 0.0000 + 0.0211i \quad 0.0000 - 0.0005i \quad 0.0000 - 0.0220i$$

$$bb^{T} = 1.0e-03 * \begin{matrix} -0.4452 & 0.0106 & 0.4642 \\ 0.0106 & -0.0003 & -0.0110 \\ 0.4642 & -0.0110 & -0.4840 \end{matrix}$$

$$Z_b = 0.0000 - 0.0700i$$

$$\gamma = 0.0000 + 37.1747i$$

$$\gamma bb^{T} = \begin{matrix} 0.0000 - 0.0166i & 0.0000 + 0.0004i & 0.0000 + 0.0173i \\ 0.0000 + 0.0004i & 0.0000 - 0.0000i & 0.0000 - 0.0004i \\ 0.0000 + 0.0173i & 0.0000 - 0.0004i & 0.0000 - 0.0180i \end{matrix}$$

$$Z_{bus1} = Z_{bus2} = \begin{matrix} 0.0000 + 0.1206i & 0.0000 + 0.0806i & 0.0000 + 0.0656i \\ 0.0000 + 0.0806i & 0.0000 + 0.1009i & 0.0000 + 0.0819i \\ 0.0000 + 0.0656i & 0.0000 + 0.0819i & 0.0000 + 0.1229i \end{matrix}$$

Zero phase sequence bus impedance matrix: Convert all quantities to same base (of 1000 MVA):

$$G3: X_1 = X_2 = j0.30\,\text{p.u.}; X_0 = j0.10\,\text{p.u.}$$

$$G4: X_1 = X_2 = j0.24\,\text{p.u.}; X_0 = j0.093\text{ p.u}$$

$$T3: X_1 = X_2 = X_0 = j0.24\,\text{p.u.}$$

$$T4: X_1 = X_2 = X_0 = j0.147\,\text{p.u.}$$

Zero phase sequence network (Figure 2.42):

Step 1: Write down the bus impedance matrix for the network consisting only of those branches connected directly to the reference bus:

$$Z^0_{bus} = j\begin{bmatrix} 0.10 & 0 \\ 0 & 0.10 \end{bmatrix}$$

Step 2: Add a branch Z_b from a new node 3 to node 1:
Apply

$$Z^n_{bus} = \begin{bmatrix} Z_{bus} & Z_i \\ Z_i^T & Z_b + z_{ii} \end{bmatrix}$$

$$(i = 1,\ j = 3,\ Z_b = j0.17)$$

$$Z^1_{bus} = j\begin{bmatrix} 0.10 & 0 & 0.10 \\ 0 & 0.10 & 0 \\ 0.10 & 0 & 0.27 \end{bmatrix}$$

Step 3: Add branch between nodes 2 and 3:
Apply

$$Z^n_{bus} = Z_{bus} - \gamma\, bb^T$$

$$\text{where} \quad b = Z_i - Z_j \quad \text{and} \quad \gamma = (Z_b + z_{ii} + z_{jj} - 2z_{ij})^{-1}$$

$$(i = 2,\ j = 3,\ Z_b = j0.17)$$

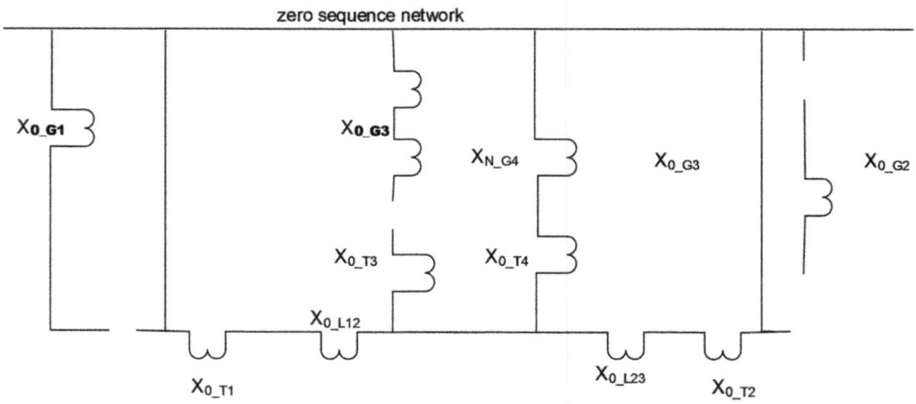

FIGURE 2.42 Zero phase sequence network.

and obtain

$$Z2 = \begin{matrix} 0 \\ 0 + 0.1000i \\ 0 \end{matrix}$$

$$Z3 = \begin{matrix} 0 + 0.1000i \\ 0 \\ 0 + 0.2700i \end{matrix}$$

$$b = \begin{matrix} 0 - 0.1000i \\ 0 + 0.1000i \\ 0 - 0.2700i \end{matrix}$$

$$b^T = \begin{matrix} 0 - 0.1000i & 0 + 0.1000i & 0 - 0.2700i \end{matrix}$$

$$Z_b = 0 + 0.1700i$$

$$\gamma = 0 - 1.8519i$$

$$\gamma bb^T = \begin{matrix} 0.0000 + 0.0185i & 0.0000 - 0.0185i & 0.0000 + 0.0500i \\ 0.0000 - 0.0185i & 0.0000 + 0.0185i & 0.0000 - 0.0500i \\ 0.0000 + 0.0500i & 0.0000 - 0.0500i & 0.0000 + 0.1350i \end{matrix}$$

$$Z_{bus0} = \begin{matrix} 0 + 0.0815i & 0 + 0.0185i & 0 + 0.0500i \\ 0 + 0.0185i & 0 + 0.0815i & 0 + 0.0500i \\ 0 + 0.0500i & 0 + 0.0500i & 0 + 0.1350i \end{matrix}$$

Hence,

$$I_f = \frac{3E}{Z_{22\_1} + Z_{22\_2} + Z_{22\_0}} = -j10.59 \text{ p.u.}$$

2.11 Computer Simulations

Power system protection and planning engineers use software tools such as Digsilent *PowerFactory* [10] for the calculation of fault currents. Various standard short circuit calculation methods are supported by these software tools. The IEC 60909/VDE 0102, ANSI and IEC 61363 short circuit calculation methods, for example, are generally used for planning purposes when the actual system operating conditions are not yet known. These methods are based on the venin theorem. An equivalent voltage source and impedance at the short circuit location is determined, and short circuit current is calculated from them. All network feeders as well as the synchronous and asynchronous machines are replaced in the calculation by their impedances (positive, negative, and zero sequence impedances). All line capacitances and the parallel admittances of nonrotating loads, except those of the zero-sequence system, are neglected. These methods use nominal values or conditions and apply correction factors for voltage and impedances to give conservative results.

However, where precise evaluation of the fault current in a specific situation is required, for example, to find out the cause of relay failure to operate for a fault, exact methods of short circuit calculation need to be applied. The exact methods calculate the short circuit current based on the existing network operating condition. That is, these methods must be preceded by a load flow calculation to establish the system operating state. These methods include the *Complete Method* that uses the superposition theorem.

2.12 Tutorial Problems

1. Explain how the method of symmetrical components may be used to represent an unbalanced three-phase system by an equivalent set of balanced phasors.
2. A solid three-phase fault occurs on the terminals of a generator. Describe the effect of this fault on the current output of the generator.
3. Explain why it is necessary to calculate short circuit currents in large electrical systems.
4. Sketch three methods of connecting busbar reactors at a generating station.
5. Discuss the behaviour of the following types of motors under short circuit fault conditions: (a) synchronous motor, (b) induction motor.
6. A section of a 400 kV power system consisting of three busbars is shown in Figure 2.43. The fault levels at busbars 1 and 2 are as shown in the figure. The reactances of the lines TL12, TL13 and TL23 are equal to $j5\ \Omega$ (each). Calculate the fault level at bus 3. (24640 MVA)
7. Using a 20 MVA base, calculate the fault current (in amperes) for a three-phase short circuit at busbar 3 of the network given in Figure 2.44. The line reactances are 32 Ω (each) and all other reactances are given (on the figure) referred to a base of 20 MVA. (1.21 kA)
8. Equipment ratings for the three-bus power system shown in Figure 2.48 are as follows:

   Generator G1: 1000 MVA, 13.8 kV, $x = j0.25$ p.u.

   Generator G2: 1000 MVA, 17.5 kV, $x = j0.20$ p.u.

   Transformer T1: 1000 MVA, 13.8/400 kV, $x = j0.15$ p.u.

   Transformer T2: 1000 MVA, 17.5/400 kV, $x = j0.15$ p.u.

   Line 12: $x = j40\ \Omega$

   Line 23: $x = j60\ \Omega$

   Line 14: $x = j40\ \Omega$

   Line 24: $x = j40\ \Omega$

   Line 34: $x = j60\ \Omega$

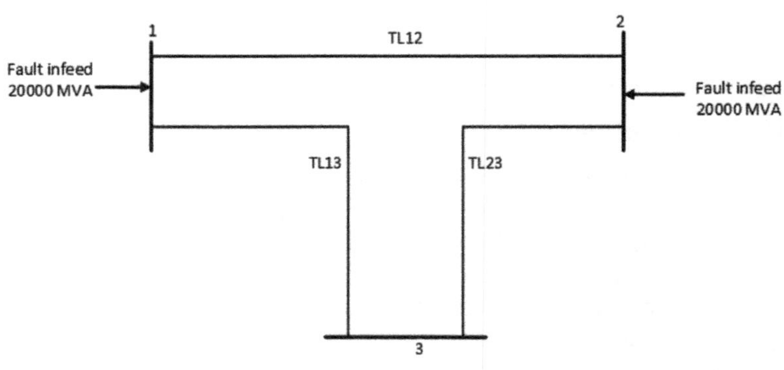

FIGURE 2.43 Section of a 400 kV power system.

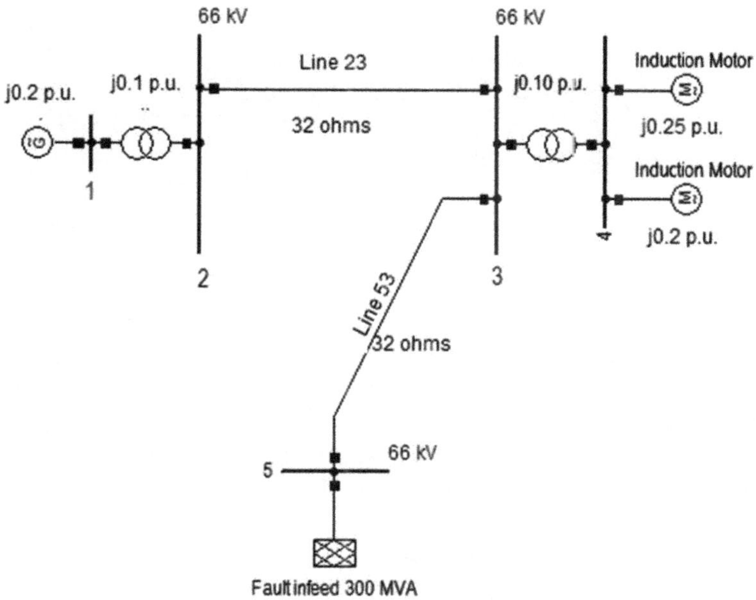

FIGURE 2.44 One-line diagram of power network with motor loads.

Using a 1000 MVA base, calculate the fault current (in amperes) for a three-phase short circuit at busbar 2 (Figure 2.45). (6.87 kA)

9. An 11 kV synchronous generator is connected to a 11/66 kV transformer that feeds a 66/11/3.3 kV three-winding transformer through a short feeder of negligible impedance, as shown in Figure 2.46. Calculate the fault current (in amperes) when a single-phase-to-earth fault occurs on a terminal of the 11 kV winding of the three-winding transformer. The relevant data for the system are as follows:

   Generator: $X_1 = j0.15$ p.u., $X_2 = j0.10$ p.u., $X_0 = j0.03$ p.u., all on a 10 MVA base; star point of winding grounded through a 3 Ω resistor.

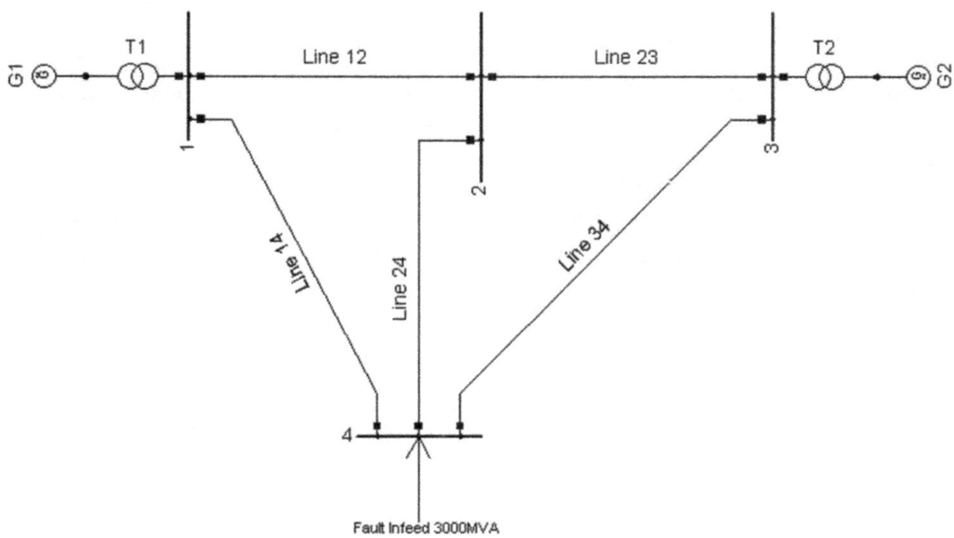

FIGURE 2.45 One-line diagram of a four-bus power network.

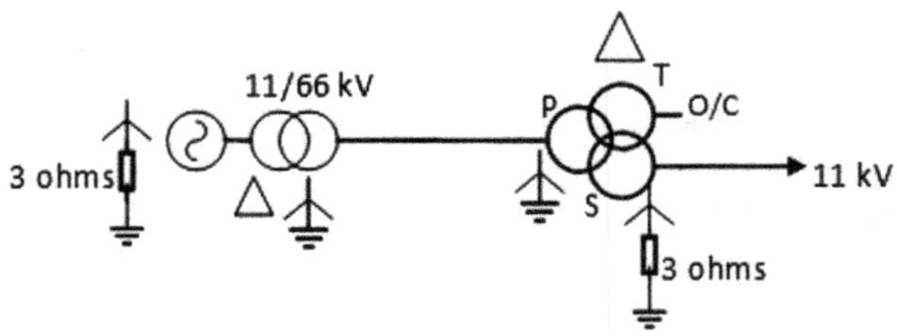

FIGURE 2.46 Simple power network.

11/66 kV Transformer: $X_1 = X_2 = X_0 = j0.10$ p.u. on a 10 MVA base; 11 kV winding delta connected and the 66 kV winding star connected with the star point solidly grounded.

Three-winding transformer: A 66 kV winding, star connected, star point solidly grounded; 11 kV winding, star connected, star-point grounded through a 3 Ω resistor; 3.3 kV winding, delta connected; the three windings of an equivalent star connection to represent the transformer have sequence impedances:

66 kV winding $X_1 = X_2 = X_0 = j0.04$ p.u.

11 kV winding $X_1 = X_2 = X_0 = j0.03$ p.u.

3.3 kV winding $X_1 = X_2 = X_0 = j0.05$ p.u.

all on a 10 MVA base. Resistance may be neglected throughout. (1.58 kA)

10. A 25 MVA, 11 kV synchronous generator with sub-transient reactance $Xd'' = 0.2$ p.u. is supplying three identical synchronous motors through a step-up and step-down transformer, as shown in Figure 2.47. Each motor has a sub-transient reactance $Xd'' = 0.25$ p.u. on its 5 MVA, 6.6 kV base. Both the 11/66 kV step-up and 66/6.6 kV step-down transformers are rated at 25 MVA and have the same leakage reactance of 0.1 p.u. on their respective base. Assuming the voltage at the motor terminals to be 6.6 kV when a three-phase fault occurs at point *F*, calculate the magnitude of the fault current in kA. Choose the system base as 25 MVA.
11. A *Y*-connected unbalanced load draws currents in RYB sequence of $I_R = 10\angle 10°$ A, $I_Y = 10\angle 100°$ A, $I_B = 20\angle 150°$ A. Evaluate the positive, negative and zero sequence components of the *R*-phase current.

$$(7.82\angle 113.10°; 6.74\angle 15.23°; 9.66\angle -48.36°)$$

12. A three-phase 10 MVA, 11 kV synchronous generator with a solidly grounded neutral point supplies a feeder at the same voltage level. The positive, negative and zero sequence impedances of the generator and the feeder are shown in Table 2.4. For a phase-to-ground fault at the far end of the feeder, calculate the fault current in kA. (2.54 kA)
13. Using the *Z*-building procedure, calculate the current (in per unit) feeding into a three-phase fault at the busbar labeled "East" of the three-bus system shown in Figure 2.48. The system data is given in Table 2.5. (5.98 p.u.)

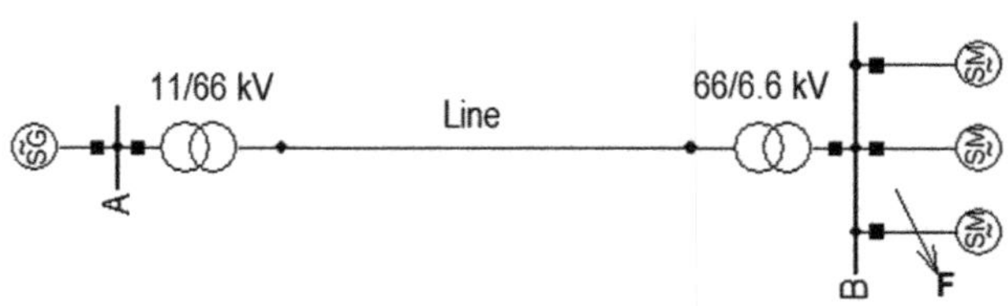

FIGURE 2.47 Two-bus power network with motor loads.

TABLE 2.4

Network Data for the Generator and Feeder

	Generator	Feeder
Positive sequence, $Z1$	$j1.2\ \Omega$	$j1.0\ \Omega$
Negative sequence, $Z2$	$j0.9\ \Omega$	$j1.0\ \Omega$
Zero sequence, $Z0$	$j0.4\ \Omega$	$j3.0\ \Omega$

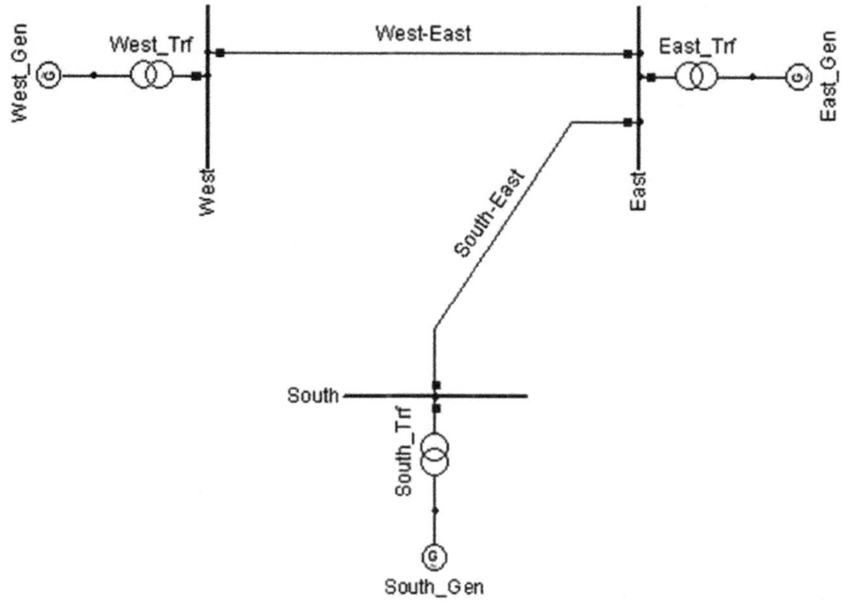

FIGURE 2.48 Three-bus power network for Tutorial Problem 13.

TABLE 2.5

Network Impedances

Item	Impedances
West_Gen	$X = j0.15$ p.u.
East_Gen	$X = j0.20$ p.u.
South_Gen	$X = j0.20$ p.u.
West_Trf	$X = j0.30$ p.u.
East_Trf	$X = j0.20$ p.u.
South_Trf	$X = j0.25$ p.u.
West-East	$X = j0.10$ p.u.
South-East	$X = j0.15$ p.u.

14. Using the indirect bus impedance matrix method, calculate the current (in per unit) feeding into a single phase to earth fault at bus 2 of the system shown in Figure 2.49. The system data is given in Table 2.6. A single phase to earth fault occurs at the secondary (load side) terminals of transformer $T3$. Will any current feed into this fault? Explain. (9.047 p.u.)
15. A simple power network consisting of two substations connected by a single 132 kV transmission line has the following positive sequence bus impedance matrix:

$$Z_{bus\_1} = j\begin{bmatrix} 0.2520 & 0.1540 \\ 0.1540 & 0.3080 \end{bmatrix} \text{(per unit)}$$

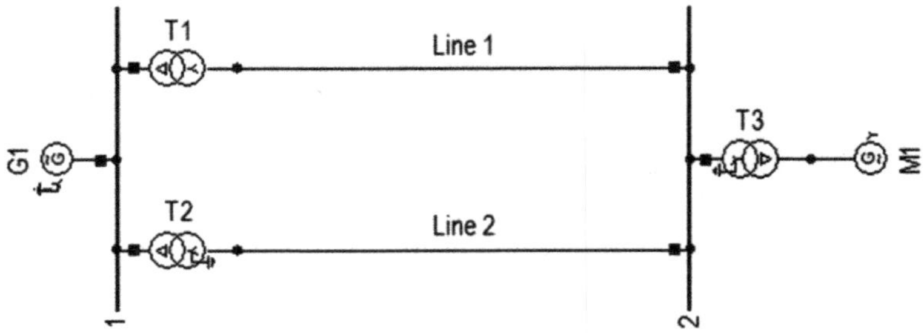

FIGURE 2.49 A double circuit power system.

TABLE 2.6

Sequence Reactances for the Double Circuit Power System

Item	Sequence Reactance (in p.u. Referred to Same Base)
G1	$x_1=0.18$, $x_2=0.18$, $x_0=0.10$ p.u., earthing reactance $=j0.05$
M1	$x_1=0.08$, $x_2=0.08$, $x_0=0.16$ p.u.
T1	$x_1=0.15$, $x_2=0.15$ p.u., $x_0=0.18$
T2	$x_1=0.12$, $x_2=0.12$, $x_0=0.12$ p.u.
T3	$x_1=0.10$, $x_2=0.10$, $x_0=0.10$ p.u.
Line 1	$x_1=x_2=0.20$, $x_0=0.40$ p.u.
Line 2	$x_1=x_2=0.30$, $x_0=0.70$ p.u.

Using the *Z*-building procedure, determine the current flowing into a three-phase fault at the midpoint of the transmission line (in amperes), given that the positive sequence impedance of the transmission line is $j0.35$ p.u. Assume a 400 MVA base. (5.74 kA)

16. Explain how the method of symmetrical components may be used to represent an unbalanced three-phase system by an equivalent set of balanced phasors.
17. A solid three-phase fault occurs on the terminals of a generator. Describe the effect of this fault on the current output of the generator.
18. Explain why it is necessary to calculate short circuit currents in large electrical systems.

2.13 Conclusion

Fault analysis is an important component of power system design. This chapter presented various methods and techniques for the calculation of different types of faults. Network reduction and bus impedance matrix methods were discussed, including application of the concept of symmetrical components for the analysis of unbalanced faults. Detailed examples were given to enhance student's understanding of the various methods of fault analysis. Last, a brief discussion of software tools used by engineers to facilitate system planning and protection calculations was given.

REFERENCES

1. B.M. Weedy, B.J. Cory, N. Jenkins, J.B. Ekanayake and G. Strbac, *Electric Power Systems*, 5th ed., John Wiley & Sons, 2012.
2. J.D. Glover, M.S. Sarma and T.J. Overbye, *Power System Analysis and Design*, 4th ed., Toronto, Canada: Thomson, 2008.
3. L.G. Hewitson, M. Brown and R. Balakrishnan, *Practical Power System Protection*, Oxford, UK: Newnes, 2004.

4. B.A. Oza, N.C. Nair, R.P Mehta and V.H Makwana, *Power System Protection and Switchgear*, New York: McGraw-Hill, 2010.
5. The Electricity Training Association, *Power System Protection*, 2nd ed., London, UK: The Institution of Electrical Engineers, 1995.
6. C.L. Wadwa, *Electric Power Systems*, 6th ed., New Delhi, India: New Age, 2009.
7. J. Schlabbach, *Short-Circuit Currents*, London, UK: The Institution of Engineering and Technology, 2005.
8. J.J. Grainger and W.D. Stevenson, *Power System Analysis*, New York: McGraw-Hill, 1994.
9. Alstom Grid, *Network Protection & Automation Guide*, 2nd ed., Stafford, UK: Alstom Grid, 2011.
10. DIgSILENT GmbH (2015, May), DIgSILENT PowerFactory, User Manual.

3

Fuses and Circuit Breakers

Abhishek Chauhan, Padmanabh Thakur, and Ramesh Bansal

CONTENTS

3.1 Introduction

Switchgear holds a comprehensive range of switching, controlling and current interrupting equipment (gear) in the power system that is responsible for necessary action during normal and abnormal conditions. A switchgear system consists of the following:

1. Relays
2. Circuit breakers
3. Switches
4. Fuses
5. Isolators
6. Lightning arrestor
7. Current transformers (CTs)
8. Potential transformers (PTs)

A power system is divided into many sections, and every section has its own zone of protection with protective scheme, circuit breakers and relays, which are accountable for the segregation of the faulty part from the healthy part of the section under fault. Protective relays detect the fault location and issue a command to the circuit breaker to disconnect the faulty region. Relays continuously monitor electrical quantities, as they are likely to be changed during faulty conditions. So, on the basis of change in electrical quantities like current, voltage, frequency and phase angle, relays are able to distinguish between healthy and faulty conditions [1–8]. In power systems, abnormal conditions not only signify short circuit current transients but also overspeeding of generators and electrical drives, underfrequency, loss of excitation, temperature rise of alternator, overvoltage, etc.

3.1.1 Zone of Protection

Generators, transformers, busbars, transmission lines and distribution lines are the building blocks of power system infrastructure. Each element of a power system has its own protective scheme, so it is evident that a power system is distributed into several zones of protection [1–8]. The power system is completely covered by these protective zones, and no elements are left unprotected. Figure 3.1 show the zone of protection of a power system; note that different zones overlap with the adjacent zones. Overlapping of protective zones enables the protection of sections that lie on the boundaries of two zones and of the sections that may not lie in any of these zones [1–4,6–8].

3.1.2 Functional Characteristics of Switchgear

The functions of switchgear are as follows:

- Switchgear is liable for the detection of faults and enables the automatic disconnection of faulty parts from the healthy system.
- Intentional isolation of circuit section from main lines for maintenance.
- Reallocation of loads.
- Deterrence of the occurrence of abnormal conditions in exception case of gas actuated Buchholz relay.

3.1.3 Properties of Protection System

The protection system should have several properties that make it capable of limiting transient conditions (short circuit or faulty) conditions before they cause harm to the power system. The basic properties of a protective system are the following:

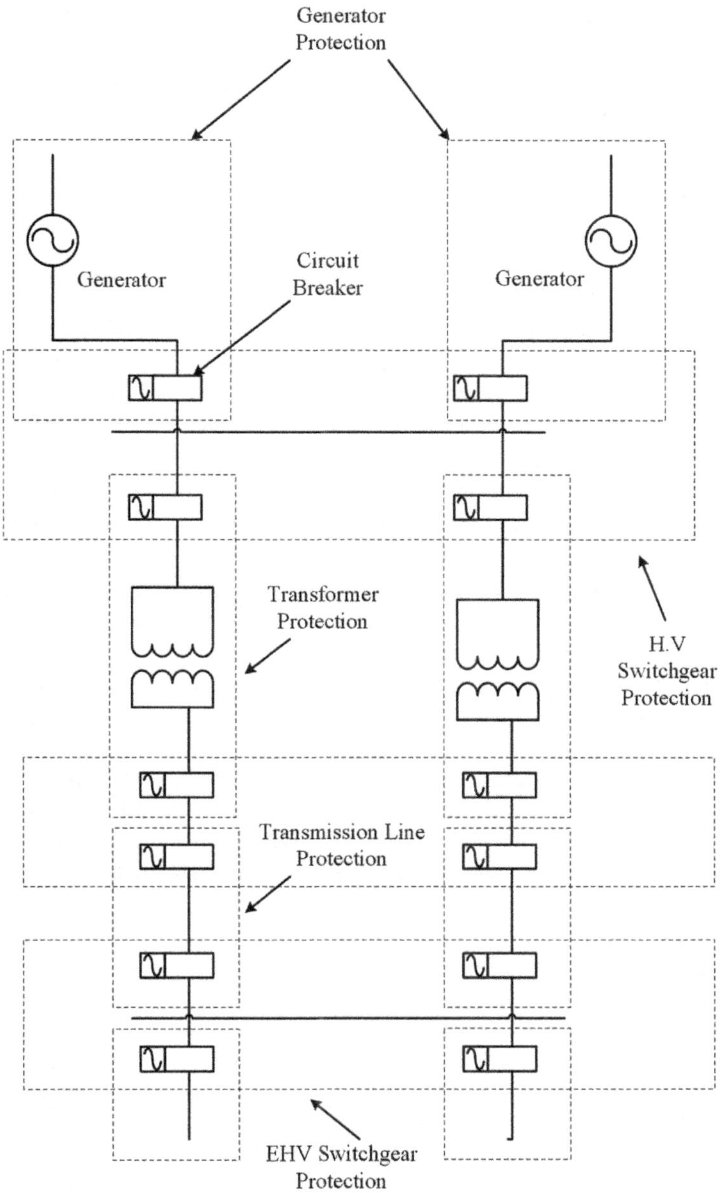

FIGURE 3.1 Zones of protection in power system.

1. *Fast and reliable operation*: A protective system should be reliable and act instantly when transients are in its zone of protection. In order to obtain high reliability, all the peripherals of the protection system should be maintained and meticulously checked at consistent intervals of time [1–4]. A protective scheme should have a reliability value of about 95%.
2. *Selectivity*: The protective peripheral (relay) should be able to differentiate a faulty from a normal condition. Moreover, it should precisely identify whether the fault lies in its zone of protection. For instance, when a parallel connected alternator loses its synchronism, there is a transient current that starts flowing through lines (i.e., a power surge) that are similar to the short circuit conditions [1–4]. Hence, the protective system should have the capability to identify such conditions for necessary action to be introduced.

3. *Sensitivity*: The protective system should be sensitive towards its pick value, i.e., preset values of electrical quantity above which the protective system comes into action [1–4].
4. *Manual control option*: The manual control option should be implemented if there is any malfunction during automatic operation.

3.1.4 Classification of Switchgear

Switchgear systems are broadly classified on the basis of the following:

1. *Actuating quantity*: Different types of switchgear protection use different electrical quantities for their actuation. For instance, an overcurrent protection system operates for levels that are above the pickup current (threshold value of actuating quantity). Similarly, distance protection identifies the impedance, reactance and change in phase angle as the actuating quantity, whereas a differential protection system operates in response to the difference between two actuating quantities [1–4,8].
2. *Insulating medium*: The insulating medium between fixed and moving contacts is responsible for arc quenching. As gas (SF_6), oil, vacuum and CO_2 are the modes for providing insulation and inhibit restriking of arc [1–4].
3. *Application*: Power systems involve different sections that are broadly classified as generation, transmission and distribution. Each section has its own challenges towards protection. Hence, switchgears are employed according to the applications [1–4,6–8].
4. *Construction*: A switchgear system is classified on the basis of its construction, which includes ingress protection marking that explains the degree of protection against intrusion. Ingress protection is published by International Electrotechnical Commission (IEC). Moreover, metal-enclosed and metal-clad systems are the two similar constructional types of switchgear systems. Metal-enclosed systems are enclosed by metal sheets from all sides. Metal-clad systems are those metal enclosures where all live wires are safely separated by different compartments that are well insulated and have mechanical interlocks that minimize operating mistakes [1–4,6–8].

3.2 Fuses

A fuse is a basic device used for circuit protection against overloads or short circuits. Fuse have less efficiency and reliability. They were devised in 1890 by Edison. They are the cheapest and most elementary form of protection system and operate by melting the conductor link (fuse wire or fuse element) between the contacts and thus break the circuit [1–4,6–8].

Fuses have an inverse-time characteristic: The fuse blows more quickly with an increase in magnitude of current. It means that time of fuse blowing is a magnitude of current dependent parameter. Fuses are used in applications where repeated operations are not expected, for instance distribution transformers, small or medium size motors, electronic circuits, extension boards and car electrical fittings [1–4].

3.2.1 Fuse Element Materials

Selection of fuse material is based on its application and its current bearing capacity.

- Lead, tin and zinc are the most commonly used materials for fuse elements as they have low melting point. Table 3.1 shows that, along with a low melting point, these materials have high specific resistance. Hence, for a particular current interruption application, a fuse element of low melting point and high specific resistance leads to a greater diameter, resulting in a large

TABLE 3.1

Fuse Element Materials (Thermal and Resistance Specifications)

S. No.	Metal	Melting Point (°C)	Specific Resistance (μΩ)-cm
1	Tin	230	11.2
2	Lead	328	21.5
3	Zinc	419	6.1
4	Aluminium	670	2.85
5	Silver	960	1.64
6	Copper	1090	1.72

Source: Rajput, R.K., *Power System Engineering*, 2nd ed., Laxmi Publications, 2015.

amount of oxidized material during operation [1–4]. Therefore, fuse elements of lead, tin and zinc are limited to low current interruption, i.e., 10 A.

- For currents above 15 A, silver is used as fuse material, due to low resistance or high conductivity. A lesser amount of silver is used for a given current interruption when compared with lead, tin and zinc [1–4].
- The modern high rupturing capacity (HRC) fuses use silver element surrounded by chalk, plaster of Paris, quartz, and marble dust, which act as the arch quenching medium and allow cooling. HRC is capable for voltage ratings up to 66 kV [1–4,6–11].

3.2.2 Terms and Relation for Fuses

For proper understanding of fuses, several technical terms related to fuses are discussed below:

1. *Fusing current*: It is defined as the least value of operating current at which the fuse element melts. Fusing current is more than the rated value of the fuse element. Fuse law identifies the fusing current for an element of particular diameter and is expressed as:

$$I = kD^{3/2} \tag{3.1}$$

where fuse constant k depends on the metal of wire, and D is the diameter of the wire. Table 3.2 gives the value of k for a particular fuse element metal.

TABLE 3.2

Value of Constant "*K*" for Different Fuse Material

S. No.	Metal	Value of Fuse Constant, "k" for "D" in mm
1	Copper	80
2	Aluminium	59
3	Tin	12.8
4	Lead	10.8

Source: Rajput, R.K., *Power System Engineering*, 2nd ed., Laxmi Publications, 2015.

2. *Fuse element current rating*: Rated current for fuse element is expressed as the current that a fuse wire will be able to carry without being blown off or melted.
3. *Fusing factor*: The ratio of minimum fusing current and the rated current of fuse element is defined as the fusing factor:

$$\text{Fusing Factor} = \frac{\text{Minimum fusing current}}{\text{Rated fuse element current}} = \frac{i_{f(\min)}}{i_r} \tag{3.2}$$

   A smaller fusing factor signifies greater trouble in evading deterioration due to overheating and oxidation at rated element current. Time-current characteristics determine the fusing factor of a fuse.
4. *Prospective current*: It is the root mean square (r.m.s). current that should have flowed through circuit during the fault, if it is presumed that the fuse link is traded by a resistance having negligible ohmic value [1–4]. In Figure 3.2, the dotted symmetrical wave indicates the prospective current (I_p), where I_{pp} is the peak prospective current.
5. *Cut-off current*: It is the current at which the arc is initiated and the fuse contact tends to melt, labelled as I_{cc} in Figure 3.2. The fuse contact melts before reaching prospective current, as sufficient heat generated by I_{cc} results in melting of the fuse element [6].
6. *Pre-arcing time*: It is the time from the instant of instigation of the fault current to the instant when current reaches to the cut-off current [3,6] (denoted by $T_{\text{p.arc}}$, i.e., *ox* in Figure 3.2).
7. *Arcing time*: This is the time between the cut-off time and the time at which the arc is extinguished and current reaches zero permanently [3,6,9]. Refer to T_{arc} in Figure 3.2, i.e., *xy*.
8. *Operating time or total time:* It is the summation of pre-arcing time and arc time, i.e., *oy* in Figure 3.2.
9. *Rupturing capacity* (*breaking capacity*): It is defined as the Volt Ampere (VA) rating of the fuse corresponding to the peak prospective current at which the fuse is able to rupture at system voltage. A fuse never bears the actual current equivalent to its rupturing or breaking capacity, as the fuse element melts earlier at cut-off current [3,6].

3.2.3 Characteristic of Fuse (Inverse-Time)

Fuses traces inverse time characteristic as shown in Figure 3.3, it concludes that with the increase in prospective current there is a decrease in time of operation. Fuse characteristic is usually defined in terms of

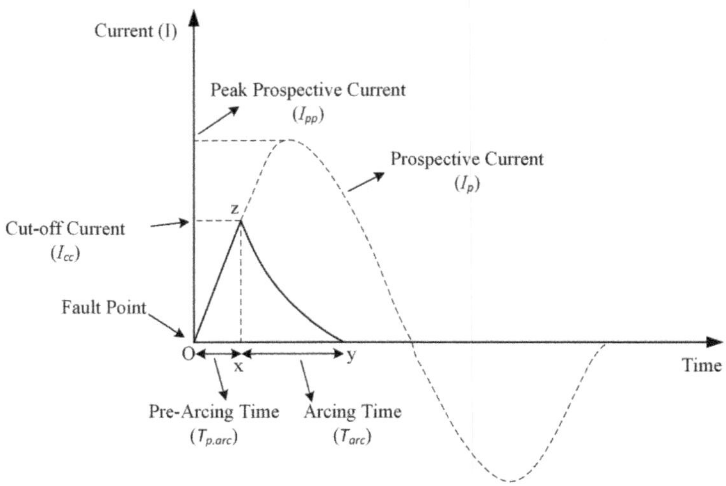

FIGURE 3.2 Cut-off characteristic (fuse).

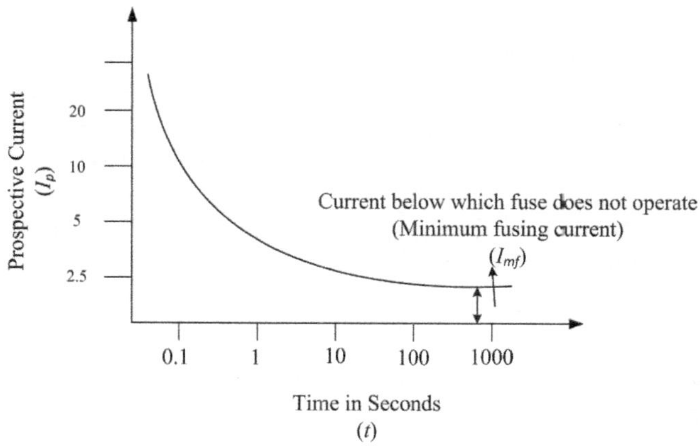

FIGURE 3.3 Time current characteristic (fuse).

pre-arcing time and r.m.s. value prospective current. Figure 3.3 asserts that the minimum current below which the fuse does not operate is known as minimum fusing current [3,6].

3.2.4 Classification of Fuses

On the basis of voltage levels fuses are classified as described in the following subsections.

3.2.4.1 Low Voltage Fuses

Low voltage fuses are designed for low voltage operations or low voltage applications. They are further classified as follows:

1. *Rewireable*: As the name suggested, rewireable fuses can rewire the fuse every time it blows off during unstable condition (fault). Rewireable fuses are popularly known as *kit-kat* fuses (shown in Figure 3.4). A rewireable fuse consists of two sections, namely, fixed section or fuse base (i.e., fastened on wall or board and carry incoming and outgoing connections, as shown in Figure 3.4a) and a removable section or fuse carrier (see Figure 3.4b) made of porcelain that

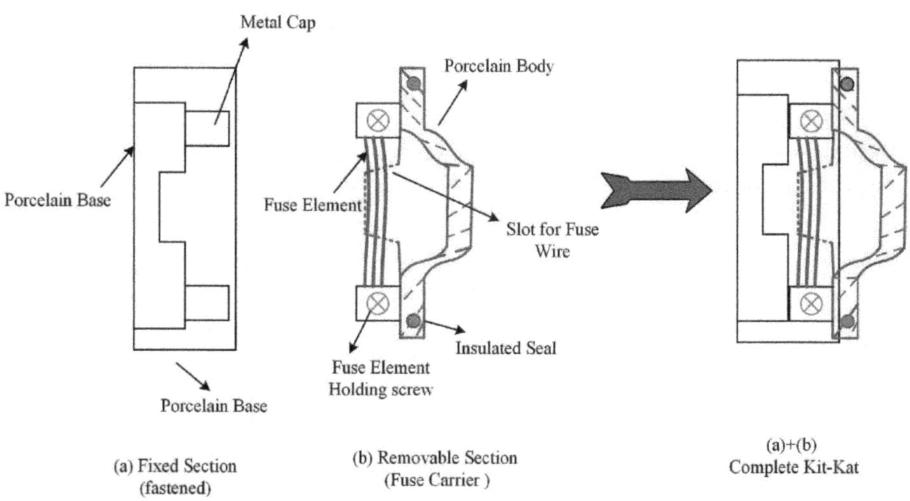

FIGURE 3.4 Rewireable fuse (kit-kat fuse).

holds the fuse element in order to replace the fuse element. Fuse carrier can be taken out and placed again after replacing fuse element [1–4,6].

a. Fuse wire may be of tinned copper, lead or an alloy of tin-lead. Rewireable fuses are affected by ambient temperature, as they are exposed to the atmosphere, which results in no-uniform time-current characteristics. Hence, rewireable fuses are used for domestic and lighting loads.
b. Fuse wires of different diameters are used for different current ratings and are standardized for rated current up to 200 A [3,6].

Advantages of rewireable fuses:

- Cheap and simple to install.
- Due to rewireable feature, it can be used for long period of time until the contacts are in good condition.
- Low maintenance required.

Disadvantages of rewireable fuse:

- Exact estimation of wire diameter is a challenge at times.
- Less reliable operation, as it is affected by ambient temperature.
- Due to low-current breaking capacity, it cannot be employed with the circuits having the probability of large currents.
- Continuous heating of fuse element results in oxidation, which causes the fuse to blow at a current below its rated current.

2. *High rupturing capacity (HRC) cartridge fuse*: The HRC fuse holds the fuse element that is surrounded by a closed heat resisting medium of ceramic with good mechanical strength. In contrast, Figure 3.5 showcases HRC without a tripping device [3,6].
   a. The whole structure is filled with arc extinguishing agent, commonly quartz powder and packed with brass end caps having fuse link contact. Fuse link is the unit in which fuse element is enclosed and is to be replaced when it blows off [1–4].
   b. Long cylindrical shaped fuse wire is replaced by bimetallic sections joined together with tin joint, as long cylindrical shaped fuse after melting forms molten droplets. The arc will collide between each droplet, which evaporates at a later stage and results in a long arc [1–4].
   c. Tin has a lower melting point than silver. Therefore, tin melts more rapidly than silver element and this can rectify the problem of long arc [1–4].
   d. During arcing period, the chemical reaction between the metal vapour (fuse element vapour) and quartz powder provides a high resistance substance, which assists in arc quenching [3–6].

HRCs are widely employed in low-voltage distribution systems against overload and short circuit condition and for protection of motors, cables, backup protection, etc. Low voltage HRCs are applicable for the breaking capacity of 16,000 to 30,000 A at the voltage level of 440 V.

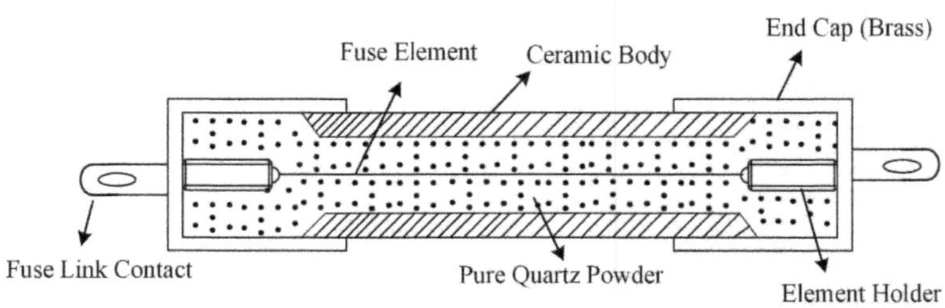

FIGURE 3.5 High rupturing capacity (HRC) fuse.

Advantages of HRC fuse:

- Due to packed construction, it is safe, as probability of arc leakage is too low.
- No maintenance required.
- Quartz-powder-based arc quenching makes the arc quenching approach stronger.
- Operating speed is high.
- Capable of clearing high and low magnitude currents.

Disadvantages of HRC fuse:

- The fuse should be replaced by a new one for each operation.
- It produces overheating of adjacent contacts.
- It is not possible to interlock HRC.

HRC has undergone several advancements with over time, including the addition of tripping circuit with simple HRC to make it more reliable. Figure 3.6 shows an HRC fuse with tripping feature. When faulty condition come into existence, silver fuse elements are the first line of protection to blow off, which is followed by fusing of weak link in series with the tungsten wire. This results in the chemical charge being ignited. Consequently, the plunger is pushed out with a force to activate the circuit breaker but remains attached with the fuse body [3].

Advantages of HRC fuse with tripping function:

- When HRC with tripping circuit is used for three-phase system and when the circuit faces a fault in any one of the three phases, the plunger stimulates the tripping mechanism to trip all three phases.
- It evades the obligation for replacing the fuse during low current operation. Fuse-tripped circuit breaker is capable of handling small fault current itself except the high rated currents for which it is designed.

3.2.4.2 High Voltage Fuses

High voltage fuses are designed for low voltage operations, and further classified as:

1. *Liquid type HV fuses*: These fuses minimize the corona, and the fuse elements are surrounded by a liquid known as carbon tetra chloride. These types of fuses have a wide range of designs and are employed for transformer protection or as a circuit with operating current of 400 A at 132 kV or above. These fuses have the breaking capacity of a current range of 6100 A at power of 350 MVA with 33 kV [11].

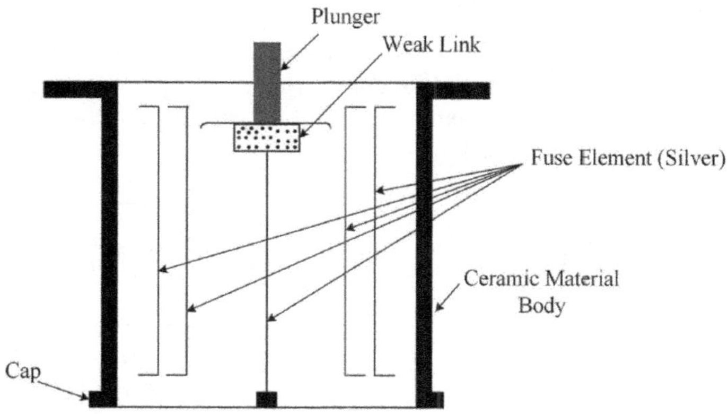

FIGURE 3.6 HRC fuse along with tripping function.

2. *Cartridge type HV fuse*: In some designs of cartridge type fuses, two fuse elements are employed in parallel, one with high resistance and the other with low resistance. During a fault the low-resistance fuse element blows off, and the high resistance element diminishes the short-circuit current and disrupts the circuit. In order to minimize the effect of the corona, the element is wound in the shape of helix. These are employed for voltage levels up to 33 kV with a breaking capacity of 8700 A [11].
3. *Metal clad type HV fuse*: These fuses are oil submerged type fuses and act as a substitute of oil circuit breaker [3–6].

3.3 Circuit Breaker

Circuit breakers (CBs) are mechanical system designed to close or open an electrical circuit under normal or abnormal condition [12]. A relay senses the abnormal condition and sends a tripping signal to the circuit breaker for operation. The maximum value of current that a circuit breaker can interrupt without any damage is known as rupturing or interrupting current. A circuit breaker has two contacts: fixed contact and moving contact. During normal conditions the contacts are in closed position, whereas during abnormal conditions the contacts move to interrupt the circuit that results in an arc between the contact, as shown in Figure 3.7 [1–4,6–12]. The circuit breakers are rated in MVA, and heavy duty circuit breakers interrupt several thousand amperes of high short circuit currents.

3.3.1 Arc Phenomenon

Electrodes are the current carrying contacts in circuit breaker. They remain closed under normal conditions and separate to interrupt the circuit during abnormal conditions. When the current carrying contacts separates, an arc forms. When the electrodes move the area of contact decreases. The high magnitude fault current causes an increase in current density which in turn increase the temperature between electrodes. The heat between the contacts have the ability to strike the electrons from air molecules or ionise the air surrounding the electrodes. This avalanche of free electrons in the medium is sufficient to carry charges and results in arc strikes. The arc has low resistance path; therefore the current in the circuit remains continuous as long as the arc is sustained [12,13].

3.3.2 Arc Interruption and Its Techniques

Till the arc is sustained between electrodes the circuit remains energized and the fault current continues to flow into the circuit, in other words the circuit remained uninterrupted. Therefore, in order to interrupt the circuit under faulty condition or for fast operation of circuit breaker, the arc should be extinguished completely as soon as possible. Some of the arc quenching techniques are explained below.

3.3.2.1 High Resistance Interruption

In this type of interruption, the resistance of an arc increases so that the current tends to zero and is unable to restrike again. This method is not suitable for high voltage because if the potential between the circuit breakers is more than the withstand capacity of the gap, the arc might restrike [1–4,6,9,11,12]. The resistance of the arc can be increased by:

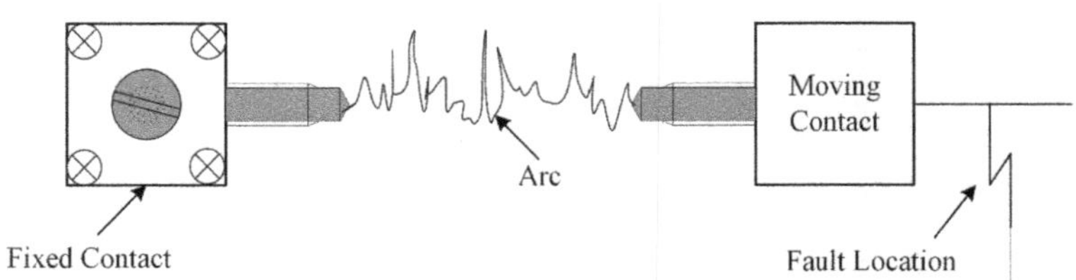

FIGURE 3.7 Circuit breaker contacts.

1. *Cooling the arc*: As discussed previously the heat due to the arc results in striking out of electrons from air molecules, which decreases the resistance of the medium (the medium has an ample number of charge carriers). If the arc is cooled, the rate of ionization of air decreases, which in turn increases the resistance of arc.
2. *Lengthening the arc*: As per the relation,

$$R = \frac{\rho l}{A} \tag{3.3}$$

   resistance can be increased by increasing the length. Thus, in circuit breakers the length of the arc can be increased by increasing the gap length of the contacts, but it is not possible to draw out the contact by such a distance that the voltage available to maintain the arc is disabled.
3. *Splitting the arc*: The arc is split into several small arcs in series, which results in an increase in arc resistance. Each section of the arc experiences the effect of lengthening and cooling [1–4,3–6,11–13].
4. *Constraining the arc*: The arc is constrained into a very narrow channel, which results in a decrease in the voltage necessary to maintain the arc [3,6].

3.3.2.2 Current Zero Interruption

This arc interruption method is used in AC systems where natural current zero is observed twice in one cycle (or we can say 100 zero/sec in a supply of 50 Hz). It is also known as a low resistance method, as the arc resistance keeps low till the current reaches natural zero. The arc becomes extinct and does not restrike after current zero [1–4,13].

- *Restriking voltage*: At normal current zero condition the voltage across the breaker terminals does not normalise instantaneously and behaves as a transient. Hence, the transient oscillations that appear across the breaker contacts during arc extinction are known as restriking voltage and are shown in Figure 3.8 [3,6,9,11–13].
- *Recovery voltage*: The voltage that appears across the contacts after the arc is extinguished and the transient voltage dies out is known as recovery voltage and is presented in Figure 3.8 [11–13].

Two current zero theories are:

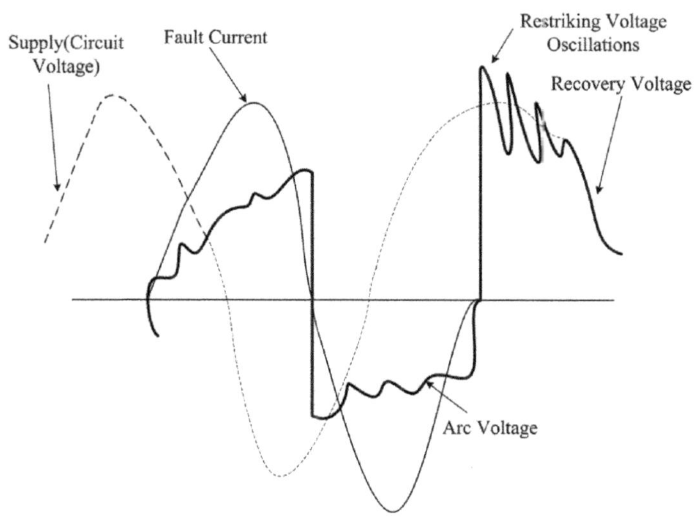

FIGURE 3.8 Restriking and recovery voltage.

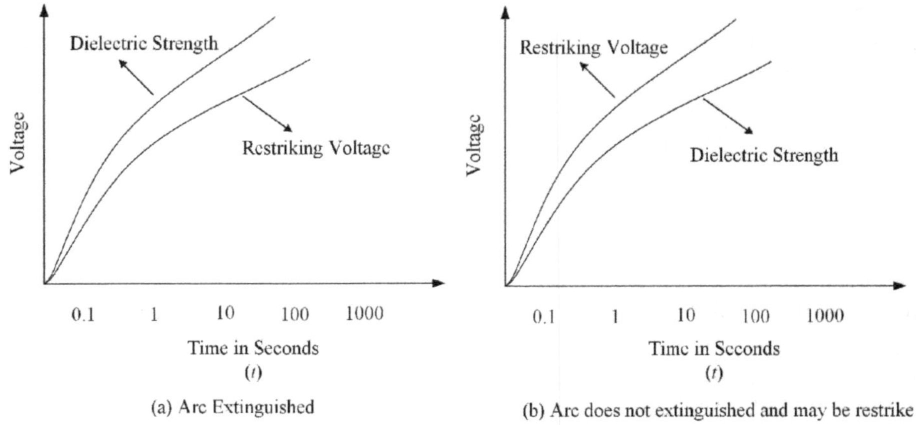

FIGURE 3.9 Relationship between arc voltage and dielectric strength.

1. *Recovery rate theory* (*Slepain's theory*): To extinguish the arc, it is important to eliminate the electrons and ions from the gap instantly after current zero. Removal of ions and electrons can be attained by converting them into neutral molecules through recombination or sweeping them by inserting gas or a liquid medium into the gap [3,6,9].
   a. Recovery rate theory states that the arc in the gap can only be quenched when the rate of removal of ions from the gap is faster than the rate of ionisation [3,6].
   b. Hence, this theory underlines the approach of comparing the rate of rise of restriking voltage with the rate of recovery of the dielectric strength of the gap. As shown in Figure 3.9a and b, if the dielectric strength increases more promptly than the restriking voltage, the arc becomes extinct, and vice versa [1–4,6,9].

   If $\frac{d(Ds)}{dt} > \frac{d(R_V)}{dt}$, then the arc will be extinguished; if $\frac{d(Ds)}{dt} < \frac{d(Rv)}{dt}$, then the arc restrikes and will not extinguished at current zero condition. D_s is the dielectric strength of the medium, and R_v is the restriking voltage.
2. *Energy balance theory* (*Cassie's theory*): According to this theory, there is a dissipation of energy in the form of heat between the contacts. Figure 3.10 depicts that an arc is initiated just after $t = 0$, where power is zero, and reaches its maximum value attaining maximum power. Then it tends to zero power at current zero, where r reaches infinity. At these instants the power increases and then decreases, finally reaches its zero [1–4,6,11]. The circuit breaker should be

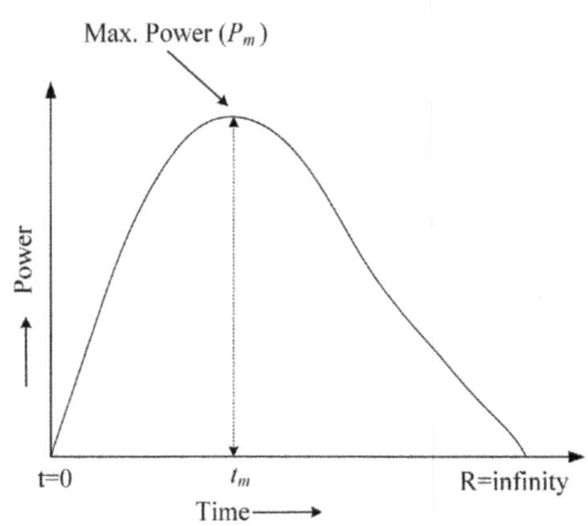

FIGURE 3.10 Energy balance theory.

designed so that it removes the heat from the contact as soon as possible and cools down the contacts immediately. This in turn will increase the resistance between contact gaps. Hence, if the rate of exclusion of heat is faster than the rate of dissipation of heat, the arc is extinguished.

3.3.3 Rate of Rise of Restriking Voltage (RRRV)

Rate of rise of restriking voltage (RRRV) identifies the rate at which the oscillations of restriking voltage changes; the estimation of RRRV identifies the probable voltage transients at current zero [3,6]. In order to express RRRV, assume that a fault occurs on a feeder; the equivalent circuit holds inductance (L_{pp}) and capacitance (C_{pp}) per phase of the system measured at breaker location with resistance R and current i (Figure 3.11).

Restriking voltage oscillates with the natural frequency given by:

$$f_n = \frac{1}{2\pi}\frac{1}{\sqrt{L_{pp}C_{pp}}} \tag{3.4}$$

The voltage across the capacitor is the restriking voltage; therefore, it can expressed in terms of L_{pp}, C_{pp}, f_n, and E_s (system voltage at the instant of arc interruption), as per the circuit:

$$L_{pp}\frac{di}{dt} + \frac{1}{C_{pp}}\int i dt = E_s \tag{3.5}$$

$$i = \frac{dq}{dt} = \frac{d(C_{pp}v_{cp})}{dt} \tag{3.6}$$

$$\frac{di}{dt} = \frac{d^2(C_{pp}v_{cp})}{dt^2} = C_{pp}\frac{d^2(v_{cp})}{dt^2} \tag{3.7}$$

where, v_{cp} is the voltage across capacitor, i.e., restriking voltage:

$$\int \frac{idt}{C_{pp}} = \frac{q}{C_{pp}} = v_{cp} \tag{3.8}$$

Substituting Equations 3.7 and 3.8 in Equation 3.5:

$$L_{pp}C_{pp}\frac{d^2(v_{cp})}{dt^2} + v_{cp} = E_s \tag{3.9}$$

By taking Laplace of Equation 3.9 it is expressed as:

$$L_{pp}C_{pp}S^2 v_{cp}(S) + v_{cp}(S) = \frac{E_s}{S} \tag{3.10}$$

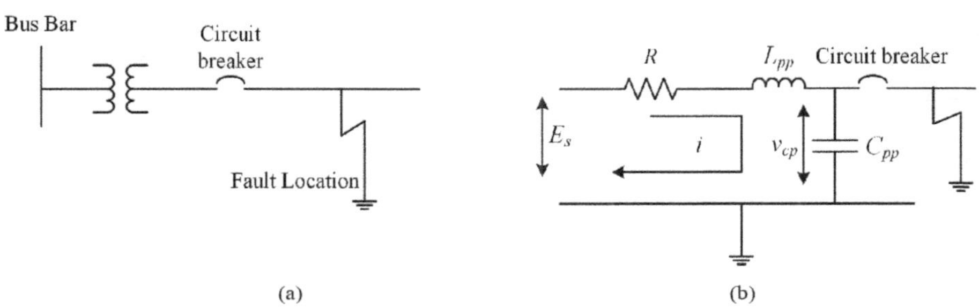

FIGURE 3.11 (a) Fault near feeder and (b) equivalent circuit for RRRV analysis.

$$v_{cp}(S)\left[L_{pp}C_{pp}S^2+1\right]=\frac{E_s}{S}$$

$$v_{cp}(S)=\frac{E_s}{S\left[L_{pp}C_{pp}S^2+1\right]}=\frac{E_s}{L_{pp}C_{pp}S\left[S^2+\frac{1}{L_{pp}C_{pp}}\right]} \tag{3.11}$$

where,

$$w_n=\frac{1}{\sqrt{L_{pp}C_{pp}}}$$

Therefore, $w_n^{\,2}=\frac{1}{L_{pp}C_{pp}}$. By substituting value of $w_n^{\,2}$ in Equation 3.11 we get:

$$v_{cp}(S)=\frac{w_n^{\,2}E_s}{S\left[S^2+w_n^{\,2}\right]}=\frac{w_nE_s}{S}\frac{w_n}{\left[S^2+w_n^{\,2}\right]} \tag{3.12}$$

By taking the inverse Laplace transform of Equation 3.12, we get:

$$v_{cp}(t)=w_nE_s\int_0^t \sin(w_nt) \tag{3.13}$$

On integrating Equation 3.13:

$$v_{cp}(t)=w_nE_s\left|\frac{-\cos w_nt}{w_n}\right|_0^t=E_s\left(1-\cos w_nt\right)=E_s\left(1-\cos\frac{1}{\sqrt{L_{pp}C_{pp}}}t\right)$$

$$v_{cp}(t)=E_s(1-\cos w_nt)=\text{Restriking voltage} \tag{3.14}$$

Equation 3.14 shows that v_{cp} is the restriking voltage expressed in terms of L and C.

Moreover, the rate of rise *(d/dt)*, of restriking voltage (RRRV) is:

$$\text{RRRV}=\frac{dv_{cp}}{dt}=\frac{dE_s}{dt}(1-\cos w_nt)$$

Hence,

$\text{RRRV}=w_nE_s(\sin w_nt)$

PROBLEM 3.1

A three-phase 220 kV undergoes a short circuit, and the breaker operates where power factor of the fault is 0.6. The recovery voltage is 0.7 times the full line voltage. The breaker current is symmetric, and the natural frequency of the restriking transient is 12 kHz. Determine RRRV if the fault that occurred is a ground fault.

Solution

Phase voltage:

$$V=\frac{220}{\sqrt{3}}\times\sqrt{2}=179.62\text{ kV}$$

Recovery voltage is $V_{\max}=0.7\times179.62=125.73\text{ kV}$

The time taken to reach the peak value is $t_p = \frac{1}{2f_n} = \frac{1}{2\times 12\times 10^3} = 41.6\ \mu\text{ sec}$
Active recovery voltage:

$$kV_{\max}\sin\phi = 1\times 125.73\times \sin(\cos^{-1}0.6) = 89.77\ \text{kV}$$

NOTE: For ground fault $k = 1$.

Peak value of restriking voltage $= 2\times 89.77 = 179.54$ kV

$$\text{RRRV} = \frac{\text{Peak restriking voltage}}{t} = \frac{179.54}{41.6} = 4.315\ \text{kV}/\mu\text{ sec}$$

PROBLEM 3.2

Calculate (a) maximum voltage across the contacts of a circuit breaker when short circuit current breaks at the instant of zero current, and (b) frequency of transient oscillations and average RRRV up to first peak of the oscillation. A 60 Hz generator having an e.m.f. to neutral is 6 kV (r.m.s.), reactance of the system and the generator is 7 Ω with a distributed capacitance to neutral is 0.02 μF.

Solution

a. So, maximum voltage across contacts:

Recovery voltage (active), $v = \sqrt{2}\times 6 = 8.48$ kV

Max. restriking voltage (Ph-N) $= 8.48\times 2 = 16.96$ kV

$$X = 2\pi fL_{pp} = 7\ \Omega,$$

b. Transient frequency, $f_n = L_{pp} = \frac{X}{2\pi f} = \frac{7}{2\times 3.14\times 60} = 0.0185\ H$

$$f_n = \frac{1}{2\pi\sqrt{L_{pp}C_{pp}}} = \frac{1}{2\pi\sqrt{0.0185\times 0.02\times 10^{-6}}} = 8.33\ kHz$$

As:

$$\cos\frac{1}{\sqrt{L_{pp}C_{pp}}}t = -1 \text{ or } \cos\frac{1}{\sqrt{L_{pp}C_{pp}}}t = \cos\pi$$

$$t = \pi\sqrt{L_{pp}C_{pp}} = \frac{\pi}{2\pi f_n} = \frac{1}{2f_n} = \frac{1}{2\times 8.33\times 10^3} = 60.02\ \mu\text{ sec}$$

Average RRRV:

$$\text{RRRV} = \frac{\text{Peak RV}}{t} = \frac{16.96}{60.02} = 0.28\ \text{kV}/\mu\text{ sec}$$

3.3.4 Current Chopping

When a low inductive current (magnetising current of a transformer, i.e., i_o) is to be interrupted, the circuit breaker exerts the same force that is required to quench high magnitude current, which is more than sufficient to quench the low magnitude inductive currents. Thus, the current is interrupted before the actual current zero. Hence, *the phenomenon of extinction of current before the actual current zero is known as current chopping.* As electromagnetic energy that is stored in the system cannot becomes zero instantly; it converts into some other forms of energy [1–4,6].

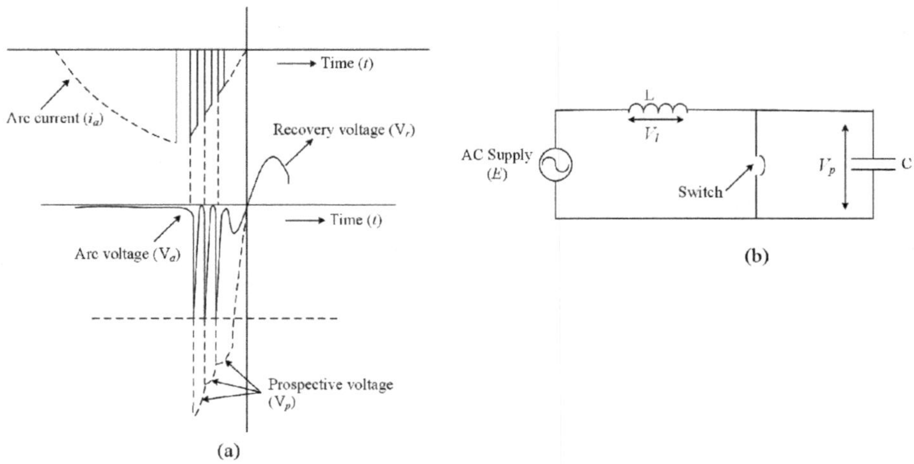

FIGURE 3.12 Current chopping.

The electromagnetic energy is converted into electrostatic energy by possible diversion of current through capacitor, as shown in Figure 3.12b. If i_{ar} is the arc current, C is the capacitance between breaker contacts and L is the inductance of the circuit, the inductive energy converted into capacitive component, is given by $\frac{1}{2}Li_{ar}^2 = \frac{1}{2}Cv_p^2$, which identifies a voltage across circuit breaker contacts, $v_p = i_{ar}\sqrt{\frac{L}{C}}$, i.e., known as *prospective voltage.*

PROBLEM 3.3

A magnetizing current of a 50 MVA transformer at 132 kV is interrupted by a circuit breaker. Magnetizing current magnitude of a transformer is 10% of the full load current. Determine the maximum voltage that appears across the contacts of the circuit breaker if the magnetizing current is interrupted at 27% of its peak value, and if the capacitance and inductance is of 2900 μF and 25 H, respectively.

Solution

Current of a full load transformer:

$$I = \frac{P}{V} = \frac{50 \times 10^6}{\sqrt{3} \times 132 \times 10^3} = 218.69 \text{ A}$$

Magnetizing current is 10% of full load current I, $\frac{10}{100} \times 218.69 \text{ A} = 21.869 \text{ A}$

As per the problem, current chopping occurs at i_{ar} $0.27 \times 21.86\sqrt{2} = 8.34$ A:

$$\frac{1}{2}Li_{ar}^2 = \frac{1}{2}Cv_p^2$$

$$\frac{1}{2} \times 25 \times (8.34)^2 = \frac{1}{2} \times 2900 \times 10^{-6} v^2$$

$$v = 0.24 \text{ kV}$$

3.3.5 Resistance Switching

Resistance switching plays a significant part in current chopping. During chopping a very high voltage appears across the circuit breaker contacts that may cause damage to the circuit breaker. Hence, in order to prevent this, a resistance is connected in parallel with the arc, as shown in Figure 3.13. In other words, to reduce restriking voltage, a resistance is connected in parallel with the arc [1–4,6].

Application of shunt resistance enables the following functions:

- As part of the arc current diverges through shunt resistance, it results in decrease in arc current that in turn is responsible for removal of ions from the arc path and an increase in the arc resistance. Decrease in arc current is responsible for the decrement of restringing voltage and RRRV itself.
- During switching of inductive and capacitive loads, the shunt resistance ensures the damping of high frequency voltage transients.

Voltage equation for the given circuit:

$$L\frac{di_l}{dt}+\frac{1}{C}\int i_{cp}dt = E_s \tag{3.15}$$

where $i_l = i_{cp} + i_r$

$$L\frac{d\left(i_{cp}+i_r\right)}{dt}+V_{cp} = E \tag{3.16}$$

On separating i_{cp} and i_r:

$$L\frac{di_{cp}}{dt}+L\frac{di_r}{dt}+V_{cp} = E \tag{3.17}$$

where

$$\frac{dq}{dt} = i_{cp} = \frac{d(CV_{cp})}{dt}$$

Hence:

$$\frac{di_{cp}}{dt} = \frac{d^2(CV_{cp})}{dt^2} = C\frac{d^2V_{cp}}{dt^2} \tag{3.18}$$

$$\frac{di_r}{dt} = \frac{d\left(V_{cp}/R_p\right)}{dt} = \frac{1}{R_p}\frac{dV_{cp}}{dt} \tag{3.19}$$

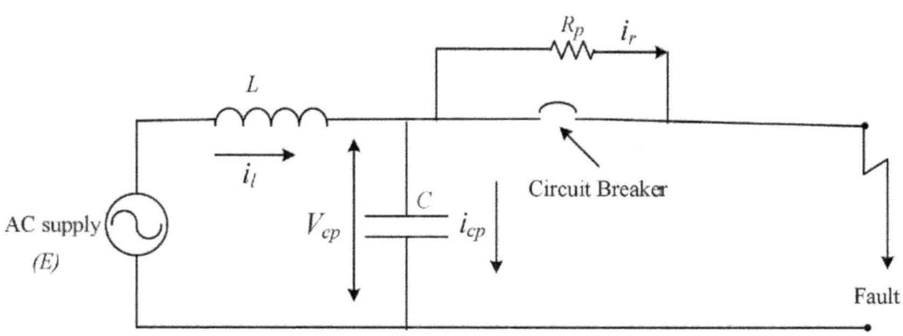

FIGURE 3.13 Equivalent circuit for resistance switching analysis.

On substituting Equations 3.18 and 3.19 in 3.17:

$$LC\frac{d^2(V_{cp})}{dt^2}+\frac{L}{R_p}\frac{dV_{cp}}{dt}+V_c=E \tag{3.20}$$

On taking Laplace of Equation 3.20:

$$LCs^2V_{cp}(s)+\frac{L}{R_p}sV_{cp}(s)+V_c(s)=\frac{E}{s} \tag{3.21}$$

at $t=0$, $V_{cp}=0$

$$V_{cp}(s)=\frac{E}{sLC\left(s^2+\frac{1}{R_pC}s+\frac{1}{LC}\right)} \tag{3.22}$$

In order to estimate the value of resistance that will be attached in parallel to the arc to minimize the restriking voltage, the roots of Equation 3.22 can be analysed. For no transients, all roots of Equation 3.22 should be real. One root is $s = 0$, which is real. The other two roots have to be obtained as follows:

$$\left[\left(\frac{1}{2R_pC}\right)^2-\frac{1}{LC}\right]\geq 0 \tag{3.23}$$

On solving, Equation 3.23:

$$R_p^{\,2}\leq\frac{LC}{4C^2} \tag{3.24}$$

or

$$R_p\leq\frac{1}{2}\sqrt{\frac{L}{C}} \tag{3.25}$$

Equation 3.25 shows that if the value of resistance attached in parallel to the circuit breaker contacts is less than or equal to the $\frac{1}{2}\sqrt{\frac{L}{C}}$, there will be no transient oscillations or restriking voltage, where $R_p=\frac{1}{2}\sqrt{\frac{L}{C}}$ is known as the critical resistance [3,6]. Figure 3.14 shows the restriking oscillations at different values of R.

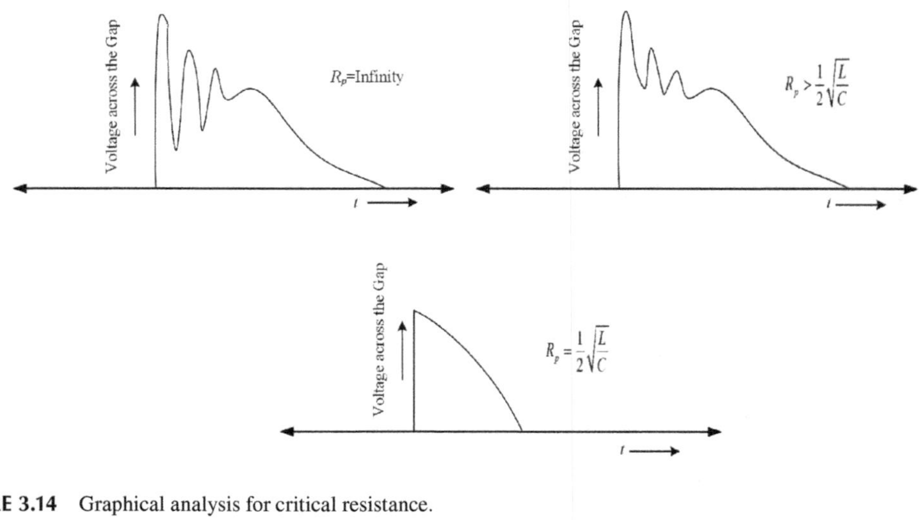

FIGURE 3.14 Graphical analysis for critical resistance.

PROBLEM 3.4

The line to ground capacitance is 0.04 μF, and inductance is measured to 5.4 H of a 66 kV, three-phase 50 Hz power system. Determine the prospective voltage when the magnetizing current of magnitude 6 A is interrupted, and determine the resistance to be connected across the contacts to eliminate the restriking voltage.

Solution

Prospective voltage: $v_p = i_{ar}\sqrt{\frac{L}{C}} = 5\times\sqrt{\frac{5.4}{0.04\times10^{-6}}} = 58.09\text{ kV}$

where, i_{ar} is the arc current or current to be interrupted. The resistance (R_p) to be connected is:

$$R_p = \frac{1}{2}\sqrt{\frac{L}{C}} = 0.5\sqrt{\frac{5.4}{0.04\times10^{-6}}} = 5.8\text{ k}\Omega$$

3.3.6 Classification of Circuit Breaker

Circuit breakers are mainly classified on the basis of voltage level of applications for which they are employed, i.e., *high voltage circuit breakers* and *low voltage circuit breakers*. Involvement of high voltage in high voltage circuit breakers is important to have an arc quenching medium that is capable of withstanding faulty condition. Therefore, transformer oil is popularly used as an insulating and arc quenching medium in oil circuit breakers that are mainly used for high voltage applications [1–4,6] (Figure 3.15).

3.3.6.1 Oil Circuit Breaker

Mineral oil exhibits better insulating properties than air. Thus, mineral oil is used in many electrical peripherals that include protection systems, i.e., circuit breakers. During arc formation, the arc energy is absorbed by oil that decomposes into gas. Oil circuit breaker ranges from 25 MVA at 2.5 kV and 5000 MVA at 250 kV. In bulk oil circuit breakers the contacts are completely submerged into the insulated oil. Plain break, self-generated oil pressure and externally generated oil pressure are the types of bulk oil circuit breakers explained below [1–8,10,11,13].

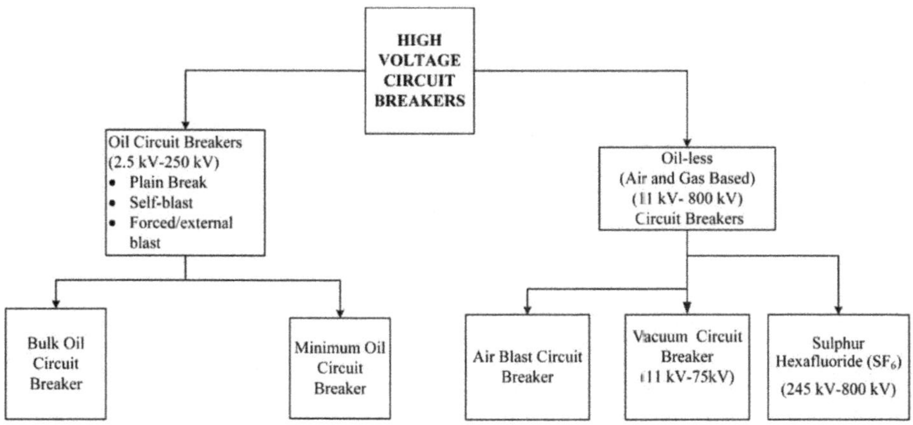

FIGURE 3.15 Classification of high voltage circuit breakers.

Advantages:

- Provide large cooling surface due to the presence of arc quenching medium in close vicinity.
- High dielectric strength.

Disadvantages:

- It is flammable and may cause a fire hazard.
- Periodic replacement and monitoring of oil level is required.
- May cause explosion when mixed with air.

3.3.6.1.1 Plain Break Oil Circuit Breaker

Plain break oil circuit breakers have fixed and moving contacts that are immersed in oil. The metal tank carrying the contact is strong, grounded, weather tight and contains oil up to a certain level along with air cushion above the oil level, as shown in Figure 3.16. During operation, contacts separate, which results in the formation of an arc [3,6]. Consequently, the oil vaporises and forms hydrogen gas. The volume of gas is about 1,000 times that of oil decomposed. After decomposition the arc is surrounded by the gas, and oil is pushed away from the arc. The air cushion provides sufficient room to generate the gas and keep the chamber safe from high pressure consequences. As the contact displaces, the arc length increases, and the arc extinguishes when the distance between fixed and moving contacts reach a critical value and the arc current passes through its zero. Figure 3.16 shows the double break plain oil circuit breaker; two series breaks have been provided for arc elongation without special fast contact. Double break provides an appropriate gap distance before arc interruption. Double break plain oil circuit breaker has the disadvantage of unequal voltage distribution among breaks [3,6,13].

Voltage ratio among breaks are expressed in Equation 3.28, where, V_{b1} and V_{b2} is the voltage across the first and second contact, respectively; C_b is the capacitance between fixed and moving contact. Circuit breaker is under stress due to fault current i_f; C_e signifies the capacitance between moving

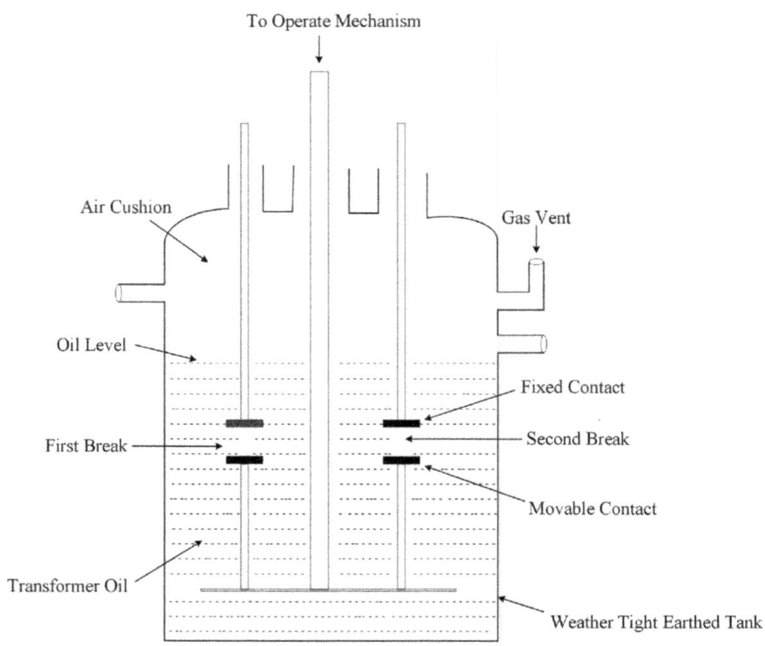

FIGURE 3.16 Plain break oil circuit breaker.

contact and earth. This circuit breaker suffers from inconsistent arcing times and low speed interruption due to this. Plain break oil circuit breakers are limited to low voltage and low current, i.e., not exceeding 150 MVA at 11 kV.

$$V_{b1} = \frac{i_f}{wC_b} \tag{3.26}$$

$$V_{b2} = \frac{i_f}{w(C_b + C_e)} \tag{3.27}$$

$$\frac{V_{b1}}{V_{b2}} = \frac{C_b + C_e}{C_b} \tag{3.28}$$

3.3.6.1.2 Blast Oil Circuit Breaker

In this type of oil circuit breaker, the pressure generated by the arc itself is used to speed up the process of arc extinction. A pressure is created by putting contacts into a pressure chamber or a pot, which results in an increase in breaking capacity of the circuit breaker and a decrease in the operating time. As high pressure is associated with high currents, the challenge is to design the pot to withstand such high pressure; in case of low currents, the pressure is low and the challenge is to produce a pressure that extinguishes the arc. Manufacturers have produced various designs of self-generated pressure oil circuit breaker [1–4,6].

1. *Plain explosion pot:* It is considered the meekest form of explosion pot; it is well labelled in Figure 3.17. During a fault, the moving contact separates and oil decomposes into gas; it generates the pressure corresponding to the arc current within the pot. The pressure forces the oil and gas around the arc to extinguish the arc. If the arc is not extinguished and the moving contact is still inside the pot, then the arc will definitely be extinguished after the moving contact leaves

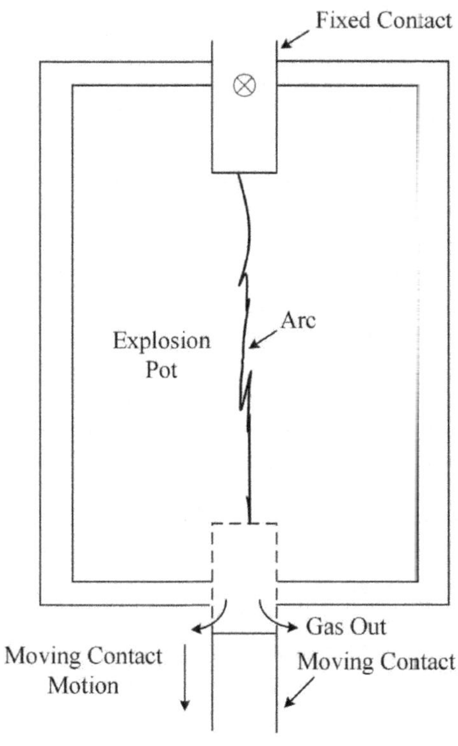

FIGURE 3.17 Plain explosion pot.

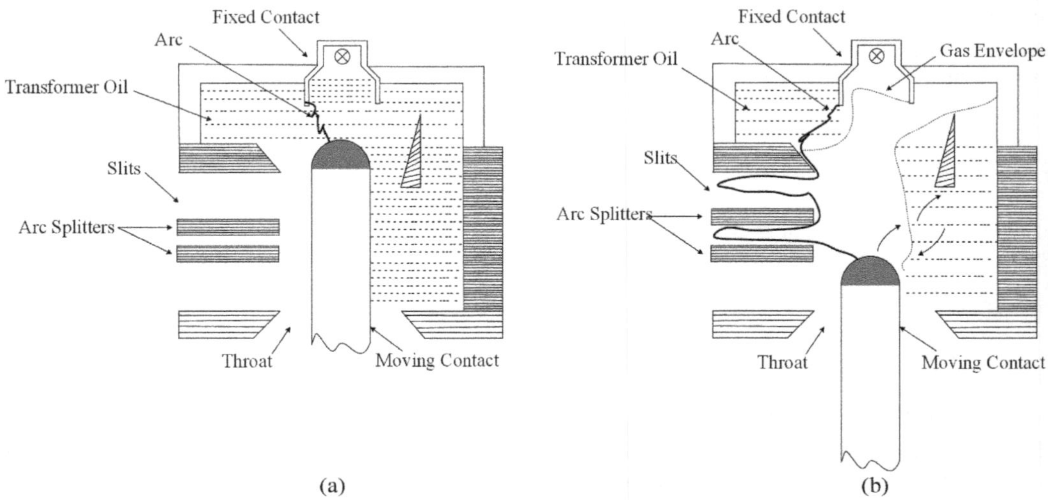

FIGURE 3.18 Cross jet explosion pot.

the pot. This results in the axial blast of the gas, which releases through the throat at the end of the pot. Due to the axial extinction of arc, the plain explosion pot is also known as an axial extinction pot. Its drawbacks are long operating time with low operating current. Moreover, high current causes high pressure and may burst the pot [3,6,13].

2. *Cross-jet explosion pot:* This pot operates on the phenomenon of cross or transverse extinction and is shown in Figure 3.18. The pot is mechanically loaded with channels known as arc splitters that are responsible for lengthening of the arc. Figure 3.18a shows that, during a fault, moving contacts are displaced and an arc is formed between fixed and moving contact. In Figure 3.18b, the gas generated from the arc starts moving the oil backward and results in the motion of the arc sideways into splitters, which causes the lengthening of the arc and arc extinction. This type of construction is suitable for high current operation at 66 kV [3,6,13].
3. *Self-compensated explosion pot*: This pot construction is the remedy for both low and high current operation limitations. It is a combination of both the cross-jet explosion pot and the plain explosion pot, and hence is suitable for both low and high current applications [3,6].

Figure 3.19 shows a self-compensated explosion pot with two chambers: an upper chamber that acts as cross-jet and a lower chamber that behaves as a plain explosion pot. During high fault current, the gas generation is high and the device operates as cross-jet explosion. Arc extinction takes place when arc is forced into slit. When the fault current is low, the rate of generation of gas is low, but pressure builds up till the moving contact reaches the tip. The pressure leakage is very low through the slits, as the oil towards these openings is restricted by the right angle bends and also by the presence of arc. The arc is extinguished when moving contact appears from the throat.

3.3.6.1.3 Externally Generated Pressure Circuit Breaker

In this type of circuit breaker, oil pressure is established by externally operated piston-cylinder arrangement instead of being created by the arc itself. The movement of the piston is coupled with the moving contact, which leads to the appropriate oil pressure for high speed interruption. This circuit breaker has the advantage of requiring less oil. Moreover, it performs well with high current as well as with low fault current applications as the pressure is independent of fault current [1–4,6].

3.3.6.1.4 Minimum Oil Circuit Breaker

As explained previously, in bulk-oil circuit breaker, the contacts are submerged in oil and act as a quenching medium, as well as it insulates the live parts from earth. Note that only a small portion of oil is used

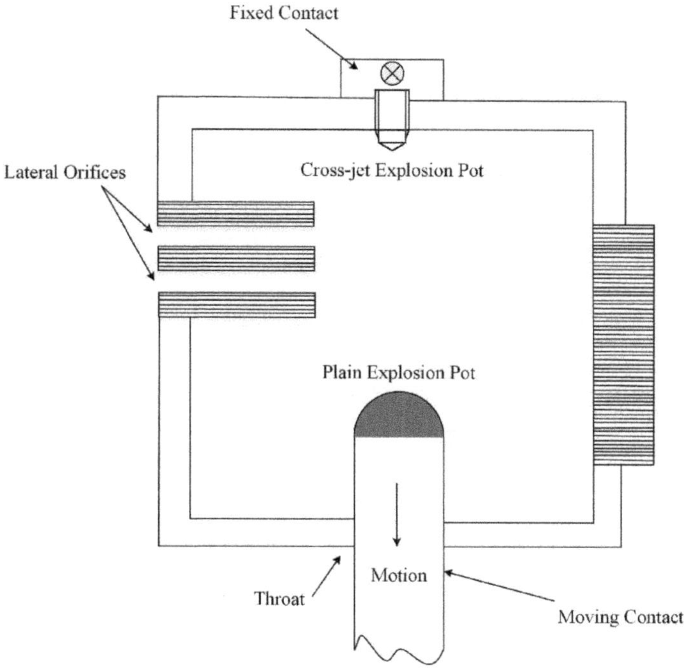

FIGURE 3.19 Self-compensated explosion pot.

for arc extinction; the rest of the oil is used for insulation purpose. With an increase in system voltage, the quantity of oil required also increases, which in turn increases the overall cost of circuit breakers that is to be employed. These disadvantages of bulk oil circuit breakers led to the introduction of minimum oil circuit breakers. Minimum oil circuit breakers employ solid insulation material, and only a small quantity of oil is used for arc extinction. It may be self-blast type, external blast type, or a combination of both. The operating principle of minimum oil circuit breakers is similar to bulk oil type circuit breakers. Glass fibre and reinforced synthetic resin are used as insulation in minimum type circuit breakers. On the basis of venting, two different designs of arcing chambers are provided: axial venting and radial venting. Minimum oil circuit breakers are available up to 8000 MVA, with a total break time of 3–5 cycles. Minimum oil circuit breakers have the disadvantage of high degree of carbonisation that results in the deterioration of dielectric strength of oil [1,4,6].

1. *Axial venting:* The arc is in longitudinal direction and it is employed for low voltage and low current interruption as it generates high gas pressures and has high dielectric strength.
2. *Radial venting:* It is employed for relatively heavy current at low voltage interruption as the gas pressure developed is low with low dielectric strength. Radial venting sweeps the arc in transverse direction.

3.3.6.2 Water Type Circuit Breaker

This circuit breaker converts water into pressurised steam by the arc. This circuit breaker chamber is divided into two parts: lower part (pressure chamber) and upper part (expansion chamber), as shown in Figure 3.20. During a fault, the moving contact starts moving and the water in the pressure chamber starts converting into high pressure steam. When the steam pressure reaches 12 bar, it lifts the lid and expands in the expansion chamber. It results in sudden fall in temperature and pressure in the pressure chamber, which extinguishes the arc. These circuit breakers are small compared to oil circuit breakers and operate in one or two half cycles [3].

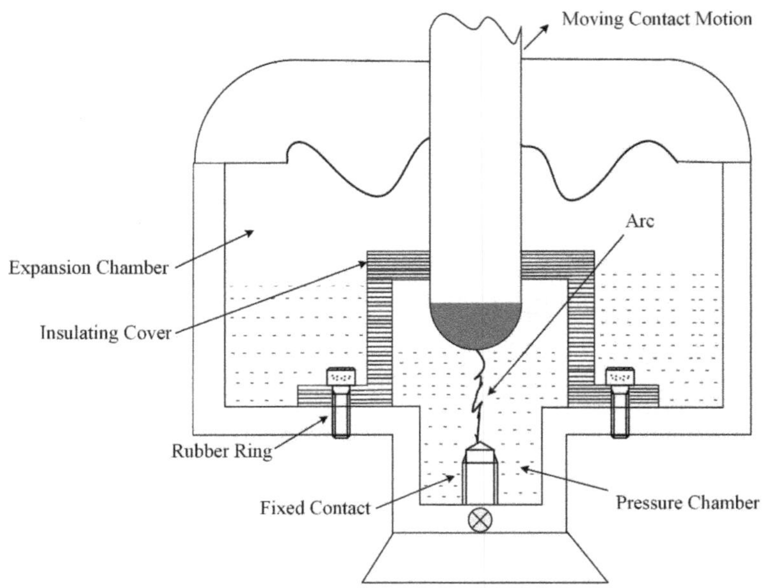

FIGURE 3.20 Water type circuit breaker.

3.3.6.3 Air Blast Circuit Breaker

Compressed air at pressure of 20–30 kg/cm^2 is used as arc quenching medium in air blast circuit breakers. They are suitable for the voltage level of 132 kV and above [3,12].

Air blast circuit breakers are conventional circuit breakers, but their drawbacks are current chopping and restriking voltage, along with noise and maintenance of the air compressor plant. They are being replaced by vacuum circuit breakers at 11–33 kV level and by SF_6 at the voltage level of 132 kV and above.

Air blast circuit breaker have several advantages over oil circuit breakers: They eliminate the fire hazard and are suitable for frequent operation. Air blast circuit breakers are classified into cross-blast and axial blast [12].

1. *Cross blast circuit breaker*: In cross blast air circuit breakers, the blast of air is perpendicular to the arc in order to extinguish the arc. Compressed air from an air compressor is fed into the circuit breaker through suitable chute, as shown in Figure 3.21. A perpendicular blast of air tends towards lengthening of arc through arc splitters, which in turn increases arc resistance and the arc is extinguished. Cross-blast circuit breakers are suitable for current interruption on the order 100 kA at low voltages [3,6,12].
2. *Axial blast circuit breaker*: Axial blast air circuit breaker is further classified into single blast and double blast, as shown in Figure 3.22a and b, respectively. In the axial blast circuit breaker, air is directed axially, i.e., in line with the arc. As the interrupting chambers are fully enclosed in porcelain tubes, these can be employed for extra high voltage (EHV) application, whereas resistive switching is employed to minimize transient overvoltage [3,6,12].
3. *Air-break circuit breaker*: Oil circuit breaker faces a problem for interrupting high currents at low voltage due to carbonization of oil. Hence, an air-break circuit breaker can be employed for applications of high arc current at low voltage. The air-break circuit breaker consist of two pairs of contacts: main contact and arcing contacts, as shown in Figure 3.23. Main contact has low resistance and carries current during closed position. During fault, when the contacts open, the main contact separates first, whereas the arcing contact remain closed and separates later, which results in an arc drawn between them. The resistance of the arc is increased by splitting, cooling and lengthening of arc in air break circuit breaker. As the arc moves upward due to electromagnetic and thermal effect, the length of the arc increases by being pushed into arc

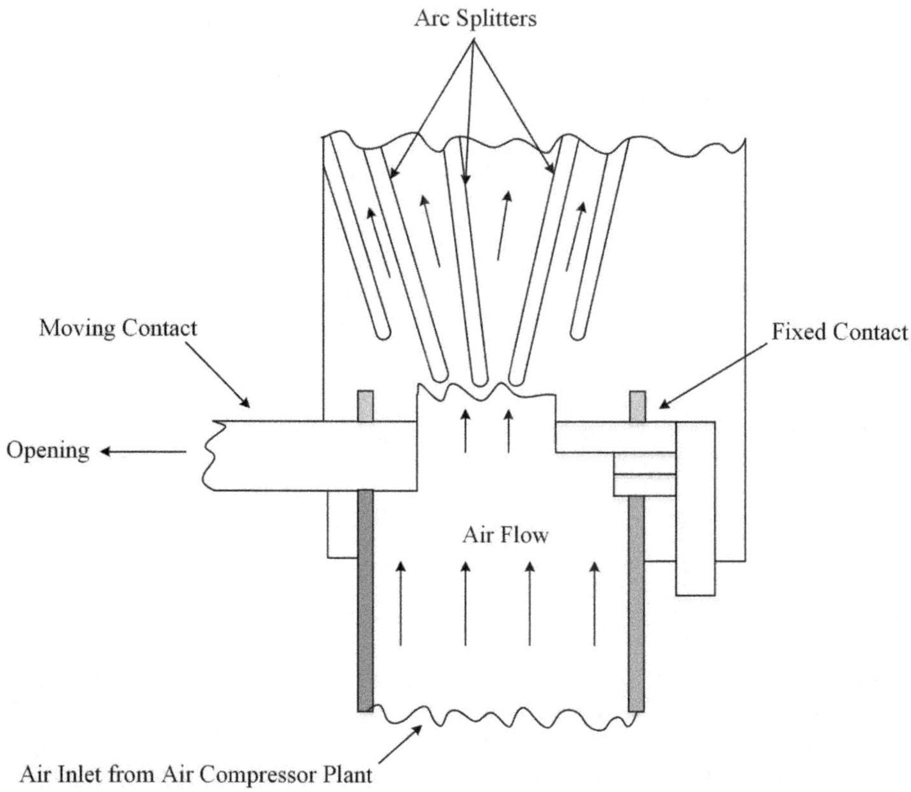

FIGURE 3.21 Cross blast air circuit breaker.

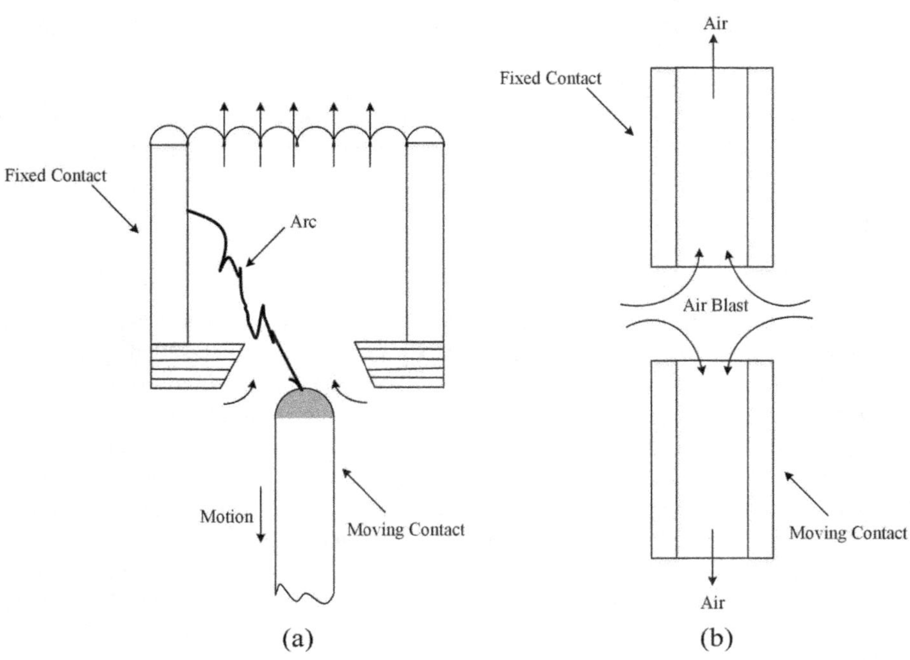

FIGURE 3.22 (a) Single blast type axial blast circuit breaker and (b) double blast type.

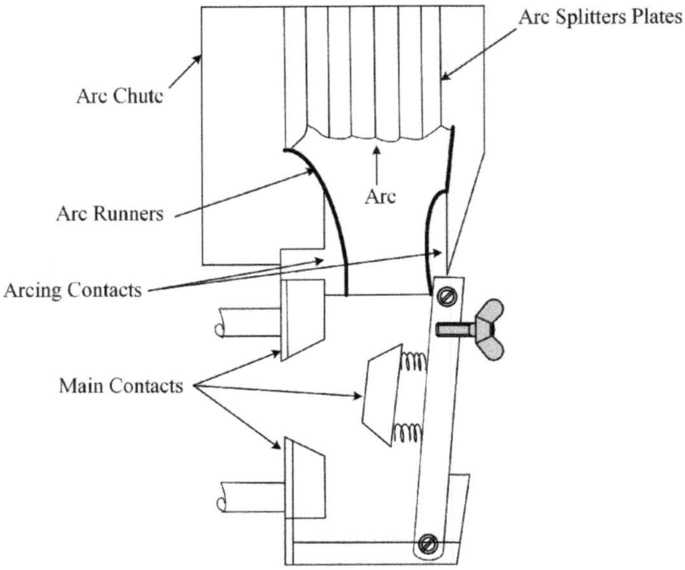

FIGURE 3.23 Air break circuit breaker.

runners and arc chutes [12]. Magnetic field set up by the blown-out coil exerts a force to the arc to speed up the arc movement into arc splitters. The arc is interrupted at current zero. Air-break circuit breaker is available for the voltage range of 400 V to 12 kV. Air-break circuit breakers can be employed in applications that are prone to fire, for example, in an electric furnace with large motor having frequent starting [12].

3.3.6.4 Sulphur Hexafluoride Circuit Breaker (SF_6)

The SF_6 is a combination of sulphur and fluorine. In 1953, the Americans discovered its properties for arc extinction [1–4,14–17]. Properties of SF_6 that make it suitable for arc extinction applications in power systems:

- SF_6 has a very high dielectric strength (2.4 times of air at atmospheric pressure and equal to transformer oil at 3 atmospheric pressure), with excellent arc extinction properties.
- Inert, nontoxic, heavy and nonflammable gas in nature.
- Highly electronegative.
- Outstanding property of recombination after the removal of the source that energises the arc.

Due to the high electronegative property of SF_6, it has higher affinity for electrons, which means that when free electrons collide with SF_6 gas molecule, the free electrons are absorbed by the gas to form negative ions. The negative ions are heavy and keep the gas from being ionised, because negative ion do not attain sufficient energy to ionise. Under normal operating conditions SF_6 remains inert, but at high temperatures, for instance, at 1000°C, it decomposes into agents like sulphur tetrafluoride (SF_4), sulphur dimmer (S_2) and fluoride (F_2). Recombination time of these products is approximately 1 microsecond. It is significant to keep the gas free from moisture as moisture results in the deterioration of contacts, rubber sealing and other sensitive parts of the circuit breaker. Removal of decomposed products from the arc chamber is necessary because these products act as a source of free electrons and may increase the arc quenching time. Therefore, a mixture of soda lime and activated alumina can be placed in the arcing chamber. SF_6 circuit breakers are preferably employed for the current level up to 60,000 A and voltage range of 132 kV to 765 kV [1–4,14–17].

3.3.6.4.1 Constructional and Working Features of SF_6

Figure 3.24 illustrates the structure of SF_6 circuit breaker. The contact faces are coated with copper-tungsten and are housed in an arc extinction chamber. When the breaker contact opens during fault, the valve mechanism allows SF_6 to flow inside the chamber from the gas reservoir. Fixed contacts and moving contact are hollow in construction. Fixed contacts are fitted with rectangular arc horn, and the moving contact consists of rectangular holes in the side to allow the SF_6 gas to be let out after flowing along and across arc. Under normal conditions, the contacts are surrounded by the SF_6 at 2.8 bar pressure; during fault, the moving contact moves, which stimulates the valve and permits the flow of SF_6 at pressure of 14 bar inside the chamber. The high pressure SF_6 absorbs the free charge carriers and forms immobile negative ions. These immobile negative ions result in high dielectric strength of the medium that cause extinction of arc [14–17].

Advantages:

- Silent operation when compared to air blast circuit breakers.
- Can be employed for high current levels as it provides high dielectric strength.
- Free from fire hazard and can thus be employed where explosion hazards persist, for example, in coal and oil fields.
- Short arcing time.

Disadvantages:

- SF_6 gas condenses at low temperatures, pressure level decides the pressure at which the gas condenses into liquid. SF_6 condenses at 15 atm. pressure at a temperature of 10°C. Therefore, to counter low temperature problems, SF_6 should be equipped with thermostatically controlled heaters [3,6,14–17].
- High cost associated with SF_6 increases the overall cost of the circuit breaker.

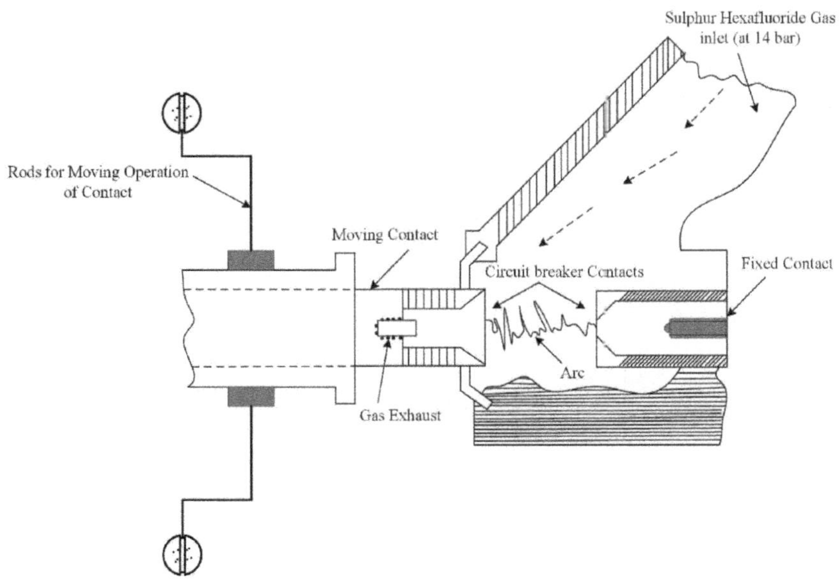

FIGURE 3.24 Sulphur hexafluoride circuit breaker.

3.3.6.5 Vacuum Circuit Breaker (VCB)

Vacuum is defined as the pressure below atmospheric pressure, i.e., 760 mm of Hg. A vacuum circuit breaker holds a low pressure on the order of 10^{-5} to 10^{-7} mm of Hg. Mean free path of air is small, and ionisation results in electron avalanche [3,6]. Under high vacuum (range 10^{-5} mm of Hg) the mean free path of the gas molecules become large (few meters); hence, during a fault, when the contacts move a few mm, the electrons pass through the gap without collision. This inhibits the avalanche of free electrons in between the contact gap, and this is the reason for the remarkable dielectric property of vacuum. In a vacuum circuit breaker, the arc forms due to the evaporation of surface material of electrodes. Arc extinction depends on the arc length and the surface condition of the electrodes. Silver-lead, silver-bismuth, copper-lead, and copper-bismuth are usually used for making contacts in circuit breakers [18–25].

The surface of the electrodes is not smooth and has many micro projections that form the last point of contact, and the current concentrates in these micro projections when contact separates. Due to the small cross section area, the resistance at micro projection contact increases, which results in the resistive heating that leads to the formation of hot spots. These hot spots radiate electrons and act as cathode; as a consequence, an arc forms. Hence, it is important to keep the surface of the contacts smooth and polished.

3.3.6.5.1 Constructional and Working Features of VCB

Figure 3.25 shows the labelled elementary structure of VCB. The vacuum circuit breaker is compounded in a glass envelope sealed with metallic end lids. The transparency of the glass allows easy operation monitoring. A milky white colour of the glass indicates the leakage of vacuum. A spit shield made of stainless steel prevents the metal vapours from reaching the envelope. The shield consists of moving and fixed contacts; moving contact move a distance of 5–10 mm on the basis of operating voltage. Bellows made of stainless steel are used to move the lower contact [1–4,18–25]. One side of the end lid is fixed with fixed contact, whereas moving contact comes out of the chamber for external connection. Vacuum circuit is suitably employed for repetitive operations because it advantages include lower maintenance,

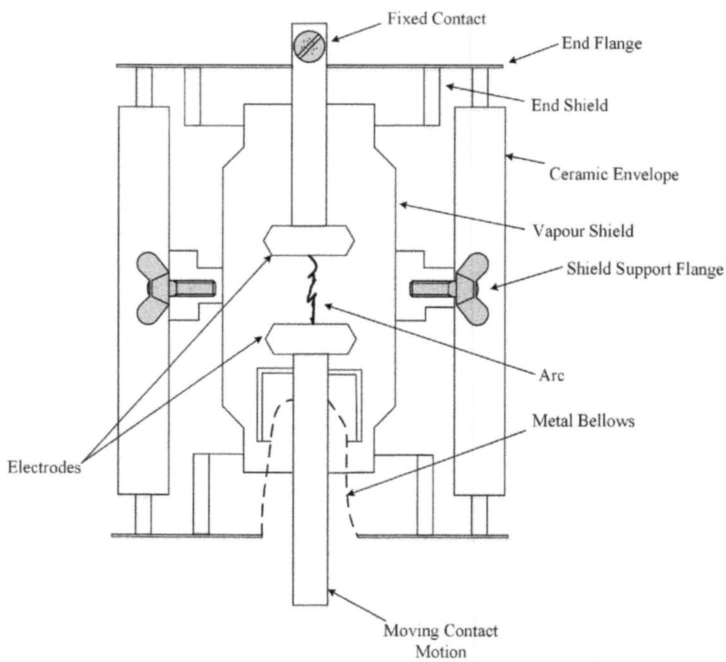

FIGURE 3.25 Vacuum circuit breaker.

silent operation and long life. Arc current increase the vapour emission; near current zero the emission rate of vapour particles tends to zero. Hence, immediately after current zero the vapours condense due to the increase in the dielectric strength of the medium [18–25].

3.3.7 Ratings of Circuit Breaker

Rating categorises the circuit breakers according to the levels for which they are to be employed. In other words, the ratings of circuit breakers depend on their intended duties. Circuit breaker are related on the basis of the functions discussed in the following subsections.

3.3.7.1 Breaking Capacity

Breaking capacity is the ability to open the contacts to isolate the faulty portion of the circuit. Breaking capacity can also be explained as the r.m.s. value of a fault current that a circuit breaker breaks (current for which it can operate) at a given recovery voltage with specified power factor and RRRV. The r.m.s. value of the fault current is estimated at the instant of contact separation. During this time, the current become asymmetrical as it holds both AC and DC components [3,6,13]; the DC component dies swiftly, as shown in Figure 3.26.

Symmetrical breaking current or r.m.s. value of AC component $= i_{sym} = \dfrac{x_{ac}}{\sqrt{2}}$

where:

x_{ac} is the maximum value of AC component
y_{dc} is the DC component

Asymmetrical breaking current holds both AC and DC component, $i_{asym} = \sqrt{\left(\dfrac{x_{ac}}{\sqrt{2}}\right)^2 + \left(y_{dc}\right)^2}$.

The breaking capacity of a circuit breaker is always expressed in MVA. For a three-phase system:

$$BC_{sym} = \sqrt{3} \times \text{ratedvoltage kV} \times i_{sym} \text{ in kA}$$

$$BC_{asym} = \sqrt{3} \times \text{ratedvoltage kV} \times i_{asym} \text{ in kA}$$

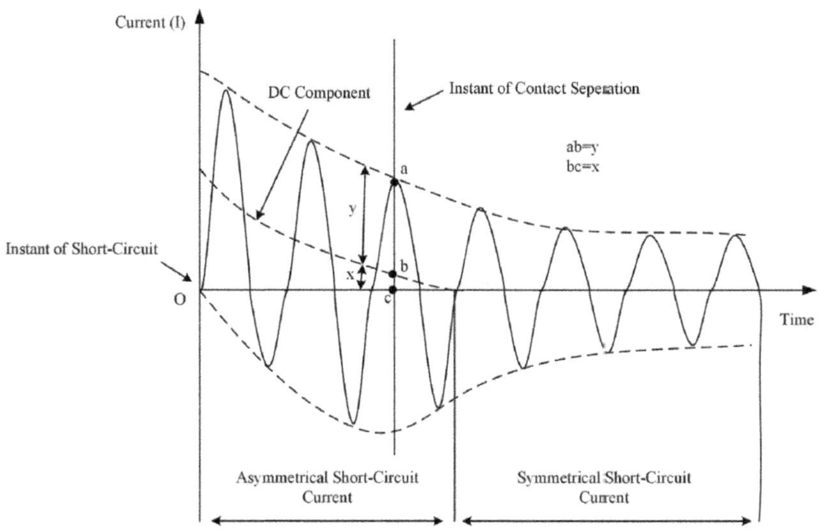

FIGURE 3.26 Short circuit current and instant of contact separation.

3.3.7.2 Making Capacity

Making capacity is defined as the peak value of current during the first cycle of the current wave at which a circuit breaker can be closed on a short circuit. It is also defined as the ability of the circuit breaker to withstand the electromagnetic forces that determine the capability of the circuit breaker to be closed during short circuit [3,6].

Making capacity = $\sqrt{2} \times 1.8 \times$ symmetrical breaking capacity

Making capacity is expressed in peak current. Hence $\sqrt{2}$ converts r.m.s. into peak; 1.8 is used as doubling effect of maximum asymmetry.

3.3.7.3 Short-Time Capability

Short-time capability is the ability of the circuit breaker to carry short circuit current safely until another circuit breaker clears the fault. Short-time is of 3 sec if the ratio of symmetrical breaking current to the rated normal does not exceed 40, and it is 1 sec if the ratio exceeds 40 [13].

3.3.7.4 Rated Voltage, Current and Frequency

The voltage levels are floating in nature and change per the operating conditions. Hence, it is important to specify the level of voltages for which the circuit breaker can operate safely. Rated voltage is defined as the voltage at which the operation of the circuit breaker is guaranteed. Rated current is the r.m.s. value of the operating current that a circuit breaker carries without any rise in temperature or by keeping the temperature with its specified limits. Rated frequency is the operating frequency for the circuit breaker. Change in frequency might affect the performance of circuit breaker, so any change in frequency should be considered [3,6].

3.3.8 Testing of Circuit Breaker

From the previous sections, it is evident that the circuit breakers play a vital role as the line of defence for the power system. Hence, it become necessary to test circuit breakers according to the standards before employing them in the field. There are several methods for testing circuit breakers, and testing is broadly categorised into routine testing and type testing. Routine testing has been done on every piece of circuit breaker manufactured by the firm, whereas type testing has been done on randomly selected circuit breaker units that are tested in a high voltage laboratory as per international standards for breaking capacity, making capacity, duty cycle and short time current test [1–4,26–28]. Furthermore, testing of circuit breakers is classified into several testing modes, as shown in Figure 3.27.

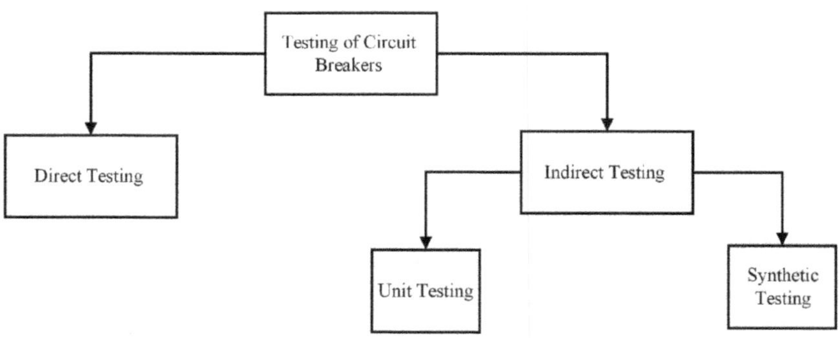

FIGURE 3.27 Block diagram for circuit breaker testing.

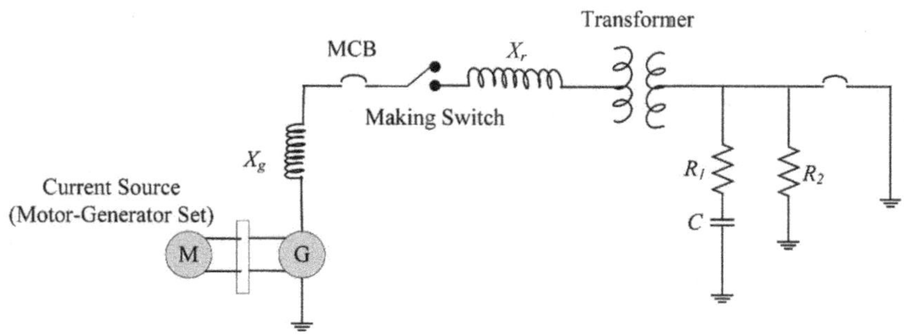

FIGURE 3.28 Direct testing.

3.3.8.1 Direct Testing

Direct testing is the testing of circuit breakers under the condition existing in the power system. Short-circuit tests are performed under direct test, using short circuit testing stations. Circuit breakers are subjected to transient recovery voltage (TRV), which is expected in practical situations [1,3]. Figure 3.28 describes direct testing; reactor X_r is responsible for controlling short-circuit current, whereas C, R_1 and R_2 adjust the transient condition of restriking voltage [26–28].

1. *Breaking capacity testing*: Both the master circuit breaker (MCB) and the circuit breaker under test are closed. Then short circuit current is interrupted by opening the breaker under test at the desired moment of operation. Symmetrical breaking current, recovery voltage, frequency of oscillation and asymmetrical breaking current are measured. The circuit breaker is capable of breaking all the currents up to rated capacity, but testing for all current levels is not possible. Hence, the breaking capacity of circuit breaker is tested for different fractions of rated breaking current [1–4].
2. *Making capacity testing*: The MCB and circuit breaker under test are initially closed. A short circuit is stimulated by closing the circuit breaker under test. Rated making current is estimated by peak value of the first major loop of the short circuit wave [3].

3.3.8.2 Indirect Testing

With the increase in electrical power transmission levels, circuit breakers with increasing breaking capacity are important. For instance, circuit breakers for 245–1100 kV power system face short circuit currents up to 120 kA. Very large testing stations are required to test the circuit breaker of such capacity, whereas increasing the testing capacity of a testing station is not economical [26–28].

The Keuring van Elektrotechnische Materialente Arnhem (KEMA) is the largest test facility in the world, with a short circuit power of 8400 MVA at 145 kV having 31.5 kA, three-phase direct test setup. Presently, the SF_6 circuit breaker consists of a single interrupting chamber with an interrupting power level of 10 GVA. Even the KEMA laboratory does not have the capacity to verify such levels through direct testing. In India, Central Power Research Institute (CPRI) testing facility is responsible for testing of power equipment. CPRI testing facilities are graded up to 2500 MVA at 36/72.65 kV for three-phase and 1400 MVA at 245 kV for single-phase testing. Hence, limited power for testing of high voltage and extra high voltage circuit breakers and the cost associated with the installation of testing stations are the challenges for direct testing. Therefore, indirect testing is needed to quantify the limitations with direct testing. Indirect testing is used for the testing of large circuit breaker, and indirect methods of testing are unit and synthetic testing [6,26–28].

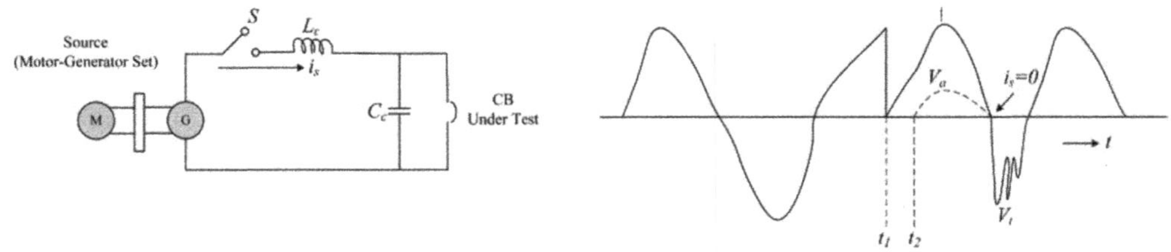

FIGURE 3.29 Equivalent circuit and voltage waveform for test circuit breaker.

1. *Unit testing*: The IEC defines unit testing as the test made on a making or breaking unit or group of units at the making current or the breaking current, specified for the test on the complete pole of a circuit breaker and at the appropriate fraction of the applied voltage, or the recovery voltage, specified for the test on the complete pole of the circuit breaker. High voltage circuit breakers have several arc interrupter units connected in series. In unit testing approach one unit is tested separately; on the basis of testing results, the capacity of whole circuit breaker is estimated [1–4,26–28].
2. *Synthetic testing*: In synthetic testing, two power supply sources are employed, i.e., current source and voltage source. Current sources have high current and low voltage configuration that supplies short circuit current during test. Whereas voltage source have high voltage with low current supplies. According to Figure 3.29, a short circuit current i_s flows through CB under test when switch S is closed. During this time, the circuit breaker contacts begin to open and arc voltage V_a appears across the contacts of the circuit breaker [26–28]. Arc extinguishes during current zero instant, and a transient voltage V_t appears across the circuit breaker. Circuit constants L_c and C_c along with the generator characteristic are responsible to identify the transient voltage. In order to clear the fault, the breaker has to withstand the transient recovery voltage. It is also important to know that during the period of short circuit current flows, there is a small arc voltage across the breaker contacts, whereas much less current flows through the circuit breaker during the period of transient voltage [26–28].

Hence, it can be summarised that, in order to simulate the arc voltage condition, there is no need to use a single high MVA power source, as the tested current can be supplied by low voltage sources. The arc voltage is only 1%–3% of the rated voltage of the circuit breaker. On the other hand, the transient recovery voltage can be simulated by applying low energy, high voltage source with low current rating.

Advantages:

- Test voltage and current can be varied independently.
- Test plant capable to test the breaker with test capacity five times that of the test plant.
- The current supplied to the circuit breaker by short-circuit generator is supplied at relatively low voltage.
- The circuit breaker can be analysed or tested for desired TRV and RRRV.

3.3.8.2.1 Types of Synthetic Testing

Synthetic testing methods are classified as:

- Current injection
- Voltage injection

In general, the on sequence of voltage source classifies the type of synthetic testing, as switched on of voltage source before current zero is known as current injection, whereas the on condition of voltage source after current zero is identified as voltage injection [26–28]. Further classifications are shown in Figure 3.30.

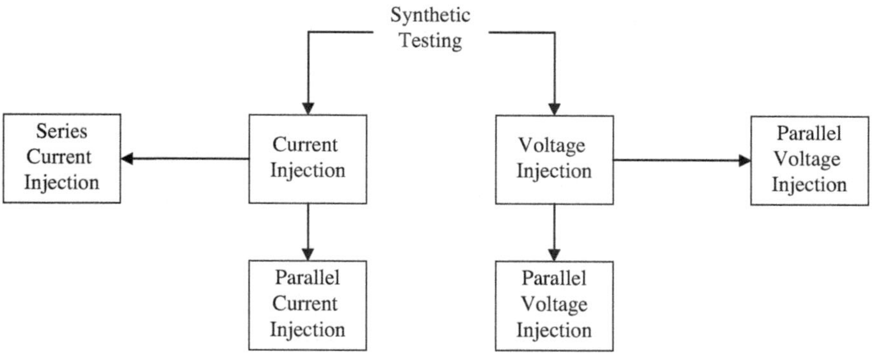

FIGURE 3.30 Block diagram for synthetic testing.

1. *Current Injection Methods as per IEC 62271-101*: As prescribed in Figure 3.30, the current injection methods are further classified as parallel and series current injection methods of circuit breaker testing. In parallel current injection, the voltage circuit is in parallel with the breaker to be tested; in series current injection, the test circuit breaker is in series with the voltage circuit [26–28].
   a. *Parallel current injection method*: This method of circuit breaker testing is explained in Figure 3.31. A short circuit current i_s is injected at low voltage in the circuit through a high current source generator. L_{cs} controls the short circuit current (i_s) from the generator, MCB and CB under test trips, before i_s reaches natural zero. The MCB and CB are fully opened at t_o, MCB is us ed as backup circuit breaker if CB under test fails. Closing of the make switch (MS) allows the flow of short circuit current at the desired moment during test. Control circuit is employed to trigger the spark gap at the desired moment when capacitor C_r charges to give require recovery voltage, where L_c, R_c, and C_c control recovery voltage and RRRV. Magnitude and frequency of transient recovery voltage (TRV) depend on the circuit parameters and the voltage to which the capacitor bank C_r is charged. An automatic controller is used for closing and opening of the CB [1–4,6,26–28]. The steps for the parallel current injection method of circuit breaker testing (in contrast to Figures 3.31 and 3.32) [6,26–28] are as follows:
      i. The test is stimulated by closing the MS, which in turn initiates the current i_s, from low voltage and high current source generator operated through CB under test.
      ii. As the i_s tends to zero, the spark gap is triggered by the controlling circuit at time t_{st}, and injected current i_{in} begins to flow.
      iii. The current i_s+i_{in} flows through the CB under test until the time reaches t_{in}.

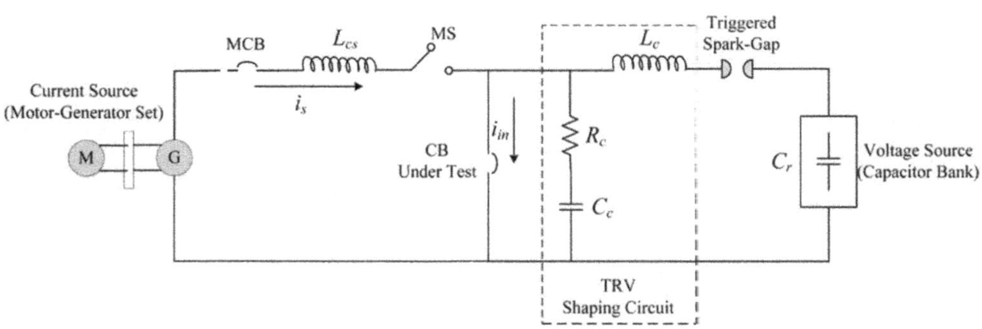

FIGURE 3.31 Parallel current injection method.

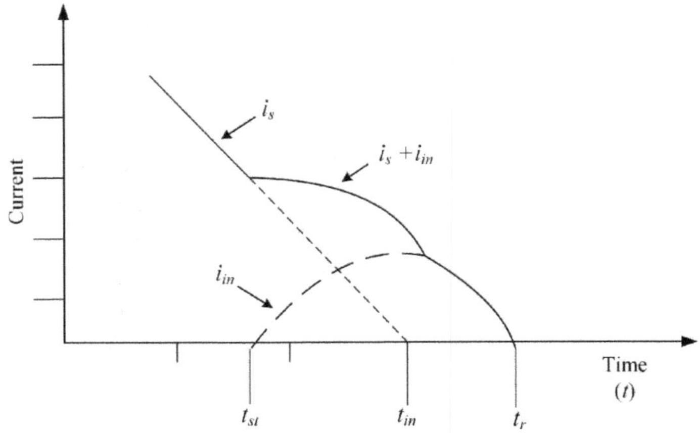

FIGURE 3.32 Waveform for parallel current injection testing method.

iv. At t_{in} the i_s goes to zero, whereas at t_r, the isolation breaker separates the two sources of power. This results in the interruption of injected current, which in turn supplies the high voltage by the voltage source (capacitor bank) that provides the desired TRV that subsequently appears across the terminals of CB under test.

b. *Series current injection method*: Voltage circuit is in series with CB under test in series current injection method or in parallel with auxiliary CB and source capacitance. The voltage circuit is connected to the current circuit in series before i_s reaches current zero [1–4,26–28].

Figures 3.33 and 3.34 show that the capacitor C_r is charged with opposite polarity to the current, and hence injected current i_{in} is opposite to i_s and subtract from i_s. As voltage and current circuit are in series; therefore, it is difficult to estimate the circuit parameters to suit both current and voltage circuits to give TRV at the same time. As i_s and i_{in} are equal and opposite at t_{in}, the current in the isolating breaker is interrupted and the current flowing from instant t_{in} to t_r through the CB under test is i_{tst}.

2. *Voltage injection methods as per IEC 62271-101*: Voltage injection methods are classified as parallel and series voltage injection methods of circuit breaker testing. In this type of testing, the voltage circuit is activated after current zero condition, circuit provides entire current i_s and the first part of TRV to the CB under test after current zero. In order to estimate parameters

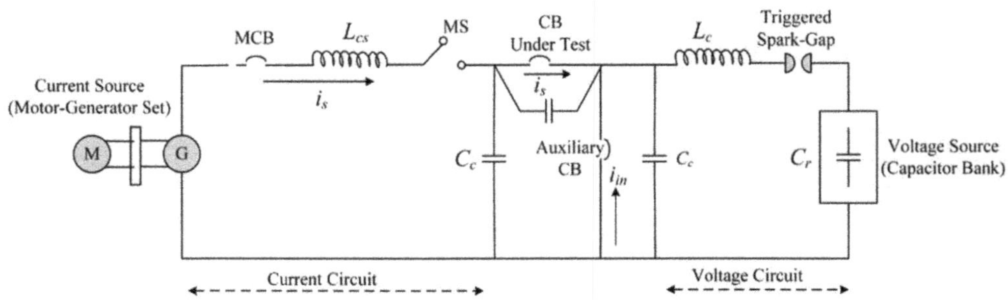

FIGURE 3.33 Series current injection method.

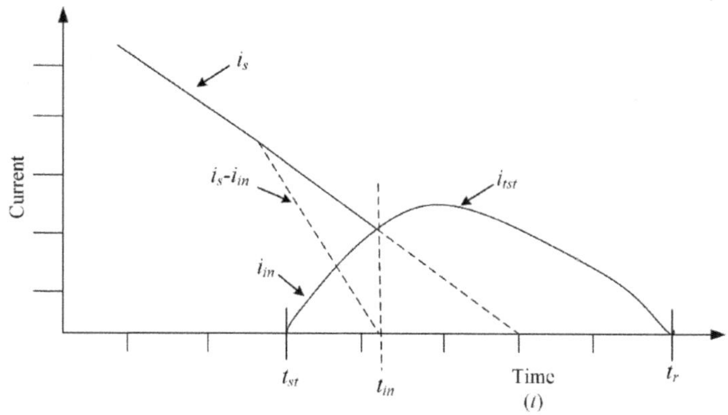

FIGURE 3.34 Waveform for series current injection testing method.

like power factor, current and first part of TRV, choice of correct value of voltage and natural frequency is essential. The voltage circuit is switched on with the help of voltage-controlled circuit so that a quantified transient recovery voltage is sustained at the instant of first peak of transient recovery voltage of current circuit. Hence, delay between the current stress and the voltage stress is eliminated [26–28].

a. *Series voltage injection method of circuit breaker testing*: In this, the voltage circuit is in series with the CB under test (we can also say that it is in parallel with the auxiliary CB). Entire i_s is supplied by the current circuit, whereas a suitable value capacitor is connected in shunt with the auxiliary circuit breaker. This capacitor is responsible for the transmission of the entire transient recovery voltage of the current circuit to the test circuit breaker after current zero of the power-frequency short circuit current. During this, necessary energy is passed for the post-arc current [3,6].
b. *Parallel voltage injection method of circuit breaker testing*: In this method of synthetic testing, the voltage circuit is in parallel with the CB under test; the rest is similar to the series voltage injection method. This method of testing is not commonly used [3].

3.3.9 Concept of High Voltage DC Circuit Breaker

HVDC found its relevance over AC due to the advantage of less corona, less radio interference and low transmission cost. Hence, HVDC is preferred over HVAC for point-to-point transmission over long distances. Power-electronics-based circuits are employed for current controlling, whereas switching operation are controlled by the circuit breaker on AC side. Circuit breakers for HVDC are still in the research phase where more advancement is required to achieve the goal of high power system stability and safety. HVDC circuit breaker face the challenge of achieving the current zero instant. In AC circuit breaker it is easy to quench the arc at its current zero instant, but in case of HVDC the current zero is not available. If the current is suppressed abruptly in DC, a high transient voltage appears at the contact of the circuit breaker. Hence, artificial current zero circuits are employed with the HVDC circuit breakers [6].

Figure 3.35 shows the HVDC having MCB with artificial current zero circuit along with the transient voltage suppression circuit. The combination of R_v and C_v in parallel with the MCB reduces *dv/dt* after final current zero instant. In order to reduce *di/dt*, a saturation reactor L_i is placed in series with MCB, whereas parallel *LC* circuit is responsible for artificial current zero. Switch *S* is triggered immediately after the separation of contacts. The capacitor C_{pre} is pre-charged with the given polarity, discharged through the main MCB, and sends the current opposite to the circuit current that suppress the main current zero with few oscillations.

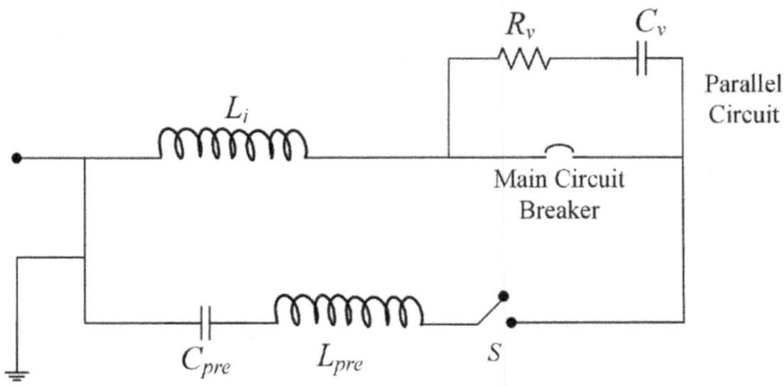

FIGURE 3.35 HVDC circuit breaker.

3.4 Tutorial Problems

1. Explain the term *abnormal conditions* in power systems.
2. Why it is important to protect the power lines and other power peripherals from overvoltage conditions?
3. Explain resistance switching and current chopping.
4. List the factors that are responsible for increasing the resistance of arc path.
5. What are the reasons for sustaining an arc?
6. Explain restriking voltage and rate of restriking voltage, and derive the expression for RRRV.
7. What is pickup value?
8. List the advantages of HRC fuse over rewireable fuses.
9. What are the ratings of circuit breakers?
10. What are circuit breakers and how can you classify them on the basis of their medium of arc extinction?
11. List the conditions that are important for the efficient operation of vacuum circuit breaker.
12. Compare the oil and SF_6 based arc circuit breakers.
13. What are the advantages of vacuum circuit breakers over air blast circuit breakers?
14. Explain the importance of testing of circuit breakers.
15. Classify the modes of testing of a circuit breakers?
16. What are the challenges to designing circuit breakers for DC-based power systems?
17. The line to ground capacitance is 0.09 μF and inductance is measured to 6.3 H of a 33 kV, three phase 60 Hz power system. Determine the prospective voltage when the magnetizing current of magnitude 8 A is interrupted and the resistance to be connected across the contacts has to eliminate the restriking voltage.
18. Calculate the breaking current of a 1900 A, 500 MVA, 33 kV, three-phase, 2 sec rated circuit breaker.
19. Calculate the maximum voltage across the contacts of avcircuit breaker when short circuit current breaks at the instant of zero current. Assume that there is a 50 Hz generator with an e.m.f. to neutral is 8 kV (r.m.s.), reactance of the system and the generator is 6 Ω, and the distributed capacitance to neutral is 0.03 μF.
20. A three-phase 132 kV undergoes a short circuit, and the breaker operates where power factor of the fault is 0.7, and the recovery voltage is 0.6 times the full line voltage. The breaker current is symmetric and natural frequency of the restriking transient is 13 kHz. Determine RRRV if the fault that occurred is a ground fault.

3.5 Conclusion

Protection of power systems is the major concern among researchers. As the power applications change, protection peripherals undergo various modifications, for instance, from basic wire fuse to HRC fuse. This chapter covers various principles and concepts of circuit breakers that are important to understand for a power engineer. Moreover, testing of circuit breakers highlights the prominence of testing circuit breakers under different spans of time for reliable operations throughout their tenancy. Selection of appropriate circuit breakers along with relays leads to the design of reliable protection scheme.

REFERENCES

1. Electricity Council, *Power System Protection*, vol. I, II and III, 1st ed., London, UK: Macdonald & Co. Ltd., 1969.
2. B.D. Russel and M.E. Council, *Power System Control and Protection*, New York, USA: Academic Press, 1978.
3. R.K. Rajput, *Power System Engineering*, 2nd ed., New Delhi, India: Laxmi Publications, 2015.
4. S.S. Rao, *Switchgear and Protection*, 16th ed., New Delhi, India: Khanna Publishers, 2016.
5. D. Robertson, *Power System Protection Reference Manual*, London, UK: Oriel Press, 1982.
6. B. Ram and D.N. Vishwakarma, *Power System Protection and Switchgear*, 34th ed., New Delhi, India: McGraw-Hill, 2010.
7. A.E. Guile and W. Paterson, *Electrical Power Systems*, 2nd ed., vol. I & II. Oxford, UK: Pergamon Press, 1977.
8. B. Ravindranath and M. Chander, *Power System Protection and Switchgear*, 1st ed., New Delhi, India: New Age International, 1977, Reprint 2005.
9. M. Murano et al., "Current zero measurement for circuit breaking phenomenon," *IEEE Trans. Power App. Syst.*, vol. 94, no. 5, pp. 1890–1900, 1975.
10. J.G. Steel and D.T. Swift-Hook, "Statistics of circuit breaker performance," *Proc. IEEE (IET)*, vol. 117, no. 7, pp. 1337–13345, 1970.
11. C.J.O. Garrard, "High voltage switchgear: A review of progress," *Proc. IEEE (IET)*, vol. 113, no. 9, pp. 1523–1539, 1966.
12. Siemens, *High voltage Circuit Breakers: Trends and Recent Developments*, Berlin, Germany: Siemens AG Energy Sector, 2011.
13. E.B. Rietz and J.W. Beatty, "Effect of voltage recovery rates on interrupting performance of air-blast circuit breakers," *IEEE Trans. Power App. Syst.*, vol. 72, no. 2, pp. 303–311, 2008.
14. W.M. Leads et al., "The use of SF_6 for high power arc quenching," *IEEE Trans. Power App. Syst.*, vol. 76, no. 3, pp. 906–909, 1957.
15. T. Ushio et al. "Practical problems on SF_6 gas circuit breakers," *IEEE Trans. Power App. Syst.*, vol. 90, no. 5, pp. 2166–2174, 1971.
16. R.E. Friedrich and R.N. Yeckley, "A new concept in power CB design utilizing SF_6," *IEEE Trans. Power App. Syst.*, vol. 78, no. 3, pp. 695–702, 1959.
17. T. Nitta and Y. Shibuya, "Electrical breakdown of long gaps in SF_6," *IEEE Trans. Power App. Syst.*, vol. 90, no. 3, pp. 1065–1071, 1971.
18. T.V. Armstrong and P.B. Headley, "Vacuum techniques in modern circuit breakers sign in or purchase," *IEEE Trans. Electronics Power*, vol. 20, no. 5, pp. 198–201, 1974.
19. S. Sehrawat and K. Pandey, "Evolution of vacuum circuit breakers and the role of axial magnetic field (AMF) technology," in *Proceedings of 2015 International Conference on Recent Developments in Control, Automation and Power Engineering* (*RDCAPE*), Noida, India, Amity University, October 2015.
20. H. Fink, M. Heimbach and W. Shang, "Vacuum interrupters with axial magnetic field contacts," *ABB Technological Review*, Germany, pp. 59–64, 2000.
21. B. Fenski et al., "Characteristics of a vacuum switching contact based on bipolar axial magnetic design," *IEEE Trans. Nucl. Plasma Sci. Soc.*, vol. 27, no. 4, pp. 949–953, 1999.
22. A.S. Joglekar and B.N. Karekar, "Contact material for the development of vacuum circuit breaker: A review," *J. Inst. Eng. (India)*, vol. 53, 1973.

23. R.B. Shores and V.E. Philips, "High vacuum circuit breaker," *IEEE Trans. Power App. Syst.*, vol. 94, no. 5, pp. 1821–1830, 1975.
24. C. Zhiyuan et al., "Breaking capacity test of vacuum circuit breaker using synthetic test circuit with current zero forecast," *IEEE 21st International Symposium on Discharges and Electrical Insulation in Vacuum*, October 2004.
25. J. Panek, "Synthetic methods for testing of vacuum breakers," *IEEE Trans. Power App. Syst.*, vol. 97, no. 4, pp. 1328–1336, 1978.
26. J.G.P. Anderson et al., "Synthetic testing of AC circuit breaker: Part I, methods of testing and relative severity," in *Proc. IEEE*, vol. 113, no. 4, pp. 611–621, 1966.
27. J.G.P. Anderson et al., "Synthetic testing of AC circuit breaker: Part II, Methods of circuit breaker proving," in *Proc. IEEE*, vol. 115, no. 7, p. 996, 1968.
28. G. Casagrande and S. Rovelli, "Synthetic testing under short-circuit fault conditions," in *Proc. IEEE*, vol. 115, no. 1, pp. 136–140, 1968.

4

Instrument Transformers

Rajiv Singh and Asheesh Kumar Singh

CONTENTS

4.1 Introduction

Instrument transformers (ITs) are mainly used for the measurement of electrical parameters like currents, voltages, power in power system and for operation of protection devices like relays. This chapter focuses mainly on the theory, operation and use of ITs in detail. Due to limitations of dielectric materials and the current carrying capacity of the conductors, the voltmeters can be economically designed to measure the voltages in the range of 100–220 V, while the ammeters can be conveniently designed to measure currents in the range of 1–5 A [1]. These instruments with limited range of measurement are normally not suitable for measuring electrical parameters in modern power systems generally operating at high voltage and current levels. Measurements in modern power systems may involve voltage measurements in the range of 66–400 kV in primary transmission systems and distribution voltages in the range of 6.6–11 kV in primary distribution systems [2]. The problems get much more complex when transient currents up to the value of 40,000 A are required to be measured in order to operate the protective relays for disconnecting the faulty portion of the power system using circuit breakers (CBs). Though shunts and multipliers to some extent can be used for extending the range of conventional ammeters and voltmeters like d.c. measurements, they too, however, find limited applications in a.c. measurements due to the inherent mismatch between the time constant of the meter and the shunt during manufacturing [2,3].

A better illustration of the range extension of indicating instruments using shunts and multipliers is presented in the subsequent sections. ITs serve as a better alternative than shunts and multipliers for extending the range of conventional ammeters and the voltmeters without subjecting the involved humans and the instruments to the hazards of high voltages and currents. ITs are normally used in conjunction with the measuring instruments for measurement of electrical parameters; hence the name instrument transformers has been suitably coined for them. ITs referred as current transformers (CTs) are connected with ammeters for the measurement of current, as depicted in Figure 4.1. In CTs, the value of secondary current (I_s) is lowered substantially in proportion to the primary current (I_p) and theoretically has zero angle of phase difference between the two [3]. The lowered currents are fed into the ammeters from the secondary of CTs for display. Similarly, the voltage transformers (VTs), also referred as potential transformers (PTs), are used along with the voltmeters for the measurement of voltages, as depicted in Figure 4.2. They reduce the value of secondary voltage (V_s) compared to the primary input voltage (V_p), and theoretically both are in the same phase [3]. The similarity between the power transformers and CTs can be drawn on the basis of the common operating principle of electromagnetic induction; however, significant differences between the two can be established on the basis of their designs and use. The CTs used for measurement and metering purposes are generally referred as measuring CTs while those used for protection purposes with the relays are called protection CTs [2,3]. Fair accuracy requirements are desired from measuring CTs for current measurements in the range of 125% of the rated value; however, for the overcurrents beyond this range, the accuracy concerns are nominal due to the core saturation [4].

VTs are more or less the power transformer equivalents with comparatively reduced transformation errors, and they possess similar accuracy requirements as CTs for both the measurement and protection applications. In some special applications involving extended voltage measurements using VTs, additional accuracy limits are imposed depending on the value of permissible errors [3,4]. Sometimes, a residual winding connected in open delta is used with VTs for protection applications [4].

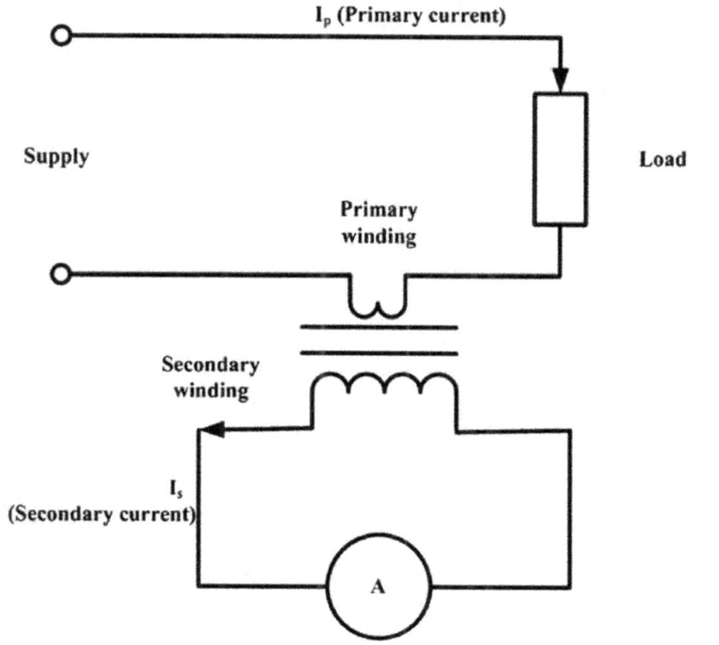

FIGURE 4.1 Connection diagram of a CT in operation.

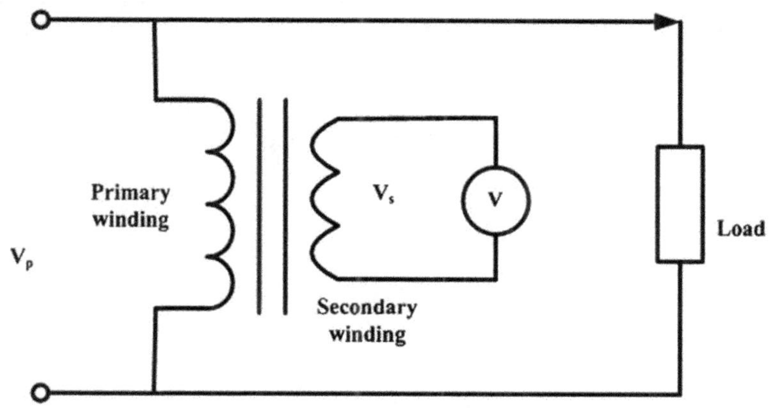

FIGURE 4.2 Connection diagram of a PT in operation.

4.2 Shunts and Multipliers for Range Extension

Low resistances generally made up of magnanin are used as shunts for extending the range of moving coil ammeters for d.c. measurements, while high resistances called multipliers are used in series with the voltmeters for the range extension. The shunts as depicted in Figure 4.3 are connected in parallel with the ammeters for diverting the major portion of the current to be measured (I_T) so that the meter current (I_m) remains within its measuring range [1,3].

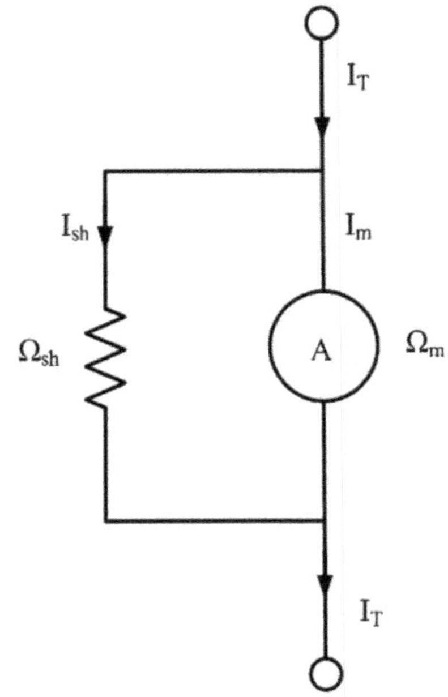

FIGURE 4.3 Shunts for ammeter range extension. (From Golding, E.W., and Widdis, F.C., *Electrical Measurements and Measuring Instruments*, 5th ed., Reem Wisdom Pages LLP, New Delhi, India, 2017; Sawhney, A.K., *A Course in Electrical and Electronic Measurements and Instrumentation*, 19th ed., Dhanpat Rai and Company, New Delhi, India, 2016.)

Low thermal e.m.f. production gives magnanin an advantage over copper for construction of shunts. They exhibit difficulties, however, in soldering and are affected by corrosion [3,4]. The measured I_T can be expressed in terms of the meter resistance (Ω_m), the shunt resistance (Ω_{sh}) and the meter current I_m by Equations (4.1) through (4.3).

$$n = \frac{I_T}{I_m} \text{ is the shunt multiplying factor} \tag{4.1}$$

Also

$$n = 1 + \frac{\Omega_m}{\Omega_{sh}} \tag{4.2}$$

$$I_T = I_m\left(1 + \frac{\Omega_m}{\Omega_{sh}}\right) \tag{4.3}$$

Multipliers are generally made up of magnanin and constantan and have very high values of resistances connected in series with the voltmeters for limiting the current through it so that the voltage across the meter lies within its full deflection range. Basically, the multipliers connected in series with the voltmeters, as depicted in Figure 4.4, serve as potential divider for the total applied voltage (V_T) to be measured. The applied voltage is divided in proportion to the values of the voltmeter resistance (Ω_V) and the multiplier resistance (Ω_{se}) according to the voltage division rule [3,4].

Here, the measurand V_T can be expressed in terms of the voltmeter resistance (Ω_v), the series resistance (Ω_{se}) and the meter voltage V_v by Equations (4.4) through (4.6), respectively.

$$M = \frac{V_T}{V_v} = \text{Series multiplying factor} \tag{4.4}$$

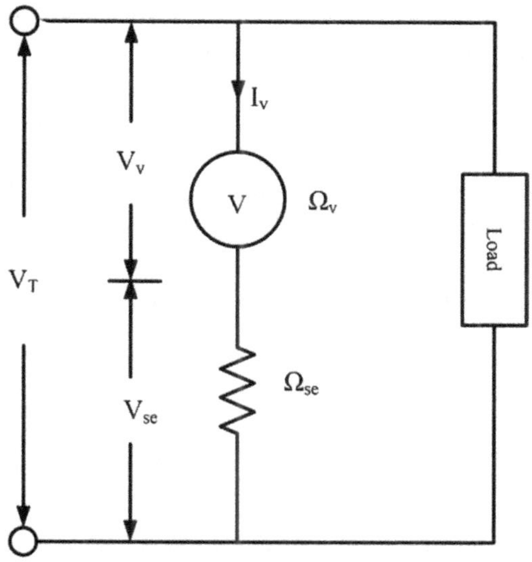

FIGURE 4.4 Multipliers for voltmeter range extension. (From Sawhney, A.K., *A Course in Electrical and Electronic Measurements and Instrumentation*, 19th ed., Dhanpat Rai and Company, New Delhi, India, 2016; Bharat Heavy Electricals Limited, *Handbook on Switch Gears*, 2nd ed., Tata-McGraw-Hill, New Delhi, India, 2005.)

Also,

$$M = 1 + \frac{\Omega_v}{\Omega_{se}} \tag{4.5}$$

$$V_T = \left(1 + \frac{\Omega_v}{\Omega_{se}}\right) V_v \tag{4.6}$$

The shunts and multipliers must be designed very carefully in view of the desired accuracy requirements [1,3]. The design requirements for both the shunts and multipliers are similar and as listed below:

1. Low variation of resistance with time
2. Minimum temperature coefficient of resistance
3. Low temperature rise of the shunts during current flow
4. Minimum production of thermal e.m.f.

4.3 Limitations of Shunts and Multipliers in Range Extension

The range extension of ammeters and voltmeters using shunts and multipliers is restricted only for the measurement of low values of currents and voltages and exhibits several disadvantages for high range a.c. measurements, as listed below:

1. In a.c. measurements the division of currents between the ammeter and the shunt depends on their time constants represented by the *X/R* ratio in a series R-L circuit. For accurate current division between the shunt and the ammeter, their time constants should match [1,3,4]. Hence, separate shunts are required to be connected with every ammeter for the measurements involving range extension. Moreover, accuracy may be compromised if the ammeter is used for

measurements in a wide frequency range with the same shunt as *X/R* ratio of the meter, and the shunt will change with the frequency variations [2,4].

2. Shunts and multipliers are mostly preferred for low current and voltage measurements in order to avoid significant power losses, which may exceed the upper limit of 7.5 W for instruments operating at high current and voltage levels [1,3].
3. Non-isolation between the power and the measuring circuits may be dangerous for the personnel and the equipment while operating at high voltage and currents.
4. Better insulation requirements for limiting the leakage currents in multipliers may lead to bulkier designs of multipliers operating at high voltage levels.

4.4 Merits of ITs

ITs are used widely for the measurement of currents and voltages as they exhibit several advantages over the shunts and multipliers, as follows:

1. Measuring circuit being isolated from power circuit is safe and secure to be handled by operators.
2. Unlike shunts and multipliers, the currents and voltages in ITs are independent of the instrument constants like resistance, inductance and capacitance. Hence, practically the same reading is produced by the ITs regardless of any instrument connected on its secondary [3].
3. Properly standardised CTs and PTs for low values of currents and voltages can be used with moderately ranged instruments, even for measuring high voltages and currents. For example, a CT standardised at 5 A having turn ratio of 1000/5 can be used for measuring currents up to 1000 A with a moderately ranged ammeter of 5 A [2,4].
4. With the standardization of CT and PT secondary windings at moderate values of currents and voltages the ammeters and voltmeters can also be standardised around the same ratings, consequently reducing the cost of ITs and the connected instruments [4].

4.5 Technical Performance Parameters of ITs

Some important technical performance parameters for ITs generally marked on their rating plates are discussed below:

1. Rated primary current: The value of maximum current that can be handled by the primary windings in ITs normally appearing on the transformer rating plates and affecting its accuracy and performance [2,4].
2. Rated secondary current: The value of maximum current that can be handled by the secondary windings in ITs normally appearing on the transformer rating plates and affecting its accuracy and performance [1,3,4].
3. Rated or total burden: It denotes the permissible volt-ampere loading of a transformer secondary without compromising the limits of accuracy classes [4].

Total secondary winding burden

$$= \frac{(\text{secondary winding induced voltage})^2}{\text{total secondary winding impedance including secondary circuit impedance}}$$

or

$$= (\text{Secondary current})^2 \times (\text{total secondary winding impedance including secondary circuit impedance})$$

$$\text{Secondary winding burden due to load} = \frac{(\text{secondary winding terminal voltage})^2}{\text{impedance of load connected on secondary winding}}$$

or

$$= (\text{Secondary current})^2 \times (\text{impedance of load connected on secondary winding})$$

4. Rated frequency: It denotes the system frequency on which an IT operates and is generally expressed in Hz [3,4].
5. Accuracy class: The maximum allowed permissible total error in percentage at rated current/voltage for a category of IT. As per IS2705 the accuracy classes for measuring CTs are 0.1, 0.2, 0.5, 1, 3 and 5, while for protection CTs the accuracy classes are 5 P, 10 P and 15 P [1,4].
6. Rated short-circuit: The primary current value, expressed in RMS, which can be sustained by the CT for a specified time with secondary winding short-circuited without damaging the transformer [3,4].
7. Rated voltage: The RMS value of voltage generally used to designate the CTs for a particular highest system voltage [4].
8. Instrument security factor (ISF): The limit primary current at which CT core starts saturating is generally referred as instrument security current (I_{ps}) for measuring CTs and accuracy limit current for protective CTs as the errors increase significantly after this value of current. The ratio of instrument security current (I_{ps}) to the rated primary current for a measuring CT is referred as ISF. In the event of fault current circulating through the CT primary, the safety of equipment connected to its secondary is maximum when ISF is minimum [2,4].
9. Accuracy limit factor (ALF): The ratio of accuracy limit current to the rated primary current for protective CTs is referred as ALF. A few ALF values based on IS2705 are 5, 10, 15, 20 and 30 [1,3,4].
10. Transformation ratio (T_R): The ratio of magnitudes of primary phasor to the secondary phasor is referred as transformation ratio given by Equations 4.7 through 4.9.

$$T_R = \frac{|\text{Primary phasor}|}{|\text{Secondary phasor}|} \tag{4.7}$$

$$= \frac{\text{Primary winding current } (I_p)}{\text{Secondary winding current } (I_s)} \text{ for CT} \tag{4.8}$$

$$= \frac{\text{Primary winding voltage } (I_p)}{\text{Secondary winding voltage } (I_s)} \text{ for PT} \tag{4.9}$$

11. Nominal ratio (N_R): It is the ratio of rated primary current (or voltage) to the rated secondary current (or voltage) given by Equations 4.10 and 4.11 [3,4].

$$N_R = \frac{\text{Rated primary current}}{\text{Rated secondary current}} \text{ for CT} \tag{4.10}$$

$$N_R = \frac{\text{Rated primary voltage}}{\text{Rated secondary voltage}} \text{ for PT} \tag{4.11}$$

12. Turn ratio: It is the ratio of the number of turns on the primary and secondary windings

$$n = \frac{\text{Secondary winding turns}}{\text{Primary winding turns}} \text{ for CT} \tag{4.12}$$

$$= \frac{\text{Primary winding turns}}{\text{Secondary winding turns}} \text{ for PT} \tag{4.13}$$

13. Ratio correction factor (RCF) is the ratio of transformation (T_R) to the nominal ratio (N_R) [1,3,4]
    Hence, Transformation ratio = Ratio correction factor × Nominal ratio
    or

$$T_R = \text{RCF} \times N_R \tag{4.14}$$

    Generally the ITs contain nominal ratio markings [5].
14. Composite error: The difference between primary winding instantaneous current and the secondary winding instantaneous current multiplied by the nominal ratio under steady state conditions denotes the composite error [4,5].
15. Rated primary voltage: It is the maximum voltage that can be sustained by the primary windings of ITs generally appearing on their rating plates and affecting their performance [3–5].
16. Rated secondary voltage: It is the maximum voltage that can be sustained by the secondary windings of ITs generally appearing on their rating plates and affecting their performance [3,5].
17. Rated voltage factor: A voltage factor determined on the basis of minimum operating voltage of the ITs along with the grounding conditions of PT primary winding and the overall power system. Standard rated voltage factors with different earthing conditions are given in Part 1, IS: 3156 [4,5].
18. Knee point: The point on the magnetization curve of CTs beyond which there is 50% increment in the exciting m.m.f. upon increasing the flux density by 10%. It is the point on the magnetization curve of CTs after which the saturation starts. The knee point is also sometimes termed as the knee point voltage, which is defined as the point on secondary winding voltage and excitation current curve beyond which just a 10% increase in the secondary voltage causes a 50% increase in the transformer excitation current. Beyond this point a large amount of primary current is required for magnetizing the core, thereby limiting the output secondary voltage [3–5].

4.6 Current Transformers

As depicted in Figure 4.5, the CTs, when either used for measurement or protection purposes, have their primary windings connected in series with the line carrying the measured current, which may change depending on the changes in the load connected with the line. Hence, the primary current is not essentially a reflection of the secondary current, unlike in power transformers, where the primary current largely depends on the load component of the secondary current. The primary of a CT normally has a much smaller number of turns compared to the number of turns in the secondary windings [2,3,5]. As the secondary in CTs are connected with ammeters, wattmeter pressure coil and the relay coils having very low resistances, it normally operates in short-circuit conditions. However, one of the terminals in secondary is earthed for the protection of personnel and the instruments in the event of insulation failure.

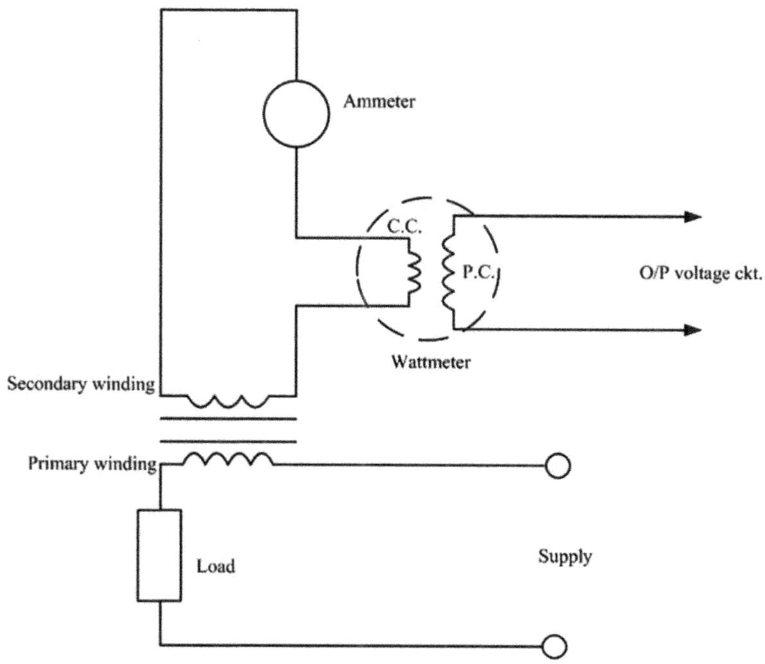

FIGURE 4.5 Connection of CT for current and power measurement.

4.6.1 Equivalent Circuit and Phasor Representation of CTs

The equivalent circuit of an electromagnetic device facilitates the application of circuit theory approach for its analysis and investigation. The device is represented in the form of resistances, inductances, capacitances, voltages and currents involved with it [6]. The equivalent circuit and phasor representation of CT is depicted in Figure 4.6a and b, respectively. The equivalent circuit is similar to any other transformer, while the phasor representation slightly differs from that of conventional transformers. The equivalent circuit parameters and the phasor quantities are also listed and explained along with the figures. On the basis of phasor representation some important parameters like the transformation ratio (T_R) and the phase angle of a CT are calculated in the subsequent sections.

The flux (ϕ_m) flowing through the core linking both the primary and the secondary sides is drawn as the reference phasor, while the induced e.m.f in the secondary winding (E_s) is shown lagging behind the flux by 90°. The voltage drops on the primary side has been neglected due to its small value because of much fewer primary winding turns. The excitation component of current I_0 is resolved into two subcomponents: I_m referred as the magnetizing current responsible for the production of flux (ϕ_m) and hence shown in phase with it and core loss component I_e supplying the hysteresis and eddy current losses in the core; δ is the phase difference between the secondary induced e.m.f (E_s) and the secondary current I_s; Δ is the phase angle generated due to the burden on the transformer [2,5]. The secondary current phasor when referred on the primary side is reversed and multiplied with the turn ratio n. The total current flowing in the primary windings (I_p) is given as the vector sum of phasors nI_s and I_0. Unlike VTs the flux density in the core of CTs changes with variations in the primary currents, while the voltage and flux density in a VT remains constant and the current varies with the load variations [3–5].

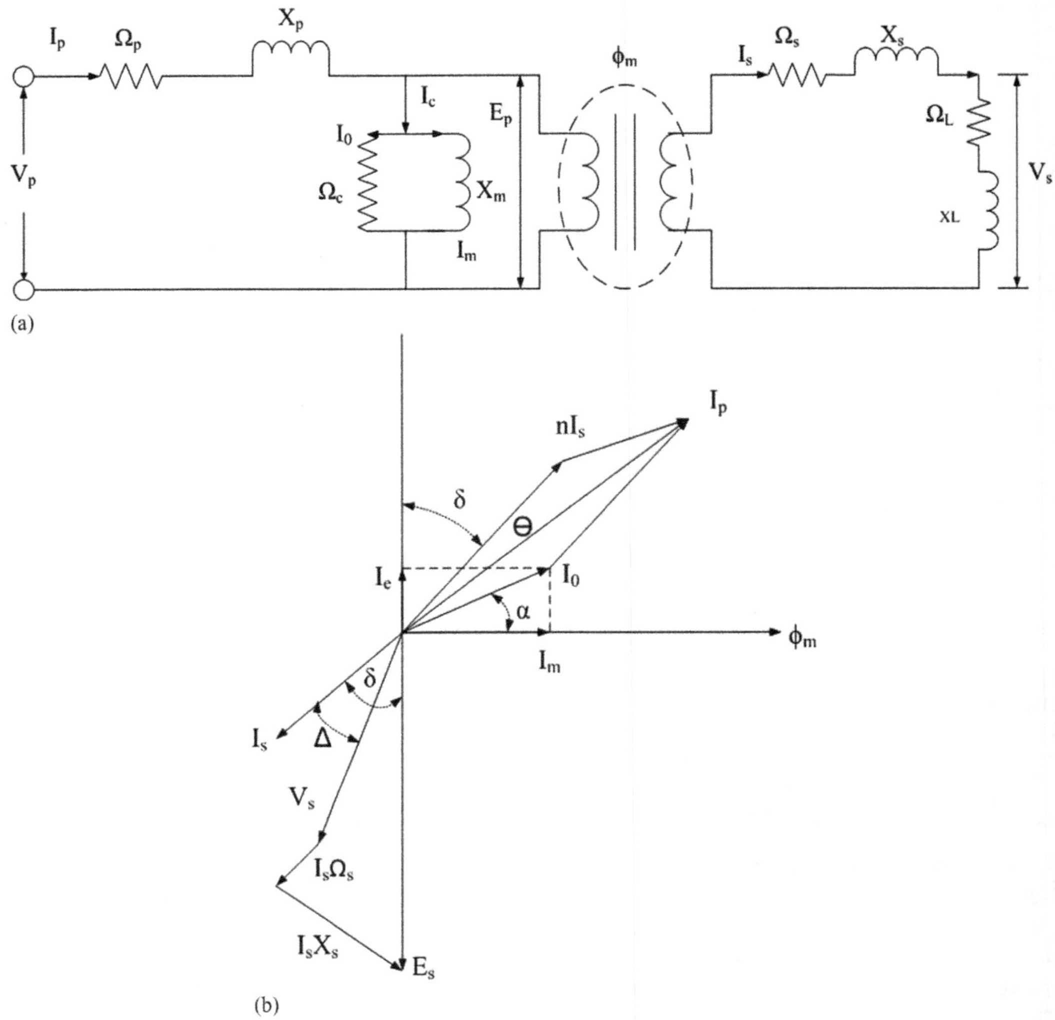

FIGURE 4.6 (a) Equivalent circuit of a CT (b) Phasor and symbol description of a CT with lagging power factor burden.

Symbol description

ϕ_m is maximum core flux
I_m is magnetizing component of current I_0
I_c is core (hysteresis and eddy current) loss component of current I_0
I_0 is exciting current
I_s is secondary winding current
n is the turn ratio
Ω_s is resistance of secondary winding
X_s is the reactance of secondary winding
V_s is open circuit secondary winding terminal voltage
α is hysteresis angle between I_0 and ϕ_m
δ is angle between secondary induced voltage (E_s) and secondary current (I_s)

In terms of total burden including the total winding impedance, it is given as:

$$\delta = \tan^{-1}\left(\frac{x_s + x_L}{\Omega_s + \Omega_L}\right)$$

Δ is the phase difference between the secondary current (I_s) and the secondary terminal voltage (V_s).

$$\Delta = \tan^{-1}\left(\frac{x_L}{\Omega_L}\right)$$

4.6.2 CT Transformation Ratio

In order to derive the formula for CT transformation ratio in terms of equivalent circuit parameters, a portion of its phasor representation, depicted in Figure 4.7, is considered.

$$\text{Angle LPC} = 90 - \delta - \alpha,\ \text{PC} = I_0,\ \text{OP} = nI_s,\ \text{OC} = I_p \tag{4.15}$$

Therefore,

$$\text{LC} = I_0 \sin(90 - \delta - \alpha) = I_0 \cos(\delta + \alpha);\ \text{PL} = I_0 \cos(90 - \delta - \alpha) = I_0 \sin(\delta + \alpha) \tag{4.16}$$

Since

$$(\text{OC})^2 = (\text{OL})^2 + (\text{LC})^2 = (\text{OP} + \text{PL})^2 + (\text{LC})^2 \tag{4.17}$$

From (4.15) through (4.17), we get:

$$I_p^2 = [nI_s + I_0 \sin(\delta + \alpha)]^2 + [I_0 \cos(\delta + \alpha)]^2 \tag{4.18}$$

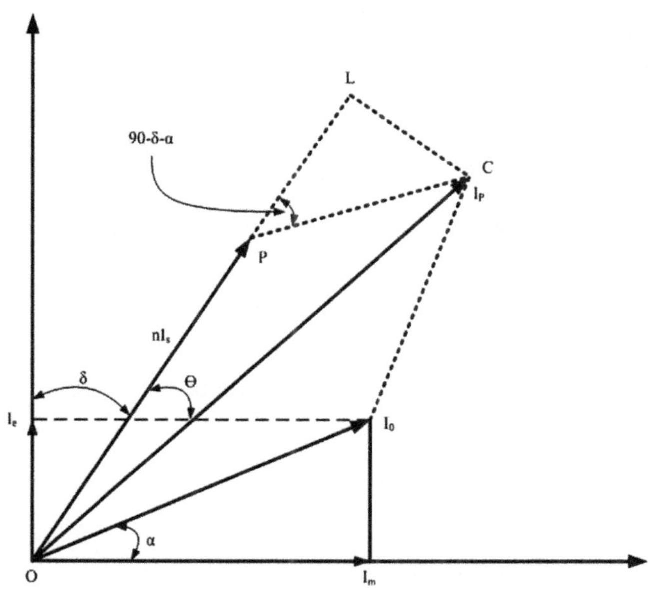

FIGURE 4.7 Primary section of CT phasor representation. (From Sawhney, A.K., *A Course in Electrical and Electronic Measurements and Instrumentation*, 19th ed., Dhanpat Rai and Company, New Delhi, India, 2016; Bharat Heavy Electricals Limited, *Handbook on Switch Gears*, 2nd ed., Tata-McGraw-Hill, New Delhi, India, 2005.)

$$I_p^2 = n^2 I_s^2 + 2nI_s I_0 \sin(\delta+\alpha) + I_0{}^2 \sin^2(\delta+\alpha) + I_0^2 \cos^2(\delta+\alpha) \tag{4.19}$$

$$I_p^2 = n^2 I_s^2 + 2nI_s I_0 \sin(\delta+\alpha) + I_0\left[\sin^2(\delta+\alpha) + \cos^2(\delta+\alpha)\right] \tag{4.20}$$

$$I_p = (n^2 I_s^2 + 2nI_s I_0 \sin(\delta+\alpha) + I_0)^{1/2} \tag{4.21}$$

So, transformation ratio:

$$T_R = \frac{I_p}{I_s} = \frac{\left\{n^2 I_s^2 + 2nI_s I_0 \sin(\delta+\alpha) + I_0\right\}^{1/2}}{I_s} \tag{4.22}$$

Since for CTs with perfect designs, $I_0 << nI_s$ and also I_0 is less than 1% of I_p, hence, $I_p \approx nI_s$. So the formula for transformation ratio in Equation (4.22) can be suitably approximated by Equation (4.23) [4,5]:

$$T_R \approx \frac{nI_s + I_0 \sin(\delta+\alpha)}{I_s} \approx n + \frac{I_0}{I_s}\sin(\delta+\alpha) \tag{4.23}$$

Expanding Equation (4.23) we obtain

$$T_R \approx n + \frac{I_0}{I_s}(\sin\delta\cos\alpha + \cos\delta\sin\alpha) \approx n + \frac{(I_m \sin\delta + I_e \cos\delta)}{I_s} \tag{4.24}$$

as $\left\{I_m = I_0\cos\delta; I_e = I_0\sin\alpha\right\}$

4.6.3 Phase Angle

The angle denoted by θ is the phase difference between the reversed secondary current phasor ($-I_s$) and the primary current phasor (I_p), as depicted in Figure 4.6b. θ also referred as the phase angle in CT is considered positive if the reversed secondary current phasor $-I_s$ leads the phasor I_p and is said to be negative if $-I_s$ phasor lags I_p [3–5].

According to Figure 4.7 we can write:

$$\tan\theta = \frac{\text{LC}}{\text{OL}} = \frac{\text{LC}}{\text{OP}+\text{PL}} = \frac{I_0\cos(\delta+\alpha)}{nI_s + I_0\sin(\delta+\alpha)} \tag{4.25}$$

Generally the value of phase angle (θ) is very small, so Equation (4.25) can be written as:

$$\theta = \frac{I_0\cos(\delta+\alpha)}{nI_s + I_0\sin(\delta+\alpha)}\text{rad} \tag{4.26}$$

Also we know that $I_0 << nI_s$; hence the term I_0 *sin* $(\delta + \alpha)$ can be neglected, so the phase angle can be approximately given by:

$$\theta \approx \frac{I_0\cos(\delta+\alpha)}{nI_s}\text{rad} \tag{4.27}$$

Expanding the term on the numerator we obtain:

$$\theta \approx \frac{I_0\cos\delta\,\cos\alpha - I_0\sin\delta\,\sin\alpha}{nI_s}\left\{\text{Since } \cos(a+b) = \cos a\cos b - \sin a\sin b\right\} \tag{4.28}$$

Also,

$$\theta \approx \frac{I_m \cos\delta - I_e \sin\delta}{nI_s} \text{rad since} \{I_m = I_0 \cos\delta; I_e = I_0 \sin\alpha\} \tag{4.29}$$

Converting Equation (4.29) in degrees we get:

$$\theta \approx \frac{(I_m \cos\delta - I_e \sin\delta)}{nI_s} * \frac{180}{\Pi} \text{degrees} \tag{4.30}$$

4.6.4 Errors in CTs

Two types of errors generally encountered in CTs are:

1. Ratio or current errors and
2. Phase angle errors

The ratio error is due to the difference between the nominal ratio (N_R) and the actual transformation ratio (T_R) given by Equation (4.31) [4,5].

$$\%\text{Ratio error} = \frac{\text{nominal ratio} - \text{actual ratio}}{\text{actual ratio}} \times 100 \tag{4.31}$$

$$= \frac{N_R - T_R}{T_R} \times 100 \tag{4.32}$$

In power transformer, the turn ratio (n) is equal to the transformation ratio (T_R), which is not the case in CTs, as observed from Equation (4.24). Transformation ratio depends on four varying parameters (a) I_m (magnetizing current), (b) I_c (core loss current), (c) I_s (secondary load current), and (d) load power factor (δ). Due to the variations in the above parameters, the secondary current (I_s) cannot be represented as a constant fraction of the primary current (I_P) like power transformers thereby introducing errors in current measurement [3,4].

Phase angle error is given by the phase difference (θ) between the reversed secondary current phasor ($-I_s$) and the primary current phasor (I_p), as depicted in Figure 4.7 [3,5,6].

$$\theta = \frac{180}{\Pi}\left[\frac{I_m \cos\delta - I_e \sin\delta}{nI_s}\right] \text{degrees}$$

The reversed secondary current phasor ($-I_s$) in CTs with inductive burden normally leads the primary current phasor (I_p) by phase angle θ and hence is considered positive. It can be observed from Equation (4.30) that θ also depends on I_e, I_m, I_s and the phase difference due to secondary burden (δ). Due to the variations in these parameters in CTs the error θ also varies [1,5].

The formula for transformation ratio and phase angle in Equations 4.24 and 4.30, respectively, can be approximated by Equations (4.33) and (4.34) because the instrument burden on CTs is largely resistive with very low inductance; hence δ is very small and positive [1,4,5]. Therefore, $\sin\delta \approx 0$ and $\cos\delta \approx 1$.

Hence,

$$T_R = n + \frac{I_e}{I_s} \tag{4.33}$$

$$\Theta \approx \left(\frac{180}{\Pi}\right)\left(\frac{I_m}{nI_s}\right) \tag{4.34}$$

From Equations (4.24) and (4.30), it can be inferred that the current and phase angle errors in CTs can be reduced by:

1. Using materials with low hysteresis and high permeability for making the core [1,5].
2. Reducing the number of core joints and the length of flux path and increasing the area of the cross-section [1,4,5].
3. Lowering the winding and leakage impedances to the minimum possible value [1,5].

4.6.5 Characteristics of CTs

CT characteristics can be analysed in the light of expressions for ratio and phase angle errors given by Equations (4.24) and (4.30) explained by the points given below:

1. ***Effect of secondary burden power factor on ratio errors***

   Delta (δ) represents the phase difference between the secondary induced e.m.f. (E_s) and the secondary current (I_s) in a CT and depends on the total burden of the CT, including the secondary winding impedance. The secondary burden power factor affects the errors in two ways:

   - For inductive burdens, δ is positive; therefore, I_s lags E_s and transformation ratio (T_R) > turn ratio (n).
   - For capacitive burdens, δ is negative; therefore, I_s leads E_s and transformation ratio (T_R) < turn ratio (n) as δ approaches *−90°*.

2. ***Effect of secondary burden power factor on phase angle errors***

   The conclusions listed below can be drawn on the basis of Equation (4.27) and visualised from phasor representation in Figure 4.6b:

   - For burdens with low inductance, the phase angle θ is +ve and the value of δ is small, thereby representing high secondary power factor.
   - For highly inductive burdens, as δ approaches 90°, the secondary power factor is low; hence phase angle θ is −ve.
   - For capacitive burdens, δ is always negative, and therefore phase angle θ is +ve.

   The above conclusions can be inferred by assuming that the magnitude of secondary impedance remains constant. The change in ratio and phase angle errors w.r.t. the variations in δ is depicted in Figure 4.8a and b, respectively [1,4,6].

3. ***Variation in the primary current (I_P)***

   The values of I_m and I_c in a transformer are generally constant; hence at low values of I_P their effect becomes significant because a major portion of I_P is taken in feeding the loss and magnetization components. The larger values of I_m and I_c as compared to I_P lead to increase ratio and phase angle errors [3,5]. At the same time as the value of I_P changes, a corresponding change in the value of I_s also takes place in accordance with the turn ratio of the transformer, thereby affecting the value of ratio and phase angle errors according to Equations (4.33) and (4.34) [1,3,5].

4. ***Variation in the burden of secondary winding circuit***

   The effect of increase in the secondary winding impedance in CTs leads to:

   a. Increase in the VA rating, thereby increasing the secondary winding induced voltage (E_s).
   b. Increase in the value of flux and flux density in the core.
   c. Consequently, there is an increase in the values of I_m and I_c.

   The above three findings suggest a resultant increase in the phase angle and ratio errors according to Equations (4.24) and (4.30), as depicted in Figure 4.9a and b, respectively.

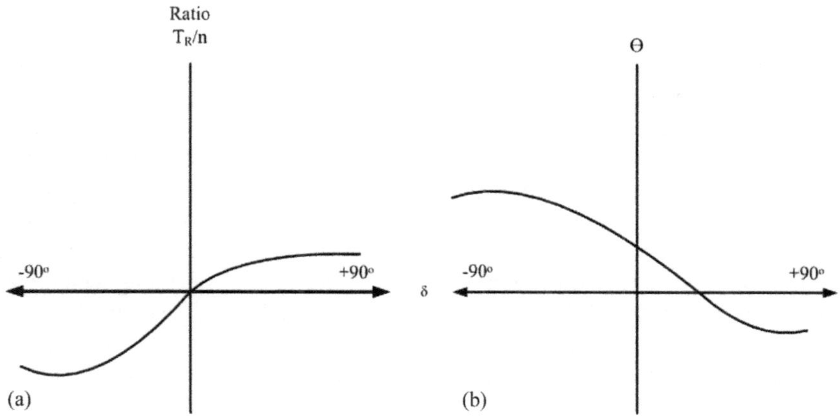

FIGURE 4.8 CT characteristics (a) T_R/n (b) θ vs δ. (From Golding, E.W., and Widdis, F.C., *Electrical Measurements and Measuring Instruments*, 5th ed., Reem Wisdom Pages LLP, New Delhi, India, 2017; Sawhney, A.K., *A Course in Electrical and Electronic Measurements and Instrumentation*, 19th ed., Dhanpat Rai and Company, New Delhi, India, 2016.)

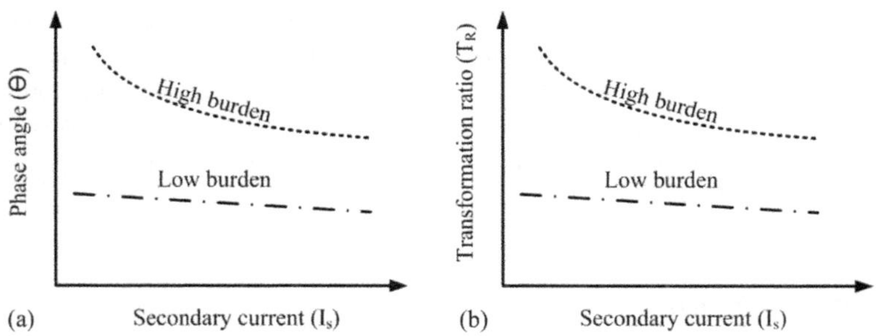

FIGURE 4.9 Variation of phase angle and ratio errors (a) θ vs I_s (b) T_R vs I_s.

5. ***Change in frequency***

   CTs normally operate at a fixed rated frequency for which they have been designed; hence considering the effect of variations in frequency is not significant. However, the increase in frequency causes a proportionate decrease in the value of flux density, thereby having similar effect as a decrease in the value of secondary impedance on CTs [1,5].

4.6.6 Design Features for Reducing Errors in CTs

Deviance from ideal behaviour in CTs is due to the inherent physical limitations of the electric and magnetic circuits used for its manufacturing, consequently leading to ratio and phase angle errors. These errors are mainly due to:

- Presence of magnetization current (I_m) required for producing the working flux.
- Presence of the loss component of current (I_e) needed to feed the hysteresis and eddy current losses in the CT core.
- Nonlinear flux density in the core causing its saturation.
- Leakage flux resulting in unsymmetrical flux linkages on the primary and secondary sides.

It is evident from Equations (4.24) and (4.30) that higher values of I_m and I_e as compared with I_p result in the ratio and phase angle errors in CTs. Hence, the magnitude of errors can be reduced by reducing the values of I_m and I_e. A few design features, as explained below, can be helpful in reducing these errors.

4.6.6.1 Core Design

The core design and materials must be selected very carefully as they serve as the path for the flow of working magnetic flux. The core should be made up of materials with high permeability and must also possess minimum length, large cross section area and minimum flux density. These features help in maximizing the core permeability, thereby reducing the values of I_m and I_e and consequently the ratio and phase angle errors. Moreover, the part of the core at the joints should be appropriately overlapped and tightly bonded to reduce the air gaps, which offer high reluctance for the flux. The preferred material for core design should possess low reluctance, low hysteresis and eddy current losses, and should also be subjected to low flux densities during normal operations [1,5]. The protective CTs operating with protective relays normally handle currents up to 20–30 times the rated value; hence, they are generally designed for operating at lower flux densities in order to achieve fair accuracy at higher currents [5]. Table 4.1 presents various types of magnetic materials, their characteristics, and the common core shapes used for CT core designs.

4.6.6.2 Effect of Rated Primary mmf

It is desirable to have high values of primary to excitation mmf ratio for CTs during operation by obtaining the high values of primary to excitation current ratio (I_p/I_0). Achieving this condition is very difficult, particularly at low values of the primary current (I_p), which results in low values of primary mmf. The performance of CTs at low values of primary currents can be improved by increasing the primary mmf and keeping it fixed at a standard value for rated primary currents. The CTs can operate satisfactorily if the values of primary mmf lie at 500 AT. Hence, the CTs with 500 A rated primary currents normally have single turns on the primary windings so that the mmf is maintained at 500 AT [4,5,7]. However, those CTs with less primary current ratings can have multi-turn primary windings, thereby facilitating reduced core size.

TABLE 4.1

Classification of Magnetic Materials for CT Core Design

S. No.	Type of Magnetic Material	Major Constituents	Important Characteristics	Core Shapes
1.	Hot rolled Si-steel	0%–4% Si	—	Ring shape, L, E and I type stampings
2.	C.R.G.O. Steel	0%–6.5% Si in the proportion Si:5Fe	Flux travels along the grains hence experiences minimum reluctance	Manufactured in all the above shapes but specially used in toroidal shaped core as clock springs
3.	Fe alloys	Mumetal (76% Ni)	• High permeability • Low loss and retentivity • Saturates at low flux density hence not fit for overload relays • Low magnetic anisotropy and magnetostriction	In all the above shapes
		Permendur (49% Co and 49% Fe)	Very high saturation density of 2–4 Wb/m^2 as compared to 0.7–0.8 Wb/m^2 for other alloys	
		Hipernik (50% Fe and 50% Ni)	• High permeability at low flux densities • High saturation density • Frequently used	

Source: Golding, E.W., and Widdis, F.C., *Electrical Measurements and Measuring Instruments*, 5th ed., Reem Wisdom Pages LLP, New Delhi, India, 2017; Sawhney, A.K., *A Course in Electrical and Electronic Measurements and Instrumentation*, 19th ed., Dhanpat Rai and Company, New Delhi, India, 2016; Purkait, P. et al., *Electrical and Electronics Measurements and Instrumentation*, 1st ed., McGraw-Hill Education, New Delhi, India, 2016.

With the recent improvements in the properties like permeability in the magnetic materials, single turn primary can be used even for the transformers of 100 A rated primary currents [6,7].

4.6.6.3 Reducing Leakage Reactance

More leakage flux results in higher leakage reactance thereby increasing the value of errors in CTs. The primary and secondary windings, if placed close together, will lead to reduced leakage reactance and consequently to errors. Moreover, the CTs having toroidal windings wound uniformly over the ring-shaped core also facilitate reduction of the errors due to reduced leakage reactance [1,5,7].

4.6.6.4 Adjusting the Turn Ratio

The errors in CTs can also be reduced by turn compensation, which involves adjusting the secondary winding turns, consequently changing the turn ratio n. The approximate actual transformation ratio in CTs is given by Equation (4.33).

Generally the nominal ratio (N_R) and the turn ratio in the transformers is equal; i.e., $N_R = n$. Hence the transformation ratio is given as:

$$T_R = N_R + \frac{I_e}{I_s} \tag{4.35}$$

It can be observed from Equation (4.35) that the transformation ratio is always greater than the nominal ratio. However, if we reduce the turn ratio the by changing the number of turns on the secondary winding, the transformation ratio will also be reduced, thereby making it equal to the nominal ratio [1,5,7]. The importance of turn compensation can be explained with the help of Example 4.1.

Example 4.1

A CT having the primary and secondary current ratio 1000/2 A and the loss component of current as 0.5% of the primary current has to be compensated for ratio errors with the help of turn compensation. The adjustment in the number of secondary turns has to be calculated.

Solution

From the data available, the nominal ratio can be calculated as:

$$N_R = \frac{1000}{2} = 500$$

The loss component of current is:

$$I_e = \frac{0.5}{100} \times 1000 = 5\text{A}$$

Let the number of primary turns is $N_p = 1$, and if the nominal ratio = the turn ratio, we have $n = 500$, and the secondary winding turns N_S is given by:

$$\because N_s = nN_p = 500 \times 1 = 500$$

The transformation ratio is:

$$T_R = n + \frac{I_e}{I_s} = 500 + \frac{5}{2} = 502.5$$

Hence, it can be observed that the nominal ratio and the transformation ratio are unequal. Now, instead of using 500 turns on the secondary, if it is adjusted to have 488 turns, the transformation ratio will be:

$$T_R = n + \frac{I_e}{I_s} = 488 + \frac{5}{2} = 500.5$$

which is nearly equal to the nominal ratio N_R. Thus, by reducing or compensating the number of turns on the secondary, the transformation ratio is obtained almost equal to the nominal ratio and the turn ratio. Though single or double turn secondary compensation is sufficient for reducing the ratio errors significantly, it has very little affect in reducing the phase angle errors. The turn compensation is accurate only for fixed values of currents and impedance [1,4,7].

4.6.6.5 *Use of Shunts*

The transformation ratio can also be reduced by reducing the value of secondary current (I_s) by using a shunt across the primary or secondary windings as an alternate path for bypassing the large currents. This method, in addition to reducing the phase angle errors, also provides exact correction to ratio errors for burdens of particular values and characteristics [1,5,6]. Turn compensation, besides being a coarse technique for ratio error reduction, has little effect on the phase angle errors. A more refined technique having a similar effect as turn compensation was suggested by S. Wilson, a General Electric engineer, for reducing both the ratio and phase angle errors. It is named after him as, the Wilson compensation method [5,6].

4.6.6.6 *Two Stage Design*

Here the CTs are designed in two stages so that the error in the secondary current of the first CT is compensated by the second transformer. This type of design is prevalent in the energy meters as they contain a secondary coil for carrying the error correcting current, unless an auxiliary transformer is used [2,3,5].

4.6.7 Types of CTs and Their Construction Features

The broad classification of CTs based on their field application is:

1. Indoor type
2. Outdoor type

The indoor CTs are usually mounted within metal cubicles insulated by tapes and cast resin like epoxy, polyurethane or polycrete [3,4].

4.6.7.1 *Indoor Type CT*

Based on their construction, the indoor type CTs are classified as wound type, bar type and slot/window/ring type [1,4,5].

The wound type, depicted in Figure 4.10a, contains more than one full primary turn wound on the core. The outward primary terminals are similar to the bar type CTs or in the form of rectangular pads [4,5,7]. These CTs normally use *NI-Fe* cold rolled core, while a hot rolled steel core uses a stack of ring type stampings. The core laminations are insulated using wrappings of elephantide and pressphan and stacked in consolidated form using end collars. The secondary windings are then put on the consolidated core with the help of a toroidal winding machine [6,7]. In addition to providing insulation, the pressphan and elephantide on the CT core also protect the secondary winding conductors from mechanical damage due to sharp corners. After placing the secondary windings on the ring core, the complete arrangement is again insulated by exterior tapes and circumferential wraps [1,3,7].

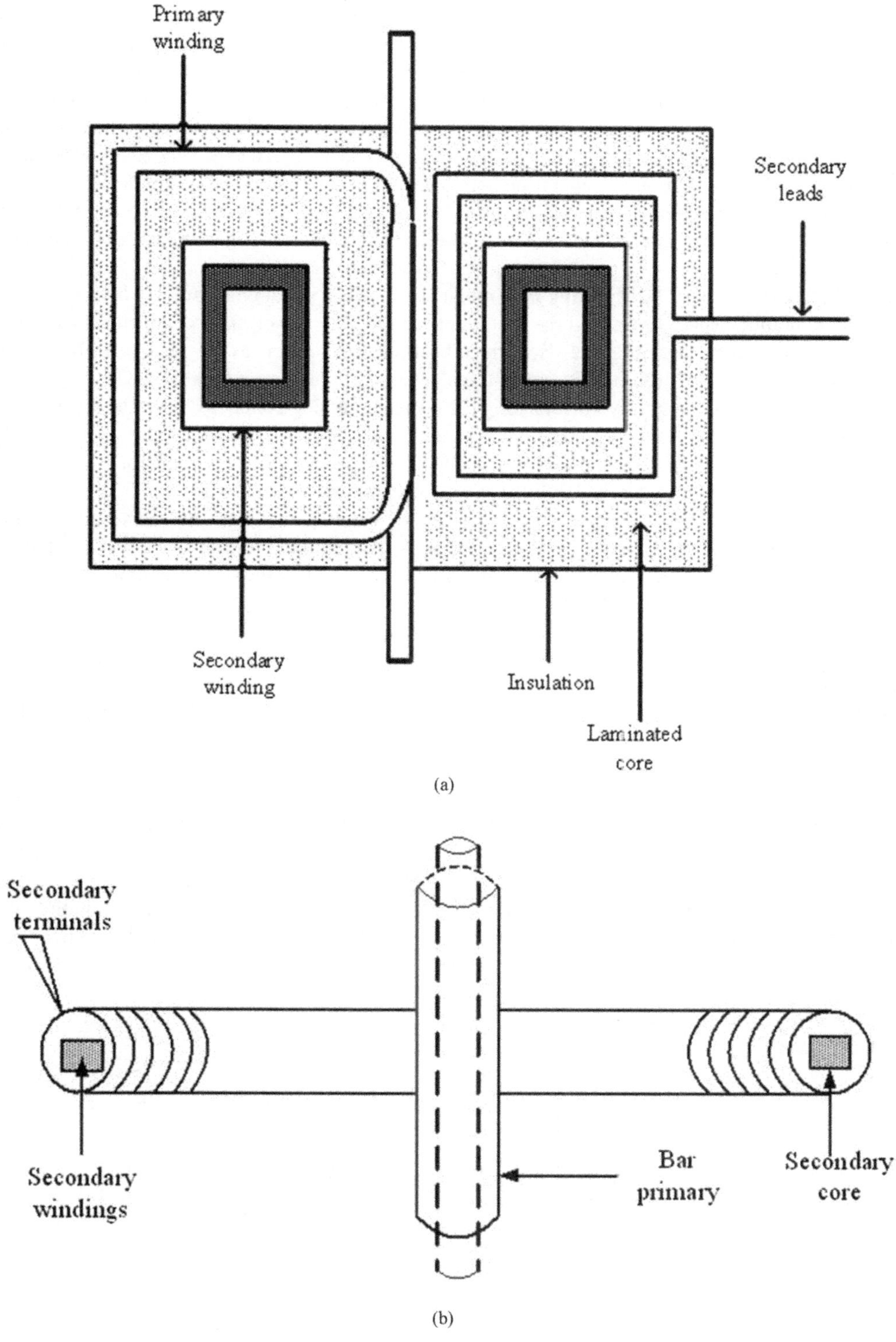

FIGURE 4.10 Construction of CT (a) wound type (b) bar type. (From Sawhney, A.K., *A Course in Electrical and Electronic Measurements and Instrumentation*, 19th ed., Dhanpat Rai and Company, New Delhi, India, 2016; Purkait, P. et al., *Electrical and Electronics Measurements and Instrumentation*, 1st ed., McGraw-Hill Education, New Delhi, India, 2016.)

The bar type CTs, on the other hand, have a primary winding in the form of a bar of suitable material and size integrated within the body of the transformer, as depicted in Figure 4.10b. All the features of the core in a bar type CT are similar to the ring type and wound transformers except an insulation coated bar forms its integral part [1,3,6]. The insulation used may be a bakelised paper tube or a resin molded directly on the bar. The secondary windings in LV CTs are wound on a former which is insulated by bakelite and a heavy primary winding is mounted concentrically over the secondary winding after applying appropriate insulation for separating both the windings. Usually the CTs are manufactured with cores having a larger cross-sectional area compared to the ordinary transformers; however, the design is such that the reluctance of interleaved corners is minimum, thereby minimizing the required magnetizing current [3,7].

Slot/window/ring type CTs have ring or circular, window or rectangular and stadium shaped core stampings containing space or a slot at the centre to accommodate the primary conductor, as depicted in Figure 4.11. Sometimes, these CTs are also termed bushing transformers as they are generally placed on the oil end of the terminal bushing of a power transformer or an oil circuit breaker (OCB).

Handling insulation problems while isolating the current carrying conductor from the measuring circuit in high voltage (HV) networks become a serious and costly issue. For circuits operating above 750 kV, the problem is handled using CTs in cascade or by using telemetry systems for transmitting the measured current signal through the RF links to receiving equipment on the ground [5,7].

In ring CTs, it is normally desired that the secondary windings are uniformly distributed over the transformer core so that the full core length is utilised by maintaining uniform space between the turns. Moreover, the insulation level between the transformer core and the windings must have the capacity to bear the voltage peaks caused when secondary winding gets open-circuited while the primary winding remains energised. This insulation problem is normally handled in case of layered secondary windings by sectioning the windings to significantly reduce the peak voltage between the layers [3,5,6].

In wound CTs the problems of overvoltages cannot be handled in the manner similar to ring CTs, as explained above. Here, however, the relative positioning of primary and secondary windings is adjusted to minimise the axial forces acting on both the windings produced due to short circuit currents in the primary [5,7].

In all the above types of CTs, like any other transformer, the primary and secondary windings are generally kept close in order to reduce the leakage reactance and consequently the ratio errors. Generally the secondary windings are made up of round copper conductors of 3 mm^2 area for rated currents up to 5 A. Cu strip primary windings of suitable dimensions depending on the value of current are used for constructing the primary windings. To cater the effects of corona, the CTs with bar type primary windings use tubes of sufficiently large diameters for directing the voltage gradient on the dielectric surface [1,6,7].

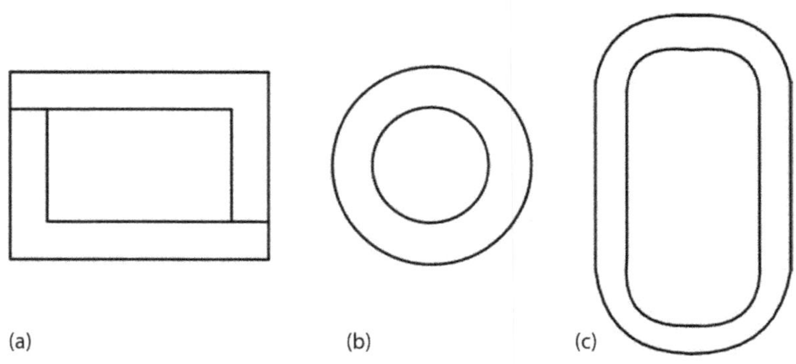

FIGURE 4.11 Shapes of stampings in ring type CTs: (a) Window (b) Ring (c) Stadium. (From Golding, E.W., and Widdis, F.C., *Electrical Measurements and Measuring Instruments*, 5th ed., Reem Wisdom Pages LLP, New Delhi, India, 2017; Sawhney, A.K., *A Course in Electrical and Electronic Measurements and Instrumentation*, 19th ed., Dhanpat Rai and Company, New Delhi, India, 2016.)

4.6.7.2 Outdoor Type CT

The CTs designed to work in outdoor environments are generally immersed in oil or any other liquid medium for insulation and cooling purposes. These CTs are hermetically sealed to isolate them from the environment and thus minimizing the atmospheric effects [4,5,7]. Outdoor CTs are further classified as live tank type and dead tank type based on the voltage level of the transformer body.

***Live tank type*:** In this configuration, the voltage level of the tank in which the CT is stored is maintained same as the system voltage. Figure 4.12 shows that these transformers may suffer from bushing damages as the tank containing CT is placed at large height [2,5,7].

***Dead tank type*:** In this configuration, the voltage level of the tank in which the CT is stored is maintained at ground potential. Figure 4.13a and b show dead tank configuration with single and double bushing. In both configurations, the CT tank is placed at a lower centre of gravity compared to the live tank type; hence the chances of damage to the bushings during transit is reduced [5,7].

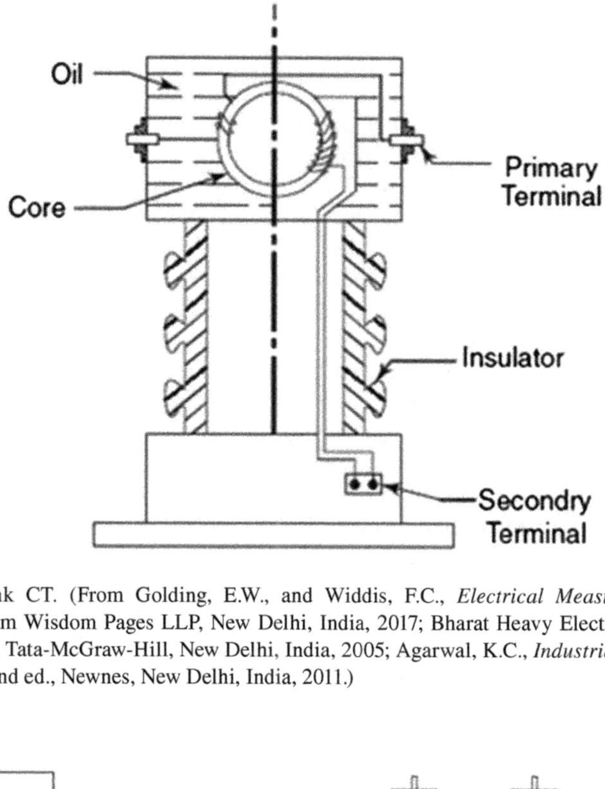

FIGURE 4.12 Live tank CT. (From Golding, E.W., and Widdis, F.C., *Electrical Measurements and Measuring Instruments*, 5th ed., Reem Wisdom Pages LLP, New Delhi, India, 2017; Bharat Heavy Electricals Limited, *Handbook on Switch Gears*, 2nd ed., Tata-McGraw-Hill, New Delhi, India, 2005; Agarwal, K.C., *Industrial Power Engineering and Application Handbook*, 2nd ed., Newnes, New Delhi, India, 2011.)

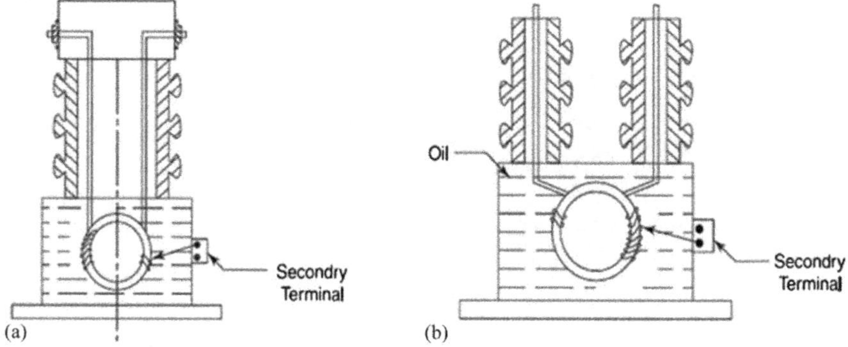

FIGURE 4.13 Dead tank CT (a) single bushing (b) double bushing.

In addition to the classifications as discussed above, CTs can also be classified on the basis of core and winding designs, as follows:

***Multi-core CT*:** CTs having several cores and secondary windings and used for metering and protection applications are called multi-core CTs [6,7].

***Multi-ratio CT*:** CTs in which it is possible to obtain several turn ratios by using tapings on the primary and secondary windings are called multi ratio CTs. For example, in 800-400-200/1 A CT, variable turn ratio can be obtained. In these transformers, normally tapings in the secondary windings are desirable [3,4,6].

***Dual-purpose CT*:** CTs finding application for both protection and measurement purposes are generally termed as dual-purpose CTs.

***Split core CT*:** CTs with split or cut core used for measuring current in a power system bus bar are called split core instrument transformers. As depicted in Figure 4.14, the groove created due to the slot in the core is used to envelop a low tension bus bar during measurement without obstructing its current flow [5,7].

4.6.8 Precautions in CTs

One of the most important precautions while operating a CT is to always keep its secondary windings connected to an ammeter, relay coil or the current coil of a wattmeter when its primary windings are energised. Open circuiting the secondary windings of a CT when its primary winding is connected to supply may have serious repercussions on the transformers and the personnel handling it [1,3,7].

The secondary winding of a power transformer is generally connected to a load, and its primary current is largely dependent on the load component of the secondary winding current. However, the primary winding in a CT is connected in series to the line containing the measurand current, and unlike power transformers, it is not at all dependent on the secondary current [5,7].

During normal operation when both the primary and secondary windings of CT contains current, their mmfs are in phase opposition and hence balance each other. As the secondary mmf is slightly less than the primary mmf and also in phase opposition to it, the resultant mmf which is responsible for production of flux in the CT core remains small but sufficient enough to cater the iron losses and the core magnetization requirements. Low value of flux and the flux density in the core therefore results in low value of induced e.m.f. in the secondary winding [4,5,7].

FIGURE 4.14 Split core CT. (From Agarwal, K.C., *Industrial Power Engineering and Application Handbook*, 2nd ed., Newnes, New Delhi, India, 2011; Purkait, P. et al., *Electrical and Electronics Measurements and Instrumentation*, 1st ed., McGraw-Hill Education, New Delhi, India, 2016.)

If the secondary winding is open circuited and the primary winding is kept energised, the neutralizing effect produced by the secondary mmf vanishes, but the primary mmf remains present. Hence, the resultant mmf I_pN_p is very large and results in a large flux in the CT core, causing a high value of induced e.m.f. in the secondary winding. The personnel handling the measuring instruments connected on the secondary side are therefore exposed to high voltage and electric fields. Under these conditions, the transformer may have increased ratio and phase angle errors, increased core losses and overheating of the transformer sometimes leading to its permanent damage and permanent core magnetization [1,7].

Many times handling CTs requires altering the secondary circuit conditions, with the primary winding being energised. Hence, short-circuit links or switches provided on the CT secondary are closed while altering the circuit conditions on the secondary side [6,7].

4.6.9 Permanent Magnetisation of CT Core

The CT core may get permanently magnetised under following conditions:

1. When secondary is open circuited with primary windings connected to the supply, very high flux density in the core may magnetise the permanent core due to residual magnetism.
2. High transient currents arising during switching may also result in residual magnetism.
3. When d.c. current is passed deliberately through the windings for finding their resistances (r_{dc}) and polarity.

The effect of permanent magnetization is normally to reduce the permeability of the CT core at its operating flux density, causing an increase in the ratio and phase angle errors. Hence, for proper operation, the permanent magnetization of the CT should be removed to restore normal conditions [2,4,7].

Two most commonly used demagnetization methods are:

- Using a motor-generator set to pass same amount of current through the CT primary windings which was passed when the secondary winding of CT was open circuit. As the motor is shut down with alternator field under excited state, the CT experiences several magnetization cycles with gradually damped amplitude, finally resting the magnetization to zero [3,5,7].
- Creating a condition similar to the opening of CT secondary by looping the secondary winding through a very high resistance and keeping the primary windings under excited state. The resistance connected on the secondary is then gradually reduced to zero in uniform steps leading to the demagnetization of the core [5,7].

4.6.10 Solved Examples on CTs

Example 4.2

A CT with bar primary has 300 turns on the secondary winding. The burden connected on the secondary windings is an ammeter of resistance 1.5Ω and a reactance of 0.5Ω, while the secondary winding resistance is of 0.2Ω and reactance of 0.3Ω; 100 AT magnetization mmf is required for magnetizing the core and 50 AT for the core loss current component.

1. Calculate the current flowing through the primary windings and ratio errors when the ammeter connected in the secondary windings shows a reading of 5 A.
2. Determine the turn compensation in the secondary windings for reducing the ratio errors to zero under the above conditions.

Solution

Total resistance in the secondary circuit = 1.5 + 0.2 = 1.7Ω
Total reactance in the secondary circuit = 0.5 + 0.3 = 0.8Ω
Hence, the secondary circuit phase angle δ is = $\tan^{-1}(0.8/1.7) = 25.196°$

$$\cos\delta = 0.90 \text{ and } \sin\delta = 0.426$$

As there is a single turn primary winding N_p in the form of a bar and $N_s = 300$, the turn ratio $n = N_s/N_p = 300$.

The magnetizing current is:

$$\mathrm{I_m} = \frac{\text{Magnetizing mmf}}{\text{Primary turns}} = \frac{100}{1} = 100\text{A}$$

$$\text{The loss component } \mathrm{I_c} = \frac{\text{Magnetizing mmf equivalent of iron loss}}{\text{Primary turns}} = \frac{50}{1} = 50 \text{ A}$$

The transformation ratio is given as:

$$T_R = n + \frac{(I_m \sin\delta + I_e \cos\delta)}{I_s} = 300 + \frac{(100 \times 0.426 + 50 \times 0.90)}{5} = 317.52$$

$$\text{Primary current} = \text{actual transformation ratio} \times \text{secondary current}$$

$$I_p = 317.52 \times 5 = 1587.6 \text{ A}$$

3. The value of transformation ratio is greater than the nominal ratio hence the ratio errors can be eliminated by reducing the number of secondary winding turns and hence by reducing the turn ratio.

Nominal ratio (N_R) = 300

The transformation ratio is:

$$T_R = n + \frac{(I_m \sin\delta + I_e \cos\delta)}{I_s}$$

For eliminating the ratio errors:

$$T_R = N_R$$

$$300 = n + \frac{(100 \times 0.426 + 50 \times 0.90)}{5} n + 17.52$$

$$n = 282.5$$

Hence the turns on secondary windings are:

$$N_s = n N_P = 282.5 \times 1 = 282.5$$

So, the required reduction in secondary winding turns is:

$$= 300 - 282.5 = 17.5 \approx 17$$

Example 4.3

A bar type CT has 300 turns on the secondary side. The secondary side impedance is $(3 + j1.5)\Omega$. Determine the values of the ratio and phase angle errors when the secondary winding draws 3 A current having magnetizing mmf at 80 AT and the iron loss is 1 watt.

Solution

As the CT has bar primary, hence $N_p = 1$. $N_s = 300$. Hence, turn ratio $n = N_s/N_p = 300$.
Total burden impedance on the secondary side:

$$Z_s = \sqrt{(3^2) + (1.5^2)} = 3.354\ \Omega$$

Secondary circuit power factor = $\cos\delta = 3/3.354 = 0.89$ and $\sin\delta = 1.5/3.354 = 0.44$
Secondary induced e.m.f. $E_s = I_s Z_s = 3 \times 3.354 = 10.062$ V
Primary induced voltage $E_p = E_s/n = 10.062/300 = 0.03354$ V
Core-loss component of current I_c = Iron loss/E_p = 1/0.03354 = 29.8 A
Magnetizing current I_m = mmf/N_p = 80/1 = 80 A.
The transformation ratio is:

$$T_R = n + \frac{(I_m \sin\delta + I_e \cos\delta)}{I_s}$$

$$T_R = 300 + \frac{(80 \times 0.44 + 29.8 \times 0.89)}{3} = 335.654$$

$$\text{Percentage error} = \frac{N_R - T_R}{T_R} \times 100 = 10.6\%$$

$$\text{Phase angle error is } \theta = \frac{180}{\Pi}\left[\frac{I_m \cos\delta - I_e \sin\delta}{nI_s}\right]$$

$$\theta = \frac{180}{\Pi}\left[\frac{80 \times 0.89 - 29.8 \times 0.44}{300 \times 3}\right] = 3.40°$$

4.7 Potential Transformers

Potential transformers (PTs), also called voltage transformers (VTs), are generally connected with a voltmeter, potential coil of a watt meter and voltage relays on the secondary side; the primary side is connected with a high voltage line, as depicted in Figure 4.2. They are in every way very similar to a common power transformer but have relatively low VA loading capacity. Hence, the secondary current in power transformers lies in the same range as the magnetizing current [1,3,8]. The secondary windings of PT are designed to conveniently handle the voltages in the range of 110–120 V, which can be measured by the voltmeters.

The major differences between the CT and PT can be listed as follows:

- Primary windings of PTs ares connected in parallel with the supply voltage, while in CTs they are connected in series with the power line. However, the secondary windings in PTs are connected across a voltmeter, virtually imitating open circuit conditions. In CTs secondary windings are generally connected through an ammeter, virtually imitating short circuit conditions. Hence the PTs are also sometimes termed as parallel transformers, and CTs are sometimes called series transformers [1,3,8].
- Unlike PTs, the secondary burden in CTs has no effect on the primary current.
- Primary windings in CTs experience high currents and low voltages in power system as they are connected in series, while in PTs they experience high voltages and low currents due to parallel connection.
- Unlike CTs, the flux density and hence the excitation current in PTs are nearly constant and vary in a very limited range as the line voltage feeding the primary windings exhibit restricted variations [5,7,8].

4.7.1 Equivalent Circuit and Phasor Representation of CTs

The equivalent circuit is similar to a common power transformer and CT, as depicted in Figure 4.6a. The vector diagram is presented in Figure 4.15.

All the symbols used in the phasor representation are the same as for a CT; however, on the primary side, the voltage drops due to the primary winding resistance (Ω_p) and primary leakage reactance (X_p) are added up; Δ, which represents the phase difference between secondary current I_s and secondary terminal voltage V_s, depends on the value of load resistance Ω_L and load reactance X_L connected on the secondary side according to Equation (4.36).

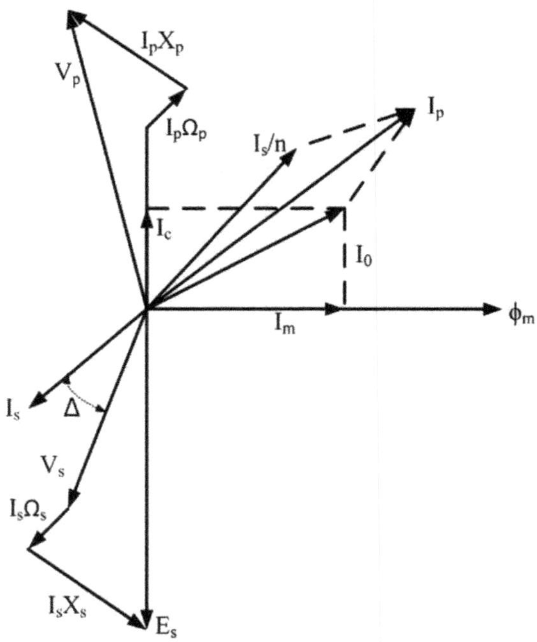

FIGURE 4.15 Phasor diagram of a PT. (From Golding, E.W., and Widdis, F.C., *Electrical Measurements and Measuring Instruments*, 5th ed., Reem Wisdom Pages LLP, New Delhi, India, 2017; ALSTOM, *Network Protection and Automation Guide*, 3rd ed., Chapter 6: Current and voltage transformers, http://rpa.energy.mn/wp-content/uploads/2016/07/network-protection-and-automation-guide-book.pdf, 2011.)

$$\Delta = \tan^{-1}\left(\frac{X_L}{\Omega_L}\right) \tag{4.36}$$

The load on the secondary side is generally a voltmeter with a very high value of resistance compared to the reactance; hence Δ is usually very small. The secondary terminal voltage V_s is obtained by subtracting the resistive and reactive drops from the induced e.m.f E_s. Also, the total applied voltage on the primary side (V_p) is the vector addition of the resistive and reactive drops $I_p(\Omega_p + X_p)$ on the primary side and the primary induced e.m.f (E_p).

4.7.2 Calculation of PT Transformation Ratio

The value of the transformation ratio in PTs can be calculated from Figure 4.16, which depicts the enlarged view of its primary side vector diagram. Secondary voltage V_s and current I_s are shown referred on the primary side; hence they are respectively multiplied and divided by turn ratio (n)

$$\text{Where, } n = \frac{\text{primary winding turns } (N_p)}{\text{secondary winding turns } (N_s)} \tag{4.37}$$

From Figure 4.16, it can be observed that:

$$od = V_p\cos\theta \tag{4.38}$$

Also:

$$od = nV_s + nI_s\Omega_s\cos\Delta + nI_sX_s\sin\Delta + I_p\Omega_p\cos\beta + I_pX_p\sin\beta \tag{4.39}$$

From (4.38) and (4.39):

$$V_p\cos\theta = nV_s + nI_s\Omega_s\cos\Delta + nI_sX_s\sin\Delta + I_p\Omega_p\cos\beta + I_pX_p\sin\beta \tag{4.40}$$

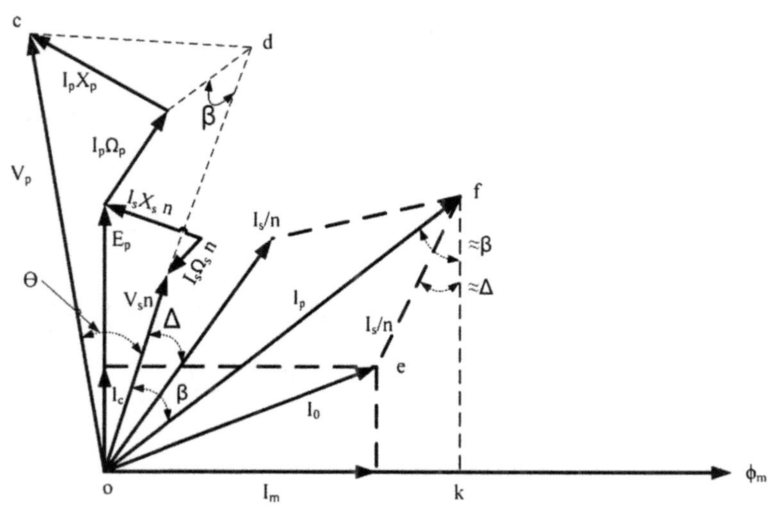

FIGURE 4.16 Primary section of PT phasor. (From Golding, E.W., and Widdis, F.C., *Electrical Measurements and Measuring Instruments*, 5th ed., Reem Wisdom Pages LLP, New Delhi, India, 2017; ALSTOM, *Network Protection and Automation Guide*, 3rd ed., Chapter 6: Current and voltage transformers, http://rpa.energy.mn/wp-content/uploads/2016/07/network-protection-and-automation-guide-book.pdf, 2011.)

$$= nV_s + nI_s\left(\Omega_s\cos\Delta + X_s\sin\Delta\right) + I_p\Omega_p\cos\beta + I_pX_p\sin\beta \tag{4.41}$$

θ is the transformer phase angle = angle between the phasors V_p and V_s
Δ is phase angle of secondary load circuit
β is the phase angle between phasors I_p and V_s

Angle θ is very small; hence, the phasors V_p and V_s reversed can be taken to be in quadrature with the phasor ϕ.

So ∟ of $k \approx \beta$ and ∟ of $k \approx \beta$

Therefore:

$$I_p\cos\beta = I_c + \frac{I_s}{n}\cos\Delta \tag{4.42}$$

$$I_p\sin\beta = I_m + \frac{I_s}{n}\sin\Delta \tag{4.43}$$

As $\theta \approx 1°$, which is very small, $\cos\theta \approx 1$ and $V_p\cos\theta \approx V_p$.

Now after suitable substitutions from Equations (4.42) and (4.43) into Equation (4.41) we get:

$$V_p = nV_s + nI_s\left(\Omega_s\cos\Delta + X_s\sin\Delta\right) + \Omega_p\left(I_c + \frac{I_s}{n}\cos\Delta\right) + X_p\left(I_m + \frac{I_s}{n}\sin\Delta\right) \tag{4.44}$$

$$nV_s + I_s\cos\Delta\left(n\Omega_s + \frac{\Omega_p}{n}\right) + I_s\sin\Delta\left(nX_s + \frac{X_p}{n}\right) + I_c\Omega_p + I_mX_p \tag{4.45}$$

The resistance and reactance are considered to be referred on the primary side, and their values are substituted in Equation 4.45:

So

$$V_p = nV_s + \frac{I_s}{n}\cos\Delta\, R_p + \frac{I_s}{n}\sin\Delta\, X_{pr} + I_c\Omega_p + I_mX_p \tag{4.46}$$

$$= nV_s + \frac{I_s}{n}\left(\cos\Delta R_p + \sin\Delta X_{pr}\right) + I_c\Omega_p + I_mX_p \tag{4.47}$$

where $R_p = n^2\Omega_s + \Omega_p$ and $X_{pr} = n^2X_s + X_p$ are the referred resistance and reactance, respectively, on the primary side of the PT. Hence the actual voltage transformation ratio is:

$$T_R = \frac{V_p}{V_s} = n + \frac{nI_s(R_p\cos\Delta + X_{pr}\sin\Delta) + I_c\Omega_p + I_mX_p}{V_s} \tag{4.48}$$

The difference between the transformation ratio (T_R) and the turn ratio (n) are given by:

$$T_R - n = \frac{nI_s\left(R_p\cos\Delta + X_{pr}\sin\Delta\right) + I_c\Omega_p}{V_s} \tag{4.49}$$

4.7.2.1 Phase Angle Calculation in PT

From the phasor depicted in Figure 4.16, we can write

$$\tan\theta = \frac{cd}{od} \tag{4.50}$$

$$\tan\theta = \frac{I_p X_p \cos\beta - I_p \Omega_p \sin\beta + nI_s X_s \cos\Delta - nI_s \Omega_s \sin\Delta}{nV_s} \tag{4.51}$$

$$= \frac{X_p\left(I_c + \frac{I_s}{n}\cos\Delta\right) - \Omega_p\left(I_m + \frac{I_s}{n}\sin\Delta\right) + nI_s X_s \cos\Delta - nI_s \Omega_s \sin\Delta}{nV_s} \tag{4.52}$$

$$= \frac{I_s\cos\Delta\left(\frac{X_p}{n} + nX_s\right) - I_s\sin\left(\frac{\Omega_p}{n} + n\Omega_s\right) + I_c X_p - I_m \Omega_p}{nV_s} \tag{4.53}$$

$$= \frac{\frac{I_s\cos\Delta}{n}\left(X_p + n^2 X_s\right) - \frac{I_s\sin\Delta}{n}\left(\Omega_p + n^2\Omega_s\right) + I_c X_p - I_m \Omega_p}{nV_s} \tag{4.54}$$

$$= \frac{\frac{I_s\cos\Delta}{n} X_{pr} - \frac{I_s\sin\Delta}{n} R_p + I_c X_p - I_m \Omega_p}{nV_s} \tag{4.55}$$

$$= \frac{\frac{I_s}{n}(X_{pr}\cos\Delta - R_p \sin\Delta) + I_c X_p - I_m \Omega_p}{nV_s} \tag{4.56}$$

Since $\theta \approx 0°$, $\tan\theta = \theta$, we have:

$$\theta \approx \frac{\frac{I_s}{n}(X_{pr}\cos\Delta - R_p \sin\Delta) + I_c X_p - I_m \Omega_p}{nV_s}\ \text{rad} \tag{4.57}$$

4.7.3 Types of Errors in PT

PTs also exhibit ratio and phase angle errors like CTs, and they can be defined as follows:

1. Ratio error is the difference between the nominal voltage ratio and the actual transformation ratio:

$$\%\ \text{ratio error} = \frac{N_R - T_R}{T_R} \times 100 \tag{4.58}$$

   where:
   N_R is the nominal ratio
   T_R is the actual transformation ratio

   Due to the presence of ratio errors in PTs, the voltage to be measured and connected on the primary windings (V_p) cannot be accurately determined simply by multiplying the turn ratio (n) and secondary voltage (V_s) measured by the voltmeter [5,7,9]. Deviation of the actual voltage ratio V_p/V_s from the turn ratio (n) is due to the presence of magnetization and core loss component of current.

2. Phase angle error (θ) is the phase difference between the primary winding voltage (V_p) and the reversed secondary voltage phasor (V_s). Ideally the value of $\theta \approx 0°$, which is not the case in actual PTs. The phase error is given by:

$$\theta \approx \frac{\frac{I_s}{n}(X_{pr}\cos\Delta - R_p\sin\Delta) + I_c X_p - I_m \Omega_p}{nV_s} \tag{4.59}$$

Conventionally θ is considered +ve when reversed V_s phasor leads the V_p phasor; otherwise, it is considered −ve. Like ratio errors the phase angle errors are also affected by magnetization and core loss component of current. The ratio errors are important while measuring the voltage, and phase angle errors are important when power is being measured [1,9].

4.7.4 Design Features for Reduction of Errors in PTs

Advancements due to research for improvement in the magnetic properties of materials have considerably reduced the errors in PTs. The errors can be further reduced, however, by adopting improved design features to meet objectives such as:

- Reduction of magnetizing component (I_m) and core loss component (I_c) of the excitation current.
- Reduction of resistance and leakage reactance.
- Turn compensation.

Unlike the values in CTs and power transformers, the value of I_m and I_c in PTs are comparable to the load current (I_s). Equations (4.47) and (4.58) suggest that the ratio and phase angle errors in PTs depend on the currents I_m and I_c. Therefore, the value of ratio and phase angle error can be reduced significantly by reducing their values. These reduction requirements can be achieved by incorporating design features like minimising the length of the magnetic path, using a high permeability core, maintaining high operational flux density without saturation and minimizing the air gaps between the core interleaves [3,5,10].

The winding resistance can be minimised by using thick winding conductors and smallest mean turn length of the coil, while the leakage reactance can be minimised by keeping the primary and secondary windings as close as possible; however, there should be sufficient space to meet out the insulation requirements [4,10].

The objective while designing a PT should be to keep the operational flux density in the core as high as possible without causing its saturation too early in order to achieve reduced area of core cross section consequently reducing the winding resistance due to lesser length of mean turn requirements. The high flux density requirements can be achieved by using *CRGO* steels, allowing a flux density of 1.5 Wb/m^2 compared to 1.2 Wb/m^2 in materials used previously.

Adopting the turn ratio adjustment, also called the turn compensation to minimise the ratio errors. Turn compensation is required when the N_R (nominal ratio) > n (turn ratio) and is achieved by reducing the primary turns (N_P) or increasing the secondary turns (N_s). Hence, the actual transformation ratio (T_R) can be made approximately equal to the nominal ratio (N_R) for a type of burden and its value. The turn compensation has negligible effect, however, on the value of phase angle error (θ) [5,7,10].

4.7.5 Construction of PTs

There is a lot of similarity between the conventional power transformers and PTs in terms of construction. However, a few vital differences exist between the two, as listed here:

1. The power transformers are designed for achieving maximum efficiency, and minimum voltage regulation and cost; on the other hand, PTs are designed with objectives of low ratio and phase angle error. These requirements give rise to comparatively larger size of PTs compared to power transformers because using reduced core and conductor size for economic designs results in large ratio and phase angle errors [1,9,11].

2. As mentioned earlier, the power handling capacity of PTs is very small compared to a power transformer. The loading capacity of PT is limited due to accuracy constraints, while in power transformer it is based on heating constraints. Unlike power transformers, the high voltage PTs are capable of handling loads up to 30 times their capacity on the basis of thermal loading.

A few important components of PTs are discussed below:

1. **Core:** The core is manufactured in the same way as in the case of power transformers by stacking thin laminations of Si-steel. Special precautions are taken for minimizing the air gaps at joints in order to minimise the core reluctance. Though both core and shell type construction is adopted, shell type construction is generally used for low voltage PTs.
2. **Windings:** Minimizing leakage reactance is a very important objective for reducing the errors in PTs and requires co-axial arrangement of the windings. Also, for reducing the insulation requirements, the LV winding is wound near the core and the secondary windings are divided into several sections in HV transformers with the objective of reducing the inter-turn insulation requirements [1,9,11].
3. **Insulation:** Insulation choice is generally based on the voltage level handled by the transformer. Cotton tapes, varnished cambrics and hard fibre separators are generally used as insulators. Oil insulation is also used in PTs operating above 7 kV voltage levels, while dry insulators like porcelain are used at voltage levels above 45 kV [10,11].
4. **Bushings:** To minimise the overall size of oil-filled PTs, oil-filled bushings are used. Single bushings are used when PTs are connected between a line and a neutral wire, and two bushings are used when none of the lines are at ground potential [10,11].

A cost saving arrangement utilizing a single bushing is used in CTs due to the very small voltage between the two terminals of the primary windings. In this arrangement both the terminals of primary windings are brought out through the same insulator [9,10]. The constructional features of two winding single-phase potential transformer are depicted in Figure 4.17.

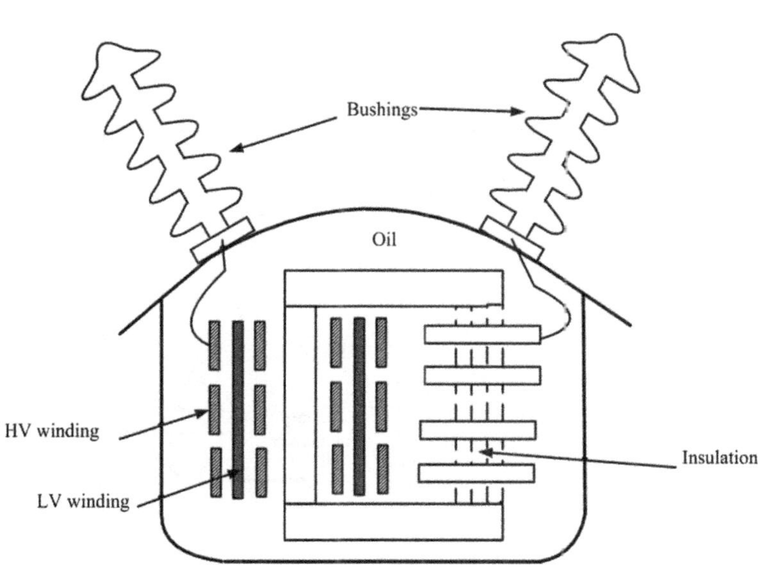

FIGURE 4.17 Single phase PT.

4.7.6 Types of PTs

Based on operating methodology, two types of PTs, normally called electromagnetic type and capacitive type, are used. Their details are described in the subsections below.

4.7.6.1 Electromagnetic Type PTs

They are also called high voltage (HV) PTs and used for voltages at 100 kV and above. They are very large and costly due to very strict insulation requirements. PTs operating at 110 kV are giants at 7.5 m height and weigh nearly 5 tonnes. Recent advancements in design process and core materials have significantly reduced the size and cost of PTs [7,8,12]. A few designs involving insulated casing, molded rubber and cascaded arrangement do not require bushings on the HV side, consequently reducing the overall cost of PT. All these arrangements are generally used to measure L-G voltages in a 3-ϕ arrangement. In an insulated casing arrangement, the transformer is put in a HV oil-filled insulator casing, resulting in economy of space and material [8,12].

In molded rubber PTs, the HV oil insulation is replaced by relatively cheaper molded rubber. In this arrangement the difficulties arising due to porcelain breakage are eliminated.

The idea behind the cascaded arrangement is to reduce the insulation and space requirements by using several stages of cascaded PTs. If N stages in the cascaded arrangement are used, the voltage to be sustained by each stage will be $1/N$, consequently reducing the insulation requirements for each stage [8,12].

4.7.6.2 Capacitive Type PTs

At above 100 kV L-G voltages, the conventional electromagnetic type PTs become very costly due to insulation requirements. Less costly capacitive type PTs have therefore been developed as a suitable alternative [9,10,12]. Here an auxiliary electromagnetic transformer is connected to a capacitive potential divider, as depicted in Figure 4.18.

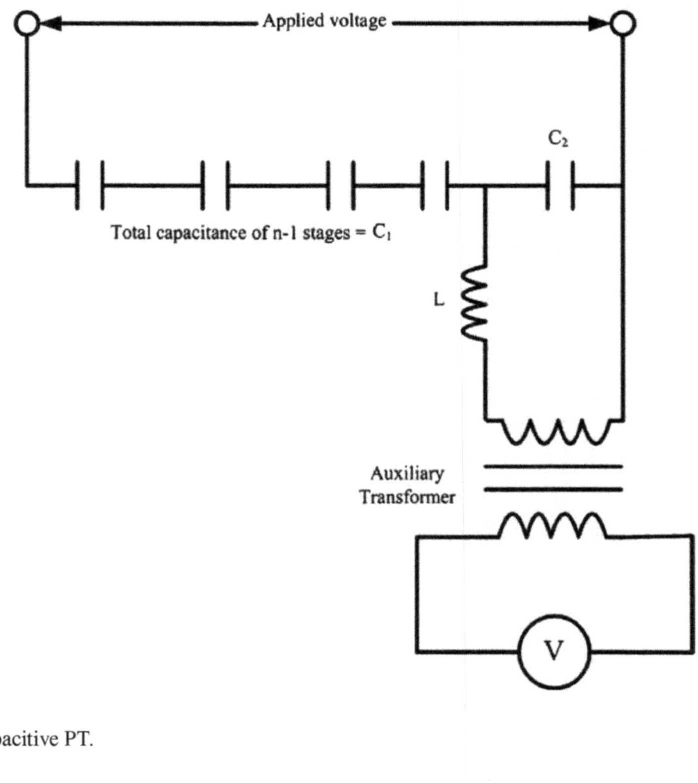

FIGURE 4.18 Capacitive PT.

The measurand line voltage is divided by each capacitive unit equally to lower voltage level. The voltage across the last stage is further stepped down and brought into the scale range of voltmeter by an auxiliary electromagnetic transformer. Total capacitance of n-1 stages is taken as C_1 and the last stage is taken as C_2. The inductance L given by Equation (4.60) largely contains the leakage inductance of the windings of auxiliary transformer. Its value is adjusted in order to compensate for the current drawn out from the divider [10,12].

$$L = \frac{1}{\omega^2 (C_1 + C_2)} \tag{4.60}$$

In this arrangement, the total transformation ratio being the product of the ratios of auxiliary transformer and capacitive divider is independent of the burden. Though the compensation in capacitive PTs is not complete owing to the losses in inductance and minute changes in the frequency, their performance is not much inferior to the conventional electromagnetic type PTs [10,12].

4.7.7 Protection of PTs

PTs may be operated at 1.2 times the rated voltage and are protected with the help of fuses connected on their primary side for avoiding the sudden short circuits on the LV (secondary) side. The fuses are also used on the secondary side for protecting it from faulty switching and defective earthing [9,10,11].

4.7.8 PT Characteristics

1. ***Impact of secondary burden***

   Increase in secondary VA increases the secondary winding current (I_s), consequently increasing the primary current (I_p) and the voltage drops across both the primary and secondary windings. Hence, the overall effect of an increase in the secondary burden is to increase the ratio errors, making it more negative. However, the graph between the secondary burden and the ratio errors is almost linear [10,12]. Here, the phase angle between V_p and V_s reverses and becomes more negative, thereby increasing the phase angle errors too with an increase in the secondary burden.

2. ***Impact of secondary burden power factor***

   The effect of a secondary burden power factor can be understood form the phasor representation of PT, depicted in Figure 4.15. Angle Δ between V_s and I_s increases as the secondary burden power factor gets reduced, thereby shifting the primary current vector (I_p) closer to I_0. Consequently, the phasors V_p and V_s try to align towards E_p and E_s, respectively, due to nearly constant voltage drops on both the primary and secondary sides. Hence, there is a relative increase in V_p compared to E_p, but since applied voltage V_p is generally constant, the resultant effect is to reduce E_p and reduce V_s in proportion to E_s. Hence, the overall impact is to increase the transformation ratio (T_R) with the reduction in secondary burden power factor [5,10,12].

3. ***Change in frequency***

   If the applied voltage is constant, there is always an indirect proportionality relationship between flux generated and supply frequency. Therefore, as the frequency increases, the values of I_m and I_c decrease due to a decrease in the value of flux; consequently, the voltage ratio also drops. However, an increase in frequency also results in an increase of leakage reactance values, and consequently the voltage drops, thereby increasing the voltage ratio. Hence, as both the effects stated above tend to neutralise one another, there is very little variation in the voltage ratio due to an increase in the supply frequency. The phase angle error increases, however, due to an increase in the frequency because with the increase in the value of X_s, V_s decreases, resulting in the increase of phase angle θ [5,10].

4. ***Variation of the voltage supply***

   As the supply voltage generally remains constant, it has very little effect on the ratio and phase angle errors [10,11,12].

4.7.9 Design Requirements for PTs

The PT design requirements can be listed under three categories: insulation requirements, short-circuit with stand capability and mechanical requirements [10,13].

The insulation requirements give an overview of the insulation type to be used by the primary windings during normal operation and various test conditions. The highest system voltage is the deciding factor for primary winding insulation requirements, on the basis of several factors listed in Table 4.2.

The short-circuit with stand capability describes the capacity of PTs to sustain mechanical and thermal stresses due to external short circuits of specified duration. It is desired that no significant damage due to mechanical and thermal stresses should occur because of external short circuits for duration of 1 second in PTs when they are energised at rated voltage.

Mechanical requirements give an overview of the total static load, including the ice and wind loading that a PT can bear for a particular highest system voltage. Two load classes (a) Class I and (b) Class II, as specified by IEC 60044-2, can be applied in any direction of primary windings during load tests [9,10,13].

4.7.9.1 Accuracy Class Requirements of PTs

The accuracy requirements for PTs are designated by accuracy classes represented by the highest permissible error in voltage and phase angle displacements when the PT is operating at rated burden and frequency. The errors are determined at the transformer terminals after considering the effects of fuses and resistors as integral parts of PT. Standard accuracy classes for PTs are:

$$0.1 - 0.2 - 0.5 - 1.0 - 3.0$$

The error tolerance limits in PTs for voltage and phase displacements are presented in Table 4.3.

4.7.10 3-Φ Connections of PTs

PTs are connected in a 3-Φ power supply system where the system voltages are very high and difficult to measure with the help of conventional instruments. Their purpose is to act as a reduced replica of the

TABLE 4.2

Deciding Factors for the Primary Winding Insulation Requirements

S. No.	Highest System Voltage (V_n) (kV)	Factors Deciding the Primary Winding Insulation Requirements
1.	0.72 or 1.2	Rated power-frequency withstand voltage
2.	3.6–300	Rated lightning impulse and power-frequency withstand voltages
3.	≤300	Rated switching and lightning impulse withstand voltages

TABLE 4.3

Voltage and Phase Displacement Error Limits for PTs

		Phase Displacements	
Accuracy Class	Voltage Ratio Error (%)	Minutes	Centi-radian
0.1	0.1	5	0.15
0.2	0.2	10	0.3
0.5	0.5	20	0.6
1	1	40	1.2
3	3	Unspecified	Unspecified

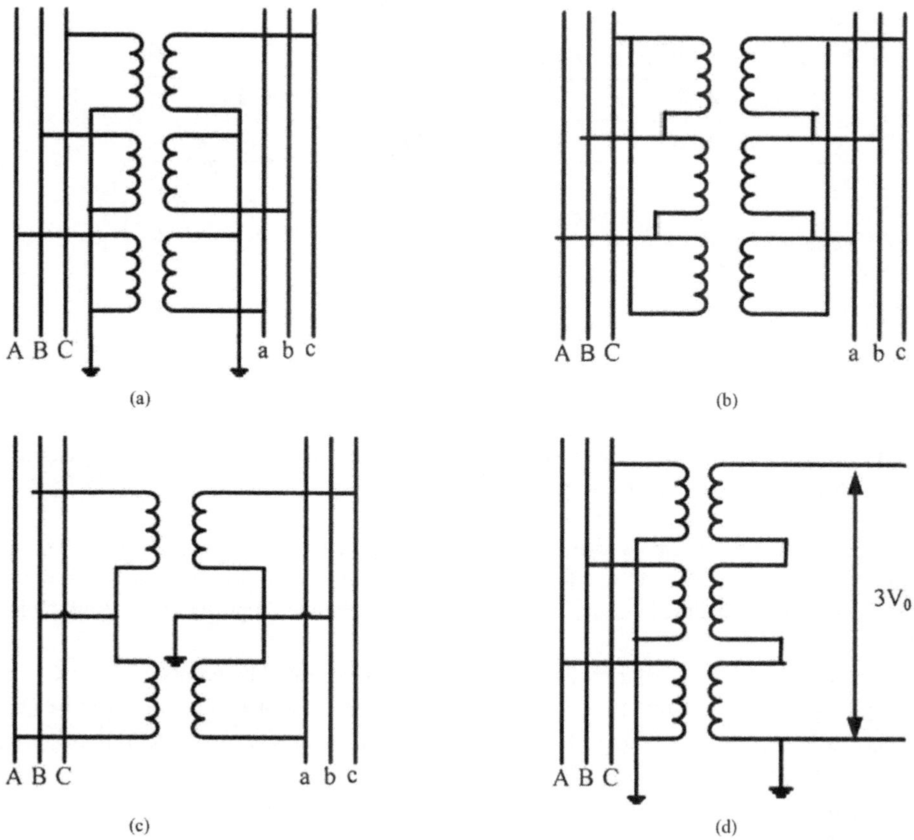

FIGURE 4.19 3-Φ connections of PTs (a) star connection (b) delta connection (c) open-delta connection (d) broken-delta connection.

voltage input to their primary windings with the desired level of accuracy [13,14]. The *PTs* in a 3-Φ system are normally connected in star, delta, open-delta and broken delta form, as depicted in Figure 4.19a–d.

4.7.11 PT Markings

Normally two types of markings, (a) rating plate markings and (b) terminal markings, are used by manufacturers for customer usage.

4.7.11.1 Rating Plate Markings

It is mandatory for manufacturers to clearly indicate the following four parameters indelibly marked either on the outer body or on a rating plate inseparably attached to the PTs:

1. Identification marks specific to the PT manufacturer: either its name or a symbol and logo should be marked [13,14].
2. Product serial number and its type should be clearly indicated.
3. VA rating and accuracy class, besides frequency and rated winding voltages, should also be mentioned [13,14].
4. The highest system voltage and rated insulation level should be mentioned on the rating plate.

TABLE 4.4

Rating Plate Description of a PT

Voltage Transformer		A-N 220000:√3V			
Manufacturer		1*a*–1*n*		(2*a*–2*n*)	d_a–d_n
Serial no..		110: √3			110:3
Type	50 Hz	VA:25	50		25
245/460/1050 kV	1.9 U_n 30 s	CI:0.5	3P		6P

A typical PT rating plate is represented in Table 4.4.

4.7.11.2 Terminal Markings

The thumb rules for terminal markings mentioned below are equally applicable to a single core 3-ϕ PT and a set of three 1-Φ PTs in 3-ϕ arrangement.

- 3-ϕ primary and secondary terminals are indicated by capital letters A, B and C and small letters a, b and c, respectively, as depicted in Figure 4.20. Terminal N (n), which is earthed, is generally weakly insulated compared to the other six terminals represented by English letters [12,14].
- If a PT contains two secondary windings, they are distinguished by numbers 1, 2, as depicted in Figure 4.21, but the terminals are represented in the same manner as explained above [12,14].
- PTs containing residual windings for supplying residual voltages have terminals represented by d_a and d_n, as depicted in Figure 4.22.
- The relative polarities of terminals represented by capital and small letters must be same at a particular time instant [12,14].

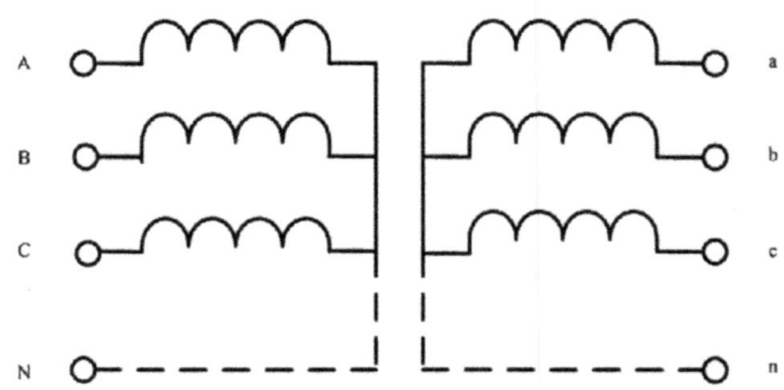

FIGURE 4.20 Terminal markings on 3-ϕ PT.

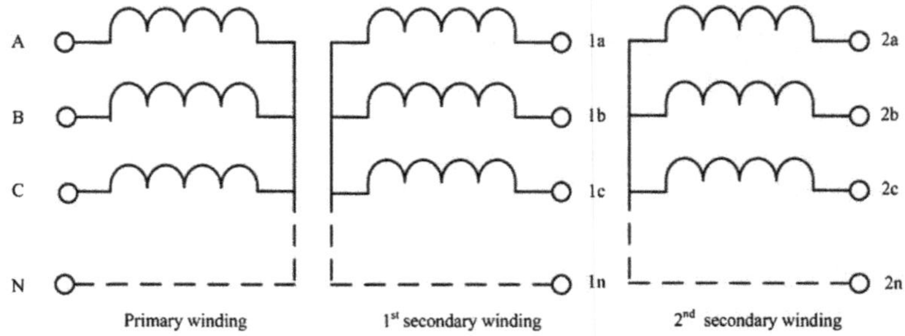

FIGURE 4.21 Terminal markings on 3-ϕ PT with two secondary windings.

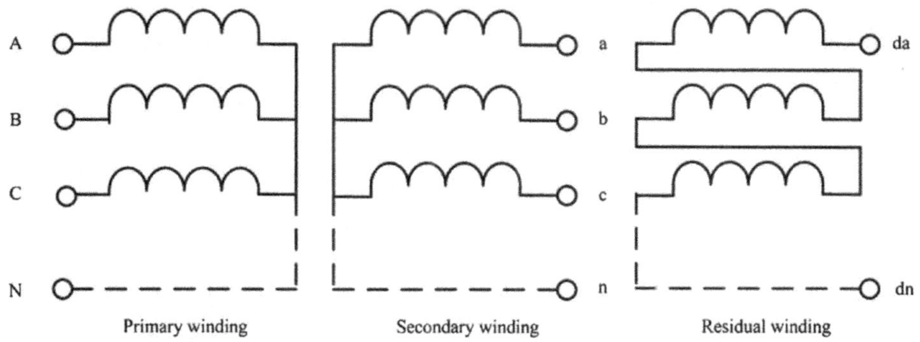

FIGURE 4.22 Terminal markings on 3-ϕ PT with residual winding.

4.7.12 Solved Examples on PTs

Example 4.4

The equivalent circuit parameters of a PT with turn ratio n = 1000/100 are: primary resistance (Ω_p) = 90.4Ω, secondary resistance (Ω_s) = 0.66Ω, primary reactance (X_p) = 60.3Ω, total equivalent reactance (X_{pr}) = 100Ω, no-load current (I_0) = 0.02 A at 0.4 power factor. Determine (1) the no-load phase angle error and (2) the burden in VA at unity power factor at which there will be zero phase angle error.

Solution

Given parameters:
(a) $n = 1000/100$, (b) $\Omega_p = 90.4\Omega$, (c) $\Omega_s = 0.66\Omega$, (d) $X_p = 60.3\Omega$, (e) $X_{pr} = 100\Omega$, (f) $I_0 = 0.02$ A, (g) $\cos\alpha = 0.4$.

1. The power factor at no-load $\cos\alpha = 0.4$, so we can calculate

$$\sin\alpha = \sqrt{(1)^2 - (0.4)^2} = 0.917.$$

Also, the phase angle error in PT is given by Equation (4.57):

$$\theta \approx \frac{\frac{I_s}{n}(X_{pr}\cos\Delta - R_p\sin\Delta) + I_c X_p - I_m \Omega_p}{nV_s} \text{ rad}$$

At no-load condition, the secondary current $I_s = 0$; hence, Equation (4.57) reduces to:

$$\theta \approx \frac{I_c X_p - I_m \Omega_p}{nV_s} \text{ rad}$$

The values of I_c and I_m can be calculated as:

$$I_c = I_0 \cos\alpha = 0.02 \times 0.4 = 0.08 \text{ A}$$

$$I_m = I_0 \sin\alpha = 0.02 \times 0.917 = 0.01348 \text{ A}$$

Substituting the values of various parameters obtained in the above steps and calculating the value of phase angle error we get:

$$\theta \approx \frac{0.008 \times 60.3 - 0.01834 \times 90.4}{10 \times 100} \approx -0.06735°$$

The no-load phase angle error is $\theta \approx -0.06735°$.

2. We know that, at unity power factor, cos $\Delta = 1$ and sin $\Delta = 0$; hence, Equation (4.57) reduces to:

$$\theta \approx \frac{\frac{I_s}{n}(X_{pr}\cos\Delta) + I_c X_p - I_m \Omega_p}{nV_s} \text{rad}$$

For zero phase angle error, we put $\theta = 0°$ and the above equation reduces to:

$$\left(\frac{I_s}{n}\right) X_{pr}\cos + I_c X_p - I_m \Omega_p = 0$$

$$I_s = \frac{n}{X_{pr}}(I_m \Omega_p - I_c X_p) = \frac{10}{100}(0.01834 \times 90.4 - 0.008 \times 60.3) = 0.11755\text{A}$$

Hence, burden $= V_s I_s = 100 \times 0.11755 = 11.755$ VA.

Example 4.5

A PT operating at 6700/100 V rated voltages has 22,300 primary turns and 375 secondary turns. The secondary open circuit current when primary is energised at its rated voltage is measured as 0.007 amperes, which lags the voltage by 73.7°. When the PT is operated at a certain secondary burden, the primary draws a current of 0.0125 amperes, lagging the voltage by 53.1°. The equivalent circuit parameters of PT are:

(a) Primary winding resistance = 1200Ω, (b) primary winding reactance = 2000Ω, (c) secondary winding resistance = 0.4Ω, (d) secondary winding reactance = 0.7Ω
From the above parameters, determine the following:

1. The terminal voltage and current for the secondary winding when the primary winding is supplied with the rated phasor voltage of 6700 + *j*0 taken as reference.
2. The actual transformation ratio and the phase angle error.
3. The required turn compensation when the actual ratio is equal to the nominal ratio under the above-mentioned conditions.

Solution

1. The nominal ratio is given by $N_R = 6700/100 = 61$, primary turns (N_p) = 22,300, secondary turns (N_s) = 375.

Hence, the turn ratio is $n = \frac{N_p}{N_s} = \frac{22300}{370} = 60.27$

No-load current (I_0) = 0.007 A
No-load power factor = cos73.7° = 0.28 and sin73.7° = 0.96
Primary current at certain burden (I_p) = 0.0125 A
Primary power factor=cos53.1° = 0.6 hence, sin53.1° = 0.8
Considering the phasor value of primary applied voltage as reference:

$$V_p = 6700 + j0$$

$$I_p = 0.0125(0.6 - j0.8) = 0.0075 - j0.2$$

$$I_0 = 0.007(0.28 - j0.96) = 0.00196 - j0.0067$$

From the phasor representation in Figure 4.15, we know that I_s/n is the phasor difference of primary current (I_p) and excitation current (I_0). Therefore:

$$I_s/n = (0.0075 - j0.2) - (0.00196 - j0.0067) = 0.00554 - j0.1933$$

Hence:

$$I_s = n \times I_s/n = 60.27\ (0.00554 - j0.1933) = 0.3334 - j11.65$$

The reversed secondary current is given as $-I_s = -(0.3334 - j11\ 65) = -0.3334 + j11.65$.

$$I_s = 11.654\text{A}$$

From the phasor representation in Figure 4.15, the induced e.m.f. in the primary winding is given by $E_p = V_p - I_pZ_p = (6700 + j0) - (0.0075 - j0.2).\ (1200 + j2000) = (6700 + j0) - (409 - j225) = 6291 + j225$ V.

So $E_p = 6295.022$ V

Secondary induced e.m.f is $E_s = E_p/n = (6291 + j225)/60.27 = (104.38 + j3.733)$

Secondary terminal voltage as per the phasor representation in Figure 4.15 is given by:

$$V_s = E_s - I_sZ_s = (104.38 + j3.733) - (-0.3334 + j11.65)\ (0.4 + j0.7)$$

$$= (104.38 + j3.733) - (-8.286 + j4.426) = (104.38 + 8.286) + j(3.733 - 4.426)$$

$$= 112.666 - j0.693$$

$$V_s = 112.6680\text{V}$$

Therefore, the secondary burden is $V_sI_s = 112.6680 \times 11.654 = 1313$ VA (approx.).

2. The actual voltage ratio = $V_p/V_s = 6700/112.6680 = 59.4$

Reversed secondary voltage phasor is $-V_s = -(112.666 - j0.693) = -112.666 + j0.693$

Angle by which the reversed secondary voltage phasor ($-V_s$) lags V_p is given by:

$$\tan^{-1}\left(-\frac{0.693}{112.666}\right) = 0.352 \text{ rad}$$

3. From the results obtained in the above two steps, it can be inferred that equalizing the turn ratio and the nominal ratio by turn compensation can be achieved by increasing the turns in the primary windings (N_p).

The new primary turns (N_p) required $\left(\frac{61}{60.27}\right) \times 59.4 \times 375 = 22544.79$

Hence, the increase in the primary turns is 22544.79 – 22300 = 224.79. But since the turns cannot be expressed in fractions, the net required increment in the secondary turns can be obtained by rounding off 224.79 to 225.

4.8 Tutorial Problems

1. Compare the shunts and multipliers with ITs for range extension of instruments.
2. With the help of schematics, explain the method of measurement of high current and voltages using ITs.
3. Explain the phasor and equivalent circuit diagrams of a CT with the help of schematic diagrams.
4. What are the essential design features for reducing the ratio and phase angle errors in CT? Explain.

5. Explain the method of turn compensation for reducing the ratio errors in CTs.
6. What is the reason behind always keeping the CT secondary windings short-circuited when its primary winding is energised?
7. Discuss the various reasons for permanent magnetization of the CT core and its effect on the errors. Suggest a few methods of core demagnetization in such situations.
8. Discuss the essential differences between wound type, bar type, clamp type and bushing CTs.
9. Enumerate the various points of differences between a CT and a PT.
10. Point out a few major differences between a PT and a regular power transformer.
11. Draw the phasor representation and equivalent diagram of a PT. Also explain the various symbols used.
12. Describe the methods for reducing the ratio and phase angle errors in a PT.
13. A ring core type CT with nominal ratio 1000/10 and a bar primary has a secondary winding resistance of 0.7Ω and very low reactance. The no-load current is 5 A at a power factor of 0.35. Full load secondary current flows when a burden of 1.5Ω pure resistance is connected on the secondary. Calculate the ratio error and phase angle error at full load. Also calculate the flux in the core at 50 Hz.
14. A bar type CT has 200 turns on the secondary windings. The total secondary impedance is (1.5 + 2.3j) Ω. With 5A current flowing in the secondary windings the magnetizing mmf is 1000 AT and the iron loss is 2 watt. Determine the ratio and phase angle errors.
15. The equivalent circuit parameters of PT having nominal ratio 1000/100 v are: primary resistance = 90Ω, secondary resistance = 0.7Ω, primary reactance = 90Ω, secondary reactance = 0.67Ω, no-load current = 0.03 A at a power factor of 0.35. Calculate the values of (a) the ratio and phase angle errors and (b) the burden in VA at unity power factor for which the phase angle error will be zero.

4.9 Conclusion

ITs are very important components for control and protection of modern power systems. Hence, their timely maintenance becomes a necessary issue. Several approved guidelines such as IEC-60044-1 and IEC-60044-2 have been issued by various agencies around the world for their installation and maintenance. ITs generally aid us in measuring high voltages and currents in power systems with moderate ranged instruments. The chapter covered almost every aspect of ITs in detail.

REFERENCES

1. E.W. Golding and F.C. Widdis, *Electrical Measurements and Measuring Instruments*, 5th ed., New Delhi, India: Reem Wisdom Pages LLP, 2017.
2. W.D. Stevenson, *Elements of Power System Analysis*, 4th ed., Singapore: McGraw-Hill, 1982.
3. A.K. Sawhney, *A Course in Electrical and Electronic Measurements and Instrumentation*, 19th ed., New Delhi, India: Dhanpat Rai and Company, 2016.
4. Bharat Heavy Electricals Limited, *Handbook on Switch Gears*, 2nd ed., New Delhi, India: Tata-McGraw-Hill, 2005.
5. K.C. Agarwal, *Industrial Power Engineering and Application Handbook*, 2nd ed., New Delhi, India: Newnes, 2011.
6. S.J. Chapman, *Electric Machinery Fundamentals*, 4th ed., Singapore: McGraw-Hill, 2005.
7. P. Purkait, B. Biswas, S. Das, C. Koley, *Electrical and Electronics Measurements and Instrumentation*, 1st ed., New Delhi, India: McGraw-Hill Education, 2016.

8. ALSTOM, *Network Protection and Automation Guide* (3rd ed.). Chapter 6: Current and voltage transformers, 2011, http://rpa.energy.mn/wp-content/uploads/2016/07/network-protection-and-automation-guide-book.pdf.
9. AREVA T&D Publication, *Network Protection & Automation Guide (NPAG)*. Chapter 6: Current and voltage transformers, 2008, www.areva-td.com.
10. International Standards IEC 60044-1, *Terminal Markings-General Rule*, https://webstore.iec.ch/preview/info_iec60044-1%7Bed1.2%7Den_d.pdf.
11. ABB Publication, *ABB Outdoor Instrument Transformers: Application Guide*, Edition 2.1, Ludvika, Sweden, 2005, www.abb.com.
12. International Standards IEC 60044-2, *Rating Plate Markings*, http://www.normservis.cz/download/view/iec/info_iec60044-2%7Bed1.2%7Den_d.pdf
13. Electrical Power Engineering Reference & Applications Handbook, *Instrument and Control Transformers: Applications and Selection*, http://www.electricalengineering-book.com/pdf/chapter-327907.pdf.
14. C.R. Mason, *The Art and Science of Protective Relaying*, General Electric, http://www.gegridsolutions.com/multilin/notes/artsci/.

5

Protective Relaying System

Senthil Krishnamurthy

CONTENTS

5.1 Introduction

This chapter provides an overview of the overcurrent and differential protection functions. The main function of a protective relay is to discriminate faults within its zone of protection, energize the trip coil of the circuit breaker (CB), and provide security against false tripping for faults outside the zones. A detailed description of the commonly used protective relays in power systems are presented; they include overcurrent, differential protection functions and their characteristics. In addition, this chapter provides examples for time overcurrent, grading principles, relay coordination, and transformer differential protection configuration settings in the DigSilent simulation environment.

5.1.1 Protective Relaying System

Power system protection is a branch of electrical power engineering that deals with the protection of electrical power systems from faults through isolation of faulty parts from the rest of the electrical network. There are several textbooks on protective relaying system published from 1998 to 2014; however, this chapter is unique in comparison with the available literature [1–9] because it provides case studies on time-overcurrent protection, relay-to-relay coordination and transformer differential protection functions using DigSilent simulation models.

Protection schemes need fast and selective operation to prevent unnecessary shutdown of large networks and excessive damage to expensive primary plants. Protection devices are installed with the aim of protecting assets and ensuring continued supply of energy. The three classes of protective devices are (a) protective relays that control the tripping of the circuit breakers surrounding the faulty part of the network; (b) automatic operation, such as auto-reclosing or system restart and (c) monitoring equipment that collects data from the system for post-event analysis.

A fail-free power system is neither economically justifiable nor technically feasible. Faults can occur in any power system component, i.e., generators, transformers, buses, transmission line. Faults fall into two general categories (a) short-circuit (SC) faults and (b) open-circuit (OC) faults. Other types of faults that don't fall in these categories are unbalanced generator operation and loss of generator excitation. Short-circuit faults are the most severe kind and can lead to extensive damage to equipment, whereas open-circuit faults cause abnormal system operation and danger to personnel. Voltages tend to rise well beyond acceptable values in certain parts of the system with the possibility of insulation failure.

Faults should be instantly detected and the faulty section isolated from the rest of the system in the shortest possible time. It is obviously not possible to do this manually; therefore, and it must be accomplished automatically. Faults are detected automatically by means of relays, and the faulty section is isolated by circuit breakers connected at the boundaries of the line section, transformer or generator. The combination of relays and circuit breakers is known as a protective system.

5.1.2 Elements of the Protection System

A protection system usually has five components: namely, current and voltage transformers, protective relays, circuit breakers, batteries and communication channels. Detailed descriptions of the power system protection elements are given below:

1. Current and voltage transformers are used to step down the high voltages and currents of the electrical power system to convenient levels and then feed signals to protection relays to deal with. Usually the current and potential transformers' nominal secondary ratings are 1 or 5 amps and 110 to 120 volts, respectively. It is necessary to know that the functions of metering and protection CT and PT are not the same.
2. Protective relays are used to sense the fault and initiate a trip or disconnection signal. Relays analyse the CT and PT signals and, if an abnormal condition has occurred (in other words, high currents and voltage decay in the network), analyse whether the power system network is in danger, and find out which element of the network is faulty and how quickly it must be removed.
3. The relay energizes the close/trip coils which command circuit breakers to close or open. Relays initiate auto-recloser commands.
4. Circuit breakers are used to energize the close/trip coils of the power system based on relay and auto-recloser commands.
5. Batteries are used to provide power in case of power disconnection in the system. The batteries provide DC supply to the circuit breaker (CB) trip coil, control/protective circuits, lighting, and emergency alarms. The commonly rated station batteries are 48/125/250 V DC. AC is not reliable during fault conditions; therefore, batteries are essential for providing power for the relays and circuit breaker tripping and closing circuits. Loss of tripping supply is crucial (DC); it causes loss/destruction of expensive equipment, so the battery charger/DC alarm must be dealt with immediately.
6. Communication channels are used to allow analysis of current and voltage at remote terminals of a line and to allow remote tripping of equipment. Examples of the communication channels are pilot wire/power line carriers, microwave, digital channel (fibre optics). The applications of the communication-assisted protection schemes are drect under-reaching transfer trip (DUTT), permissive under-reaching transfer trip (PUTT), permissive over-reaching transfer trip (POTT), directional comparison blocking (DCB) and directional comparison un-blocking (DCUB).

5.1.3 Objective of Power System Protection

A power system is geographically spread over a vast area. Disturbances such as thunderstorms, cyclones, snowstorms or many such natural occurrences may result in uprooting of transmission or distribution lines, causing interruption in power transfer and delivery. During a thunderstorm, a branch of a tree can cause a short circuit between two lines. The magnitude of a short circuit current is several times the normal operating current, and the power apparatus such as generators, transformers and lines have to be protected from such a large current. The objective of power system protection is to isolate a faulty section of the electrical power system from the rest of the live system so that the live portion can function satisfactorily without any severe damage due to fault current. Circuit breakers isolate the faulty system from the rest of the healthy system and automatically open during fault condition due to the trip signal coming from the protective relay.

5.1.4 Main Reasons for Power System Protection

The need for protection is to protect human life (safety), protect expensive primary equipment, reduce outage times, trip faulty plant only and achieve fast clearance times. In addition, it is mandatory to prevent voltage dips for customers (quality of supply), minimize loss of revenue for industry and customers, prevent and/or minimize plant damage, maintain system stability (for transmission networks), maintain public and personnel safety, ensure discrimination between healthy and faulty sections, and maximize reliability.

5.1.5 Functional Requirements of Protective Relay

The main philosophy about protection is that no power system protection can prevent the flow of fault current through the system; it only can prevent the continuation of flowing of fault current by quickly disconnecting the short circuit path from the system. To make this quick disconnection, the protection relays should have following functional characteristics (a) speed, (b) selectivity, (c) reliability and (d)sensitivity.

5.1.5.1 Speed

Operation must be fast to minimize equipment damage. Faults at any point in the system must be detected and isolated in the shortest possible time. This time is on the order of 30–100 ms, depending upon the fault level of the section involved.

5.1.5.2 Selectivity

Relaying equipment must clearly discriminate between normal and abnormal system conditions so that it never operates unnecessarily and trips the minimal number of segments.

5.1.5.3 Reliability

Protective equipment must be found in healthy operating condition when called upon to act, as years might elapse between two consecutive operations of relays at a particular station. It means operate properly, or trip, when required.

5.1.5.4 Sensitivity

The relaying equipment must be sufficiently sensitive so that it can be operated reliably when the level of fault condition just crosses the predefined limit. The protection function will send the trip signals to the circuit breaker, once the measured input quantity reaches the threshold setting.

5.2 Overcurrent Relays and Characteristics

Overcurrent relays respond to the magnitude of the input current and operate when this magnitude exceeds the preset level (pickup level). When this level is exceeded, the relay closes its trip contacts and energizes the circuit breaker trip coils. A wide variety of devices are used to protect the generation equipment, transmission, and distribution networks. The particular type of protection used depends on the system element being protected and the system voltage level. There are no specific standards for overall protection of distribution networks. The devices mostly used for distribution system protection are overcurrent relays, reclosers, sectionalisers, and fuses. This section of the chapter only focuses on the overcurrent relay and its characteristics. Overcurrent relay is very well suited to distribution system protection because it is basically simple, inexpensive and very often does not need to be directional which excludes the need of potential transformer to supply the voltage. It is possible to use a set of two overcurrent relays for protection against interphase faults and a separate overcurrent relay for ground faults. It provides backup protection for the differential protection if it fails.

Overcurrent relays are the most common form of protection used to deal with excessive currents on power systems. They should not be installed purely as a means of protecting systems against overloads, which are associated with the thermal capacity of machines or lines. Overcurrent protection is primarily intended to operate only under fault conditions. However, the relay settings that are selected are often a compromise in order to cope with both overload and overcurrent conditions. The example in Figure 5.1 provides the basic understanding of the full-load, overload and short-circuit (overcurrent) conditions in the power system.

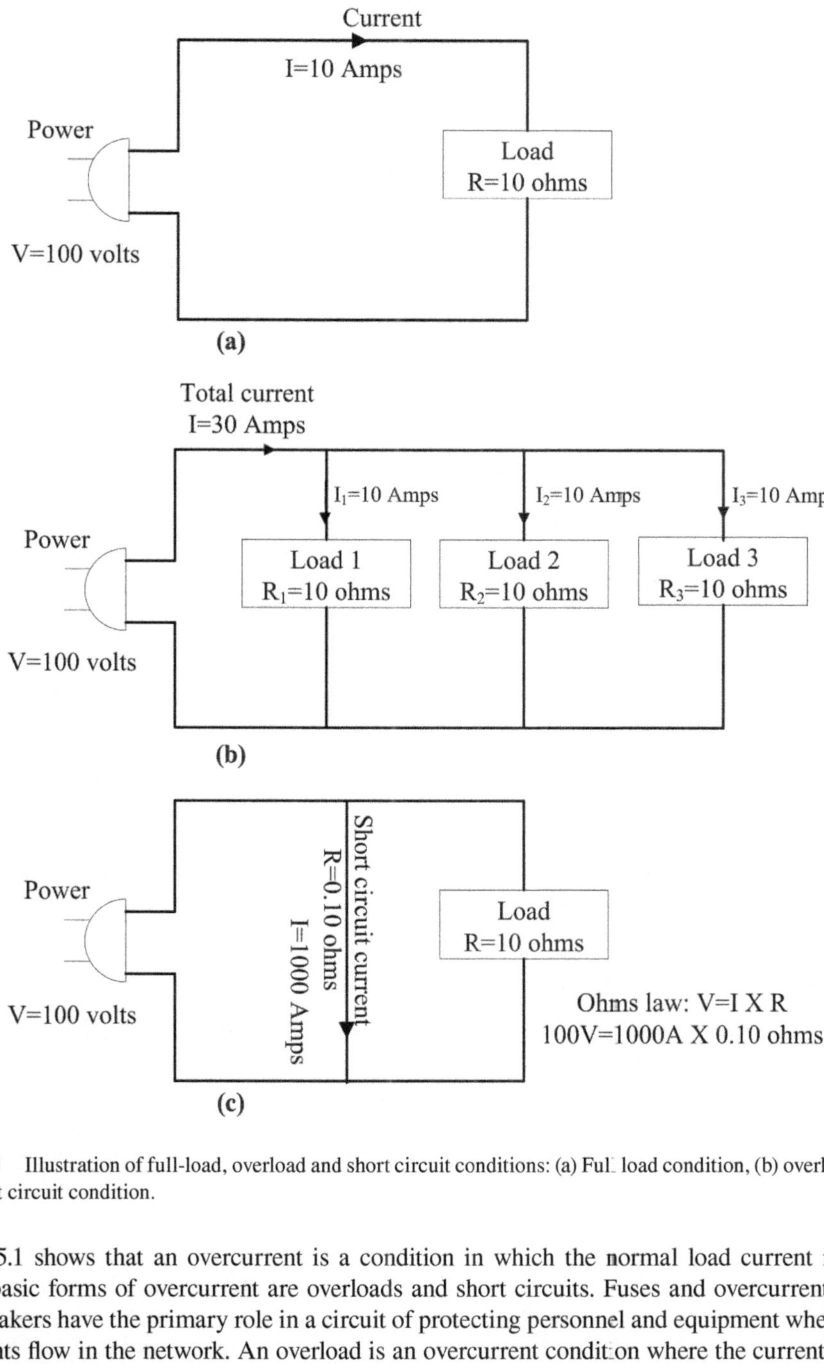

FIGURE 5.1 Illustration of full-load, overload and short circuit conditions: (a) Full load condition, (b) overload condition, and (c) short circuit condition.

Figure 5.1 shows that an overcurrent is a condition in which the normal load current is exceeded. The two basic forms of overcurrent are overloads and short circuits. Fuses and overcurrent relays with circuit breakers have the primary role in a circuit of protecting personnel and equipment when dangerous overcurrents flow in the network. An overload is an overcurrent condition where the current exceeds the normal full-load capacity of the circuit but no fault condition (short-circuit) is present. A short circuit is an overcurrent condition where an abnormal, low-resistance, circuit path is introduced into the circuit. This low-resistance path bypasses the normal load and can create extremely high currents up to 1000 times the normal current.

5.2.1 Types of Overcurrent Relays

The operating characteristics of the overcurrent relays can be classified into four groups: (a) instantaneous overcurrent relay (definite current), (b) definite time overcurrent relay, (c) inverse time overcurrent

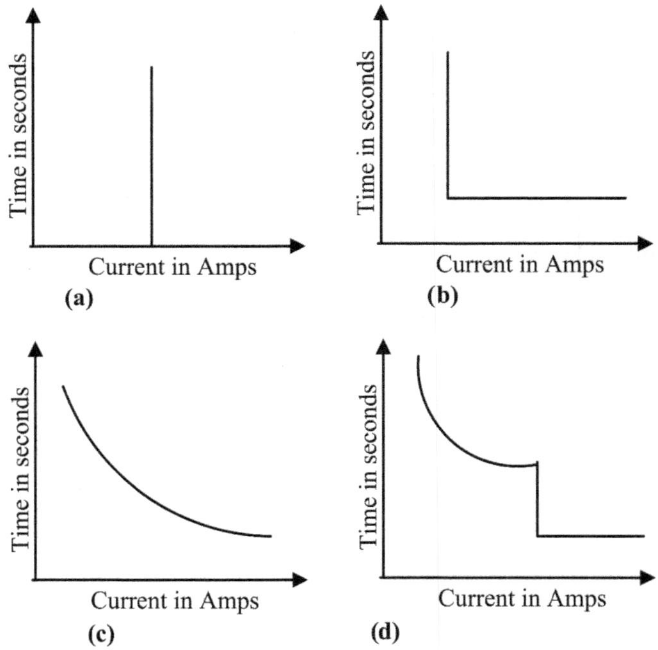

FIGURE 5.2 The operating characteristic of the overcurrent relays: (a) definite current, (b) definite time, (c) inverse time, and (d) inverse definite minimum time.

relay and (d) inverse definite minimum time (IDMT) overcurrent relay. The operating characteristic of the overcurrent relays is given in Figure 5.2.

5.2.1.1 Instantaneous Overcurrent Relay

This type of relay operates instantaneously when the current reaches a predetermined value. The setting is chosen so that, for the substation furthest away from the source, the relay will operate for a low current value and the relay operating currents are progressively increased at each substation moving towards the source. The operating characteristic curve of a definite current relay is shown in Figure 5.2a. It operates in 0.1 s or less; the operation criterion is current magnitude only.

5.2.1.2 Definite Time Relay

This type of relay enables the setting to be varied to cope with different levels of current by using different operating times. The settings can be adjusted so that the breaker nearest to the fault is tripped in the shortest time and then the remaining breakers are tripped in succession using longer time delays, moving back towards the source. The operating characteristic curve of definite time relay is shown in Figure 5.2b.

5.2.1.3 Inverse Time Overcurrent Relay

The fundamental property of these relays is that they operate in a time that is inversely proportional to the fault current and is illustrated by the characteristic curve shown in Figure 5.2c. The advantage of this type of relay over definite-time relays is that, for very high currents, much shorter tripping times can be obtained without risk to the protection selectivity.

5.2.1.4 Inverse Definite Minimum Time (IDMT) Overcurrent Relay

The IDMT overcurrent relay has a combination of definite current and inverse time characteristics to achieve discrimination. Its operating characteristic curve is shown in Figure 5.2. The inverse characteristics of the overcurrent relays are further classified as follows:

1. IEC normal inverse
2. IEC very inverse
3. IEC extremely inverse
4. IEC long-time inverse
5. IEC short-time inverse

The IEC inverse characteristic curves are shown in Figure in 5.3.

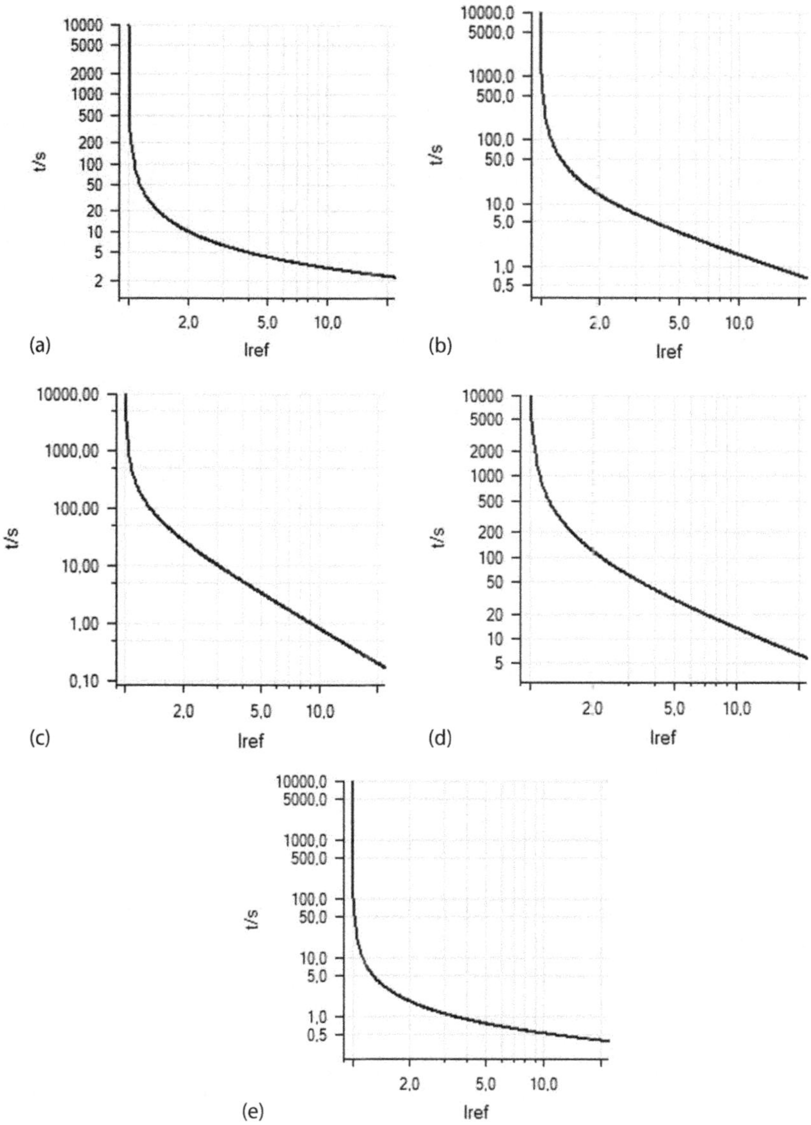

FIGURE 5.3 IEC inverse characteristic curves of the overcurrent relays: (a) IEC normal inverse, (b) IEC very inverse, (c) IEC extremely inverse, (d) IEC long-time inverse, and (e) IEC short-time inverse.

5.2.1.5 Normal or Moderately Inverse Overcurrent Relay

A moderately inverse overcurrent relay is shown in Figure 5.3a. It is used when adaptive coordination is required with other devices, namely, relays, reclosers and fuses. Sometimes it is also used for backup protection for relays on other circuits. Therefore, this type of modern programmable numerical relay allows relay engineers to implement the required characteristics for special applications such as protection coordination.

5.2.1.6 Very Inverse Time Overcurrent Relay

This type of relay provides more inverse characteristics than that of the IDMT, and is particularly suitable if the short-circuit current drops rapidly with the distance from the substation. The grading margin may be reduced to a value in the range from 0.3 to 0.4 seconds, when overcurrent relays with very inverse characteristics are used. The operating characteristic curve of a very inverse time overcurrent relay is shown in Figure 5.3b.

5.2.1.7 Extremely Inverse Overcurrent Relay

This type of relay has more inverse characteristics than that of the very inverse overcurrent relay. These relays are suitable for the protection of machines against overheating. The operating time of a time overcurrent relay with an extremely inverse time-current characteristic is approximately inversely proportional to the square of the current, as illustrated by its operating characteristics curve in Figure 5.3c. The use of extremely inverse overcurrent relays makes it possible to use a short time delay in spite of high switching-in currents.

5.2.2 Operating Time of Overcurrent Relay

The operating time of the overcurrent relay is defined by IEC 60255 and IEEE C37.112, and is given in Equation (5.1) [5]:

$$t = \frac{k(\beta)}{\left(\frac{I}{I_s}\right)^{\alpha} - 1} + L \tag{5.1}$$

where:

t is the relay operating time in seconds
k is the time dial, or time multiplier setting
I is the fault current level in secondary amperes
I_S is the tap or pickup current setting
L is the constant
α is the slope constant
β is the slope constant

Table 5.1 provides the constant values of the parameters for curves defined by IEEE C37.112 and IEC 60255, and Figures 5.2 and 5.3 illustrate these curves.

5.2.3 Directional Overcurrent Relay

When the power system is not radial (source is on one side of the line), an overcurrent relay may not be able to provide adequate protection. Directional overcurrent relay operates in one direction of current flow and blocks in the opposite direction. Three conditions must be satisfied for its operation: current magnitude, time delay and directionality. The directionality of current flow can be identified using voltage as a reference of direction.

TABLE 5.1
IEEE and IEC Constants for Standard Overcurrent Relays

IDMT Curve Description	Standard	A	B	L
Moderately inverse	IEEE	0.02	0.0515	0.114
Very inverse	IEEE	2	19.61	0.491
Extremely inverse	IEEE	2	28.2	0.1217
Inverse	US-CO8	2	5.95	0.18
Short-time inverse	US-CO2	0.02	0.02394	0.01694
Standard inverse	IEC	0.02	0.14	
Very inverse	IEC	1	13.5	
Extremely inverse	IEC	2	80.0	
Long-time inverse	IEC	1	120	

Source: ALSTOM, *Network Protection & Automation Guide*, Network Protection & Automation Guide, pp. 1–497, Cornell University, Ithaca, NY, 2010.

5.2.4 Grading of Overcurrent Relays

Grading is the adjustment of the settings of the overcurrent relays to ensure discrimination and selectivity. Consider a radial system with multiple feeders in parallel. An overcurrent protection relay is installed at every breaker location. When a fault occurs at any given point, only the relay located closest to the fault should operate. This is known as grading. A margin between two successive devices in the order of 0.2–0.4 s should be used to avoid losing selectivity. The different types of grading [14] are: (a) current grading, (b) time grading and (c) time-current grading.

5.2.4.1 Current Grading

Current grading refers to the discrimination achieved by reducing the current setting as moving from load to power source. This ensures that the relay closest to the fault trips first. The downside of this arrangement is that the fault current does not always vary with the location. Hence, it is not possible to accurately discriminate between the relays.

5.2.4.2 Time Grading

Time grading refers to the discrimination achieved by varying the time delay for the different relays. In this method, the relay farthest from the source has the shortest time delay, and the time delay increases as we move towards the source; i.e., the source breaker will have the highest time delay. This works in systems where the fault current is uniform across the system. However, this type of grading will not be sufficient in systems where the fault current varies with the location of the fault.

5.2.4.3 Time-Current Grading

The time-current grading system is the most widely used method of grading. This method uses a combination of time and current grading to achieve discrimination. In this method, the time setting varies with the fault current. A severe fault has a shorter time delay, while the delay is longer for a mild fault.

5.2.5 Example of Time-Overcurrent Protection Settings

This section provides an example of time-overcurrent protection of the simple power system network using the DigSilent power factory simulation tool [10,11]. The radial power system network is given in Figure 5.4. The input data of the power system are given in Tables 5.2 and 5.5, respectively.

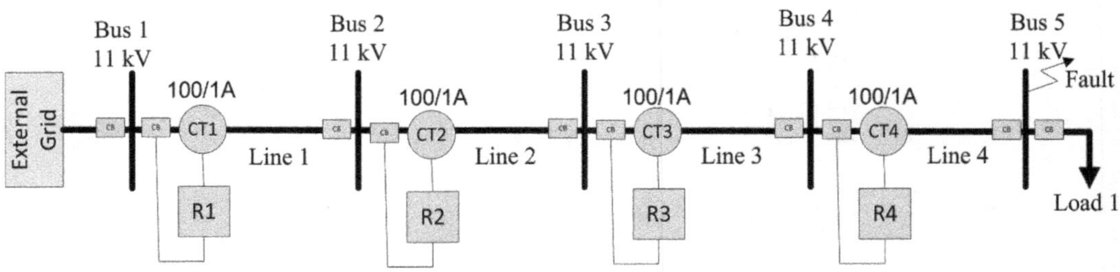

FIGURE 5.4 Radial power system networks.

TABLE 5.2

Bus Data of the Radial Power System Network

Name	Voltage in kV	System Type	Phase Technology
Bus 1	33 kV	AC	ABC
Bus 2	33 kV	AC	ABC
Bus 3	11 kV	AC	ABC
Bus 4	11 kV	AC	ABC
Bus 5	11 kV	AC	ABC

TABLE 5.3

External Grid Data

Description	Voltage in kV	Bus Type	Active Power in MW	Voltage Setpoint in p.u
External Grid	33	PV	2	1.0

External grid data of the radial power system network are given in Table 5.3. The transformer data with short-circuit voltage rating, copper losses and rated power of the transformers are given in Table 5.4.

The transmission line data and load data of the power system network are given in Tables 5.5 and 5.6, respectively. The specification of the three transmission line ratings are the same as that given in Table 5.5.

TABLE 5.4

Transformer Data

Description	Voltage in kV	Short-Circuit Voltage uk	Cu Losses in kW	Rated Power in MVA
Transformer 1 YN	33 kV (HV) 11 kV (LV)	10%	10.5	150
Transformer 2 YN	11 kV (HV) 400 V (LV)	6%	7.5	50

TABLE 5.5

Transmission Line Data

Description	Rated Voltage in kV	Rated Current in kA	Line Length in km	Type	Resistance Ω/km	Reactance Ω/km
Line 1	33	42.41	10.12	Cable	0.3454	0.9234
Line 2	33	42.41	10.12	Cable	0.3454	0.9234
Line 3	11	12.34	2.67	Cable	0.1204	0.2304

TABLE 5.6

Load Data

Description	Voltage in p.u.	Power Factor (Inductive)	Active Power in MW
General Load 1	0.9534	0.96	0.78
General Load 2	0.9565	0.95	0.65
General Load 3	0.9123	0.92	0.45

TABLE 5.7

Current Transformer Data

Relay Model	1	Transformer 1	Schweitzer	SEL 751-1A	Current Transformer	CT	200A/1A
					Current Transformer	CoreCT	200A/1A

The instrument transformer data are given in Table 5.7, and the overcurrent relay setting is given in Table 5.8. The current transformer (CT) uses three-phase, ABC phase technology and star-connected secondary terminals.

In Figure 5.4, the feeder protection overcurrent relay SEL 751A device [12] is available close to the circuit breaker terminal at bus 3. Table 5.8a and b provides the phase and ground element settings, respectively, of the overcurrent relay.

The three-phase short-circuit and single-line-ground faults are applied at bus 3, and the simulation results are given in Figure 5.5. Note that three-phase short-circuit occurs in the network at 0.1 s and is cleared in 0.2 s, whereas the single-line-ground fault occurs in the network at 0.3 s and is cleared in 0.4 s. The three-phase voltage and current waveform simulation results are calculated according to the IEC60909 standard. Note that 10.874 kA and 12.065 kA are shown, respectively, for three-phase short-circuit and single-line-ground faults at bus 3 in Figure 5.5.

TABLE 5.8a

Overcurrent Setting of Phase Elements

Protection Device	Location	Branch	Manufacturer	Model	Stage (Phase)	Current [pri. A]	Current [sec. A]	Current [p.u.]	Time	Characteristic	Directional
Relay Model	1	Transformer 1	Schweitzer	SEL 751-1A	51P1	640	3.20	3.20	0.10	C1 – IEC Class A (Standard Inverse)	None
					50P1	2000	10.00	10.00	0.01	Definite	None

TABLE 5.8b

Overcurrent Setting of Ground Elements

Protection Device	Location	Branch	Manufacturer	Model	Stage (Earth)	Current [pri. A]	Current [sec. A]	Current [p.u.]	Time	Characteristic	Directional
Relay Model	1	Transformer 1	Schweitzer	SEL 751-1A	51G1	200	1.00	1.00	0.10	C1 – IEC Class A (Standard Inverse)	None
					50G1	1000	5.00	5.00	0.10	Definite	None

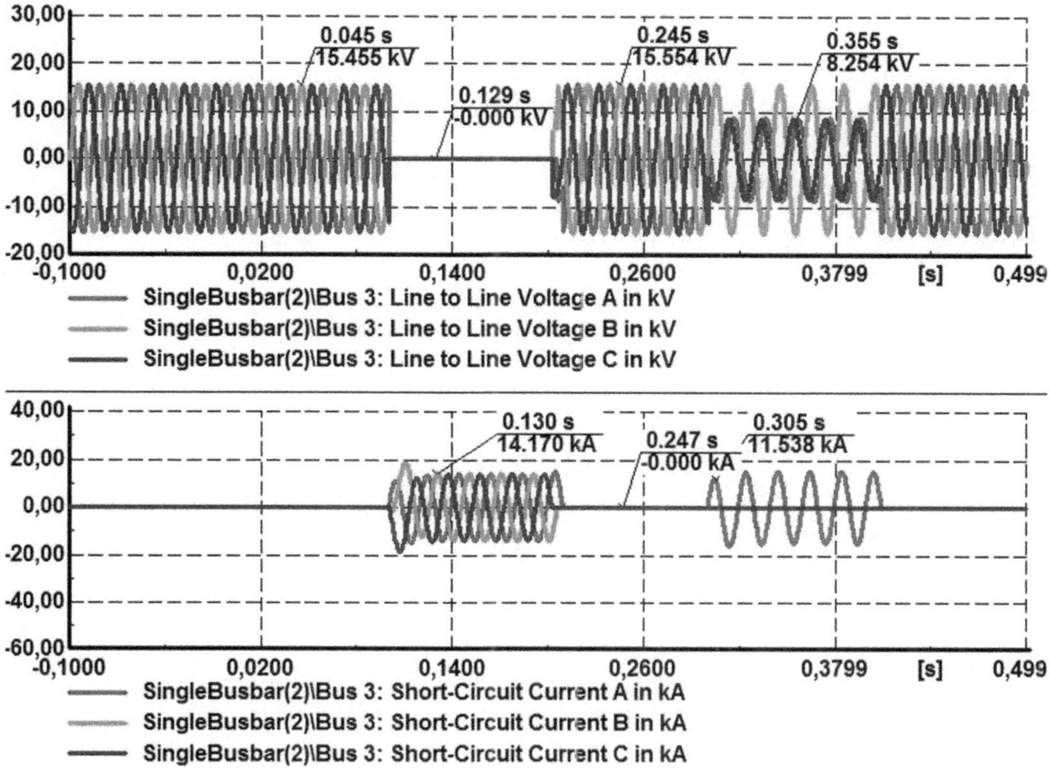

FIGURE 5.5 Simulation results of three-phase short-circuit and single-line-ground faults at bus 3.

The hands-on calculation for three-phase short circuit and single-line-to-ground fault using the symmetrical component method [13]; these results are compared with DigSilent [10] simulation results. The following procedure is used to calculate the fault analysis of the power system network

Step 1: Calculate the per unit of all the power system components given in Figure 5.4 using a base of 150 MVA and 33 kV on the primary side of the transformer 1:

$$S_b = 150\,\text{MVA}$$

$$Z_b = \frac{V^2{}_b}{S_b} = \frac{(33\times10^3)^2}{150\times10^6} = 7.2600\,\Omega$$

$$C_{factor} = 1.1$$

$$Z_{g(actual)} = \frac{C_{factor}\times V_{rated}}{\sqrt{3}\times I_{line}}$$

$$I_{line} = I_{line1} + I_{line2} = 42.41\,\text{kA} + 42.41\,\text{kA} = 84.82\,\text{kA}$$

$$Z_{g(actual)} = \frac{1.1\times(33\times10^3)}{\sqrt{3}\times(84.82\times10^3)} = j0.2471\,\Omega$$

$$Z_{g(\text{p.u.})} = \frac{Z_{g(actual)}}{Z_b} = \frac{j0.2471}{7.2600} = j0.0340_{\text{p.u.}}$$

Per unit of the transmission line 1 is calculated using a base of 150 MVA and 33 kV:

$$S_b = 150\,\text{MVA}$$

$$V_b = 33\,\text{kV}$$

$$Z_{actual} = (0.3454 + j0.9234) \times 10.12 = 9.9772\angle 69.4916°$$

$$Z_{pu} = \frac{Z_{actual}}{Z_b} = \frac{9.9772\angle 69.4916°}{7.2600} = 1.3743\angle 69.4916°$$

$$Z_{\text{p.u.}} = \sqrt{0.4815^2 - 1.2827^2} = j1.1889_{\text{p.u.}}$$

$$Z_{\text{p.u.}(line1)} = Z_{\text{p.u.}(line2)} = Z_{\text{p.u.}} = j1.1889_{\text{p.u.}}$$

Per unit of the transmission line 2 is calculated using a base of 150 MVA and 33 kV:

$$S_b = 150\,\text{MVA}$$

$$V_b = 33\,\text{kV}$$

$$Z_b = 7.2600\,\Omega$$

$$R_{\text{p.u.}} = \frac{P_{Cu.loss}}{S_{rated}} = \frac{10.5 \times 10^3}{150 \times 10^6} = 0.00007_{\text{p.u.}}$$

$R_{\text{p.u.}}$ is a small value and is neglected compared with $X_{\text{p.u.}}$. Therefore $Z_{T1\text{p.u.}} = X_{T1\text{p.u.}} = j0.1_{\text{p.u.}}$
Per unit of the transmission line 3 is calculated using a base of 150 MVA and 11 kV:

$$S_b = 150\,\text{MVA}$$

$$V_b = 11\,\text{kV}$$

$$Z_b = \frac{V_b^2}{S_b} = \frac{(11 \times 10^3)^2}{150 \times 10^6} = 0.8067$$

$$Z_{actual} = (0.1204 + j0.2304) \times 2.67 = 0.3215 + j0.6152\,\Omega$$

$$Z_{\text{p.u.}} = \frac{Z_{actual}}{Z_b} = \frac{0.3215 + j0.6152\,\Omega}{0.8067} = 0.3985 + j0.7626_{\text{p.u.}}$$

$$Z_{\text{p.u.}} = \sqrt{0.3985^2 - 0.7626^2} = j0.6502_{\text{p.u.}}$$

Per unit of the transformer 2 is calculated using a base of 150 MVA and 11 kV:

$$S_b = 150\,\text{MVA}$$

$$V_b = 11\,\text{kV}$$

$$X_{\text{p.u.}} = 0.06 \times \frac{150 \times 10^6}{50 \times 10^6} = j0.1800_{\text{p.u.}}$$

$$P_{loss} = 7.5\,\text{MW}$$

$$R_{\text{p.u.}} = \frac{P}{S_{rated}} = \frac{7.5 \times 10^3}{50 \times 10^6} = 0.0002_{\text{p.u.}}$$

$$R_{\text{p.u}new} = R_{\text{p.u.}} \times \frac{S_{bnew}}{S_{b\,old}} \times \left(\frac{V_{b\,old}}{V_{b\,new}}\right)^2$$

$$= 0.0002_{\text{p.u.}} \times \frac{150 \times 10^6}{50 \times 10^6} \times 1 = 0.0006_{\text{p.u.}}$$

$R_{\text{p.u.}}$ is too small compared with $X_{\text{p.u.}}$. Therefore $Z_{T2} = X_{T2} = j0.1800_{\text{p.u.}}$

Step 2: Draw the positive, negative and zero sequence impedance diagrams with regard to the fault location at bus 3. The positive sequence impedance diagram is shown in Figure 5.6 with respect to the fault location at bus 3. The negative sequence impedance diagram is shown in Figure 5.7 with respect to the fault location at bus 3. The zero sequence impedance diagram is shown in Figure 5.8 with respect to the fault location at bus 3.

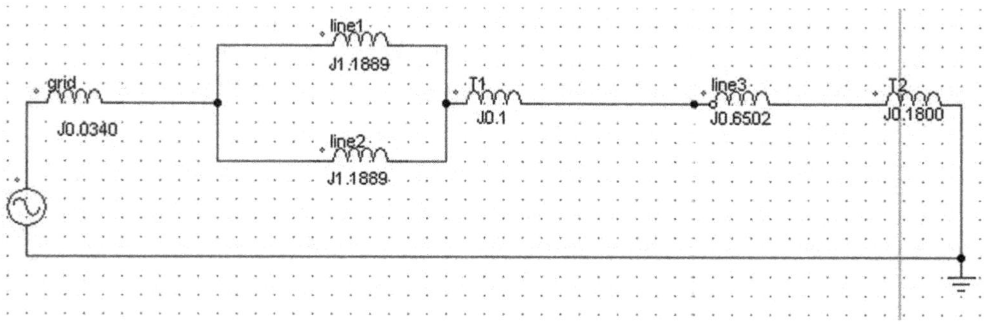

FIGURE 5.6 Positive sequence diagram for short-circuit at bus 3.

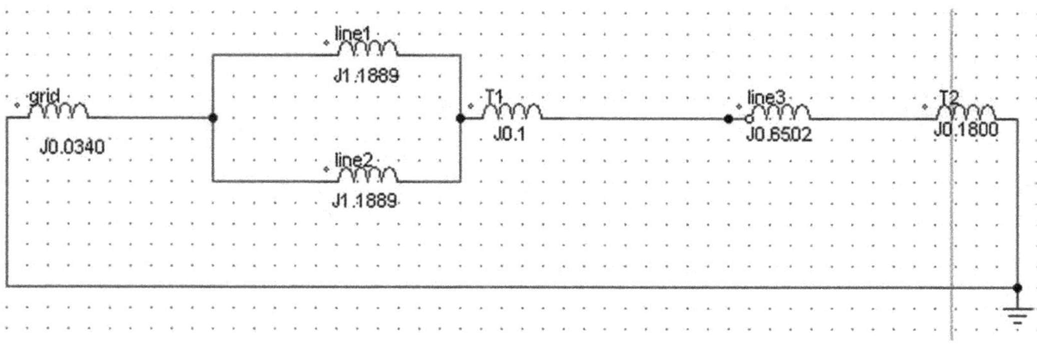

FIGURE 5.7 Negative sequence diagram for short-circuit at bus 3.

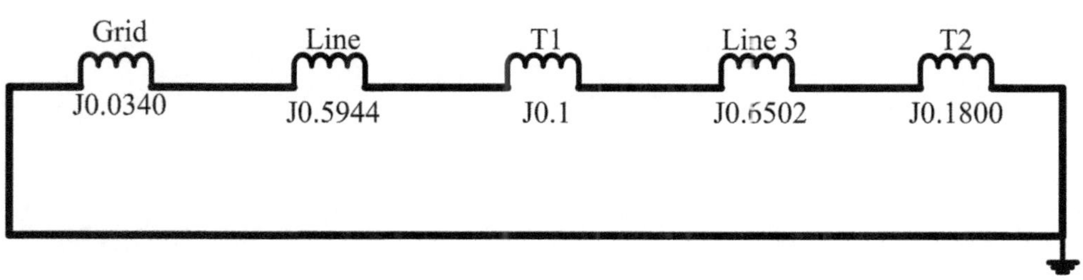

FIGURE 5.8 Zero sequence diagram for short-circuit at bus 3.

Step 3: Find the positive, negative and zero sequence equivalent impedances as given in Equations (5.2) through (5.4) with respect to the fault location at bus 3.

The positive sequence equivalent impedance is calculated with respect to the fault location at bus 3 as given in Equation (5.2):

$$\begin{aligned} X_1 &= X_g + (X_{L1} // X_{L2}) + X_{T1} \\ &= j0.0340 + \frac{j1.1889}{2} + j0.1 \\ &= j0.7284_{\text{p.u.}} \end{aligned} \tag{5.2}$$

The negative sequence equivalent impedance is calculated with respect to the fault location at bus 3 as given in Equation (5.3):

$$\begin{aligned} X_2 &= X_g + (X_{L1} // X_{L2}) + X_{T1} \\ &= j0.0340 + \frac{j1.1889}{2} + j0.1 \\ &= j0.7284_{\text{p.u.}} \end{aligned} \tag{5.3}$$

The zero sequence equivalent impedance is calculated with respect to the fault location at bus 3 as in Equation (5.4):

$$\begin{aligned} X_0 &= X_g + (X_{L1} // X_{L2}) + X_{T1} \\ &= j0.0340 + \frac{j1.1889}{2} + j0.1 \\ &= j0.7284_{\text{p.u.}} \end{aligned} \tag{5.4}$$

Step 4: Find the three-phase short circuit and single-line-to-ground fault currents at bus 3 as given in Equations (5.5) and (5.6), respectively. Per unit of the external grid is calculated using a base of 150 MVA and 33 kV.

The three-phase fault current is calculated at bus 3 as given in Equation (5.5):

$$X_1 = j0.7284$$

$$\text{MVA}_{short\ circuit} = \frac{S_b}{X_1} = \frac{150 \times 10^6}{0.7284} = 205.9308\,\text{MVA}$$

$$I_{a(short\ circuit)} = \frac{\text{MVA}_{short\ circuit}}{\sqrt{3} \times V_{bus3}} = \frac{205.9308 \times 10^6}{\sqrt{3} \times 11 \times 10^3} = 10.8086\,\text{kA} \tag{5.5}$$

The single-line-to-ground fault current is calculated at bus 3 as given in Equation (5.6):

$$\begin{aligned} I_{a\ short\ circuit} &= \frac{3 \times 1\angle 0}{X_1 + X_2 + X_0} = \frac{3 \times 1\angle 0}{j0.7284 + j0.7284 + j0.7284} \\ &= 1.3729\angle -90 \end{aligned}$$

$$I_b = \frac{S_b}{\sqrt{3} \times V_b} = \frac{150 \times 10^6}{\sqrt{3} \times 11 \times 10^3} = 7872.9582\,\text{A}$$

$$\begin{aligned} I_{a\ actual} &= I_b \times I_{a\ short\ circuit\ \text{p.u.}} = 7872.9582 \times 1.3729\angle -90 \\ &= 10.8086\angle -90\ \text{kA} \end{aligned} \tag{5.6}$$

The three-phase short circuit and line-to-ground faults are applied on bus 3 to find the short-circuit rating capability of the network. The summary of the simulation results are given in Table 5.9.

It is clear that hands-on calculated results for the three-phase short-circuit are close to the DigSilent short-circuit simulation results according to the IEC 60909 and VDE 0102 calculation methods given in Table 5.9. However, there is a significant difference between the hands-on calculated value for the single-phase-to-ground fault and DigSilent results according to the IEC 60909 and VDE 0102 methods. On the other hand, the calculation for the single-phase-to-ground fault according to the ANSI method was found to be much more similar to the value obtained by hands-on calculations. The relay tripping time and results of the IDMT characteristics curve are given in the time overcurrent plot in Figure 5.9.

According to IEC 60909 method, the three-phase fault short-circuit current of 10.87 kA at bus 3 trips the circuit breaker at 0.030 s and 0.240 s for definite and inverse time overcurrent phase elements 51P1 and 50P1, respectively, as shown in Figure 5.9. Single-line-to-ground on phase A has a short-circuit current of 12.07 kA at bus 3 and trips the circuit breaker at 0.120 s and 0.199 s for definite and inverse time overcurrent ground elements 51G1 and 50G1, respectively, as shown in Figure 5.9.

TABLE 5.9

Comparison of the Short-Circuit Analysis Results for Hands-on Calculation with Simulation Results

		DigSilent Simulation Results		
Fault Type	**Results of Hands-on Calculations**	**Simulation Results According to IEC 60909**	**Simulation Results According to ANSI**	**Simulation Results According to VDE 0102**
3-Phase short-circuit	10.80 kA	10.874 kA	9.869 kA	10.874 kA
Single-phase-ground	10.80 kA	12.065 kA	10.953 kA	12.065 kA

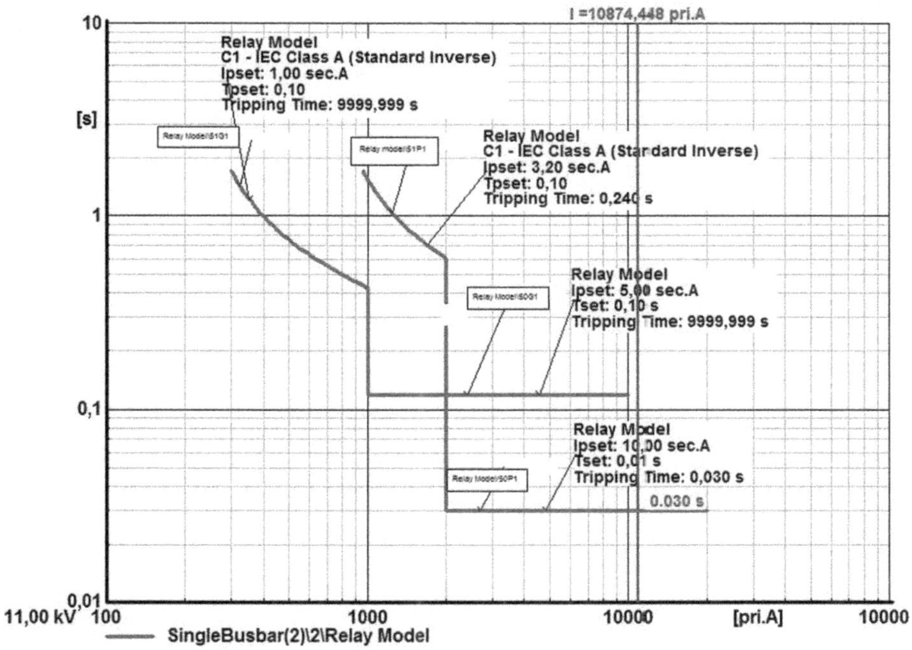

FIGURE 5.9 Simulation results of time-overcurrent plot for phase and ground elements for three-phase short-circuit and single-line-to-ground fault at bus 3.

5.2.6 Relay-to-Relay Coordination

The purpose of the relay-to-relay coordination is to make sure that the relay close to the fault operates first; however, if the relay close to the fault fails to trip, then the adjacent relay should trip and act as a backup protection. The principle of relay-to-relay coordination is explained with an example in Section 5.2.7.

5.2.7 Example of Relay-to-Relay Coordination

The principle of relay-to-relay coordination [14] is explained Figure 5.10. Protection devices exist at different zones of the power system network with its associated inverse-time overcurrent relays. The relay operating curves are selected for each of the relays, plotted as a function of fault location. Observe that the magnitude of the fault current decreases as the fault moves away from the source.

The input data of the power system are given in Tables 5.10 through 5.12, respectively. The specifications of all the four 11 kV overhead lines are the same. Tables 5.14 and 5.15 provide the instrument transformers and SEL751A overcurrent relays setting, respectively. The external grid and bus data of the distribution network are given in Tables 5.10 and 5.11, respectively.

The transmission line data and load data of the distribution network are given in Table 5.12a and b, respectively. The specifications of the four transmission lines rating are the same and are given in Table 5.12a.

The specification of the current transformer is given in Table 5.13, and the overcurrent relay setting is given in Table 5.14. The current transformer (CT) uses three-phase, ABC phase technology and star-connected secondary terminals.

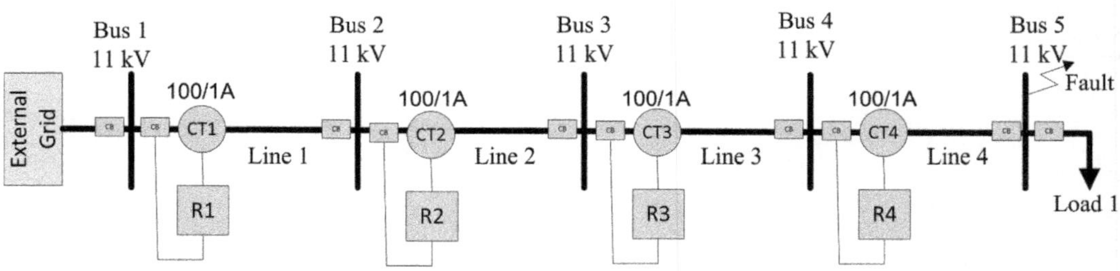

FIGURE 5.10 Distribution network with protection devices.

TABLE 5.10

External Grid Data

Supply	Bus Type	Set Point	Angle	Voltage Set Point
Bus 1	SL	Local	0	1.00

TABLE 5.11

Bus Data of the Radial Power System Network

Bus	Voltage Rating in kV	Technology	System Type
Bus 1–6	11	ABC	AC

TABLE 5.12a

Transmission Line Data

Description	Rated Voltage in kV	Rated Current in kA	Line Length in km	Type	AC-Resistance R|(20°C) Ω/km	AC-Reactance R0| Ω/km
Line 1	11	0.37	10	Overhead line	0.3242	0.471
Description	**Reactance X| Ω/km**	**Reactance X0 Ω/km**	**Susceptance B| uS/km**	**Susceptance B0 uS/km**	**Conductor material**	**Max. Operating Temp. deg C**
Line 1	0.3621	1.817	3.582	1.324	Aluminium	80

TABLE 5.12b

Load Data

Description	Voltage in p.u.	Technology	Reactive Power in Mvar	Active Power in MW
General Load	1.0	3PH PH-E	0.70	1.1

TABLE 5.13

Current Transformer Data

No.	Protection Device	Location	Branch	Manufacturer	Model	CT	Slot	Ratio [pri. A/sec. A]
1	Relay 1	Bus 1	Line 1	Alstom	KCGG142-1A	CT 1	Ct-3p	100A/1A
						CT 1	Ct-3I0	100A/1A
2	Relay 2	Bus 2	Line 2	Alstom	KCGG142-1A	CT 2	Ct-3p	100A/1A
						CT 2	Ct-3I0	100A/1A
3	Relay 3	Bus 3	Line 3	Alstom	KCGG142-1A	CT 3	Ct-3p	100A/1A
						CT 3	Ct-3I0	100A/1A
4	Relay 4	Bus 4	Line 4	Alstom	KCGG142-1A	CT 4	Ct-3p	100A/1A
						CT 4	Ct-3I0	100A/1A

TABLE 5.14
Overcurrent Setting of Phase Elements

No.	Protection Device	Location	Branch	Manufacturer	Model	Stage (Phase)	Current [pri. A]	Current [sec. A]	Current [p.u.]	Time in sec.	Characteristic	Directional
1	Relay 1	Bus 1	Line 1	Alstom	KCGG142-1A	I>/t>	100	1.00	1.00	0.25	No. 1 – SI30xDT (Standard Inverse)	None
						I>>/t>>	1015	10.15	10.15	0.38	Definite	None
2	Relay 2	Bus 2	Line 2	Alstom	KCGG142-1A	I>/t>	100	1.00	1.00	0.20	No. 1 – SI30xDT (Standard Inverse)	None
						I>>/t>>	576	5.76	5.76	0.38	Definite	None
3	Relay 3	Bus 3	Line 3	Alstom	KCGG142-1A	I>/t>	100	1.00	1.00	0.13	No. 1 – SI30xDT (Standard Inverse)	None
						I>>/t>>	400	4.00	4.00	0.38	Definite	None
4	Relay 4	Bus 4	Line 4	Alstom	KCGG142-1A	I>/t>	100	1.00	1.00	0.08	No. 1 – SI30xDT (Standard Inverse)	None
						I>>/t>>	307	3.07	3.07	0.38	Definite	None

In Figure 5.10, the distribution network protection device has five time-overcurrent relays SEL751A, which are located at zones of the network. The next step is to explore the knowledge of relay-to-relay coordination by selecting the suitable current, time and grading margin settings of the time-overcurrent relay. Table 5.14 provides the phase element settings of the overcurrent relay.

Figure 5.11 shows the relay-to-relay coordination characteristics curve for a fault downstream from the network. A fault downstream (Bus 5) with reference to Figure 5.10 is seen by all the time-overcurrent relays upstream from the network. The purpose of the relay-to-relay coordination is to make sure that the relay close to the fault operates first; however, if the relay close to the fault fails to trip, then the relay adjacent to the relay that failed should trip (act as a backup). In Figure 5.11, the relay R4 operates at 0.47s, R3 operates at 0.78s, R2 operates at 1.26s and R1 operates at 1.57s. These trip times are for a three-phase short-circuit at bus 5 with a fault current magnitude of 301.49 A.

For fault F1 at the end farthest from the generating source, relay R4, CB tripping breaker (4), operates first; relay R3 at breaker (3) has a higher time level setting that includes a coordinating time delay S of 0.3s to let breaker (4) trip if it can; similarly, relay R2, at breaker (2), coordinates with the relay at breaker (3) by having a still longer time delay (including the same coordinating time S); and finally, relay R1 at breaker (1) has the longest time delay and will not trip unless none of the other breakers trip, provided it can see the fault (i.e., provided the fault current is greater than its pickup setting).

Should a fault occur between breakers (3) and (4), relay R4 receives no current and therefore does not operate; relay R3 will trip, since its operating time is faster than that of relay R2. For the settings shown in Figure 5.10, relay R1 will not see this fault. Relay R2 must still provide backup relaying for this fault, as discussed above. The relay tripping time and grading between the relays is shown in the time-overcurrent plot given in Figure 5.11.

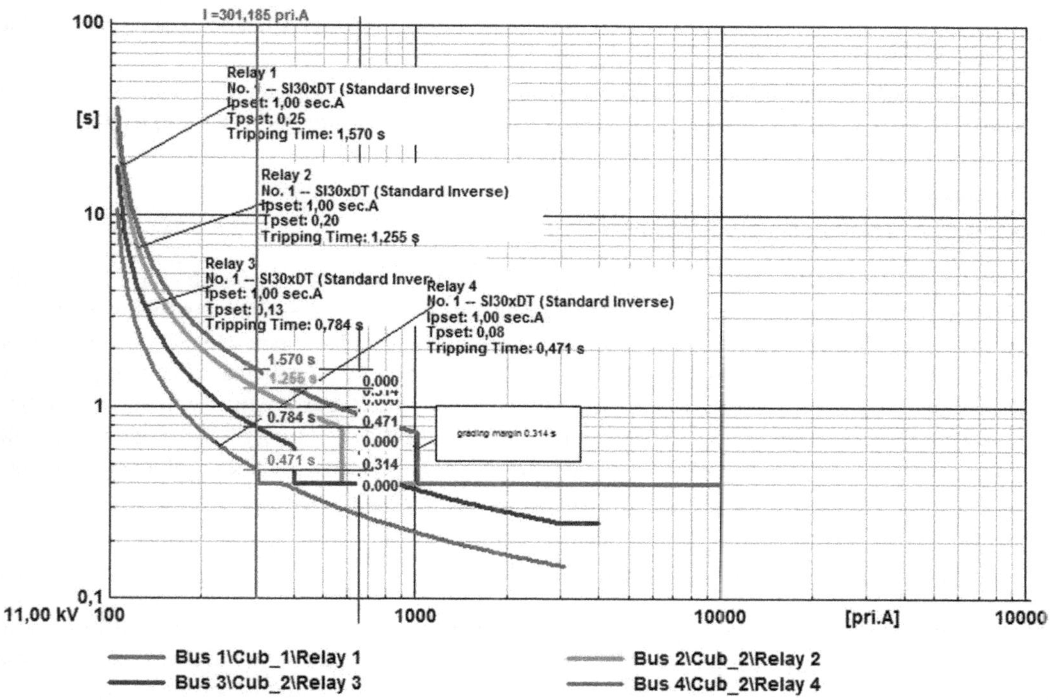

FIGURE 5.11 Relay-to-relay coordination with 0.3 s grading margin.

TABLE 5.15

Relay-to-Relay Coordination and its Tripping Time for 3-phase Unbalanced Events at Different Locations of the Network

3phase Unbalanced Fault Location	Fault Current	Relay Tripping Time in Seconds			
		R4	R3	R2	R1
Bus 5	301.49 A	0.47 s	0.78 s	1.26 s	1.57 s
Bus 4	388.33 A	9999.99 s	0.64 s	1.02 s	1.27 s
Bus 3	544.91 A	9999.99 s	9999.99 s	0.81 s	1.01 s
Bus 2	908.16 A	9999.99 s	9999.99 s	9999.99 s	0.78 s
Bus 1	2450.78 A	9999.99 s	9999.99 s	9999.99 s	9999.99 s

Note: The time 9999.99 s means relay does not send trip signal to the circuit breaker in DigSilent simulation software environment.

Relay-to-relay coordination is explained with an example in this section. A three-phase unbalanced fault is applied at bus 5 in order to explore the relay operating times along with the grading margin of 0.3 s. These results are given in Table in 5.15.

From the results given in Table 5.15, it is understood that the relay farthest from the power source has minimum fault current value. In summary, this section provides the theoretical overview of the protection coordination for the time-overcurrent relays at different locations in the network. The protection coordination and grading between the relays is explained by applying a three-phase unbalanced fault at different locations of the network using the Alstom KCGG142 overcurrent relay model. The time-overcurrent protection coordination setting is explained with illustration of the time-overcurrent plot given in Figure 5.11, which shows the required fault current magnitude, relay operating time and grading margin settings of the considered power system protection devices.

A distance protection scheme for the transmission lines is covered in Chapter 8. The next section of this chapter provides the description of the differential protection scheme with an example.

5.3 Differential Relays and Their Characteristics

Power systems are divided into zones of protection in which the differential scheme protects a specific area of the system, which can be a transformer, transmission line, capacitor, generator, motor, or bus bar. Protection systems applied to these may be broadly classified as unit and non-unit protection systems. Unit systems are bounded by current transformer locations. The major advantages of unit protection over non-unit protection are selectivity and speed. The most positive and effective method of obtaining selectivity is differential protection. For less important installations, selectivity may be obtained, at the expense of speed of operation, using time-graded protection. This section provides the types and characteristics of the differential protection relays, schemes and their operation.

5.3.1 Principles of Differential Relay

The principle of unit differential protection scheme was initially established by Merz and Price, who were the creators of the fundamental differential protection scheme. The differential protection scheme is based on Kirchhoff's current law; that is, the algebraic sum of the currents entering and leaving the node is zero. Any deviation from this must indicate an abnormal current path. In addition, any disturbance or operating condition outside the area of interest is totally ignored, and the differential protection must be designed to be stable above the maximum possible fault current that could flow through the protected area.

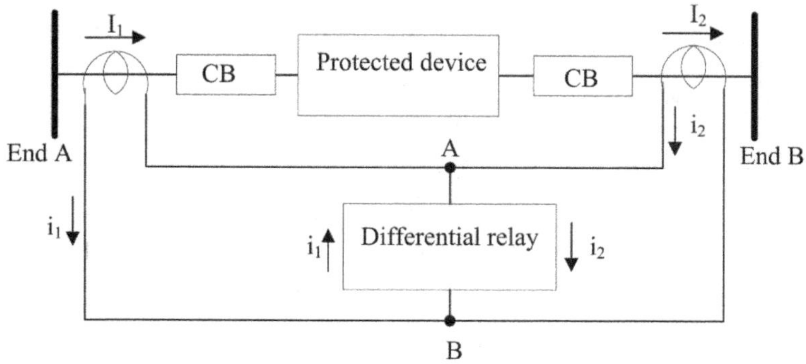

FIGURE 5.12 Balanced current differential protection scheme.

The differential protection scheme is given in Figure 5.12, where the CTs are connected in series and the secondary current circulates between them. The currents are measured by the current transformers on both sides of the protected device.

The relay is connected across the midpoint; thus, the voltage across the relay is theoretically nil. Therefore no current goes through the relay and hence there is no operation for any faults outside the protected zone. Similarly under normal conditions the currents leaving zone A and B are equal, making the relay inactive by the current balance. If a differential current exists, the relay produces a trip signal to simultaneously trip both circuit breakers on the both sides of the protected device, as shown in Figure 5.12.

5.3.2 Differential Protection Categories

Differential protection compares the currents entering and leaving the protected zone and operates when the difference between these currents exceeds a predetermined magnitude. This type of protection can be divided into two types: (i) balanced current and (ii) balanced voltage. The balance current differential scheme is further classified into (a) high impedance type, called unbiased differential protection, and its objective is to ensure stability under fault conditions, and (b) low impedance type, called biased differential protection. The application of the high impedance/unbiased differential protection includes restricted earth fault (REF) protection, bus zone protection, generator and motor protection. Low impedance/biased differential protection includes solkor protection and protection of a transformer, generator, motor and feeder. The description of the balanced current and voltage differential scheme is explained below.

5.3.2.1 Principles of Balanced Current Differential Protection Scheme

The balanced current differential protection scheme is illustrated with the aid of Figure 5.12. At normal condition, the incoming current I_1 is equal to the outgoing current I_2 ($I_1 = I_2$) and by virtue of current transformer connections, the vector sum of I_1 and I_2 is zero through the relay $\left(I_{diff} = |I_1 - I_2| = 0\right)$. Therefore, in circulating current differential protection, the secondary currents appear to circulate in the CT secondaries only, and no relay current is implied. The voltage across the relay terminals is zero $\left(V_{AB} = 0\right)$, since the relay is located at the electrical midpoint, as shown in Figure 5.12.

5.3.2.2 Principles of Balanced Voltage Differential Protection Scheme

In the balanced voltage differential scheme, it is necessary to create a balanced voltage across the relays in end A and end B under healthy and out-of-zone fault conditions, as shown in Figure 5.13.

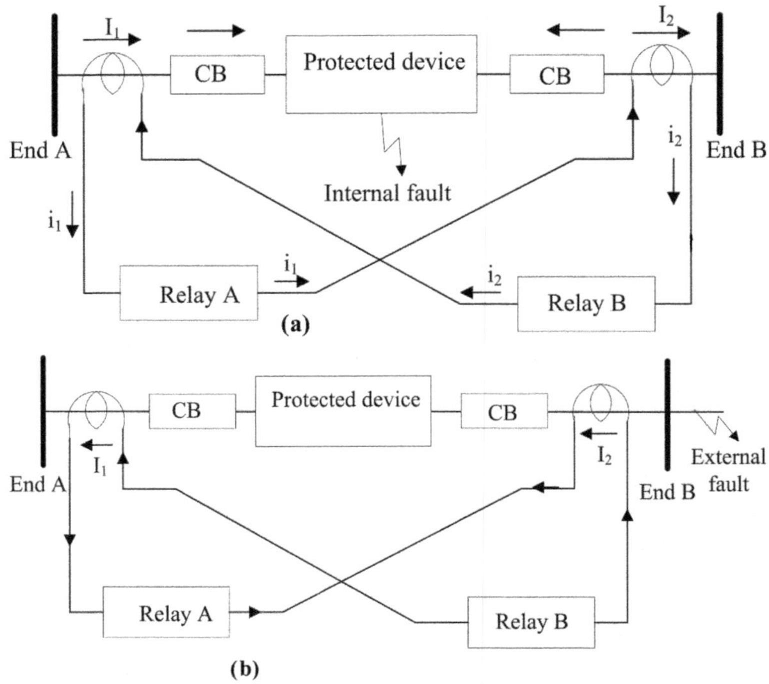

FIGURE 5.13 Balanced voltage differential protection scheme: (a) internal fault and (b) external fault.

In this arrangement, the CTs are connected to oppose each other, and the voltages produced by the secondary currents are equal and opposite. Therefore, no currents flow in the pilots or relays during external-fault conditions, and under internal fault conditions the electromagnetic forces are no longer balanced. Hence current starts flowing through the relay coil to operate and trips the circuit breaker. The disadvantage in the voltage balance differential relay is that the multi-tap transformer construction is required to accurately balance between current transformer pairs. This type of protection is suitable for protection of cables of relatively short length.

5.3.3 Principles of Percentage Restraint Differential Relay

Percentage restrained differential relays measure the individual branch currents and quantify the through current in the zone of protection. This relay has restraining coils in addition to the operating coil of the relay, as shown in Figure 5.14. The restraining coils produce torque opposite to the operating torque. Under normal and external fault conditions, the restraining torque is greater than the operating torque, and the relay remains inactive. The restraining torque is in the contact opening direction and is proportional to the vector sum of the incoming and outgoing currents.

On an external fault, this contact-opening torque is strong and tends to prevent false tripping due to the current caused by saturation effects on the current transformers. It is particularly suitable where heavy saturation currents are encountered. The burden of current transformers used in differential relaying schemes is important in maintaining the proper relationship between the two sets of current transformers. The CTs should not saturate when carrying the maximum external symmetrical fault current.

On internal faults, most of the current in the restraining coil is in the opposite direction so that the total resistant torque is much less than in the case of the external fault. The relay will trip when the operating torque is greater than the restraining torque, that is, when the operating current is higher than a certain percentage of the smaller or larger of the two restraining currents, depending on the type of relay used. The calculation method for restraint current depends on the relay manufacturer and is given in Table 5.16. Some relays are designed to trip when a constant percentage of unbalance exists between the restraining currents.

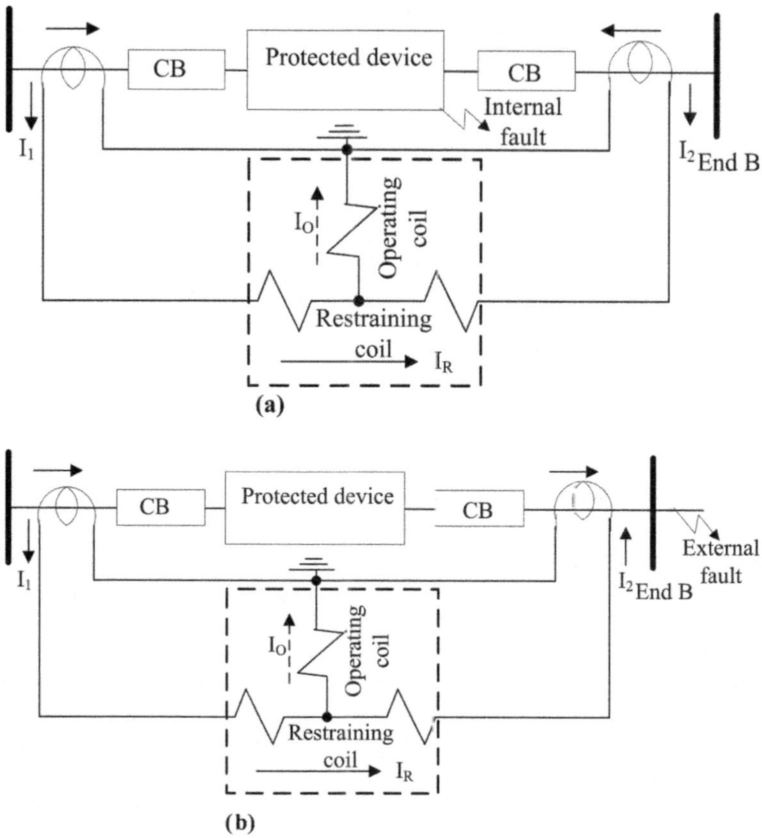

FIGURE 5.14 Percentage restraint differential relay: (a) internal fault and (b) external fault.

TABLE 5.16

Calculation Method for Restraining Currents According to the Relay Manufacturer

Calculation Method for Restraining Currents	Relay Manufacturer
$I_R = \frac{(\lvert I_1\rvert + \lvert I_2\rvert)}{K_1}$	Siemens ($K_1 = 1$), e.g., 7UT5x/7UT6x GEC ($K_1 = 1$), e.g., Series KBCH SEL ($K_1 = 2$), e.g., SEL587 SEL ($K_1 = 1$), e.g., SEL487E Alstom/Areva ($K_1 = 2$)
$I_R = \frac{(\lvert I_1\rvert - \lvert I_2\rvert)}{K_1}$	Alstom/Areva ($K_1 = 2$) e.g., PQ7x, P6x Conventional electromechanical relays
$I_R = \max(I_1, I_2)$	Elin/VATECH, e.g., DRS GE Multilin SR745
$I_R = \min(I_1, I_2)$	SEG/Woodward Conventional electromechanical relays
$I_R = \sqrt{(\lvert I_1\rvert \times \lvert I_2\rvert \times \cos(\alpha))}$	ABB

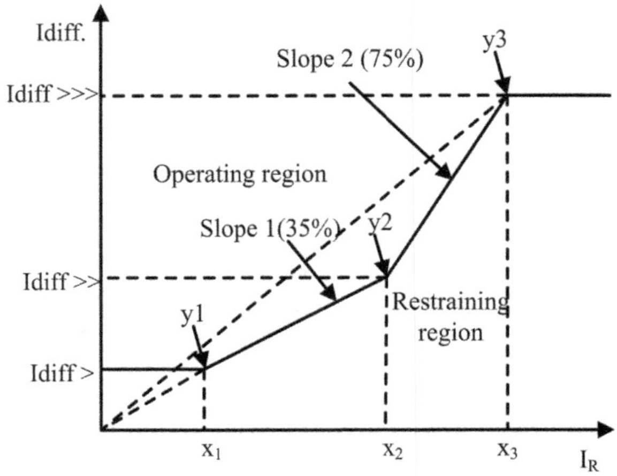

FIGURE 5.15 Dual slope characteristics of percentage restraint differential relay.

Other relays operate over a variable range of different currents. These relays have "variable percentage" characteristics and as the magnitude of the restraining current increases, a greater amount of operating, or differential current is required to trip the relay. The dual slope characteristic of the percentage restraint differential relay is given in Figure 5.15. Where the restraint currents are x_1, x_2, and x_3 and the differential/break point currents are y_1, y_2, y_3, I_{diff}>, I_{diff}>> and I_{diff}>>> represents the differential currents at points y_1, y_2, and y_3 respectively.

The differential characteristic curve shown in Figure 5.15 is defined by the following settings (a) minimum pickup, I_{diff}> in p.u.; (b) slope 1 in %; (c) break point 1, y_1 in p.u.; (d) slope 2 in %; (e) break point 2, y_2 in p.u.; and (f) break point 3, y_3 in p.u. Slope 1 is the slope setting from pickup to break point 1 and is based on CT errors during normal load currents and tap changes. Slope 2 identifies the slope where the CT saturation is likely and maximum restraint is required. The slope 1 and 2 settings express the slope of the operating characteristic as a function of differential current (I_{diff}) and restraint current (I_R), as shown in Figure 5.15.

The equation of a straight line is given in Equation (5.7):

$$y = mx + c \tag{5.7}$$

where:

y is the differential current
m is the slope of the differential characteristic
x is the restraint current
c is the constant which represents starting point on y-axis

The slope of the differential characteristic can be found using Equation (5.8):

$$m = \frac{I_{diff}}{I_R} \text{ or } \frac{(y_2 - y_1)}{(x_2 - x_1)} \tag{5.8}$$

The variable percentage relay is more sensitive than a constant percentage relay on light internal faults, but it is less sensitive on heavy external faults due to variable percentage characteristic.

5.3.4 Example of Differential Protection Scheme for Power Transformer

The principle of a differential protection scheme is illustrated in Figure 5.16. The input data of the power system are given in Tables 5.2 through 5.5 respectively. The external grid data of the radial power system network are given in Table 5.3. The transformer data with short-circuit voltage rating, copper losses and

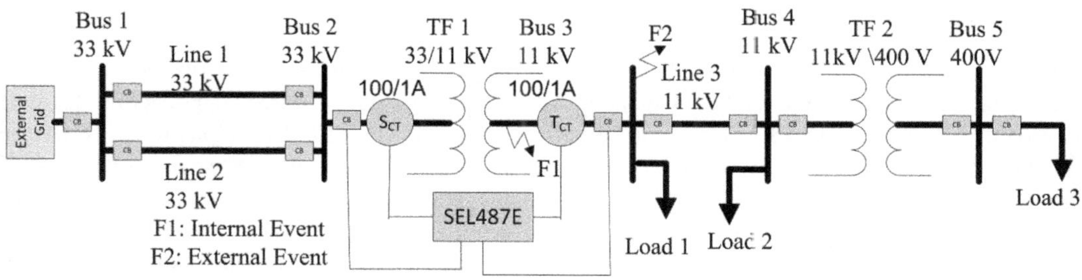

FIGURE 5.16 Differential protection setting on power transformer.

rated power of the transformers are given in Table 5.4. The transmission line data and load data of the power system network are given in Tables 5.5 and 5.6, respectively. The specifications of the three transmission line ratings are the same as given in Table 5.5.

The specification of the current transformer is given in Table 5.17, and the differential relay setting is given in Table 5.18. The current transformer (CT) uses three-phase, ABC phase technology and star-connected secondary terminals. The CT type name is 25 VA Class 10 P 20 (i.e., apparent power = 25 VA, accuracy class = 10 and accuracy limit factor = 20). The accuracy parameters are set according to the IEC apparent power. The VT connection type is YN for both primary and secondary sides. Model type of the VT is ideal transformer.

TABLE 5.17

Instrument Transformer Settings of the Differential Protection Scheme

Protection Device	Location	Branch	Manufacturer	Model	CT	Slot	Ratio [pri. A/sec. A]
Relay Model	Substation 2/ Substation 2		Schweitzer	SEL 487E-1A	33 kV CT @ SS 2	S Ct	100A/1A
					33 kV CT @ SS 2	S Ct-3I0	100A/1A
					Current Transformer	T Ct	100A/1A
					33 kV CTL @ SS 2	T Ct-3I0	100A/1A

TABLE 5.18a

Differential Element and Transformer Tap Settings of the SEL487E Differential Relay Model

Description	Release Threshold	Restraint 1st Slope Threshold	Restraint 1st Slope Threshold	Restraint 1st Slope	Restraint 2nd Slope	Unrestrained Differential Threshold
Differential element setting	0.50 p.u	0- tap	0.01 tap	35%	75%	8.0 p.u
Transformer tap settings	**Tap 1**	**Tap 2**	**Tap 3**	**Tap 4**	**Tap 5**	**Max Rated Power**
	5.8 A	10.0 A	32.0 A	32.0 A	32.0A	20.0 MVA

TABLE 5.18b

Measurement Winding Adapter Setting of the SEL487E Differential Relay Model

Description	Current Transformer Ratio	Nominal Terminal Line-Line Voltage	Current Transformer Connection	Vector Group
SMeas→Wd Adapter	100	33 kV	Y	—
TMeas→Wd Adapter	100	11 kV	Y	0

5.3.4.1 Setting of the Differential Relay

This section describes the settings of the differential relay, which includes differential elements, transformer tap settings, and measurement winding adapter settings. The differential protection example is provided using the SEL487E [15] transformer relay model. The SEL487E transformer protection relay has five sets of current coils/windings, namely, S, T, U, W and X. This example uses only S and T winding parameter settings, as given in Table 5.18. The rest of the windings are not configured in this case. The S and T windings are connected between the primary and secondary of the 33/11 kV power transformer to measure the current on both sides of the transformer. The differential current measured between the S and T windings are used as the setpoint in the differential element on the SEL487E relay model. The parameter setting of the SEL487E differential relay model in the DigSilent Power Factory is given in Table 5.18.

The SEL487E transformer differential relay is located at the breaker close to the primary side of the 33/11 kV power transformer at substation 2, as shown in Figure 5.16. The differential relay settings are provided in Table 5.18, and a current differential characteristics curve is used to analyse faults in the transformer.

The relay operating characteristic is analysed for three-phase faults at internal and external zones of protection at 33/11 kV power transformer, and is given in Figures 5.17 and 5.18, respectively. The differential relay operates for the internal fault at 3.6 kA at 15 ms, and the relay does not operate for the fault at external zone of protection, as given in Figures 5.17 and 5.18, respectively.

In summary this section provided a brief overview of the differential protection scheme and slope settings. The balanced voltage and current differential protection scheme was discussed in detail. An application example of the transformer differential protection scheme was provided, and the operating characteristics of the differential relay were analysed for the three-phase internal and external faults.

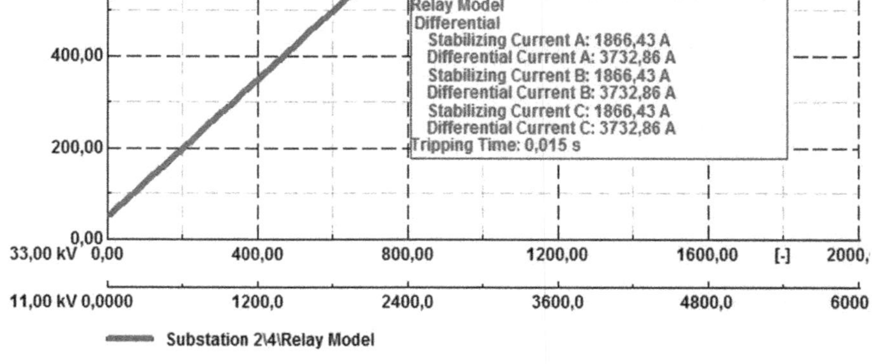

FIGURE 5.17 DigSilent simulation results of the three-phase fault at internal zones of protection of the power transformer.

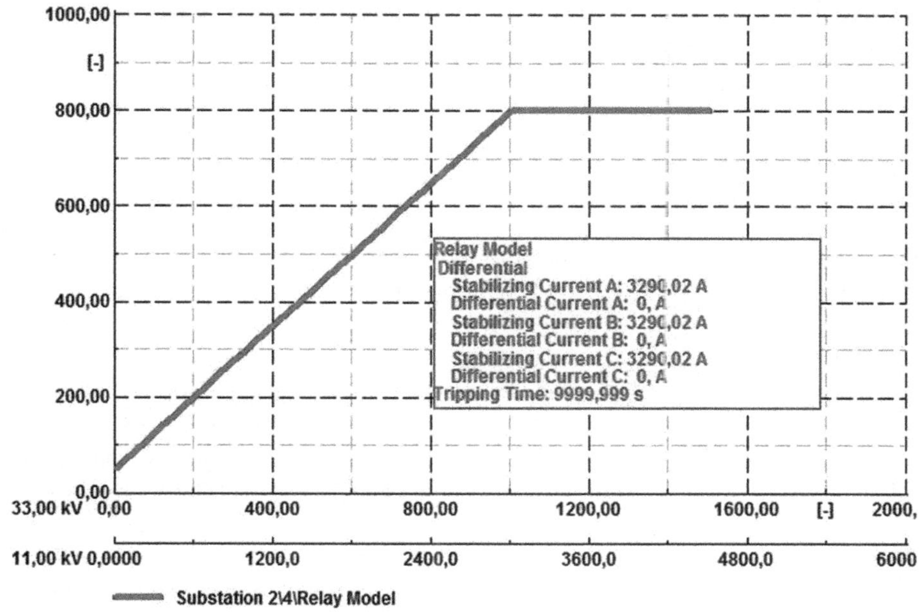

FIGURE 5.18 DigSilent simulation results of the three-phase fault at external zones of protection of the power transformer.

5.4 Solved and Unsolved Problems on Protective Relaying System

5.4.1 Solved Problems on Protective Relaying System

PROBLEM 5.1

The current transformer rating is 1000/1 A and the relay pickup current $I_s = 1.4$ A. Find the operating time of the IEC very inverse time overcurrent relays for the following scenarios:

a. If the fault current $I = 2.1$ A and time multiplier $k = 0.2$ s
b. If the fault current $I = 3.211$ A and time multiplier $k = 0.2$ s
c. If the fault current $I = 3.985$ A and time multiplier $k = 0.2$ s

Solution

The operating time of the overcurrent relay is given in Equation (5.1), and the slope constants for IEC very inverse characteristics are given in Table 5.1, which has $\beta = 13.5$ and $\alpha = 1$:

a. $$t = \frac{k(\beta)}{\left(\frac{I}{I_s}\right)^{\alpha} - 1} = \frac{0.2(13.5)}{\left(\frac{2.1}{1.4}\right)^{1} - 1} = 5.4\,\text{s}$$

b. $t = \dfrac{0.2(13.5)}{\left(\dfrac{3.211}{1.4}\right)^{1} - 1} = 2.08\,\text{s}$

c. $t = \dfrac{0.2(13.5)}{\left(\dfrac{3.985}{1.4}\right)^{1} - 1} = 1.46\,\text{s}$

From the above calculations, it is understood that the higher the fault current value, the less operating time for the same pickup current and time multiplier settings.

PROBLEM 5.2

Find the operating time of the IEC standard inverse time overcurrent relays for the typical relay-to-relay coordination example given in Section 5.2.7. The three-phase unbalanced fault at bus 5 has a fault current of $I = 301.49$ A, as given in Table 5.15, and the current transformer rating is 100/1 A. The time multiplier settings of the IEC standard inverse time overcurrent relay are given in Table 5.19.

Solution

The operating time of the overcurrent relay is given in Equation (5.1), and the slope constants for IEC standard inverse characteristics are given in Table 5.1, which has $\beta = 0.14$ and $\alpha = 0.02$:

$$t_1 = \frac{k(\beta)}{\left(\dfrac{I}{I_s}\right)^{\alpha} - 1} = \frac{0.25(0.14)}{\left(\dfrac{3.01}{1.00}\right)^{0.02} - 1} = 1.57\,\text{s}$$

$$t_2 = \frac{0.20(0.14)}{\left(\dfrac{3.01}{1.00}\right)^{0.02} - 1} = 1.25\ \text{s};\quad t_3 = \frac{0.13(0.14)}{\left(\dfrac{3.01}{1.00}\right)^{0.02} - 1} = 0.81\ \text{s}$$

$$t_4 = \frac{0.08(0.14)}{\left(\dfrac{3.01}{1.00}\right)^{0.02} - 1} = 0.50\,\text{s}$$

Therefore, the operating time of the IEC standard inverse overcurrent relays, R1 to R4, are $t_1 = 1.57$ s, $t_2 = 1.25$ s, $t_3 = 0.81$ s, $t_4 = 0.50$ s. This typical relay-relay coordination example uses a grading margin time delay of 0.3 s.

PROBLEM 5.3

Calculate the operating time of an overcurrent relay with a short circuit current $I = 15$ A, pickup current $I_s = 1.25$ A and time multiplier setting $k = 1$ for the following time overcurrent characteristics: (a) IEC standard inverse, (b) IEC extremely inverse, (c) IEEE moderately inverse and (d) IEEE extremely inverse.

TABLE 5.19

Time Multiplier Settings of the Time Overcurrent Relays Used in the Typical Relay-to-Relay Coordination Example

IEC Standard Inverse Time Overcurrent Relays	R1	R2	R3	R4
Time multiplier, *k*	0.25 s	0.20 s	0.13 s	0.08 s

Solution

The operating time of the overcurrent relay is given in Equation (5.1), and the slope constants β and α for an IEC and IEEE standard inverse characteristic are given in Table 5.1:

a. IEC standard inverse: $t_1 = \dfrac{k(\beta)}{\left(\dfrac{I}{I_s}\right)^{\alpha} - 1} = \dfrac{1(0.14)}{\left(\dfrac{15}{1.25}\right)^{0.02} - 1} = 2.75\,\text{s}$

b. IEC extremely inverse: $t_2 = \dfrac{1.00(80)}{\left(\dfrac{15}{1.25}\right)^{2} - 1} = 0.56\,\text{s}$

c. IEEE moderately inverse: $t_3 = \dfrac{k(\beta)}{\left(\dfrac{I}{I_s}\right)^{\alpha} - 1} + L = \dfrac{1.00(0.0515)}{\left(\dfrac{15}{1.25}\right)^{0.02} - 1} + 0.114 = 1.12\,\text{s}$

d. IEEE extremely inverse: $t_4 = \dfrac{1.00(28.2)}{\left(\dfrac{15}{1.25}\right)^{2} - 1} + 0.1217 = 0.32\,\text{s}$

From the above calculations, the comparison of the IEEE and IEC operating times of the time-overcurrent relays are obtained. The results conclude that the IEEE extremely inverse relay operates very fast in comparison to the IEC extremely inverse relay. In addition, it is observed that the IEC standard inverse relay responds very slowly in comparison to all other types of IEEE and IEC overcurrent relays.

PROBLEM 5.4

Calculate the operating time of an overcurrent relay with a short circuit current $I = 4$ A, pickup current $I_s = 1.25$ A and time multiplier setting $k = 0.025$ for the following time overcurrent characteristics: (a) IEC standard inverse, (b) IEC extremely inverse, (c) IEEE moderately inverse, and (d) IEEE extremely inverse.

Solutions

a. IEC standard inverse: $t_1 = \dfrac{k(\beta)}{\left(\dfrac{I}{I_s}\right)^{\alpha} - 1} = \dfrac{0.025(0.14)}{\left(\dfrac{4}{1.25}\right)^{0.02} - 1} = 0.15\,\text{s}$

b. IEC extremely inverse: $t_2 = \dfrac{0.025(80)}{\left(\dfrac{4}{1.25}\right)^{2} - 1} = 0.22\,\text{s}$

c. IEEE moderately inverse: $t_3 = \dfrac{k(\beta)}{\left(\dfrac{I}{I_s}\right)^{\alpha} - 1} + L = \dfrac{0.025(0.0515)}{\left(\dfrac{4}{1.25}\right)^{0.02} - 1} + 0.114 = 0.17\,\text{s}$

d. IEEE extremely inverse: $t_4 = \dfrac{0.025(28.2)}{\left(\dfrac{4}{1.25}\right)^{2} - 1} + 0.1217 = 0.20\,\text{s}$

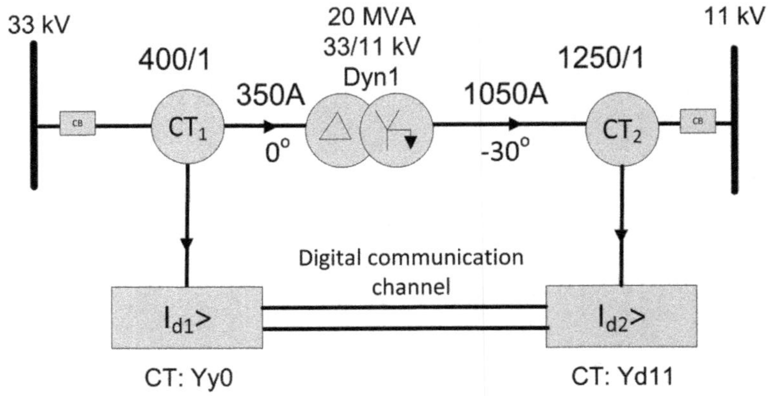

FIGURE 5.19 A power transformer differential protection scheme.

PROBLEM 5.5

A power transformer differential protection scheme is given in Figure 5.19. The current transformers (CTs) are star-connected on both sides. CT ratios are not a perfect match for the transformer winding ratio. Therefore, CT phase shift is compensated by calculating the correction factors on the given differential protection settings.

a. Find the CT correction factors based on current ratings of the transformer.
b. Find the currents I_{d1}, I_{d2}.

Solutions

The CT ratios are not a perfect match for the transformer winding ratio; therefore, correction factors for CTs can be chosen based on current ratings of transformer, and it is calculated as follows:

Correction factors for CTs on HV side is 400/350 = 1.14
Correction factors for CTs on LV side is 1250/1050 = 1.18

The turn's ratio of a power transformer:

$$N = \frac{V_1}{V_2} = \frac{11\left(10^3\right)}{33\left(10^3\right)} = 0.3333 \tag{5.9}$$

The next step is to find the turn's ratio of a current transformer:

$$N = \frac{V_1}{V_2} = \frac{400}{1} = \frac{x}{1} \tag{5.10}$$

where x represents the CT ratio on the secondary side of the power transformer.

Equating (5.9) and (5.10) in order to find the correct CT ratio, we obtained: $0.33x = 400$,

Therefore, $x = \dfrac{400}{0.3333} = 1200$

The new correction factor for CTs on the LV side is 1200/1050 = 1.14.

The differential currents are calculated as follows:

$$I_{d1} = \frac{I_{primary}}{\text{CT}_1\text{ ratio}} = \frac{350}{400} = 0.88\,\text{A}$$

$$I_{d2} = \frac{I_{Secondary}}{\text{CT}_2\,\text{ratio}} = \frac{1050}{1250} = 0.84\,\text{A}$$

Recalculate I_{d2} using the CT ratio 1200, $I_{d2} = \frac{I_{Secondary}}{\text{CT}_2\,\text{ratio}} = \frac{1050}{1200} = 0.88\,\text{A}$

The phase shift can be compensated in a new differential relay; therefore, the CTs on both sides can be star-connected.

5.4.2 Unsolved Problems on Protective Relaying System

PROBLEM 5.6

Calculate the operation time for an overcurrent relay with a short-circuit current measurement of 5 A (CT secondary), pickup of 1.5 A (CT secondary), and TMS of 0.03 s for the following overcurrent characteristics:

1. IEC standard inverse
2. IEC extremely inverse
3. IEEE moderately inverse
4. IEEE extremely inverse

PROBLEM 5.7

The current transformer rating is 100/1 A, and the relay pickup current I_s = 1.5 A. Find the operating time of the IEC very inverse time overcurrent relays for the following scenarios:

1. If the fault current I = 3.245 A and time multiplier k = 0.25 s
2. If the fault current I = 4.313 A and time multiplier k = 0.25 s
3. If the fault current I = 4.527 A and time multiplier k = 0.25 s

5.5 Conclusion

This chapter provided the theoretical principles, types and operating characteristics of the overcurrent and differential protection schemes. The industrial application of the time-current grading was provided in detail. In addition, an industrial application of the differential protection scheme was presented, and its operating characteristics were analysed for the three-phase internal and external faults at different zones of the power system protection.

ACKNOWLEDGMENT

The author gratefully acknowledges the authorities of Cape Peninsula University of Technology (CPUT), South Africa, for the facilities offered to complete this chapter. The software and hardware equipment used in the simulation study is supported by the Center for Substation Automation and Energy Management Systems (CSAEMS) with the Department of Electrical, Electronic and Computer Engineering at CPUT.

REFERENCES

1. P. M. Anderson, *Power System Protection*, pp. 1–1330, Hoboken, NJ: Wiley-IEEE Press, 1998.
2. M. Kezunovic, "Fundamentals of power system protection," W.-K. Chen, Ed., *The Electrical Engineering Handbook*, Chapter on electric power systems, pp. 787–804, St. Louis, MO: Elsevier Academic Press, 2005.

3. S. H. Horowitz and A. G. Phadke, *Power System Relaying*, 3rd ed., Chichester, UK: Wiley, 2008.
4. A. G. Phadke and J. S. Thorp, *Computer Relaying for Power Systems*, 2nd ed., Chichester, UK: John Wiley & Sons, 2009.
5. ALSTOM, *Network Protection & Automation Guide*, pp. 1–497, Ithaca, NY: Cornell University, 2010.
6. H. J. A. Ferrer and E. O. Schweitzer (Eds.), *Modern Solutions for Protection, Control and Monitoring of Electric Power Systems*, pp. 1–361, Pullman, WA: Schweitzer Engineering Laboratories, 2010.
7. J. M. Gers and E. J. Holmes, *Protection of Electricity Distribution Networks*, pp. 1–368, Stevenage, UK: The Institution of Engineering and Technology, 2011.
8. J. L. Blackburn and T. J. Domin, *Protective Relaying Principles and Applications*, pp. 1–695, Boca Raton, FL: CRC Press, 2014.
9. H. Bevrani, M. Watanabe, and Y. Mitani, *Power System Monitoring and Control*, 1st ed., pp. 1–288. Wiley-IEEE Press, 2014.
10. B. S. Features, C. Functions, and A. Functions, PowerFactory V14, *Optimization*, pp. 1–94, Cham, Switzerland: Springer, 2010.
11. F. M. Gonzalez-Longatt and J. L. R. Torres, *Power Factory Applications for Power System Analysis*, Cham, Switzerland: Springer International Publishing, 2014.
12. Schweitzer Engineering Laboratories, SEL-751A Feeder Protection Relay Instruction Manual, 20120903, 2012.
13. J. L. Blackburn, *Symmetrical Components for Power Systems Engineering (Electrical and Computer Engineering)*, 1st ed., pp. 1–443, London, UK: CRC Press, 1993.
14. J. V. H. Sanderson, *IEEE Recommended Practice for Protection and Co-ordination of Industrial and Commercial Power Systems*, vol. 3, no. 2, 1989.
15. Schweitzer Engineering Laboratories, *SEL-487E Current Differential and Voltage Protection Relay Instruction Manual*, 20121126, 2012.

Section II

Transmission Line Protection

6

Medium Voltage Phase Overcurrent Feeder Protection

Martin J. Slabbert, Raj Naidoo, and Ramesh Bansal

CONTENTS

6.1 Introduction

Medium voltage (MV) is defined by IEEE as a voltage ranging from 1 kV (including) up to a voltage of 100 kV (less than) [1]. This voltage category is defined by the IEC as a voltage ranging from 1 kV (including) up to a voltage of 35 kV (not exceeding) [2]. Traditionally, overhead medium voltage networks are designed to be radial [3]. Designing a radial network is relatively uncomplicated, starting with the source and then meeting the load requirements further down at various positions on this feeder. The protection philosophy and technology is also uncomplicated [3]. Circuit breakers (with relays), auto-reclosers and fuses are used to protect these networks. As fault current only flows in one direction, there is no need for directionality and, as such, grading a network (or then feeder) is fairly uncomplicated.

In this chapter, the philosophy of protecting an overhead medium voltage network is discussed. This protection is related to phase overcurrent protection. Concepts that are used for the phase overcurrent protection can also be applied to earth fault protection, for example, the grading methods. The concept of developing a protection philosophy can be understood well by using radial networks. This philosophy includes objectives that should be achieved by the feeder protection, for example the protection of life. These objectives are then used in conjunction with the protection philosophy building blocks, such as the protection sensitivity, to create a protection philosophy for a radial medium voltage feeder.

The various items that form part of this philosophy are then explored and includes items such as relay technology, network layout, protection equipment, available protection functions, operating curves, determination of pickup values, different grading methods, etc. The focus is also placed on let-through energy (LTE) protection, calculation of the conductor LTE limit and the application of this concept for both radial and then finally interconnected networks. Interconnected networks can consist of a connection to an adjacent feeder or embedded generation on the same feeder. Finally, the impact on the protection of a medium voltage feeder is discussed when changing a traditional radial feeder to an interconnected feeder or multisource feeder.

6.2 Protection Philosophy

A network cannot be designed that is immune to faults [4], hence there is a need for network protection. Protection philosophy forms the basis for how the complete protection system should perform for a certain physical section in the network, type of network or equipment-specific protection. The main protection philosophy building blocks are speed, sensitivity, selectivity, reliability and security [5,6].

These criteria have to be defined for the protection philosophy that is to be applied in the network. Speed refers to how quickly the protection system can clear a fault in the network. Generally, a protection system should clear a fault as quickly as possible [5,7]. The speed of the protection system is influenced by a multitude of factors: the protection philosophy, type of relay technology, type of protection element, operating curve, actual settings of the protection element, magnitude of fault current and fault resistance. The sensitivity of the protection system indicates at what current level the protection system operation will be initiated, in other words, how sensitive or insensitive is the specific protection element to the detection of a fault. The protection should be sensitive to faults [8]. The sensitivity of the protection is influenced by the protection philosophy (network contingencies and backing up other protection elements), the relay technology, magnitude of fault current and fault resistance. If the protection system is selective, protective devices will be graded with each other to allow the device closest to the fault to try and clear the fault first. If this device (closest) fails, the upstream device will then clear the fault. This selectivity ensures that the smallest portion of the network will be interrupted [4]. The reason for keeping the interrupted part of the network to the smallest possible portion is to limit the number of customers being interrupted, which may have a negative impact on energy sales. An interruption (and frequent interruptions) can influence utility image and this in turn can influence the utility capability to secure capital for future growth. This can even negatively impact future investments in that country. Interruptions can influence customer processes (manufacturing) and the ability of the utility's customers to provide services to their customers in the retail environment. Generally electricity interruptions are experienced as a nuisance (customer's perspective) for residential customers.

The reliability of the protection system can be defined as every time the protection system is required to operate, it should operate. Protection systems are human-made and as such they will fail at some stage. Regular maintenance of the power system protection equipment can prolong the period between these failures, but it will not completely prevent them. The maintenance strategy for these devices can be scheduled-based maintenance or condition-based maintenance. The security of the protection system can be defined as the protection system should not operate when it is not called on to operate. In simple terms, the protection relay should not just trip randomly (no fault present). The security of the protection system can be influenced by the aging of protection equipment, fault levels and changes in the network configuration. Reliability can be seen as a combination of dependability and security. Dependability refers to the ability of the protections scheme to operate when it is asked to do so [5,9].

When developing the philosophy for protecting the network, many of these philosophy building blocks have to be weighed against each other. The protection philosophy cannot be such that 100% of all the building blocks are present. If sensitivity is the main priority, then setting the overcurrent pickup to a minimum value means that the protection element will be able to detect any current flowing in the protected area. This will compromise the security of the protection, as the protection can trip for current that is not due to a network fault. A low pickup can also compromise selectivity (between series devices) when current grading is considered. The overcurrent pickup can be set to a maximum. This will make the device secure in that it will not trip, but then it will also be insensitive to faults. In general, protection systems are designed to be more dependable than secure [9]. If the protection speed is the main priority, then the device can be set to operate instantaneously, but then selectivity is again compromised. From this discussion, it can be seen that the protection philosophy forms the basis for any protection system as it will govern which functions should be active, how they should be set and which protection elements should operate to clear or isolate the fault. It can also influence the protection scheme design and choice of required hardware (e.g., type of protection relay).

The main objectives of protection in medium voltage feeders (in terms of priority) are to preserve life, protect equipment, reduce the risk of fire, promote the reliability and quality of supply and reduce lifecycle cost [6]. The first and most important objective is to safeguard life. This includes not only human life but life in general. Animals can get electrocuted if they make contact with low hanging conductors. It is not always possible to save the first life that makes contact with live apparatus, but a protection system enables one to avoid further incidents and avoid subsequent loss of life. The protection system should protect the equipment installed in the network. Failures of equipment can lead to lengthy outages and difficult fault-finding exercises in large networks. In resistively grounded networks such as those used in Australia and South Africa, a fault (i.e. clashing conductors) can lead to heated pieces of conductor being thrown around the faulted position. If the conditions are conducive to the development of veldt fires (dry, windy, presence of vegetation, etc.), a fault on a conductor in a resistively grounded network can lead to the ignition of veldt fires. This can again endanger life, damage the environment and have financial implications. The way in which the protection is set can affect the behaviour of the protection during network faults. It can unnecessarily affect the quality of supply, for example, with excessive voltage dips (also called voltage sags) when trying to clear a fault [10]. Similarly, if the selectivity is not maintained, it can lead to larger parts of the network being without supply for a single fault. Then finally, by reducing the time and frequency of equipment fault exposure, the life expectancy of the equipment can be prolonged.

The protection systems are human-made, and it is not a question of if it will fail but rather a question of when. To combat these failures and to improve the protection system's dependability, the concept of backup protection is introduced. The individual relay's dependability is not improved; rather, the protection system as a whole is improved. Two concepts are thus introduced: main and backup protection. Main protection is faster than backup protection and is normally functional in a smaller or equal area (sensitivity to faults) in the network. In a classic high voltage feeder, there would be main protection in the form of impedance or differential protection installed. This protection responds quickly to faults on the feeder as the fault position is known (e.g., on the feeder) and which protection is required to clear the fault. The backup protection to this main protection is overcurrent relays for both phase and earth faults. These relays provide protection in case the main protection fails to detect the fault. Examples are a loss of potential (voltage transformer input) for an impedance protection scheme or the loss of a communication medium for differential protection scheme. The backup protection will be sensitive to faults beyond the reach of the main protection and as such can lead to a loss of selectivity for faults beyond the protected feeder. Because of this, they are generally set slower than the main protection to allow the main protection on and beyond the feeder to clear the fault first. In MV feeders, there is only current-based protection; hence they have to be set in a certain way to provide this backup functionality to the system. The word *set* refers to their physical position on the feeder and also their applied protection settings in software. It is not required for main and backup protection to be installed at the same physical location in the network [9].

The protection settings applied to protective devices are based on the philosophy as determined above. The philosophy should be such that it is robust, meaning that if the network conditions change, the applied protection settings are still valid. Normally the relay does not sense if the fault levels in a network changed due to a transformer being faulted or being out for maintenance, or if a downstream circuit breaker operated, thus reducing the intended reach of the protective devices. This status update can be created by means of an automation system in a smart grid environment. Based on this, the protection system can then adjust itself to meet the reconfigured network. Even when the system is then reconfigured, the new applied protection settings should still reflect the protection philosophy. To achieve robustness, the protection functions have to be set under minimum and maximum network conditions [4]. Minimum network conditions refer to a network contingency that will result in the minimum fault current passing through the device that is being set for a specific fault. Maximum network conditions refer to the maximum fault current that will pass through the device being set for a specific fault. For radial feeders, minimum and maximum network conditions are achieved by changing the source impedance such as switching in and out source transformers [4].

Maximum network conditions are normally easy to define with all source transformers in service. Minimum network conditions can be complicated. Think of a scenario with five identical transformers in parallel at the source supplying three feeders on the secondary side. For maximum network conditions, place all the transformers in parallel. If an absolute minimum for a feeder was required, then switch out of service four of the five transformers. This network condition might not be a feasible network condition as the loading might be so high on the feeders that the network can never be operated with less than three transformers in operation. In such a case, it might be advisable to place two transformers out of service. This would then be an acceptable minimum, making provision for the load to reduce and one more transformer to trip due to a fault. The protection is then still valid for the defined minimum network condition. If the minimum or maximum network condition is too extreme, it can hamper the way in which the protection functions are set and the expected performance from the feeder.

A number of network performance indicators can be used to measure the performance of the medium voltage feeder and then the network in general. When developing a protection philosophy, these factors should also be considered as they are indicators of when the performance of a feeder is not up to standard. The protection devices are the equipment responsible for interrupting the supply once a fault has occurred. If they respond in an unfavourable way to the fault, it can have a negative impact on the network performance, such as excessive auto-reclosing on the feeder. Reliability indices that are used on medium voltage feeders can include the following [11]:

- System average interruption frequency index (SAIFI)
- System average interruption duration index (SAIDI)
- Customer average interruption duration index (CAIDI)
- Customer average interruption frequency index (CAIFI)
- Average service availability index (ASAI)
- Average system interruption frequency index (ASIFI)
- Average system interruption duration index (ASIDI)
- Momentary average interruption frequency index (MAIFI)
- Momentary average interruption event frequency index (MAIFIE)

The average indices are only recorded after a specified time duration such as 3 minutes, and it is thus beneficial to keep the total clearing time for the protection at a value less than this. The momentary indices can be beneficial as they have a shorter time and this can provide guidance towards the protection performance.

6.3 MV Network Layout and Components

Figure 6.1 shows a radial MV feeder with substation-based feeder circuit breaker (CB) auto-reclosers (AR), fuses and a number of loads on this feeder. Both the feeder CB and the AR have a cross for a symbol as both of them are circuit breakers. The use of inline fuses on MV feeders is not recommended [6] because it can lead to bad network availability (long outages), many customers being affected by one backbone fuse operation, extra labour hours for replacing the fuses when they are blown (and they can be replaced with the wrong type of fuse, which is a human error), they are not visible to a control centre (SCADA system), fault finding in the network is difficult and the risk to utility personnel increases as they have to go out to the actual fuse installation to replace the fuse [9]. Fuses can have a negative impact on protection performance indicators such as the system average interruption duration index (SAIDI) [11].

A sectionaliser can be described as an intelligent isolator or disconnector. A disconnector cannot break load or fault current; thus, it can only be opened when there is no current flowing through the disconnector. The same applies to a sectionaliser [12]. A sectionaliser is thus used in conjunction with a normal circuit breaker upstream from its position [4,13]. This circuit breaker (with its own protection relay)

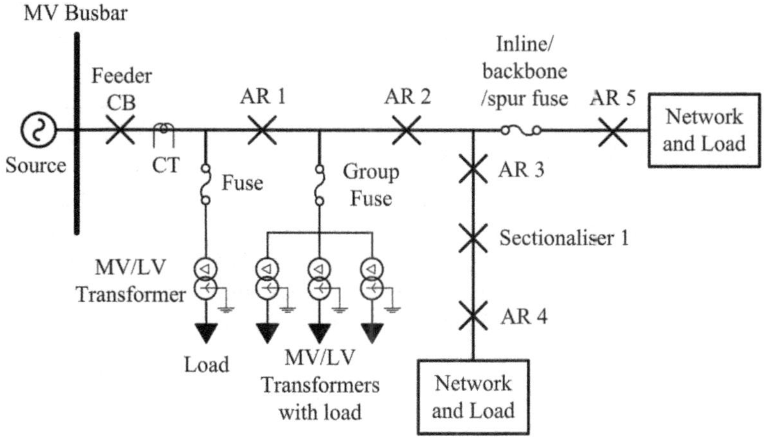

FIGURE 6.1 A radial medium voltage feeder showing different protective devices.

is used to break the fault current. During the upstream circuit breaker dead time (circuit breaker is open) the downstream sectionaliser can open, thus isolating a fault if it was beyond the sectionaliser. The circuit breaker will close again and re-energise the remaining network. For this circuit breaker and sectionaliser combination to work, the auto-reclose philosophy of the upstream circuit breaker should always be set to two or more trips to lockout [13]. The sectionaliser is seen as an extension of the upstream circuit breaker that allows smaller portions of the network to be sectionalised.

For the protection philosophy covered in this chapter, inline fuses and sectionalisers are not going to be considered as part of the network. The philosophy can be extended, however, towards including these protective devices if required.

Fuses are used at the MV/LV transformers to improve the network availability. They are not installed to protect the MV/LV transformer. This means that if there is a fault on the MV/LV transformer, its own protection should protect the unit and the fuse should remove the unit from the network, thus promoting the continuity of supply to the rest of the nonfaulty feeder. In certain cases where the transformers are relatively small, a fuse can be used to remove a group of transformers at a certain position (group fusing). This will influence a larger number of customers. The origin of group fusing comes from the 1930s, a large number of fuses had to be replaced by field staff (all spurs or t-off and transformer locations) after a storm, which increased the outage time [14]. By having group fusing (removing some of the spur and transformer fuses),s restoration time can be decreased. These fuses are sized to be insensitive to transients such as transformer inrush and lightning, and also the continuous current rating (load current estimate) of the downstream MV/LV transformer.

At the substation, the feeder protection can be a dedicated protection scheme or an auto-recloser. This raises the question, what is the difference between an auto-recloser and a protection scheme? A protection scheme is generally a collection of protection relays, switches, control knobs, current and voltage test points, and status indication displays that are housed in a protection panel. This panel in turn is usually housed in a control room, but in some applications the actual protection relay may be housed in the high voltage yard. The scheme is supplied with a backup DC supply that is also used for other secondary plant equipment. The interface with the high voltage equipment is via separate current and voltage transformers that are housed in the high voltage yard. Control and visibility of the primary plant equipment is normally provided via separate tele-control or then supervisory control and data acquisition (SCADA) system. This SCADA system can also interface with the protection scheme. An auto-recloser is an all-in-one unit, meaning it has DC backup, protection functions, operational logs, SCADA capabilities and metering functions housed in one cabinet that can be mounted at a remote location. The current transformers used for auto-reclosers are generally situated next to the control panel. The drawback of these auto-reclosers is that they have limited functions and then generally predefined input and output capabilities. However, they are common in MV networks to provide protection, automation, control and network visibility. There can be many auto-reclosers installed on an MV feeder, but it is recommended not to install more than five devices in series. This includes the feeder circuit breaker. The reasons for this are that it becomes more difficult to grade (maintain selectivity) for the network and the protection slows down at the source busbar to allow for this grading with the addition of auto-reclosers.

6.4 Protection Elements and Functions

In the previous section, various types of protection equipment and their typical positions on a radial medium voltage feeder were introduced. In this section the operating characteristics and functions of these components are discussed. These operating characteristics refer to how they can be set to isolate faults in the network. All of the functions are not always enabled and it depends on the type of protective device, the network layout, fault levels and protection philosophy that are applied.

6.4.1 Relay Technology

Various types of relay technology are present in power system networks. They can be placed into four categories [5]: electromechanical relays, static relays, digital relays, and numerical relays. Electromechanical

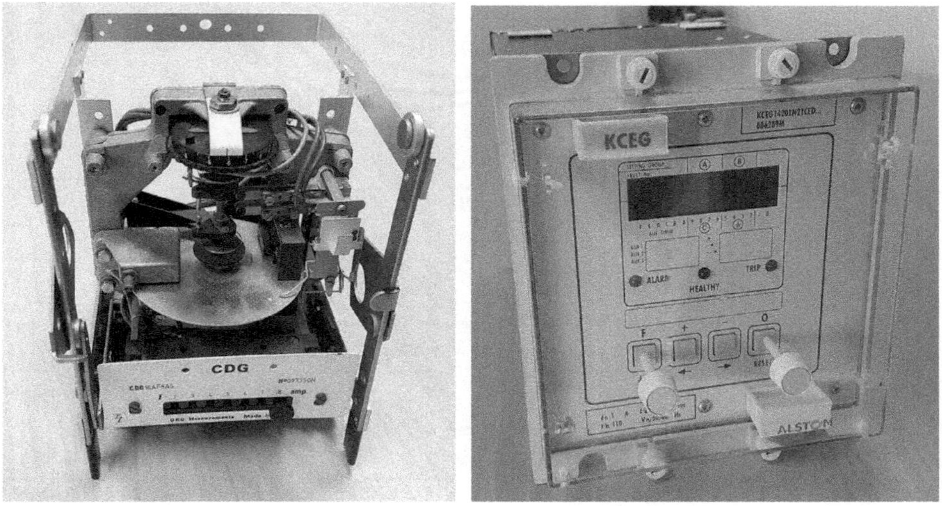

(a) (b)

FIGURE 6.2 (a) An induction disk electromechanical relay, and (b) a numeric relay.

relays, like the one shown in Figure 6.2a, make use of an induction disk that rotates at a certain speed. The speed depends on the magnitude of current that is passed through the relay. As the disk rotates, it reaches a certain position where a contact will close that will enable the trip signal to be passed to the circuit breaker. These relays are self-powering (do not require an auxiliary supply). Static, digital and numeric relays require an auxiliary supply to bias the electronic components [9]. Static, digital and numeric relays are also more susceptible to temperature damage, large overvoltages and currents [9]. A numeric relay is shown in Figure 6.2b. The life expectancy of electromechanical relays is more than those of the static, digital and numeric relays [9]. Static, digital and numeric relays do provide greater accuracy, smaller tolerances and a greater number of functions when considering the protection functions [9]. In terms of overcurrent-based relays they normally consist of one or perhaps two functions on one relay, such as an inverse definite minimum time (IDMT) element and an instantaneous current element. The relay footprint in a protection panel is also reduced due to the number of functions present in one digital relay compared to the electromechanical relay [9].

Static relays were the next generation of relays (from around 1960). The name static refers to the fact that there are no moving parts inside the relay [5]. Static relays use discrete electronic components which later included some integrated circuits [5]. This then includes the sensing of analogue values, signal filters, amplifiers and level detectors. In Figure 6.3, the primary input to a relay, primary output from the relay and an instantaneous relay are shown for one phase on a three-phase feeder. A current transformer (CT) and circuit breaker (CB) are shown for one phase of the three-phase system. There will also be a CT and circuit breaker on the other two phases. The CT will be the current input to the protection relay. The CT provides insulation between the high voltage of the feeder and the lower voltage on the secondary side and earth. It lowers the high currents on the primary side of the feeder to lower values on the secondary side, and it allows for the equipment used on the secondary side of the CT (e.g., protection relay) to use standardised secondary current values as an input to the device. The standardised secondary values that are used are 1 and 5 A secondary. Rogowski coils are also used in many of the auto-reclosers. Some of the modern auto-reclosers also sense the line voltage for the purpose of determining if the line is live and to provide directional protection on the auto-recloser or then the protection relay driving the circuit breaker at a substation. The circuit breaker is used to break the current once a trip signal is received from the protection relay. There are many types of circuit breakers, including oil, air-blast and SF6 gas circuit breakers [5]. A basic model for an electronic instantaneous protection relay is shown in Figure 6.3 [9]. The current received from the CT passes through a resistor that will bias the diodes in the rectifying bridge, and the rectifying bridge will change the AC signal to a DC signal. An example of the expected signals is shown below the relay blocks after each stage. This DC signal is dependent on the magnitude of fault

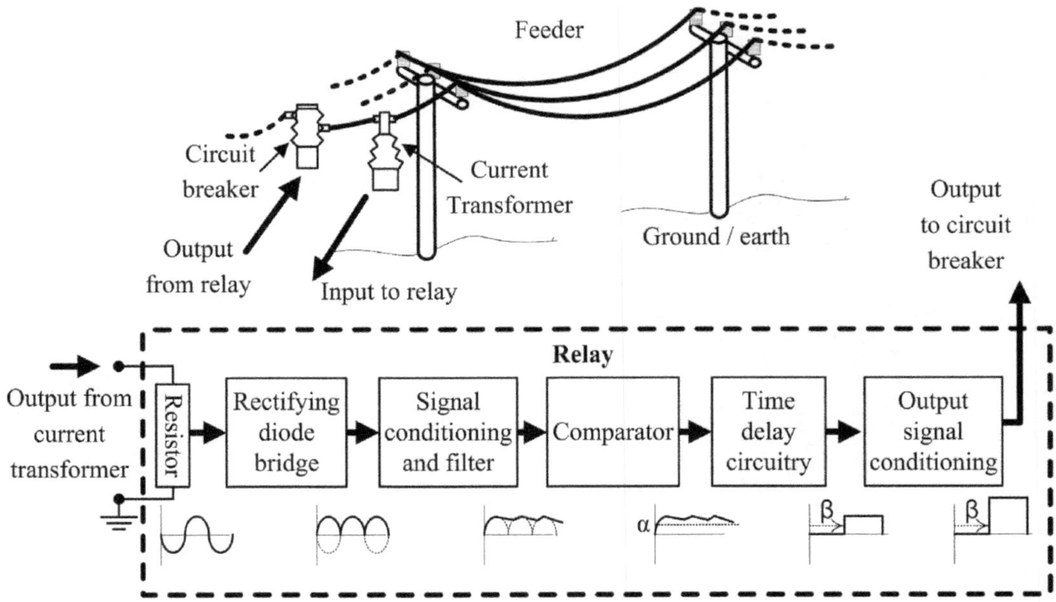

FIGURE 6.3 An electronic instantaneous relay concept.

current flowing on the actual feeder. The higher the current, the higher the DC voltage in the signal used. The signal conditioning and filtering circuit adjusts the magnitude of the signal to fit the next input state and reduces the ripple (smoothes the signal) of the DC. This filtering also slows down the response of the relay. If an R-C filter is used, the voltage would take time to change due to the capacitor characteristics. The smoothed DC signal is then passed through a comparator circuit. This can be an operational amplifier circuit with an adjustable threshold voltage (α). This adjustable voltage threshold is the pickup of the relay function. If the input voltage is above the threshold voltage, the comparator will output a logic 1; if it is below the threshold voltage, it will output a logic 0. The next step is to have a certain time delay (β) introduced for the purpose of selectivity and network security. This time delay is adjustable. Once the trip output (logic 1) from the comparator is received, it initiates the time delay circuit. Once the timer has completed a timing cycle, it will give an output signal to the final stage where the output signal is adjusted to meet the required output magnitude and driving magnitude for acceptance by the circuit breaker. The functionality available on the static relays was still fairly limited to a couple of settings such as having different IDMT operating curves available on one relay when compared to a single curve of electromechanical relays [5].

Digital relays were introduced in the 1980s. These relays have replaced the analogue circuits with digital microcontroller and microprocessor circuits [5]. The analogue quantities are sampled and then converted to digital quantities. This relay technology has improved the accuracy of the relay and provided much more protection functionality. Another advantage of these relays was the ability of the relay to report its status (self-diagnosis) and events to a SCADA system. These digital relays were slower than the static relays due to the limited processor capability (8 or 16 bit) [5]. The concept of a computer-based (digital and numerical) relay is shown in Figure 6.4 [9]. For a computer-based relay, the input filter consists of a surge and anti-aliasing filter. The surge filter is there to protect the relay from voltage spikes, and the anti-aliasing filter is a low pass filter that is used to limit the frequencies that are sampled (Nyquist theorem). There is an analogue to digital converter with a sampling clock used to convert the analogue signals to a digital version used by the controller or processor. RAM, ROM and EEPROM memory are associated with the processor for storage of data and program code. The output signal from the processor is then conditioned to meet the magnitude and driving current requirements of the output circuitry. Finally there will be some isolation filters to protect the relay from damage from the circuit it is driving. The difference between a digital and numeric relay is in its processing capability. Numeric relays make use of digital signal processors (DSPs). These processors are optimised towards signal processing and

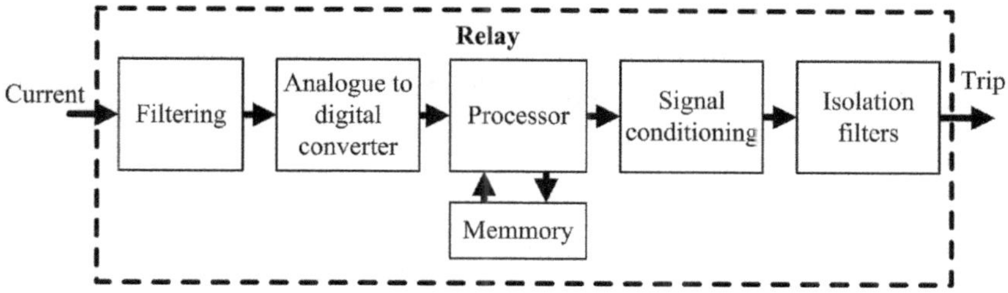

FIGURE 6.4 The concept for a computer-based relay.

this has improved the speed of the relay [5]. Numeric relays allow for the creation of customised protection logic inside the relay. With numeric relays it is easier to increase the number of input and output contacts by marshalling the contacts internally or adding extra contact cards.

6.4.2 Overcurrent Protection Relay Operating Curves

6.4.2.1 *Type of Curves and Equations*

IDMT operating curves are well suited for use in a power system as the protection will trip quickly during high fault currents and slowly during low fault currents [15]. This is excellent for coordinating multiple protection devices in series, but can result in long operating times [15]. The operation of current based protection relays are governed by selection of a time current operating characteristic or curve. Depending on the relay technology (as discussed in the previous section) the relay might only have one operating curve or it can have multiple. In both cases there will be some parameters that can be changed (or then set) to move the characteristic around in the time-current plane. A large group of curves are available; in some relays the curves can be combined and in others the curves can be customised. The latter is, of course, more applicable to newer numeric type relays. Equation (6.1) is the actual relay operating curve as derived from the basic differential equation for an electromechanical relay (induction disk) shown in Figure 6.2 [16].

$$K_I I^2 = m\frac{d^2\theta}{dt^2} + K_d\frac{d\theta}{dt} + \frac{t_F - t_S}{\theta_{max}}\theta + t_S \tag{6.1}$$

where K_I is a constant relating torque to current, I is the input current, m is the moment of inertia of the disk, K_d is the drag magnet damping factor, θ is the disk travel, t_F is the spring torque at maximum travel, t_S is the initial spring torque and θ_{max} is the maximum disk travel. Equation (6.1) can be simplified by neglecting the effect of the inertia of the disk (see the IEEE Std. C37.112 for a complete derivation) [16,17]. The simplified time-current equation is then shown in Equation (6.2) [16,18].

$$\int_{T_i}^{T_0} \frac{1}{t(I)}dt = 1 \tag{6.2}$$

where T_0 is the final time or trip time of the relay (s), T_i is the initial time of the RMS current signal (s) and $t(I)$ is the time-current equation. There are various time-current equations available. The IEEE (IEEE std. C37.112-1996) and IEC (IEC 60255-3 1998) define a set of equations for different standard IDMT curves [5,16,19]. The curves are named normal inverse (NI), very inverse (VI), extremely inverse (EI) and long-time inverse (LTI) curves. The constant time adder has been omitted from the equations below, but it can be added if required.

NI curve:

$$\text{OT} = \frac{0.14 \cdot \text{TM}}{\left(\frac{If}{Ipu}\right)^{0.02} - 1} \tag{6.3}$$

VI curve:

$$\mathrm{OT} = \frac{13.5 \cdot \mathrm{TM}}{\left(\dfrac{If}{Ipu}\right) - 1} \tag{6.4}$$

EI curve:

$$\mathrm{OT} = \frac{80 \cdot \mathrm{TM}}{\left(\dfrac{If}{Ipu}\right)^2 - 1} \tag{6.5}$$

LTI curve:

$$\mathrm{OT} = \frac{120 \cdot \mathrm{TM}}{\left(\dfrac{If}{Ipu}\right) - 1} \tag{6.6}$$

For Equations (6.3) through (6.6), *If* refers to the fault current (A), *Ipu* is the relay pickup current (A), TM is the time multiplier (unit less) and OT is the relay operating time (s). This relay operating time is the programmed delay time based on the applied protection settings. When this time is reached, the relay will issue a trip signal. The TM and *Ipu* values are user selectable settings for each curve. By increasing and decreasing the pickup current, the curve is moved left and right in the time-current plane. By increasing and decreasing the TM, the operating curve is moved up and down in the time-current plane. All of the above curves are inverse curves. More specifically, these are classified as IDMT curves. The inverse part of the explanation refers to the operating time increasing as the fault current reduces and then decreasing as the fault current increases. At a certain fault current value (*If* for the curve), the relay will reach a straight line in terms of operating time. This means that irrespective of fault current, as long as it is more than this value, the relay will have a constant trip time. The rationale behind this is based on the relay limitations. As an example, for an electromechanical relay the disk can only turn that fast; for a numeric relay, the algorithm execution time takes a certain time or the closing of contacts to energise the trip circuit, to name some factors. This can also be seen as a saturation level for the curve based on the magnitude of the current passing through the relay. This part of the curve is called the minimum definite time section of the curve. For an electromechanical device, this occurs from roughly 20 times the rated relay input current; for other relay types, this value is higher. A definite time (DT) curve can also be selected as an operating curve. This DT curve equation is shown in Equation (6.7).

DT curve:

$$\mathrm{OT} = \begin{cases} \mathrm{TD}, & If > Ipu \\ \infty, & If \le Ipu \end{cases} \tag{6.7}$$

where OT is the operating time (s), TD is the time delay of the curve (s), *If* is the fault current (A) and *Ipu* is the pickup current setting (A) of the curve. The DT curve is well suited for networks where there is a significant change in source impedance that will void grading when IDMT relays are used [5]. In Figure 6.5, the operating curves are shown on graphs using logarithmic and semi-log axis. This is for time versus current values.

Figure 6.5a shows the saturation effect (definite time effect) of the inverse curves. This is for an electromechanical relay on an NI IDMT curve. If the relay was ideal, it would be possible to obtain a trip time of 0.39 s at 10 kA with the applied protection settings and current transformer ratio. But due to the saturation of the relay from a plug setting multiple (PSM) of 20, the fastest definite time operating time that can be achieved is only 0.45 s from 6000 A. This curve was for a pickup current of 300 A primary

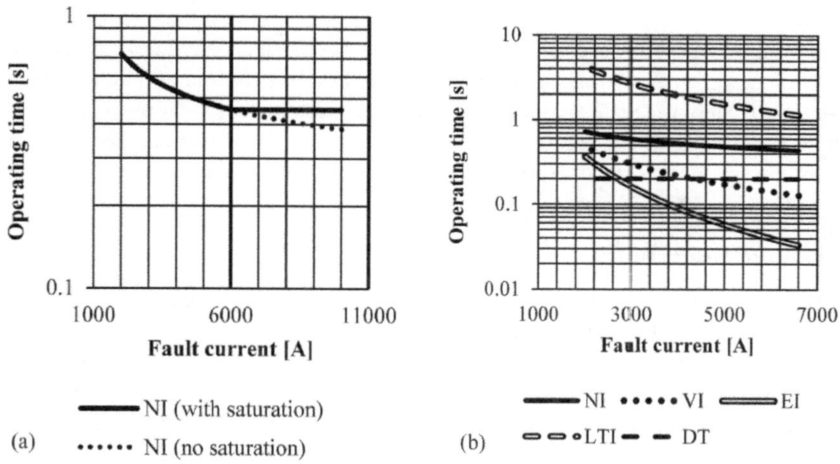

FIGURE 6.5 Time-current curves for standard IEC operating curves (a) illustrating the saturation effect and (b) showing the different IEC curves.

and a TM of 0.2. If the TM is changed, the curve can be moved up and down in the time-current plane. By changing the current transformer ratio (for electromechanical relay) and relay pickup current, the saturation level can be moved left or right. This saturation level can result in a loss of coordination if it was not accounted for during the grading purpose.

Figure 6.5b shows all the IEC curves. For these curves, they all have the same pickup current of 300 A with a TM of 0.2 and no saturation is considered. It can be seen that the more inverse the curve is, the quicker the curve operates at high fault currents. This can be seen when comparing the operating time of the NI IDMT curve at 6600 A with the VI and EI curves. When comparing the EI curve to the NI curve, it can be seen that there is a benefit in using the EI curve on feeders where there is high fault levels as the operating time will be reduced. In turn this will improve power quality (voltage dip), and reduce the damage and risk at the fault position [10]. An example of damage can be damage to the conductor where pieces of molten conductor can be expelled from the conductor at the fault position (mechanical damage) [20]. This welding effect of the conductor material is a function of the current magnitude and exposure time (I-t effect) [20]. Risk refers to the probability of starting a veldt fire or burning down a plantation due to the incandescent conductor particles at the fault position [20,21]. This then becomes an environmental incident as well. The fast fault clearing will also assist with reducing the damage due to the thermal effect of the fault current that the equipment needs to carry [5,22]. The drawback of using a curve that is more inverse (compared to NI) is that the rate at which the operating time increases down the feeder is also increased. If both these curves were set to operate in 1 s for a fault current of 4000 A on a pickup current of 300 A, the NI curve TM would be 0.4 and the EI curve TM would be 2.21. To achieve the same operating time, the EI curve requires a much greater TM value. If the fault current reduces by half (still the same settings), the NI curve will operate in 1.45 s and the EI curve will operate in 4.07 s. A drawback of inverse curves is that as the ratio of fault current to pickup current (*Ipu*) gets closer to a value of 1, the operating time can become infinitely high [23]. The opposite is also true: if both these curves are set to operate at the same time for a fault level downstream (lower fault level), then the EI curve will operate faster than the NI curve for a higher fault level close to the source.

6.4.2.2 Maximum Time Function

When IDMT curves are applied on long feeders where the fault level reduces to a small value at the end of the feeder compared to the relay pickup, the operating time can become excessively high. To counter this effect, a DT curve can be applied to the relay with the same pickup current value as the IDMT curve but with a time delay set to the desired maximum operating time. It can be seen in Figure 6.5b that the DT curve will operate in the same time, irrespective of the fault current it measures. The measured fault current simply has to be bigger than the relay set pickup current to initiate the tripping sequence (time delay). This second DT curve is normally not available on electromechanical devices (can add a second relay), but only on newer microprocessor and numerical relays. It is thus good practice to apply this DT curve with an IDMT curve to create an upper bound for the operating curve. Some manufactures do have specific functions on the relay (or auto-recloser) that do perform this function. The LTI curve is not a curve that is used very often as the operating time can be excessively long.

6.4.2.3 Constant Time Adder

Many of the IDMT protection elements in modern numeric relays also have a function called a constant time adder. This function adds a constant time to the existing operating time of the relay. The total operating time is thus the IDMT operating time and the constant time adder delay added together. This function is not used often, but it does allow for selectivity to be created when different operating curves are used for protective devices in series or where it is difficult to obtain selectivity at high fault currents. As an example, set an NI IDMT curve with a pickup current of 300 A and a TM of 0.2. At 4000 A the protection will operate in 0.527 s, and at 500 A it will operate in 2.727 s. If a constant time adder is enabled with a time delay of 150 ms, the operating time 4000 A is 0.757 s and at 500 A it is 2.877 s. If the applied constant time adder is divided by the operating time of the IDMT curve without the adder added, the effect of the adder can be evaluated. At the fault current of 4000 A we can see that the effect calculates to 28.5% and at 500 A it calculates to 6.5%. Due to the curve being inverse in nature, the constant time adder influence is greater at high fault currents than low fault currents even though the same time is added to the complete curve.

6.4.2.4 Reset Curve

If the relay is using an IDMT element that picked up, the reset of the curve of the picked-up element can also be programmed. This is for a protective device that has not issued a trip signal. An electromechanical device will take a certain time to rotate the disk back to its original starting position if the fault cleared before a trip was issued. On modern numeric relays, the reset curve can be set [4]. This can simply be a register inside the relay that is cleared, or it can be set to reset on a certain curve such as to emulate an electromechanical device [13]. This is required to ensure that the selectivity between relays is maintained. If the relay technology type is similar, all the reset curves on the relays can be set to reset instantaneously. Equation 6.8 is an electromechanical disk reset curve [16].

General form:

$$\text{OTcurve} = \frac{tr}{\left(\frac{If}{Ipu}\right)^2 - 1} \tag{6.8}$$

where *If* refers to the fault current (A), *Ipu* is the relay pickup current (A), *tr* is the relay reset time (s) and OTcurve is the relay operating time curve (s). This means that if an NI IDMT curve was used, the OTcurve will be replaced by Equation (6.3). By using this form of Equation (6.8) with Equation (6.3) substituted, the maximum reset time can be calculated. If a specific time is required, then the OTcurve time can be replaced with the time that the fault was present on the network.

6.4.2.5 Instantaneous or High-Set Element

An instantaneous element or high-set element resembles a DT curve with a zero second time delay. On numeric relay technology the trip signal can be delayed by a certain time. The benefit of using a high-set element is that the time delay is either zero seconds or very low; this reduces the damage at the fault point significantly [5,22]. It also assists with reducing the thermal damage such as LTE damage to equipment due to the fault current the equipment should carry [5,22,24]. With the reduction in operating time, it is difficult to maintain selectivity between series protective devices. For this reason it is normally set to not overreach the downstream device. Abnormal overload conditions such as cold load pickup can cause the high-set element to pick up if this is not considered when determining the high-set element pickup. Modern relays do have cold load pickup algorithms that can desensitise the overcurrent elements for a certain time period [4,25]. Inrush current can also be above the instantaneous element pickup. Inrush current can be between 8 and 30 times the rated transformer current [4]. A value of 6 times the rated current is used very often [25]. Another estimate for the inrush current magnitude and duration is 10 times the rated current for 100 ms [4]. During energisation there can be high quantities of second, third and fifth order harmonics [5]. Higher order harmonics can also be present but in reduced values. Modern relays do have the ability to block second order harmonics as they are high when transformers are energised [5,25]. Current transformer and relay errors should also be considered; a value of 20% is normally used. There are instances where it is required to overreach, but then a time delay will have to be introduced to maintain selectivity, and additional high-set elements have to be applied to downstream devices. A high-set is an additional curve to the IDMT or even DT element on a protective relay. This high-set element can be set to initiate an auto-reclose cycle. It is recommended to initiate auto-reclosing as this improves the availability of the network [5,13]. If no auto-reclose cycle is required, a high current lockout function should be used. This high current lockout element operates on a similar curve to an instantaneous element but with a zero second time delay, and no auto-reclosing will be initiated. When applying a high-set function it can also assist in creating grading margins for downstream protective devices; thus more devices can be placed in series without losing selectivity.

This is achieved by grading at the high-set pickup instead of the downstream device fault level. This concept is illustrated in Figure 6.6a, where the upstream device (A) can now grade at the high-set pickup value (C) of device (B) [23]. Normally it would have to grade at the fault current value of (B). When grading the device at (B) with the device at (D), the fault current at (D) should be used as there is no instantaneous element applied at (D). This is also illustrated on the grading graph of Figure 6.6b.

The pickup of a high-set element should always be determined in maximum network conditions. If the pickup is determined is this way, the setting is more secure. With the setting determined this way, if the network status changes from a maximum to minimum network contingency, the high-set pickup will appear to move closer to the source end of the feeder. This concept is illustrated in Figure 6.7.

In Figure 6.7, the pickup of the high-set or instantaneous (Inst.) element is determined in both maximum and minimum network conditions to thus explain why it should be determined in maximum network conditions. The fault levels for both minimum and maximum conditions are shown at Relay A and Relay B. For this example, the relay technology is numeric and the pickup is set to 130% of the downstream device fault level. This is applied to both the minimum and maximum network conditions. In Figure 6.7, the k indicates the current difference between the fault level at Relay A and the value of the instantaneous element pickup (*Ipu*). The symbol m indicates the difference between the instantaneous element pickup and the downstream protection Relay B.

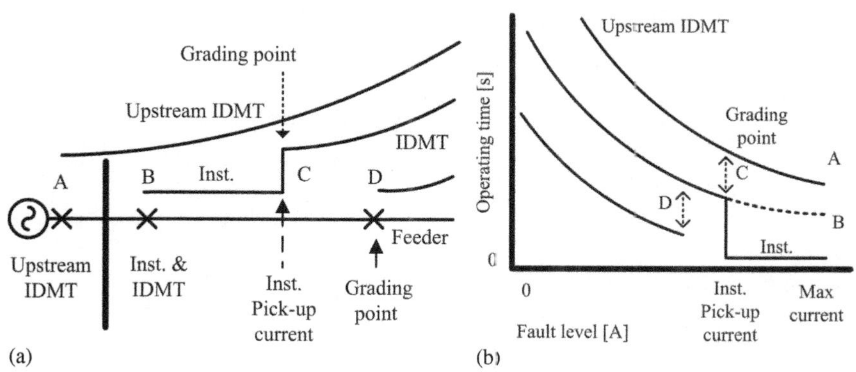

FIGURE 6.6 (a) A feeder with protective devices is shown where an instantaneous (inst.) or high-set element has been applied, (b) the associated grading graphs.

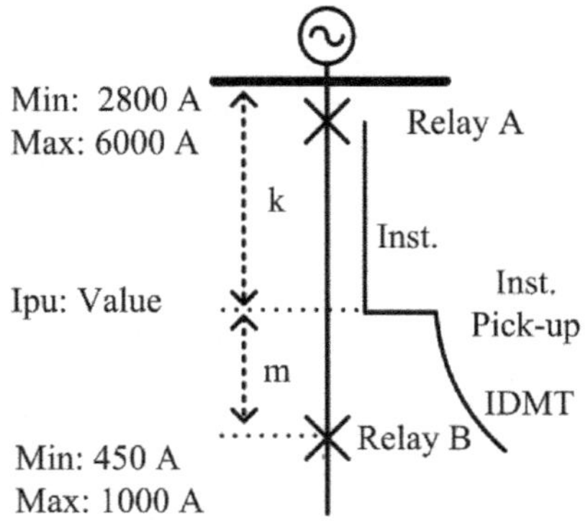

Ipu Inst. Set using Maximum network conditions Ipu: 1300 A	Ipu Inst. Set using Minimum network conditions Ipu: 585 A
Maximum conditions k: 4700 A m: 300 A	Maximum conditions k: 5415 A m: 135 A
Minimum conditions k: 1500 A m: 850 A	Minimum conditions k: 2215 A m: -415 A

FIGURE 6.7 Determining the high-set or instantaneous (inst.) element pickup in minimum and maximum network conditions.

When the instantaneous element pickup is determined using maximum network conditions, it calculates to a 1300 A. This means that under maximum network conditions, the current range for which the Inst. element will operate is 4700 A; beyond that, the IDMT element will take over and operate. If the network contingency changes to minimum conditions (Inst. pickup determined under maximum conditions), the range of k reduces to 1500 A and the range of m increases to 850 A. It appears as if the instantaneous element pickup moves closer to the source as the fault levels on the feeder decrease.

If the instantaneous element pickup was determined under minimum network conditions, it would calculate to 585 A using 130% of the downstream fault level. This means that in minimum network conditions the value of k is 5415 A, and the value of m is 135 A. Under this minimum network contingency, there is selectivity between Relay A and Relay B. However, if the network contingency changes to maximum conditions with the instantaneous element pickup determined under minimum network conditions, the values of k and m change to 2215 A and −415 A, respectively. This means that the instantaneous element will be sensitive to faults beyond the Relay B position and this can lead to a loss of selectivity. Thus, a high-set or instantaneous element pickup should be determined using maximum network conditions.

Figure 6.8 shows two possible applications of an instantaneous or high-set element [26]. First, whenever possible, a high-set element should be enabled as the reduction in operating time minimizes network damage and risk [5,14]. For both applications, numeric relays are assumed where a minimum pickup of 130% is used to ensure selectivity. For the type 1 application the fault level drops fast enough to allow the instantaneous element to be enabled without overreaching the next auto-recloser (Relay B).

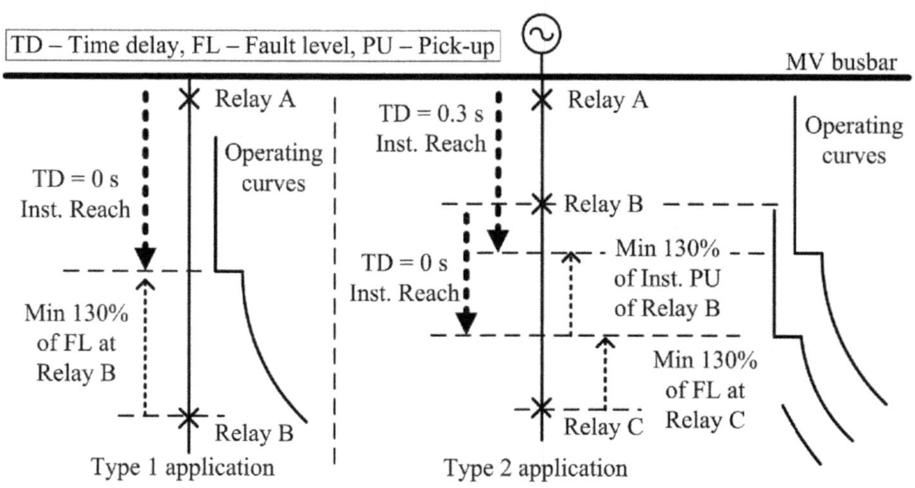

FIGURE 6.8 Two application examples for an instantaneous (inst.) or high-set element.

The instantaneous element pickup for Relay A is based on 130% of the downstream fault level at Relay B. Due to this device not overreaching Relay B, the time delay associated with this instantaneous element can be set to zero seconds.

In the type 2 application, the instantaneous element is required to overreach Relay B. This can be a result of high fault currents and the LTE associated with this. For this type of application, another instantaneous element will have to be applied on Relay B with a zero second time delay. The Relay A instantaneous element will be set to a pickup of 130% of the Relay B instantaneous element pickup current for current selectivity. A time delay of 300 ms will also be applied to the Relay A instantaneous element to ensure time selectivity between Relay A and Relay B. The instantaneous element applied to Relay B is again a type 1 application where the pickup is based on the fault level of the downstream device (Relay C). When looking at the grading graph (or operating curves), it can be seen which elements (Inst. and IDMT) are grading at which points in the network.

Depending on the protection philosophy are applied and the priority assigned to certain of these elements, an instantaneous element can be applied and it can be forced to overreach the downstream device with no time delay applied, but it will be set to initiate an auto-reclose cycle. In such a case, the priority of protecting the network with network availability is higher than that of selectivity. If the high set was disabled due to it overreaching the downstream device (loss of selectivity) and the risk of damaging the network equipment is there, then selectivity has a higher priority than reducing damage in the network. This is an example of why the philosophy elements and their priority are important. It aids in making judgement calls when faced with conflicting choices.

6.4.2.6 *Percentage Overreach*

The pickup for the high-set function is normally set higher than the downstream device fault level. This is because there can be a loss of selectivity between series protective devices due to the reduction in operating time of the upstream device, the nature of the fault and then the relay technology applied. A fault that is beyond the downstream protective device will appear closer to the source than what it actually is due to the DC offset [23]. This can cause the upstream protection to operate incorrectly. To create speed in operation, some of the signal filtering can be bypassed in the relay. This makes the relay susceptible to higher currents during transient fault periods. The relay security is thus affected by reducing the filtering component. Relays that do filter away the higher frequency component can be set more to be sensitive than the ones that do not have filtering. The percentage transient overreach can be calculated using Equation (6.9) [23,25,26].

$$\text{Percentage overreach} = 100 \cdot \left(\frac{Is - Ir}{Ir} \right) \tag{6.9}$$

where Is is the steady state pickup current (A) and Ir is the steady state RMS current that will pick up the relay with a full DC offset (A). With transient overreach and the relay error considerations, the minimum setting that can be applied to an electromechanical device is 192% of the downstream fault level. For numerical devices, this can be reduced to 126% [25,26].

6.4.2.7 *Fuses*

Fuses and their operation are well covered in the literature. In this section the common type of fuses used in medium voltage networks are introduced, and then some common fuse terms are shared. Two types of fuses can be used in medium voltage power systems: the first being a current limiting fuse and the second being an expulsion fuse [4]. Expulsion type fuses break the current at a current zero, and hence the circuit is exposed to at least one half cycle of fault current [4]. This is achieved by the gasses that are created inside the fuse chamber, and these gasses are then expelled to remove ionized air inside the fuse (extinguishing the arc). Current limiting fuses are generally more expensive than expulsion type fuses. A current limiting fuse quenches the arc by creating a fulgurite in the sand surrounding the fuse element.

A current limiting fuse will stop the fault current before it reaches its peak value. Fuses are relatively inexpensive to apply, but they are normally not visible to a control centre and cannot be remotely operated. This is not ideal for a smart grid where visibility of network components (status of components) and the control of these components are central to an intelligent network. Normally, the utilities rely on customers at the supply point to indicate that they are without supply and thus notify the network control centre of a fault beyond the fuse element. Smart metering can change this with the customer meter being visible to a central control system.

A fuse has two operating curves creating an operating area rather than just a single line operating curve (such as a protection relay). The first is a minimum melting time (mmt) curve, and the second is a total clearing time (tct) curve [4]. These curves are illustrated in Figure 6.9a for a 15 K type fuse. Also in Figure 6.9a is an EI and an NI operating curve with now saturation. Both the EI and NI curves are set to a pickup of 100 A and a TM of 0.2. If a protective device is set upstream from the fuse, then the grading margin has to be determined between the upstream device time and the total clearing time curve of the fuse. Figure 6.9a shows an NI curve, and it can be seen that the fuse curve grades well with this NI curve. If a protective device such as an auto-recloser is placed below the fuse, a good curve to use for grading purposes is an EI curve [5,23]. This EI curve will have to be faster than the mmt curve of the fuse. Figure 6.9a shows an EI curve, and it can be seen that the general form or shape of the EI curve is similar to that of the fuse.

Figure 6.9b shows a range of minimum melting time type-K fuse curves starting from 8 K up to 40 K. The higher the fuse number (e.g., 8 K versus 40 K), the greater the continuous current of the fuse. The 40 K fuse is also less sensitive than the 8 K fuse. The 40 K fuse will take longer to operate than the 8 K fuse for the same fault current. The indicator K is an indication of the type of fuse. A K type fuse is a fast acting fuse, and a type T is a slow acting fuse. This means that the type T fuse curve is not as inverse as the type K fuse and can thus sustain an overcurrent for a longer period before starting to operate. At 90 A of fault current, a 15 K fuse will operate in roughly 650 ms, and a 15 T type fuse will operate in roughly 1.9 s. The number in the type of fuse, e.g., 40 K, is indicative of the continuous current rating of the fuse. Table 6.1 provides the continuous current rating of various fuses. When grading auto-reclosers in a medium voltage network, the fuses are not always considered as the operating time of the fuse for three phase, and phase-to-phase faults are much faster than what can be achieved with the auto-reclosers at these fault levels. If the grading is determined between fuses and auto-reclosers, a grading margin of 350 ms is recommended [4].

When a fuse is applied to remove a transformer from the network in the event of a fault, the continuous rating of the fuse is considered as are a number of other factors such as through fault sensitivity on the transformer. When a transformer is energised, there is inrush current present because most circuit breakers close at a voltage zero, and this corresponds to maximum flux in the transformer core. Due to the transformer being energised, the flux is zero and thus there is a transient created, and the flux that is generated by the flow of current has to be corrected in terms of magnitude and angle. To ensure that the fuse is insensitive to this transient, the 10–100 ms guideline can be used [27]; this guideline states that

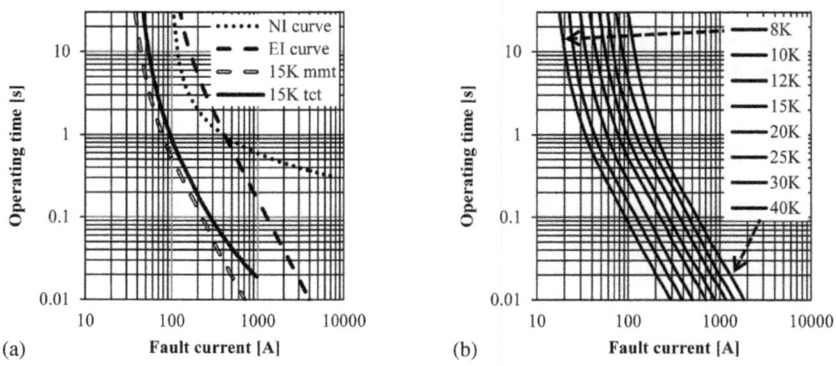

FIGURE 6.9 (a) A 15 K total clearing and minimum melting time fuse curve with extremely inverse and normal inverse relay operating curves, (b) a range of type-K minimum melting time fuse curves starting from 8 K up to 40 K.

TABLE 6.1

Continuous Current Ratings of Expulsion Type Fuse Elements

Fuse Type	Ratting [A]	Fuse Type	Ratting [A]
8 K and 8 T	8	20 K and 20 T	20
10 K and 10 T	10	25 K and 25 T	25
12 K and 12 T	12	30 K and 30 T	30
15 K and 15 T	15	40 K and 40 T	40

the fuse should operate slower than 10 ms at 25 times the rated current of the transformer and slower than 100 ms at 12 times the rated current of the transformer.

As an example, if a 500 kVA transformer is installed below a fuse, this fuse will have to be sized for transient conditions. The rated current of the transformer at 11 kV can be calculated as 26.24 A. 25 times this rated current, which is 656 A, and 12 times this rated current is 315 A. From the minimum melting time, type K fuse curves in Figure 6.9b, it can be seen that the smallest fuse that will meet both criteria in the 10–100 ms guideline is a 25 K fuse.

To enable different fuses to be compared with each other, the speed ratio of the fuse can be used. The speed ratio is the ratio of current that will melt the fuse in 0.1 s to the current that will melt the fuse in 300 s. For fuses with a rating greater than 100 A, the ratio of current melting the fuse in 0.1 s to the current melting the fuse in 600 s is used.

6.4.3 Auto-Reclosing

The auto-reclose function allows the circuit breaker to close automatically once a trip signal was sent to the circuit breaker and it opened. The main elements that will have to be set on the auto-reclose function for medium voltage feeder applications are the number of shots, dead time and reclaim time. The operating curves associated with each attempt can also be set depending on the relay technology applied.

6.4.3.1 The Number of Attempts

Two main categories for faults are transient and permanent faults [4,13]. A broken conductor is a good example of a permanent fault [4]. A permanent fault requires some human intervention to repair the network [28]. Isolating the permanent fault (faulty part of the network) and then reconfiguring the network can be automated. The most common type fault is a transient type fault. An example of this is lightning striking the feeder where the operation of the circuit breaker (open and then close) will re-establish the dielectric integrity of the feeder [23,29]. Except for lightning, a transient fault can also be caused by animals, wind and trees [4,5]. Another fault classification type is a semi-permanent fault [5]. A semi-permanent fault example is a tree branch touching the bare overhead conductor [5,28]. The philosophy that is applied to try and clear this fault is to burn away the fault with longer fault duration [5]. This will then not result in the lockout condition of a permanent fault. When considering the protection philosophy principles and objectives stated early in this chapter, increasing the trip time will increase the risk of veld fires [21]. The literature differs to the exact occurrence percentage of the three fault categories, but there is overwhelming consensus in that transient faults account for the majority of the faults [3–5,8,10,12,14,23,29].

A study done [30] evaluated 905 protection event records (2044 individual trips) detailing 547 months of data in radial medium voltage overhead lines that are resistively earthed (limited to 360 A per source transformer). In this study it was found that single phase-to-ground faults accounted for 44% of all faults, phase-to-phase faults accounted for 34% of all faults, and three-phase faults accounted for 22% of all faults. These values are dependent on factors such as tower design and dominant weather phenomenon. The 905 protection events were classified into the three main fault categories of transient, semi-permanent and permanent. Transient faults accounted for 83.5% of all faults, and this corresponds well to existing literature on this topic [5]. Permanent faults accounted for 9.5%, and semi-permanent faults accounted for 7% of all the faults. The probability of successful auto-reclose (ARC) attempts decrease with the increase in ARC attempts.

A possible auto-reclose philosophy can be to disable the auto-reclose function completely, but based on the study above, it can be seen that by enabling the auto-reclose function, the availability of the network can be improved due to the majority of the faults being transient in nature [5]. Another philosophy can be that the maximum number of ARC attempts should be applied, but this is detrimental to the long-term reliability of the network. An example of this is that the source transformers get exposed to cumulative damage every time there is a fault of sufficient magnitude (or then an ARC attempt) [13]. With the probability of having a successful ARC diminishing and the possible damage to the source transformers increasing with an increase in ARC attempts, a maximum number of ARC attempts are not advisable. If the maximum number is used, the circuit breaker duty cycle (number of open and close attempts) has to be considered to ensure that the circuit breaker can actually sustain this number of attempts. This is especially true for substation-based circuit breakers where the fault level is generally high and slow operating times can be expected. When developing an ARC philosophy, the basic concept is to maximize network availability by enabling the ARC function, but to minimize the unnecessary closure on permanent faults. Also, an ARC cycle should not be the cause of equipment damage [14]. When considering the classification percentages in the previous paragraph and then the decline in ARC success with an increase in ARC attempts, a two trip to lockout philosophy is recommended [13,14,30]. The first trip is then for the predominant fault type of transient. The second then assumes that the fault is permanent and the protection relay goes into a lockout condition. Lockout means the protection relay will not issue another close command unless it is instructed to do so either via a SCADA interface or then manual closure. If provision were to be made for semi-permanent faults, the fault will have to be left on the network for a longer period. By leaving the fault on the network for a longer period, the risk in the network increases and this goes against the initial objectives (stated at the beginning of this chapter) set for the medium voltage feeder protection philosophy. If those objectives are changed, then the ARC philosophy can be changed. When considering a cable feeder, most faults are permanent, and applying a circuit breaker to reclose it will only increase the damage at the fault position [4,5]. For a network consisting of overhead lines and cable networks, the auto-reclose function can be enabled as transient faults can occur on the overhead line sections.

The study in [30] identified that some faults do clear if the circuit breaker open time was more than 10 minutes. This accounted for 3.3% of the 905 incidents. Based on this, the control centre can try to close the circuit breaker once if it is visible to the control centre via a SCADA system. If it is not, then the number of ARC attempts can be increased from two trips to lockout to three trips to lockout, to try and improve the network availability and avoid sending field personnel to investigate unnecessarily.

6.4.3.2 Dead and Reclaim Time

Dead time refers to the time that the circuit breaker is open, but it is actually measured from the time that the trip signal is sent to the circuit breaker [5]. The items that influence the dead time are the deionisation of the air in the fault path, the required reset time of the protection relay, circuit breaker characteristics, type of loads being supplied beyond the circuit breaker, and then system stability and synchronism [5]. When a fault occurs, the air gets ionised, thus creating a path for the flow of current. The dead time should allow sufficient time for the ionised air to disperse after the fault; otherwise, the fault will restrike through the same ionised air path. The voltage level and weather conditions do influence how quickly the air disperses. For medium voltage networks a value not less than 100 ms is sufficient [5]. Dead times greater than 10 s are required to allow conductors to settle and birds to fall clear if small conductors are used on the feeder [14]. The protection relay reset time should also be considered. As stated earlier, the electromechanical relays do not reset instantaneously, whereas numeric devices can be set to reset instantaneously. If the fault is beyond the downstream numeric device and both the upstream electromechanical and the downstream numeric device detects the fault, both will start to time on its operating curve. If the numeric device trips, the fault current will drop off, and the disk on the electromechanical device will start to return to its starting position. If the numeric device closes again before the electromechanical relay's disk is at its starting position, selectivity between the two devices can be lost. Thus, when different relay technologies such as electromechanical and numeric are used, this can be compensated for in the dead time on the numeric device. Dead time settings of at least 10 s are recommended [5]. Another rule of thumb is to set the downstream device reset time to a value greater than or equal to 10 times the TM of the electromechanical device and then add 2 s to this. This means that the minimum dead time on the numeric device should be 5.5 s if the electromechanical device is set to a TM of 0.35. The characteristics of the circuit breaker should also be considered [5,13]. If more than one ARC cycle is applied, then a minimum dead time of 15 s is recommended to ensure that the circuit breaker interrupting capability does not have to be de-rated [13]. Some of the circuit breakers can take up to 300 ms to close. Modern vacuum circuit breakers can close in less than 100 ms. Industrial customers are more likely to be affected by the dead time. If motors and systems are running, sufficient time should be allocated for the system to shut down or to reset. Residential customers are not that affected by the dead time. Traditionally system stability and synchronising was not a concern for medium voltage networks as it generally supplies load from a single source. Modern medium voltage networks do have distributed generation or then embedded generation installations and bidirectional power flow can occur on these feeders. To accommodate generators on these networks, dead times should be kept to a minimum.

The reclaim timer is measured from the time the closing command is sent to the circuit breaker. The reclaim timer ensures that the duty cycle of the circuit breaker is not exceeded (number of open close cycles in a certain time period) [5]. When the timer times out, it resets the ARC trip number. If another fault is detected before the reclaim timer has timed out, then the circuit breaker will trip and not close automatically again if the maximum number of ARC cycles are reached (lockout condition). The minimum reclaim time that should be used is 30 s and the maximum is 3 minutes [14]. If the reclaim time is too long, then the availability of the network will be negatively affected as a new fault will be perceived to be part of the previous fault. An example of this might be a thunderstorm where there can be multiple transient faults in a short span of time. If the reclaim time is too short, then the fault can be on the network permanently. The type of protection (e.g., IDMT or DT operating curves) and the spring rewind time can also influence the reclaim time. Reclaim times can be as much as 180 s. Figure 6.10 shows a successful ARC sequence of events.

In the actual event record evaluation done in [30] the reclaim time was optimized based on the 905 fault record evaluations. It was recommended there to use a reclaim time of 75 s for circuit breakers that can be controlled via a SCADA system (with one ARC cycle). If a three trip to lockout ARC philosophy is used for a circuit breaker that cannot be closed via a SCADA system, a reclaim time of 90 s is recommended for medium voltage feeders.

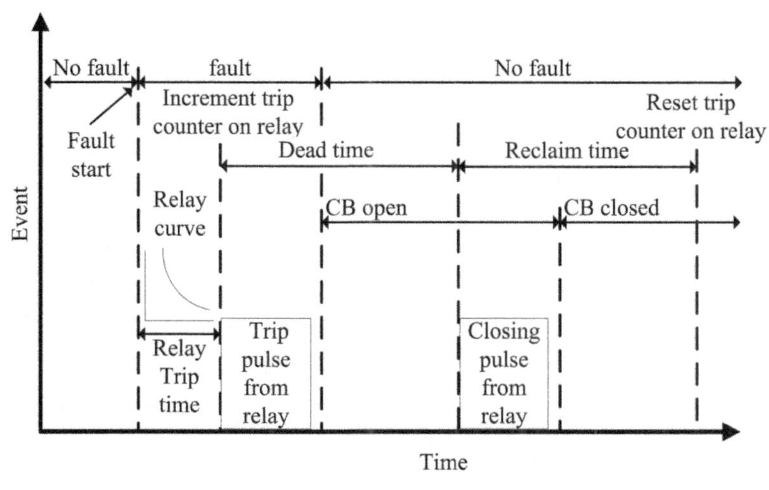

FIGURE 6.10 A successful ARC cycle on a circuit breaker.

6.4.3.3 Zone Sequence Coordination Function

The zone sequence coordination function allows an upstream relay to advance its trip counter without sending a trip command to the local circuit breaker. This allows different operating curves to be used on every trip in an ARC cycle and then to maintain coordination between the operating curves. Figure 6.11 shows different zone sequence coordination scenarios for the network diagram in the same figure.

The network in Figure 6.11 shows an auto-recloser (AR) at the substation and then another auto-recloser further downstream of AR 1. Both AR 1 and 2 are set to lock out after three trips (two ARC cycles). The curves applied to each trip at each auto-recloser can be different. For example, let us define two tripping curves, the first a slow (S) and the second a fast (F) operating curve (referencing to their operating times). The slow curve of AR 2 is set to grade with the slow curve of AR 1. The fast curve of AR 2 is set to grade with the fast and slow curve of AR 1, but the fast curve of AR 1 will not grade with the slow curve of AR 2. In Scenario 1, all the tripping curves are set to operate using slow curves. With the fault at position A, it does not matter what trip number the downstream or upstream device is at as there will always be selectivity. Zone sequence coordination can be switched on for Scenario 1. This can lead to a lockout condition if there is a thunderstorm in the area. The fault scenario is for the first fault at position A with both AR 2 and AR 1 on its second trip number and then a second transient fault occurs at position B. AR 1 will then trip and go into its lockout state. If AR 1 is closed after the storm, it will stay closed as if no fault was present. A spurious trip number increase and then ultimately a lockout condition was also identified in the practical 905 event records due to repeated transient pickups before the fault reach a stable period [30].

In Scenario 2 the upstream device is set to operate on slow curves only, and the downstream device is set to operate on a fast trip and then two slow trips. Again, similar to Scenario 1, zone sequence coordination is not required for this as there will always be grading between the upstream and downstream auto-reclosers. It can be applied with the same benefits and drawbacks as Scenario 1. For Scenario 3, zone sequence coordination has to be applied if grading is to be maintained between the upstream (AR 1) and downstream device (AR 2). In this scenario the upstream trip counter will be increased as the downstream device trips.

In general, it might not always be required to use different curves. Using only NI or then EI curves can be adequate for the feeder and network being protected. If a fuse saving philosophy is applied with a fast trip and then a slow trip, zone sequence coordination may be required. Using the zone sequence coordination function does improve the selectivity on the feeder, but it decreases the security of the protection system.

6.4.3.4 Fuse Save and Fuse Blow Philosophy

Most of the faults in an overhead network are transient in nature. If the transient fault is beyond the fuse and the fuse is slow enough to allow the upstream auto-recloser to clear the fault before the fuse on the first trip. If the fault is still present after the first trip, the protection relay will advance to its second slow

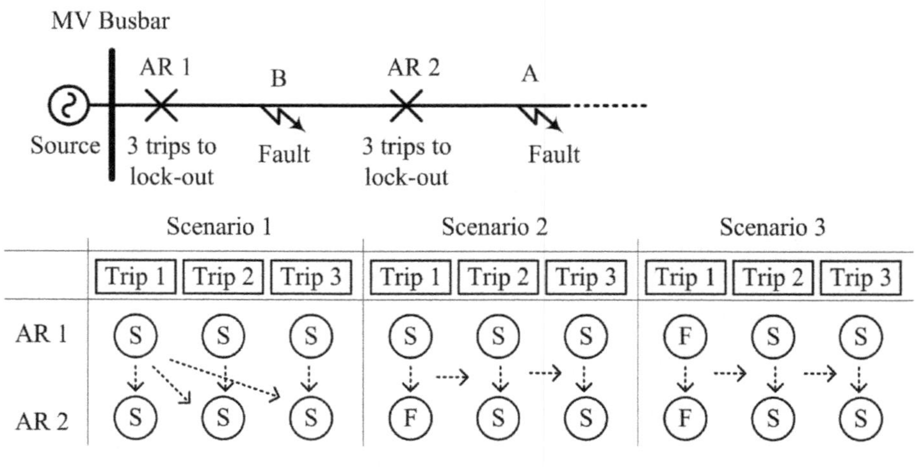

FIGURE 6.11 Different zone sequence coordination scenarios.

trip curve, allowing the fuse then to blow if the fault is still present. This is a fuse save philosophy. It is dependent on the fault magnitude, the type of fuse used and the protection settings on the upstream relay. In a fuse-saving philosophy, the number of permanent interruptions by fuses will decrease for transient faults, while the number of momentary interruptions will increase [4]. In a fuse blow philosophy, the fuse will always clear the fault before the protection relay, and the opposite will happen in terms of the type of interruptions. Some of the literature indicates that the fast or recloser curves should be used, but these curves are not always available. Instead, the EI curve can be used as the mathematical equation is defined making grading calculations possible (no need for a graphical program). The EI curve also grades well in terms of its shape with fuse curves (see Figure 6.9a). The EI curve is thus well suited for medium voltage distribution networks [4,23].

6.4.3.5 Breaker Fail Functionality

The circuit breaker fail function is not really applicable to conventional medium voltage networks where there is no automation system. One of the reasons is due to the cost of the extra relay functionality [4]. The circuit breaker fail function will issue a trip to an upstream breaker (can be multiple) when a trip command was sent to the local circuit breaker and it failed to clear the fault. This failure can be detected by either the fault current still being present or the status contacts on the circuit breaker not changing state (or both). In the case of auto-reclosers that are installed in a conventional network (not smart grid), there is no means to trip the upstream circuit breakers, and hence a breaker fail can only result in a protection not healthy alarm on the protection device and at the control centre if there is a SCADA system. If there is an automation system or smart grid system on the feeder and it detects the breaker fail condition, it can send trip signals to all the circuit breakers upstream to the failed circuit breaker. But this system does require a communication infrastructure and an automation algorithm on a substation-based computer or server. A circuit breaker fail function can be implemented at the feeder circuit breaker stationed at the substation as it is easy to create a blocking signal for the upstream protection at the substation while the protection on the feeder is healthy. This can be done either by conventional hardwire or serial communication systems. The newer IEC61850 systems can also be utilised for this type of substation automation on a breaker fail alarm.

6.5 Let-Through Energy (LTE)

In this part of the chapter, the concept of LTE is defined for radial feeders. A method is shared on how to calculate the conductor limits, and then the application concept of LTE protection to radial feeders is discussed in terms of minimum and maximum network conditions, curve selection and ARC application.

6.5.1 LTE Definition

6.5.1.1 Conductor Limit and Damage

All conductors have resistance (R) associated with them. The power (P) that will be dissipated over a conductor is equal to the square of the current (I) passing through the conductor times this resistance of the conductor [31,32]. For alternating current the root mean square (RMS) value will be used [33]. If we multiply a time component with this, the power will be changed into energy [32–34]. Based on the aforementioned we can then write Equation (6.10) for energy dissipated in a resistor.

$$\text{Energy} = I^2Rt \tag{6.10}$$

where energy is the energy calculated in watt-hour, I is the fault current (A), R is resistance (ohm) and t is time (s). Joule can also be used to represent the energy as 1 watt is equal to 1 Joule-per-second and 1 W·s is thus equal to 1 Joule. As the current passes through the conductor, heat is generated in the conductor based on the resistance of the conductor. The higher the current, the more heat is generated.

TABLE 6.2

Conductor Short Time and LTE Ratings

Conductor	1 s Short Time Rating kA	LTE MA²s
Fox	3.50	12.25
Gopher	2.50	6.27
Hare	10.02	100.39
Mink	6.03	36.31
Oak	9.52	90.68
Rabbit	5.04	26.38
Squirrel	2.00	4.00

The assumption is that all the heat that is generated by the flow of current is contained within the conductor during a fault [35–39]. Thus, this is an adiabatic process with no heat that is lost to the environment due to convection to the surrounding environment, conduction to clamps, etc. This allows for the calculation of a conductor limit (more information on this is given in the next section). This limit specifies how much current the conductor can carry for a specific time period before it gets damaged. Generally, a current value and a time value of 1 s are used to specify this energy limit. Some conductor short time ratings and LTE ratings are shown in Table 6.2.

From Table 6.2, we see that the Mink conductor can carry 6030 A for 1 s. If we look at this in terms of energy, it is an I^2t energy or then Joule per ohm (dividing the energy by the resistance in Equation [6.1]). Because the assumption is made that this an adiabatic process we can set the energy rating for one instance equal to the energy for another instance. This is shown in Equation (6.11):

$$I_1^2 t_1 = I_2^2 t_2 \tag{6.11}$$

$$t_2 = \left(\frac{I_1}{I_2}\right)^2 t_1$$

where I_1 is the fault current for instance one, I_2 is the fault current for instance two, t_1 is the fault current withstand time for instance one and t_2 is the fault current withstand time for instance two. By using Equation (6.11) we can then determine how long the conductor can sustain a fault current before it gets damaged. As an example, if the Mink conductor is exposed to a fault current of 3500 A, it can sustain that current for 2.97 s before it gets damaged. If the fault current is equal to 7000 A, the conductor can sustain this current for 0.742 s before it is damaged. If this Mink conductor is used and the times are calculated over a range of fault currents, a damage curve can be created for the Mink conductor. But if we consider this as an I^2t energy, it will result in a straight line instead of a curve. As an example, the energy rating for Mink conductor is equal to 36.31 MA²s (6030 A for 1 s). Let us call this the LTE rating of the conductor. It is easier to work with the energy rating than the short time rating as the energy rating is constant, whereas the short time rating varies with the fault current. In Figure 6.12 the short time and LTE rating for this Mink conductor is shown over a current range. Both of these are damage curves for the conductor: one is in terms of time and the other in terms of LTE.

Conductor damage in the context of this work refers to the annealing effect on the conductor material due to the heat generated by the flow of current in the conductor during a fault. [4]. If the change in the conductor material stress is proportional to the change in the strain, the material (or then conductor) is said to undergo elastic deformation [40]. The slope of the stress-strain graph for the material is called the modulus of elasticity. Thus for an elastic deformation, the modulus of elasticity of the material will be constant [40]. For an elastic deformation, the material will return to its original profile if the stress is removed. The time it takes for the material to return to its original profile is called the anelasticity of the material. For metals, this time is most often neglected due to it being small [40].

If the change in the stress and strain quantities is not proportional, the material (or then conductor) is said to undergo plastic deformation [40]. This means that the modulus of elasticity is not a constant

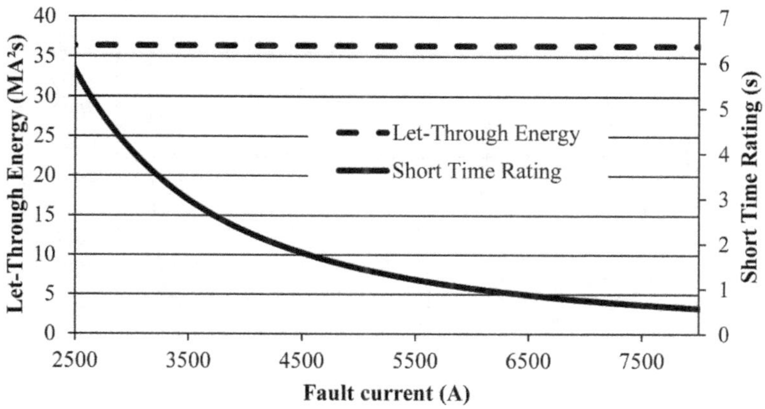

FIGURE 6.12 LTE and short time damage curves for a Mink conductor.

any more. The electrical resistivity of the conducting material will increase when the conductor has undergone plastic deformation [40]. For plastic deformation there is a small bit of recovery back to the elastic deformation region, but the material will not recover back to its original profile [40]. If we apply this concept to an overhead conductor, the conductor will encroach on its minimum clearance from the ground [4]. This means the conductor sags or, in other words, the length of the conductor increases between the supporting towers. Then, because the profile of the faulted phase conductor in a three-phase system might be at a different profile, it will swing more when exposed to cross winds and this can result in an increase in network faults such as phase-to-phase and phase-to-earth faults. The damaged conductor can stay in service, but in time the damaged conductor material can result in a failure [4]. On an actual feeder, the whole conductor will carry the same current up to the fault point. This means that the entire conductor up to the fault point is subjected to the same potential LTE damage.

It is recommended that a safety margin be applied to the material to ensure that plastic deformation does not occur. This margin is with reference to the yield strength of the material. It is recommended to use a safety margin of 1.2–4 times for the yield strength. If the safety margin is too big, it will result in overdesign and then an increase in cost of the material [40]. For this work, a safety margin of 120% will be applied. If a conductor on a feeder is exposed to excessive faults, it can lead to a reduction of the life expectancy of the conductor due to fatigue [40]. If the conductor is exposed to higher stress levels, the number of cycles required to cause severe conductor damage is reduced [40].

6.5.1.2 Let-Through Energy

As stated in the beginning of this section, the I^2Rt-energy is used to ensure that the conductor is protected [36]. The resistance is incorporated in the calculation of the conductor limit, and hence we are only looking at the I^2t part of the equation. The current used for the I^2t calculation is the fault current. The time used in the I^2t calculation is the fault clearing time [36]. This fault clearing time consists of the protection operating time and then the circuit breaker operating time [10,19]. When a fault is cleared on the IDMT protection element further down the feeder, the circuit breaker operating time influence is minimal. However, close to the source where the high fault current region is, the circuit breaker operating time influence can be significant. This is even when an instantaneous curve (or high-set curve) is applied. The LTE equation is shown in Equation (6.12). From a protection perspective, the fault current cannot be changed as this is dependent on the source impedance to the fault. The time the conductor is exposed to the fault current can be changed by changing the protection settings. This adjustment can be in the form of changes to the operating curve, pickup current, time multiplier (or time delay), the ARC philosophy or even by adding additional operating curves.

$$\text{LTE} = I^2 t \tag{6.12}$$

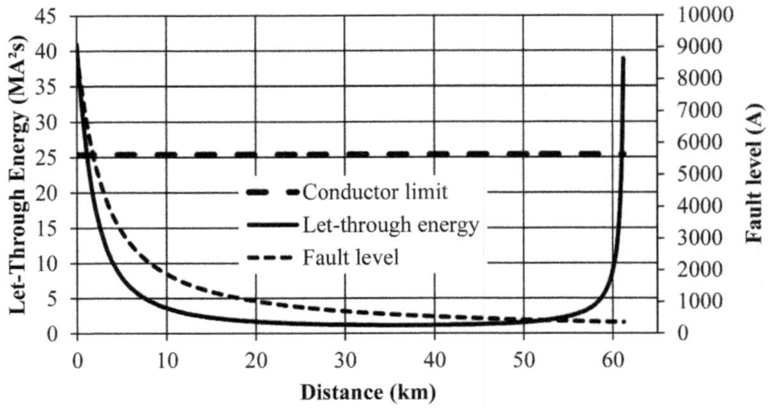

FIGURE 6.13 Conductor damage curve and the LTE exposure for a radial feeder.

where LTE is the let-through energy (A^2s), I is the fault current (A) and t is the fault clearing time (s). To ensure that the conductor is protected, the LTE exposure should be less than the conductor limit. In Figure 6.13 the LTE exposure for a rabbit conductor (26.38 MA^2s) in a radial network is shown. Radom protection settings are used to illustrate the concept of LTE. For the graph in Figure 6.13, the distance from the busbar is used instead of the fault current. This approach is applied as it provides a better feel for where on the feeder (position) damage is most likely to occur. The fault level is also shown, however, as this helps with the calculations and to visually determine possible setting values such as a high-set pickup. The LTE was calculated by applying Equation (6.12).

The conductor I^2t exposure should be less than its limit to ensure it is protected [24]. In Figure 6.13 it can be seen that the conductor limit is exceeded for the first 1.17 km of the feeder. This means that any fault within this area will result in conductor damage with the current protection settings. This is also the high fault current region of the conductor. Close to the source, the current will be the dominant factor in the LTE calculation. This is for IDMT protection operating curves such as those shown in Equations (6.3) through (6.6). As the fault level reduces (moves further away from the substation) the ratio of the fault level (If) to the pickup current (Ipu) in Equation (6.3) will get closer to 1. As this happens, the time part of the LTE calculation will become the dominant factor. This is the second time the damage curve is exceeded and occurs from about 61 km onwards. Thus even though the fault current is low in this area, the time the conductor is exposed to this fault current will cause the conductor damage.

6.5.1.3 Maximum and Minimum Network Conditions

The current and time parts of the LTE equation (I^2t) have to be evaluated under different network contingencies. As was shown in Figure 6.13, there are two energy peaks when considering the LTE. As the first peak in the radial network application is dominated by the current, this should be evaluated under a high current or then a maximum network contingency. In a radial network only the source impedance has to be changed. For an interconnected network or a network where there is additional generation active, a greater study is required. This is because the current through the conductor section under study has to be maximised. This means all the generation should be in service, and then the normal open points will have to be evaluated to determine their effect on the fault level at the study point. Some generation types can be ignored based on their fault current contribution and the time this fault current is present during a fault. This chapter is detailing overcurrent protection for medium voltage feeders; hence this work is not aimed at earth fault protection. To obtain the maximum fault current when the network is correctly switched (maximum network contingency), a bolted three-phase fault should be used.

In Figure 6.13 there is a second LTE peak further down the feeder at a low fault level. This fault level is low with reference to the protection pickup current value (a protection setting). This second energy peak area should be evaluated under minimum network conditions. Generally, for a radial medium voltage feeder, only the source transformers will have to be switched as they have the largest impact on the source impedance. Similar to maximum network conditions, if the feeder is interconnected or there are multiple points of generation on it, additional evaluations are required to find the minimum network condition. For minimum network conditions a phase-to-phase fault will be used as this will result in the minimum fault current where there is no earth path involved. In addition to using minimum network conditions with a phase-to-phase fault, suitable fault resistance has to be included [23]. Fault resistance values used for these evaluations range from 0 to 40 Ω [4]. When fault resistance is included in the fault level calculation, it will reduce the fault level. For a phase-to-phase fault, this fault resistance is associated with the arc that can exist between the faulted phases. The arc can increase in length over time and as such it is recommended to use the minimum distance between phases or then phase to ground when determining a suitable distance for the arc [23].

6.5.2 Conductor Let-Through Energy Limit Calculations

6.5.2.1 Background

Resistance in terms of Ohm's law is defined as voltage (V) divided by current [31]. Ohm's law take a holistic view on a conducting material. This means that the current passing through a conducting material is dependent on the potential difference between the two ends of the conducting material and then the resistance of this conducting material between the two ends. If the length of the material increases, the resistance will increase. The resistance is thus proportional to the length (L) of the conducting material. When the effect of current within a material conducting the current is considered, it can be defined in terms of resistivity (ρ) [31]. The resistivity of a conducting material can be calculated by dividing the electric field (E) with the current density in a vector format [31]. Current density has units of ampere per square meter. Thus, if the area (A) of the conducting material is increased, the resistivity will decrease if the electric field is kept constant. Conductivity (σ) is a measure of how good the conducting material is (e.g., wood vs. gold as a conductor) and this is the reciprocal of the resistivity [31]. Resistivity is dependent on the type of material used in the conductor and resistance is a property of the conductor itself [31]. The link between resistance and resistivity is defined by Equation (6.13) [31,33,34].

$$R = \rho \frac{L}{A} \tag{6.13}$$

where:

R is the resistance (ohm)
ρ is the resistivity of the material (Ω·m)
L is the length of conducting material (m)
A is the cross-sectional area of conducting material (m^2)

The temperature coefficient of resistance (α) is a measure of how much the resistance of the conductor will change when there is heating in the material [31]. The relationship between the temperature coefficient of resistance and the resistivity of the material for a change in temperature (T) is shown in Equation (6.14) [31,33]. The resistivity of a metal such as aluminium and copper will increase if the temperature increases [31].

$$\alpha = \frac{1}{\rho} \cdot \frac{d\rho}{dT} \tag{6.14}$$

where:

α is the temperature coefficient of resistance (per °C)
ρ is the resistivity of the material (Ω·m)
$d\rho$ is the change in the resistivity of the material (Ω·m)
dT is the change in the temperature of the material (°C)

Equation 6.14 can be rewritten into the form shown in Equation (6.15). In this form, the equation is shown with the initial resistivity (ρ_0), final resistivity (ρ_1), initial temperature (T_0) and final temperature (T_1). If (6.15) is multiplied on both sides with the length (L) and is divided by the area (A), it will result in Equation (6.16). In this equation it can be seen that the resistance of the material will change with a change in temperature [33,34].

$$\rho_1 = \rho_0\left[1+\alpha\left(T_1 - T_0\right)\right] \tag{6.15}$$

$$R_1 = R_0\left[1+\alpha\left(T_1 - T_0\right)\right] \tag{6.16}$$

6.5.2.2 Conductor Short Time Rating and LTE Limit

To determine the conductor short time rating, the heat energy generated by the flow of current has to be set equal to the heat energy absorbed in the conductor. If it is assumed to be an adiabatic process, Equation (6.17) can be used to calculate the short time rating [35].

$$I^2R\cdot t = \left(\frac{W\cdot S}{\alpha\cdot 1000}\right)\log_e\left(1+\alpha\left(T_f - T_i\right)\right) \tag{6.17}$$

where:

- I is the fault current (kA)
- t is the fault duration (s)
- W is the mass of conducting material (kg/km)
- S is the heat capacity of conducting material (J/°C·g)
- R is the resistance of conductor at Ti (Ω/km)
- α is the temperature coefficient of resistance (per °C)
- T_i is the conductor temperature before the fault (°C)
- T_f is the conductor temperature after the fault (°C)

The heat capacity of a material (S) is a measure of the energy required to increase the temperature of a unit mass of the material by 1° [33]. In Equation (6.17), by moving the resistance (R) from the left side of the equation to the right side of the equation, we have the LTE rating of that conductor. From Equation (6.13), we see that, by increasing the conductor size (increase the conducting area), the resistance of the conductor will be reduced. This will reduce the heat being generated in the conductor by the flow of current and hence the short time rating of the conductor will increase. By increasing the short time rating, the permissible LTE rating at that part of the network will be increased. It is thus beneficial to have larger conductor types in high LTE regions on the feeder.

Conductors can consist of one or multiple types of materials. An overhead conductor consisting of only aluminium is called an all aluminium conductor (AAC) [35,41]. If the conductor consists of a steel core, it is called an aluminium conductor steel reinforced (ACSR) conductor [35,41]. Another type of overhead conductor being produced is an all aluminium alloy conductor (AAAC). Both the ACSR and AAAC conductor types have better mechanical strength and can accommodate a greater span length or higher currents [35]. In Figure 6.14 Wasp (AAC) and Hare (ACSR) conductor configurations are shown [41]. It can be seen that the steel forms the core of the ACSR conductor whereas for the AAC it is only aluminium. The Wasp AAC conductor construction is specified by the code 7/4.39. This means that the conductor consists of seven strands, each with a diameter of 4.39 mm. The Hare ACSR conductor is specified by the code 6/1/4.72. The Hare conductor consists of six aluminium strands and one steel strand. All the strands have a diameter of 4.72 mm. If there are more steel strands, the conductor strand configuration can be different, but the steel strands will always form the centre or pass through the centre of the conductor [41]. The steel reinforcement not only assists with the mechanical strength of the conductor but also improves the short time rating of the conductor [35].

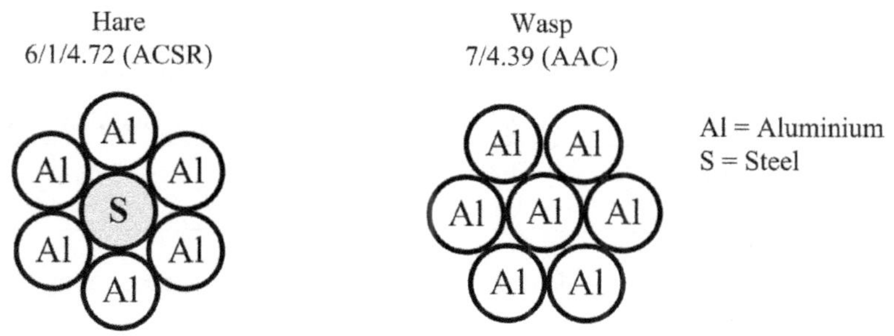

FIGURE 6.14 Hare ACSR and Wasp AAC conductor configurations.

To calculate the short time ratings of conductors, the initial and final temperatures should be known. For the short time ratings of Table 6.2, a prefault (or initial) temperature of 75°C and a final temperature of 200°C are used. The method that is recommended in [35] has made use of an initial temperature of 50°C for AAC and an initial temperature of 65°C for ACSR conductor. The final temperature was set to 200°C for both conductor types [35]. Figure 6.15 illustrates the effect of initial and final temperature choices on the conductor 1 s short time rating. The conductor 1 s short time ratings were calculated using Equation (6.17) on Hare conductor.

For the initial temperature graph of Figure 6.15, the choice of initial temperature was varied from 30°C to 90°C while keeping the choice of final temperature constant at 200°C. It can be seen that the higher the choice of initial temperature, the lower the 1 s short time rating of the conductor (fault current for 1 s). To get the worst case scenario in terms of choice of initial temperature, the highest operating temperature (for load current) of the conductor should be used. For the final temperature graph of Figure 6.15, the choice of final temperature was varied from 170°C up to 230°C while keeping the choice of initial temperature constant at 50°C. It can be seen that the higher the choice of final temperature, the higher the 1 s short time rating of the conductor.

The worst case in terms of protection for the choice of final temperature would be to choose a low value. This final value is governed by the conductor material [35]. By choosing the highest initial temperature and the lowest final temperature, the temperature delta ($T_f - T_i$ of Equation [6.17]) is small and that results in the low 1 s short time rating.

To calculate the conductor 1 s short time ratings, the values listed in Table 6.3 are used for the different material types [35,41]. The resistance of steel conductor used for the calculations is 192 Ω·mm²/km at 20°C [35]. This steel resistance will be divided by the steel cross-sectional area. From the manufacturers datasheets the following information is obtained: conductor stranding configuration, conductor diameter over steel, overall conductor diameter, aluminium area, steel area, the mass of each material type per km, ultimate tensile strength, coefficient of linear expansion, initial modulus of elasticity, final modulus of elasticity, the DC resistance per km and the continuous current rating [41].

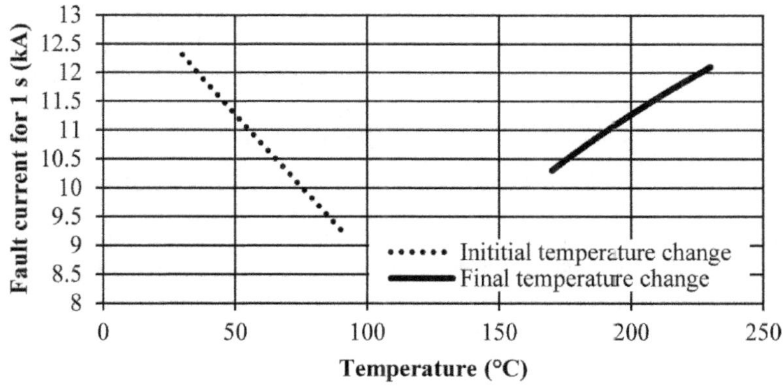

FIGURE 6.15 The effect on the 1 s short time rating of the Hare conductor for choices of initial and final temperatures.

TABLE 6.3

Conductor Material Constants Used for Calculating the Conductor 1 s Short Time Rating

	Copper	Aluminium	Aluminium Alloy	Galvanised Steel
Temperature coefficient of resistance (per °C)	0.00393	0.00403	0.0036	0.00537
Heat capacity of conducting material (J/°C·g)	0.394	0.904	0.904	0.488

All of this information is not used for calculating the conductor 1 s short time rating but is mentioned here to show what is available.

If the conductor consists of only one material type, the conductor material constants can be used, as in for Equation (6.17). But if there is more than one material type used in the construction of the conductor, some of the elements in Equation (6.17) have to be averaged or combined [35]. An example of a multi-material type conductor is the ACSR Hare conductor. The quantities that will have to be averaged or combined are the total weight of the conductor, the heat capacity of the conducting material, the temperature coefficient of resistance and the resistance of the conducting material. Equations (6.18) through (6.21) shows how this can be done for two material types in a conductor [35]. The subscript of 1 and 2 for the variables in Equations (6.18) through (6.21) are for the two different materials that the conductor consists of.

$$W = W_1 + W_2 \tag{6.18}$$

$$S = \frac{W_1 S_1 + W_2 S_2}{W_1 + W_2} \tag{6.19}$$

$$\alpha = \frac{A_1 \alpha_1 + A_2 \alpha_2}{A_1 + A_2} \tag{6.20}$$

$$R = \left(\frac{1}{R_1} + \frac{1}{R_2} \right)^{-1} \tag{6.21}$$

where:

W is the mass of conducting material (kg/km)
S is the heat capacity of conducting material (J/°C·g)
R is the resistance of conductor at T_i (Ω/km)
α is the temperature coefficient of resistance (per °C)
A is the cross-sectional area of conductor (mm^2)

In Equation (6.18) the total weight of the conductor has to be determined and this is simply the sum of each material type used in the conductor. The heat capacity of the conducting material (S) is averaged by the ratio of the mass of each type of material to the total mass of the conductor. In Equation (6.16) it was shown that the resistance is a function of temperature. In Equation (6.13) it can be seen that the resistance is influenced by the conductor cross-sectional area. The temperature coefficient of resistance is averaged by using the ratio of each material type's area to the total cross-sectional area of the conductor. The conductor materials can be considered as resistors in parallel as each material type is parallel to each other along the length of the conductor. Thus the resistance of the total conductor is determined by means of the resistors in parallel calculation in Equation (6.21). The averaging techniques shown in Equations (6.18) through (6.21) can be expanded to more than two conductor material types. Table 6.4 shows the conductor information that is used for the calculation of the 1 s short time rating of ACSR Hare conductor.

By applying Equation (6.18) to Equation (6.21) the following results are obtained. The total mass of the conducting material (W) calculates to 427 kg/km. The heat capacity of the conducting material calculates to 0.7696 J/°C·g. The temperature coefficient of resistance calculates to 0.00422 per °C. The resistance at 20°C calculates to 0.2667 Ω/km. This resistance is now moved to the initial temperature of 75°C by

TABLE 6.4

Hare Conductor Information from the Manufacturers Datasheet

	Area	Mass	Temperature Coefficient of Resistance	Heat Capacity	Resistance at 20°C
	mm²	kg/km	per °C	J/°C·g	Ω/km
Aluminium	104.98	289	0.00403	0.904	0.2733
Steel	17.5	138	0.00537[a]	0.488[a]	10.971[a]

Source: Babu, P. S. et al., *Int. J. Energy Sci.*, 1, 72–77, 2011.

[a] These values were obtained from [35].

using Equation (6.16) with the average temperature coefficient of resistance. The resistance value calculates to 0.3286 Ω/km at 75°C. All of these values are now applied to Equation (6.17) to calculate the conductor short time rating when the conductor temperature increases from 75°C to 200°C. This calculation results in an LTE rating for ACSR Hare conductor of 100.39 MA^2s. To calculate the 1 s short time rating for ACSR Hare conductor, the square root of the LTE rating has to be divided by 1 s. Thus, the 1 s short time rating for ACSR Hare conductor calculates to 10.02 kA. An example of calculating the 1 s short time rating is shown at the end of the chapter.

6.5.3 Network Layout

There are two network layouts or configurations to be considered. The first being a radial medium voltage (MV) feeder and the second being a multisource MV feeder. In this part of the chapter a radial network will be considered; later in this chapter the concept will be applied to a multisource feeder.

6.5.3.1 Radial Medium Voltage Network

Traditionally, MV networks are designed to be radial in nature. Radial means that there is only one source active at a time and the current will flow (or then radiate) from this source to the various loads on the feeder. There can be multiple or even just a single load on the MV feeder. Due to the current flow (fault current) only being in one direction from the source to the fault, the overcurrent protection applied to these types of feeders is usually nondirectional current based protection. To isolate a fault, only one circuit breaker has to operate. This radial network with current-based protection is a simplistic and relatively inexpensive design [23]. Figure 6.16 shows a simplified radial MV feeder with a number of

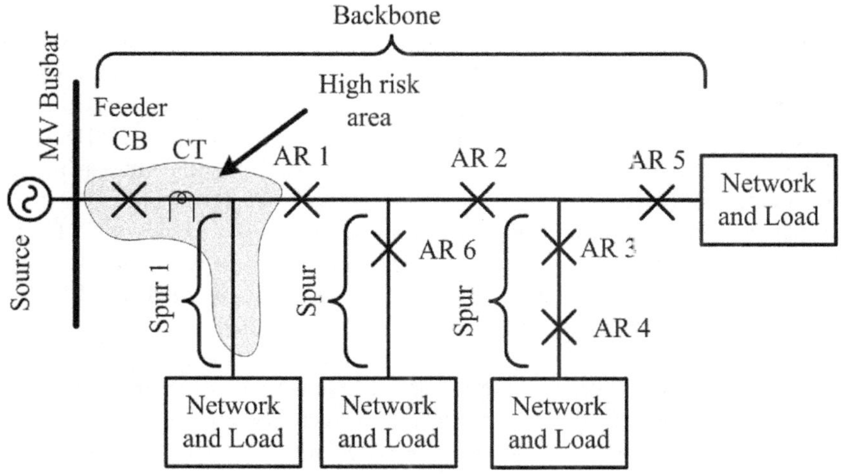

FIGURE 6.16 A simplified radial medium voltage feeder.

auto-reclosers (AR) on this feeder. All of the loads are not shown. The low voltage (LV) customers are normally supplied via Dyn-vector group transformers that are earthed on the secondary side.

For the MV feeder shown in Figure 6.16, there is a substation-based feeder circuit breaker (CB) with a substation-based current transformer (CT). This can be replaced by an AR. For a radial feeder, there is no need to have a large backbone conductor throughout the length of the feeder as bidirectional power flow and fault current flow are not possible. A large backbone conductor will only assist with reducing losses and for future expansion on the feeder. However, a larger conductor is required close to the source where the conductor can get damaged. This is a high risk area. The conductor damage is due to high fault current and slow protection operating times close to the source for the benefit of coordinating protective devices [42]. This is evident in the high LTE peak close to the source as shown in Figure 6.13. From the backbone conductor there are spurs radiating from the conductor to the various loads. These conductors can be much smaller in size based on the load they carry. If the spur conductor is located in a high-risk part of the network (such as close to the source), then the spur conductor size should be increased as this conductor can also get damaged by the LTE. If we consider the first spur from the source side in Figure 6.16, this falls within the high-risk area. If there are faults on this spur, the feeder breaker is required to clear or isolate the fault and this will influence all the customers on this feeder.

The high-risk area in terms of LTE damage can also be downstream in the network towards the end of each circuit breakers reach. In Figure 6.16 this can be towards AR 4 and AR 5. At these positions, the fault level to pickup ratio might result in slow operating times (also shown in Figure 6.13). The high-risk area is not only for the conductors in this region, but for all the conductors that carry the fault current up to the fault position. The high-risk area defines the region where a fault can cause conductor damage and not where the conductor will get damaged necessarily. For a radial feeder, the operating time of the protection can be calculated using Equation (6.3) and then the circuit breaker operating time can be added to the operating time [10]. The result is the total fault current exposure time.

6.5.4 Curve Selection

The advantage of IDMT curves are that they trip fast during high fault currents and slow during low fault currents, but the operating time can still be too long and result in thermal damage and mechanical stress to equipment [15]. Choosing a good curve can minimize these effects. As discussed earlier in this chapter, LTE consist of a current and time component. From a protection perspective, the exposure time is the only part that can be influenced without switching a working plant out of service. The fault clearing time is dependent on the protection operating time (if we neglect circuit breaker operating time). The time the protection relay takes to issue a trip signal to the circuit breaker is dependent on the fault current, the pickup current, the operating curve and the time settings related to the operating curve.

The NI, VI, EI, and DT curves are shown in Figure 6.17 (similar to Figure 6.5b with different settings) over the same current range. In the figure all of them are illustrated with the same pickup current (350 A) and time multiplier (0.25) applied. The DT curve has the same pickup current but with a time delay of 0.25 s. In Figure 6.17a it can be seen that the EI curve results in the fastest operating time for high fault currents and the slowest for low fault currents. These low and high currents are with reference to the pickup current of the protection element. The opposite is happening to the NI IDMT curve. Thus, the more inverse curve (compared to a NI curve) will result in faster operating times at high currents and then slower operating times at low currents.

In Figure 6.17b the LTE curves are shown for those operating curves in Figure 6.17a. The LTE limit or damage curve for Rabbit conductor is also shown. The best curve to use where the fault current is high is the EI curve. Not only does it operate the fastest, but it also keeps the conductor LTE exposure to a minimum. Another advantage of the EI curve is that it is almost parallel to the Rabbit conductor damage curve. Both curves are horizontal for the better part of the fault current range. Thus, it grades well with the damage curve due to it following an I^2t profile [5]. The drawback of the EI curve, however, is that there is a rapid increase in the operating time as the fault current is decreasing. This increase in time can be used when a number of devices are to be time graded that are electrically (impedance) close to each other. The NI curve does not grade as well as the EI curve with the conductor damage curve. But the NI curve does provide a horizontal profile when the operating time is considered (as the fault current drops).

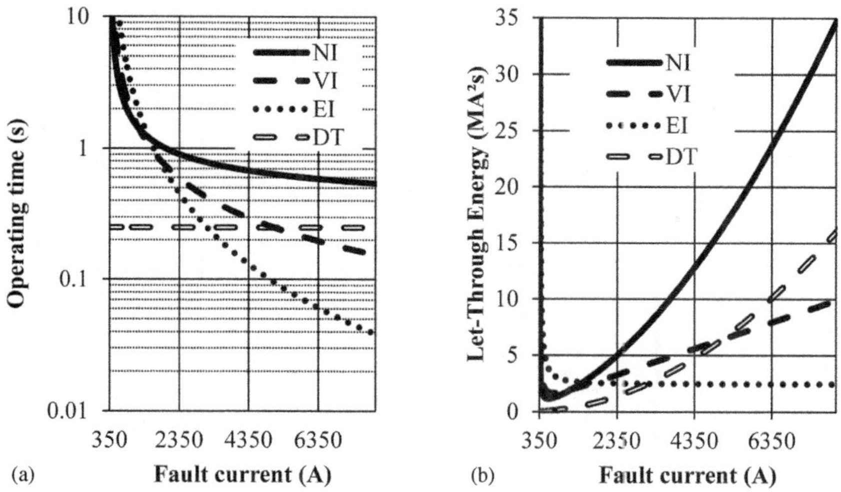

FIGURE 6.17 (a) Standard IEC IDMT operating curves and (b) LTE curves using different IEC IDMT operating curves.

There is a sudden and fast change in the LTE as for all the IDMT curves when the ratio of fault level to pickup current approaches the value of 1. From this, the curve that provides the best performance in terms of LTE protection is the EI IDMT curve [22]. It might not always be possible to use this curve when considering the grading requirements and the maximum time a fault is on the feeder (low fault currents).

The DT curve operating time is immune to the drop in fault level. Its operating time stays constant for as long as the fault current is larger than the pickup current. This can also be seen when looking at the LTE curves of Figure 6.17a as the operating time does not increase when the ratio of fault level to pickup current approaches the value of 1. The DT curve can be applied as an instantaneous element or high-set element if the time delay is set to 0. Under very high fault currents, even this element might not be able to protect the conductor from damage as the circuit breaker operating time has to be considered. The change in LTE for the DT curve is due to the change in fault current across the feeder.

For combination curves, the most commonly used and best example is a high-set curve on top of an IDMT curve. This high-set curve will reduce the LTE dramatically in the high fault current region close to the source busbar. Care should be exercised when choosing the high-set pickup value as this element is susceptible to transient overreach [5,25,26].

6.5.5 The Effect of the Auto-Reclose Function

There are many theories on how many ARC cycles should be applied. If the objective is to keep the LTE exposure to a minimum, then no ARC cycles should be applied. If the objective is to maximise the availability of the feeder, then the maximum number of ARC cycles should be applied. This shows that some scientific method should be used to determine the number of cycles to apply. The probability of having a successful ARC attempt decrease with the increase in ARC cycles.

6.5.6 A Radial Network

The approach is that during the first ARC trip, the conductor will be heated by the flow of fault current. With the dead time being short, the assumption is made that the conductor does not lose significant heat energy [22]. This means that for the second shot, the LTE is added to that of the first ARC trip. To ensure that the conductor does not get damaged, this total energy should be less than the conductor LTE limit. In Figure 6.18 the LTE is shown for a one and two trip to lockout ARC philosophy on a radial network (using Rabbit conductor). The protection settings that were applied to the relay are a NI IDMT curve, pickup of 270 A and a TM of 0.42. This results in an operating time

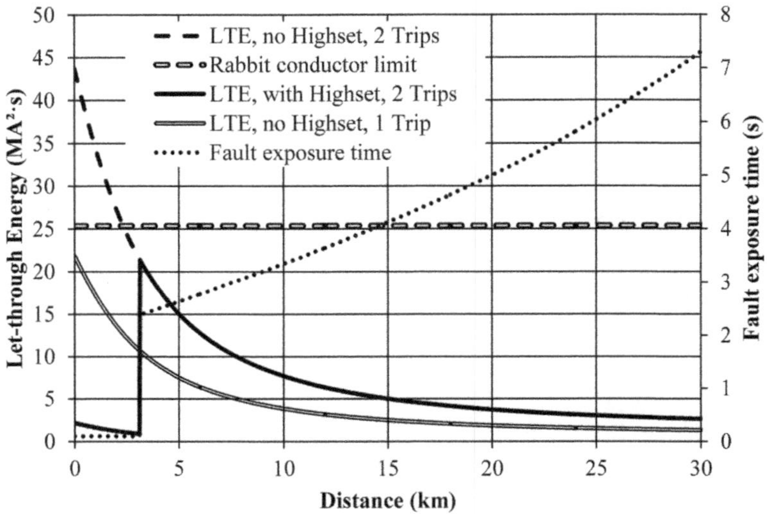

FIGURE 6.18 The LTE exposure on a radial feeder with a one and two trips to lockout ARC philosophy.

of roughly 1 s for the IDMT element for a fault close to the busbar. A high-set element with a pickup current of 3000 A and a circuit breaker operating time of 0.05 s was also applied to the NI curve with two trips to lockout (one ARC cycle).

For the LTE curve where there is only one trip to lockout (LTE, no high-set, 1 trip), it can be seen that the LTE exposure never exceeds the Rabbit conductor limit (conductor damage curve). When the protection is set to apply one ARC cycle (LTE, no high-set, 2 trips) it can be seen that the conductor limit will be exceeded for a fault within the first three kilometres from the busbar. It is recommended to use the ARC function as this improves the network availability. Figure 6.18 shows that the extra trip (due to ARC) will result in a conductor being damaged in this case. The application of the high-set function will reduce the operating time and improve LTE exposure for a fault close to the busbar. It is thus recommended to always try and apply a high-set function where possible and set this high-set function to initiate auto-reclosing of the circuit breaker. A high-set function that initiates ARC will reduce the time exposure of the conductor to the fault, keep the LTE exposure low in the high current region, improve the network availability for transient faults, improve power quality (voltage dip) and maintain selectivity if the pickup is chosen correctly [22,26]. For the LTE curve in Figure 6.18 it can be seen that the high-set reduced the LTE exposure peak from 45 MA^2s to 21 MA^2s.

If the fault level is very big, even the high-set element with auto-reclosing applied might not be able to protect the feeder conductor as the circuit breaker does still take some time to operate. In such a case the high-set should still be applied, but another function can be applied on top of this high-set function, and it is called a high current lockout function. It is similar to the high-set function except that it will not initiate an ARC cycle. This high current lockout function has a higher priority than the high-set and will trip rather than the high-set if the fault current is big enough to initiate the tripping sequence of both elements.

The LTE exposure at the end of the feeder should also be evaluated. This should generally be done using a phase-to-phase fault with some fault resistance included. Using this fault type with fault resistance will result in a minimum amount of fault current. This will lead to a maximum time exposure and LTE exposure for the conductor at low fault currents.

6.5.7 Energy-Area Value

It is difficult to quantify the effect of small protection settings on the effect of LTE. As an example, one can change the TM from 0.3 to 0.31, but the effect of the change in operating time might not be known. To assist in quantifying this change, the area under the LTE curve is determined [22]. This is calculated by applying a Riemann calculation to the LTE curve [43]. Figure 6.19a shows this energy-area

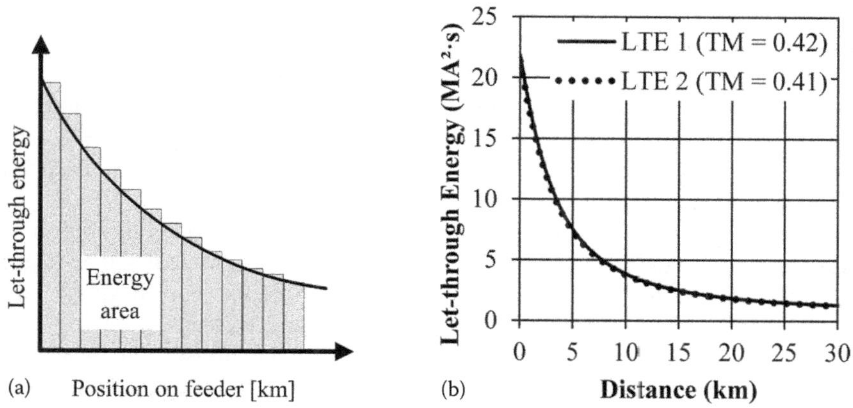

FIGURE 6.19 (a) The energy-area concept illustrated and (b) an actual application of the energy-area calculation on two LTE curves.

calculation; in Figure 6.19b it is applied to an actual LTE curve. For the two LTE curves in Figure 6.19b the pickup is the same. LTE 1 has a TM of 0.42 and LTE 2 has a TM of 0.41. The two curves are almost on top of each other. When applying the energy-area calculation to both LTE curves, the difference between the curves can be quantified. LTE 1 calculates an energy area of 133.69 $MA^2s{\cdot}km$ and LTE 2 calculates an energy-area of 130.51 $MA^2s{\cdot}km$. This illustrates that the risk is reduced when applying the lower TM in LTE 2. Using the Riemann calculation does introduce some error in the calculation, but for more complex networks such as interconnected or multisource networks the curve might not resemble a simple polynomial, especially when different curves with different ARC settings are introduced.

6.6 Grading

Grading refers to the intentional introduction of additional time and current steps to series protective devices to ensure that they are trip selective. This process of grading the protective devices does help to promote the continuity of supply in that the smallest portion of the network (faulted section) is removed and the remainder of the network can then be kept live. If there is no selectivity, a larger portion will be influenced, and this can negatively impact revenue streams for both the utility and the customers, impact the image of the utility, allow for network equipment to be stolen or vandalised while deenergised, increase fault finding time as a bigger area is considered and increase the risk to personnel and the public due to a prolonged fault-finding period. A graded network can also have negative aspects, especially if there are a large number of devices that must be graded. The sensitivity of the protection elements can be impacted with the pickup becoming too sensitive or insensitive depending on how the pickup currents are determined. With a large number of devices in series, the protection can become too slow [24], thus failing to protect the network equipment from damage and increasing the risk in the network due to long exposure times.

Grading should be done in both the current and time domain. Current grading is required to ensure that two devices maintain selectivity as a result of relay errors and then the relay pickup/drop-off ratio [5]. When it comes to the error, the upstream device can have a negative error (pickup at lower value) and the downstream device a positive error (pickup at higher error). This means that if the two relays were set to the same pickup current, the upstream device can detect the fault and the downstream device can fail to detect the fault, leading to a loss in coordination (selectivity). The pickup/drop-off ratio refers to the current value at which the relay is picking up and then at which current level the relay stops being picked up if the fault current drops prior to a trip being issued. As an example, the relay can be set to pick up at 100 A, but might only drop off at 98 A. Again, this can lead to a loss of coordination between series protective devices [14]. To accommodate the relay errors (all technology types) and the pickup/drop-off ratio, a current grading margin of 10% between series protective devices is recommended [5,6].

For time grading, provision should be made for circuit breaker operation, current and voltage transformer errors, relay measuring and timing errors, and relay overshoot [5]. Relay overshoot refers to the time it will take before the relay responds when the fault current has dropped below the pickup value (dropped off). For a numeric relay, it can be the result of capacitance that has to discharge; for an electromechanical relay, it can be the result of kinetic energy stored in the rotating disk [5]. To accommodate errors a grading margin of 0.4 s is recommended [5]. For static relays this can be reduced to 0.35 s, and then digital and numeric relays a value of 0.3 s can be applied [5].

To determine the minimum grading margin between two IDMT curves, the golden rule of grading can be used. This rule will provide the current where the minimum grading margin will be. The golden rule of grading states: Use the current where both devices will measure the same maximum current or the upstream device measures more. Figure 6.20 shows the golden rule of grading.

Let's consider the first part: use the current where both devices will measure the same maximum current [4,14,23]. This is illustrated in Figure 6.20a with two relays in series (Relay A and Relay B) and two possible fault positions (X and Y). When the fault is placed at position X, the current that both devices will measure will be a maximum and they are also measuring the same maximum. This means that the upstream relay (Relay A) will operate at its fastest. When looking at the grading curves for this scenario (Figure 6.20a), the minimum time difference (grading margin) between the curves will be at the high current end of the grading graph. As the fault current reduces, the grading margin will increase. This is illustrated in Figure 6.20a by the 2 A and 1 A fault levels. Thus when evaluating the grading, it should be done at this maximum current and not at position Y in Figure 6.20a.

The second part of the golden rule of grading (Figure 6.20b) states: use the current where the upstream device measures more. The downstream Relays D and E should grade with the upstream Relay C. The minimum grading margin will be obtained when the network contingency is with both Relay D and Relay E in service. The upstream Relay C will measure a current that is more than 1 A passing through Relay D and Relay E. This means that Relay C will operate fast and Relay D and E will operate slow, thus resulting in a minimum grading margin.

There are various methods of grading a network. Four methods are available:

- A top-down method (Figure 6.23a and b).
- A bottom-up method (Figure 6.26a and b).
- A course grading method (Figure 6.27).
- Natural busbar grading with infeeds (Figure 6.22).

Figure 6.21 shows four protective relays in series after the source protection (Relay S) at the busbar. For all the grading methods the first step is to determine the pickup of the protection element. Once this is done, the time grading can be started. The basic principle of the top-down grading method is to start

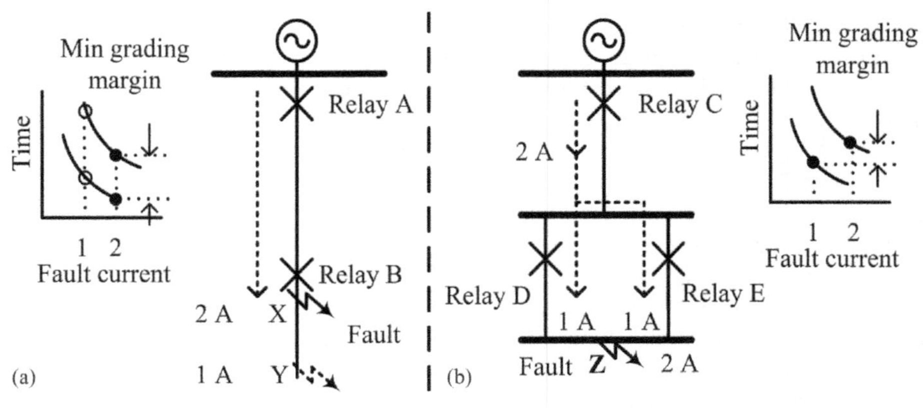

FIGURE 6.20 (a) The golden rule of grading with both devices measuring the same maximum current and (b) the golden rule of grading with the upstream device measuring more current.

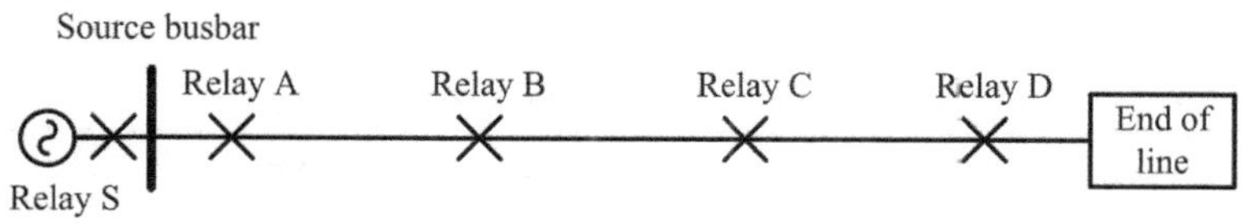

FIGURE 6.21 A radial network used illustrating the grading methods.

at Relay A and then ensure that it is grading with the source relay by placing the fault at the Relay A position (in front of Relay A). Determine the operating time of Relay S for this fault and then subtract the grading margin. This is the required operating time for Relay A. Once Relay A is set, then the same concept can be applied when grading Relay B and Relay A with the fault placed at the Relay B position.

The bottom-up grading method starts at Relay D by setting the operating time of Relay D to a specific value for a fault in front of or at Relay D's position [23,24]. The upstream relay (Relay C) is then be set to operate slower than Relay D for a fault at Relay D's position. The operating time of Relay C is thus set to that of Relay D, and added to this is a grading margin. Now the same principle can be applied when determining the settings of Relay B, up to Relay A, provided that Relay A can grade with the source relay. The operating time is thus set (chosen) at the remote end of the feeder, and the protection device will trip faster as the fault is getting closer to protective device being set. For the top-down method, the grading is set at the device, which results in slower tripping times.

The course grading method starts at Relay A and then moves downstream in the network. The course grading method is the same as the top-down method for Relay A where the upstream operating time is determined for a fault at Relay A and a grading margin is then subtracted. For the devices downstream from Relay A it is different. For Relays B, C and D, the fault level at Relay A is used. The operating time of Relay B is set by using the fault level of Relay A. The same concept is applied between Relay B and Relay C, but still with the fault level at Relay A. This means that due to the actual fault level dropping as you move further down the feeder, the actual grading margin between the downstream devices (from Relay A) will always be more than the set value. The same concept of Relay B and Relay C is then applied between Relay C and Relay D with the fault level at Relay A. This course grading method is suited for new protective devices that are added to an existing network; thus the device will be sensitive to faults as the original feeder protection (Relay A) was set using the required criteria. For this course grading method, the high-set element cannot be applied as the actual fault levels in the network are required to ensure the high-set element is selective and secure.

The natural busbar grading with infeed's method is a method that is applied to transmission networks where there are multiple sources and the network is interconnected. The concept will be explained but this method is not used very often on medium voltage radial networks. As the medium voltage grid is developing into an interconnected network with distributed generation, this method can become useful. The network applicable to this concept is illustrated in Figure 6.22. On each of the circuits, there is a contribution of 100 A to a busbar fault (100 A for illustration); thus the total busbar fault current is 500 A. A fault in front is defined as a fault on the equipment that the protection is normally protecting (or in that direction). If we consider Relay A, a fault in front is at position Y, and a fault behind is at position X. If the fault is at Y, Relay A will operate very fast as it will operate on all the contributions from the other

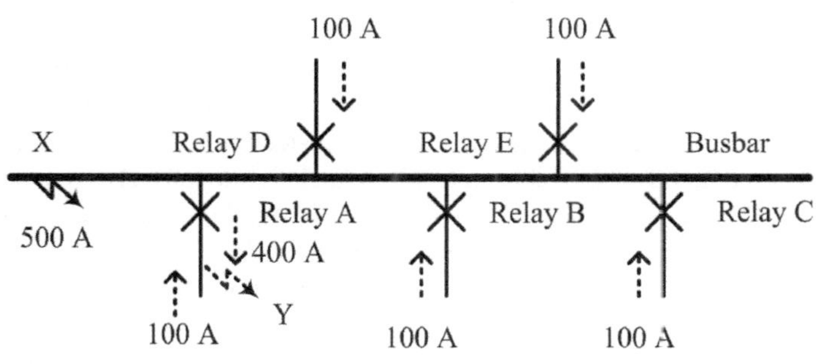

FIGURE 6.22 The natural busbar grading method with infeeds applied to transmission networks.

TABLE 6.5

General Comparison on Grading Methods for Medium Voltage Feeders

Field for Comparison	Top-Down	Bottom-Up	Course Method
Speed in clearing a fault	Average	Fast	Slow
Ease of adding additional protective devices	Fair (depends on position)	Tedious	Easy
Biggest grading margin position	At the last device	At the feeder circuit breaker – good for selectivity	Devices beyond the feeder circuit breaker
High-set element	Yes	Yes	No
Check for maximum operating time exceedance	Yes	Yes	No
Time required to do grading	Large	Large	Small
Complexity in grading	Complex	Complex	Simple

relays, which totals then 400 A for this example. All the other circuits will operate slowly because they see the fault behind them and they will only operate on their own contribution to the fault, this being 100 A. This means that by default there will be grading on the busbar due to the infeeds from the other circuits. Grading between different substations is not required due to the fast main protection (impedance or differential), but this can be done if required. This grading method will not be used for medium voltage.

A general comparison of the grading methods is shown in Table 6.5.

6.6.1 Top-Down Grading Method

The top-down grading method is shown in Figure 6.23. These steps should be taken, and they can be programmed onto a computer to generate protection settings for protection relays. The step numbers are shown in circles in the flow diagram.

In step 1 the IDMT operating curve should be selected. It is preferable to select curves that are similar to the ones that should be graded [5,23]. The operating curves were covered earlier in this chapter. In step 2 the IDMT element pickup should be determined. There are three items to consider for this [6]:

- It should be set above load current.
- It should be sensitive to faults.
- It should allow for current grading with the upstream device.

First, the protection should be set above load current [4,5,8,14,18,23]. Load current resembles positive sequence current; if the phase overcurrent protection pickup is set less than this, the protection can operate for load current. A method of estimating the load current is to make use of the conductor ratings. If errors such as modelling, measuring and relay errors are considered, a 20% value can be added to this conductor rating. This also assists with not exceeding the conductor continuous current rating. Second, it should be sensitive to faults. In medium voltage networks there is only overcurrent protection devices due to its simplistic nature and because it is relatively inexpensive. Thus, there is only one overcurrent relay (or auto-recloser) at each position on the feeder. To provide backup to downstream devices, the upstream device should back up all the immediate downstream devices (overlap protection) [5,44]. This should be determined in minimum network conditions for a phase-to-phase fault [4,23,45] because the protection device can fail and require maintenance (device out of service then). The pickup under this criterion should not be set greater than 80% of this phase-to-phase fault to allow for errors. It should be noted that there can be a conflict between setting the pickup above load and being sensitive to a fault when determining the pickup current [4]. Negative sequence current element can assist in such a case to improve sensitivity to phase-to-phase faults. Determining the sensitivity of protection relays for a medium voltage network is illustrated in Figure 6.24.

The backup concept is that the upstream relay should provide backup protection to all immediate downstream relays on the feeder [4,23]. For Relay A the immediate downstream device is only Relay B,

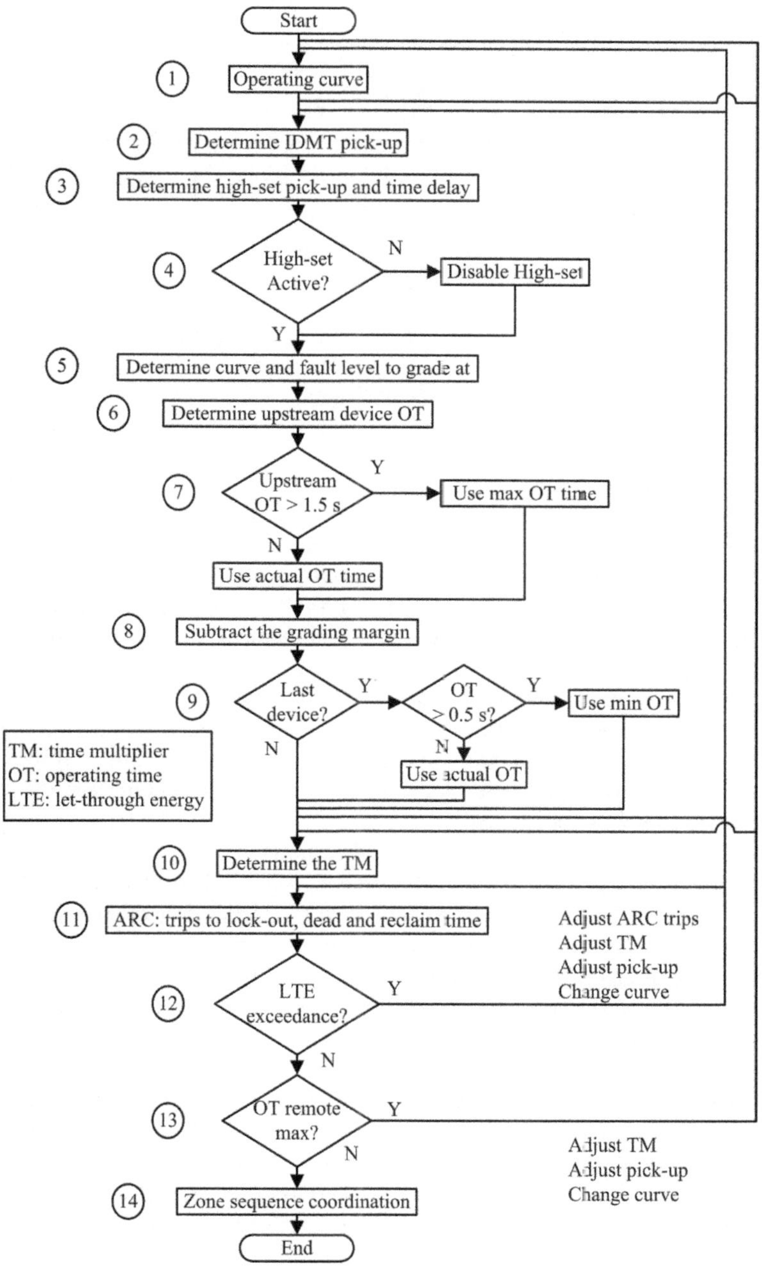

FIGURE 6.23 The top-town grading method (steps in circles).

and this relay will be bypassed when determining the sensitivity (to provide backup). Relay A is not required to be sensitive to faults beyond Relay C and Relay E. The sensitivity criteria calculates to 80% of the 2500 A fault level at Relay E. There is a practice where Relay A is set to be sensitive to faults up to the lowest phase-to-phase fault level on the feeder. This approach improves the dependability of the protection system towards faults as the feeder relay (Relay A) is able to detect (and is sensitive to) all the faults on the network irrespective of the number of auto-reclosers in service downstream of the relay. Thus, the probability of the fault not being cleared is reduced. The drawback of this approach is that it impacts the security of the protection system. The probability of the protection operating for load current increases, especially if this is a long feeder with many protection devices in series. This is because each device is

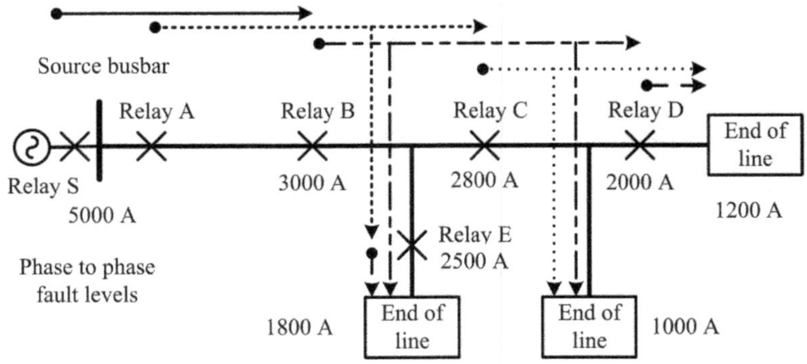

FIGURE 6.24 Determining the sensitivity of the overcurrent pickup.

set to 90% of the upstream device pickup current (maximum) to obtain current selectivity. By applying a one-over philosophy, backup of downstream devices is achieved for a maximum of one device (N-1) on bypass or out of service. Then the protection is not too sensitive to load but is still sensitive to fault current. Relay B has to provide backup to both Relay C and Relay D; thus both of them will be bypassed. To determine the pickup sensitivity, the 1800, 2000, and 1000 A positions is considered. The lowest one of these is used; thus all the other are then also covered. This same concept is also applied to Relay C and then Relay D. Finally, to determine the pickup current, the lowest value of the three criteria should be chosen. In step 3, the high-set settings are determined. As the high-set resembles a DT operating curve, it requires a pickup and a time delay setting (see high-set section). Normally the time delay is set to zero seconds. For step 4, the validity of the proposed high-set element is determined. If the high-set element pickup is greater than the maximum fault level at the relay position (one being set), then the high-set element can be disabled as it will not operate.

The curve and fault level at which to grade at is determined in step 5. If the IDMT element is only active (no high-set), then the golden rule of grading can be applied. Two applications of this is discussed. First, grading between the source transformers and then the feeder protection at the substation will be explained. With two or more transformers at the source, the same maximum current that both devices (source and then feeder protection) will measure is with the minimum number of transformers in service at the source. This increases the current per transformer, thus speeding up the protection operating time. If grading is done using minimum transformers at the source, then the grading margin increases if more transformers are added with the settings applied that was determined with minimum number of transformers in service. Thus, selectivity between the transformers and the feeder protection is maintained. If this was done incorrectly by using all the transformers in service, the grading margin decreases if one or more transformers trip, thus losing selectivity. The second application of the golden rule of grading is when two devices are graded on the feeder. An example is grading Relay B with Relay A in Figure 6.24. The same maximum current that the devices will measure is when all the source transformers are in service, and then the grading margin will be at its minimum. If one transformer is lost or out of service at the source, the grading margin between Relay A and Relay B increases, thus maintaining selectivity.

When a high-set is also applied to the IDMT curve, there is the benefit of grading at the high-set pickup and not the IDMT element pickup. This helps in creating more grading time for more devices to be placed in series. When determining the grading current, the complete operating curve has to be considered to determine which element (or curve) will be active during the required network condition. Figure 6.25 illustrates this for a curve where an IDMT and high-set element are graded at the substation medium voltage busbar. This should be done with minimum transformers in service. For the network in Figure 6.25 one of the two transformers are taken out of service. Once this is done, the fault current that the feeder relay will measure has to be compared to the high-set pickup current. If the fault current is greater than the high-set (B-side), then grading can be done at the high-set pickup. If the measured fault current is less than the pickup, grading will be done at the normal fault level on the IDMT curve (A-side).

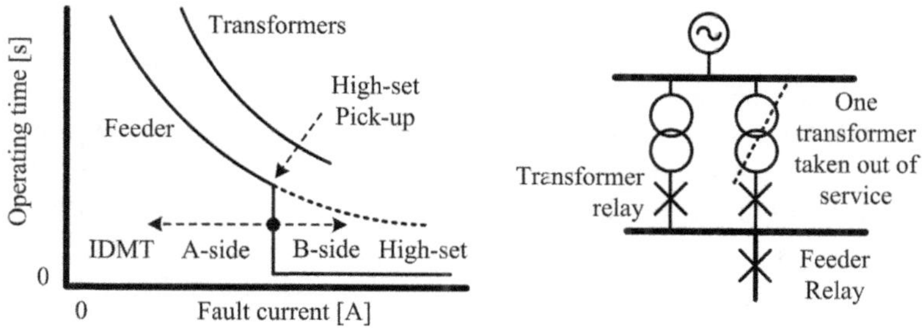

FIGURE 6.25 The network and operating curves for grading the source transformers and feeder.

The upstream protection operating time has to be determined next in step 6. This is done using the grading current from step 5 and the protection settings of the upstream protection elements. This operating time will be used in step 7 to test if it is above or below a preset value. In the flow diagram of Figure 6.23 a value of 1.5 s is used, but this can be any value. The reason for this test is to keep the operating time of the upstream device to a required minimum, thus improving the speed in clearing a fault and reducing the risk in the network. The upstream protection can reach an operating time of infinity as the ratio of fault current to pickup current approaches a value of 1. As an example, if the upstream operating time calculated to 15 s and the grading margin of 400 ms was subtracted from this, the required operating time for the downstream device would be 14.6 s. This would leave the fault on the network for too long a period. Thus, testing for an acceptable limit such as 1.5 s would result in a maximum operating time of 1.1 s for the downstream device. The actual value to be used can be determined by the user (1.5 s is a good value used as an example). This is for a fault at the downstream device position. In step 8, the grading margin is subtracted from the upstream protection operating time. In step 9, a test is done to determine the minimum operating time for the last device. For the opt down grading method, the largest grading margin would be potentially at the last device due to the operating curves being slotted in underneath each other starting at the source. The same argument as in step 7 is used where it is not required to leave the fault on the network for extended period. For this example, a value of 0.5 s is used to explain this concept. This value can be reduced or increased based on specific applications. Once the last device maximum operating time is determined in step 8, it can be compared to this user definable maximum time (0.5 s in this case). If the operating time is more than 0.5 s, then use the value of 0.5 s, else use the maximum time as was determined in step 8. The 0.5 s used here allows another protective device to be added below this last device if required later (with some grading margin). The complete feeder setting does not have to be revised then for this new last protection device. The required TM for the IDMT operating curve can then be determined in step 10. The required ARC cycles with the associated dead and reclaim times are set in step 11. At this point most of the key settings for the current-based protection elements are set, and the settings can be evaluated to ensure that the equipment such as the conductors is actually protected in step 12. Thus, an LTE evaluation has to be conducted next to ensure the conductor energy exposure is less than the conductor limit. If the conductor LTE exposure is exceeded, the following items can be changed to reduce the exposure: The TM can be reduced, the pickup can be decreased, the operating curve can be changed and the number of ARC cycles can be adjusted.

The fault current can also be reduced if operating instructions are put in place that reduces the number of transformers that are allowed to operate in series, as an example. This is not the best solution, however, and should be the last resort. In step 13, the maximum time a fault is on the network is determined. This is to ensure that the fault is not present for extended periods as this increases the risk in the network. Risk includes fire risk, damage to equipment, the potential to destroy life etc. A value of 5 s is used as the maximum time in this example [36]. A utility can determine a maximum time value based on its protection philosophy. This time should be determined in minimum network conditions with a phase-to-phase fault (for phase overcurrent protection) to increase the operating time of the IDMT elements when evaluating this step. If the operating time exceeds the maximum time, a combination of the following can be changed: The TM can be reduced, the pickup current can be decreased, and the operating curve can be changed.

A maximum time element (DT curve) can also be applied to this IDMT element to cap the protection operating time for remote faults. This method will also assist for faults where fault resistance can increase the operating time and does not require the normal IDMT settings to be revised. Finally, in step 14 the zone sequence coordination function should be enabled or disabled depending on the type of curves used for each of the ARC trips between different protective devices. If different curves are used for each of the ARC cycles, then grading on the other curves should also be done following the same flow diagram in Figure 6.23.

6.6.2 Bottom-Up Grading Method

The bottom-up grading method is shown in Figure 6.26a and b. Steps 1 to 4 in the bottom-up grading method are the same as those in the top-down method. In these steps the operating curves are selected, the pickup is determined and the settings related to the high-set element are determined. These settings are not related to the grading method as the sensitivity, security and current selectivity are set in these steps.

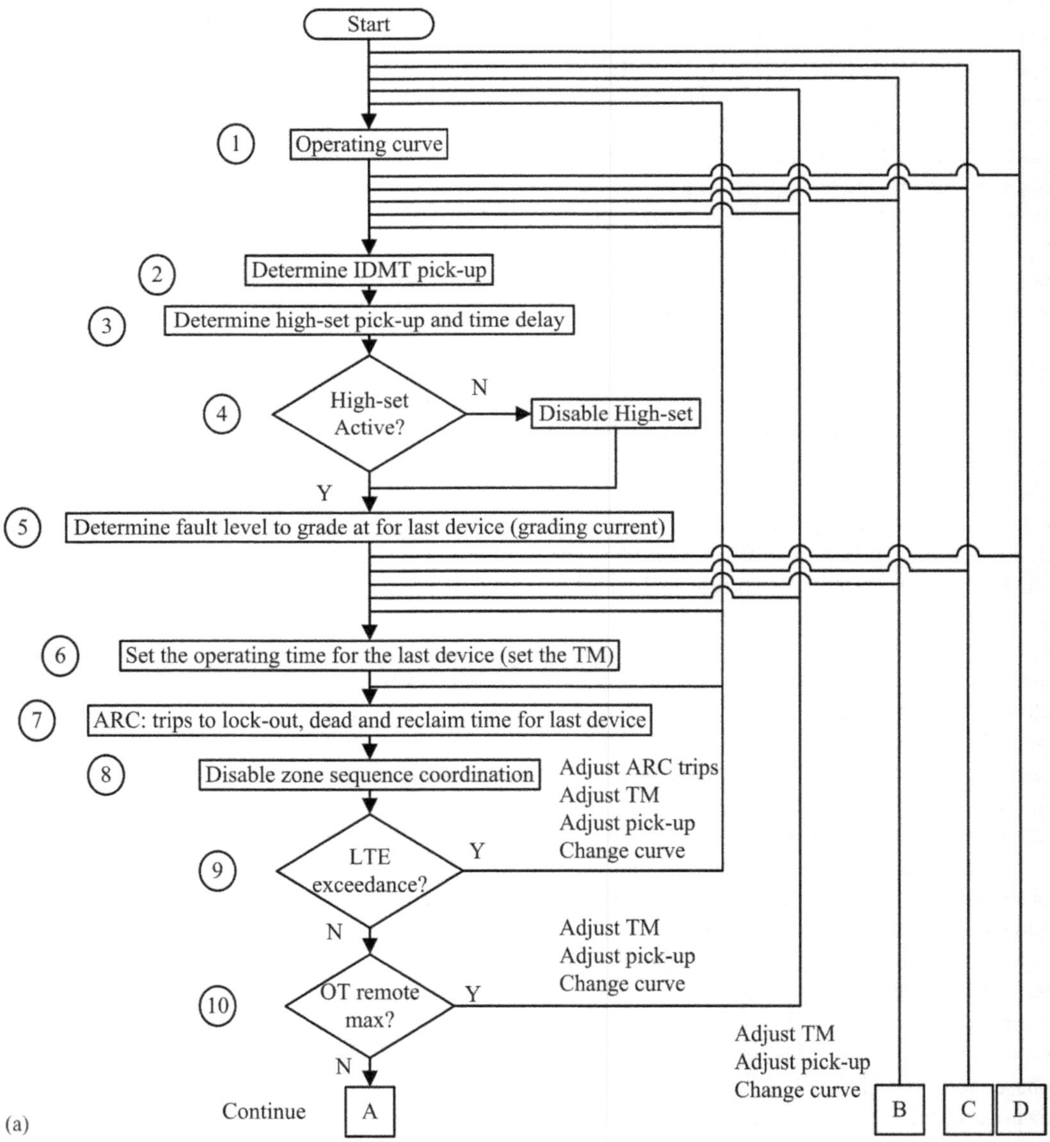

FIGURE 6.26 (a) The bottom-up grading method (steps in circles). *(Continued)*

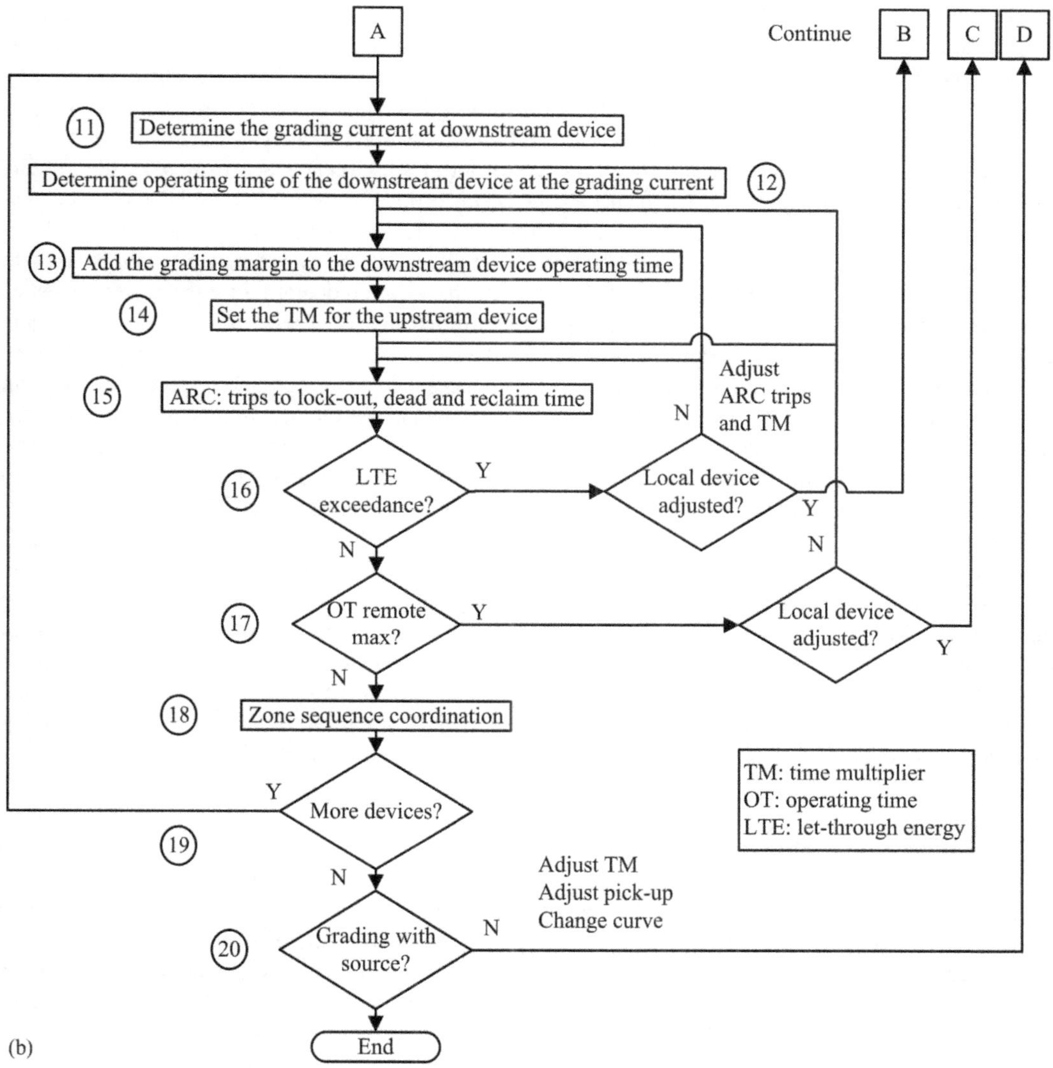

FIGURE 6.26 (Continued) (b) The bottom-up grading method (continued).

The grading current has to be determined in step 5. For this method, time grading is done starting at the last. This last device should be for the greatest number of protective relays in series. For the last device, the grading current used can be either the fault current at the end of the relay's reach or at the relay's position. If the current at the end of the reach is used, then the protection will be very fast for a close-up fault; if the current at the device position is used, then the protection will slow down as you move further into the network to the end of the feeder. By setting the operating at the last device position, it might be possible to accommodate another device downstream from this last device if the network expands in the future. Once the grading current is determined, the TM can be determined in step 6 to set the required time delay for the last device. For the last device an operating time of not more 0.5 s is a good design value. This allows for other devices to be added downstream from this device at a later stage. The operating time can be set faster than this. In step 7, the ARC cycles and dead and reclaim time can be set. Depending on the curves used for the ARC cycles, the zone sequence coordination function is enabled or disabled in step 8. At this stage the protective relay will grade with the upstream protection and will be sensitive to faults, but it might leave the fault on the network for a long time and conductor damage can still occur. Thus, in step 9 the LTE evaluation is done to ensure that the conductor is not damaged. If the conductor LTE exposure is more than its limit, the TM can be decreased, the pickup can be decreased, the operating curve can be changed or the number of ARC cycles can be reduced. In step 10, the operating time for a fault at the end of the feeder should be evaluated against the maximum time a fault is allowed to be on the feeder (user can determine limit). This is similar to step 13 for the top-down grading method. To change the operating time, the TM can be reduced, the pickup can be lowered or the operating curve can be changed. A combination of these options can also be applied.

The protection settings for the last device (first device for grading) is now set. In step 11, the grading current for the next upstream device can be determined. This is done by using fault level at the next downstream device with the highest fault level. In step 12, the operating time for the downstream device is determined at the fault level of step 11. In step 13, the time selectivity is created by adding the grading margin to the operating time of the downstream device. This will be the required operating for the upstream device being set. Determine the required TM for this operating time in step 14. The number of ARC cycles and the dead and reclaim time is set in step 15. In step 16, the LTE exposure is evaluated for this device. If there is an exceedance, then the settings at this device can be changed first to determine if the exceedance can be negated. If not, then the downstream device settings can be changed to accommodate the exceedance at the upstream device. If the downstream device settings are changed, then an evaluation of the LTE and maximum time limit should be done for those devices again. Once the LTE analysis is done, the maximum time the fault is on the network for this upstream device can be evaluated in step 17. Similar to the LTE evaluation of step 16, the changes will be made to the local upstream device first. If there is still an exceedance, the downstream device settings are adjusted. Once the settings are evaluated and no exceedances are found, the zone sequence coordination function should be enabled or disabled in step 18 depending on the curves used for each of the ARC attempts. If there are more upstream devices (for step 19), this process will repeat itself starting at step 11. Finally, if there are no more devices in series, the grading between the source protection and the feeder protection can be evaluated in step 20. This is to ensure there is selectivity. If there is no selectivity, then the complete process will have to be restarted from the last device, changing the required settings such as the TM, pickup current and operating curves at this last device.

6.6.3 Course Grading Method

The course grading method is shown in Figure 6.27. The course grading method is applicable to a network where there is existing grading and a new device is added to this network, normally at the end of the feeder or on a T-off (spur). In Figure 6.20, the settings at Relay A are already determined to meet the requirements of selectivity, sensitivity, LTE criterion and maximum time criterion. Relay B or more relays are then added below Relay A without any change to the settings on Relay A. This is a very fast way of grading, but it is not optimised. In step 1, the operating curve is chosen for the first downstream device. It is preferable to keep the operating curve similar to that of the first relay (Relay A in Figure 6.20). The pickup is determined in the step 2. As Relay A is already sensitive to faults up to the required position on the feeder, the next downstream device only has to grade (current) with the upstream device. Thus, a value of 90% of the upstream device pickup can be used. The grading current is determined in step 3. For this, use the fault level of the device that is already set to protect the complete feeder prior to starting the grading (thus Relay A). By using this fault level for all the devices, they will be graded as if they were positioned at the physical position of the upstream known device (Relay A). This means that by setting the grading margin at this current, the actual time grading margin will be more than this due to the fault level reducing further down into the network (devices operating slower).

In step 4, the operating time of the upstream device is determined at the grading current. Compare the operating time to a set limit in step 5. If this operating time is more than the set limit (user defined), then the maximum limit can be used (e.g., 1.5 s); otherwise, use the operating time as determined at the grading current. Once the upstream operating time is determined, then the grading margin can be subtracted in step 6. Test to see if this is the last device in step 7. If this is the last device and the operating time is more than the maximum user defined limit for the last device, then use the maximum value. Otherwise, the calculated operating time can be used to determine the TM in step 8. In step 9, the number of ARC attempts, the dead time and the reclaim time can be set. It is recommended to set them similar to the

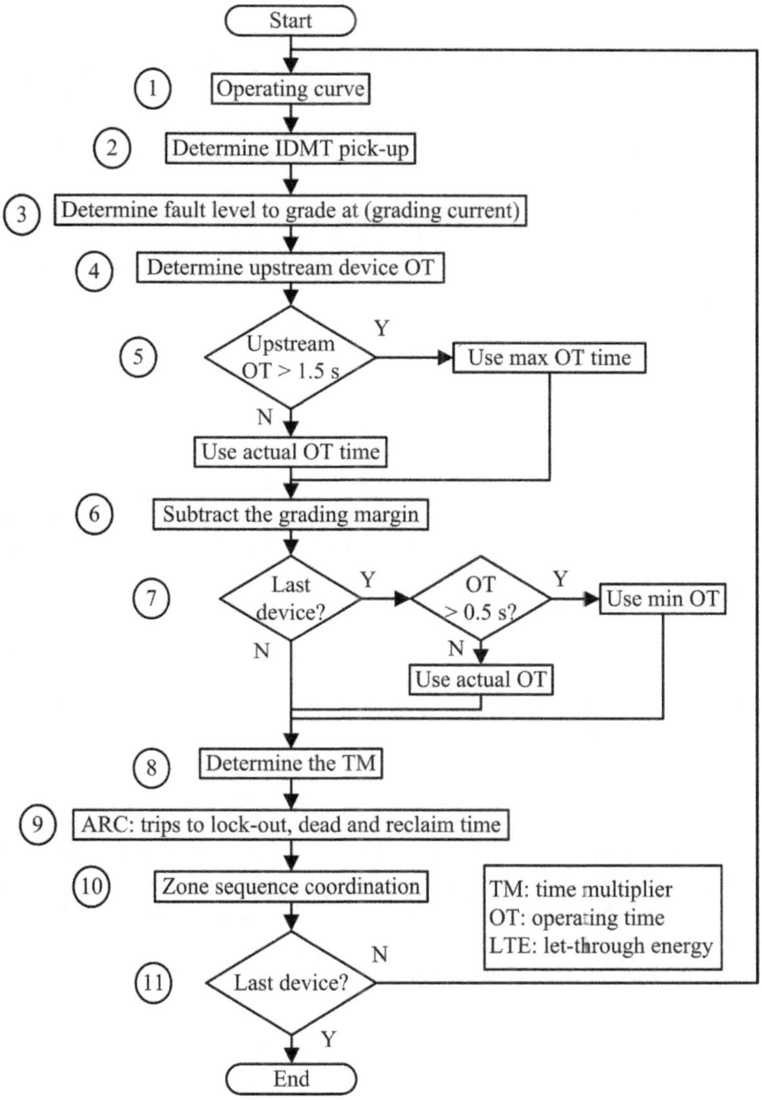

FIGURE 6.27 The course grading method (steps in circles).

upstream device on the feeder (Relay A). Enable or disable zone sequence coordination as required in step 10. Finally, if this is the last device, the grading will end, else the process will repeat itself again until the last device is reached. If different types of curves are used for each ARC trip, then the process has to be repeated for each curve type.

6.7 Settings Example (Top-Down Method)

In this section, an example showing the top-down grading method is applied to the radial feeder in the network of Figure 6.28. The network three-phase and phase-to-phase fault levels are provided in Table 6.6. The aim is to determine the overcurrent settings for the protection relay at poles 1, 50, 100,

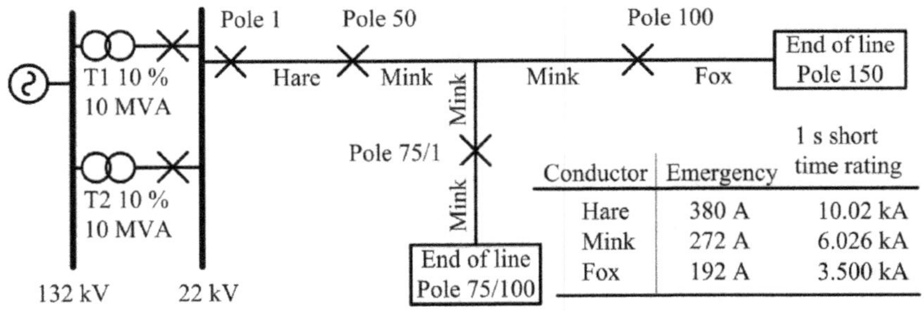

FIGURE 6.28 The protection network diagram for the top-down grading example.

TABLE 6.6

The Fault Levels for the Network in Figure 6.28

Position	Three-Phase 1 Transformer	Three-Phase 2 Transformers	Phase-to-Phase 1 Transformer	Phase-to-Phase 2 Transformers
Pole 1 (busbar)	5000 A	7000 A	4330 A	6062 A
Pole 50	2500 A	3200 A	2165 A	2771.2 A
Pole 75/1	1800 A	2900 A	1558.8 A	2511.4 A
Pole 75/100	900 A	1100 A	779.4 A	952.6 A
Pole 100	1300 A	1800 A	1126.8 A	1558.8 A
Pole 150	1000 A	1400 A	866 A	1212.4 A

and 75/1. Apply a grading margin of 400 ms. All the protection devices can accommodate a pickup with step sizes of 1 A with no decimal places. The pickup is set in primary amperes. All the protection devices can accommodate TM settings with step sizes of 0.01 A and two decimal places. Both transformers are set to pick up at 66 A on the 132 kV side of the transformer using a NI curve with a TM of 0.51. The conductor ratings used for this example are provided in Figure 6.28.

6.7.1 Pole 1

Step 1: A NI curve is chosen to grade optimally with the upstream protection device. This curve will be applied to all the ARC attempts.

Step 2: The pickup current is determined by considering three criteria and choosing the lowest one of the three.

1. Ninety percent of the upstream device pickup calculates to 356.4 A. This was calculated by taking the transformer pickup current of 66 A, referring it to the 22 kV side of the transformer and multiplying it with 90%. The value of one transformer is used as the minimum network condition for the feeder. Current selectivity is still required under this network condition (robustness of setting).
2. Eighty percent of the phase-to-phase fault level at the intended reach in minimum network conditions calculates to 900.64 A. This was calculated by bypassing all the immediate downstream devices (not more than two in series) and then searching for the lowest fault level with one transformer in service at the source.
3. One hundred twenty percent of the conductor ratings within the intended protective reach are determined last. The care conductor rating calculates to 456 A and the Mink rating calculates to 326.4 A.

Based on the above criteria, the lowest value of the three is the rating for Mink conductor. When considering the relay limitations in terms of step size, the pickup is set to 326 A (primary current value).

Step 3: The high-set pickup is determined by taking 150% of the fault level at the next device with the highest three-phase fault level in maximum network conditions. This calculates to 4800 A with a chosen time delay of 0 seconds. The 150% value is used as this is a numeric device which can be set to a minimum value of 130%.

Step 4: To determine if the high-set from step 3 is valid, the high-set pickup is compared to the fault level at the protective device position in maximum network conditions. It is found that the pickup of 4800 A is less than the 7000 A at the device position; thus the high-set element is enabled (is valid).

Step 5: The high-set pickup of 4800 A is less than the fault level of 5000 A; thus the high-set is still active when one transformer is in service. Use a grading current of 4800 A to benefit from the high-set pickup in terms of grading and operating time.

Step 6: The upstream device operating time at the grading current calculates to 1.395 s with a pickup of 396 A (at 22 kV side), a TM of 0.51 on a NI curve.

Step 7: This upstream protection operating time is less than the user defined maximum limit of 1.5 s; thus the operating time used for this device is 1.395 s. If it was more, then the maximum defined value (by user) would have been used.

Step 8: The grading margin of 400 ms is subtracted from the operating time determined in step 7. This calculates to 0.995 s. This is the desired operating time of the device being set.

Step 9: There are two grading paths on this feeder. The first path is Pole 1, Pole 50 and then Pole 75/1. The second path is Pole 1, Pole 50, and then Pole 100. The two last devices are the devices at Pole 75/1 and Pole 100. The device at Pole 1 is not the last device on any of the grading paths. This operating time of the device at Pole 1 is not adjusted (last device) and the value of 0.995 s determined in step 8 is used.

Step 10: The TM required to achieve the operating time of 0.995 s at 4800 A with a pickup of 326 A on a NI curve is 0.392. The relay can be set in two decimal places with a step size of 0.01. This value is rounded down for the top-down grading method to avoid infringing on the 400 ms grading margin. The TM is set to 0.39.

Step 11: Two trips to lockout (one ARC cycle) will be applied to the protective device with a dead time of 15 s and a reclaim time of 75 s. This reclaim time is greater than the minimum limit imposed by the circuit breaker based on its duty cycle.

Step 12: To determine if the conductor is actually protected, an LTE analysis is done on the settings. To do an accurate evaluation, the energy curves should be drawn for the complete feeder (which will be shown at the end of this example). To do a quick evaluation, the LTE value can be determined at specific positions on the feeder, and these can be compared to the conductor limit. The results are shown in Table 6.7.

TABLE 6.7

The LTE Evaluation Results for the Protective Device at Pole 1

Position	Fault Current [A]	Protection Operating Time [s]	LTE [MA^2s]	Conductor Type	Conductor I^2t Limit [MA^2s]	Exceedance
Pole 1	7000	0.863	84.57	Hare	100.4	No
Pole 1*	4800	0.988	46.53	Hare	100.4	No
Pole 50	2165	1.415	13.26	Hare	100.4	No
Pole 50	3200	1.168	23.92	Mink	36.3	No
Pole 75/1	1558.8	1.718	8.35	Mink	36.3	No
Pole 100	1126.8	2.176	6.52	Mink	36.3	No

* This is at the pickup current of the high-set element (step 3).

The LTE has to be evaluated for all the conductor types within the required protective reach of the device at Pole 1. This reach is thus from Pole 1 up to Pole 75/1, and Pole 100. The LTE was calculated by making use of the fault current and the protection operating time (I^2t) and then multiplying this with the number of trips to lockout (ARC). As the LTE evaluation consists of a current and time portion, the evaluation is done at the start (current portion) and the end (time portion) of each conductor type within the protective reach. The start position is evaluated in maximum network conditions (three-phase fault) and the end in minimum network conditions (phase-to-phase fault). To determine the conductor limit (short time rating known) Equation (6.11) is applied and then multiplied with the square of the fault current at that pole position. It can be seen that the Hare and Mink conductors are protected as the LTE exposure is less than the conductor limit at all positions.

Step 13: The maximum time the protection will take to clear a fault with all the immediate downstream protection devices on bypass in minimum network conditions (phase-to-phase fault) has to be evaluated. This is compared against the maximum limit of 5 s that is used for this example. There are two positions that should be evaluated: Pole 75/1 and Pole 100. The protection at Pole 1 will take 1.72 s to clear a fault at Pole 75/1 (1558.8 A) and 2.18 s to clear a fault at Pole 100 (1126.8 A). Both of these times are less than 5 s; thus the settings are valid.

Step 14: As only NI curves are used, the zone sequence coordination function is not required. If other curves were used, then the function will be switched on and settings will have to be determined for those curves as well using the same process in Figure 6.23.

6.7.2 Pole 50, Pole 75/1, and Pole 100

The results for Pole 1 and the remaining poles are summarised in Table 6.8. The same steps were followed as described in Figure 6.23 and applied to Pole 1.

6.7.3 Grading Graphs and LTE Graphs

Figure 6.29 shows the grading graph for the top-down grading example. It can be seen that the various curves are grading with each other and the high-set that is applied to the protective device at Pole 1. This graph is excellent for determining selectivity between devices, but it does not show if the conductors are protected. Figure 6.30 shows the LTE evaluation graph. It can be seen that, while the fault level to pickup ratio is large, the conductor LTE limit is not exceeded. The LTE limit for the Hare conductor is not shown as this is 100.4 MA^2s. The values for the Mink and Fox conductors are 36.01 MA^2s and 12.25 MA^2s, respectively. Both these limits are indicated in Figure 6.30. The largest current that the Fox conductor is exposed to is 1800 A, and the largest current for the Mink conductor is 3200 A. Both these conductors are protected while there is decent fault current. Apply a DT curve to the IDMT curve as well to ensure that none of the conductor limits are exceeded when the fault level to pickup current ratio approaches 1. The DT curve can be applied on the same IDMT protection curve with a pickup equal to the IDMT element pickup at each protective device. The time delay is set to 5 s for all of the devices. With this applied to all the devices, the maximum energy that each device will be exposed to can be determined by determining the current at which the operating time equals 5 s. The LTE can then be calculated by using the square of this current with a time of 5 s. The results of this are that Pole 1 will be limited to 1.57 MA^2s, Pole 50 to 0.59 MA^2s, Pole 75/1 to 0.32 MA^2s, and Pole 100 to 0.33 MA^2s. The benefit of this DT curve is evident in the reduction of the LTE for high-resistance faults (long operating times). When evaluating the high-set element that is applied to the protection device at Pole 1, it can be seen that there is a significant reduction in the LTE exposure for close-up faults. The energy peak is reduced and moved further into the network from the busbar.

TABLE 6.8

The Protections Settings Calculation Results for the Top-Down Method

Step	Pole 1	Pole 50	Pole 75/1	Pole 100
1	NI for all trips	NI for all trips	NI for all trips	NI for all trips
2	1. 356.4 A 2. 900.64 A 3. 326.4 A Set to 326 A	1. 293.4 A 2. 623.52 A at Pole 75/100 3. Fox 230.4 A Set to 230 A	1. 207 A 2. 623.52 A at Pole 75/100 3. Mink 326.4 A Set to 207 A	1. 207 A 2. 692.8 A at Pole 150 3. Fox 230.4 A Set to 207 A
3	Pickup 4800 A TD is 0 s	150% of 2900 A Pickup 4350 A TD is 0 s	Last device, choose not to apply, can apply if required	Last device, choose not to apply, can apply if required
4	4800 A > 7000 A? No, apply	4350 A > 3200 A? Yes, Disable	Not applicable as high-set is disabled	Not applicable as high-set is disabled
5	4800 A > 5000 A? Yes, grade at 4800 A	High-set disabled, Golden rule of grading, 3200 A	High-set disabled, Golden rule of grading, 2900 A	High-set disabled, Golden rule of grading, 1800 A
6	Upstream operating time is 1.395 s	Upstream operating time is 1.168 s	Upstream operating time is 0.781 s	Upstream operating time is 0.966 s
7	Is 1.395 s > 1.5 s? No, use 1.395 s	Is 1.168 s > 1.5 s? No, use 1.168 s	Is 0.781 s > 1.5 s? No, use 0.781 s	Is 0.966 s > 1.5 s? No, use 0.966 s
8	1.395−0.4 = 0.995 s	1.168−0.4 = 0.768 s	0.781−0.4 = 0.381 s	0.966−0.4 = 0.566 s
9	Not last device, use 0.995 s	Not last device, use 0.768 s	Is 0.381 s > 0.5 s? No, use 0.381 s	Is 0.566 s > 0.5 s? Yes, use 0.5 s
10	TM required 0.392 Set to 0.39	TM required 0.297 Set to 0.29	TM required 0.148 Set to 0.14	TM required 0.158 Set to 0.15
11	2 trips to lock-out Dead time 15 s Reclaim Time 75 s	2 trips to lock-out Dead time 15 s Reclaim Time 75 s	2 trips to lock-out Dead time 15 s Reclaim Time 75 s	2 trips to lock-out Dead time 15 s Reclaim Time 75 s
12	Pass Refer to evaluation	Pass Refer to LTE graph in following section	Pass Refer to LTE graph in following section	Pass Refer to LTE graph in following section
13	Pole 75/1 is 1.72 s Pole 100 is 2.18 s Pass, less than 5 s	Pole 75/100 is 1.64 s Pole 150 is 1.51 s Pass, less than 5 s	Pole 75/100 is 0.73 s Pass, less than 5 s	Pole 150 is 0.723 s Pass, less than 5 s
14	Disable zone sequence coordination	Disable zone sequence coordination	Not required, last device	Not required, last device

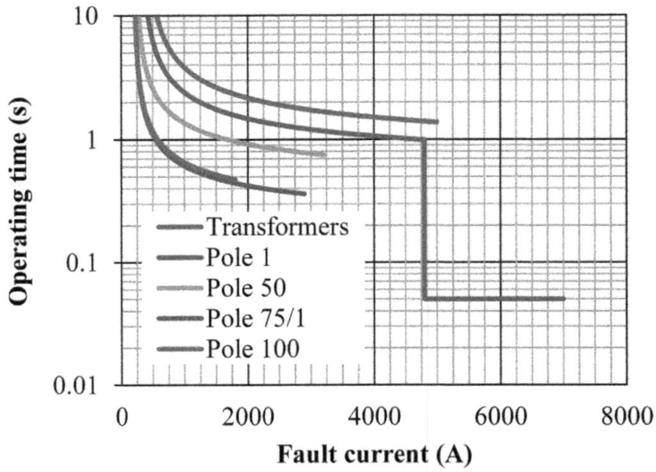

FIGURE 6.29 The grading graph for the top-down grading example.

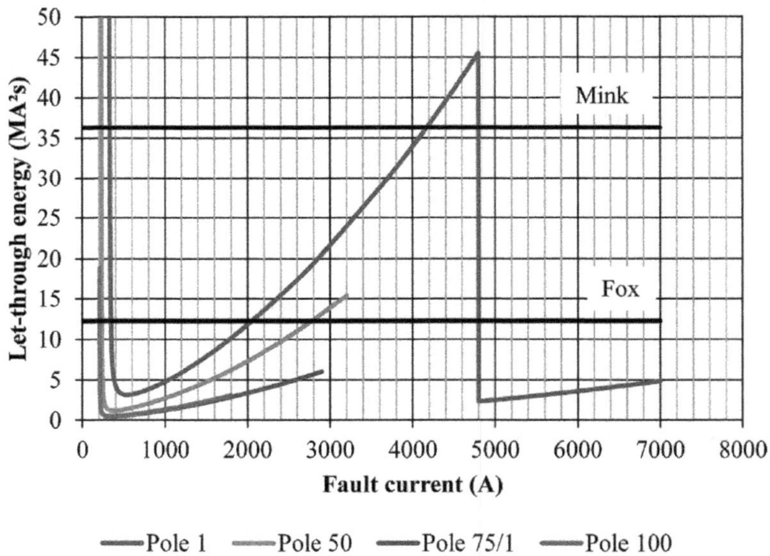

FIGURE 6.30 The LTE evaluation graph for the top-down grading example.

6.8 Interconnected Network Strategy

When there are multiple supply points on a medium voltage network with current-based protection, protection becomes more complicated. Fault current can change magnitude and even direction. Relay models become more complicated to accompany these changes. The need for intelligent protection schemes is emphasised.

6.8.1 Multisource Medium Voltage Network

Multisource MV networks refer to MV networks where there are two or more possible points of supply. These supply points can be the result of normal open points (NOPs) closing or the distributed generation that might be installed in the feeder. A distinction has to be made between the two scenarios. For the NOP, there is still only one source in the network, but there are two possible directions of current flow in the network due to the physical position of the source changing. The source impedance can then also be different when the NOP is supplying the feeder. This means that the fault current contribution will be different for a fault. Still, to isolate the fault, only one CB is required to operate on the feeder between the source and the fault. The protection operating time for a fault when the NOP is closed can still be determined using Equation (6.3).

With distributed generation installed on the feeder, there are more possible simultaneous points of supply on the feeder. This means that not only is the utility supplying the load on the feeder, but other generation points are supplying at the same time. To isolate the fault, a minimum of two different circuit breakers are required to operate on either side of the fault. Some of these generators contribute to the fault current and some of them do not. To explain the effect of additional sources on the protection of a feeder, a network with two sources (Source A and Source B) is considered. This network is shown in Figure 6.31.

For a fault on Feeder X in Figure 6.31, fault current will be supplied from both ends of the feeder. These fault currents will be limited by the source impedance, the impedance of the feeder and the fault resistance. Feeder Y shows that the sources are not independent. This means that the fault current will change from each end of the faulted feeder when there are circuit breaker operations.

For the faulty Feeder X in Figure 6.31, fault current is supplied from both ends of the feeder. This means that the conductor from Busbar A up to the fault is exposed to fault current associated with Source A. The conductor from Busbar B up to the fault is exposed to the fault current from Source B. At the fault position an arc forms between the faulty phases. At the fault position plasma forms [46–48]. In the plasma fault streamers move from the conductor [46–48]. The assumption is made that the fault current does not thus combine on the feeder. The conductor thus is not exposed to the combined fault current from both ends of the feeder. The plasma and streamers are illustrated in Figure 6.32 for the fault shown in Figure 6.31.

One of the advantages of having a distributed generation installed on a feeder is that it reduces the network investment and operation cost [42,49]. This is due to the load being closer to the generation. The losses associated with transporting power to loads at the end of the feeder are reduced. The requirement for large conductors to meet the load demand for the total feeder from the main utility source can be minimised. This is under the assumption that the distributed generation has reasonable availability on the feeder. The complexity of protecting and controlling the network does increase.

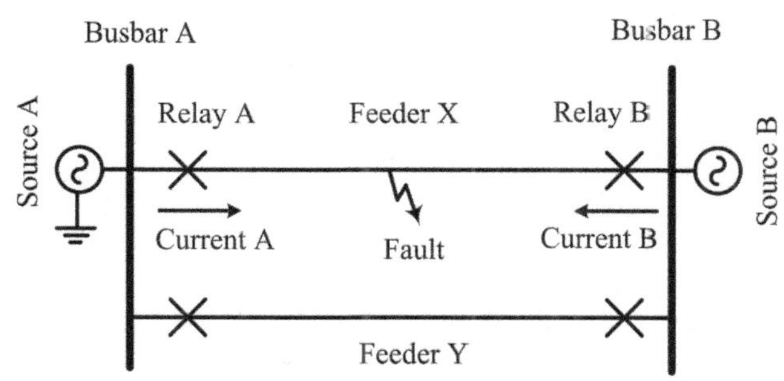

FIGURE 6.31 A feeder with two sources used to explain the effect of multiple sources on protection.

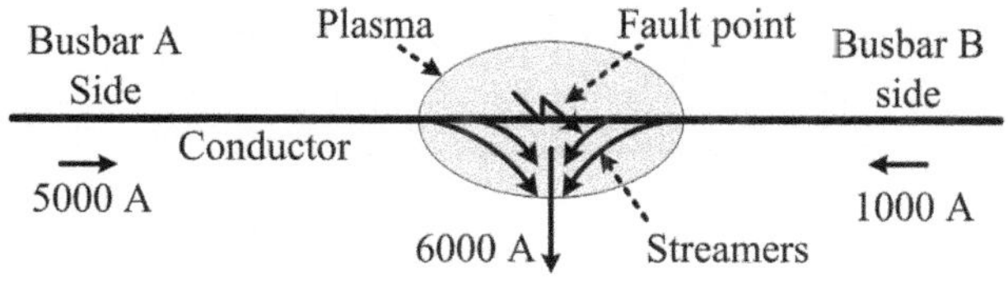

FIGURE 6.32 A plasma and streamer illustration at the fault position.

6.8.2 Overcurrent Relay Model

The protection operating time for multisource interconnected networks was not discussed in the previous section as a different relay model has to be used to accommodate this type of network. For a radial feeder, the fault current flows in only one direction, and the RMS value is constant over time (neglect higher frequency transient and subtransient part). This is after the filtering of the input current for the fundamental 50 Hz component. For this type of radial application, Equation (6.3) can be used to calculate the protection operating time. Figure 6.33 shows an RMS primary current signal. This is the current that the relay measures for a fault in the network. It can be seen that the initial fault current at 1 s is equal to 4892 A. At 1.35 s, the fault current through the relay increases to 5882 A. If we were to apply a pickup current of 550 A and a TM of 0.31 to Equation (6.3), the following trip times can be expected if the RMS fault current stayed constant. For a fault current of 4892 A, the trip time would be 0.971 s; for a fault current of 5882 A, the trip time would be 0.894 s. If we examine the current signal, we expect to see a trip time that is faster than 0.971 s and slower than 0.894 s. Equation (6.3) is a simplification of the actual overcurrent relay equation of Equation (6.1) for electromechanical devices.

To calculate the operating time, the RMS fault current in Figure 6.33 can be applied to the integral version of the equation. The application of Equation (6.2) to the current signal of Figure 6.33 is provided in Equation (6.22). This equation has been set up in terms of two changes in time. More time changes can be added. The time current characteristic is defined by Equation (6.3). The trip time calculates to 0.922 s, which confirms the earlier expectation.

$$1 = \int_{0}^{T_1} \frac{1}{t(I)} dt + \int_{T_1}^{T_2} \frac{1}{t(I)} dt \tag{6.22}$$

where:

T_1 is the end of first event time (s)
T_2 is the end of second event time (s)
$t(I)$ is the time current characteristic

Another method that can be used to evaluate the trip time is to write a script in programming code to simulate an overcurrent relay. The main equation that can be programmed is shown in Equation (6.23) [18]. This is essentially the same equation as that of Equation (6.22) except that it makes used of discrete (or sampled) values. For this simulation, a time step (and sample step) of 1 ms is used. The RMS input signal resembles that of Figure 6.33. The relay issued a trip signal after 922 samples, indicating that the trip time is 0.922 s.

$$1 = \sum_{n=1}^{n=X} \left(\frac{\text{time step}}{t(I)} \right) \tag{6.23}$$

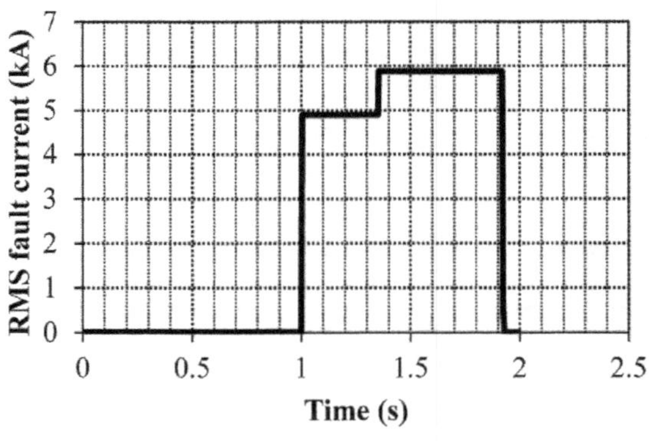

FIGURE 6.33 The RMS fault current that a relay is exposed to during a fault.

Another overcurrent relay model that can be used is based on the average disk speed of an electromechanical relay. For an electromechanical relay, the physical distance that the disk needs to travel is proportional to the TM setting of the relay. This means that the linear distance can be set equal to the TM. An assumption is made that the angular speed of the disk is constant if the input RMS current is constant. The actual distance does not matter as the speed of the disk compensates if the distance is more or less. The total distance is constant as the TM does not change while the relay is timing for a fault. Since the input current is constant Equation (6.3) (NI IDMT) can be used to determine the relay trip time, and Equation (6.24) is an average speed equation [31]. By using Equation (6.24) the speed of the disk required to travel the TM distance in the calculated trip time can be determined.

$$v = \frac{s}{t} \tag{6.24}$$

where:
v is the speed of disk (m/s)
s is the distance (time multiplier) (m)
t is the trip time (s)

For every change in the RMS current, a new trip time and disk speed can be calculated. Since the time is known when there was a change in current, the distance the disk travelled can be determined when applying Equation (6.24). This means that the disk will then travel at a different speed from the point where the current changed, and the time for a trip can be determined by determining how long the disk takes to complete the distance (TM value). If there is another current change before the trip elapsed, then the same process is followed. This model is well suited for software implementation. A flow diagram illustrating this average speed overcurrent relay model and the process is shown in Figure 6.34.

If this model is applied to the input signal of Figure 6.33, a trip time of 0.922 s is expected. The protection settings that are applied to the relay have a pickup current of 550 A and a TM of 0.31. The distance the disk has to travel is set equal to the TM, thus 0.31 m. Using Equation (6.3) the trip time at a fault current of 4892 A and then 5882 A is calculated as. 0.971 s and 0.894 s, respectively. Now the speed of the disk can be calculated for each of these RMS input currents using Equation (6.24). For the current of 4892 A, a linear disk speed of 0.3192 m/s is required and for 5882 A a linear disk speed of 0.3468 m/s is required to trip in the respective times. In Figure 6.33 there is a change in current after 0.35 s of the fault inception time at 1 s. The distance the disk travelled in 0.35 s at the speed of 0.3192 m/s is equal to 0.112 m. This means the disk should still travel 0.198 m before the relay will trip. At the new disk speed of 0.3468 m/s with the remainder of the disk distance, the relay will issue a trip signal in 0.572 s. The first time of 0.35 s can now be added to the second time of 0.572 s, and a trip time of 0.922 s is determined.

This average speed overcurrent relay model can also be applied to other IDMT curves such as an EI operating curve if the required disk speed is determined with the trip time of the EI curve for a constant RMS input current. It can also be applied to the reset curve of the relay if the reset is not instantaneous. The reset curve allows for calculation of the time the disk will take to return back to its original starting position [16]. In most modern numeric relays, the reset curve can be chosen, which means that the numeric relay can also emulate an electromechanical disk reset curve.

6.8.3 A Multisource Interconnected Network

The application of the ARC function on radial networks is relatively uncomplicated when it is compared to a multisource application. For the radial application, it was simply adding the LTE from each ARC attempt together, and the only factor that was influencing it was the protection setting of the operating element [22]. For a multisource network, it is not only the settings but also the change in fault levels, the sequencing of protection elements and the ARC settings (not only the number of ARC attempts) that influence the LTE exposure.

To explain the effect of LTE analysis on an interconnected multisource network, a single feeder with supply from both ends of the feeder will be used. The network resembles the one shown in Figure 6.31 for Feeder X.

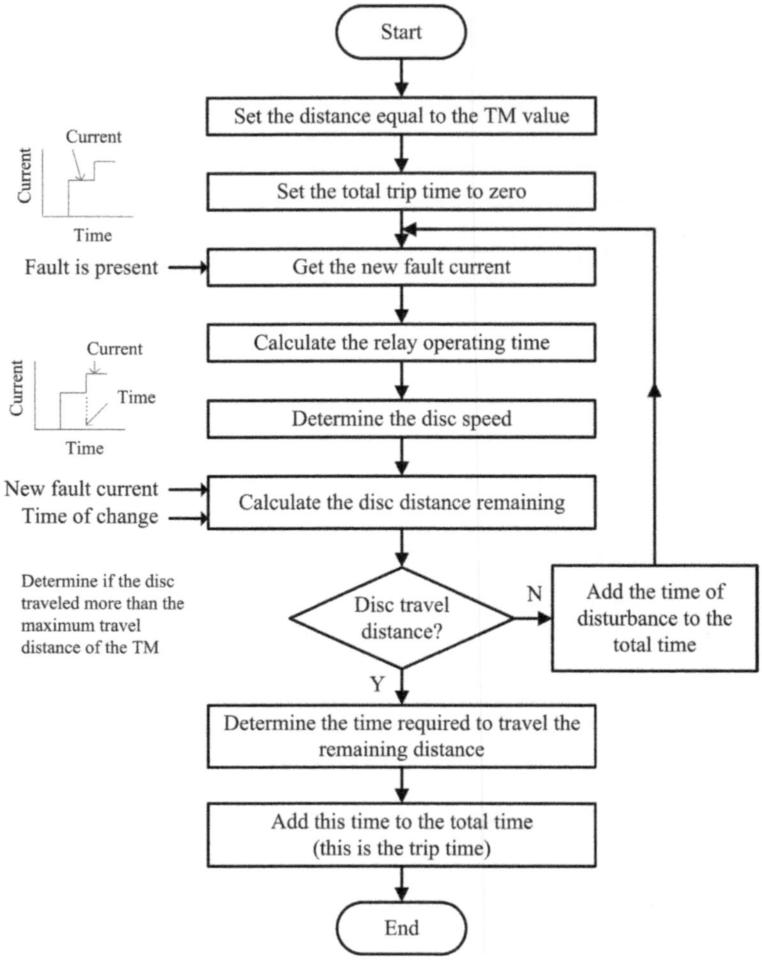

FIGURE 6.34 A flow diagram showing the average speed overcurrent relay model for software applications.

The fault levels over a 30 km 22 kV feeder (Feeder X in Figure 6.31) are shown in Figure 6.35. The fault levels shown are for when the feeder is supplied from busbar A only (A), then busbar B only (B) and for when the feeder is supplied from both ends of the feeder. When both ends are supplying, each individual end's contribution is shown [A+B (A) and A+B (B)]. In Figure 6.35, it seems as if the A+B (A) and only A fault currents are exactly the same, but they are in fact slightly different. The fault level at busbar A for A+B (A) is 4654.056 A and for A it is 4668.772 A. This difference can be attributed to the network configuration and source impedances. This is also for the fault current contribution of A+B (B) at Busbar B. which is equal to 380.747 A, and for B it is 403.683 A.

To clear the fault on Feeder X in Figure 6.31, both Relay A and Relay B have to detect the fault and trip their respective circuit breakers. Thus, for the multi-source Feeder X, a total of two circuit breakers are required to operate and thus clear the fault. Figure 6.35 shows that the fault current contribution, first, is different from each side of the feeder and, second, that if any circuit breaker opens, the fault current contribution changes. To add to this, the protection settings in each relay are also different. This means that there will be a certain sequence of events to clear the fault.

Figure 6.36 shows two possible sequences of events. In both sequences A and B the fault starts at step 1. Initially both sides of the feeder supply fault current to the fault. In Figure 6.36, this is shown as (A+B). For both sequences A and B, the B side trips first based on the protection settings and the

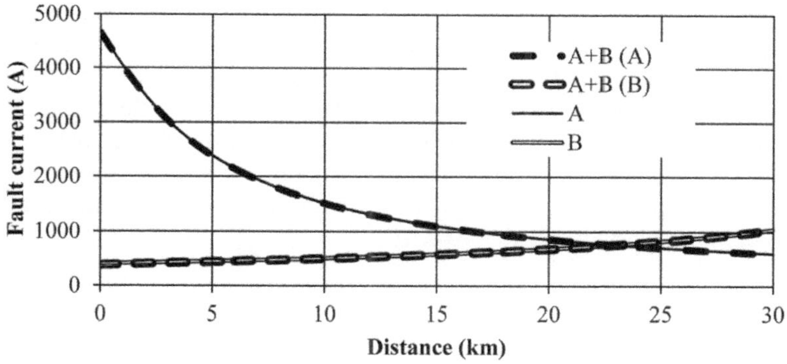

FIGURE 6.35 The three-phase fault levels across the feeder for different network contingencies.

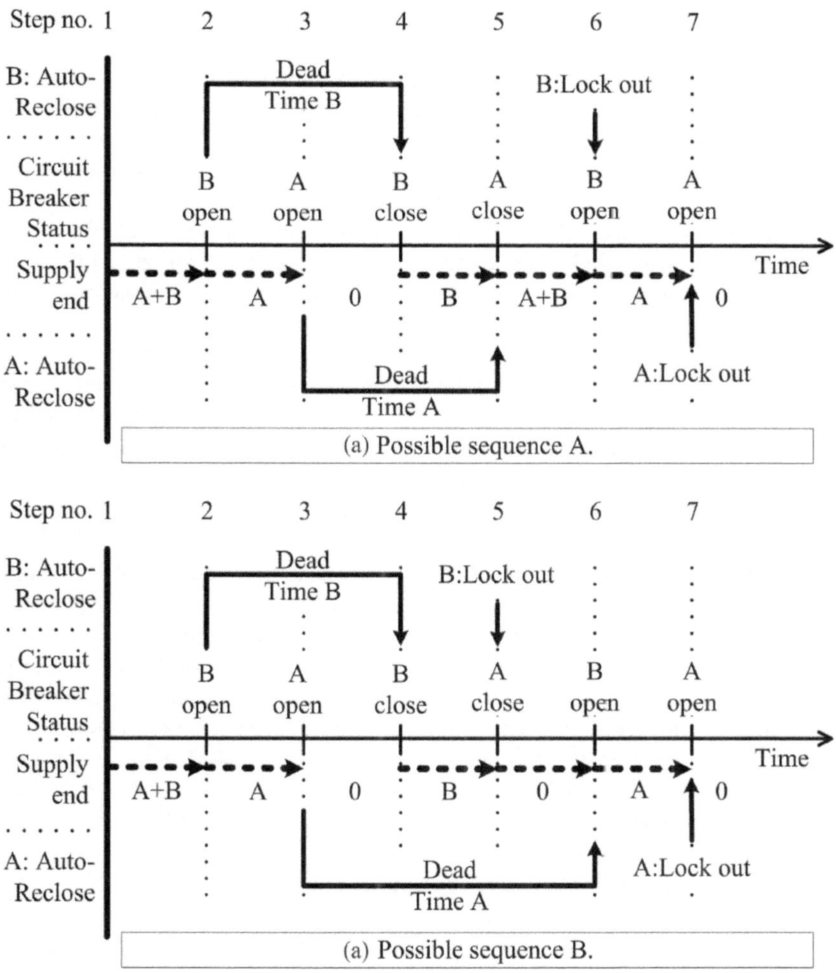

FIGURE 6.36 Two possible sequence of events are shown to clear a fault on an interconnected feeder when it is supplied from both ends.

measured fault current. At step 2, the trip signal is sent to the circuit breaker, and it is assumed that the circuit breaker opens instantaneously. If the circuit breaker did not open instantaneously, the circuit breaker operating time would have to be included and the current can be seen to flow for that time period still. With the issue of the trip signal by the relay, it also starts its dead time timer. The circuit breaker at Busbar B on Feeder X is now open. The current flowing from Busbar A changes (increases or decreases), and this is now the only side feeding the fault. The relay trip time for relay A now changes (increases or decreases) as there was a change in fault current. This trip time can be determined using the relay model as described earlier in this chapter.

1. Possible sequence A
2. Possible sequence B.

If the dead time of Relay B is longer than the remaining time required to trip by Relay A, the relay at Busbar A of Feeder X will issue a trip signal to the circuit breaker. Again, the circuit breaker time is taken as instantaneous to reduce the complexity. Hence, as the trip signal is sent to the circuit breaker, the dead time timer of Relay A is started, and the fault current stops immediately. This means that both circuit breaker A and circuit breaker B on Feeder X are open, and zero fault current is flowing into the fault. If the dead time for relay B is too small, the fault current is not interrupted. This results in a different sequence of events again from the two that are shown in Figure 6.36.

In both sequence A and sequence B, circuit breaker B is closed at step 4. In sequence A the dead time of Relay A is shorter than the trip time (or fault clearing time) of Relay B. From step 5, the fault is supplied from both ends again. This means the fault current that the B-end is measuring again is now changing, and a new trip time has to be calculated using the relay model (e.g., average disk speed model). From step 4 in sequence B, relay B trips before the dead timer times out at relay A. This means the relay at B trips and goes into a lockout condition before the circuit breaker at the A-end of the feeder closes at step 5. Thus, there is now two periods of no current flowing on the feeder for one ARC cycle.

At step 6, in sequence A, the relay at the B-end trips and goes into a lockout condition. From this point onwards, only the A-end of the feeder is supplying the fault current. This fault current that the relay at the A-end is measuring is again changing from the value in the previous step, and hence the relay operating time has to be calculated using an appropriate relay model. Finally, the circuit breaker at the A-end trips and the relay moves into a lockout condition at step 7. The fault is now isolated from both ends for sequence A. The circuit breaker at the A-end closes at step 6 again in sequence B. This re-energises the feeder, and the relay at the A-end issues a trip to the circuit breaker after the relay has timed. At step 7, the relay at the A-end issues a trip signal to the circuit breaker and moves into a lockout condition. The fault is now isolated for sequence B.

From the two sequences shown in Figure 6.36 and the associated explanation, it can be seen that multiple possible sequences can occur. It is recommended to use a software application to generate results for evaluation of LTE in a multisource network. This application would require fault levels for all possible network contingencies in terms of circuit breaker status at the two ends of the feeder. An IDMT relay model will have to be applied that can adapt its calculation of the trip time based on a change in the measured fault current. An example of this is shown in Figure 6.37 for the network of Figure 6.31 (Feeder X). The relay at busbar A has a pickup of 270 A and a TM of 0.42 on an NI IDMT curve. The relay at busbar B has a pickup of 100 A and a TM of 0.05 on an NI IDMT curve. Both Relay A and Relay B are set to initiate one ARC cycle and both hava dead time of 3 s. The resulting LTE and fault exposure time for the Rabbit conductor are shown in Figure 6.37.

Figure 6.37 shows that there is still an energy peak after the high-set curve, but the curve has another step change at 26 km due to the settings fault current from the B-end of this busbar with its associated protection settings. In Figure 6.38 the dead time is changed from 3 s to 1.5 s to illustrate that the dead time influences the LTE curve and the fault exposure time. When Figure 6.38 is compared to Figure 6.37, it is seen that the maximum energy peak is still similar, but the step change that was at 26 km is now moved to 10.5 km. The exposure time to a fault is changed where there are now two peaks of long exposure times.

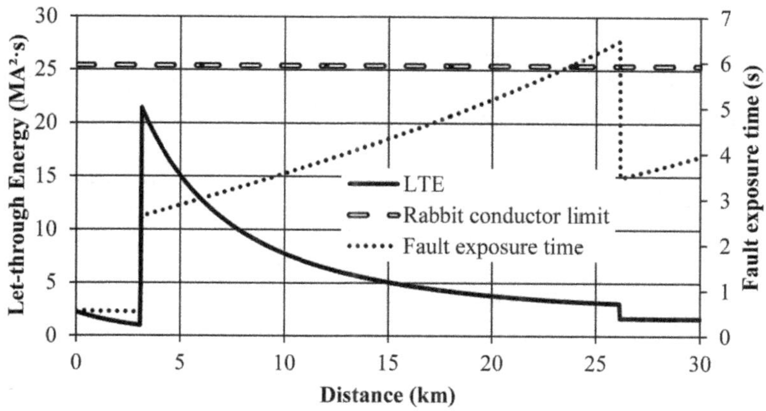

FIGURE 6.37 The LTE exposure for a multisource feeder with one ARC cycle and a dead time of 3 s.

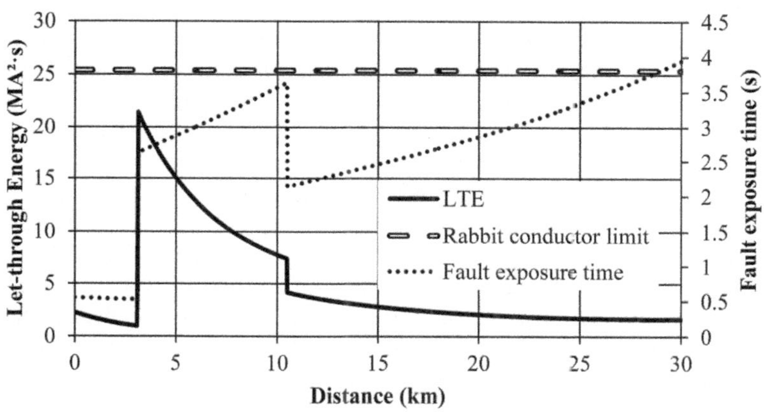

FIGURE 6.38 The LTE exposure for a multisource feeder with one ARC cycle and a dead time of 1.5 s.

If the source impedance was similar on both ends of the feeder with similar settings applied, then there would be two energy peaks at either end of the feeder. Thus, with reference to Figure 6.38, the first part of the curve (0–10 km) would be reflected on the other end of the feeder, from 30 to 20 km, that is, if a high-set element was applied on both ends. If the high-set element was only applied on the one end, the other end can experience a LTE peak that is much greater than the one shown in Figure 6.38.

6.9 Adaptive Protection Requirement

In most medium voltage networks, the power flow starts at the utility connection (e.g., medium voltage busbar) and is then distributed via the radial medium voltage feeders to the load points. This is set to change as the cost of building the network escalates and the cost of distributed generation becomes justified [44]. A multisource medium voltage network is shown in Figure 6.39. To promote the continuity of supply and thus network availability, normal open points can be introduced into the feeders. These normal open points can require manual opening via a disconnector (NO 1 in Figure 6.39) or it can be done via an auto-recloser that is installed at the normal open point (NO 2 in Figure 6.39) [4]. The benefit of the auto-recloser is that it can be done by the network controller remotely, and it does have protection and automation functions available. By having these normal open points, the network is more flexible, but it is also more complex to control and protect. Distributed generation (DG) can

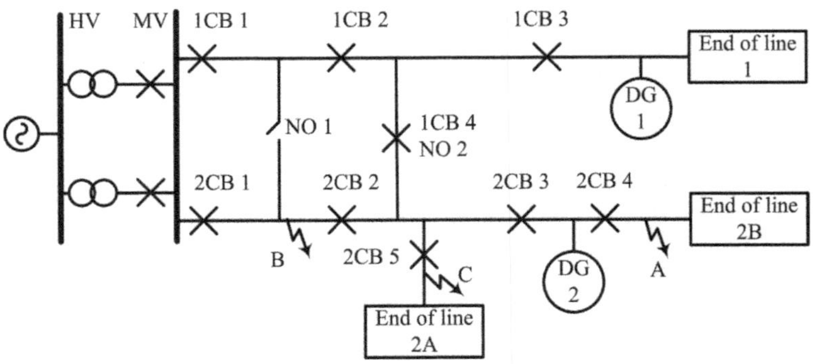

FIGURE 6.39 A multisource medium voltage network.

also be added to this network. The disadvantage of distributed generation is that it can feed a fault or cause an islanded network if the penetration is big enough. An islanded network refers to a network where the load is supplied by the distributed generation and that part of the network is not connected to the utility. If a network can operate in an islanded network condition, it does improve the reliability of the network, but system coordination is difficult [50]. Additional protection functionality such as synchronising relays may be required in these networks that are allowed to island [4]. Allowing networks to island is not common practice [4]. Adaptive protection can assist with some of the problems associated with distributed generation [7]. A rate of change of frequency function can be used to detect islanded situations to either trip the relay or change the protection settings [50]. Synchronous generators and microturbines have the biggest impact on fault current contribution from distributed generation [50]. Small capacity solar generators can be excluded from the fault analysis as their contribution is almost negligible [45]. With the distributed generation being able to supply fault current, they can influence the sensitivity of protection devices towards faults [15,51]. Protection devices are reactive in that they have to be able to detect a fault to initiate an operation. The protective device cannot predict the fault. Figure 6.39 shows two distributed generators installed in the network (DG 1 and DG 2). When normal open points are closed (or opened) or distributed generators are removed (or added) from the network, this influences the fault levels and direction of fault current [42,44,45,52]. With the protection being current based, this can result in a loss of selectivity in the network [53]. The sensitivity and speed (decrease) of the protection can also be compromised with a change in the topology or connected generation [15]. The speed of the protection devices can be influenced as well if IDMT protection is used and the fault level changes.

For a fault at position A in Figure 6.39, only the operation of circuit breaker 2CB 4 is required to isolate the fault. This will not require any of the normal open points to close and all the remaining protection devices are still adequate. If a fault is placed at position B, then 2CB 1 and 2CB 2 have to operate to clear the fault. This is with the fault current contribution from the utility via the MV busbar and then from DG 2. If NO 2 is closed, there is a contribution via this normal open point as well. This means that the protection at 2CB 2 has to be sensitive in both a forward and a reverse direction; thus directional protection is required [4,44]. It will also have to be sensitive to faults for different minimum network contingencies. If the fault was placed at position C, then only 2CB 5 is required to clear the fault. Current flow will always be in one direction for 2CB 5, and thus nondirection protection is required, but this nondirection protection has to be set under various network contingencies to ensure selectivity and sensitivity. From this explanation, it can be seen that directional and adaptive protection is required [52]. This is moving the protection and now automation of the network into the smart grid realm. If the protective devices are not capable of directional protection, all the distributed generators have to disconnected from the network, thus creating a radial feeder again [44].

The European Technology Platform [49] defines the smart grid as follows: "A Smart Grid is an electricity network that can intelligently integrate the actions of all users connected to it—generators, consumers and those that do both—in order to efficiently deliver sustainable, economic and secure electricity supplies." In [45] it is defined as follows: "The smart grid uses advanced information and communication technologies (land line, radio, internet, etc.) to improve power system operations."

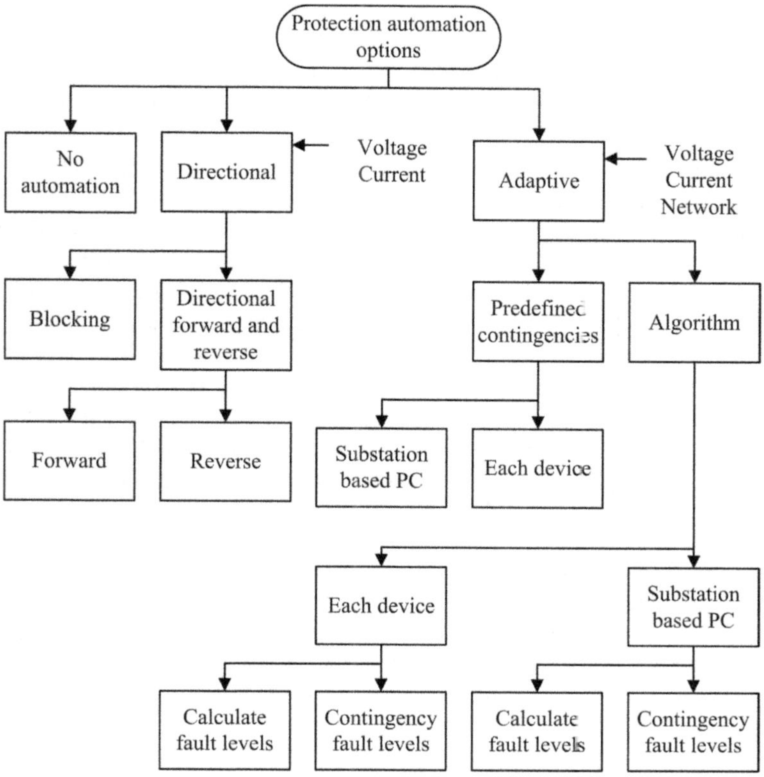

FIGURE 6.40 Protection automation options for medium voltage networks.

The options available for protection automation are shown in Figure 6.40. The first option is not to apply any protection automation. This is the normal nondirectional application of protection. Second, the protection can be directional. To achieve directional protection, voltage and current information are required at the position of the protective device. Communication between devices is not a requirement. If the device is directional, it can be set to block in the forward or then the reverse direction. It can also be set to operate on specific protection settings in the forward and the reverse directions, depending on the direction of the fault current measured at the device position.

The protection devices can be made adaptive [54]. To achieve this adaptive protection, the voltage and current measurements at the device position should be known. In addition, the status of the network should be known. As and example of status, it should be known if a normal open point is open or closed, if a circuit breaker is open or closed or if a distributed generator (contribute fault current) is in service or not. Knowing the status of the network equipment (relevant to protection) means that the protection at a specific position can adjust its own settings. This can be done either by having protection settings that are calculated based on predefined network contingencies or by having a protection algorithm that calculates new protection settings. The method of determining the network status can be achieved by either having the information centralised in a substation based personal computer (PC) or ensuring that each protection device knows the status of all other devices. The latter might not be feasible on current available auto-reclosers. For a substation-based PC, the status of the network can dictate which precalculated settings group should be activated on which protective devices [4,7,45,50]. This can be communicated to the devices (including current auto-reclosers). Communication is a key element in adaptive protection [44,52]. The actual settings are stored on the protective device. These predefined settings in settings groups are a very practical and feasible way of having adaptive protection [45]. An algorithm can be used to calculate new protection settings based on the status of the network [54]. This algorithm can be embedded in each protective device, or it can be embedded in the substation-based PC [51].

The protection can still be set directional to improve selectivity in the network. For the protection to be fully adaptive, the algorithm has to calculate new protection settings based on the network status. This means that the network information should also be available to the algorithm. This network information can be in the form of precalculated fault levels for specific network contingencies [54]. If the algorithm calculates new fault levels based on the current network status, then the total network configuration (conductor impedances etc.) should be known. The new protection settings are then based on this fault level information. If the fault levels are calculated beforehand, then it can be stored in the substation personal computer or protective device. Based on the network status, it will then use that set of fault levels. With the use of an algorithm in a smart grid, the actual load current can also be measured, and the pickup of the IDMT element can be reduced to a value just above the load current based at specific times in the day [15,51]. The conductor rating can be used to determine the maximum possible load, but these values are generally very conservative [7]. If the weather conditions are different from the extreme used to estimate the conductor rating, the conductor can sustain a larger current. With the use of new protection and then algorithms to continuously calculate and adjust the conductor rating (load estimate), more current can be passed through a conductor at certain times without compromising the protection [7]. When adaptive protection is applied, the dependability and security of the protection system can be moved [7].

Protection can also be made adaptive in terms of sensitivity towards specific weather conditions such as lightning and rain [10]. During these conditions there is a bigger likelihood of interruptions [55]. These interruptions are temporary and can result in excessive voltage sags for momentary interruptions. The protection can be adjusted to lockout for a period of 30 minutes to give the storm (with lightning) to pass. A local weather sensor must be incorporated into the automation scheme or it can be activated by the network controllers if prevailing weather conditions are known.

6.10 Worked Examples

Example 6.1: Operating Time

Calculate the operating time for each of the following IDMT operating curves and discuss the differences. All the curves are set to use a pickup of 300 A and a time multiplier of 0.2. The definite time curve makes use of a time delay of 200 ms. Calculate these operating times for a fault current of 4000 A and then 305 A. (a) NI, (b) VI, (c) EI, (d) DT.

Solution

a.

$$4000\text{ A: OT} = \frac{0.14\cdot 0.2}{\left(\frac{4000}{300}\right)^{0.02}-1} \qquad 305\text{ A: OT} = \frac{0.14\cdot 0.2}{\left(\frac{305}{300}\right)^{0.02}-1}$$

$$\text{OT} = 0.527\text{ s} \qquad \text{OT} = 84.684\text{ s}$$

b.

$$4000\text{ A: OT} = \frac{13.5\cdot 0.2}{\left(\frac{4000}{300}\right)-1} \qquad 305\text{ A: OT} = \frac{13.5\cdot 0.2}{\left(\frac{305}{300}\right)-1}$$

$$\text{OT} = 0.219\text{ s} \qquad \text{OT} = 162\text{ s}$$

c.

$$4000\text{ A: OT} = \frac{80\cdot 0.2}{\left(\frac{4000}{300}\right)^{2}-1} \qquad 305\text{ A: OT} = \frac{80\cdot 0.2}{\left(\frac{305}{300}\right)^{2}-1}$$

$$\text{OT} = 0.091\text{ s} \qquad \text{OT} = 476.033\text{ s}$$

d.

$$4000\text{ A: OT} = \begin{cases} 0.2, & If > Ipu \\ \infty, & If \le Ipu \end{cases} \qquad 305\text{ A: OT} = \begin{cases} 0.2, & If > Ipu \\ \infty, & If \le Ipu \end{cases}$$

$$\text{OT} = 0.2\text{ s} \qquad \text{OT} = 0.2\text{ s}$$

Use the NI curve results as a reference. As the curve becomes more inverse, the operating time of the curve is reduced under high fault currents. This is beneficial for reducing the LTE exposure and then the damage to equipment. The rate at which the operating time increases as the fault level drops increased as the curve is more inverse. This can have both a positive and a negative impact. For a positive impact, you can fit another device close to this device and obtain selectivity while tripping fast for high fault currents. The drawback is that it can lose selectivity with upstream devices if they use a different curve and the operating time can reach its maximum time (such as 5 s) quickly. The fault level does not influence the DT curve operating time if the fault current is above the pickup of the DT curve.

Example 6.2: Operating Time

Calculate the time multiplier for each of the following curves to achieve an operating time of 1 s for a fault current of 3500 A. All the curves are set to use a pickup current of 250 A. (a) NI, (b) VI, (c) EI

Solution

a.
$$OT = \frac{0.14 \cdot TM}{\left(\frac{3500}{250}\right)^{0.02} - 1}$$
$$TM = 0.387$$

b.
$$OT = \frac{13.5 \cdot TM}{\left(\frac{3500}{250}\right) - 1}$$
$$TM = 0.962$$

c.
$$OT = \frac{80 \cdot TM}{\left(\frac{3500}{250}\right)^{2} - 1}$$
$$TM = 2.437$$

These results show that the more inverse the curve, the larger time multiplier is required to slow down the curve.

Example 6.3: Grading Different Curves

Determine the minimum grading margin between the two operating curves. This is for a radial feeder with a feeder circuit breaker and then a downstream auto-recloser at Pole 20. The applied protection settings for each scenario are provided in Table 6.9. Draw the grading curves for each scenario. The fault level at the busbar is 360 A.

a. Feeder protection using an NI curve and the auto-recloser at Pole 20 using a DT curve
b. Feeder protection using a DT curve and the auto-recloser at Pole 20 using an NI curve

TABLE 6.9

Protection Settings for Example 6.2(a) and (b)

Device	Example 6.2(a)			Example 6.2(b)		
	Curve	Pickup	TM/TD	Curve	Pickup	TM/TD
Feeder	NI	80 A	0.2	DT	80 A	1 s
Pole 20	DT	40 A	0.5 s	NI	40 A	0.2

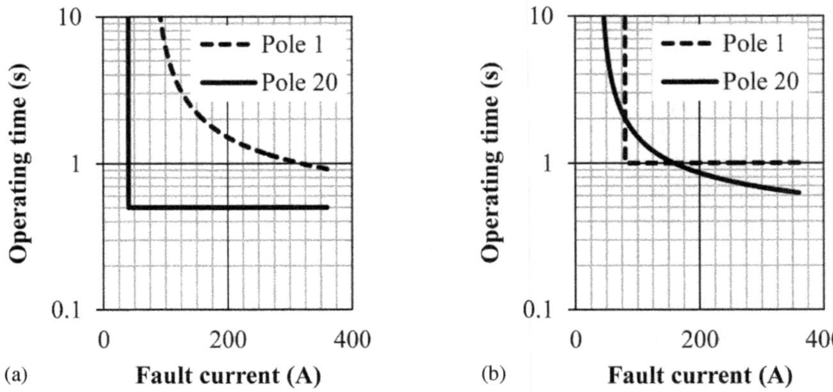

FIGURE 6.41 The grading graphs for Examples 6.2(a) and (b).

Solution

The grading graphs for (a) and (b) are shown in Figure 6.41.

a. Feeder protection using an NI curve and the auto-recloser at Pole 20 using a DT curve: It can be seen that the minimum grading margin is at the maximum current of 360 A.

   Pole 1 operating time at 360 A: 0.917 s

   Pole 20 operating time at 360 A: 0.5 s

   Grading margin is 0.917 – 0.5 = 0.417 s
b. Feeder protection using a DT curve and the auto-recloser at Pole 20 using an NI curve: Initially it appears as if the minimum grading margin is at the maximum current of 360 A, but there is actually no grading between the curves as they cross at a fault level of 159 A. when different curves are graded, the grading must be checked at various positions. It helps to draw the grading graphs as this will illustrate any potential grading issues.

Example 6.4: Minimum and Maximum Network Conditions

For the network diagrams in Figure 6.42, define how the networks below should be switched for minimum and maximum network conditions when setting the protection at the feeder (Pole 1).

Solution

Network (a):

Minimum network conditions. Switch one of the transformers out of service.

Maximum network conditions. Switch all the transformers in service.

Network (b):

Minimum network conditions. Switch transformers with an impedance of 10% and 11% out of service. This will result in the minimum network condition. With more transformers, the network has to be evaluated to see if the minimum network condition is realistic. With 5 transformers at the substation, it might be valid to switch out only three transformers.

Maximum network conditions. Switch all the transformers in service.

Example 6.5: LTE Protection

Determine if the section of the feeder shown in Figure 6.43 is protected by the protection settings applied to the feeder circuit breaker relay. The circuit breaker operating time is 50 ms and the conductor has a short time rating of 5400 A for 1 s. *Tip:* Choose four current points to evaluate. Draw the energy curves.

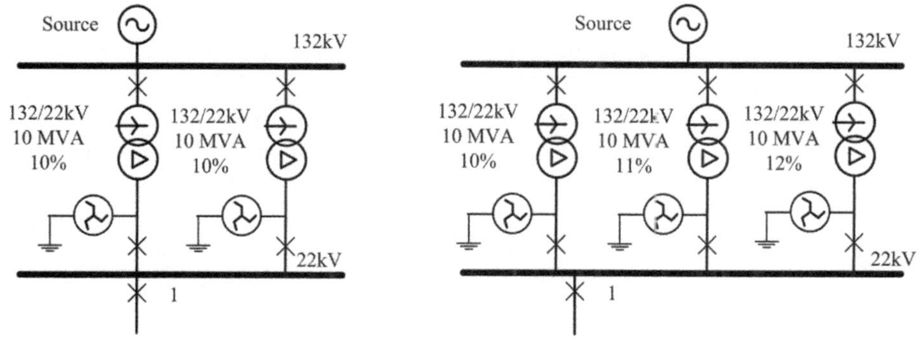

FIGURE 6.42 Two networks to be used for defining minimum and maximum conditions.

Feeder circuit breaker settings

Curve	NI
Pick-up	326 A
TM	0.25
ARC	1 trips to lock-out

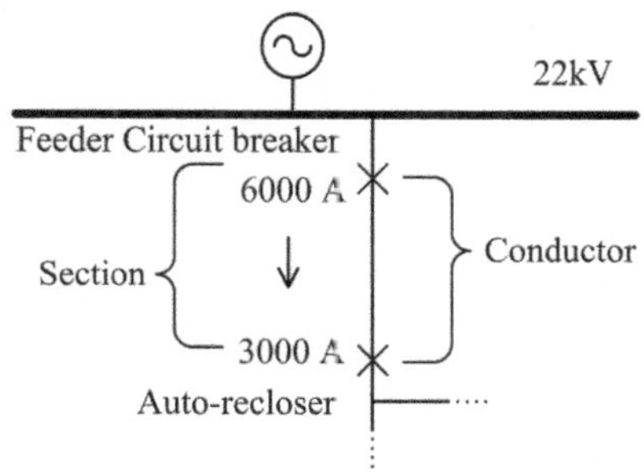

FIGURE 6.43 The network and applied protection settings for Example 6.4.

TABLE 6.10
Conductor LTE Limit Calculations

	Conductor Fault Current Withstand Time	Conductor LTE Limit
6000 A	$t_2 = \left(\frac{5400}{6000}\right)^2 \cdot 1$ $t_2 = 0.81\text{ s}$	LTE limit $= (5000)^2(0.81)$ LTE limit $= 29.16\text{ MA}^2\text{s}$
5000 A	$t_2 = 1.1664\text{ s}$	LTE limit $= 29.16\text{ MA}^2\text{s}$
4000 A	$t_2 = 1.8225\text{ s}$	LTE limit $= 29.16\text{ MA}^2\text{s}$
3000 A	$t_2 = 3.24\text{ s}$	LTE limit $= 29.16\text{ MA}^2\text{s}$

Solution

Determine the conductor energy limit across the line section: In Table 6.10 we can see that the conductor LTE limit stays constant as is expected.

Determine the LTE exposure based on the protection settings (Table 6.11).

Drawing the LTE exposure graph (Figure 6.44): In Figure 6.44 it can be seen that the LTE exposure is less than the conductor limit and, as such, the conductor is protected. The total exposure time is also shown in this graph, which shows that the time does increase as the fault current drops, but the current is still the dominant factor in this part of the curve.

TABLE 6.11

Protection Settings Related LTE Exposure Calculations

	Fault Exposure Time	LTE
6000 A	$ET = \frac{0.14 \cdot 0.25}{\left(\frac{6000}{326}\right)^{0.02} - 1} + 0.05$ $ET = 0.633\ s$	$LTE = (6000)^2(0.633)$ $LTE = 22.81\ MA^2s$
5000 A	$ET = 0.674\ s$	$LTE = 16.84\ MA^2s$
4000 A	$ET = 0.731\ s$	$LTE = 11.69\ MA^2s$
3000 A	$ET = 0.821\ s$	$LTE = 7.39\ MA^2s$

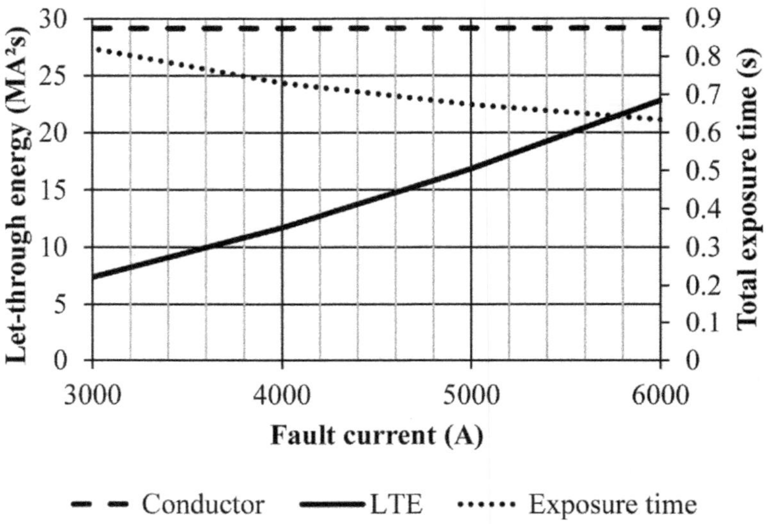

FIGURE 6.44 The conductor LTE limit and LTE exposure graph.

Example 6.6: Pickup Sensitivity

Determine the overcurrent protection element pickup in terms of sensitivity only. This is for every circuit breaker in the network shown in Figure 6.45. The three-phase fault levels are provided at the various positions in Figure 6.45.

Solution

To determine the pickup (PU) in terms of sensitivity, the criterion that is used is 80% of the phase-to-phase fault at the intended reach in minimum network conditions. Remembered that the upstream device should provide backup to all the immediate downstream reclosers. This means that the immediate downstream devices have to be placed on bypass, and then the lowest fault level should be used. It is not required to reach beyond two protective devices in series. This is only one of the three criteria used to determine the pickup. When all three are considered, the lowest of the three should be used (Table 6.12).

Another solution to the sensitivity is to set all devices sensitive to the lowest fault current for a fault in front of the device. This method is good for sensitivity and then dependability. The drawback is that the protection can be too sensitive when current grading is also considered (90% of upstream device pickup current). This can then have an impact on the security of the protection settings in that the protection can be too sensitive to load current.

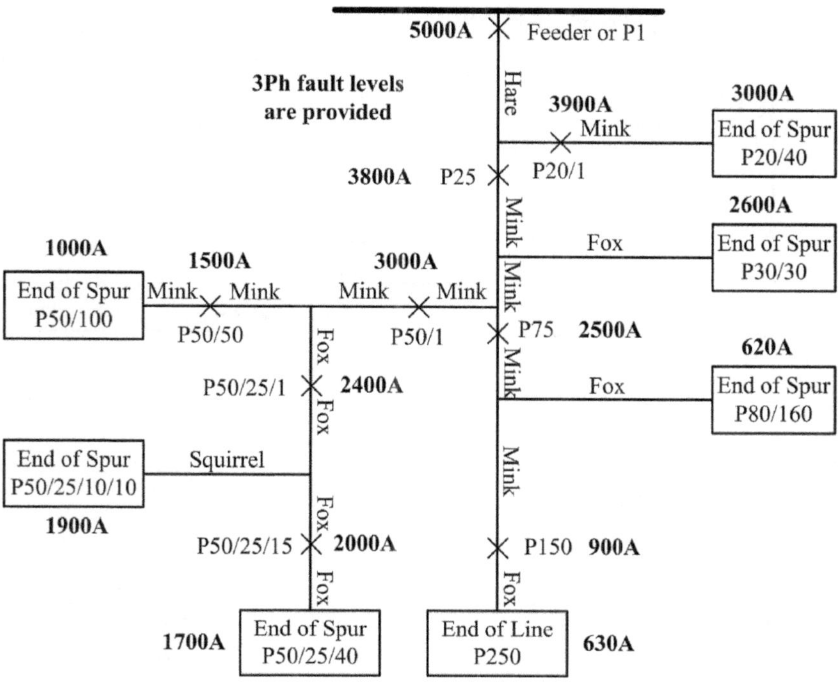

FIGURE 6.45 The radial feeder network for the sensitivity example.

TABLE 6.12

Device Pickup Current in Terms of Sensitivity Alone

Device	Bypassed	Fault Position and Levels Considered	Sensitivity Pickup
P1	P20/1 P25	3000 A P20/40 2600 A P30/30 2500 A P75 3000 A P50/1	$PU = 80\% \cdot \left(\frac{\sqrt{3}}{2}\right) \cdot 3\ ph$ $PU = 0.8 \cdot 0.866 \cdot 2500$ $PU = 1732\ A$
P20/1	None	3000 A P20/40	$PU = 0.8 \cdot 0.866 \cdot 3000$ $PU = 2078.4\ A$
P25	P75 P50/1	2600 A P30/30 620 A P80/160 900 A P150 2400 A P50/25/1 1500 A P50/50	$PU = 0.8 \cdot 0.866 \cdot 620$ $PU = 429.536\ A$
P75	P150	620 A P80/160 630 A P250	$PU = 0.8 \cdot 0.866 \cdot 620$ $PU = 429.536\ A$
P150	None	630 A P250	$PU = 0.8 \cdot 0.866 \cdot 630$ $PU = 436.464\ A$
P50/1	P50/50 P50/25/1	1000 A P50/100 1900 A P50/25/10/10 2000 A P50/25/15	$PU = 0.8 \cdot 0.866 \cdot 1000$ $PU = 692.8\ A$
P50/50	None	1000 A P50/100	$PU = 0.8 \cdot 0.866 \cdot 1000$ $PU = 692.8\ A$
P50/25/1	P50/25/15	1900 A P50/25/10/10 1700 A P50/25/40	$PU = 0.8 \cdot 0.866 \cdot 1700$ $PU = 1177.76\ A$
P50/25/40	None	1700 A P50/25/40	$PU = 0.8 \cdot 0.866 \cdot 1700$ $PU = 1177.76\ A$

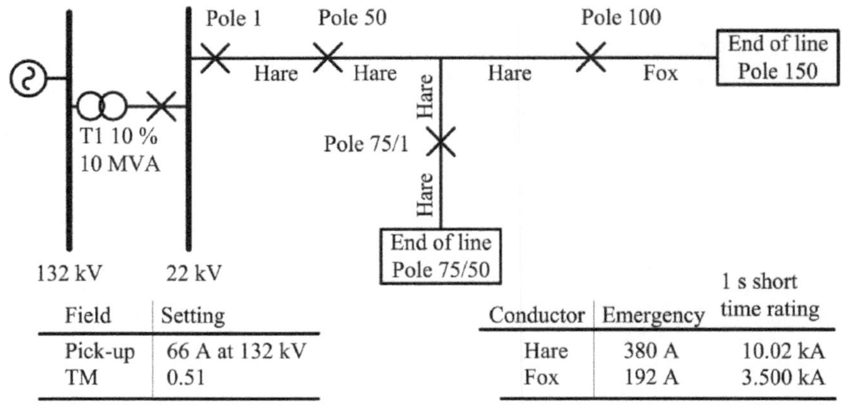

FIGURE 6.46 A single source transformer radial network.

TABLE 6.13

Fault Levels in the Single Transformer Radial Network

Position	Three-Phase Fault Level	Position	Three-Phase Fault Level
Pole 1	4000 A	Pole 75/50	1200 A
Pole 50	2000 A	Pole 100	1400 A
Pole 75/1	1600 A	Pole 150	1000 A

Example 6.7: Single Transformer Grading

Calculate protection settings for all the circuit breakers (including auto-reclosers) in the network shown in Figure 6.46. The fault levels are provided in Table 6.13. The conductor ratings and applied transformer settings (NI curve) are provided in the network diagram. Apply a minimum grading margin of 400 ms. The relay pickup can be set with no decimal places and 1 A steps in primary current values. The time multipliers and delays can be set with two decimal places and a step size of 0.01. The settings have to be calculated for two grading methods using an NI curve for all devices. It is not required to apply a high-set function for this example, and the circuit breaker time can be neglected.

a. Apply a top-down grading method.
b. Apply a bottom-up grading method.

Use a 500 ms operating time for the last device. This is for a fault at the device position.

Solution

a. Apply a top-down grading method: To determine the overcurrent NI element pickup (step 2), the tree criteria is used that was set out in Section 6.6.1. This included grading with the upstream pickup (refer to 22 kV side), setting the sensitivity and using the conductor rating. The high-set is not required for this example; those steps are marked accordingly. In this example, the minimum and maximum network conditions are the same as there is only one transformer at the source (Table 6.14).

   The LTE evaluation for step 12 is shown in the Table 6.15. The protection operating time is for one trip, and the LTE is multiplied with the number of ARC attempts. For the remote end, the three-phase fault levels are converted to phase-to-phase levels (0.866 the three-phase level) and then used.
b. Apply a bottom-up grading method: The pickup would be exactly the same as that of the top-down method. Start at the top (Pole 1) and then move down to the last device. The grading current is the same as for the top-down method; thus apply the golden rule of grading. Both Pole 75/1 and Pole 100 are the last devices and furthest away

TABLE 6.14

The Protections Settings Calculation Results for the Top-Down Method

Step	Pole 1	Pole 50	Pole 75/1	Pole 100
1	NI for all trips	NI for all trips	NI for all trips	NI for all trips
2	1. 356.4 A 2. 969.92 A at Pole 100 3. Hare 456 A Set to 356 A	1. 320.4 A 2. 692.8 A at Pole 150 3. Fox 230.4 A Set to 230 A	1. 207 A 2. 831.36 A at Pole 75/50 3. Hare 456 A Set to 207 A	1. 207 A 2. 692.8 A at Pole 150 3. Fox 230.4 A Set to 207 A
3	Not required for this example	Not required for this example	Not required for this example	Not required for this example
4	Not applicable as high-set is disabled	Not applicable as high-set is disabled	Not applicable as high-set is disabled	Not applicable as high-set is disabled
5	High-set disabled, Golden rule of grading, 4000 A	High-set disabled, Golden rule of grading, 2000 A	High-set disabled, Golden rule of grading, 1600 A	High-set disabled, Golden rule of grading, 1400 A
6	Upstream operating time is 1.508 s	Upstream operating time is 1.515 s	Upstream operating time is 1.203 s	Upstream operating time is 1.294 s
7	Is 1.508 s > 1.5 s? Yes, use 1.5 s	Is 1.515 s > 1.5 s? Yes, use 1.1 s	Is 1.203 s > 1.5 s? No, use 1.203 s	Is 1.294 s > 1.5 s? No, use 1.294 s
8	1.5 − 0.4 = 1.1 s	1.5 − 0.4 = 1.1 s	1.203 − 0.4 = 0.803 s	1.294 − 0.4 = 0.894 s
9	Not last device, use 1.1 s	Not last device, use 1.1 s	Is 0.803 s > 0.5 s? Yes, use 0.5 s	Is 0.894 s > 0.5 s? Yes, use 0.5 s
10	TM required 0.389 Set to 0.38	TM required 0.347 Set to 0.34	TM required 0.149 Set to 0.14	TM required 0.139 Set to 0.13
11	2 trips to lockout Dead time 15 s Reclaim Time 75 s	2 trips to lockout Dead time 15 s Reclaim Time 75 s	2 trips to lockout Dead time 15 s Reclaim Time 75 s	2 trips to lockout Dead time 15 s Reclaim Time 75 s
12	Pass Refer to evaluation below	Pass Refer to evaluation below	Pass Refer to evaluation below	Pass Refer to evaluation below
13	Pole 75/1 is 1.93 s Pole 100 is 2.14 s Pass, less than 5 s	Pole 75/50 is 1.83 s Pole 150 is 2.13 s Pass, less than 5 s	Pole 75/50 is 0.7 s Pass, less than 5 s	Pole 150 is 0.74 s Pass, less than 5 s
14	Disable zone sequence coordination	Disable zone sequence coordination	Not required, last device	Not required, last device

TABLE 6.15

The LTE Evaluation Results for the Protective Device at Pole 1

Device Pole	Position Pole	Fault Current [A]	Protection Operating Time [s]	LTE [MA^2s]	Conductor Type	Conductor I^2t Limit [MA^2s]	Exceed?
1	1	4000	1.073	34.34	Hare	100.4	No
	1	969.92	2.144	6.303	Hare	100.4	No
50	50	2000	1.076	8.614	Hare	100.4	No
	75/50	1200	2.04	4.08	Hare	100.4	No
	100	1400	1.294	5.072	Fox	12.25	No
	150	1000	1.025	2.049	Fox	12.25	No
75/1	75/1	1600	0.469	2.404	Hare	100.4	No
	75/50	831.36	0.695	0.96	Hare	100.4	No
100	100	1400	0.467	1.83	Fox	12.25	No
	150	692.8	0.744	0.71	Fox	12.25	No

from the source. Both will be set to an operating time of 500 ms. The first part of the procedure is applicable to the last device, and the second part is applicable to upstream devices. As Pole 75/1 has the higher fault level, it will have the greatest impact on the grading margin. For the last device, the fault is placed at the device position, but this can also be placed at the remote end of the device reach and this speeds up the protection even more (Table 6.16).

Example 6.8: Course Grading Example

A new auto-recloser is added to the network of Example 6.6. This auto-recloser is placed at Pole 75/25 (between Pole 75/1 and Pole 75/50). Use the course grading method to calculate protection settings for this device. Apply a minimum grading margin of 200 ms and use an NI curve. The relay pickup can be set with no decimal places and 1 A steps in primary current values. The time multipliers and delays can be set with two decimal places and a step size of 0.01.

Solution

For the course grading method, it is not required to determine the new fault levels. Only the fault level at the upstream device is required. As this is now the new last device, the grading margin is reduced to 200 ms. The upstream device was set to an operating time of 500 ms to accommodate one new device if required, such as in this case. Again, the pickup is determined using the three criteria that were set out in Section 6.6.1 (step 2), but only grading with the upstream device has to be calculated. The other two criteria are covered by the upstream device. The settings determined using the course grading method are not optimal, but the method does produce very quick settings that are sensitive to faults and allow for selectivity with upstream devices. The upstream protection at Pole 75/1 is set to a pickup of 207 A, a TM of 0.14 and an NI curve (Table 6.17).

Example 6.9: Grading Graphs

Draw the following grading graphs and include the transformer curve as well for all of them.

a. Grading graph for the top-down grading method of Example 6.7(a).
b. Grading graph for the bottom-up grading method of Example 6.7(b).
c. Draw a grading graph with only the circuit breaker at Pole 1 for both Examples 6.7(a) and (b) on the same graph.

TABLE 6.16

The Protections Settings Calculation Results for the Bottom-Up Method

Step	Pole 75/1	Pole 100	Pole 50	Pole 1
1	NI for all trips	NI for all trips	NI for all trips	NI for all trips
2	1. 207 A 2. 831.36 A at Pole 75/50 3. Hare 456 A Set to 207 A	1. 207 A 2. 692.8 A at Pole 150 3. Fox 230.4 A Set to 207 A	1. 320.4 A 2. 692.8 A at Pole 150 3. Fox 230.4 A Set to 230 A	1. 356.4 A 2. 969.92 A at Pole 100 3. Hare 456 A Set to 356 A
3	Not required for this example	Not required for this example	Not required for this example	Not required for this example
4	Not applicable as high-set is disabled	Not applicable as high-set is disabled	Not applicable as high-set is disabled	Not applicable as high-set is disabled
5	Device fault level 1600 A	Device fault level 1400 A	Not applicable here as this is not a last device	Not applicable here as this is not a last device
6	Require an operating time 0.5 s Set the TM to 0.14	Require an operating time 0.5 s Set the TM to 0.13	Not applicable as this is not a last device	Not applicable as this is not a last device
7	2 trips to lock-out Dead time 15 s Reclaim Time 75 s	2 trips to lock-out Dead time 15 s Reclaim Time 75 s	Not applicable here as this is not a last device	Not applicable here as this is not a last device
8	Disable zone sequence coordination	Disable zone sequence coordination	Not applicable here as this is not a last device	Not applicable here as this is not a last device
9	Pass This evaluation can be done. If the top down pass, the bottom up will also pass.	Pass	Not applicable here as this is not a last device	Not applicable here as this is not a last device
10	Pole 75/50 is 0.7 s Pass, less than 5 s	Pole 150 is 0.74 s Pass, less than 5 s	Not applicable here as this is not a last device	Not applicable here as this is not a last device
11	Golden rule of grading, 1600 A	Golden rule of grading, 1400 A	Golden rule of grading, 1600 A	Golden rule of grading, 2000 A
12	Operating time is 0.469 s	Operating time is 0.467 s	Slowest downstream device P75/1 operating time 0.469 s	Slowest downstream device P50 operating time 0.792 s
13	Not applicable	Not applicable	0.469 + 0.4 = 0.869 s	0.792 + 0.4 = 1.192 s
14	Not applicable	Not applicable	TM required 0.246 Set to 0.25	TM required 0.246 Set to 0.3
15	Already set	Already set	2 trips to lock-out Dead time 15 s Reclaim Time 75 s	2 trips to lock-out Dead time 15 s Reclaim Time 75 s
16	Pass	Pass	Pass	Pass
17	Already checked	Already checked	Pole 75/50 is 1.34 s Pole 150 is 1.57 s Pass, less than 5 s	Pole 75/1 is 1.83 s Pole 100 is 2.07 s Pass, less than 5 s
18	Already set	Already set	Disable zone sequence coordination	Disable zone sequence coordination
19	Yes	Yes	Yes	No
20	Not applicable as not at the source	Not applicable as not at the source	Not applicable as not at the source	At 4000 A Source operating time = 1.508 s Pole 1 operating time = 0.847 s Grading, yes

TABLE 6.17

The Protections Settings Calculation Results for the Course Grading Method

Step	Pole 75/1
1	NI for all trips
2	1. 186.3 A 2. Covered by upstream 3. Covered by upstream Set to 186 A
3	The upstream device (Pole 75/1) fault level is 1600 A
4	Operating time is 0.469 s
5	Is 0.469 s > 1.5 s? No, use 0.469 s
6	0.469 − 0.2 = 0.269 s
7	Is 0.269 s > 0.5 s? No, use 0.269 s
8	TM required 0.085. Set to 0.08
9	2 trips to lock-out. Dead time 15 s. Reclaim Time 75 s
10	Disable zone sequence coordination
11	Yes last device

Solution

a. Grading graph for the top-down grading method of Example 6.7(a) (Figure 6.47): It can be seen that grading between the curves is maintained throughout the current ranges. The biggest grading margins are between the last devices and the associated upstream device. An example is between Pole 75/1 and Pole 100.

b. Grading graph for the bottom-up grading method of Example 6.7(b) (Figure 6.48): For the bottom-up grading method, it can be seen that there is grading between all the required curves across the current range. The biggest grading margin is at the source between the transformer and the device at Pole 1.

c. Draw a grading graph with only the circuit breaker at Pole 1 (Figure 6.49): From this graph, it can be seen that the bottom up method results in faster protection times.

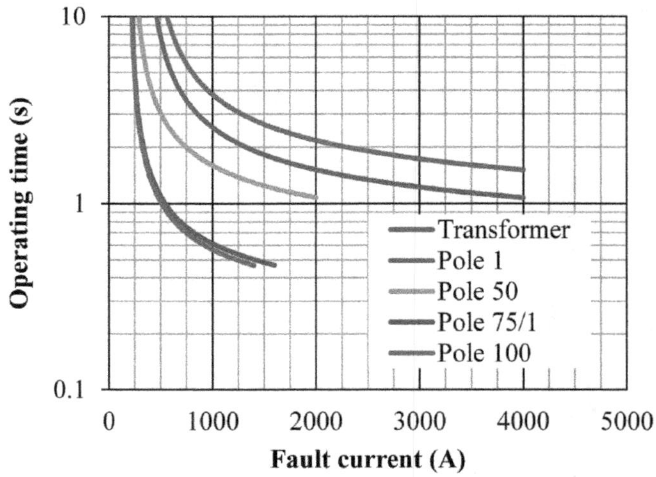

FIGURE 6.47 The grading graphs associated with the top-down method.

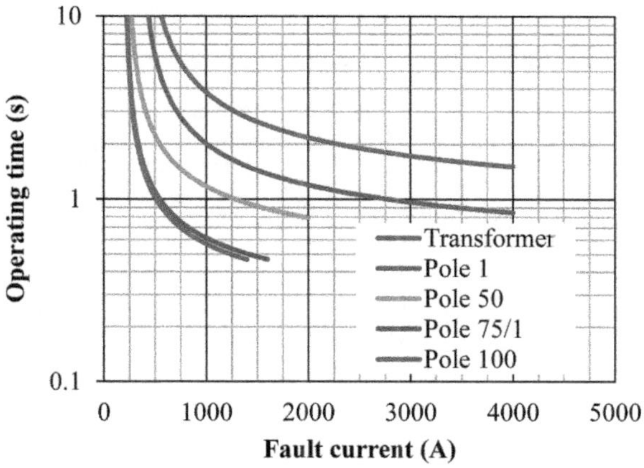

FIGURE 6.48 The grading graphs associated with the bottom-up method.

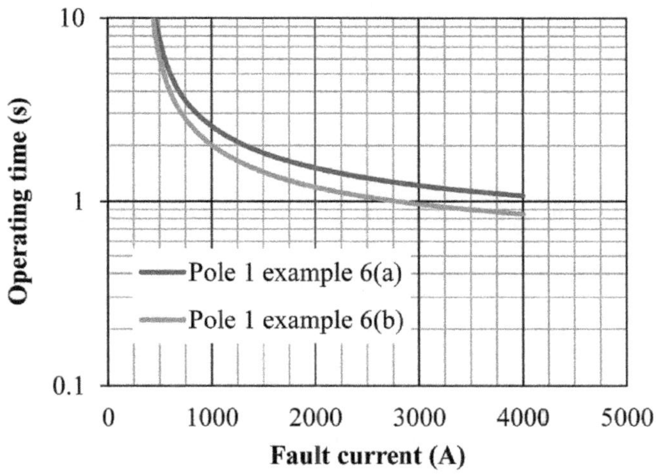

FIGURE 6.49 The operating curves of the circuit breaker for the top-down and bottom-up grading methods.

Example 6.10: Conductor LTE Limit

Calculate the short time rating for the Mink conductor. This conductor consists of two material types and is classified as an aluminium conductor steel reinforced (ACSR) conductor. Use an initial conductor temperature of 75°C and a final temperature of 200°C. The conductor material properties are provided in the Table 6.18.

Solution

The method used is obtained from the work done in [35]. The first step is to average the material parameters. Use Equation (6.18) to calculate the total mass (W) of the conductor:

$$W = W_1 + W_2$$

$$W = 174 + 82.8 \tag{6.25}$$

$$W = 256.8 \text{ kg/km}$$

TABLE 6.18

Mink Conductor Information from the Manufacturer's Datasheet

	Area	Mass	Temperature Coefficient of Resistance	Heat Capacity	Resistance at 20°C
	mm^2	kg/km	per °C	J/°C·g	Ω/km
Aluminium	63.13	174	0.00403	0.904	0.4546
Steel	10.52	82.8	0.00537[a]	0.488[a]	18.251[a]

Source: Aberdare, Overhead aluminium conductors, https://www.aberdare.co.za/product-categories/overhead-conductors/, January 2008.

[a] These values were obtained from [35].

The average heat capacity (S) of the conducting material can be determined by means of Equation (6.19):

$$S = \frac{W_1 S_1 + W_2 S_2}{W_1 + W_2}$$

$$S = \frac{174 \cdot 0.904 + 82.8 \cdot 0.488}{174 + 82.8} \tag{6.26}$$

$$S = 0.7699 \, \text{J/°C} \cdot \text{g}$$

Use Equation (6.20) to average the temperature coefficient of resistance (α) for the conductor:

$$\alpha = \frac{A_1 \alpha_1 + A_2 \alpha_2}{A_1 + A_2}$$

$$\alpha = \frac{63.13 \cdot 0.00403 + 10.52 \cdot 0.00537}{63.13 + 10.52} \tag{6.27}$$

$$\alpha = 0.00422 \, \text{per °C}$$

Now the resistance for the conductor can be calculated using Equation (6.21):

$$R = \left(\frac{1}{R_1} + \frac{1}{R_2} \right)^{-1}$$

$$R = \left(\frac{1}{0.4546} + \frac{1}{18.251} \right)^{-1} \tag{6.28}$$

$$R = 0.44355 \, \Omega/\text{km}$$

The resistance calculated with Equation (6.28) is at 20°C. This can now be moved to the initial temperature of 75°C with Equation (6.16):

$$R_1 = R_0 \left[1 + \alpha \left(T_1 - T_0 \right) \right]$$

$$R_1 = 0.44355 \left[1 + 0.00422 \left(75 - 20 \right) \right] \tag{6.29}$$

$$R_1 = 0.5465 \, \Omega/\text{km}$$

Use Equation (6.17) to calculate the LTE limit for the Mink ACSR conductor:

$$I^2R\cdot t=\left(\frac{W\cdot S}{\alpha\cdot 1000}\right)\log_e\left(1+\alpha\left(T_f-T_i\right)\right)$$

$$I^2t=\left(\frac{256.8\cdot 0.7699}{0.00422\cdot 0.5465\cdot 1000}\right)\log_e\left(1+0.00422(200-75)\right) \tag{6.30}$$

$$I^2t=36.317\,\text{kA}^2\text{s}$$

The 1 s short time rating for the Mink ACSR conductor can now be calculated using Equation 6.31:

$$I=\frac{\sqrt{I^2t}}{\text{required time}}$$

$$I=\frac{\sqrt{36.317}}{1} \tag{6.31}$$

$$I=6.026\,\text{kA}$$

6.11 Tutorial Problems

1. Protection philosophy:
   a. Explain how determining the pickup current relates to the protection philosophy building blocks.
   b. How do the protection objectives (as set in this chapter) relate to the protection philosophy building blocks?
2. Protection element operating curves:
   a. What considerations can be used when determining the type of operating curve to implement on a protection relay on an MV feeder for overcurrent protection? The type of curve refers to an NI or EI operating curve (not limited to these) as an example.
   b. What type of IDMT curve is well suited to minimize damage in high fault level areas?
   c. What are the drawbacks and advantages of using an IDMT curve that is more inverse than an NI curve?
   d. What can be done to ensure that the protection operating time does not reach infinity during low fault levels at the end of a feeder (at its intended protective reach)?
3. Grading:
   a. Why should we grade series protective devices?
   b. Why should two devices be graded at the maximum current that both devices will measure?
   c. What do minimum and maximum network conditions refer too?
   d. What will happen to the grading margin if grading is done in minimum network conditions? Explain your answer.
4. Operating time:

   Answer the question related to the grading graph in Figure 6.50. There are four protective devices in series on this radial feeder. These are the source transformers, the feeder circuit breaker (Pole 1), the auto-reclosers at Pole 40 and then Pole 120.

   a. What is the operating time of the auto-recloser at Pole 40 for a fault current of 2000 A?
   b. What is the grading margin between the auto-recloser at Pole 40 and Pole 120 for a fault current of 500 A?

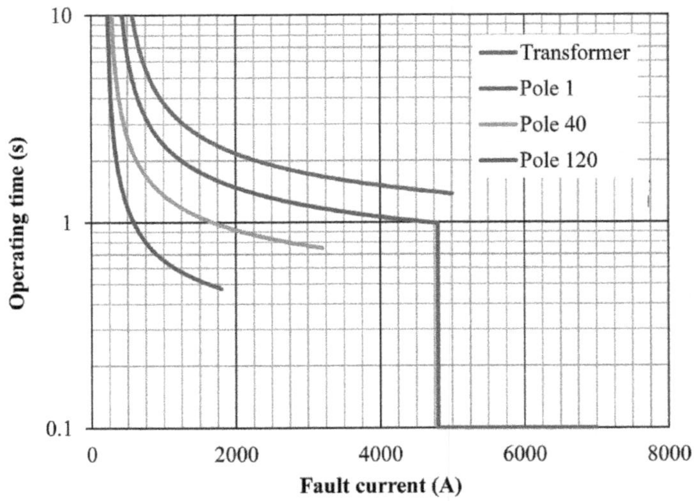

FIGURE 6.50 The grading graphs for question 1 in Section 6.11.

c. What is the minimum grading margin between the auto-recloser at Pole 40 and Pole 120? Also provide the location of the fault level at this minimum grading margin.
d. What is the high-set pickup set to for the feeder circuit breaker?
e. With one transformer operational, the maximum fault level at the busbar is equal to 5000 A. What is the minimum grading margin between the transformer and the feeder circuit breaker? Also provide the fault level.
f. With two similar transformers operational at the substation, the busbar fault level is 7000 A. Determine the minimum grading margin between the transformers and the feeder circuit breaker. Also provide the current that each device is measuring.
g. There is a fault on the network beyond the auto-recloser at Pole 40. The fault current is 2600 A at the fault position. Determine the operating time of the protective devices at Pole 40, the feeder and then also the transformers (2 transformers operational).

5. LTE protection:

Determine if the section of the feeder is protected by the protection settings applied to the feeder circuit breaker relay in Figure 6.51. The circuit breaker operating time is 50 ms. There are two conductor types in the first section of the conductor. The first conductor type has a short time rating of 7000 A for 1 s and the second conductor has a rating of 5400 A for 1 s. *Tip:* Choose current points to evaluate. Draw the energy curves.

6. Multi-transformer grading:

Calculate protection settings for all the circuit breakers (including auto-reclosers) in the network below. The fault levels are provided in Table 6.19 for both one (1 Trfr) and then when both (2 Trfr's) transformers are operational. This is the current passing through the device at that position. The conductor ratings and applied transformer settings (NI curve) are provided in the network diagram. The same settings are applied to both transformers. Apply a minimum grading margin of 400 ms. The relay pickup can be set with no decimal places and 1 A steps in primary current values. The time multipliers and delays can be set with two decimal places and a step size of 0.01. The settings have to be calculated for two grading methods using an NI curve for all devices. It is not required to apply a high-set function for this example, and the circuit breaker time can be neglected.

a. Apply a top-down grading method.
b. Apply a bottom-up grading method. Use a 500 ms operating time for the last device. This is for fault at the device position (Figure 6.52).

Feeder circuit breaker settings

Curve	NI
Pick-up	326 A
TM	0.25
High-set	5000 A
TD	0 s
ARC	2 trips to lock-out (both IDMT and high-set)

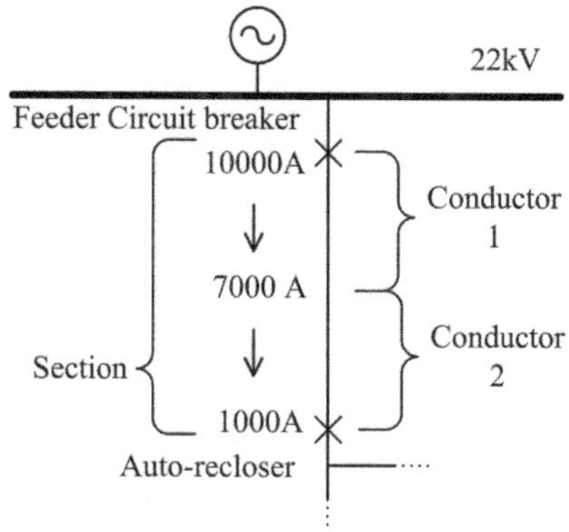

FIGURE 6.51 The network and applied protection settings for question 2 in Section 6.11.

TABLE 6.19

Fault Levels in the Double Transformer Radial Network

	Three-Phase Fault Level			Three-Phase Fault Level	
Position	**1 Trfr**	**2 Trfr's**	**Position**	**1 Trfr**	**2 Trfr's**
Pole 1	4000 A	6000 A	Pole 75/50	1200 A	1800 A
Pole 50	2000 A	3500 A	Pole 100	1400 A	2000 A
Pole 75/1	1600 A	2800 A	Pole 150	1000 A	1400 A

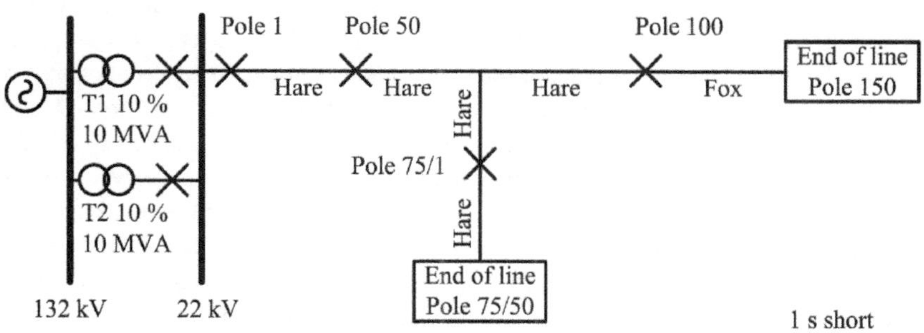

Field	Setting
Pick-up	66 A at 132 kV
TM	0.51

Conductor	Emergency	1 s short time rating
Hare	380 A	10.02 kA
Fox	192 A	3.500 kA

FIGURE 6.52 A radial network consisting of multiple source transformers.

7. LTE graphs:

   Draw the following LTE graphs.

   a. LTE graph for the top-down grading method of question 6(a) in Section 6.11.
   b. Let-through graph for the bottom-up grading method of question 6(b) in Section 6.11.
   c. Draw an LTE graph with only the circuit breaker at Pole 1 for questions 3(a) and (b) in Section 6.11 on the same graph.

TABLE 6.20

Applied Protection Settings to Device A and B

Device	Operating Curve	Pickup	TM
Device A	NI	380 A	0.4
Device B	EI	350 A	0.4

8. Relay operating time in an interconnected network:

   Two protection relays are placed in series with each other in a multisource network. Device A is at the source, and Device B is further down in the network. The applied protection settings for each protection device are provided in Table 6.20. A fault occurs in the network and both devices measure the same current. The fault happened at 1 s. For period one (1 s to 1.25 s), the RMS fault current was equal to 2200 A. For period two (1.25 s to 1.75 s), the RMS fault current was equal to 2000 A. Finally, for period three (1.75 s to trip time s), the RMS fault current was equal to 3000 A.

   a. Illustrate the RMS fault current signal over time (the fault described above).
   b. Determine how long Devices A and B will take to issue a trip signal for the fault current signal described.
   c. Draw the grading graphs for these two devices up to the maximum current of 3000 A.
   d. From the grading graph, is selectivity maintained?
   e. Is selectivity maintained between the devices for the fault described above?
   f. Can the grading graph be used to evaluate the selectivity if the fault current change before a trip signal is issued by the protective device?

6.12 Conclusion

This chapter focused on the application of protection settings to medium voltage networks that are protected using overcurrent protection. Protection philosophy concepts were discussed and applied to medium voltage networks. This application included the considerations towards relay technology and protection device functionality. Different operating curves were discussed, together with decisions on choosing which curve to use by means of the effect it will have on operating time and its reaction to the change in fault levels across the feeder. For the auto-reclose function, determining the number of cycles, dead and reclaim time influence were discussed. We also looked at the effects of the auto-reclose function on LTE exposure. Traditionally a large focus is placed on selectivity and security (not tripping for load). This chapter showed how to determine if the applied protection setting will result in conductor damage by using the LTE concept. A method was shared to determine conductor LTE limits. Different grading methods with flow diagrams were applied to feeders to illustrate how to determine protection settings for a medium voltage network using overcurrent-based protection. These grading methods were then based on the philosophy principles and objectives set out in the initial part of the chapter. Finally, the application of overcurrent-based protection on interconnected or multisource networks were discussed. This discussion showed that the normal IDMT equations cannot be used in their current form as the current that the relay is measuring can change before a trip is issued to the circuit breaker. This current change influences the protection operating time. Different relay models are shared to provide a solution to determining the operating time.

REFERENCES

1. IEEE Recommended Practice for Electric Power Distribution for Industrial Plants, IEEE Std. 141, 1994.
2. IEC Standard Voltages, IEC 60038, 2002.
3. S. Javadian and M. Haghifam, "Maintaining the recloser-fuse coordination in distribution systems in presence of DG by determining DG's size," *Proceedings of the 9th International Conference Institute of Engineering and Technology Development in Power System Protection*, Glasgow, UK, pp. 124–129, 2008.

4. IEEE Guide for protective relay applications to distribution lines, IEEE Std. C37.230, 2007.
5. AREVA T&D, "Fundamentals of protection practice", "Relay technology", "Overcurrent protection for phase and earth faults", "Auto-reclosing", and "Transformer and transformer feeder protection," in *Network Protection and Automation Guide*, AREVA T&D, Stafford, UK, 2011, Ch. 2, 7, 9, 14, and 16.
6. Protection settings philosophy for medium voltage distribution networks, Eskom 240-76628317, 2015.
7. V. Calderaro, "Adaptive relays for overhead line protection," *Elect. Power Syst. Res.*, vol. 77, pp. 1552–1559, 2007.
8. H. B. Funmilayo, J. A. Silva and K. L. Butler-Purry, "Overcurrent protection for the IEEE 34-node radial test feeder," *IEEE Trans. Power Del.*, vol. 27, no. 2, pp. 459–468, 2012.
9. S. H. Horowitz and A. G. Phadke, "Introduction to protective relaying", and "Relay operating principles," in *Power System Relaying*, 3rd ed., West Sussex, UK: John Wiley & Sons, 2008, Ch. 1 and 2.
10. J. C. Gomez and M. M. Morcos, "Voltage sag mitigation using overcurrent protection devices," *Electr. Power Compon. Syst.*, vol. 29, no. 1, pp. 71–81, 2010.
11. IEEE Guide for Electric Power Distribution Reliability Indices, IEEE Std. 1366, 2012.
12. S. Jamali and H. Shateri, "Optimal application of reclosers and sectionalisers to reduce non-distributed energy in distribution networks," *Presented at the 18th International CIRED Conference*, Turin, Italy, June 6–9, 2005.
13. IEEE Guide for Automatic Reclosing of Line Circuit Breakers for AC Distribution and Transmission Lines, IEEE Std. C37.104, 2002.
14. S. H. Money and J. Harris, "Autoreclosing switchgear in distribution practice," *Proc. Inst. Electr. Eng.*, vol. 115, no. 2, pp. 288–300, 1968.
15. A. Conde and E. Vazquez, "Application of a proposed overcurrent relay in radial distribution networks," *Elect. Power Syst. Res.*, vol. 81, pp. 570–579, 2011.
16. IEEE Standard Inverse-Time Characteristic Equations for Overcurrent Relays, IEEE Std. C37.112, 1996.
17. E. Sorrentino, "Behavior of induction disc overcurrent relays as a function of the frequency," *Electr. Power Syst. Res.*, vol. 143, pp. 474–481, 2017.
18. A. C. Enrizquez, E. V. Martinez and H. J. Altuve, "Time overcurrent adaptive relay," *Electr. Power Energy Syst.*, vol. 25, pp. 841–847, 2003.
19. C.-R. Chen, C.-H. Lee and C.-J. Chang, "Optimal overcurrent relays coordination in power distribution system using a new approach," *Electr. Power Energy Syst.*, vol. 45, pp. 217–222, 2013.
20. J. Meppelink and M. Benzin, "Lightning protection of metal roofs", in *28th International Conference on Lightning Protection*, Kanazawa, Japan, pp. 1298–1303, 2006.
21. A. D. Stokes, "Fire ignition by electrically produced incandescent particles," *Aust. J. Electr. Electron. Eng.*, vol. 10, no. 3, pp. 175–187, 1990.
22. M. J. Slabbert, S. J. van Zyl, R. Naidoo and R. C. Bansal, "Evaluating phase over-current protection philosophies for medium-voltage feeders applying let-through energy and voltage dip minimization," *Electr. Power Compon. Syst.*, vol. 44, no. 2, pp. 206–218, 2016.
23. C. R. Mason, "Line protection with overcurrent relays," in *The Art and Science of Protective Relaying*, Wiley, 1956, pp. 259–295.
24. T. Keil and J. Jager, "Advanced coordination method for overcurrent protection relays using nonstandard tripping characteristics," *IEEE Trans. Power Del.*, vol. 23, no. 1, pp. 52–57, 2008.
25. Guide for the application and setting of phase instantaneous overcurrent protection, Eskom 34-994, 2008.
26. M. J. Slabbert, S. J. van Zyl and R. Naidoo, "Using let-through energy to determine the application of a high current lock-out function on MV feeders," in *Southern African Power System Protection Conference*, Johannesburg, South Africa, November 12–14, 2014.
27. A. Wright and P. G. Newbery, "Application of fuses," in *Electric Fuses*, 3rd ed., Stevenage, UK: The Institution of Engineering and Technology, 2004, p. 161.
28. A. E. D. C. Tio, I. B. N. C. Cruz, B. M. Malquisto and R. D. del Mundo, "A binary programming model for reliability optimization considering fuse-blow and fuse-save schemes," *TENCON 2012 IEEE Region 10 Conference*, November 19–22, pp. 1–6, 2012.
29. *Electrical Transmission and Distribution Reference Book* (formerly the Westinghouse Electrical Transmission and Distribution Reference Book), ABB Power T&D Company, Inc., pp. 666–808, 1997.
30. S. van Zyl and A. Lubbe, "Characterisation of MV overhead feeder faults using auto-recloser event data," in *Southern African Power System Protection Conference*, Johannesburg, South Africa, November 12–14, 2014.

31. D. Halliday, R. Resnick and J. Walker, "Motion along a straight line", "Current and resistance" and "Conduction of electricity," in *Fundamentals of Physics*, 6th ed., New York: John Wiley & Sons, 2001, Ch. 2, 27, and 42.
32. J. W. Nilsson and S. A. Riedel, "Circuit elements," in *Electric Circuits*, 6th ed., Upper Saddle River, NJ: Prentice Hall, 2001, pp. 31–33.
33. J. D. Cutnell and K. W. Johnson, "Temperature and heat", and "Electric circuits," in *Physics*, 4th ed., New York: John Wiley & Sons, 1998, Ch. 12 and 20.
34. G. M. Masters, "Basic electric and magnetic circuits," in *Renewable and Efficient Electric Power Systems*, Hoboken, NJ: John Wiley & Sons, pp. 8–17, 2004.
35. B. Wareing, "Conductor characteristics and selection," in *Wood Pole Overhead Lines*, 1st ed., vol. 48, London, UK: The Institution of Engineering and Technology, pp. 114–140, 2008.
36. G. Parise and M. Adduce, "Conductor protection against short circuit current: Available I^2t evaluation," *Conference Record—IEEE IAS Annual Meeting*, pp. 2336–2341, 1998.
37. E. S. Thomas, "Bonding requirements for conductive poles," *Presented at IEEE Rural Electric Power Conference (REPC)*, Milwaukee, WI, April 15–17, pp. A4-1–A4-6, 2012.
38. M. Mitolo and M. Tartaglia, "An analytical evaluation of the factor k^2 for protective conductors," *IEEE Trans. Ind. Appl.*, vol. 48, no. 1, pp. 211–217, 2012.
39. M. Tartaglia and M. Mitolo, "An analytical evaluation of the prospective I^2t to assess short-circuit capabilities of cables and busways," *IEEE Trans. Power Del.*, vol. 25, no. 3, pp. 1334–1339, 2010.
40. W. J. Callister, Jr., "Mechanical properties of metals", "Failure", and "Electrical properties," in *Materials Science and Engineering an Introduction*, 6th ed., New York: John Wiley & Sons, 2003, Ch. 6, 8, and 18.
41. Aberdare. January 2008. Overhead aluminium conductors, https://www.aberdare.co.za/product-categories/overhead-conductors/.
42. A. H. Osman, M. S. Hassan and M. Sulaiman, "Communication-based adaptive protection for distribution systems penetrated with distributed generators," *Electric Power Components and Systems*, vol. 43, no. 5, pp. 556–565, 2015.
43. J. Stewart, "Integrals," in *Calculus Early Transcendentals*, 4th ed., Pacific Grove, CA: Brooks/Cole Publishing, pp. 378–388, 1999.
44. P. S. Babu, S. V. J. Kumar and P. R. K. Chaitanya, "Digital relay based adaptive protection for distributed systems with distributed generation," *Int. J. Energy Sci.*, vol. 1, no. 2, pp. 72–77, 2011.
45. M. Yen Shih and A. Conde, "Implementation of directional over-current relay coordination approaches in electrical networks," *Electr. Power Compon. Syst.*, vol. 43, no. 19, pp. 2131–2145, 2015.
46. H. Schmidt and G. Speckhofer, "Experimental and theoretical investigation of high-pressure arcs-Part I: The cylindrical arc column (two-dimensional modelling)," *IEEE Trans. Plasma. Sci.*, vol. 24, no. 4, pp. 1229–1238, 1996.
47. E. Nasser, "Some physical properties of electrical discharges on contaminated surfaces," *IEEE Trans. Power. App. Syst.*, vol. PAS-87, no. 4, pp. 957–963, 1968.
48. M. Arrayas and J. L. Trueba, "Investigations of pre-breakdown phenomena: Streamer discharges," *Contemp. Phys.*, vol. 46, no. 4, pp. 265–276, 2005.
49. J. Ekanayake, K. Liyanage, J. Wu, A. Yokoyama and N. Jenkins, "The smart grid," in *Smart Grid Technology and Applications*, 1st ed., West Sussex, UK: John Wiley & Sons, 2012, Ch. 1 and 2.
50. M. Farsadi, A. Yazdani Nejadi and A. Esmaeilynasab, "Reducing over-current relays operating times in adaptive protection of distribution networks considering DG penetration," *2015 9th International Conference on Electrical and Electronics Engineering (ELECO)*, Bursa, Turkey, pp. 463–468, 2015.
51. A. C. Enríquez, and E. V. Martínez, "Sensitivity improvement of time overcurrent relays," *Electr. Power Syst. Res.*, 2006.
52. W. Sheng, X. Meng and J. Ma, "A new adaptive current protection scheme for distributed systems with distributed generation," *Presented at theInternational Conference on Power, Electronics and Materials Engineering*, Dalian, China, May 16–17, 2015.
53. X. Lin, Y. Lu and L. Wang, "Agent-based multi-source wide-area current protection for systems with distributed generators," *Aust. J. Electr. Electron. Eng.*, vol. 6, no. 1, pp. 33–44, 2015.
54. A. Abdelaziz, H. Talaat, A. Nosseir and A. A. Hajjar, "An adaptive protection scheme for optimal coordination of overcurrent relays," *Electr. Power Syst. Res.*, vol. 61, no. 1, pp. 1–9, 2002.
55. M. J. Slabbert and S. J. van Zyl, "Analysis of MV feeder protection performance using autorecloser event data," *Presented atEskom Protection Workshop*, Johannesburg, South Africa, November 9–10, 2011.

7

Bus-Bar Protection

Arvind R. Singh, Ranjay Singh, Abhishek Kumar, Raj Naidoo, and Ramesh Bansal

CONTENTS

7.1 Introduction

Bus-bar is an essential part of the electrical power system; it connects various types of equipment to each other, forming a node. The bus connects most of the circuits, and it is also called thes nodes of a power system [1]. The word *bus* is defined as a distinct set of conductors (bars) carrying signals to which various pieces of equipment are connected in parallel. Bus-bar has incoming lines and outgoing lines (transmission or distribution lines). If there are N number of incoming lines and M number of outgoing lines, then

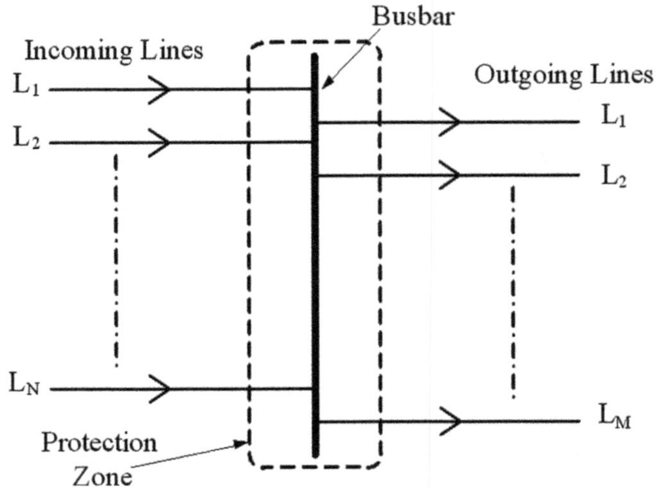

FIGURE 7.1 Equivalent representation of busbar.

the incoming current is always equal to outgoing current, as depicted in Equation (7.1). Figure 7.1 shows the equivalent representation of bus-bar having incoming N number of incoming lines and M number of outgoing lines, with its protection zone covered by its protective relay.

$$\sum_{i=1}^{N} I_i = \sum_{j=1}^{M} I_j \tag{7.1}$$

There are various types of bus-bar arrangements available, but depending upon its installation location, it has mainly two categories as an outdoor bus-bar and the indoor bus-bar. The bus-bar carries the bulk of the current through it; hence it acts as a sink for short-circuit current. It is very rare to have a fault on bus-bar, but if it does occur, it causes severe damage and in some cases complete shutdown of the power system. Bus-bars are located in switchyards at substations, making bays with the main bus-bar and auxiliary bus-bar; in gas-insulated substations, bus-bars are located inside the substation. The outdoor bus-bars are well shielded from lighting strikes but subjected to harsh weather conditions.

7.2 Conventional Bus-Bar Arrangements

There are different types of the bus-bar arrangements and it is required to choose the arrangement. This depends on various factors such as

1. System voltage.
2. Position of a substation in the system.
3. Reliability of supply.
4. Flexibility.
5. Cost.
6. Availability of alternative arrangements if an outage at any of the pieces of equipment happens.
7. Bus-bar arrangement should be simple and easy to maintain.
8. In case of load growth, there should be a possibility to extend the system to meet the load requirements.
9. The installation should be as economical as possible, keeping in view the needs and continuity of supply.

Types of bus-bar arrangements

1. Single bus-bar arrangement.
2. The single bus-bar arrangement with bus sectionalized.
3. Double bus arrangement.
4. Double bus double breaker arrangement.
5. One and a half breaker arrangement.
6. Main and transfer bus arrangement.
7. Double bus system with bypass isolators.
8. Ring main arrangement.
9. Mesh arrangement.

7.2.1 Single Bus-Bar Arrangement

This single bus-Bar arrangement consists of only one bus-bar, and all the incoming feeders and outgoing distributors are connected to this bus-bar only. All the fuses, circuit breakers, generators, and transformers are connected to this, as shown in Figure 7.2.

Advantages:

1. It is smooth in operation.
2. Initial cost is less.
3. Requires less maintenance.

Disadvantages:

1. When damage occurs, then there will be complete interruption of power supply.
2. Flexibility and immunity are reduced.

7.2.2 Single Bus Arrangement with Bus Sectionalized

In the single bus-bar arrangement with bus sectionalized we divide a single bus-bar into two sections with the help of a circuit breaker and isolator switches, and the load is distributed equally among both sections, as shown in Figure 7.3.

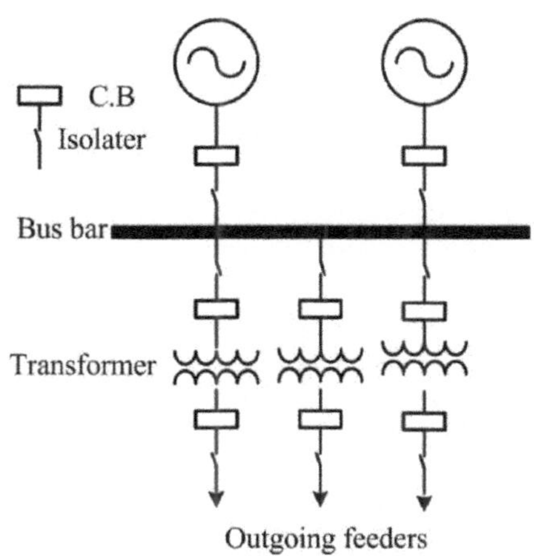

FIGURE 7.2 Single bus-bar arrangement.

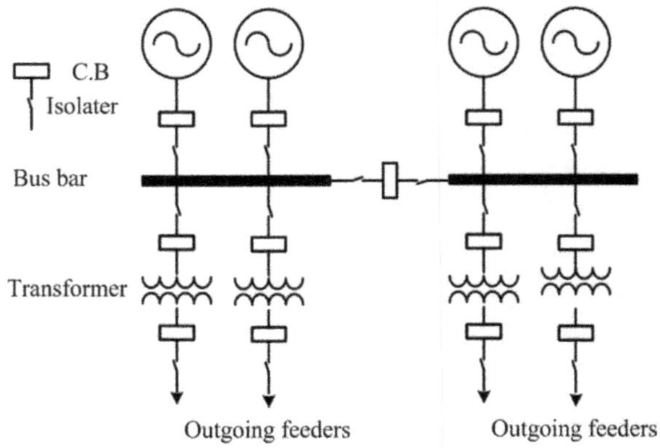

FIGURE 7.3 Single bus-bar arrangement with bus sectionalized.

Advantages of sectionalized single bus-bar arrangement:

1. As we are using the circuit breaker to divide a bus-bar into two sections, a fault on one section will not affect the other sections. Only a few loads will have lack of power supply.
2. The fault level can be reduced by adding a current limiting reactor.

Disadvantages of sectionalized single bus-bar arrangement:

1. It requires extra isolators and circuit breakers, which increases the cost.

7.2.3 Double Bus Arrangement

Double bus arrangement has two bus-bars, and the incoming feeders and outgoing feeders are connected in parallel to both buses with the help of isolators. By closing the isolator switch, we can connect the feeders to either bus-bar 1 or bus-bar 2. We can divide the load among two buses with the help of isolator switches. By closing the isolator switch connected to bus-bar 1 and feeder, the load can be connected to bus-bar 1, and by closing the isolator switch connected to bus-bar 2 and feeder, the load gets connected to bus-bar 2, as shown in Figure 7.4.

A bus coupler breaker is used for bus transfer operation. When we need to transfer load from one bus to another, we need to close the bus coupler first, then close isolators of the associated bus to which load is to be connected, open the isolator switch coupled to the fault bus, and then open the bus coupler breaker.

Advantages:

1. It has greater flexibility.
2. During fault conditions, there is no interruption of power supply to the load.

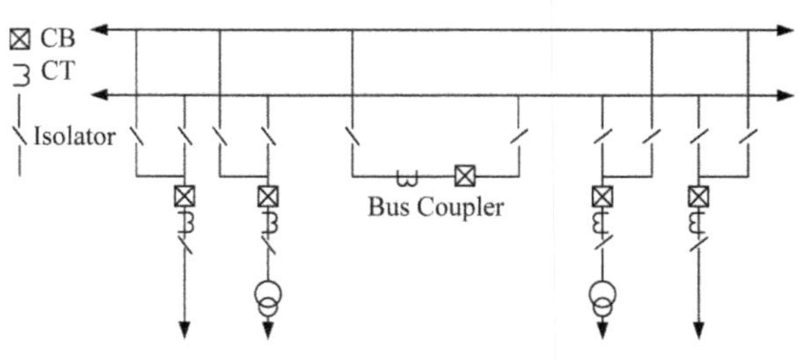

FIGURE 7.4 Double bus-bar arrangement.

Disadvantages:

1. We cannot transfer load from one bus to another without interruption of power supply for few minutes.

7.2.4 Double Bus with Double Breaker Arrangement

In a double bus with double breaker arrangement, we connect a feeder in parallel to both buses with the help of two circuit breakers and isolator switches instead of a bus coupler, as shown in Figure 7.5.

Both feeders are energized and divided between feeders among both the buses, but it can connect the desired feeder to the desired bus at any time For this purpose, an isolator needs to be closed with the associated circuit breaker of the required bus-bar, and later open the circuit breaker with isolator from which it has to be disconnected.

Advantages:

1. During fault conditions, the load can be transferred from one bus to another to prevent interruption in power supply.
2. Here we are not using a bus coupler, so there will not be much delay in power supply while closing the circuit breaker to transfer load from one bus to another.
3. High flexibility.

Disadvantages:

1. The high number of circuit breakers used increases the cost.
2. Maintenance cost will also be high.

Thus, this type of arrangement is very rarely used.

7.2.5 One and a Half Breaker Arrangement

With the high cost of the double bus with double breaker arrangement, the one and a half breaker arrangement is preferred. In this arrangement, two feeders are fed through their corresponding bus-bars, and these two feeders are coupled by a third circuit breaker called a tiebreaker, as shown in Figure 7.6.

During normal conditions, all three circuit breakers are closed, both circuits operate in parallel and power is fed to feeders from the two bus-bars. If a fault occurs on one bus-bar, then, with the help of the second bus-bar, the feeder circuit breaker and tiebreaker power are fed to feeders. This means each feeder breaker has to be rated to feed both feeders which are coupled by a tiebreaker.

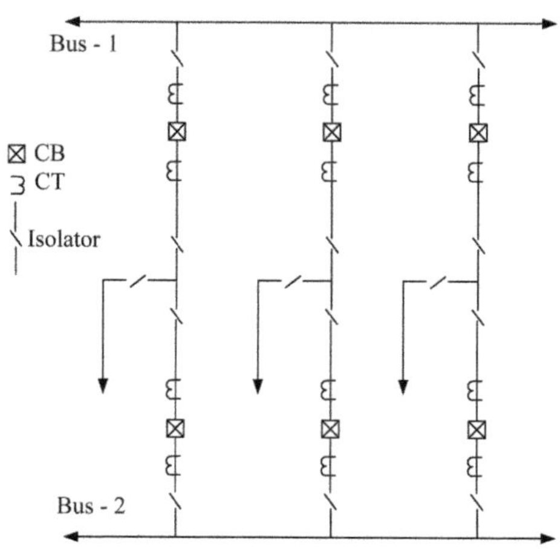

FIGURE 7.5 Double bus with double breaker arrangement.

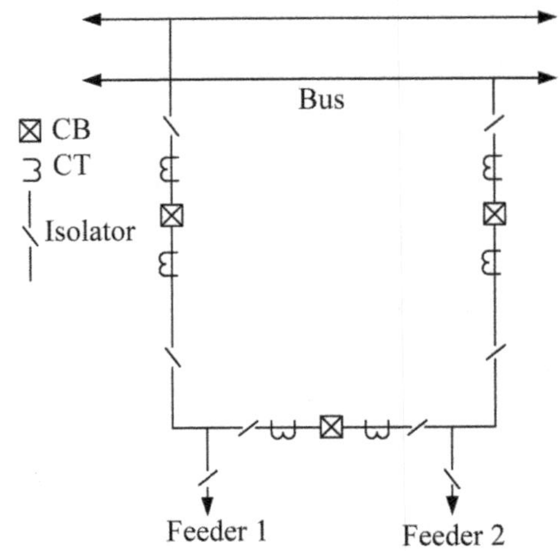

FIGURE 7.6 Double bus with one and half breaker arrangement.

Advantages:

1. There will be no interruption of power in case of a fault because all the feeders can be transferred to the working bus immediately.
2. Additional circuits can be easily added to the system.
3. Cost is less compared to the double bus double breaker arrangement.

Disadvantages:

1. This arrangement of fault isolation is complicated because, during a fault, two circuit breakers are to be opened.
2. Maintenance cost is high.

7.2.6 Main and Transfer Bus Arrangement

The main and transfer bus arrangement has two buses: one is the main bus, and the other is the transfer bus. With the help of isolator switches, it is connected to the transfer bus, which is called bypass isolators and with the help of circuit breakers and isolator switches, it is connected to the main bus. A bus coupler is also required in this arrangement, as shown in Figure 7.7.

In normal conditions, the feeders are fed through the main bus, but during fault conditions, load is transferred to the transfer bus. To transfer load from the main bus to the transfer bus, we need to close the

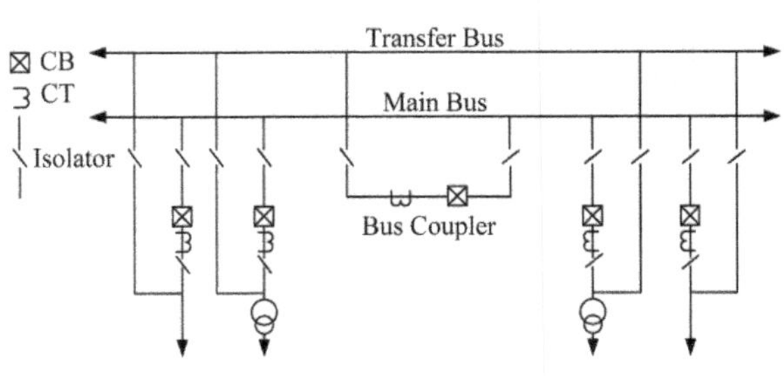

FIGURE 7.7 Main and transfer bus arrangement.

bus coupler first and then close bypass isolators of the feeder to be connected to the transfer bus. Then we must open the isolator switch of the feeder coupled to the main bus and then open the bus coupler breaker.

Advantages:

1. No interruption of power supply because, in case of a fault, load can be shifted to the transfer bus.
2. The load can be divided into two groups since it can be feed from either of the buses.

Disadvantages:

1. Two bus-bars are used, which increases the cost.

7.2.7 Double Bus Arrangement with Bypass Isolators

A double bus arrangement with bypass isolators is the combination of the double bus system and main and transfer bus system. Here the feeders are connected to both buses with the help of isolators. During fault condition, loads can be transferred to the healthy bus by closing isolators of feeders associated with the healthy bus and opening isolators of feeders associated with the faulty bus. The connection diagram is given Figure 7.8.

A bus coupler is also provided for transferring the load to the healthy bus. The bus coupler closed the breaker first and then the isolator of the feeder to which it has to be transferred, and then it opened the isolator switch of the feeder coupled to the fault bus with the bus coupler breaker.

Advantages:

1. No interruption of power supply because, in case of a fault, load can be shifted to the transfer bus.
2. The load can be divided into two groups since they can be fed from either of the buses.

Disadvantages:

1. Cost is high as we are using two bus-bars and additional isolator switches.
2. The arrangement is complex.

7.2.8 Ring Main Arrangement

The ring main arrangement provides double feed to each feeder circuit. In ring main arrangement, the ends of the bus-bar are returned upon themselves to form a ring; hence the name main ring arrangement. This arrangement is shown in Figure 7.9.

In this arrangement, if one circuit breaker is damaged, it is opened and the feeder can be supplied from the other circuit breaker near to it.

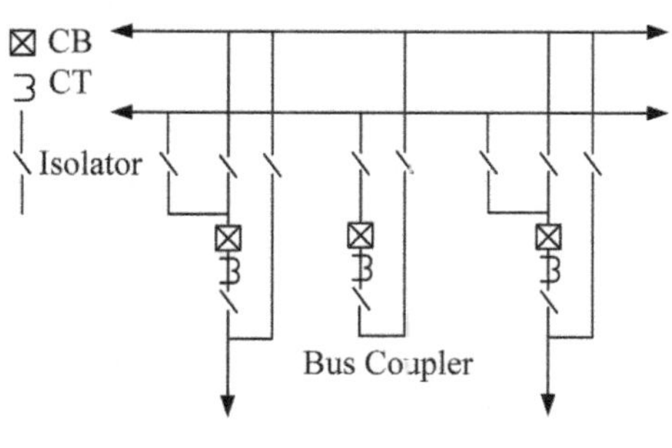

FIGURE 7.8 Double bus arrangement with bypass isolators.

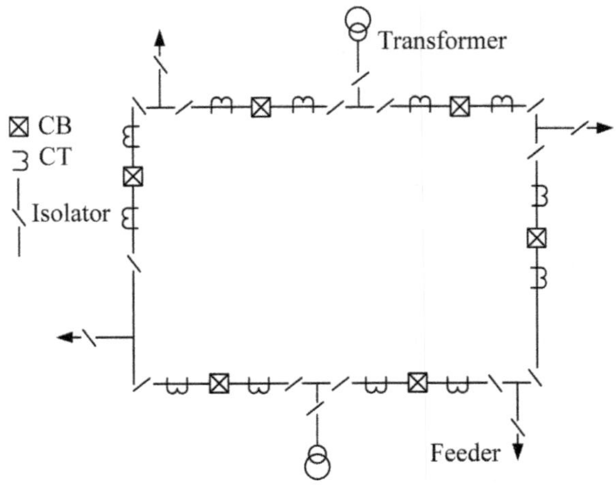

FIGURE 7.9 Ring main arrangement.

Advantages:

1. Each feeder is fed from two circuit breakers; even if a fault occurs in one system, the feeder can be fed from another system so there will be no interruption of the power supply under fault conditions.
2. The effect of the fault is localized to that section alone, and the rest of the section continues to operate normally.

Disadvantages:

1. It is difficult to add any new circuit in the ring.
2. Overloading problems may occur.

7.2.9 Mesh Arrangement

In the mesh arrangement, circuit breakers are installed between the mesh formed by bus-bars, as shown in Figure 7.10.

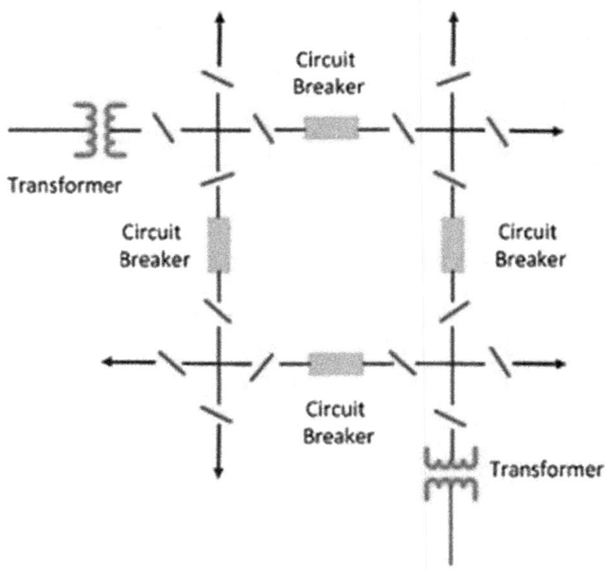

FIGURE 7.10 Mesh arrangement.

From the node point of the mesh, the circuit is tapped. We need to open two circuit breakers when the fault occurs so that protection can be obtained, but switching is not possible.

Advantages:

1. Provides protection against the fault.
2. For substations having a large number of circuits, this arrangement is suitable.

Disadvantages:

1. It doesn't provide switching facility.
2. Not suitable for all type of substations.

These are the different substation layouts or bus-bar arrangements.

7.3 Bus Faults

Bus faults are relatively rare in comparison to overhead line faults. Bus fault accounts for 6%–7% of the all faults in power system, whereas line faults comprise 60%–80% [2]. The leading causes of bus-bar faults are weakening of insulation due to aging, corrosion, insulation breakdown due to transient overvoltages, and destruction by animals (snakes, lizards, rodents, etc.) or fine particle deposits (charged particles), etc. Most of the bus faults are not reported; however, bus faults statistics, in percentages, are shown in Table 7.1, to have a comparative understanding of various types of faults. It is notable that most of the bus faults are single line to ground faults. The most substantial numbers of failure are due to flashovers and insulation failures, which often occur due to severe weather conditions. The statistics shown in Table 7.1 are for outdoor substation bus-bar arrangements and not for indoor gas-insulated substation bus-bar arrangements.

Bus-bar protection is also called bus zone protection due to the fact that it protects all the equipment connected to it. However, this terminology is more descriptive and not used by protection engineers; today they refer to it as bus protection, which is more straightforward. There are many types of

TABLE 7.1

Bus Faults Statistic

	Types and Number of Faults					
Reported Causes of Faults	**1LG**	**2LG**	**3LG**	**3PH**	**Unknown**	**Total %**
Insulation flashover	20	6	1	–	–	21.0
Insulation failure	16	2	2	–	–	15.5
Switchgear insulation failure	19	2	–	–	1	17.0
Other insulation failure	4	1	1	3	–	7.0
Current transformer failure	3	–	–	–	–	2.3
Disconnection (open) or grounded	8	1	5	1	–	11.6
Safety grounds lefts on	6	1	8	–	–	11.6
Accidental contact	5	–	2	–	–	5.4
Debris falling	4	1	–	1	–	4.7
Miscellaneous/unknown	2	1	–	1	1	3.9
% of each type	67.4	11.6	14.7	4.7	1.6	100.0

Source: Hewitson, L.G. et al., 16 switchgear (busbar) protection, in *Practical Power System Protection*, Newnes, Oxford, UK, pp. 233–243, 2005.

protection equipment; the choice depends on where it is applied and which component of power system it is protecting. For bus protection, the fundamental concept is to use differential protection which measures the sum of all currents at the bus. In the normal operating condition of the bus-bar, the measured sum of currents should be zero for a considered bus section.

7.4 Bus Protection Requirements

The relay selection is a very important aspect of bus protection system. The relays must have very accurate discriminative characteristics to restrain from tripping for close-in faults near the bus. It is necessary to protect the buses which have sufficient benefits of its implementation. The basic requirements of bus protection are speed and security, which must be achieved. High speed is necessary to limit the damage due to faults on a bus in high voltage buses. Bus fault clearing should be faster compared to other connected transmission line backup protection. Most of the high voltage bus protection relays provide faster clearing of faults, on the order of one cycle. Fast bus fault clearing is also required to provide safety for the substation and other connected equipments.

Security is also very important for bus protection since many lines can be tripped by tripping a single bus, which may result in a number of line outages. There are various causes for security failures: mainly failures of relay circuits, lack of selectivity, incorrect settings and maintenance personnel mistakes. Successful bus protection can be achieved with relays complying with the following [4]:

1. Selectivity
    - Trip only for the faulty equipment
    - Important for bus-bar with divided zones
2. Speed
    - Limited damage at fault point
    - Limited effect on security
3. Security
    - Restraint for tripping for outside zone faults
    - Security must be guaranteed
4. Cause for loss of security
    - Interruption in CT circuits-imbalance, saturation
    - Accidental operation during maintenance

7.5 Bus Protection

The main components of bus protection are current transformer (CT) and relay which detect the faults occurring on the bus zone of protection. Relays with input from the CTs detect faults and issue the tripping decision based on the setting [5]. The CT turns ratio selection and connection are important aspects to be considered while installing it on the bus-bar. In following sections, the CTs turns ratio selection and its different connections will be explained with examples. The bus differential protection is the most common method of bus protection schemes. There are various types of protection schemes based on the issues in bus protection. The differential protection scheme works on Kirchhoff's current law at the node. Most bus differential protection uses sum of all currents (entering and leaving) at the bus, which should be zero. Figure 7.11 shows the bus differential protection overcurrent (OC) relay arrangement with CT ratios for one phase.

The CT ratio is selected based on the maximum primary current of a feeder for all feeders, as shown in the figure. This is one of the most accurate and flexible turn ratio selection methodology and is followed

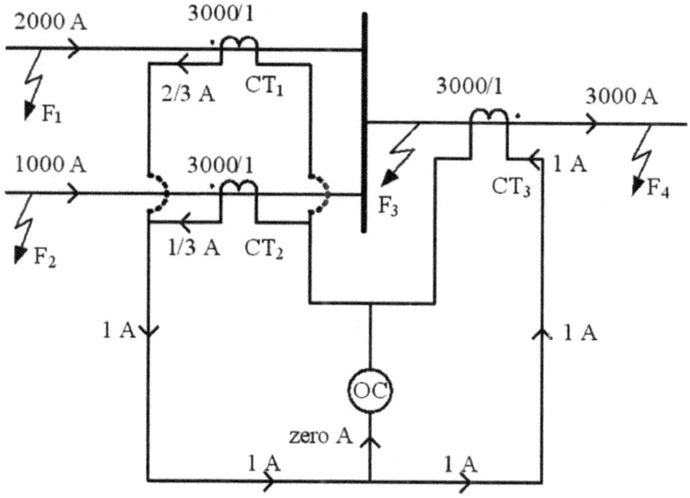

FIGURE 7.11 Bus differential overcurrent relay with CTs ratios.

by most of the utilities. Using the ratio as 3000/1 for all CTs, the primary and secondary currents of all the CTs are shown in Figure 7.11. With the total secondary current as per the arrangement shown for bus differential protection with OC relay, no current (ideally) will flow though the OC relay [1].

7.5.1 Bus Differential Protection Response to External Faults

Consider faults on the feeder or transmission lines F_1, F_2, and F_4 shown in Figure 7.11. Let's assume that the external fault F_1 occurs on the feeder and that the maximum current will go to the fault point, say ten times of the currents shown in all the feeders. So all the feeders will contribute ten times their normal currents; hence the summation of currents in CTs secondaries (ideally) will remain zero and no current will flow in OC relay (ideally). Thus, no action will be issued for this fault. Similar observation can be drawn for faults F_2 and F_4 as these faults are also treated as external fault by the bus differential protection scheme.

It was assumed that all CTs will reflect the primary current without any error in the secondary current, but in practice it is not possible to have ideal CT condition. Figure 7.12 shows the response of a bus differential protection scheme for external fault F_4. The CT_3 will be subjected to saturation due to high current in the primary and it will not able to reflect it ideally. The secondary current of CTs are also shown; it can be seen that the CT_3 is subjected to saturation, only giving output 4 A in the secondary instead of 10 A (ideally). From Figure 7.12 it is observed that 6 A current is flowing through bus differential overcurrent relay and it will issue a trip decision that is highly unexpected; the relay malfunctions, issuing the wrong trip decision. Depending upon the various conditions of external faults, the OC relay is configured with impedance or resistance in series with the relay to avoid erroneous tripping. The relay settings and other configuration parameters related with the bus differential relaying will be discussed in the following section, with examples.

7.5.2 Bus Differential Protection Response to Internal Faults

Figure 7.13 shows the bus differential protection response for internal fault F_3 with primary as well as secondary currents. It is assumed that the source is available on the left side or one can neglect the infeed effect on the right side. Hence the current flowing in the CT_3 secondary is zero. The position of the fault point is slightly shifted to the left, and its effect is entirely different compare to F_4 fault condition. As shown in Figure 7.13, 10 A current will flow for this internal fault and the bus differential OC relay issues the correct trip decision.

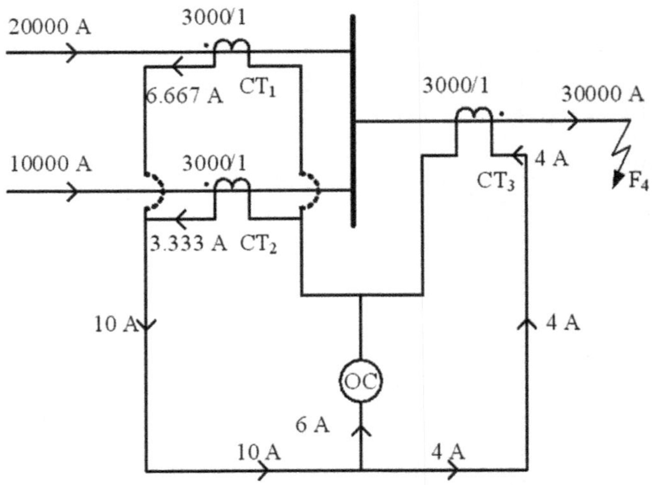

FIGURE 7.12 Bus differential protection response for external fault.

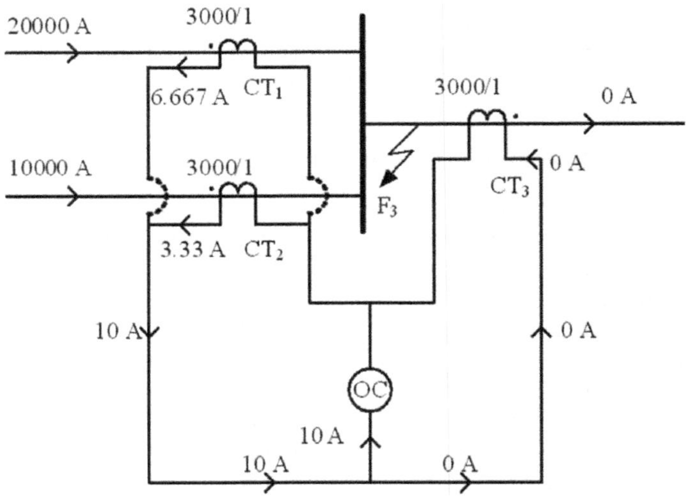

FIGURE 7.13 Bus differential protection response for internal fault.

From the discussion of external and internal fault responses of bus differential protection, it is evident that the CT plays quite an important role in sensing the primary current and its response to the nature of the current it is subjected to. Depending on the nature of the DC offset present in the primary current of the CT, the nature of the secondary current will change and it will add delay in detection of faults. Hence understanding and selection of various properties of CT are very important considerations: for example, its core, type of material, winding type, magnetic characteristics, burden, etc.

7.6 CTs for Differential Protection

The discussion in the previous section about the bus differential protection response for internal and external faults showed that CTs unequal saturation is one of the major problems in bus protection. It is due to the fact that the fault current magnitude is not same for all CTs, and large differences in spill current exist in the relay circuit. Also, the effect of DC offset in the fault current saturates the CT, and its reflected secondary current is not a replica of primary current in magnitude as well as in wave shape [6].

1. Bushing type CT is most widely used due to its lower installation cost. It has tapped secondary windings on a circular magnetic core. The magnetic core encircles the high voltage conductor to form one turn primary winding of the CT. The secondary windings of the CT are distributed equally around the circular magnetic core to minimize the leakage reactance of CT. The tapping is provided on the secondary of the bushing CT, and its ratios and connection are provided by the manufacturer. The nominal ratios mostly used are 600:5, 1200:5, 2000:5, and 3000:5 with only two IEEE relay accuracy class of C400 and C800. Due to the availability of tappings on the secondary mostly as X_1 to X_5, one can achieve the desired CT ratio. For example, in nominal CT ratio of 600:5, it is possible to get 50:5 ratio by using the tapping connection of the secondary. Due to this feature, it is also called a multi-ratio bushing CT.
2. Window type CTs have a circular magnetic core with an opening in the centre. It is also called a window-through CT because the core has an opening through which conductor carrying current is passed. The current carrying conductor makes primary winding of CT, and secondary winding is wound on the circular magnetic core in distributed manner.
3. Wound type CT is similar in construction to a power transformer: both primary and secondary winding are wound on the same magnetic core. These types of CTs have high accuracy and higher burden capacity. To achieve the high efficiency, usually wound type of CTs is designed with toroidal cores. Today these types of CTs are mostly not in use due to the fact that primary winding has to carry full load current through it and the insulation requirement is high; hence cost is also high.
4. For CT accuracy per the IEEE standard C37.110-2007 for relaying, the ratio correction should not exceed 10% of rating [6]. CTs are classified in *C*, *K*, or *T*. In *C* or *K* type CTs the core leakage flux does not have an appreciable effect on ratio within the current limit and burden. *K* type CTs should have at least 70% knee-point voltage of the secondary voltage rating. *T* type CTs have a significant effect on core flux leakage on the ratio within current limit and burden. More detailed information of the secondary terminal voltage, CT burden, thermal current rating, short-term and continuous current ratings, etc., are given in IEEE standard C37.110-2007 for relaying applications [6].

7.7 Bus Differential Protection

In previous sections, discussion about the bus differential relay was based on the assumption that all the CTs are ideal and currents in the secondary are an exact replica of the primary line current, which is not true in real-world applications. The fault F_4, shown in Figure 7.12, is high current flow to the fault and the iron core CT_3 is likely to get saturated. The CT will experience saturation of varying degree depending on the total current. The primary current measurement has two components: decaying DC component and symmetrical AC component. The DC component may start at maximum value depending on the fault inception time and decays exponentially. The equation for the primary current measurement is given as:

$$i = I_m\left[\sin(\omega t + \theta - \phi) + e^{-Rt/L}\sin(\theta - \phi)\right] \tag{7.2}$$

where

$\phi = \tan^{-1}\left(\frac{\omega L}{R}\right)$

θ = fault inception angle after voltage zero

The first component in Equation (7.2) is a symmetrical AC component, and the second term is an exponential DC decaying component with a time constant of L/R, where L and R are the primary circuit inductance and resistance between bus and fault point. The DC offset is an important factor in primary current measurement. If the fault inception angle or fault occurrence time is close to voltage zero crossing, DC offset is maximum. For high voltage systems, the magnitude of the DC offset is high and it causes the CT magnetizing flux to give rise to high error or even saturation of CT core in both components.

7.8 Bus-Bar Differential Protection with High Impedance

The equivalent circuit of a CT is basically a transformer. At low values of primary current, lower secondary current and voltage are induced, which leads to lower induced flux in the CT. Hence the CT nearly works ideally with $I_s = I_p/N$. For a higher value of primary current, as shown in the case of CT_3 in Figure 7.12, higher secondary current and voltage are induced, leading to a higher working flux requirement of the CT. The transformer needs to draw more magnetising current as flux increases, and due to the nonliners nature of B–H curve, the I_s is no longer proportional to I_p and finally reaches a constant value at which the transformer is said to be saturated. Under this condition the equitant circuit of the transformer core can be represented as a short circuit with a current flowing through it equal to saturation current, as shown in Figure 7.14. The factors on which CT saturation depends are as follow [1–5]:

- Core cross-sectional area
- CT ratio
- Amount of residual flux
- Connected burden
- Amount and direction of DC offset current
- Core steel saturation flux density

7.8.1 Requirement of High Impedance Bus-Bar

For the condition of external fault of 30,000 A shown in Figure 7.12, equivalent secondary circuit of CTs is shown in Figure 7.14, with CT_3 assuming to be saturated. It is assumed that all the CTs are identical with a turn ratio of N.

For the scenario shown in Figure 7.15, the CT_1 and CT_2 currents are $(I_1/N) - I_{o1}$ and $(I_2/N) - I_{o2}$, respectively. The summation of these currents have two paths to flow, one is through the OC relay and the other through the saturated CT_3. Thus, the fault current flows though the OC relay even for external faults, leading to false tripping. Hence, the OC relay is restrained from tripping for external faults with

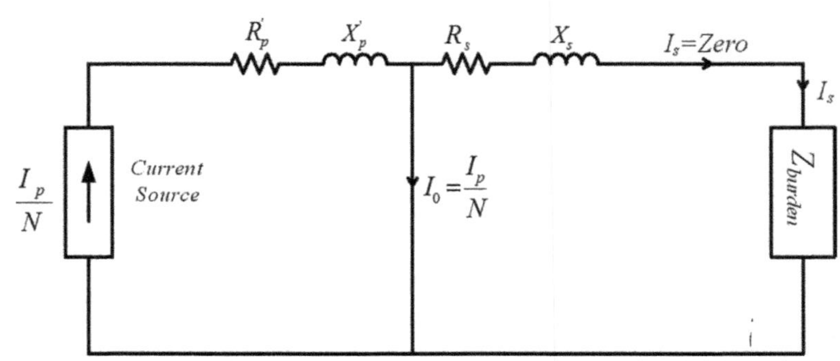

FIGURE 7.14 Equivalent circuit diagram for a saturated CT referred to the secondary.

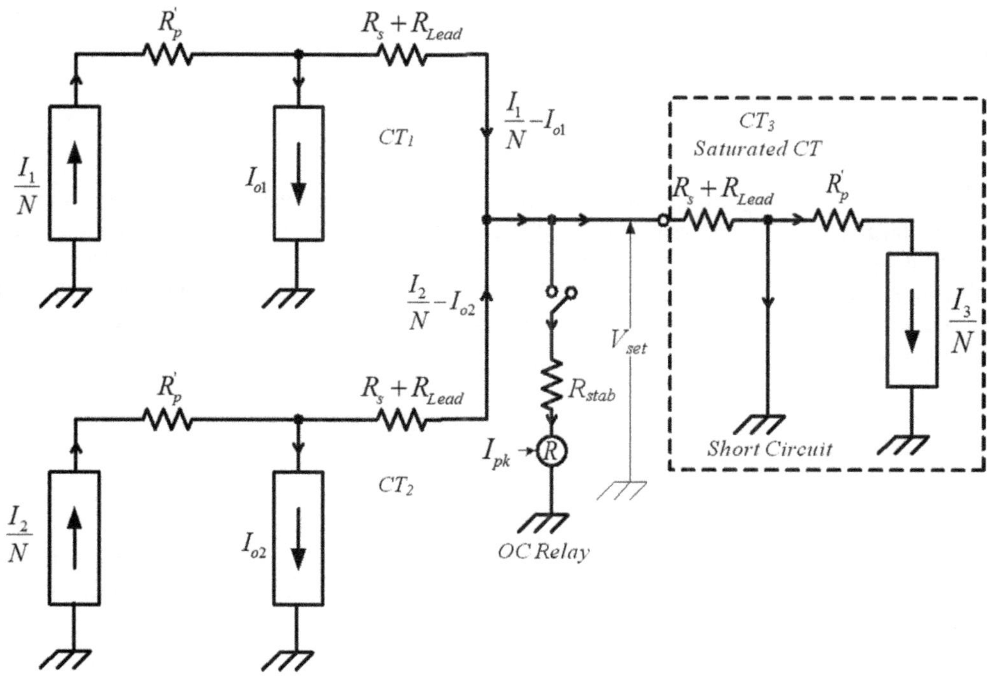

FIGURE 7.15 Equivalent circuit for external fault with CT_1 saturated.

one CT saturated using stabilizing resistance. The stabilizing resistance value should be such that, in the worst case scenario of one CT being saturated and maximum external fault, the current passing thought the OC relay should be less than its pick value I_{pk}.

7.8.2 High Impedance Bus-Bar Differential Protection Technique

In order to find the stabilizing resistance, we need to find out the voltage that will be developed across it with the OC relay not connected. Then with the OC relay connected, the value of stabilizing resistance can be such that the current flowing through the OC relay is less than its pickup value I_{pk}. The voltage V_{set} developed with the OC relay disconnected is given by Equation (7.3):

$$V_{set} = \left[\left(\frac{I_1}{N} + \frac{I_2}{N}\right) - \left(I_{o1} + I_{02}\right)\right] Z_s \tag{7.3}$$

where, Z_s represents the secondary impedance of the saturated CT, which includes the secondary resistance and lead wire resistance. For the stabilizing resistance to work properly, it should satisfy the following condition:

$$R_{stab} \geq \frac{V_{set}}{I_{pk}} \tag{7.4}$$

The generalized formulation of V_{set} for an n bus system with identical turn ratio and magnetizing current is given by Equation (7.5):

$$V_{set} = \left[\left(\sum_{b=1}^{n-1} \frac{I_b}{N}\right) - I_o(n-1)\right] Z_s \tag{7.5}$$

The maximum external fault current for which the high impedance differential protection techniques remains stable can be found using Equation (7.6):

$$I_{f,max,ext} = \frac{NV_{set}}{Z_s} \tag{7.6}$$

In order to find the minimum internal fault current that can be evaluated using this technique, it is assumed that all the CTs in the system are unsaturated (since the currents will be low), with identical turn ratio and magnetizing current. From the high impedance technique, it is known that an OC relay has a stabilizing resistance of R_{stab} and pickup current of I_{pk}, and needs a minimum voltage of V_{set}. Therefore, the minimum fault current on the secondary side of the CT that develops the required voltage V_{set} is given by Equation (7.7):

$$I_{f,min.int,sec} = \frac{V_{set}}{R_{stab}} = I_{pk} = \left(\frac{\sum_{b=1}^{n} I_b}{N} - NI_o \right) \tag{7.7}$$

The quality of the differential protection is measured by its stability ratio, which is a dimensionless quantity defined as the ratio of the maximum external fault current to the minimum internal fault current for which the operation of high impedance differential bus-bar technique is stable.

$$S = \frac{I_{f,max,ext}}{I_{f,min.int}} \tag{7.8}$$

The higher the value of the stability ratio, the better is the performance of the differential relay. Stability ratio for EHV lines is usually in the order of few tens.

Precaution has to taken not to connect any other device in the CT circuit used in this technique because the added burden may lead to CT saturation or may lead to settings that are not in the range of high impedance relay. The advantage of the high impedance protection technique is that it allows junction point of all CTs to be in the yard, which minimizes the wire requirement and leads to lower settings thus making the protection system more sensitive.

7.8.3 High Impedance Bus-Bar Differential Protection Technique Numerical Example

A bus-bar having three incoming lines and four outgoing lines is set at 132 kV. The switchgear capacity at 132 kV is 3000 MVA, and its parameters are as follows:

- CT secondary resistance $R_s = 0.6\ \Omega$
- Relay resistance = 1.2 Ω
- Lead wire resistance = 2.0 Ω
- Maximum full load current = 450 A
- CT magnetizing current up to 125 V = 0.25 mA/V (linear)
- CT saturation voltage $V_{knee} > 125$ V

For a secondary current setting of CT equal to 1.0 A and voltage setting of $V_{set} = 100$ V, find the following:

1. Maximum external fault current detected by the protection technique.
2. For the switchgear used, is the maximum fault current sustainable?

3. Minimum internal fault current detected by the protection scheme.
4. What is the current pickup setting for detecting minimum internal fault current of 450 A?
5. What is the value of the stabilizing resistance?
6. What is the stability ratio for the scheme that is sustainable for switchgear used?

Solution

1. As mentioned in the question, the CT ratio for all CTs is chosen as 450:1 (N:1).

$$I_{f,max,ext} = \frac{NV_{set}}{Z_s}$$

$$\frac{450 \times 100}{2 + 0.6} = 17307.7 \text{ A}$$

2. Maximum current handling capacity for the given switchgear is:

$$\frac{3000 \times 10^6}{\sqrt{3} \times 132 \times 10^3} = 13121.59 \text{A}$$

The maximum current handling capacity of the switch gear is 13121.59 A is less than the protection scheme maximum external current 17907.7 for the current voltage setting. Therefore new V_{set} is found:

$$V_{set} = \frac{13121 \times (2 + 0.6)}{450} = 75.8 \text{ V}$$

For the V_{set} of 75.8 V, the maximum external fault current will be 13121.59 A, which is sustainable for the switchgear used.

3. In order to find the minimum internal fault current for which the protection technique is stable, we need to find the magnetizing current for the CT's. Since the V_{set} is below the knee voltage, the linear curve of the magnetizing current is considered with a slope of 0.25:

$$I_o = (0.25 \text{ mA/V})(75.8 \text{ V}) \text{ mA} = 0.01895 \text{ A}$$

For the given magnetizing current, the minimum internal fault current is calculated as follows:

$$I_{f,min,int,pri} = N\left(I_{pk} + nI_o\right)$$

$$I_{f,min,int,pri} = 450\left(1 + (7 \times 0.01895)\right) = 509.69 \text{ A}$$

4. The pick value for minimum internal fault of 450 A:

$$I_{pk} = \frac{I_{f,min,int,pri}}{N} - nI_o = \frac{450}{450} - (7 \times 0.01895) = 0.8673 \text{ A}$$

5. Net resistance of the relay branch is:

$$R_{stab} + R_{relay} = \frac{V_{set}}{I_{pk}} = \frac{75.8}{0.8673} = 87.39\ \Omega$$

Since the overcurrent relay resistance is 1.2 Ω, the required stabilizing resistance is 86.19 Ω.

6. Stability ratio for which the protection scheme is stable:

$$S = \frac{I_{f,max,ext}}{I_{f,min,int,sec}} = \frac{13121}{(509.69/450)} = 11584.4$$

7.9 Percentage Restrained Differential Relay

The percentage restrained differential relay relies on the error in the current of the differential circuit. There are two coils in the relay circuit, as shown in Figure 7.16. One is an operating coil, and the other is a restraining coil. The current in the restraining coil provides restraining torque that helps in preventing malfunctions in the relay.

The relay operation is based on the graph shown in Figure 7.17. The relay operates when the ratio, I_{op} (the operating current) and I_{rest} (the restraining current) are greater than the characteristic slope. A theoretical characteristic slope is shown in Figure 7.17; the slope is dependent on the restrain setting.

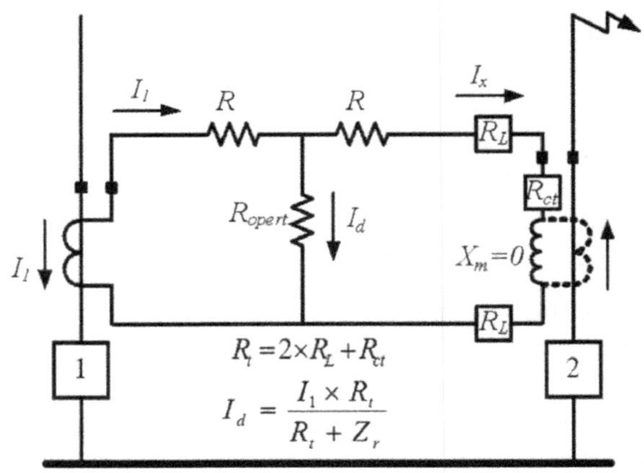

FIGURE 7.16 Busbar percentage differential protection relay.

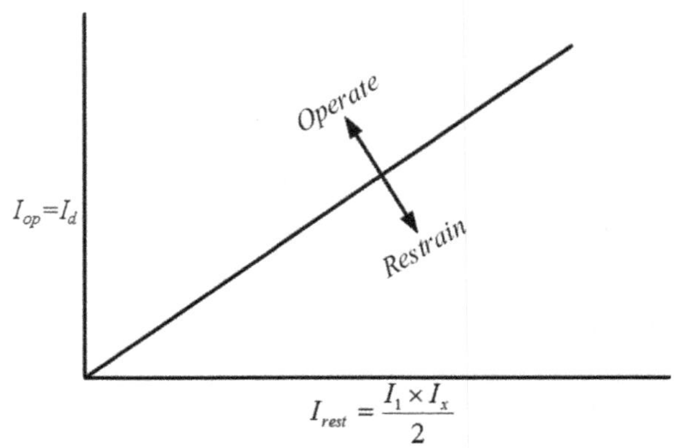

FIGURE 7.17 Operating characteristic of percentage restrained differential relay.

A stabilizing resistor is also chosen so that the operating current does not exceed a set value even when total saturation occurs in the CT circuit having maximum resistance. The selection of a stabilizing resistor is based on:

- The slope of the characteristic setting
- The sensitivity of the relay
- Largest total resistance in the CT secondary circuits measured from CT to the relay

The advantages of the percentage differential relay technique in comparison to the high impedance relay are:

- Operating time is 1–2 ms in comparison to 20–25 ms in high impedance technique.
- CT ratios of all CTs need not be same.
- Inclusion of auxiliary CT in CT circuits introduces ratio error to which this technique is immune.
- It is stable against through faults even for infinite MVA, unlike the high impedance technique where an increase in the fault level leads to an increase in the voltage setting, which leads to saturation.

7.10 Percentage Differential Bus-Bar Protection Technique Numerical Example

For the bus-bar percentage differential protection technique shown in Figure 7.18, the percentage relay setting has a slope of 25% and a pickup current of 2 A. Determine the following:

1. For an external fault F_1 leading to primary pilot current of 24,000 A and secondary pilot wire current of 1150 A, will the relay operate?
2. If the relay slope is charged to 60% with the relay currents remaining the same, what will be the operational difference?
3. For an internal fault F_2, if the secondary pilot current changes to 2020 A and the slope of the relay is 25%, will the relay operate?

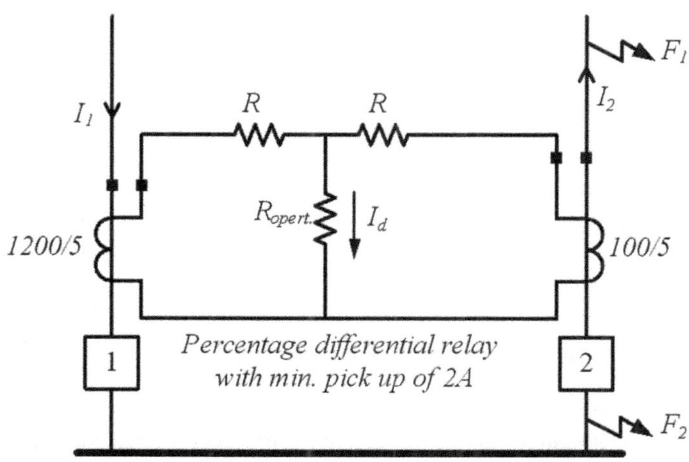

FIGURE 7.18 Equivalent connection diagram for bus percentage differential protection relay.

Solution

1. Primary pilot current:

$$I_1 = \frac{24000 \times 5}{1200} = 100 \text{ A}$$

Secondary pilot current:

$$I_2 = \frac{1150 \times 5}{100} = 57.5 \text{ A}$$

Operating current:

$$I_{op} = I_1 - I_2 = 100 - 57.5 = 42.5 \text{A}$$

Restraining current:

$$I_{rest} = \frac{I_1 + I_2}{2} = \frac{100 + 57.5}{2} = 78.75 \text{ A}$$

Slope:

$$\text{Slope} = \frac{I_{op}}{I_{rest}} = \frac{42.5}{78.75} = 0.5396$$

Since the slope is greater than 25%, the the point lies in the operating region of the curve, as shown in Figure 7.17. Hence the relay will operate.

2. For a slope setting of 60%, the slope point of 0.5396 lies in the restraining region. Therefore the relay will no longer operate.
3. Primary pilot current:

$$I_1 = \frac{24000 \times 5}{1200} = 100 \text{ A}$$

Secondary pilot current:

$$I_2 = \frac{2020 \times 5}{100} = 101 \text{ A}$$

Operating current:

$$I_{op} = I_1 - I_2 = 100 - 101 = 1 \text{ A}$$

Since the operating current is less than the minimum pick current of 2 A, the the relay will not operate.

7.11 Partial Differential Protection

Partial differential protection is commonly used on distribution buses, as is often discussed in the literature and shown in Figure 7.19. Simple overcurrent relays (time and instantaneous) are most often used in these applications. The partial differential relays, Rb1 and Rb2, operate for respective bus faults, but they must be set to coordinate with the downstream devices.

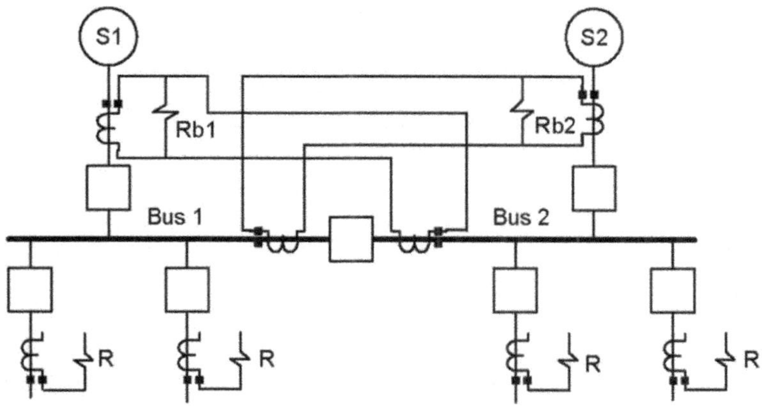

FIGURE 7.19 Partial differential relaying.

7.12 Directional Comparison Bus Protection

Another type of protection that is available but not extensively used is bus protection by directional comparison, as shown in Figure 7.20. This type of protection requires a directional function (overcurrent or distance) on each of the lines, with each function connected so that each function is looking into the bus. A trip output is produced when all of the directional functions operate. This will occur only in the case of a bus fault. Electromechanical implementation of the scheme requires all the contacts to be connected in series, which in effect forms an AND function that is used to implement the analog version of the system. This system can be applied with current transformers of different ratios, but it does require a source potential to provide directional action of the functions.

7.13 Tutorial Questions

1. What is a bus-bar?
2. What is the necessity of bus-bars?
3. What are the advantages of bus-bars?
4. What are the materials used for bus-bars?
5. What are the different types of bus-bar arrangements generally used in the system?
6. What are the advantages of a single bus-bar arrangement?

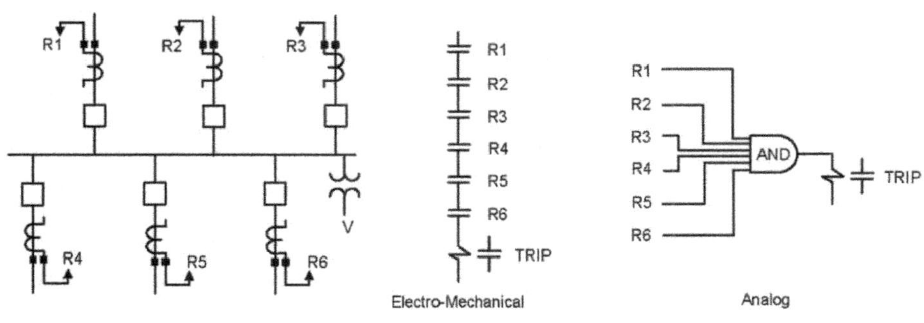

FIGURE 7.20 Directional comparison bus protection.

7. What are the advantages of the duplicate bus-bar arrangement over the single bus-bar arrangement?
8. What are the advantages of the double bus-bar arrangement?
9. What are the advantages of sectional arrangement of bus-bars?
10. Where and why is the one and a half breaker arrangement used?
11. What are the advantages of the ring bus-bar arrangement?

7.14 Conclusion

This chapter summarises the different arrangements in which bus-bars can be connected. Each arrangement has its own merits and disadvantages; therefore the system engineer should make a practical choice based on complexity, cost and reliability of various bus-bar arrangements. Since the bus-bar acts as node of power system to which several types of equipment are connected, bus-bar protection becomes very essential, and different protection schemes have been discussed. High impedance relays have been used to provide adequate, low-cost bus protection for many years, but they have limitations because of complex bus arrangements and arrangements involving multi-ratio CTs. Traditional low impedance current differential relays generally cannot be used because of the problems associated with CT saturation. A digitally implemented differential relay overcomes the effects of CT saturation and is currently the most preferred scheme for bus-bar protection.

REFERENCES

1. Y. G. Paithankar and S. R. Bhide, *Fundamentals of Power System Protection*. New Delhi, India: PHI Learning, 2011.
2. GEC Alsthom Measurements Limited, *Protective Relays Application Guide*. Stafford, UK: GEC Alsthom, 1987.
3. L. G. Hewitson, M. Brown, and R. Balakrishnan, "16 switchgear (busbar) protection," in *Practical Power System Protection*. Oxford, UK: Newnes, 2005, pp. 233–243.
4. L. C. W. Frerk, "Busbar protection," in *Power System Protection Application*, Edited by Electricity Training Association. London, UK: Institution of Electrical Engineers. pp. 81–148, 1995.
5. S. H. Horowitz and A. G. Phadke, "Bus, reactor and capacitor protection," in *Power System Relaying*. Chichester, UK: John Wiley & Sons, 2008, pp. 225–241.
6. "IEEE guide for the application of current transformers used for protective relaying purposes—Redline," *IEEE Std C37.110-2007 (Revision of IEEE Std C37.110-1996) - Redline*, pp. 1–83, 2008.

8

Distance Protective Relaying System for Long Transmission Lines

Senthil Krishnamurthy

CONTENTS

8.1 Introduction

The transmission and subtransmission lines play an essential role in the distribution of electrical power to the end users. They provide connections between the generation and distribution system. Transmission lines operate at voltage levels from 66 to 765 kV and are tightly interconnected for reliable operation. The key challenges of the transmission and subtransmission line protection schemes are reliable detection and isolation of the faults from the healthy part of the system and provision of security to the power system. Factors like deregulated market environment, economics, high-speed fault clearing and green energy requirements have pushed the electrical utilities to operate transmission line voltages close to their operating limits. The chapter provides an introduction in Section 8.1, and the distance relay and its characteristics, and communication-assisted protection schemes in Sections 8.2 and 8.3, respectively. References [1–4] provide the traditional protective relaying textbooks used by utilities vendors, end users, consultants, academics and graduate students. The unique feature of this chapter, in comparison with the traditional textbook, is that it provides distance protection setting on DigSilent Power Factory simulation tool and Schweitzer Engineering Laboratories (SEL) numerical relay settings in Sections 8.4 and 8.5, respectively; solved examples in Section 8.6; and a conclusion in Section 8.7.

8.2 Distance Relays and Characteristics

Electrical protection plays an important role in the transmission of electricity. Distance protection, which forms part of electrical protection, is one of the most important types of protection to ensure the continuity of supply to the end users. This section describes the operation of the distance relay, where the impedance of a transmission line is proportional to the line length, and uses voltage and current to calculate the apparent impedance. The distance relay compares and operates when the apparent impedance is less than the set reach point impedance. This type of protective device operates for faults between the relay and the reach point.

8.2.1 Principles of Distance Relay

Transmission lines play an essential role in the distribution of electrical power to the customers, and they can provide a connection between the generation and the load. To ensure the disturbance does not affect the healthy sections on the transmission line, it is significant to provide the main distance protection and backup overcurrent protection of the power transmission lines. Transmission lines operate at voltage levels from 66 to 765 kV and should be tightly interconnected for reliable operation. The protection schemes used in transmission lines are designed to identify the location of faults and isolate only the faulted section. The key requirements for the transmission line protection schemes are to be reliable in detecting and isolating the faults and to provide security to the power system.

Distance relays respond to the voltage and current, i.e., the impedance, at the relay location. The impedance per mile is fairly constant in impedance-based distance relay. Therefore. impedance characteristics of the distance relay is analysed between the relay and the fault locations, respectively. As the power systems become more complex and the fault current varies with changes in generation and system configuration, directional overcurrent relays become difficult to apply and to set for all contingencies, whereas the distance relay setting is constant for a wide variety of changes external to the protected line. In a three-phase power system, 10 types of faults are possible: three single phase-to-ground, three phase-to-phases, three double-phase-to-ground, and one three-phase fault. It is essential that the relays provided have the same setting regardless of the type of fault.

8.2.2 Types of Distance Relay

There are three general distance relay types; each is distinguished by its application and its operating characteristic. They are: (a) impedance relay, (b) admittance relay, and (c) reactance relay. The descriptions of these types are explained in the following sections.

8.2.2.1 Impedance Relay

The impedance relay has a circular characteristic centred at the origin of the R–X diagram. It is nondirectional and is used primarily as a fault detector.

8.2.2.2 Admittance Relay

The admittance relay is the most commonly used distance relay. It is the tripping relay in pilot schemes and is used as backup relay in step distance schemes.

8.2.2.3 Reactance Relay

The reactance relay is a straight-line characteristic that responds only to the reactance of the protected line. It is nondirectional and is used to supplement the admittance relay as a tripping relay to make the overall protection independent of resistance. It is particularly useful on short lines where the fault arc resistance is the same order of magnitude as the line length.

8.2.3 Zones of Protection

The impedance of a transmission line is proportional to the line length. The distance protection relays are usually provided with five to six zones, but the majority of the distance relays have only three zones of protection: zone 1 covers 80%–90% of the protected line (0.8–0.9 Z_{AB}), zone 2 covers 120% of Line 1 (100% of Line 1 + 20% of Line 2) and zone 3 covers 240% of Line 1 (100% of Line 1 + 100% of Line 2 + 40% of Line 3), as shown in Figure 8.1. The detailed explanation of three zones of distance protection is given below with reference to the Figure 8.1. The set time delay for the measuring zone must decrease with the fault impedance measured inside the zone characteristic. In general, zone 1 has no time delay ("instantaneous") and all other zones have time delays.

Zone 1: The first step of distance protection is set to reach up to 80% of the length of the line section. This is instantaneous protection; i.e., there is no intentional delay.

Zone 2: The second zone is required in order to provide primary protection to the remaining 10%–20% of the line and to cover up to 20% of the next line section. The operating time of this zone is 15–30 cycles (typically 200 or 400 ms) delayed so as to be selective with zone 1.

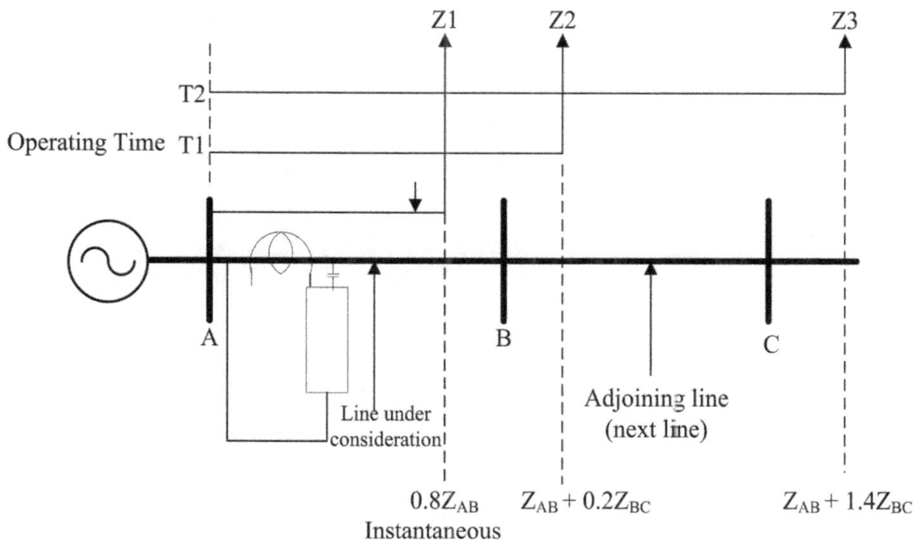

FIGURE 8.1 Time-distance diagram of the three-step distance protection.

Zone 3: The third zone is provided with an intention to give full backup to adjoining line section. It covers the line of the section plus 100% of the next line section and 40% of the third line (240%). The motivation behind the extended reach of this step is to provide full backup to the next line section. Its operating time is slightly more than 30 cycles (typically 400 or 800 ms) that of zone 2.

8.2.4 Distance Relay Characteristics

All electromechanical relays respond to one or more of the conventional torque-producing input quantities: (a) voltage; (b) current; (c) product of voltage, current and the angle θ between them; and (d) a physical or design force such as a control spring. Similar considerations hold for solid-state relays as well. It is difficult to analyse the response of an intelligent electronic device (IED) for all system conditions due to variation in the voltages for each fault, or with same fault level but with different system conditions. Therefore, it is common practice to analyse and visualise the relay response using (a) mho, (b) lens/tomato, and (c) line cartesian characteristic. The line cartesian type can be changed to line polar, arc cartesian, and arc polar type. Figure 8.2 shows the mho characteristic of a distance protection scheme; it has five zones and it is optional to set zone 3 and zone 5 to be reversed.

The reverse zone 3 and zone 5 elements at the one-line terminal must be set to overreach the forward zone 2 and zone 4 elements, respectively, at the other line terminal. In most of the cases, zone 2 is set forward by default (hard-coded in the relay). There is an additional significant advantage in R–X diagram of the mho characteristic, which allows the user to represent both the relay and the system on the same diagram.

Table 8.1 provides the forward reach, offset value and angle settings of the mho, lens and line cartesian characteristic of the distance relay. With reference to the distance relay settings given in Table 8.1, the individual and combined characteristic of the mho, lens and line cartesian are given in Figure 8.3.

1. Mho characteristic with 75° positive sequence angle.
2. Lens/tomato characteristic.
3. Line cartesian characteristic.
4. Combination of mho, lens and line cartesian characteristics.

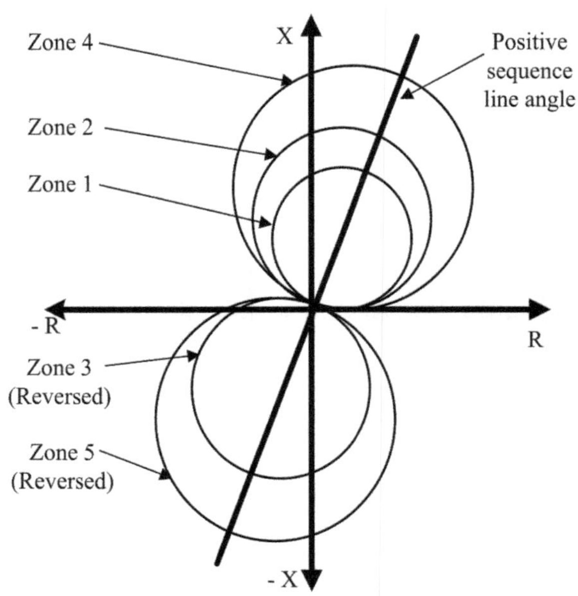

FIGURE 8.2 Mho characteristic of the distance relay.

TABLE 8.1

Mho, Lens and Line Cartesian Characteristic Setting of the Distance Relay

Mho		Lens/Tomato		Line Cartesian				
Element list	1	Element list	1	Element list	1	2	3	4
Forward reach	1.00 Ω	Forward reach	1.00 Ω	R	0.00 Ω	1.00 Ω	0.00 Ω	0.00 Ω
Offset	0.00 Ω	Offset	0.00 Ω	X	0.00 Ω	0.00 Ω	1.00 Ω	0.00 Ω
		Angle	75.00°					
Angle	75.00°	Width	800.00 mΩ	Angle	−15.00°	90.00°	180.00°	135.00°
		A/B	0.800					

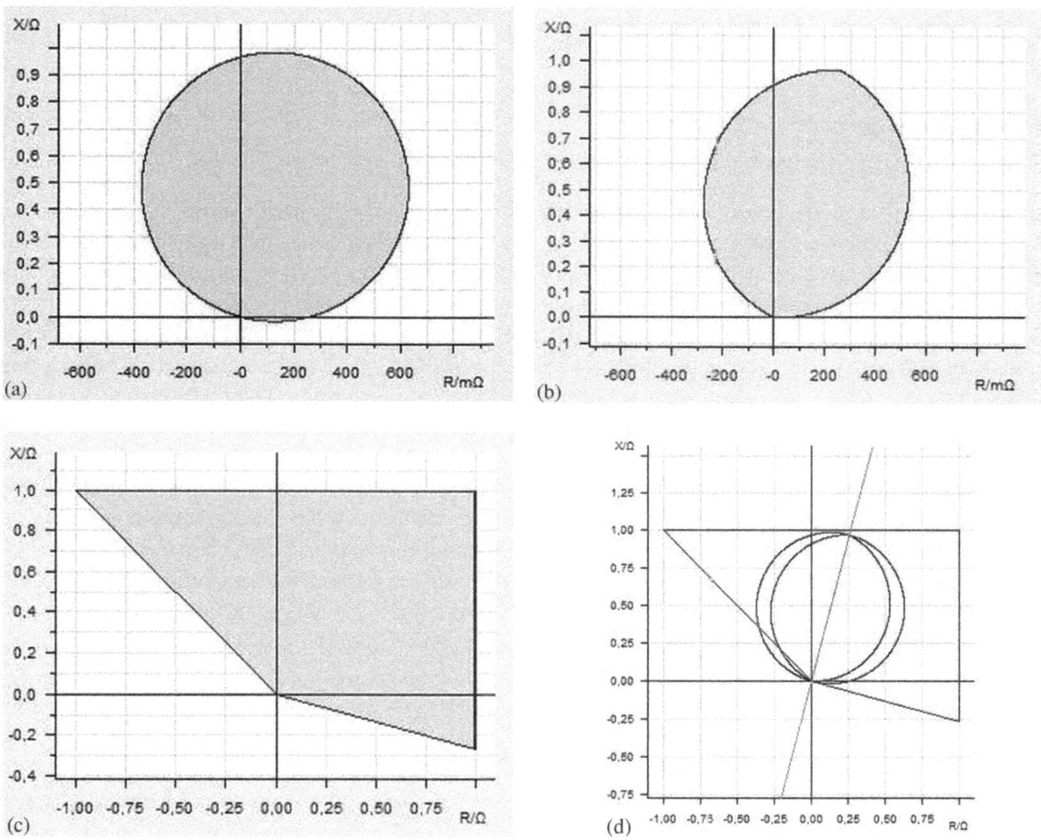

FIGURE 8.3 Mho, lens and line cartesian characteristic of the distance relay: (a) mho characteristic with 75° positive sequence angle, (b) lens/tomato characteristic, (c) line cartesian characteristic, and (d) Combination of mho, lens and line cartesian characteristics.

Mho and lens characteristics are shown in Figure 8.3a and b with reference to a 75° positive sequence angle and forward reach setting impedance at 1 Ω. The line cartesian characteristic is shown in Figure 8.3c, which has four elements. Element 1 has a −15° angle setting; elements 2 and 3 have resistance and reactance of 1 Ω reach setting, respectively, and a 90° positive sequence angle; element 4 has a 135° angle setting. In addition to the individual Mho, lens and line cartesian characteristics shown in Figure 8.3a–c, the combination of mho, lens and line cartesian characteristics is given in Figure 8.3d. Therefore, these relay characteristics are used to analyse and visualise the distance relay response for either different fault levels or the same fault level with different system conditions.

The next section of the chapter provides an overview of the communication assisted protection schemes. To understand the communication assisted protection schemes, it necessary to know the definition of the following terms: (1) protection signalling, (2) intertripping, (3) communication message, (4) blocking, and (5) tripping

1. *Protection signalling* refers to communication between the relays located remotely from each other.
2. *Intertripping* refers to the communications facilities required when remote operation of a circuit breaker is required as a result of a local event.
3. *Communication messages* involve instructions for the receiving device to take some defined action, either trip or block. It may be the passing of measured data from one device to another.
4. *Blocking mode* describes the presence of a transmitted signal that prevents tripping of a circuit breaker.
5. *Tripping mode* describes the condition when the signal initiates tripping a circuit breaker.

8.3 Communication-Assisted Protection Schemes

Communication-assisted protection schemes are used to speed up the fault clearing times for remote faults. These faults would normally be cleared by the conventional step distance protection scheme after the zone 2 time delay. During this time the fault conditions could lead to system instability or cascading blackout. For all communication-assisted trip schemes, zone 2 and 3, is enabled, as shown in Figure 8.4.

Zone 2 is always set forward by default (hard-coded in the relay). Zone 3 must be set to reverse direction in the software of the relay configuration as DIR3: = R. Zones 1, 4, and 5 are optional. At least three zones must be enabled, but zone 1 can be turned off in the reach settings. The reverse zone 3 elements at one line terminal must be set to overreach the forward zone 2 elements at the other line terminal; this is shown in Figure 8.4.

Communication-assisted protection schemes use communication between the local and remote relays by transmitting relevant information from one end of the line to the other. The communication-assisted pilot protection scheme shown in Figure 8.5 requires a communications channel between each terminal on the line to achieve high-speed fault clearing anywhere on the protected line. Communications-based schemes allow high-speed fault clearing at both ends of the line for any type of fault and fault location. Information about the status of each terminal on the line is transmitted to the remote end(s). As a result, all terminals can make tripping decisions based on more complete data.

8.3.1 Direct Under-Reaching Transfer Trip (DUTT)

The direct under-reaching transfer trip (DUTT) scheme uses an instantaneous zone 1 element to trip the local circuit breaker and initiate a transfer trip to the remote end, as shown in Figure 8.6. The remote end trips immediately on receipt of the transfer trip signal, without any additional qualification. This scheme is extremely simple but is susceptible to faulty operation if channel noise keys the direct trip signal.

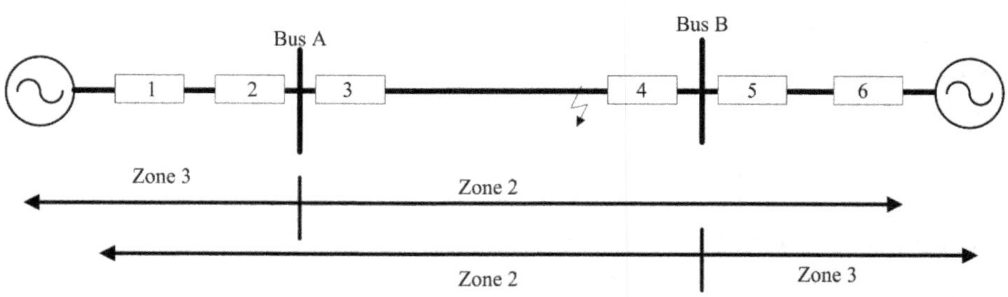

FIGURE 8.4 Communications-assisted trip scheme with zones 2 and 3 enabled.

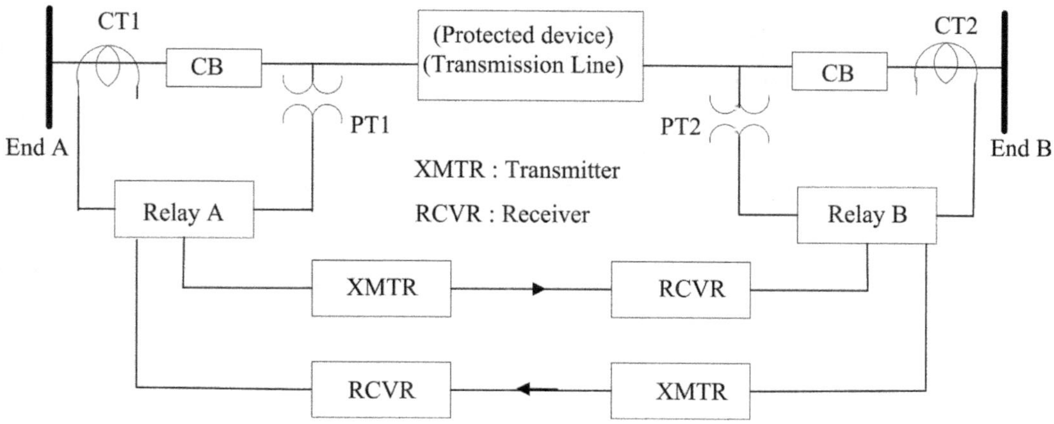

FIGURE 8.5 Pilot protection scheme.

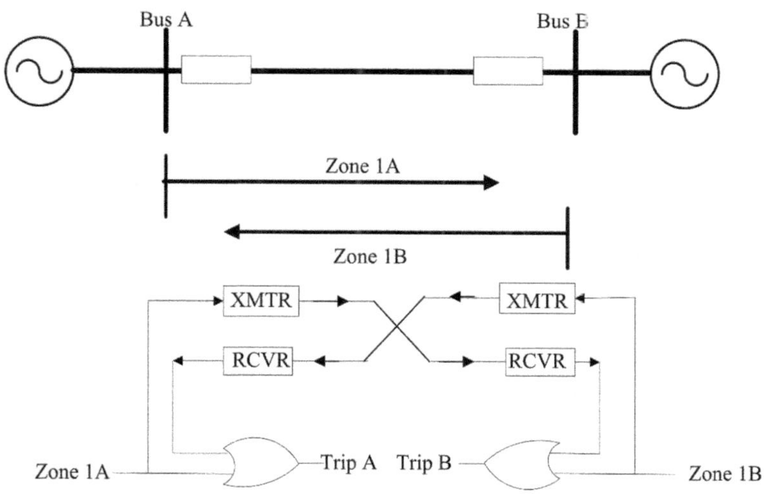

FIGURE 8.6 Direct under-reaching transfer trip (DUTT).

8.3.2 Permissive Under-Reaching Transfer Trip (PUTT)

The permissive under-reaching transfer trip (PUTT) scheme uses zone 1 to trip the local breaker and sends a permissive trip signal to the remote end, as shown in Figure 8.7. The remote end breaker trips when it receives the permissive signal if its zone 2 element is detecting a fault. By using the zone 2 element to supervise tripping on receipt of the permissive signal, this scheme is less susceptible to faulty operation under noisy channel conditions than the DUTT scheme.

Because the scheme uses an under-reaching element to send permission, PUTT does not send a permissive signal for out-of-section faults. PUTT schemes do not require additional supervisory logic to maintain security under current reversal conditions on parallel lines.

8.3.3 Permissive Over-Reaching Transfer Trip (POTT)

The permissive over-reaching transfer trip scheme uses an over-reaching zone 2 element to send a permissive trip signal to the remote end. The remote end breaker trips when it receives the permissive signal if its zone 2 element is detecting a fault, as shown in Figure 8.8. Because the scheme uses an over-reaching element to send permission, the POTT scheme needs an additional supervisory logic to maintain security under current reversal conditions on parallel lines.

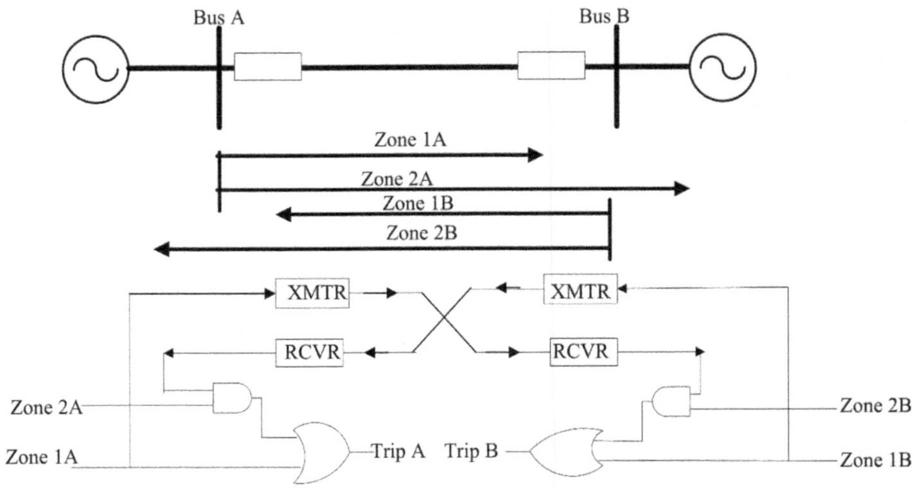

FIGURE 8.7 Permissive under-reaching transfer trip (PUTT).

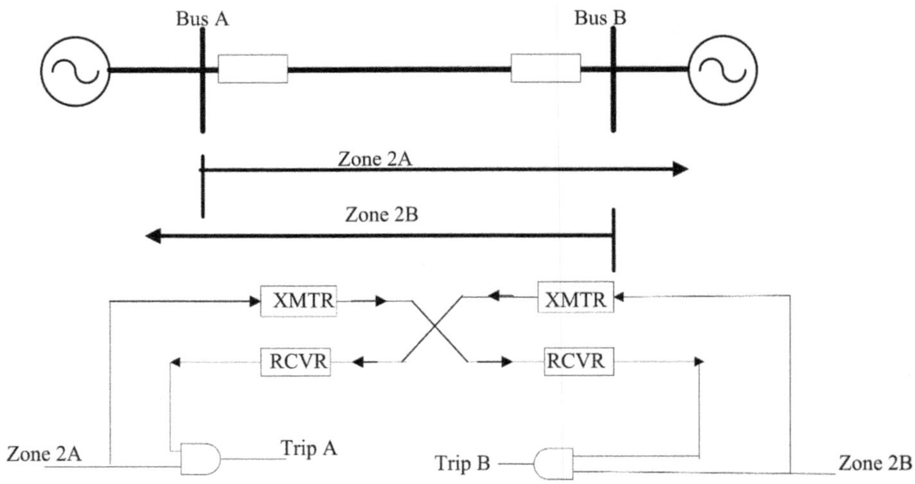

FIGURE 8.8 Permissive over-reaching transfer trip scheme.

8.3.4 Directional Comparison Blocking (DCB)

Unlike the schemes described above, which send a signal when a fault is detected in the forward direction, the directional comparison blocking (DCB) scheme sends a signal (block trip) when a fault is detected in the reverse direction, as shown in Figure 8.9. If the local relay detects a reverse fault, it sends a block trip signal to the remote end. At the remote end, the over-reaching zone 2 elements are allowed to trip, following a short coordinating time delay, if they are not blocked by the arrival of the block trip signal. In many applications, a nondirectional element is used to send the block trip signal. In these cases, the block signal is quickly shut off if the fault is in the forward direction.

8.3.5 Directional Comparison Un-Blocking (DCUB)

In the directional comparison un-blocking (DCUB) scheme, a guard signal is continuously sent between the two ends of the transmission line. If a fault is detected by the local relay zone 2 element, the guard signal is shut off and a trip signal is sent, as shown in Figure 8.10.

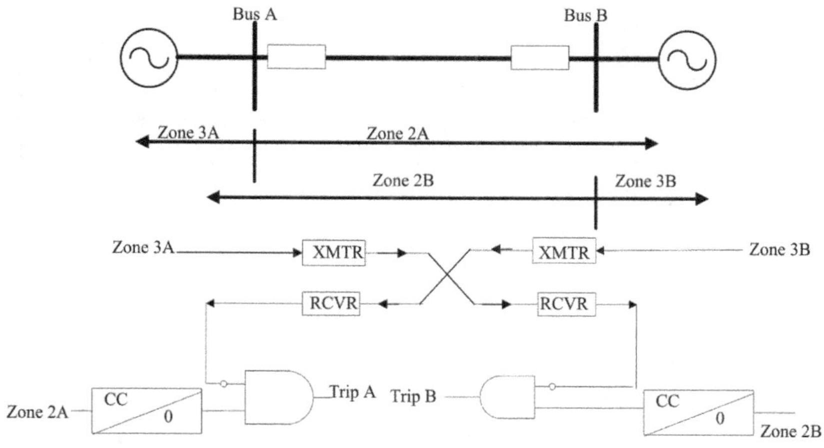

FIGURE 8.9 Directional comparison blocking.

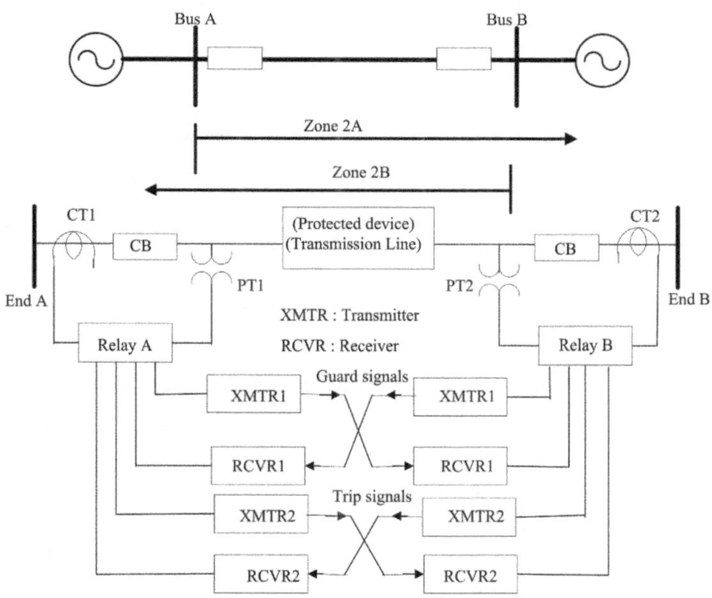

FIGURE 8.10 Directional comparison unblocking.

The remote relay detects the change in signals from guard to trip. If it also detects a fault in zone 2, it trips. DCUB schemes also use logic that permits a trip if a loss of guard is detected and a fault in zone 2 is also detected, even if a trip signal is not received.

8.4 Distance Protection Relay Configuration Setting in DigSilent Power Factory Simulation Tool

The principle of distance protection scheme is illustrated with reference to Figure 8.11. It has four 220 kV transmission lines, each 100 km in line length. The emphasis is to provide the distance protection relay (SEL421) at line 1, which covers three zones of protection and includes full protection of the main line and adjacent line (backup protection) using mho characteristic.

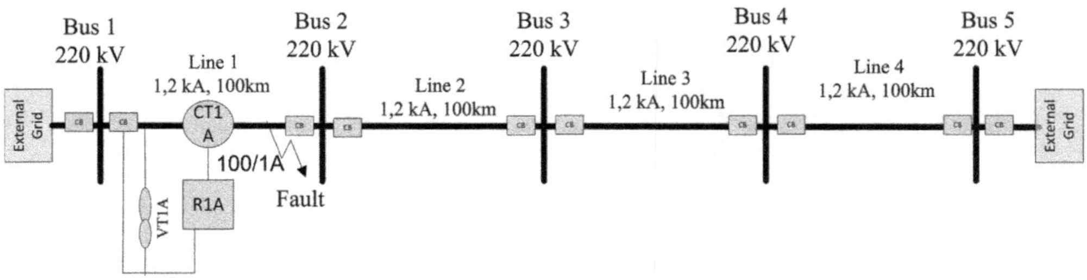

FIGURE 8.11 Single line diagram of the distance protection scheme on 220 kV transmission lines.

TABLE 8.2

External Grid Data

Description	Voltage in kV	Bus Type	Angle in Degree/ Active Power in MW	Voltage Set Point in p.u.
External Grid 1	220	SL	0	1.0
External Grid 2	220	PV	0	1.0

TABLE 8.3

Bus Data of the Radial Power System Network

Name	Voltage in kV	System Type	Phase Technology
Buses 1–5	220 kV	AC	ABC

TABLE 8.4

Transmission Line Data

Input Mode: Geometrical Parameter	Frequency HZ	No. of Earth Wires	No. of Line Circuits	Transposition	Conductor Material	Earth Conductivity uS/cm	No. of Parallel Lines
Line 1	50	1	1	Circuit wise	Aluminium	14.28571	1

The input data of the power system are given in Tables 8.2 through 8.4, respectively. The specifications of all four 220 kV overhead lines are the same. Tables 8.7 and 8.8 provide the instrument transformers and SEL421 distance relay setting, respectively. The external grid and bus data of the power system network are given in Tables 8.2 and 8.3, respectively.

The transmission line data of the power system network are given in Table 8.4. The specifications for the four transmission line ratings are the same and are given in Tables 8.4 through 8.6, respectively. The transmission line input parameters of the line and earth conductors are given in Tables 8.5 and 8.6, respectively.

TABLE 8.5

Transmission Line Input Parameters of the Line Circuits

Description	Nominal Voltage in kV	Nominal Current in kA	Line Length in km	Line Model	No. of Sub Conductors	Bundle Spacing in m
Line 1	220	1.2	100	Lumped Parameter PI	2	0.38

Sub Conductor	DC-Resistance (20°C) Ω/km	DC-Resistance (80°C) Ω/km	GMR (Equivalent radius) mm	Outer Diameter in mm	Max. Operating Temperature °C
Line 1	0.1189	0.1189	7.85	20.7 mm	80

TABLE 8.6

Transmission Line Input Parameters of the Earth Conductors

Description	Nominal Voltage in kV	Nominal Current in kA	Line Length in km	Conductor Mode	Number of Subconductors
Line 1	220	1	100	Solid conductor	1
Subconductor	**DC-Resistance (20°C) Ω/km**	**DC-Resistance (80°C) Ω/km**	**GMR (Equivalent radius) mm**	**Outer Diameter in mm**	**Max. Operating Temperature °C**
Line 1	1.85	1.85	4.998	13.2	80

TABLE 8.7

Current Transformer Data

Protection Device	Location	Branch	Manufacturer	Model	CT	Slot	Ratio [pri. A/sec. A]	VT	Slot	Ratio [pri. V/sec. V]
Relay Model	Bus 1	Line 1	Schweitzer	SEL 421-1A	CT 1A	CT	100A/1A	VT 1A	Vt	22000V/ 110V

The specifications of the current transformer are given in Table 8.7, and the distance relay setting is given in Table 8.8. The current transformer (CT) used three phases, ABC phase technology and star-connected secondary terminals. The CT type is label as 50 VA Class 10 P 20 (i.e., apparent power = 50 VA, accuracy class = 10 and accuracy limit factor = 20). The accuracy parameters are set according to the IEC apparent power. The VT connection type is YN for both primary and secondary sides. The model type of the VT is an ideal transformer.

The 220 kV power system network in Figure 8.4 is protected using a distance relay located at line 1. The next step is to set the distance relay zone impedances, positive sequence angle and time delay settings for the respective zones, as given in Table 8.8 for the phase and ground elements, respectively.

8.4.1 Setting the Distance Relay

This section describes the settings of the distance protection relay in the DigSilent Power factory simulation tool [5] environment. It includes the polarizing element, starting element, zone impedance settings of the phase and ground element with its time delay.

8.4.2 Description of the Residual Compensation ($k0$) Factor, Polarizing Angle and Its Setting

The residual compensation ($k0$) and polarizing angle must be calculated for the polarizing unit of the distance relay. It compensates the phase current for the mutual coupling between the faulted phase and the other two nonfaulty phases. In most cases, the compensation factor $k0$ of overhead transmission lines is a real number and varies between 1.0 and 2.5. In this specific example, the compensation factor is auto-calculated by the DigSilent software and set at $k0 = 1.072$ with a polarizing angle of −4.62°.

8.4.3 Description of the Pickup Element and Its Setting

Next, we must set the starting element, which consists of earth fault and overcurrent elements. It is important that these elements are set sensitively enough to pick up all faults within the zones of protection. To determine this sensitivity, it is necessary to calculate the three-phase fault on zone 3 of line 1, that is, applying a three-phase unbalance fault at bus 4 using complete calculation method with a fault impedance of $Z_f = 10 + j0\ \Omega$. The simulated results provide a fault current of 2195.43 primary amps and this value is used to set the starting element 50PP.

TABLE 8.8

Distance Relay Element Settings of a 220 kV Transmission Line

(a) Phase Element

Protection Device	Location	Branch	Manufacturer	Model	Stage (Phase)	Impedance [pri. Ohm]	Impedance [sec. Ohm]	Angle [°]	Time	Directional
Relay model	Bus 1	Line 1	Schweitzer	SEL 421-1A	Z1MP	26	1.30	70	0.00	Forward
					Z2MP	48	2.43	70	0.30	Forward
					Z3MP	80	4.06	70	0.60	Forward

(b) Ground Element

Protection Device	Location	Branch	Manufacturer	Model	Stage (Earth)	Impedance [pri. Ohm]	Impedance [sec. Ohm]	Angle [°]	Time	Directional
Relay model	Bus 1	Line 1	Schweitzer	SEL 421-1A	Z1MG	26	1.30	70	0.00	Forward
					Z2MG	48	2.43	70	0.30	Forward
					Z3MG	80	4.06	70	0.60	Forward

8.4.4 Description of the Zone Impedance Settings of the Phase and Ground Element and Its Time Delay

Terminology such as *replica impedance* and *transactor* is defined in this section before the distance relay zone impedance settings are discussed. The distance relay uses a magnetic circuit called a transactor to develop transmission line replica impedance. A transactor is an iron core reactor with an air gap, and it produces an output voltage proportional to the input current.

This section describes the zone impedance settings of the phase and ground elements. An option in the DigSilent software calculates the secondary ohm impedance value for line 1. The assumption is to set the line impedance (PGZ1) of zone 1 to 80%. Therefore the replica impedance of zone 1 is set to 1.30 sec. ohms (1.625% × 80%) and the relay angle is set to 70°.

The zone 2 reach is set to cover the protected line plus 50% of the shortest adjacent line or 120% of the protected line, whichever is greater. In this example, zone 2 is set to cover the protected line plus 50% of the shortest adjacent line. Therefore, the replica impedance of zone 2 (PGZ2) is set to 2.43 sec.ohms (1.625% × 150%) and the relay angle is set to 70°. The operating time of zone 2 (Z2PD) is set to 15 cycles, i.e., 300 ms selective delay with zone 1.

The zone 3 reach is set to cover the protected line plus adjacent line plus 50% of the next line (line 1 + line 2 + 50% of line 3). Therefore, the replica impedance of zone 3 (PGZ3) is set to 4.06 sec.ohms (1.625% × 250%) and the relay angle is set to 70°. The operating time of zone 3 (Z3PD) is set to 30 cycles, i.e., 600 ms selective delay with zone 1.

The impedance angle is set to 90° to maintain a circular tripping characteristic and the offset impedance is set to 2.2 sec.ohms. The zone impedance and time delay settings of the ground elements are the same as the phase elements. It is necessary to set the logic element to define which circuit breaker to trip for the fault on specific zones of the protection. In this example the trip element logic unit is set to default. However, it is necessary to set the trip logic unit if it is needed to trip a different breaker than the one in the same cubicle as the relay.

The next step is to apply the three-phase unbalance event on line 1 at 30% of line length and analyse the operating characteristics of the distance relay through an R–X diagram, as shown in Figure 8.12.

The simulation result of the R–X diagram shows that three-phase fault at zone 1 trips the circuit breaker at 15 ms, and the subsequence zones are tripped at 30 and 60 ms, respectively. In addition, it provides the replica zone impedance and the impedance relay angle. The zones of protection using a time-distance diagram are given in Figure 8.13.

The time distance zones of protection diagram in Figure 8.13 shows that three zones of protection are enabled in this example. Zone 1 covers 80% of the protected line and is an instantaneous unit without

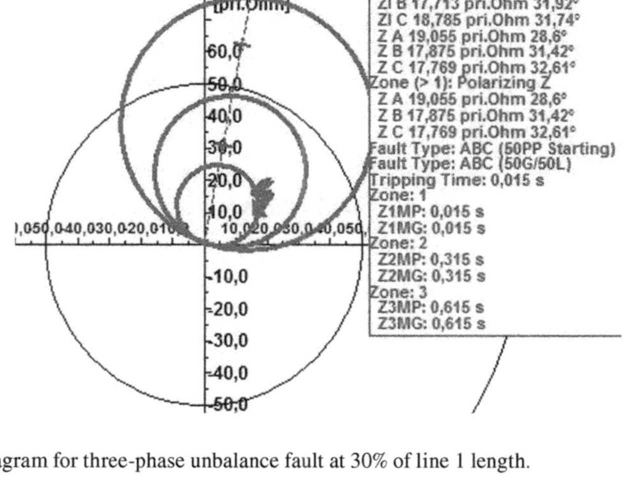

FIGURE 8.12 R–X diagram for three-phase unbalance fault at 30% of line 1 length.

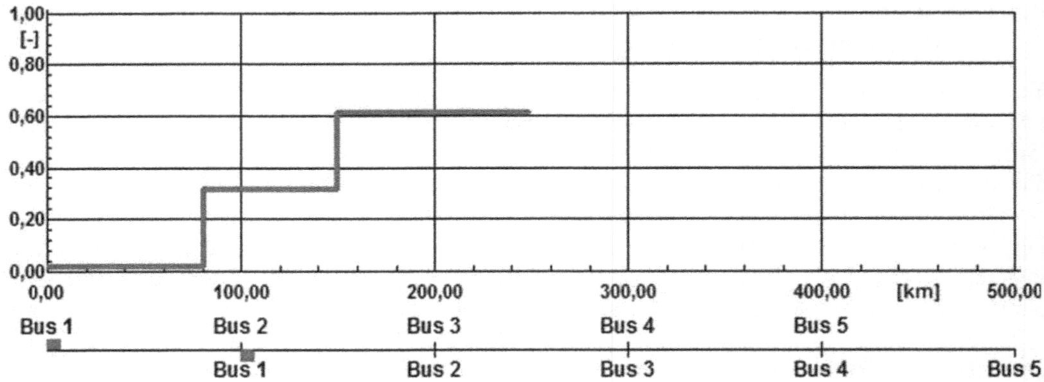

FIGURE 8.13 Zones of protection of the distance relay using time-distance diagram.

any time delay. Zones 2 and 3 cover 150% and 250% of the protected line with selective time delays of 30 and 60 ms, respectively.

In summary, this section provides the distance protection operating principles, types of distance relays, zones of protection, R–X diagram, time-distance diagram, and application of the distance protection scheme using an example. The next section of this chapter provides a description of the distance protection setting on numerical relays.

8.5 Distance Protection Configuration Setting in Numerical Relay

This section provides the distance protection setting on SEL421 numerical relays [6] and omicron test injection device [7]. The distance protection test bench setup is shown in Figure 8.14. The test bench setup has a distance protection intelligent electronic device (SEL421), omicron test injection device (CMC356) and personal computer loaded with the relay configuration tools (AcSELerator Quickset and Test universe softwares). This equipment is connected in the static IP network using the Ruggedcom Ethernet Switch.

This test bench setup provides the lab scale illustration on how the physical power transmission lines are protected using SEL421 IED, as shown in Figure 8.15. The transmission line's high voltage and current signals are produced and injected into the voltage and current channels of the SEL421 IED using the omicron test injection device. Both ends of the transmission line are connected with the switching device (circuit breakers). The trip signal of the circuit breaker is represented using the binary signal connected to the output port 101 of the SEL421 IED, which is mapped to the binary input 1 of the test injection device. For in-zone events, the SEL421 IED sends a trip signal to the binary contact of the test injection device.

The input data for the distance protection setting are given in Tables 8.9 through 8.11, respectively. The general information of the SEL421 device is given in Table 8.9, distance protection zone reach settings are given in Table 8.10 and the zone timers are given in Table 8.11. The zone timers given in Table 8.11 are represented for the 50 Hz system.

The next step is to establish communication between the SEL421 IED, the personal computer and the omicron test injection device, and is described in Section 8.5.1.

8.5.1 Communication Setting of the SEL421 IED

Establish communication between the SEL421 IED and the personal computer (PC), on which is installed with the relay configuration tools (AcSELerator quickset and Test Universe softwares). Check the IP address of the SEL421 IED (192.168.1.4) on the front panel, and then set the IP address of the PC as 192.168.1.50 to

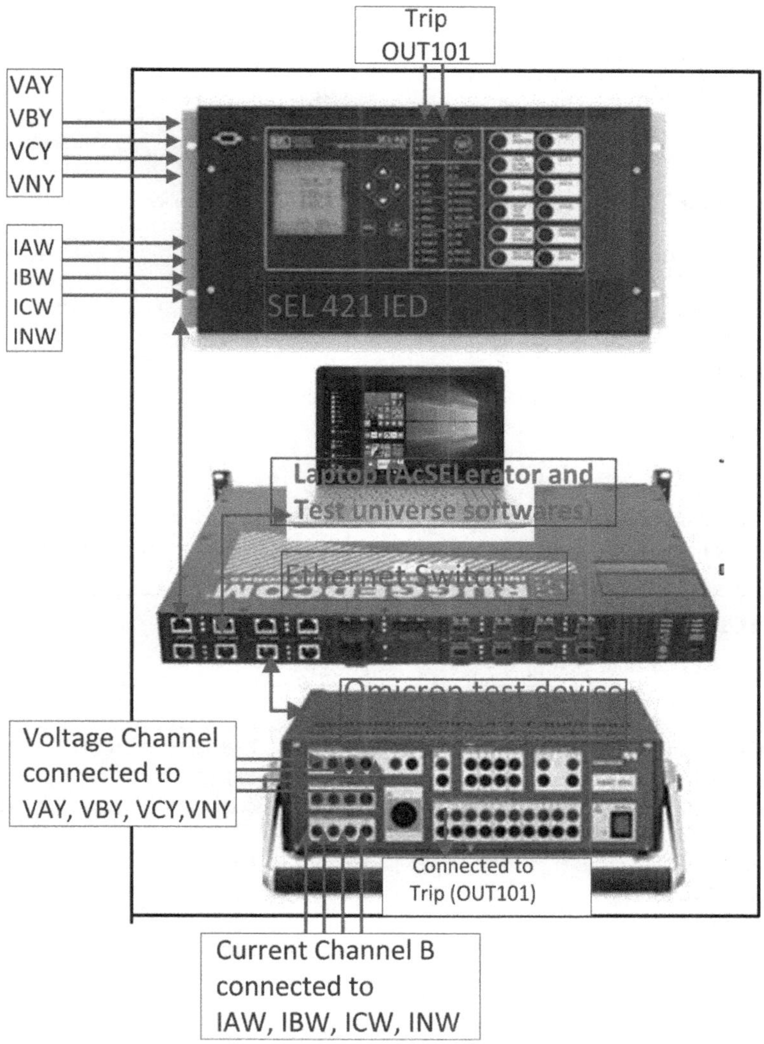

FIGURE 8.14 Distance protection test bench setup.

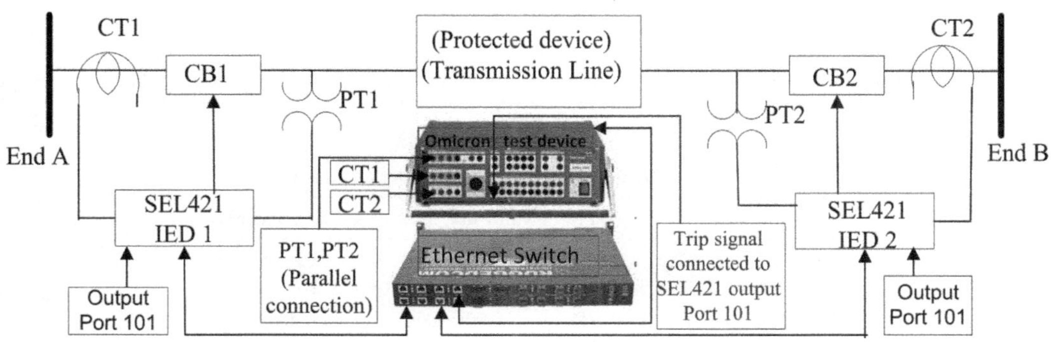

FIGURE 8.15 Distance protection single line diagram of the test bench setup.

TABLE 8.9

SEL421 Device General Information

Description	Name	Value
Substation name	Substation	Substation
Feeder name	Feeder	Feeder
Relay FID and part number	Part number	FID=SEL-421-5-R314-V0-Z017013-D20130222",''0904 PARTNO=0421561 1XC0 × 3H31XXXXX",''07CF
Secondary voltage, L-L in V	V Nominal	110.00 V
Secondary rating of CT, i.e 1 or 5 A	I Nominal	1.00 A
Nominal frequency of power system	Frequency	50.00 Hz
Positive sequence magnitude of protected object	Z1MAG	20.00 Ω
Positive sequence angle of protected object	Z1ANG	70.00°
Voltage ratio, (Primary/Secondary)	PTR	1200.00 i.e 132,000/110 V
Current ratio, (Primary/Secondary)	CTR	800.00
Voltage transformer position, i.e. busbar or line side of circuit breaker	VT position	Line
Current Transformer star point connection, i.e. towards busbar or towards line	CT star point direction	Towards line

TABLE 8.10

SEL421 Device Zone Settings

Description	Name	Value
Earthfault characteristic (Quad or Mho)	Characteristic for LN	Mho
Number of Mho characteristic zones for LL faults, 0 = none	PMHOZ	4
Number of Mho characteristic zones for LN faults, 0 = none	GMHOZ	4
Direction for zone 1	DIR1	F
Direction for zone 2	DIR2	F
Direction for zone 3	DIR3	R
Direction for zone 4	DIR4	F
Zone 1 LL reach	Z1P	16.00 Ω
Zone 2 LL reach	Z2P	24.00 Ω
Zone 3 LL reach	Z3P	16.00 Ω
Zone 4 LL reach	Z4P	36.00 Ω
Zone 1 LN reach	Z1MG	16.00 Ω
Zone 2 LN reach	Z2MG	24.00 Ω
Zone 3 LN reach	Z3MG	16.00 Ω
Zone 4 LN reach	Z4MG	36.00 Ω
Compensation factor magnitude	k01M	0.80
Compensation factor angle	k01A	−1.00°

TABLE 8.11

SEL421 Device Zone Timers (Trip Time in Cycles or Seconds)

Description	Name	Value in Cycles	Value in Seconds
Zone 1 LL time delay	Z1PD	1.00 cycle	0.02 s
Zone 1 LN time delay	Z1GD	1.00 cycle	0.02 s
Zone 2 LL time delay	Z2PD	20.00 cycles	0.40 s
Zone 2 LN time delay	Z2GD	20.00 cycles	0.40 s
Zone 3 LL time delay	Z3PD	100.00 cycles	2 s
Zone 3 LN time delay	Z3GD	100.00 cycles	2 s
Zone 4 LL time delay	Z4PD	50.00 cycles	1 s

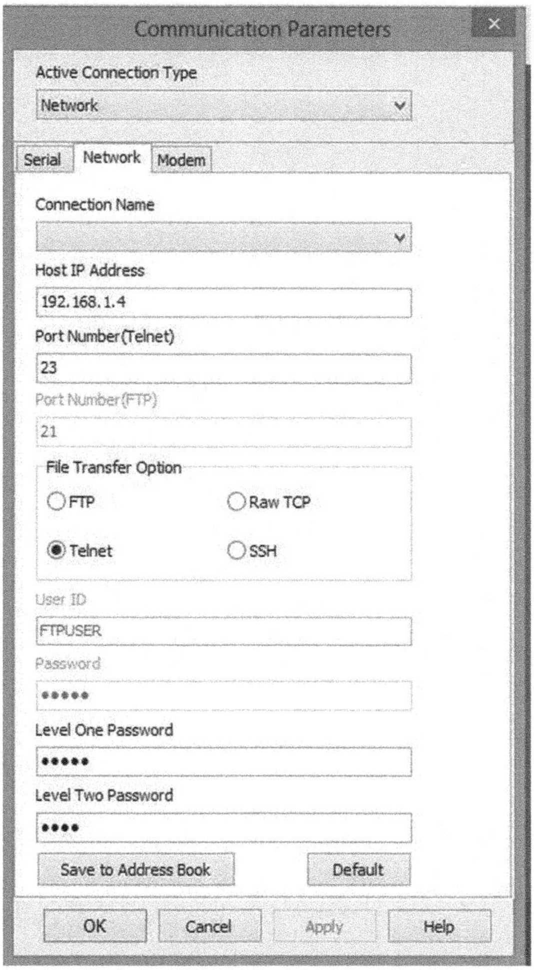

FIGURE 8.16 Communication parameter setting of the SEL421 IED.

correspond to the domain address of the SEL421 IED. Establish communication between the SEL421 IED and PC, where the quickset accelerator is installed, and check the communication parameters such as IP address, port number and file transfer option of the SEL421 IED, as shown in Figure 8.16.

The Telnet file transfer option in the AcSELerator software uses the userID as FTPUSER and password as OTTER to communicate between the SEL421 IED and the laptop, whereas FTP file transfer option uses the userID as 2AC and password as TAIL. To check the communication setup, open the command window and type 192.168.1.4 IP address of the SEL421 IED, then test whether the communication is established between the IED, laptop and omicron 356 test device using the static IP network connected using the Ethernet switch and RJ45 cables.

Alternately check the communication of the SEL421 IED in the terminal window on the AcSELerator quickset software, and then type terminal command ID, which provides the FID that is the part number of the relay connected in the network, as shown in Figure 8.17.

8.5.2 Mho Distance Protection Configuration Setting of the SEL421 IED Using the AcSELerator Quickset Software

AcSELerator quickset software is used to provide engineering configuration on the Schweitzer Engineering Laboratory (SEL) IEDs (SEL421), as shown in Figure 8.18.

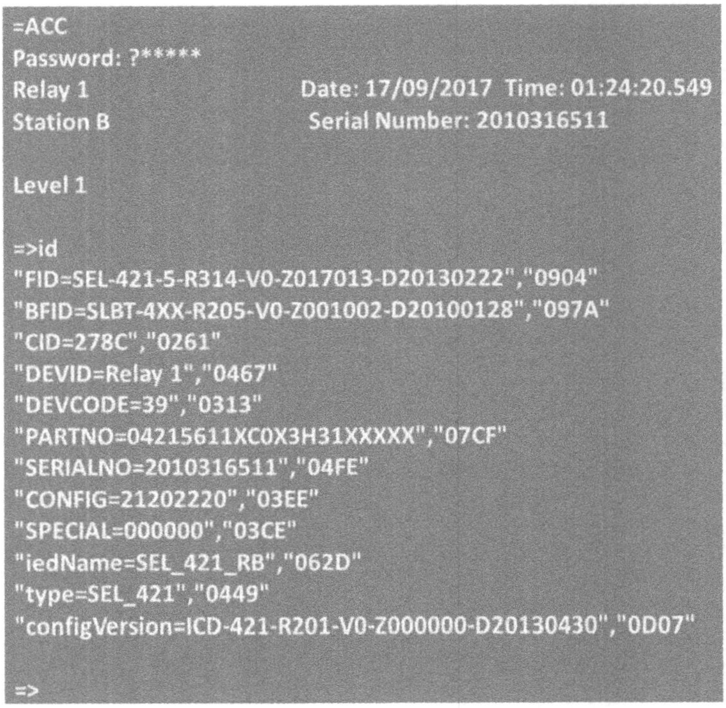

FIGURE 8.17 Terminal command (ID) information of the SEL421 IED.

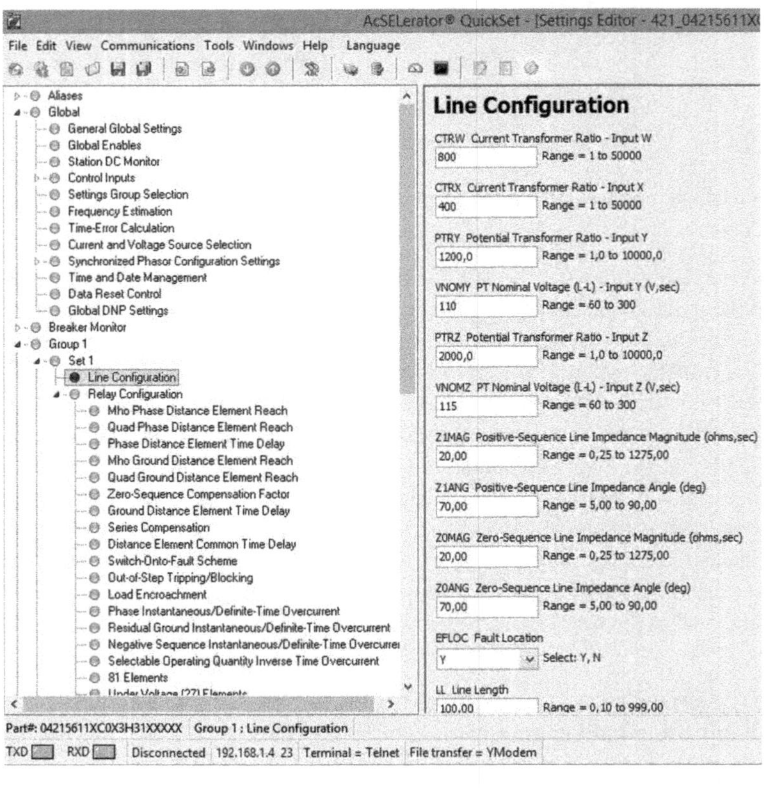

FIGURE 8.18 Line configuration setting of the SEL421 device.

SEL421 IED supports up to six group settings, and each group has five zones for mho and quadrilateral (phase and ground) distance elements. By factory default (mandatory), zones 1 and 2 of the SEL421 distance protection IED are set in the forward direction, and it is optional to set either forward or reverse direction for zones 3 to 5. SEL421 is a multifunctional IED that supports switch-onto-fault, out-of-step, load encroachment, frequency, over-/undervoltage, communication schemes (DCB, POTT, and DCUB), reclosing, and breaker failure and loss-of-potential functions. The SEL421 is shown in Figure 8.18.

However, this example only provides the mho characteristics–based distance protection configuration setting for the phase and ground elements. The general global settings provide the nominal system frequency (NFREQ), system phase rotation (PHROT) and fault condition equation (FAULT), as given in Table 8.12.

The transmission line configuration settings provide the current transformer ratio (CTRW), potential transformer ratio (PTRY), potential transformer nominal voltage (VNOMY), positive (Z1MAG) and zero sequence (Z0MAG) line impedance magnitude, positive (Z1ANG) and zero sequence (Z0ANG) line impedance angle, line length (LL). These settings are given in Table 8.13.

The relay configuration of the SEL421 IED provides Mho phase and ground distance element reach and the time delay setting. The zero-sequence compensation factor, zone/level direction and trip logic setting are given in Table 8.14.

The SEL421 supports two philosophies of zone timing (a) independent timing, in which the phase and ground-distance elements drive separate timers for each zone, and (b) common timing, in which the phase and ground-distance elements both drive a common timer. Zone 3 is reverse, and zone 4 is forward directional control.

For phase-to-phase faults, the distance between the relay and the fault can be calculated from the loop impedance by just using the line impedance. However, for phase-to-ground faults, the ground impedance is also needed. To take the ground impedance into account, a zero sequence compensation factor (grounding factor) is used. The definition of this factor depends on the manufacturer of the relay. The SEL421 device uses the formula $\overline{K_L} = \overline{Z_G}/\overline{Z_{line}} = \left(\overline{Z_0} - \overline{Z_L}/3\overline{Z_L}\right)$ to calculate the zero-sequence compensation factor.

Relay word bit is a single relay element or logic result. A relay word bit can equal either logical 1 or logical 0. Logical 1 represents a true logic condition, picked-up element, or asserted control input or

TABLE 8.12

General Global Setting of the SEL421 Device

Abbreviation	Description	Value
NFREQ	Nominal system frequency	50 Hz
PHROT	System phase rotation	ABC
FAULT	Fault condition equation (SELogic Equation)	50P1 OR 51S1 OR M2P OR Z2G OR M3P OR Z3G OR M4P OR Z4G

TABLE 8.13

Line Configuration Setting of the SEL421 Device

Abbreviation	Description	Value
CTRW	Current transformer ratio of W channel input	800
PTRY	Potential transformer ratio of Y channel input	13,2000/110 = 1200
VNOMY	Potential transformer nominal voltage of Y channel input	110 V
Z1MAG	Positive sequence line impedance magnitude	20 Ω
Z0MAG	Zero sequence line impedance magnitude	20 Ω
Z1ANG	Positive sequence line impedance angle	70°
Z0ANG	Zero sequence line impedance angle	70°
LL	Line length	100 km

TABLE 8.14

Relay Configuration Setting of the SEL421 Device

Abbreviation	Description	Value
Z1MP	Zone 1 Mho phase distance element reach	16 Ω, sec
Z2MP	Zone 2 Mho phase distance element reach	24 Ω, sec
Z3MP	Zone 3 Mho phase distance element reach	16 Ω, sec
Z4MP	Zone 4 Mho phase distance element reach	36 Ω, sec
Z1PD	Zone 1 phase distance time delay	1 cycle or 0.02 s
Z2PD	Zone 2 phase distance time delay	20 cycles or 0.40 s
Z3PD	Zone 3 phase distance time delay	100 cycles or 2 s
Z4PD	Zone 4 phase distance time delay	50 cycles or 1 s
Z1MG	Zone 1 Mho phase distance element reach	16 Ω, sec
Z2MG	Zone 2 Mho phase distance element reach	24 Ω, sec
Z3MG	Zone 3 Mho phase distance element reach	16 Ω, sec
Z4MG	Zone 4 Mho phase distance element reach	36 Ω, sec
Z1GD	Zone 1 phase distance time delay	1 cycle or 0.02 s
Z2GD	Zone 2 phase distance time delay	20 cycles or 0.40 s
Z3GD	Zone 3 phase distance time delay	100 cycles or 2 s
Z4GD	Zone 4 phase distance time delay	50 cycles or 1 s
k0M1	Zone 1 zero-sequence compensation factor magnitude	0.80
k0A1	Zone 1 zero-sequence compensation factor angle	−1.00°
DIR3	Zone/level 3 directional control	R (Reverse)
DIR4	Zone/level 4 directional control	F (Forward)
TR OUT101	Trip equation (SELogic) mainboard output OUT101 (SELogic)	M1P OR Z1G OR M2PT OR Z2GT OR M3PT OR Z3GT OR M4PT OR Z4GT

control output. Logical 0 represents a false logic condition, dropped out element, or deasserted control input or control output. Schweitzer Engineering Laboratory (SEL) IEDs use relay word bits in SELOGIC control equations. The relay word bits for the four zone-enabled distance protection phase and ground elements are given in Table 8.15. The relay word bit is used in the trip equation and the SELogic of the output port OUT101 is given in the last row of Table 8.14.

The functions such as switch-onto-fault, out-of-step, load encroachment, frequency, over-/undervoltage, communication schemes (DCB, POTT and DCUB), reclosing, breaker failure and loss-of-potential are not considered in this example.

The next section of the distance protection setting on numerical relay provides the configuration setting of the CMC356 test injection device using the Omicron Test Universe software tool. The SEL421 IED receives the transmission line high voltage and current signals generated by the omicron test device. The distance protection test bench setup is given in Figure 8.15, where the Omicron CMC356, SEL421IED and Ethernet Switch are connected in a static IP network to test the Mho distance protection scheme in the lab scale environment.

8.5.3 Omicron Test Universe Configuration Setting for the Mho Phase and Ground Distance Protection Elements

Omicron has designed the advanced distance test module for testing distance protection functions [7] and is available on the start screen of the Omicron Test Universe software, or it can be inserted into an Omicron Control Center (OCC) File. It is necessary to provide the configuration settings for the following three modules: (1) test object, (2) hardware configuration, and (3) advanced distance modules.

TABLE 8.15

Relay Word Bits of the Four Zone Enabled Phase and Ground Distance Elements of the SEL421 Device

Abbreviation	Description
Common relay word bits for both phase and ground elements for both Mho and quadrilateral distance protective relay	
Z1T	Zone 1 phase or ground distance, time-delayed
Z2T	Zone 2 phase or ground distance, time-delayed
Z3T	Zone 3 phase or ground distance, time-delayed
Z4T	Zone 4 phase or ground distance, time-delayed
Individual relay word bits for phase and ground elements for both Mho and quadrilateral distance protective relay	
Z1G	Zone 1 ground distance element
Z1GT	Zone 1 ground distance element, time-delayed
Z1P	Zone 1 phase distance element
Z1PT	Zone 1 phase distance element, time-delayed
Z2G	Zone 2 ground distance element
Z2GT	Zone 2 ground distance element, time-delayed
Z2P	Zone 2 phase distance element
Z2PT	Zone 2 phase distance element, time-delayed
Z3G	Zone 3 ground distance element
Z3GT	Zone 3 ground distance element, time-delayed
Z3P	Zone 3 phase distance element
Z3PT	Zone 3 phase distance element, time-delayed
Z4G	Zone 4 ground distance element
Z4GT	Zone 4 ground distance element, time-delayed
Z4P	Zone 4 phase distance element
Z4PT	Zone 4 phase distance element, time-delayed
Individual relay word bits for Mho phase and ground elements for each zones	
M1P	Zone 1 Mho phase distance element
M1PT	Zone 1 Mho phase distance element, Time-delayed
MAB1	Zone 1 Mho A-B phase element
MAG1	Zone 1 Mho A-phase-to-ground element

1. *Test object*: It has device settings and distance module (system and zone settings). The device setting given in Table 8.16 provides the general relay settings (e.g., relay type, relay ID, substation details, CT and VT parameters) that are entered in the Relay Interface by Omicron (RIO) function device. The system and zone settings of the distance protection are given in Figure 8.19. The detailed zone values of the Mho phase and ground distance elements are given Table 8.14.

   The line impedance/short-circuit angle between the fault voltage and current is a characteristic factor of a particular fault and is given in Table 8.17. This angle depends on the voltage level and the protected equipment (e.g., overhead line, cable and transformer). The example considers 132 kV, 100 km transmission line; therefore, the line impedance angle is chosen as 70° for the 132 kV line.

2. *Hardware configuration*: The global hardware configuration specifies the general input/output configuration of the CMC test set. The omicron CMC356 test set provides voltage, current inputs and binary trip signal to the SEL421 distance relay, as shown in Figure 8.20. The analog outputs of the CMC test are mapped to the voltage and current channels of the SEL421 distance relay, as shown in Figure 8.21.

TABLE 8.16

Device Settings of the SEL421 IED in the Omicron Test Universe Environment

Description	Value
Number of phases	3
Nominal frequency	50 Hz
Secondary voltage	110 V (L-L)
Primary voltage	132 kV (L-L)
Secondary current	1 A
Primary current	800 A
Residual voltage factor (VLN/VN)	1.732 V
Residual current factor (IN/Inom)	1 A
Maximum voltage	120 V (L-L)
Maximum current	12.5 A
Debounce time	3 ms
Deglitch time	0 sec
Overload detection sensitivity	50 ms

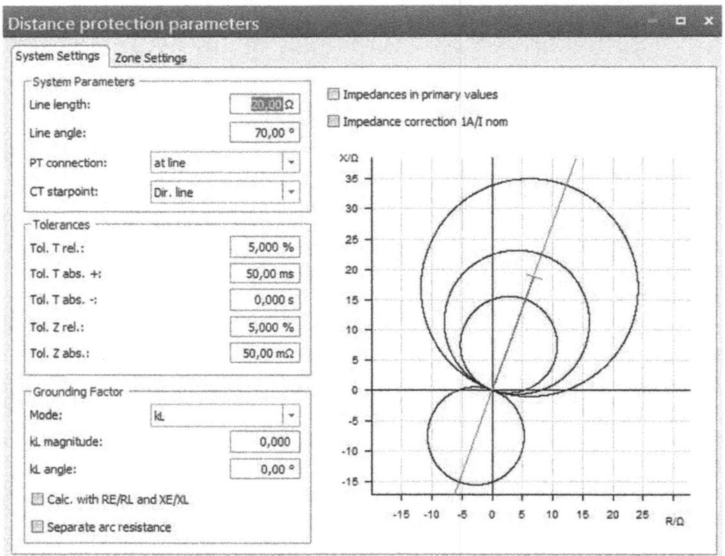

FIGURE 8.19 System and zone settings for the Test Universe distance protection function.

TABLE 8.17

Line Impedance/Short-Circuit Angle Versus Voltage Level of the Protected Equipment

Voltage level	380 kV	220 kV	110 kV	10–30 kV	Electric arc
Short circuit angle	Approx. 85°	Approx. 80°	Approx. 72°	30°–50°	Approx. 0°

3. *Advanced distance relay module*: The distance relay module has three test functions: (1) pickup test, (2) trip time test, and (3) zone reach test. The trip time test and the zone reach test are performed with the Advanced Distance test module. The description of these tests follow:
   1. *Pickup test*: The respective pickup function (e.g., overcurrent starting) is tested.
   2. *Trip time test*: The trip times of the distance protection function are verified.

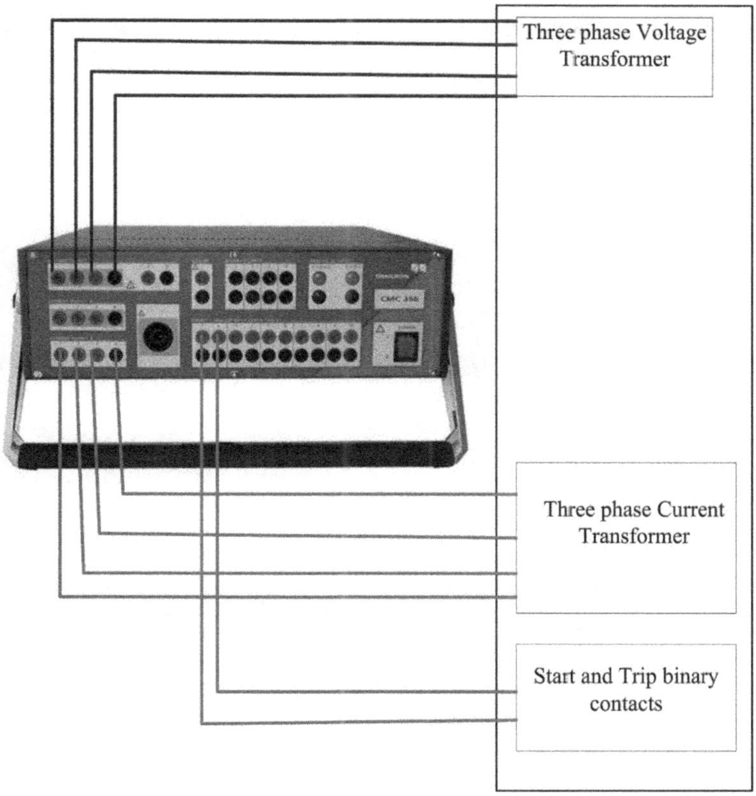

FIGURE 8.20 Global hardware configuration setting for the Test Universe distance protection function.

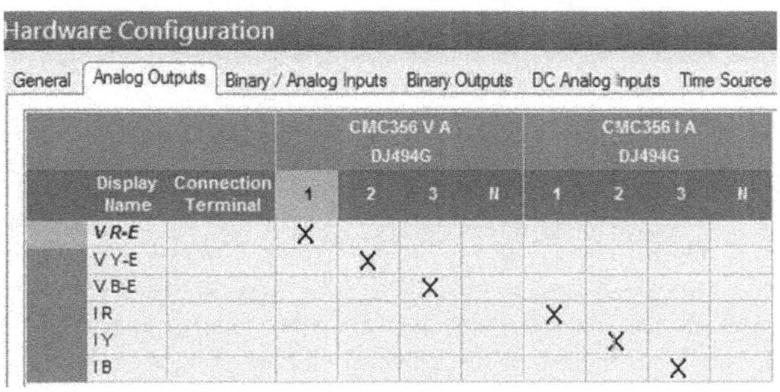

Hardware Configuration

General | Analog Outputs | Binary / Analog Inputs | Binary Outputs | DC Analog Inputs | Time Source

		CMC356 V A DJ494G				CMC356 I A DJ494G			
Display Name	Connection Terminal	1	2	3	N	1	2	3	N
V R-E		X							
V Y-E			X						
V B-E				X					
I R						X			
I Y							X		
I B								X	

FIGURE 8.21 Analog outputs of the CMC356 test set.

3. *Zone reach test*: The zone reaches of the distance protection are verified. Search lines can be defined either by dragging a line in the impedance plane or by setting a start point, an angle and the length. For the zone reach test, the search test is used. With this tab, the user can define search lines. The zone reach test results for the LL and LG events are shown in Figures 8.22 and 8.23, respectively, and the corresponding Mho phase and ground distance relay characteristics for LL and LG events are given in Figures 8.24 and 8.25, respectively.

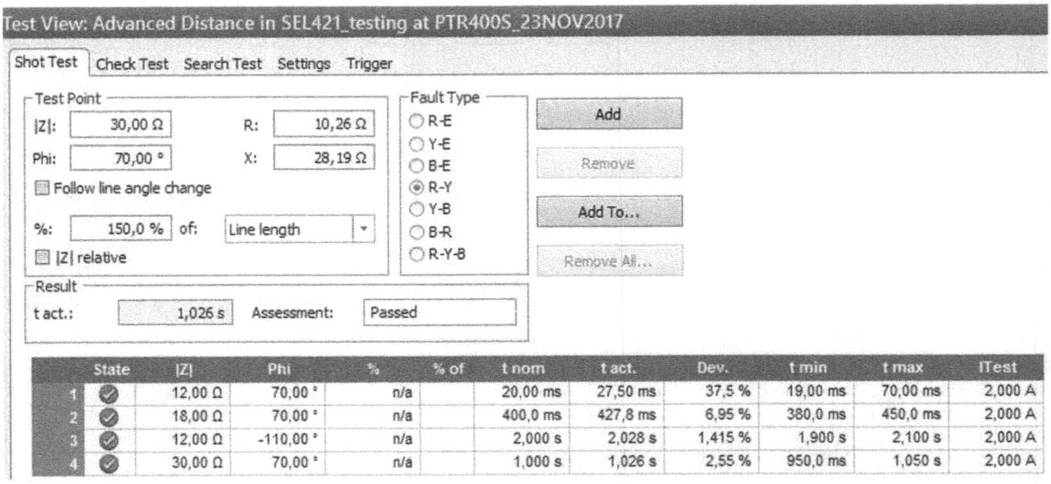

FIGURE 8.22 Zone reach test results for the LL fault.

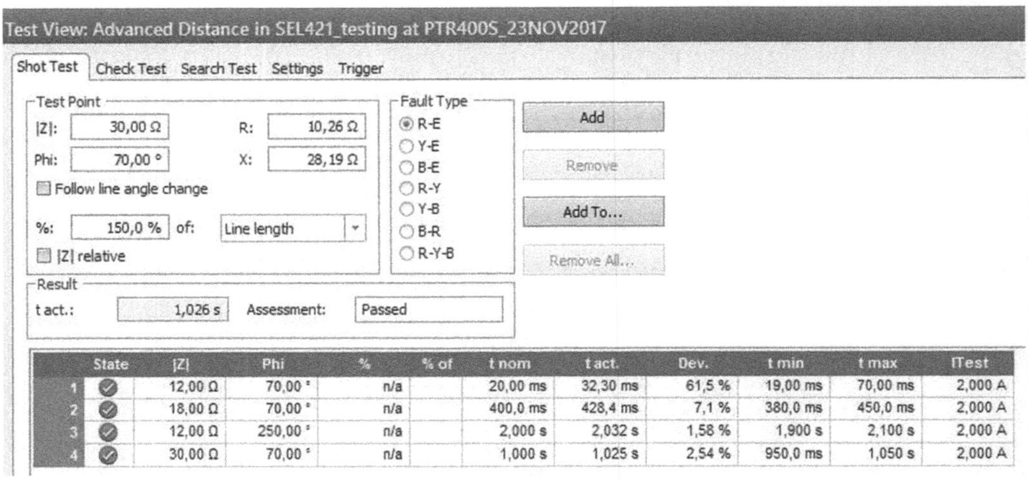

FIGURE 8.23 Zone reach test results for the LG fault.

Figures 8.24 and 8.25 show all four zones of the Mho phase and ground distance protection elements. In the zone reach test, zones 1–4 are set to 60%, 90%, 60% (reverse direction) and 150%, respectively, of the protected line length. Zones 1, 2, and 4 have forward directionality at an impedance angle of 70°. The short circuit angle of zone 3 is either −110° or 250°, which corresponds to the reverse directionality. The test module applies test shots along these lines to search for the reach of each zone. However, time test shots are placed along these lines outside the tolerance bands. This confirms if the zone reach is within the defined tolerances.

8.5.4 Analysing the Test Results of the Mho Phase and Ground Distance Protection Elements

The test results of the Mho phase and ground distance protection elements are analysed using the Omicron Test Universe and SEL event report tool. In the Omicron Test Universe Tool, the distance

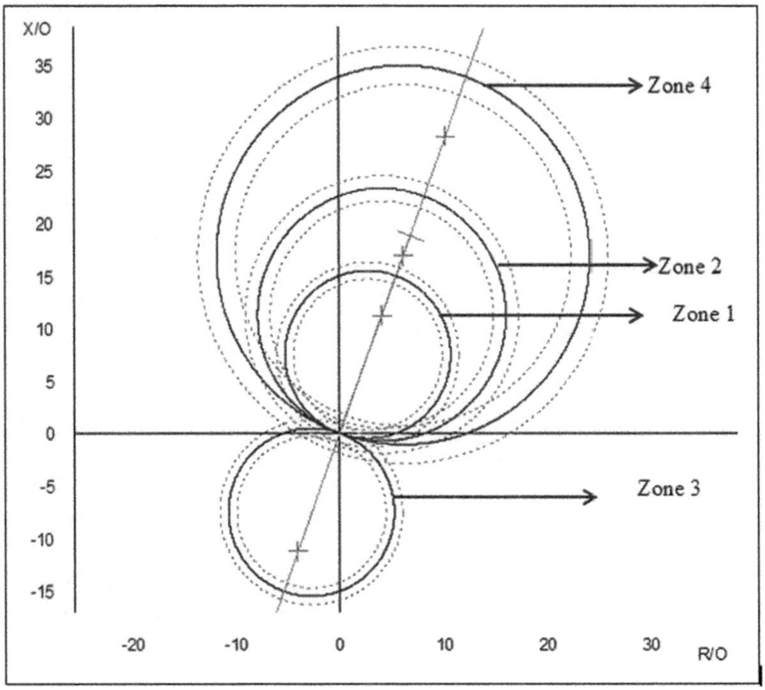

FIGURE 8.24 Mho phase distance relay characteristics for LL event.

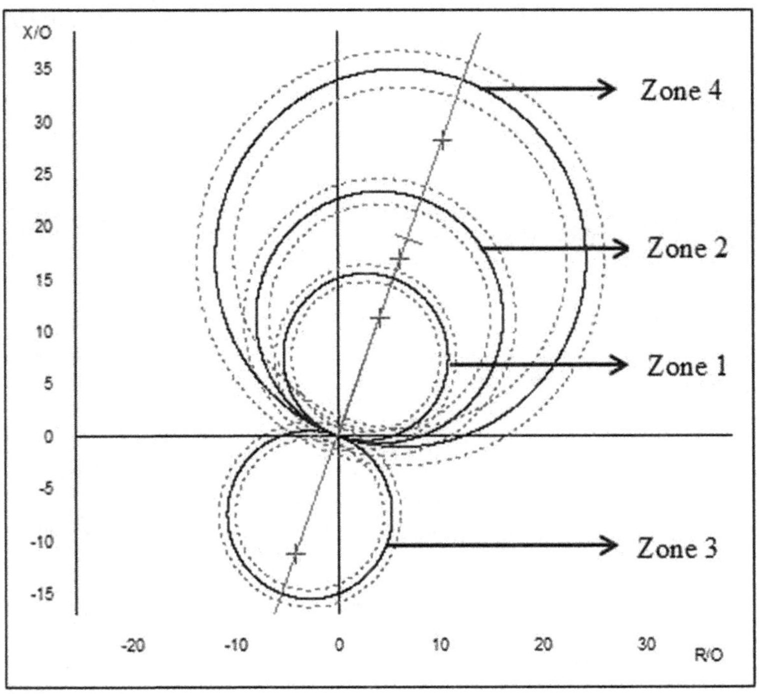

FIGURE 8.25 Mho ground distance relay characteristics for LL event.

protection test results are analysed using the (1) impedance view, (2) time signal view, (3) phasor view, (4) zone diagram, and (5) report view.

1. The *impedance view* shows the four zones of the Mho phase and ground distance relay characteristics for LL and LG events and is given in Figures 8.24 and 8.25, respectively. The test module applies test shots along these lines at 70° (forward) and −110° (reverse) directions to search for the reach of each zone. The time test shots are placed along these lines outside the tolerance bands. This confirms if the zone reach is within the defined tolerances.
2. The *time signal view* shows the time period between fault and trip signals along the voltage and current signals during the prefault, fault and post-fault conditions. Figure 8.26 shows the time signal view of the zone reach test results for the LG event on phase R in zone 1 at 60% of the line length in the forward direction. The SEL421 distance relay is tripped for this LG event at 32.30 ms.
3. The *phasor view* provides information about the line-to-ground voltage and current signals and the positive, negative and zero sequence current and voltage signals along with their phasor representation, as shown in Figure 8.27. The phasor view provides magnitude, phase, real and imaginary quantities of the voltage and current signals for both LL and LG events. LG events in zone 1 at 60% of line length produce a current magnitude of 2A at −70° with reference to the voltage signal V_{RE} at 60 V, which is equal to the line-line voltage of $\left[(60)(\sqrt{3}) = 103.92V_{LL}\right]$, and the nonfaulty phases V_{YE} and V_{BE} maintain the steady state secondary voltage of 63.51 V, which is equal to the line-line voltage of $\left[(63.51)(\sqrt{3}) = 110V_{LL}\right]$, as shown in Figure 8.27.
4. The *zone diagram* shows the line impedance of the zones versus the time delay. Figure 8.28 shows the zone diagram of all four zones for LL event. Zone 3 is in the reverse direction, and all other zones are in the forward direction with an impedance angle of 70°. Figure 8.28 provides detailed information about the line impedance reach and its time delay settings as follows: zone 1 is at 12 Ω line impedance (60% of line length) with time delay of 0.02 s or 20 ms, zone 2 is at 18 Ω line impedance (90% of line length) with time delay of 0.40 s or 400 ms, zone 3 is at 12 Ω line impedance (60% of line length in reverse direction) with time delay of 1 s, and zone 4 is at 30 Ω line impedance (150% of line length) with time delay of 2 s.

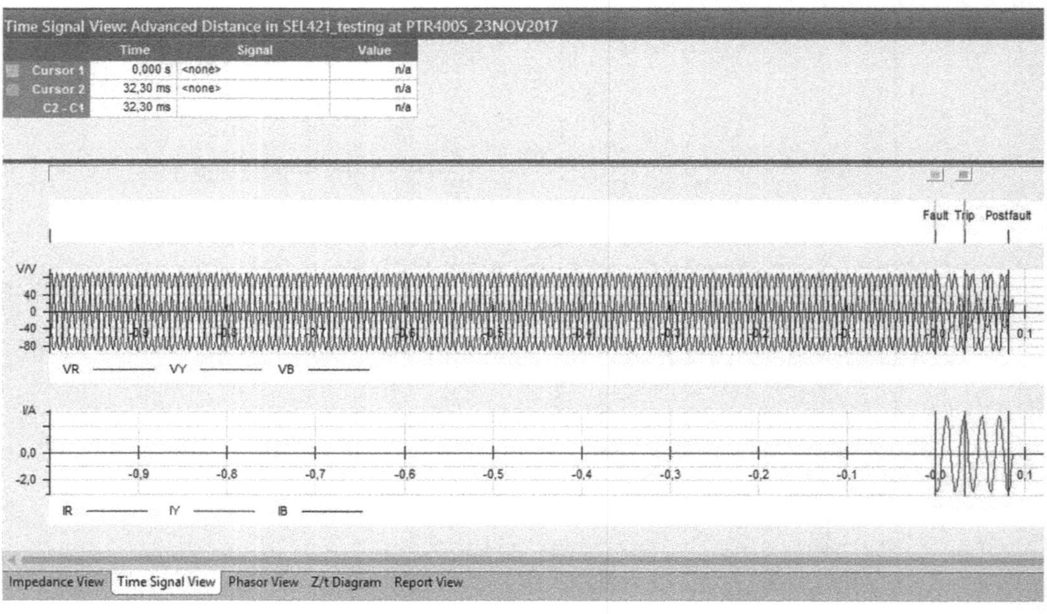

FIGURE 8.26 Time signal view of the zone reach test results for the LG event.

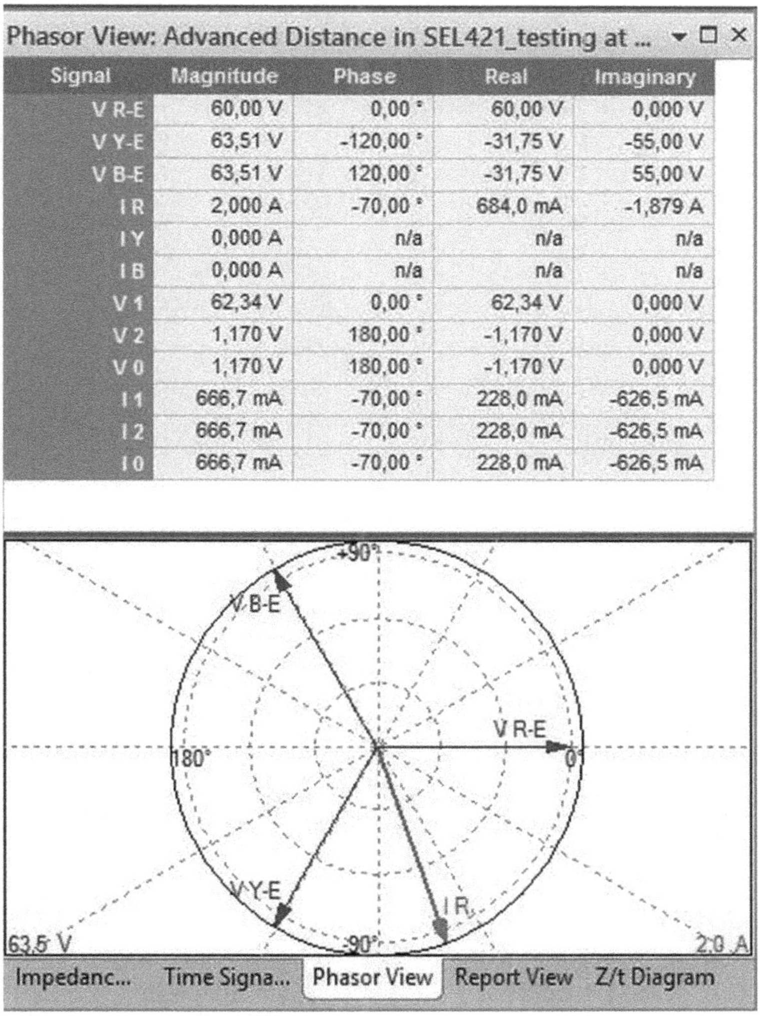

Phasor View: Advanced Distance in SEL421_testing at ...

Signal	Magnitude	Phase	Real	Imaginary
V R-E	60,00 V	0,00 °	60,00 V	0,000 V
V Y-E	63,51 V	-120,00 °	-31,75 V	-55,00 V
V B-E	63,51 V	120,00 °	-31,75 V	55,00 V
I R	2,000 A	-70,00 °	684,0 mA	-1,879 A
I Y	0,000 A	n/a	n/a	n/a
I B	0,000 A	n/a	n/a	n/a
V 1	62,34 V	0,00 °	62,34 V	0,000 V
V 2	1,170 V	180,00 °	-1,170 V	0,000 V
V 0	1,170 V	180,00 °	-1,170 V	0,000 V
I 1	666,7 mA	-70,00 °	228,0 mA	-626,5 mA
I 2	666,7 mA	-70,00 °	228,0 mA	-626,5 mA
I 0	666,7 mA	-70,00 °	228,0 mA	-626,5 mA

FIGURE 8.27 Phasor view of the zone reach test results for the LG event.

5. The *report view* summarises the test results, which includes the test object device settings, distance protection settings, test results and test state. Figure 8.29 provides the report view of the distance protection relay. It shows only part of the device settings, which include the substation, bay and device information, nominal values, residual voltage and current factors, maximum voltage and current limits, debounce and deglitch filter, overload and CB configuration information. The report view of the distance protection provides system parameters, tolerances, grounding factor, and zone settings. The test results of the report view provide information about the test module, test settings and search and shot test results for all types of events.

8.5.5 Protective Relay Event Report Analysis Tool

This section analyses the test results of the Mho phase and ground distance protection elements using the following SEL event analysis tools: (1) SynchrnoWAVe Event and (2) AcSELerator Analytic Assistant tools.

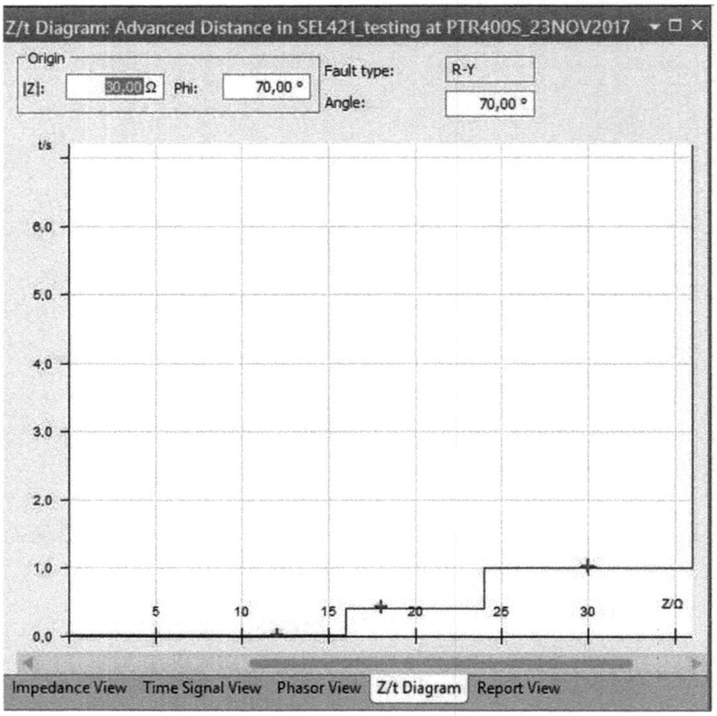

FIGURE 8.28 Zone diagram of the Mho phase distance relay for LL event.

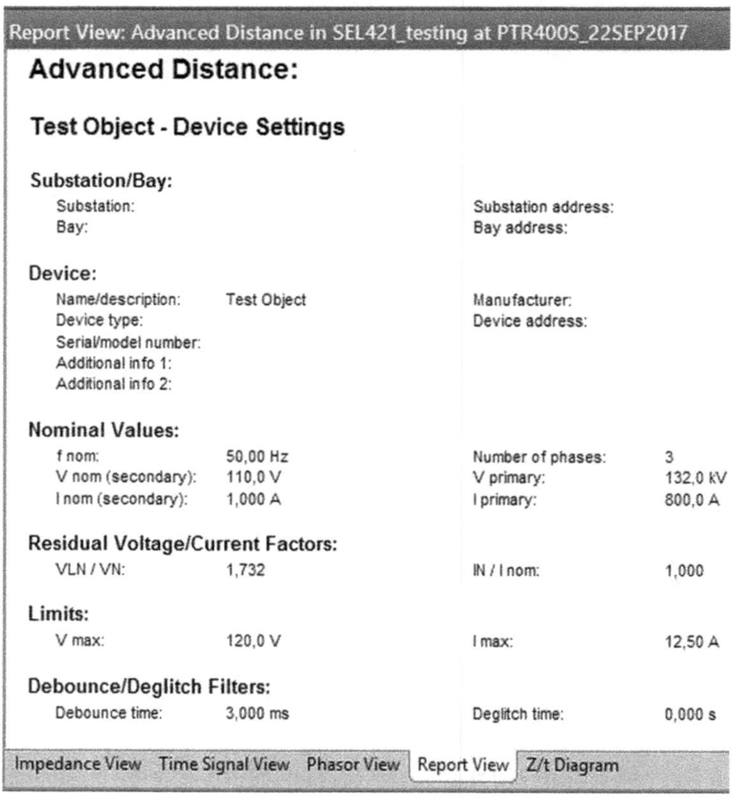
Report View: Advanced Distance in SEL421_testing at PTR400S_22SEP2017

Advanced Distance:

Test Object - Device Settings

Substation/Bay:

Substation:		Substation address:	
Bay:		Bay address:	

Device:

Name/description:	Test Object	Manufacturer:	
Device type:		Device address:	
Serial/model number:			
Additional info 1:			
Additional info 2:			

Nominal Values:

f nom:	50,00 Hz	Number of phases:	3
V nom (secondary):	110,0 V	V primary:	132,0 kV
I nom (secondary):	1,000 A	I primary:	800,0 A

Residual Voltage/Current Factors:

VLN / VN:	1,732	IN / I nom:	1,000

Limits:

V max:	120,0 V	I max:	12,50 A

Debounce/Deglitch Filters:

Debounce time:	3,000 ms	Deglitch time:	0,000 s

Impedance View | Time Signal View | Phasor View | Report View | Z/t Diagram

FIGURE 8.29 Report view of the distance relay.

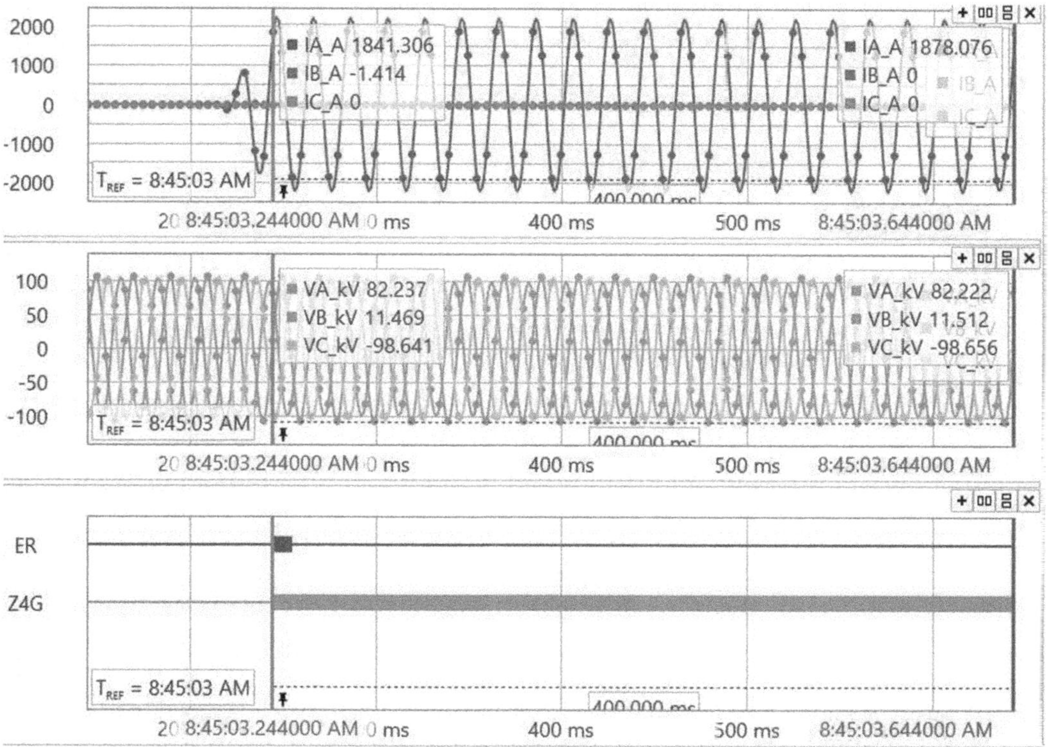

FIGURE 8.30 SynchrnoWAVe event display for LG fault at zone 4 Mho ground distance (Z4G) protective relay.

1. The *SynchrnoWAVe Event tool* is used to analyse the SEL relay event reports. It is a useful tool to diagnose distance protective relay behaviour during an event condition. It provides analog waveforms of the current and voltage signals, and digital signals of the start and trip relay word bits during the event condition. Figures 8.30 and 8.31 show the SynchrnoWAVe event display for LG fault at zone 4 Mho ground distance (Z4G) protective relay and zone 4 Mho ground distance, time-delayed (Z4GT) protective relay, respectively.

   LG fault occurs at 150% of the line length, which asserts zone 4 ground distance element (Z4G) and zone 4 ground distances, time-delayed elements. From Figures 8.31 and 8.32, the vertical line on the left side provides the timestamp of the pickup signal (Z4G) at 3.24 ms and trip signal (Z4GT) at 4.24 ms. Therefore total time elapsed between Z4G and Z4GT signal is approximately 1 ms and is the time delay setting for the zone 4 Mho phase and ground distance relay.
2. The *AcSELerator Analytic Assistant tool* is used to analyse event reports. Each time the relay issues a trip signal, an event report is recorded. The event report shows exactly what the relay saw and how it responded. The analysis of event reports and sequential event record (SER) data can provide the root cause of protection system operations, which is used to identify whether equipment can quickly be restored or testing and repairs are required. The event reports are frequently analysed to improve the protection and maintenance of the equipment. The SEL relay captures 15 or 64 cycles of event data and creates an event report that includes four sections: (a) analog values of current and voltage (translated into waveforms), (b) digital states of the protection and control elements and digital inputs/outputs, (c) event summary (trip type, date, and time), and (d) settings in service at the time of the event.

Figures 8.32 and 8.33 show the AcSELerator Analytic Assistant event display for LG fault at zone 4 Mho ground distance (Z4G) protective relay and zone 4 Mho ground distance, time-delayed (Z4GT) protective

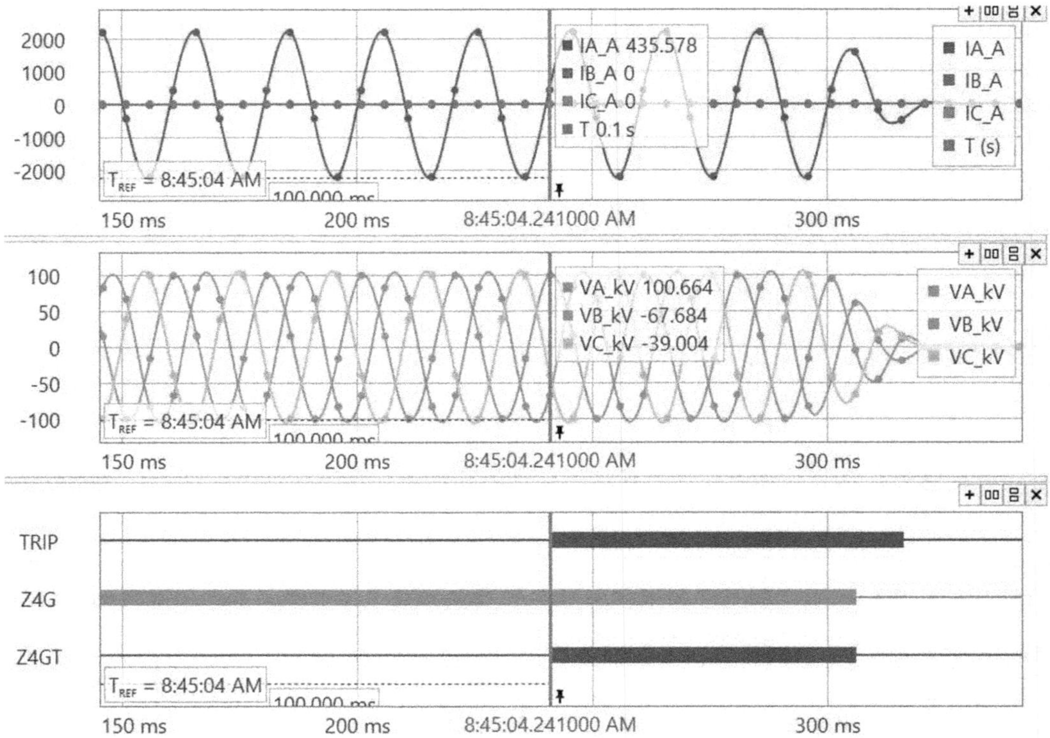

FIGURE 8.31 SynchrnoWAVe event display for LG fault at zone 4 Mho ground distance, time-delayed (Z4GT) protective relay.

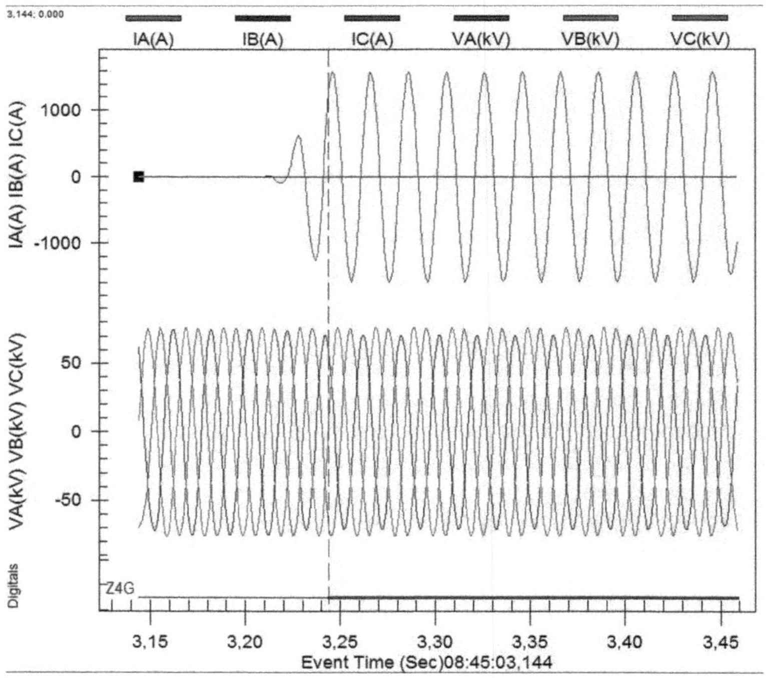

FIGURE 8.32 AcSELerator Analytic Assistant event display for LG fault at zone 4 Mho ground distance (Z4G) protective relay.

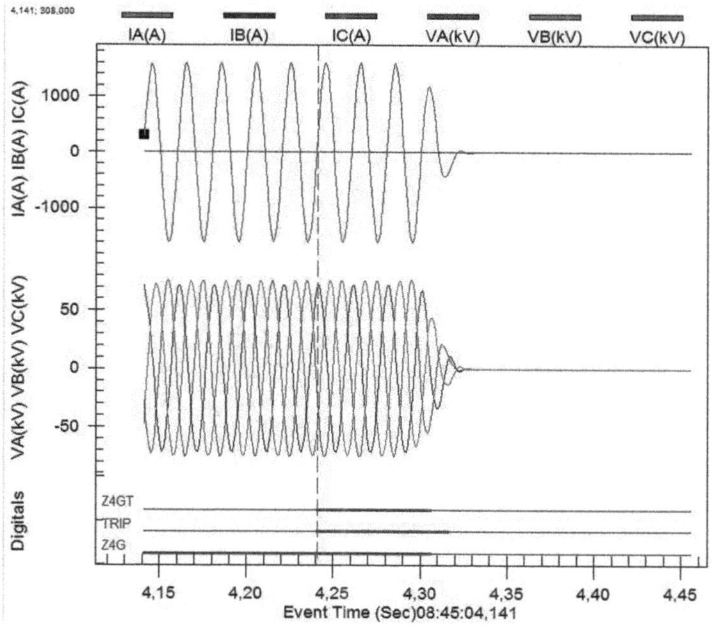

FIGURE 8.33 AcSELerator Analytic Assistant event display for LG fault at zone 4 Mho ground distance, time-delayed (Z4GT) protective relay.

relay, respectively. LG fault occurs at 150% of the line length, which asserts zone 4 ground distance element (Z4G) and zone 4 ground distances, time-delayed elements. Figures 8.32 and 8.33 show a vertical dashed line that provides the timestamp of the pickup signal (Z4G) at 3.24 ms and trip signal (Z4GT) at 4.24 ms. Therefore, the total time elapsed between Z4G and Z4GT signal is approximately 1 ms and is the time delay setting for zone 4 Mho phase and ground distance relay.

8.6 Solved Problems on Distance Protective Relaying System

PROBLEM 8.1

Three-zone mho relays are used for the protection of the power system shown in Figure 8.34. The rated voltage at bus 1 is 500 kV, and the CT and PT ratios are 1500: 5 and 4500:1, respectively. In a three-zone scheme for B3, zone 1 (Z_1) protects 80% of line 1–2, zone 2 (Z_2) protects 120% of line 1–2 and zone 3 (Z_3) protects 100% of line 1–2 and 120% of line 2–3.

The positive sequence impedances are given in the Table 8.18.

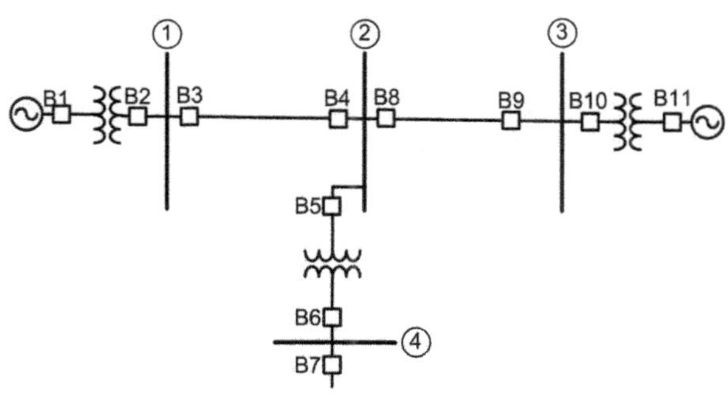

FIGURE 8.34 Mho relay with three zones.

TABLE 8.18
Line Impedance

Line	Positive Sequence Impedance in Ω
1–2	$6 + j60$
2–3	$5 + j50$

1. Determine the settings of zone-1 (Z_1), zone-2 (Z_2) and zone-3 (Z_3).
2. The maximum current through line 1-2 during emergency loading condition is 1400 A at 0.9 power factor lagging. Will any of the relays trip during this condition?

Solutions

1. The line impedance is calculated using the CT and PT ratios:

$$Z = \frac{V_{1LN}}{I_{12}} = \frac{4500/1}{1500/5} = 15$$

Now to calculate the setting for zone 1 of B3 relay for 80% of line 1–2 impedance:

$$Z_1 = 0.80\left(\frac{6+j60}{15}\right) = 0.32 + j3.2\ \Omega$$

The setting for zone 2 for B3 relay, with a reach of 120%, is calculated as follows:

$$Z_2 = 1.20\left(\frac{6+j60}{15}\right) = 0.48 + j4.8\ \Omega$$

The zone 3 is set for 100% of line 1–2 and 120% of line 2–3 is calculated as follows:

$$Z_3 = 1.00\left(\frac{6+j60}{15}\right) + 1.2\left(\frac{5+j50}{15}\right) = 0.8 + j8.0\ \Omega$$

2. The bus voltage at bus 1 is 500 kV, and the maximum current for an emergency loading condition is 1400 A. Therefore:

$$Z = \frac{Z}{15} = \frac{1}{15} \times \frac{500\times 10^3/\sqrt{3}}{1400\angle -25.8^\circ} = 13.7\angle 25.8^\circ$$

Since this impedance exceeds the zone 3 trip setting, the impedance during the emergency loading condition is outside the trip settings of any of the zones. Therefore none of the relays will trip. Moreover, the impedance during a normal loading condition will be even less, and hence it will be further away from the trip regions.

PROBLEM 8.2

Calculate the zero sequence compensation factor for a transmission line with positive and negative sequence impedance of $Z_1 = Z_2 = 0.0 + j0.6$ Ω/km and zero sequence impedance of $Z_0 = 0.0 + j.2$ Ω/km.

Solution

The zero sequence compensation factor (K_0) is calculated as follows:

$$K_0 = \frac{|Z_0 - Z_1|}{3Z_1}$$

$$K_0 = \frac{|0.2 - 0.6|}{3(0.6)}$$

$$K_0 = 0.2222$$

PROBLEM 8.3

Distance protection of a 100 km transmission line has a current transformer ratio (CTR) = 800, potential transformer ratio (PTR) = 132000/110 = 1200, positive and zero sequence line impedance magnitude of $Z_{1,0\,MAG}$ = 20 Ω.sec and positive and zero sequence line impedance angle of $Z_{1,0\,ANG}$ = 70°. Zones 1, 2, and 4 are in the forward direction, and zone 3 is in the reverse direction. Zones 1 to 4 mho phase distance element reach settings are 16, 24, 16 and 36 Ω.sec, respectively. Find the mho phase distance elements zone reach setting as a percentage.

Solution

Zone 1 = (zone 1 reach setting/ Z_{1MAG})*100
Zone 1 = (16 Ω.sec /20 Ω.sec)*100
Zone 1 = 80% impedance of the line

Zone 2 = (zone 2 reach setting/ Z_{1MAG})*100
Zone 2 = (24 Ω.sec /20 Ω.sec)*100
Zone 2 = 120% impedance of the line

Zone 3 = (zone 3 reach setting/ Z_{1MAG})*100
Zone 3 = (16 Ω.sec /20 Ω.sec)*100
Zone 3 = 80% impedance of the line (at 250° looking at reverse direction)

Zone 4 = (zone 4 reach setting/ Z_{1MAG})*100
Zone 4 = (36 Ω.sec /20 Ω.sec)*100
Zone 4 = 180% impedance of the line

PROBLEM 8.4

Transmission line data are given in Table 8.19, and an Mho phase distance relay configuration is given in Table 8.20. Draw the zone diagram and time-distance diagram for the given mho phase distance relay configuration.

Solution

The zone diagram and time-distance diagram for the given mho phase distance relay configuration are given in Figures 8.35 and 8.36, respectively.

PROBLEM 8.5

Distance protection zones 1 to 3 are set to the forward direction at a relay angle of 70°. Find the relay angle for zone 4 in the reverse direction.

Solution

Relay angle in the forward direction: $\alpha = 70°$

Relay angle in the reverse direction: β = full cycle – (fourth quadrant +20°)

Relay angle in the reverse direction: $\beta = 360° - (90° + 20°)$

$$\beta = 360° - 110° = 250°$$

TABLE 8.19

Transmission Line Data

Description	Rated Voltage in kV	Rated Current in kA	Line Length in km	Type	Line Model	Positive and Negative Sequence Impedances per Length in Ω/km	Zero Impedances per Length in Ω/km
Line 1	220	20	160	Overhead line	Lumped parameter	$0.0 + j0.40$	$0 + j0.40$

TABLE 8.20

Mho Distance Relay Phase Element Configuration

Stage (Phase)	Impedance [pri. Ohm]	Impedance Reach Setting in Percentage of Line Length	Angle [°]	Time [sec]	Directional
Zone 1 mho phase element (Z1MP)	48	80%	70	0.20	Forward
Zone 2 mho phase element (Z2MP)	60	100%	70	0.40	Forward
Zone 3 mho phase element (Z3MP)	48	80%	250	0.80	Reverse
Zone 4 mho phase element (Z4MP)	72	120%	70	1.20	Forward

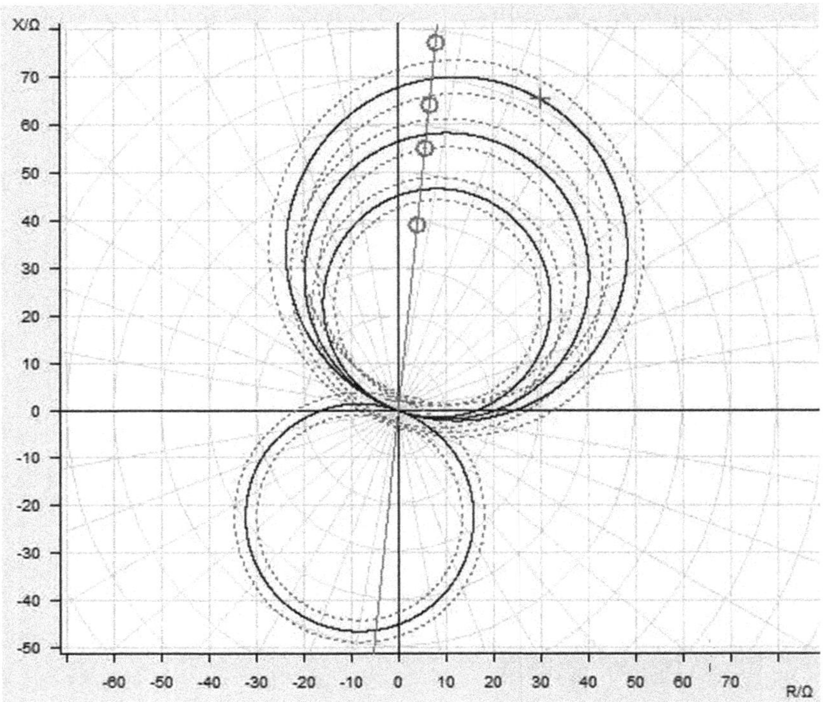

FIGURE 8.35 Zone diagram for the Mho phase distance relay.

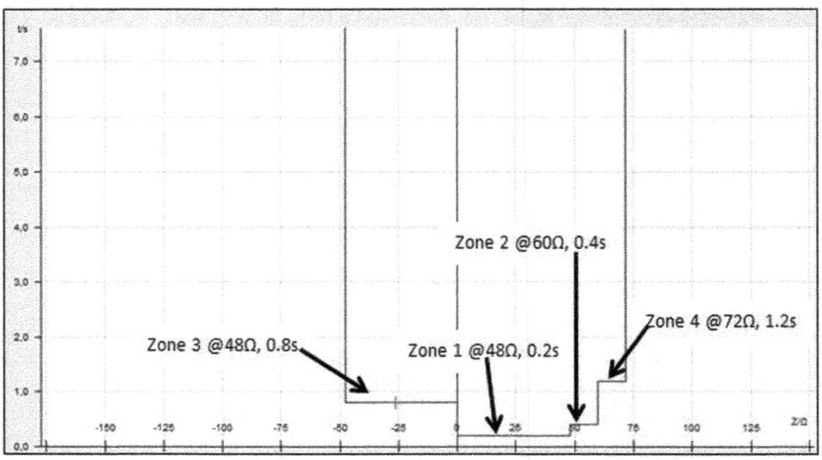

FIGURE 8.36 Time-distance diagram for the Mho phase distance relay.

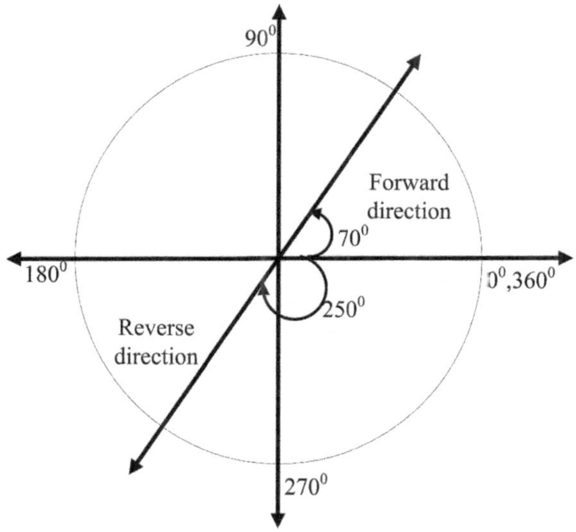

FIGURE 8.37 Mho relay angles for forward and reverse directions.

Figure 8.37 shows the mho relay angles for the forward and reverse directions.

8.7 Conclusion

Distance protective relaying systems are used to protect the transmission lines. This chapter provides an overview of the distance protective relaying system, communication-assisted distance protection schemes, distance protection settings on DigSilent Power Factory tool and numerical relays.

ACKNOWLEDGMENTS

The author gratefully acknowledges the authorities of Cape Peninsula University of Technology (CPUT), South Africa, for the facilities offered while completing this chapter. The software and hardware equipment used in the simulation study is supported by the Center for Substation Automation and Energy Management Systems (CSAEMS) with the Department of Electrical, Electronic and Computer Engineering at CPUT.

REFERENCES

1. J. L. Blackburn and T. J. Domin, *Protective Relaying Principles and Applications*, pp. 1–695, vol. 67, no. 4. Boca Raton, FL: CRC Press, 2014.
2. S. H. Horowitz and A. G. Phadke, *Power System Relaying: Third Edition*. Hoboken, NJ: Wiley, 2008.
3. ALSTOM, *Network Protection & Automation Guide*, pp. 1–497. Ithaca, NY: Cornell University, ALSTOM, 2010.
4. H. J. Altuve Ferrer and E. O. Schweitzer, Eds., *Modern Solutions for Protection, Control and Monitoring of Electric Power Systems* pp. 1–361. Pullman, WA: Schweitzer Engineering Laboratories, 2010.
5. DiGSilent Power Factory, *User Manual*, Gomaringen, Germany: DigSilent GmbH, 2015.
6. Schweitzer Engineering Laboratories, *SEL-421-4, -5 Relay, Protection and Automation System Instruction Manual*, 20170820, 2017.
7. OMICRON electronics GmbH, *Testing Distance Protection*, Test Universe Manual, 2015.

9

Protection of Reactors and FACTS Devices

K. A. Nzeba, J. J. Justo, Aishwarya Biju, and Ramesh Bansal

CONTENTS

9.1 Introduction

During the last two decades, application of flexible alternating current transmission (FACTS) devices in power systems to increase power transfer and provide optimum utilization of the designed system capability through pushing the power systems to their limits has been of worldwide interest. FACTS devices enhance the controllability and stability of the power systems. Literature reviews indicate that FACTS devices introduce new power system dynamics that must be carefully analysed by the system protection engineer [1,2]. By definition, a FACTS device refers to *a combination of power electronics switching components with traditional power system components.* This configuration improves the power system reliability, transient and dynamic stability, and voltage regulation [3]. In fact, installation of FACTS devices is one of the solutions for transmission expansion planning to meet expected increase in power demand. This is true as the constructions of new transmission line infrastructure with higher power transferring capacities faces stiff resistance from an economic point of view and because of environment concerns. Alternatively, FACTS devices cause efficient utilisation of existing transmission system network by controlling active and reactive power flow, and contribute to enhance system stability by increasing transient stability and power system oscillation damping [4,5]. Taking a static Var compensator (SVC) as one of the members of the FACTS family, controlling the thyristor valve that is the heart of the SVC system implies reactive power regulation, which in turn enables the dynamic voltage control at the point of common coupling (PCC) with the grid network. Because of its fast response, the SVC is highly suitable for fulfilling functions like: (i) steady-state as well as dynamic voltage stabilisation, which increases the power transfer capability; (ii) reducing voltage variations; (iii) reduces flicker at industrial arc furnaces; (iv) confines power flow to designated directions; (v) secures loading of the transmission lines close to their thermal limits; and, more important, (vi) prevents cascaded blackout by contributing to emergency control [6,7]. Therefore, SVCs are special devices that are needed the most during network disturbances. SVC makes the difference between a network collapse and successful continued operation. In that regards, it is important that SVC do not go malfunction when they are needed the most. Thus, security is the main requirement on SVC protections, as reasonable dependability is maintained [8,9].

Looking into the structural configurations of the SVC system and shunt reactors, e.g., thyristor switched reactors (TSRs) are defined as inductive loads used to absorb reactive power to reduce the overvoltages generated by line capacitance. In other words, TSRs are installed to offset the capacitive effect of transmission lines and therefore improve the voltage profiles of the transmission lines [9,10]. Specific implementations of shunt reactors may greatly differ between utilities. Technically, reactors and capacitor banks can be placed on a section of the transmission line or on the adjacent bus alongside with current transformers (CTs) that may be installed on the reactors, or the line protection devices such as surge arrestors for the reactors to rely on bus CTs. It is also good to note that each protection philosophy presents different challenges for reactor, transmission line, and/or busbar protection schemes [11,12]. Meanwhile the series reactors are used as current limiting devices to reduce fault current by increasing the impedance of the system. These are also used for impedance matching for parallel feeders. In most cases, their application in transmission and distribution networks is to ensure that fault ratings are not exceeded or, when feeders are added to a substation, the resulting fault current should not exceed the rating of the existing equipment [13].

Moreover, SVCs can be operated to provide reactive power control or closed-loop AC voltage control. In general, these devices (i.e., reactors, capacitor banks, filters and SVCs) are usually exposed to severe current and voltage transient during system disturbances. Thus, SVCs configured with reactors, capacitor banks, and filters are normally not grounded on the medium voltage (MV) busbar, which in this case the residual voltage protection is designed to detect ground faults. Apart from having their own special integrated protection scheme, a grounding transformer or an automatic reclosers scheme can be used as selective earth fault protection for SVC [14,15].

This chapter presents the different protection principles applicable to reactors, SVCs and other related reactive power compensating devices. It also includes an overview of protection methods, which covers the protection strategy, configuration approaches at substation busbars, transmission lines, impedance calculation and protective relay sizing. For the SVC itself, the SVC medium voltage busbar is covered, while for the active branches, the thyristor switched capacitors (TSCs), thyristor controlled reactor (TCR), and harmonic filters are of interest. Moreover, different kinds of protection functions and their interaction with power system as an important criterion for selection and application are covered in detail. The rest of the chapter is organized as follows: configuration and principle of operation for FACTS devices is presented in Section 9.2, while Section 9.3 describes their protection strategies. Moreover, the conclusion of the chapter is given in Section 9.5 and the details of various worked examples are given in Section 9.4.

9.2 Principle of Operation of Reactors, SVCs and STATCOM

Structurally, the SVC system comprises one or more banks of fixed or switched shunt capacitors or reactors, of which at least one bank is switched by thyristors [16]. Figure 9.1 shows a typical configuration of SVC; Figure 9.2 gives different elements that may be used to make an SVC system, which typically includes:

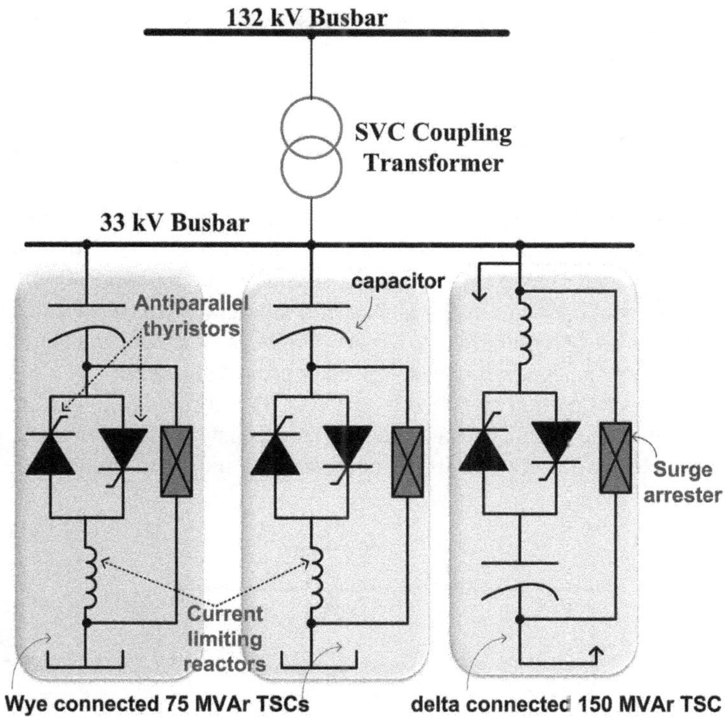

FIGURE 9.1 Typical configuration of SVC at power substation with wye and delta connected TSCs.

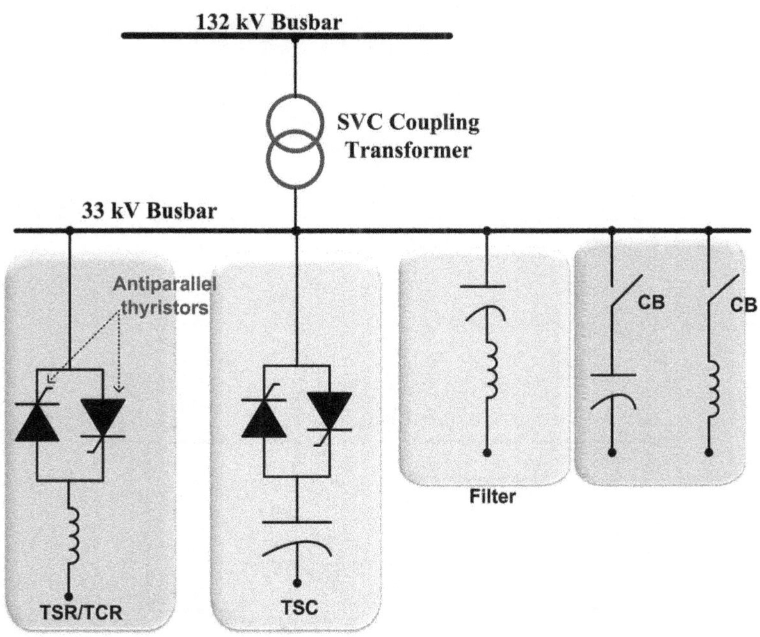

FIGURE 9.2 Shunt-connected controllers: static var compensator (SVC), static var generator (SVG), static var system (SVS) with (i) TCR, (ii) TSC, (iii) TSR, and (iv) Harmonic filter.

- TCRs, whereby the reactor may have an air or iron core and may be series connected or shunt connected.
- TSCs.
- Harmonic filter(s).
- Mechanically switched capacitors or reactors (switched by a circuit breaker).

Shunt reactors are mainly used in transmission networks. Their function is to regulate the line voltage by consuming the excess reactive power generated by overhead lines under low-load conditions and thereby stabilizing the voltage. They are quite often switched in and out on a daily basis, following the load situation in the system. The shunt reactors may either be grounded, or reactor grounded neutral. Alternatively, they may also be connected to tertiary windings of power transformers for the same purpose. The shunt reactors may have grounded, or reactor grounded neutral.

Using the phase angle modulation switched by the pair of anti-parallel thyristors, the reactor can be variably switched into the circuit and provide a continuously variable reactive power injection (or absorption) to the electrical network. In the configuration indicated by Figures 9.3 through 9.5,

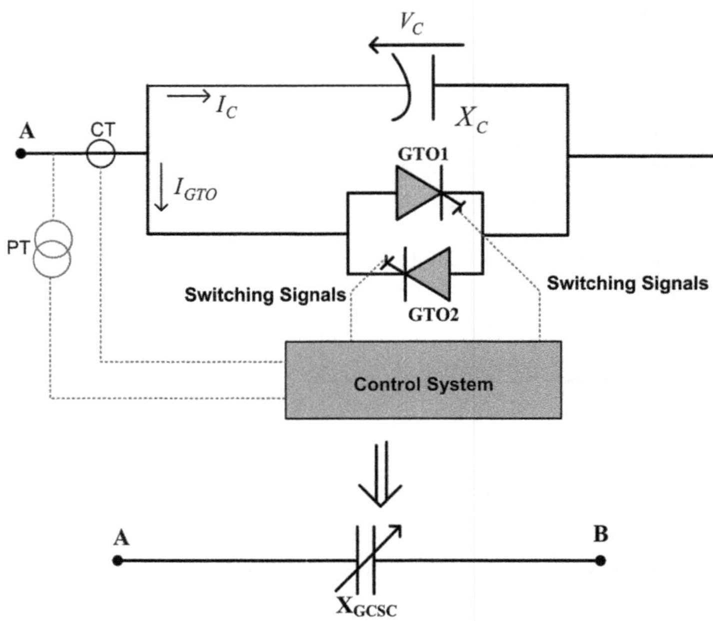

FIGURE 9.3 Transmission line in presence of GCSC.

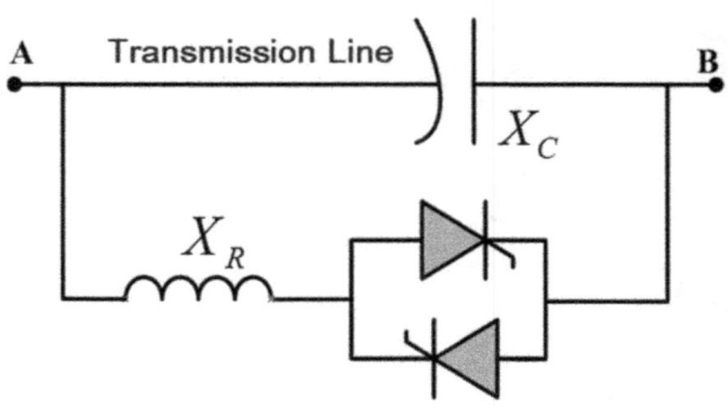

FIGURE 9.4 Typical configuration of thyristor-controlled series reactor (TCSR) with transmission line between point A and B.

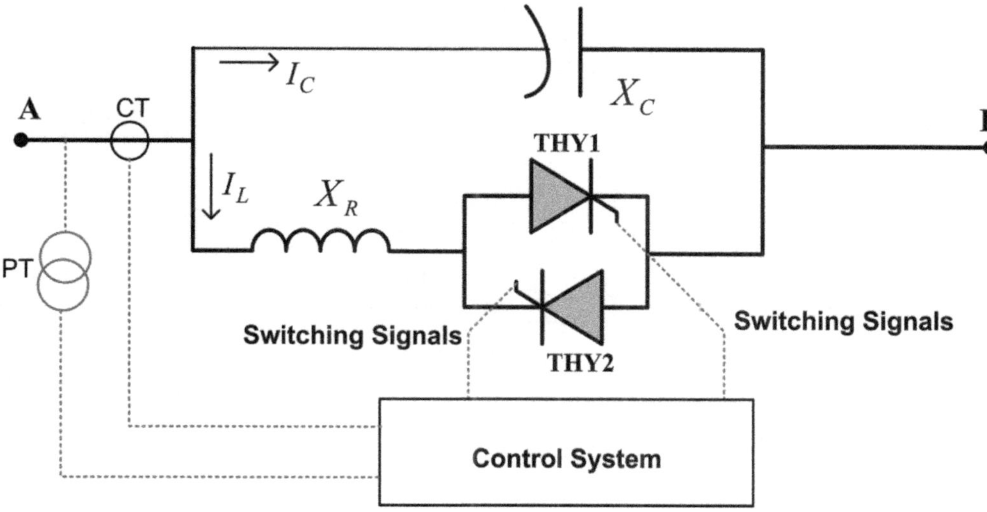

FIGURE 9.5 Transmission line in presence of TCSR.

voltage control is provided by the capacitor bank; meanwhile the TCR provides smooth control. More smooth control and flexibility can be provided with thyristor controlled capacitor (TCC) switching [17,18].

The thyristors are electronically controlled. Thyristors, like all semiconductors, generate heat, and deionized water is commonly used to cool them. Chopping reactive load into the circuit in this manner injects undesirable odd-order harmonics, and so banks of high-power filters are usually provided to smooth the waveform. Since the filters themselves are capacitive, they also export MVARs to the power system. More complex arrangements are possible where precise voltage regulation is required. Voltage regulation is provided by means of a closed-loop controller. Remote supervisory control and manual adjustment of the voltage set-point are also common today.

$$X_{GCSC}(\gamma) = X_{Cmax}\left[1-\frac{2}{\pi}\gamma-\frac{1}{\pi}\sin(2\pi)\right] \tag{9.1}$$

where

$$X_{Cmax} = \frac{1}{\omega C} \tag{9.2}$$

Figure 9.6 shows the capacitor inrush/outrush reactor used to reduce the severity of some of the transient which occurs in the line in order to minimize the dielectric stress on the breakers, capacitors, transformers, surge arrestors and associated substation electrical equipment. In this case, the reactor effectively reduces the transients associated with capacitor switching as the reactor limits the magnitude of transient current in kA, as indicated below:

$$I_{peak} = V_{LL} \times \sqrt{(C_{eq}/L_{eq})} \tag{9.3}$$

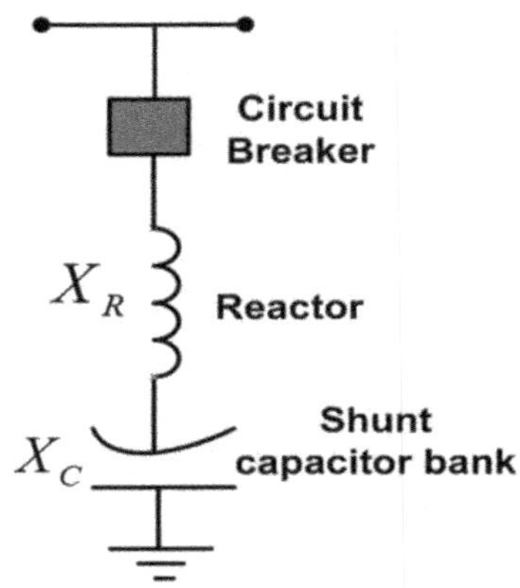

FIGURE 9.6 Typical capacitor inrush/outrush reactor connection.

$$f = \frac{1}{\left[2\pi\sqrt{(C_{eq}/L_{eq})}\right]} \tag{9.4}$$

where:

C_{eq} is the equivalent capacitance of the circuit, F
L_{eq} is the equivalent inductance of the circuit, H
V_{LL} is the system line-to-line voltage, kV

9.2.1 Static Shunt Compensators: SVC and STATCOM

A STATCOM (which is also known as an advanced static VAR compensator) is a shunt connected FACTS device. It generates a set of balanced three-phase sinusoidal voltages at the fundamental frequency, with rapidly controllable amplitude and phase angle [19]. A typical application of a STATCOM is for voltage support. In this section, the STATCOM is modelled as a GTO PWM converter with a DC-link capacitor. Figure 9.7 shows the basic structures of STATCOM.

The objective of the STATCOM is to regulate the voltage at the PCC rapidly in the desired range and keep its DC-link voltage constant. It can enhance the capability of a wind turbine to ride through transient disturbances in the grid. Figure 9.8 shows the voltage profile at the receiving end with an SVC connected and without an SVC connected.

The capability of providing dynamic reactive power compensation using SVC and STATCOM can raise the network voltage during and after fault. This indeed increases the electric torque produced by the fixed speed induction generator (FSIG) and makes generators less likely to overspeed and thus to increases system stability. It consists of a number of (2 in this example) TSCs in shunt with a TCR. The TSC provides step change of connected shunt capacitance, while the TCR provides continuous control of the equivalent shunt reactance. In this case, the SVC can be operated to provide reactive power control or closed-loop AC voltage control. For closed-loop AC voltage control, the line voltage, as measured at the point of connection, is compared to a reference value and an error signal is produced. This is passed to a PI controller to generate the required susceptance value. It is then transmitted to the nonlinear admittance characteristic to generate the firing angle for the TCR and to determine the number of TSC stages need to be switched on. The firing angle is passed to the gate pulse generator, which then generates the firing pulse for the TCR.

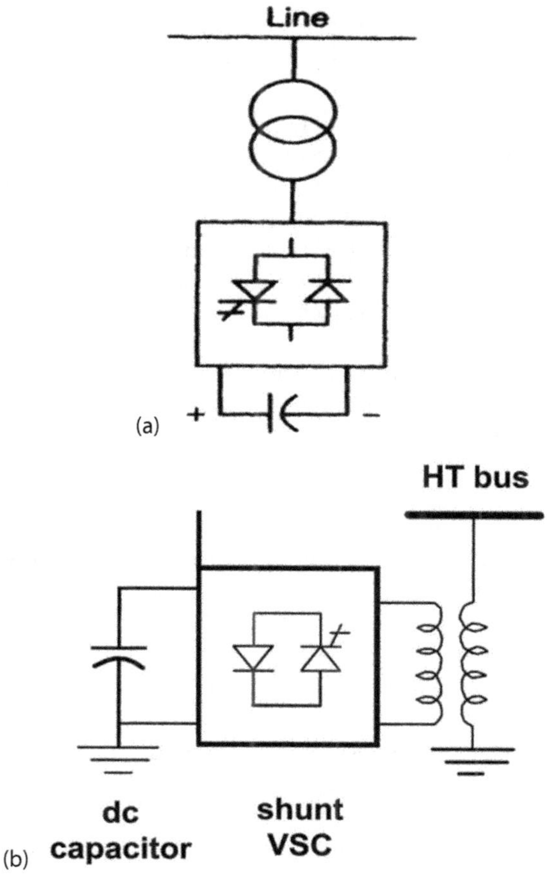

FIGURE 9.7 Shunt-connected controllers: (a) Static synchronous compensator (STATCOM) based on voltage-sourced converter (b) alternative configuration of VSC based STATCOM.

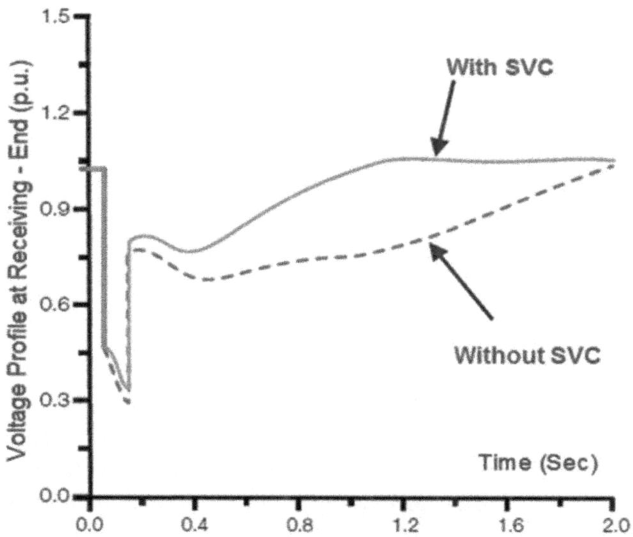

FIGURE 9.8 Voltage profile recoveries with and without SVC.

9.2.2 Static Series Compensators: GCSC, TSSC, TCSC, and SSSC

It has long been recognized that the steady-state transmittable power can be increased and the voltage profile along the line controlled by appropriate reactive shunt compensation. The purpose of this reactive compensation is to change the natural electrical characteristics of the transmission line to make it more compatible with the prevailing load demand. Thus, shunt connected, fixed or mechanically switched reactors are applied to minimize line overvoltage under light load conditions, and shunt connected, fixed or mechanically switched capacitors are applied to maintain voltage levels under heavy load conditions [20]. In this section, basic considerations to increase the transmittable power by ideal shunt-connected var compensation will be reviewed in order to provide a foundation for power electronics-based compensation and control techniques to meet specific compensation objectives. The ultimate objective of applying reactive shunt compensation in a transmission system is to increase the transmittable power. This may be required to improve the steady-state transmission characteristics as well as the stability of the system. Var compensation is thus used for voltage regulation at the midpoint (or some intermediate) to segment the transmission line and at the end of the (radial) line to prevent voltage instability, as well as for dynamic voltage control to increase transient stability and damp power oscillations.

Generally, SVCs are characterized by their ability to rapidly vary reactive power output to compensate for changing system conditions. The following are some of the typical applications for SVCs:

- Maintain voltage at a preset level by compensating the varying loads and correct voltage fluctuations caused by load rejections and outages
- Reduce some types of voltage flicker caused by rapidly changing loads such as arc furnaces or variable speed drives
- Improve power system stability by providing system voltage support
- Improve voltage profile at key locations on the network or near high voltage direct current (HVDC) converter terminals
- Suppress subsynchronous resonance (SSR)
- Minimize transmission losses by improving the power factor
- Alleviate phase unbalance if single phase control is used
- Compensate surge impedance of transmission systems

On the other hand, TCSC introduce a number of important benefits in the application of series compensation:

- Elimination of subsynchronous resonance risks
- Damping of active power oscillations
- Post-contingency stability improvement
- Dynamic power flow control

9.3 Principles of Protection Strategies

Figure 9.9 shows the types of faults within a reactor zone which can be classified based on the location of faults. As indicated in the figure, a reactor can experience phase and phase-to-ground faults on the external connection between the breaker and the high-voltage terminals of the reactor bank indicated by numbers 1 and 2, phase-to-phase faults inside the reactor (number 4), high-voltage bushing-to-ground faults (which is similar to number 2), winding-to-ground faults shown by numbers 3 and 5, and turn-to-turn faults on individual phase windings, as indicated by number 6 in the figure [13].

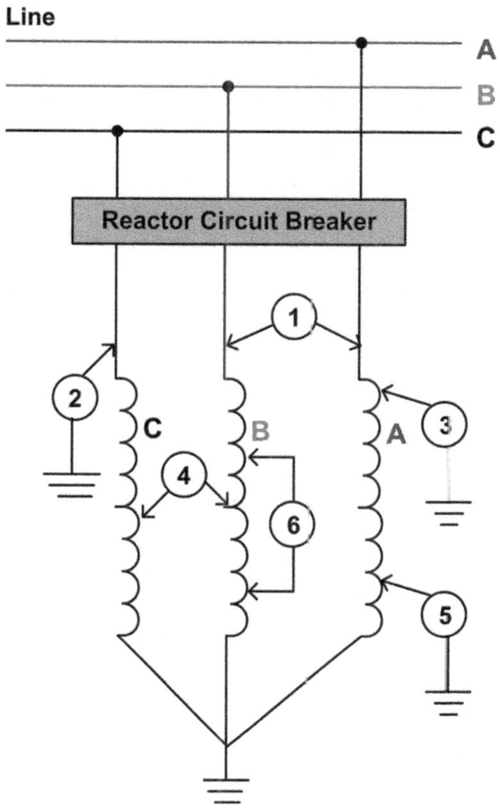

FIGURE 9.9 Types of common faults in a shunt reactor.

S/N	Types of Fault	Descriptions
1	**Phase-to-phase faults**	Phase-to-phase faults (numbers 1 and 2) generate high-magnitude fault currents, especially in the external connection between the reactor bushings and the high-voltage breaker. Internal phase-to-phase faults may not be a concern in dry-type air-core reactors because the individual units are sufficiently spaced apart from each other.
2	**Phase-to-ground faults**	Due to sufficient insulation to ground in a dry-type air-core reactor, the probability of a winding-to-ground fault is very low, unless the ground insulator is bridged by a conducting medium. In oil-immersed reactors, winding-to-ground faults can occur due to proximity to the core and tank. The magnitude of phase-to-ground faults for reactors in general depends on the location of the fault on the winding. The current is higher for ground faults near the supply side (numbers 2 and 3) because the fault currents are limited only by the source impedance and result in large unbalance currents. For ground faults near the neutral end of the reactor (number 5). the magnitude of the fault is limited by the impedance of the reactor. For a ground fault near the neutral, the faulted phase current remains nearly the same as the prefault current and results in a low unbalance current. The neutral point of the bank is in the fault loop and can experience a high current due to the autotransformer effect. The autotransformer effect is caused by the voltage in the nonfaulty turns being magnetically coupled into the faulted turns, driving current in the fault loop. High-magnitude bushing-to-ground faults can occur due to lightning impulses and reactor switching transients that create significant stress on the reactor bushings and eventually damage the bushing insulation.

(*Continued*)

S/N	Types of Fault	Descriptions
3	**Turn-to-turn faults**	Turn-to-turn faults (number 6) within the reactor result in small changes in the magnitudes of the faulted phase current. This type of fault reduces the impedance of the faulted winding and causes a corresponding increase in the phase current. This disrupts the three-phase symmetry of the terminal currents, resulting in zero-sequence unbalance currents flowing through the neutral to ground. Although the change in phase current can be very small, the current in the faulted turns increases the operating temperature and pressure inside the tank. This effect can be detected by nonelectrical protection devices such as the sudden pressure relay. These faults are more likely near the high-voltage terminals of the windings.

9.3.1 Protection of FACTS/SVC Components

Generally, there is a power transformer between the power grid and SVC medium voltage (MV) busbar. Harmonic filters, thyristor controlled reactors and capacitors are connected on this bus. In some cases, an auxiliary power transformer is also connected to this bus. These named components can be considered as FACTS/SVC components [20].

All FACTS/SVC components shall be protected against abnormal occurrences, such as faults and disturbances which can lead to power system failure, in order to limit damage and thereby enhance reliability. In this section, FACTS/SVC components are discussed in detail.

9.3.1.1 Transformer and Busbar Protection

Like generator transformer, SVC transformers are made with large turn ratios. Normally, the transformer turn ratio is 400/25 kV. This large turn ratio results in very high short circuit currents on the MV bus; they are frequently in the range of 50–90 kA. When designing the protection system, the large fault and load current should be considered. Given the fact that the connection of the SVC MV busbar to the grid is made by one single power transformer and the very large load current, protection can be ensured by the differential current relay connected directly across the power transformer. A better design is the inclusion of the SVC MV busbar in the protection zone. Sometimes the busbar can be protected by differential relays or backup transmission line relays.

Generally, the main protection used for transformers is the differential current relays. Other types of protection like overcurrent and overvoltage relays can be used as backup or supplemental protection. Note that buchholz relays will clear all bus bushing faults, but these relays are relatively slow and are designed to perform a different function than high speed, active fault clearing [21,22].

9.3.1.2 Capacitor Size Design, Calculations and Protection

For a given size and voltage rating, a capacitor bank consists of a number of series groups of individual capacitors and a number of parallel capacitors per series group. The utilisation of capacitors with the highest voltage rating results in a capacitor bank with the fewest number of series groups. This provides the simplest and most economical structure and the greatest sensitivity for unbalance detection schemes.

The number of capacitor units in parallel per series group is governed by both a minimum and a maximum limitation. The minimum number of capacitor units per group is determined by overvoltage considerations upon isolation of a capacitor unit in the group. Generally this number can be considered equal to two units, and the maximum number of capacitor units in a parallel per series group depends on the value of discharge current from the parallel unit [14,15].

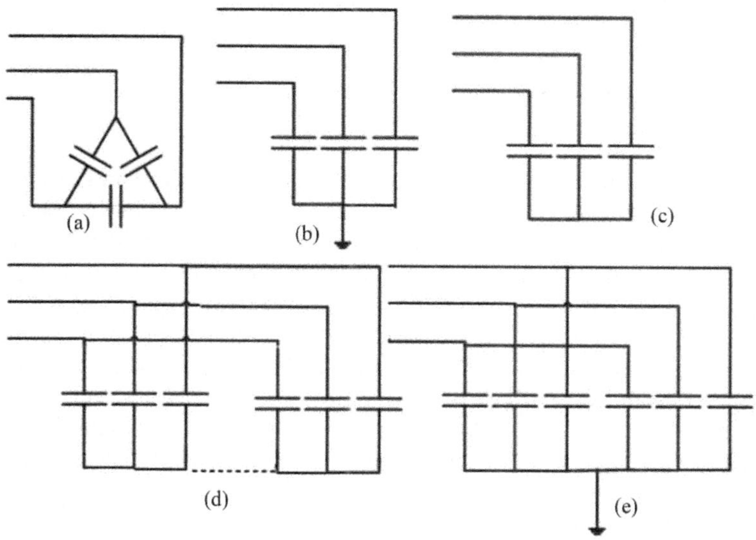

FIGURE 9.10 Common capacitor bank configurations: (a) delta, (b) grounded Y, (c) ungrounded Y, (d) ungrounded double Y-neutrals (can or cannot be tied) and (e) grounded double Y.

There are five common capacitor bank connections, as illustrated in Figure 9.10. All substation banks are connected in wye formation. However, distribution capacitor banks may be connected in wye or delta formation when capacitors are rated for line-to-line voltage. It is important to notice that the optimum configuration of capacitor banks is determined by fusing and protective relaying for a given capacitor voltage rating.

Grounded banks have the following characteristics:

- Low impedance to ground for lightning surge currents and triplen and other harmonics currents.
- A degree of protection from surge voltages.
- Reduced recovery voltages for switching equipment.

Ungrounded banks present the following characteristics:

- No path for zero sequence currents, triplen and other harmonic currents.
- No path for capacitor discharge currents during system faults.
- The neutral must be insulated to full line voltage.

The protection of shunt capacitors requires an understanding of the capabilities and limitations of the individual capacitor units and associated electrical equipment, including individual capacitor-unit fuses, bank switching devices, power fuses, and voltage or current sensing devices.

The design of the capacitor bank protective system should take into account the influence of the following seven basic conditions:

- Overcurrents due to capacitors banks faults
- System surge voltages
- Overcurrents due to individual capacitor failure

- Continuous capacitor unit overvoltages
- Discharge current from parallel capacitor units
- Inrush current due to switching
- Arc-over within the capacitor rack

Fuses are used for protecting individual capacitor units against system failure by removing the unit from service fast enough to prevent case rupture and damage to adjacent units. A proper fuse operation function of bank configuration is required in order to reduce, at minimum, the possibility of cascading failure of additional capacitor units that may, in turn, lead to a major bus fault within the capacitor bank. Therefore, surge arresters are used for the protection against system surge voltages.

Protecting against major faults, including phase-to-phase faults or phase-to-ground faults, generally requires some external protections for capacitor banks such as power fuses, circuit breakers or circuit switchers with relay circuits associated. In the case of grounded Y capacitor banks, the backup protection must respond only to high magnitude faults. However, for an ungrounded Y bank, a phase-to-earth fault results in an increase of line current. The backup protection should allow 125% or 135% of rated current to be carried continuously.

The protection of capacitor banks against phase-to-phase faults or phase-to-ground faults is ensured by using overcurrent relays or bank buses. Time overcurrent relays can be used with normal settings without encountering false operations due to inrush currents. However, instantaneous relays should be set high to override these transients.

Unbalance protection is the most effective protection against faults within a capacitor bank, including arc-over within the capacitor bank. The fast timing of the unbalance relay allows good protection for rack faults and has increased the utilisation of unbalance relays on power systems that are effectively grounded. However, the unbalance time delay should never be less than the maximum clearing time of the capacitor unit fuse for a completely shorted capacitor unit. In some SVC installations, the time overcurrent relays of the short time characteristic can be used for minimizing damage from rack faults.

The type of unbalance relays to be used depends on the configuration of the capacitor banks. The different types of unbalance relays are:

- Summation of line to neutral voltage protection method with optional line to neutral voltage protection
- Neutral voltage unbalance protection method
- Neutral current unbalance protection method
- Neutral current differential protection method
- Voltage difference protection methods used only for single grounded Y connected or double grounded Y connected capacitor banks

Finally, the capacitor bank may be subjected to overvoltages resulting from abnormal system operating conditions. Unbalance sensing with current or voltage relays for Y or double Y banks must be used. If the overvoltage is sufficient to damage the bank, overvoltage relays should be considered.

9.3.1.3 TCR Protection

A TCR or TSR branch is always connected in delta, with each phase consisting of a thyristor valve and two reactor stacks. The thyristor valve is electrically located between the reactors.

A protective zone encompasses the two reactor halves and a thyristor valve [20]. A differential protection is generally used as main protective system through the combination of one-line current transformer (CT) with two branch CTs. By permutation, three such zones are aggregated in the TCR to provide detection and clearance of interzone faults.

Time delayed overcurrent relays, with an added instantaneous step sensing the branch currents, are generally used as backup. The protection of reactors can be ensured by thermal overload relays. The split arrangement of the reactors in each phase provides extra protection to the thyristors in the event of a reactor fault; i.e., the fault current is limited and the risk for steep front voltage surges are eliminated. The valves are also protected against thermal overload by a specific function (TCR current limiter) in the SVC control system.

Differential protection can be of high or low impedance type. Short circuits between the different protective zones are mainly detected and cleaned by the differential protection relays. The protection should be unaffected by SVC energisation and any valve misfiring. Differential protection of low impedance type will have higher requirements on CTs compared to high impedance types. Differential protection of low impedance types needs to be blocked during SVC energisation, in the case where energisation is performed with fully conducting TCR valves. Due to the presence of large DC components and large time constant false differential current arise. This leads restraint criterion not to be fulfilled and protection system not to stabilize.

A negative phase sequence protection allows detecting unsymmetrical TCR operation and turn-to-turn faults. However, turn-to-turn faults are extremely difficult to detect because the small unbalances and sequence currents associated with turn-to-turn faults are generally smaller than the existing tolerable unbalances in the system; it is difficult to distinguish the intolerable from the tolerable conditions. As the turn-to-turn fault spreads to more turns, the current increases. Negative sequence relays must consider conditions mentioned above; the settings are generally high which makes the relay insensitive. The relay should be time delayed to avoid operation on system transients and external faults.

The protection of valves against thermal overload is generally ensured by a specific function (TCR current limiter) in the SVC control system. The temperature inside the reactor is simulated by the selective protection which works with the time constant of the reactors. For some installations this protection is installed in the SVC control system. Experience from working with hundreds of SVCs has also shown that it is very rare that this protection operates since the TCR current will be limited by protective control features implemented in highly reliable SVC control systems.

9.3.1.4 TSC Protection

A TSC is always connected in delta where each phase consists of a thyristor valve, a reactor and a capacitor bank. The thyristor valve is electrically located between the reactor and the capacitor bank. The capacitor bank is generally divided in two parallel halves constructed by the series/parallel combination of a number of capacitor units. As for TCRs or TSRs, TSCs are mainly protected using the differential protection relays. An overcurrent relay sensing the line currents in the TSC can be used as backup protection system. Unbalance protection function supervises the voltage across the capacitor by measuring unbalance current.

In the TSC branches the thyristors are protected by arresters across the valve. Arresters are preferably located so that a current flow in an arrester will not be seen as a transient fault by the differential relay and cause false tripping. In TSC topologies where currents are bypassed from differential CTs, extra time delays must be added to avoid false tripping.

9.3.1.5 Harmonic Filter Protection

Harmonic filters are connected for most SVC installations. Harmonic filters aim to provide reactive power generation at fundamental (grid) frequency and perform the harmonic filtering needed to take care of the harmonics generated by the TCR. Filter banks for SVC applications are generally divided into two parallel banks in Y-Y connection with ungrounded neutrals tied together. Internal fuses protect the capacitor units [20].

Harmonic filters are generally not grounded and are double wye connected. This means that strings in the capacitor bank are tied together internally. Differential protections for filter banks require CTs with a high current rating in the neutral. Harmonics generated by the system and the TCR are important when designing small capacitor banks and shall be considered in rating calculations as well as for the protection of the capacitors. The supervision of the voltage across capacitors shall be ensured by the overload protection by measuring branch currents and calculating of the resulting capacitor voltage, including the effects of harmonic frequencies. Relays designed to operate for fundamental components shall not be used. Unbalance protection function also supervises the voltage across the capacitor by measuring unbalance current.

9.3.1.6 Auxiliary Transformer Protection

Commonly, the SVC MV bus is used for one source of auxiliary power to the SVC. Given the high short circuit power on the bus, it is difficult to trip the auxiliary power transformer in case a fault occurs, which is not usually handled by fuses nor circuit breakers. Fuses can be used for maximum 40 kA short circuit power and circuit breakers for maximum 63 kA. The installation of a series reactor in front of the auxiliary transformer aims to overcome the difficulties. It should be designed to bring down the fault current below 40 kA. Current limiting fuses are the fastest and best means to minimise damage to the auxiliary power transformer. A disconnector should be used in order to be able to replace the fuses or to avoid unsymmetrical operation after a fault.

The auxiliary power transformer shall be protected by means of overcurrent relays, tripping the complete plant in case the fuse operation fails. Load current is very low, typically around 10 A. A protection scheme is needed to detect current slightly above the maximum load for overload purposes and at the same time detect short circuit current of around 40 kA. This can be done by two different overcurrent relays, one connected to a CT with a turn ratio matching the load current and a second one having a turn ratio selected for short circuit current.

9.3.2 Reactor Protection

A TCR or a TSR including its thyristors valve is usually considered a separate zone of protection. Such a zone normally overlaps with the low voltage side of the SVC bus differential and coordinates with the transformer/bus overcurrent zone of protection. In this regard, *IEEE* C37.109-1988 Guide describes the methods and configurations of protection of power system shunt reactors. It also provides information about the protection of the SVC-based reactor coils. There are two types of shunt reactors: dry-type reactor and oil immersed shunt reactor. Shunt reactors can be protected from damage due to major system faults by overcurrent, differential, or distance relays, or by a combination of these relays. However, some specific faults including turn-turn faults can be achieved through voltage unbalance relay schemes for dry-type reactor and sudden-pressure relays for oil immersed shunt reactors.

In many designs, the reactor branch is connected to the SVC bus by a relatively slow motor-operated disconnect switch. In some installations, the SVC can be put back on line without the faulty reactor branch and operated with a limited capability to handle fault conditions.

Similarly to the power transformers, HV oil immersed shunt reactors have the following built-in mechanical fault detection devices:

- Gas detection relay (i.e., Buchholz relay) with alarm and trip stage
- Sudden pressure relay
- Winding temperature contact thermometer with alarm and trip stage
- Oil temperature contact thermometer with alarm and trip stage
- Low oil level relay

These mechanical relays are an excellent complement to the electrical measuring relays. It is recommended that these mechanical relays trip reactor circuit breaker independently from electrical relays. However, signals from mechanical devices shall be connected to binary inputs of numerical relays in order to get time tagging information, disturbance recording and event reporting in case of their operation.

9.3.2.1 Dry-Type Reactors

Dry-type reactors are often connected to the delta connected tertiary of the transformer bank and switched on the supply side or on the neutral side, as shown in Figure 9.11. Lower initial and operating costs, lower weight, and lower losses are the main advantages of the dry-type reactors. However, their limitation in voltage and kVA rating are disadvantages [6,11].

Dry-type reactors are installed in the atmosphere, and their cooling is ensured by natural convection of air circulating between core layers. The reactors are mounted on insulated supports that provide standard clearances to ground, phase-to-ground faults cannot occur for dry-type reactors; thus, they are only subject to phase-phase faults and turn-to-turn faults.

Relaying protection for phase-phase faults consists of overcurrent, differential or negative sequence relays, or a combination of these relays. Figure 9.12 gives the relaying schemes for protecting dry-type reactors against phase-to-phase faults.

Turn-to-turn faults in dry-type reactors are difficult to detect. The variation of current and voltage during a turn-to-turn fault is small and remains closed to the magnitude in normal service; thus, the relays used for phase-to-phase are inadequate for such faults. Therefore, the protection of dry-type reactors can be ensured by the voltage-unbalance relay.

9.3.2.2 Oil-Immersed Shunt Reactors

This type of reactors is often connected to one or both ends of a long transmission line, as illustrated in Figure 9.13. Oil immersed shunt reactors can be connected in some installation to the substation bus and are usually wye-connected and solidly grounded.

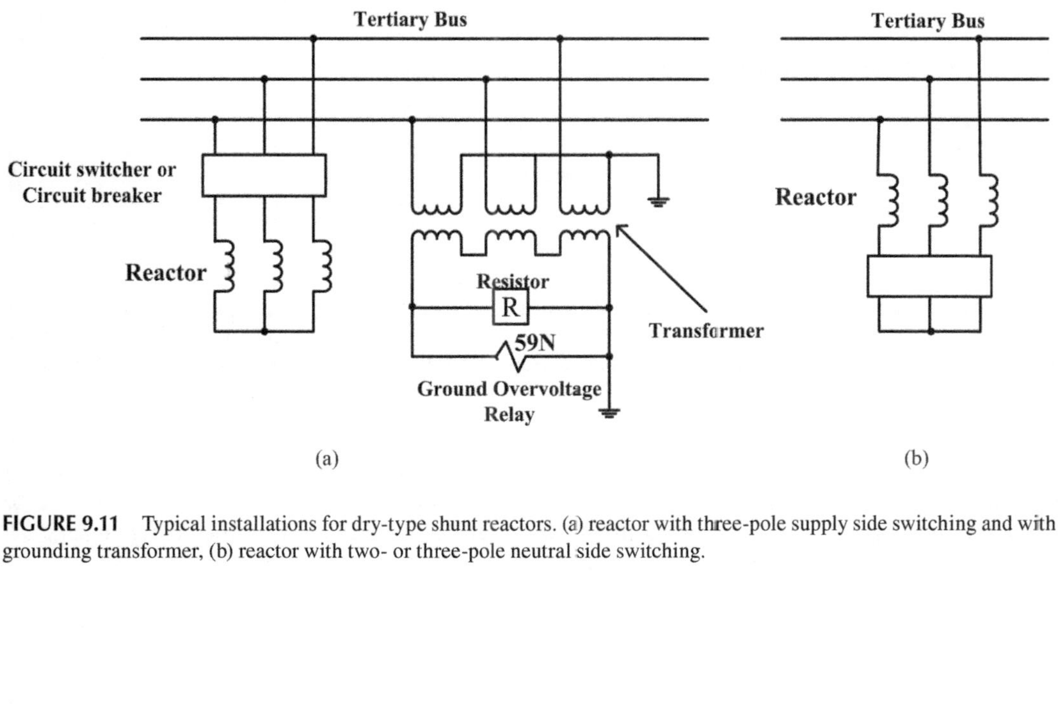

FIGURE 9.11 Typical installations for dry-type shunt reactors. (a) reactor with three-pole supply side switching and with grounding transformer, (b) reactor with two- or three-pole neutral side switching.

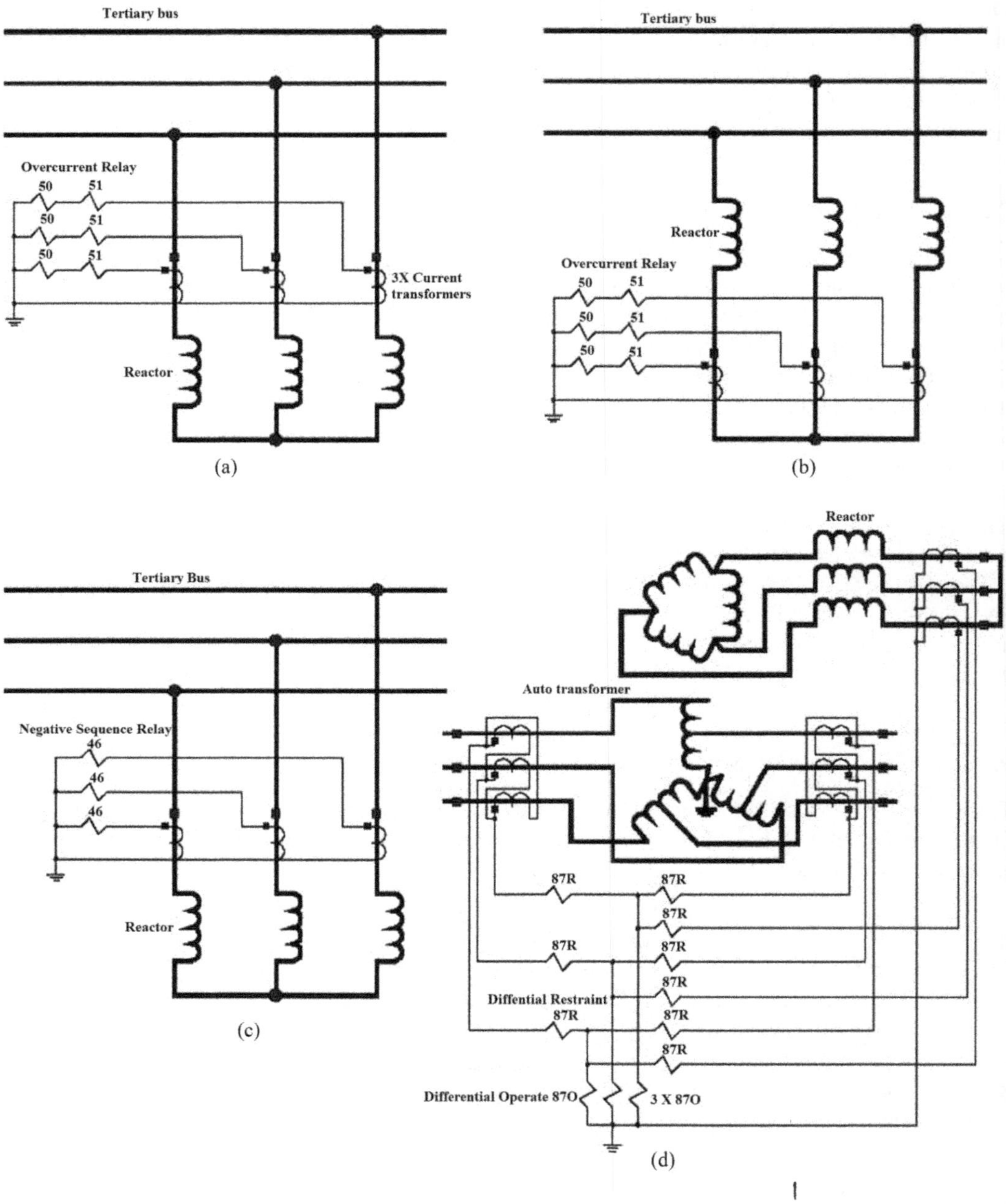

FIGURE 9.12 Protective relaying schemes for dry-type reactors. (a) Overcurrent relay protection with supply-side connected current transformers, (b) overcurrent relay protection with neutral-side connected current transformers, (c) negative-sequence relay protection and (d) differential relay protection for dry-type reactors (reactor is part of zone of auto-transformer tertiary).

Unlike the dry-type reactors, phase-to-ground faults can occur in oil immersed shunt reactors given the proximity of their windings to the core and tank. The value of current in reactors during this type of fault depends upon the location of the phase-to-ground fault or winding-to-ground fault with respect to the reactor bushing. Without winding-to-ground faults (grounding failures) resulting in high magnitude ground current, the other failure possibilities with oil-immersed reactors can be listed as follows:

1. Equipment failure: such as bushing failure, insulation failure, etc., which results in large changes in the magnitude of phase currents.

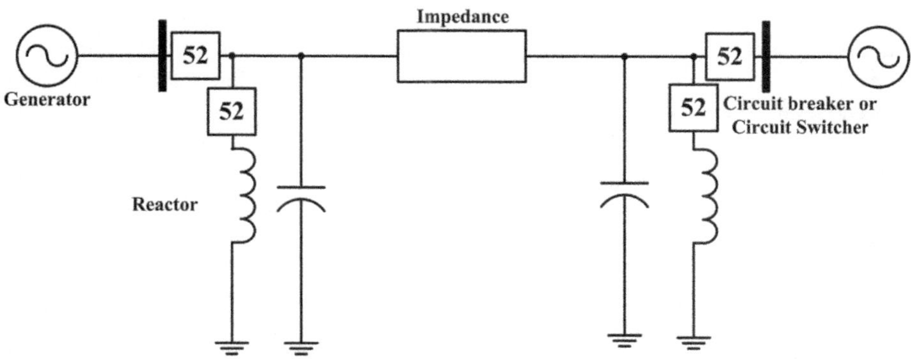

FIGURE 9.13 One-line diagram of line-connected switched shunt reactors.

2. Phase-to-phase faults which result also in high magnitude phase current.
3. Turn-to-turn faults within the reactor winding. As for dry-type reactors, these faults result in small variation of the magnitude of current.
4. Miscellaneous failure such as low oil and loss of cooling.

The protection of oil immersed shunt reactors against faults producing large increases in the magnitude of phase current consists of overcurrent, differential or distance relays. The detection of grounding failures is generally performed by the ground differential relay. Therefore, the sudden-pressure relays, gas-accumulator relays, or a combination of both can be used for detecting turn-to-turn faults within oil immersed rectors.

Oil immersed reactors can be designed with forced cooling, and the continued operation of the cooling motors is critical. Loss of cooling can be detected by monitoring oil flow with flow indicators, monitoring the AC voltage supply to the fans and oil pumps, or by monitoring the temperature with temperature relays or sensors.

Under extreme high-voltage conditions, overvoltage relays can be used to disconnect reactors, but in this case, the associated transmission line should be de-energized at the same time; otherwise disconnection of the reactors tends to aggravate the overvoltage condition on the system. This problem must be solved by opening the line circuit breaker.

9.3.3 SVC Protection

SVCs are devices which provide reactive power required by the load and control voltage at the point of connection. The fast response of SVCs makes them highly suitable for fulfilling functions such as steady-state as well as dynamic voltage stabilisation, meaning power transfer capability increases, reduced voltage variations, and flicker reduction at industrial arc furnaces. SVCs typically consist of both capacitive and inductive branches. The capacitive branch can be of fixed capacitor (FC) type or TSC type. The variable inductive or capacitive susceptance is a function of the thyristor firing angle α and can be controlled by the same angle α.

Generally, the interaction of SVC protective functions with the power system is an important criterion for the selection and application of each protection device. The protection scheme for SVCs is made up of a number of zones. In some cases, faults in a zone should shut down the entire SVC system. In other cases, the relaying can be coordinated so that only the faulty zone is cleared and the SVC is kept in operation with limited capabilities. The protection usually provided for each zone is described in the following sections. Sometimes, these protective functions can be provided as part of an integrated protective system that is supplied by the SVC manufacturer, or they can be supplied by the user. In each case, an agreement on the selection of protective devices and their settings is required between the user and the manufacturer.

Also, the SVC installation includes traditional microprocessor-based protective relays. The relays provide additional protection for the major substation equipment. In that case, the transformer is protected with differential, overcurrent, and overflux relays. The high voltage and low voltage busses are protected with differential and overcurrent relays. With the low voltage bus being ungrounded, over- and under-voltage is also used. Moreover, the TSC branches are protected with differential, thermal overload, overcurrent, and capacitor bank unbalance relays. These relays provide high speed fault protection.

9.3.3.1 SVC Protection Requirements

The protection system should be able to protect all the equipment in the static var compensator from damage due to all types of system failures, including overcurrent, overvoltage, excess reactive power loading, unbalance due to component failure, phase-to-phase faults, phase-to-ground faults, failure of cooling equipment, control malfunctions, and any other type of failure that can place undue stress on the SVC components.

As previously mentioned, the protection system should consist of two independent systems called the main and backup protection systems, or systems A and B. Both main and backup protections must perform with high reliability, speed, selectivity and sensibility.

The two systems must be different in their design and operation, and they must use different measurement transducers where possible. Note that the two systems can have the same descriptions in some cases. In this case, the equipment used to meet the protection requirements should always be different in the two systems, using hardware design or different tripping logic.

The main advantage is the possibility to remove either protective system for maintenance and repair while the second system continues to function normally. Both of them, main and backup protection systems, must perform with high reliability, speed, selectivity and sensibility.

Usually, the capacitors in the TSC are connected in parallel arrays so that the determination of the failure of individual capacitor units can be made by measuring the current in the bridge connection between the parallel arrays. During normal operation conditions, the two branches of capacitors are balanced and the current in the bridge is zero. If one capacitor unit fails, the bridge current becomes not null and can be computed based on knowledge of the capacitor array structure. Failure of a second unit in the same branch results in a greater unbalance current. The SVC should be shut down for repair if the additional failure causes an overvoltage beyond 10% by blocking the thyristors.

The control action of the thyristors allows faults to be interrupted in the SVC branches by quickly blocking the flow of current in the defective branch while allowing the nonfaulty branches to continue in operation. The tripping of circuit breaking serves as backup for fault cleaning. Thus, it is important to have coordination between thyristor blocking and circuit breaker operation. Breaker operation can be limited to faults that require complete removal of the entire SVC from the power system.

9.3.3.2 Protection of SVC Components

Apart from the components of SVCs mentioned in the previous section, transformer and the bus at which the SVC is connected are also considered as SVC components. Transformers can be effectively protected using differential relays, phase and ground overcurrent relays, gas pressure, low level oil and temperature relays. Time overcurrent relays are used for bus backup protection for phase faults even if, in many cases, the transformer overcurrent relays can provide this function. Against ground faults, the SVC buses can be protected using earth overcurrent relays. However, TSCs, FCs, and thyristor valves which are vulnerable to overvoltages can be protect using overvoltage protection relays.

In general, current differential protections (with overlapping protection zones) and overcurrent protections are used for short circuit protection for all SVC components. Time-delayed residual overvoltage protection is used as ground fault protection for the SVC bus and all SVC medium voltage components. Current differential protection is employed for the capacitor banks in TSCs and filters. Special capacitor overload protection is used on the filters to protect against low order harmonics from the system. Other protection functions requested by customers are restricted earth fault, negative sequence current protection and overexcitation.

9.3.4 STATCOM Protection

FACTS devices are normally used to enhance controllability and increase power transfer capability of the grid. Figure 9.14 shows a long transmission line with a STATCOM connected. Generally, three installation positions are considered for STATCOMs: at the near end bus, the midpoint of the line and the remote end bus. In the case of the near end bus, bus protection allows to protect STATCOM as well; at the midpoint of the line, the protections of the transmission line generally the distance relays allow to protect them.

However, in power systems grids, the presence of FACTS devices such as STATCOM usually affects the operation of some subsystems such as distance relays. It is very important that the distance relays do not malfunction under system fault conditions as this will result in the loss of stability or the security of the system. STATCOM devices have fast response times, and their functional characteristics and control system introduce dynamic changes during fault conditions in a transmission line. Thus, it is important that distance relays perform correctly irrespective of such dynamic changes introduced during fault.

The measured impedance at the relay point is the basis of the distance protection operation. Several factors affect the measured impedance at the relaying point, as presented in [1,2]. Some of these factors are related to the power system parameters prior to the fault instance, and they can be categorized into two groups. The first group includes the structural conditions, represented by the short circuit levels at the transmission line ends; the second group includes the operational conditions, represented by the line load angle and the voltage magnitude ratio at the line ends.

$$\mathbf{Z} = \mathbf{Z}_1 + \left[\left(\frac{I_{res}}{I_p}\right) \times \mathbf{Z}_{res}\right] + \left[\left(\frac{I_{mut}}{I_p}\right) \times \mathbf{Z}_{mut}\right] \quad (9.5)$$

where:

$\mathbf{Z}_1$ is the positive sequence impedance reach setting
I_p is the current in the faulted phase
$I_{res} = (I_a + I_b + I_c)$ is the residual current
I_{mut} is the residual current in the parallel line
$\mathbf{Z}_{res} = (\mathbf{Z}_0 + \mathbf{Z}_1)/3$ is the residual impedance, which includes the earth impedance
$\mathbf{Z}_{mut}$ is the mutual compensating impedance

For any fault behind the STATCOM, the STATCOM is in the front of the fault. At these conditions, STATCOM injects a current in quadrature with the line voltage feeding the fault and boosts the voltage at the midpoint, which is seen as additional impedance by the relay. This impedance can be either inductive or capacitive, depending on the mode of operation of the STATCOM prior to the fault. In this situation the relationship of Equation (9.5) does not apply, and therefore the apparent impedance calculated by the

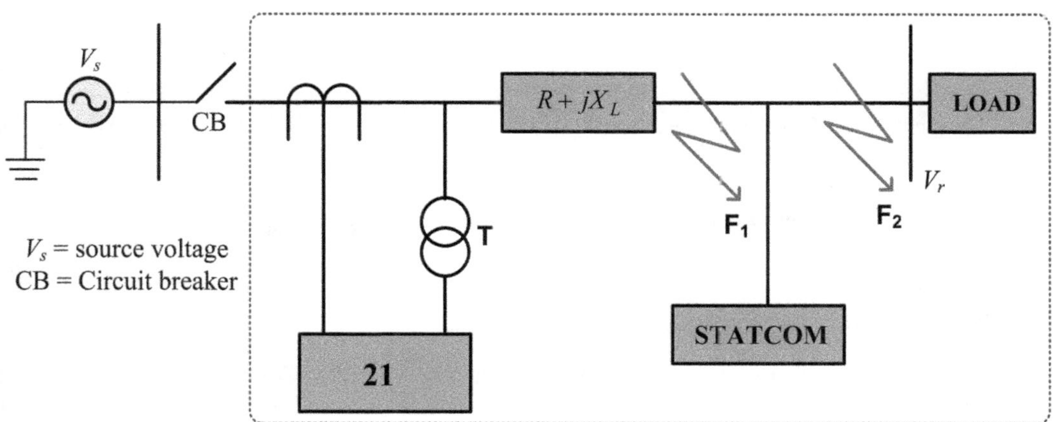

FIGURE 9.14 Distance relay in a transmission system with a STATCOM at the midpoint.

distance relay is different from the actual fault impedance. This scenario leads to possible underreach or overreach of the measuring elements of the distance relay. Hence the relay must be provided with some form of compensation to eliminate the underreach or the overreach given by Equation (9.6):

$$\mathbf{Z}^{/} = \mathbf{Z} + \mathbf{Z}_{compensating} \tag{9.6}$$

9.4 Tutorial Problems

1. List the various components of protection systems applied for FACTS devices.
2. What are various types of faults in reactors?
3. Explain the guidelines to design the protection for shunt reactors applied in transmission lines.
4. Derive the equation for selecting suitable relay current for SVC protection under- and overvoltage.
5. Explain the various principles of protection for STATCOM.
6. How many capacitor cans would be required to build a single star capacitor bank of 20 MVAr at 88kV using the can ratings of 12 kV and 186 kVAr? (At nominal system voltage, the can operating voltage is not to exceed 85% of its rated voltage.) Determine:
   - Number of cans per phase.
   - Number of series groups per phase.
   - Number of parallel cans per group.
7. The SVC consists of a 735 kV/16 kV, 333 MVA coupling transformer, one 109 MVAr TCR bank and three 94 MVAr TSC banks (TSC1 TSC2 TSC3) connected on the secondary side of the transformer. Switching the TSCs in and out allows a discrete variation of the secondary reactive power from zero to 282 MVAr capacitive (at 16 kV) by steps of 94 MVAr, whereas phase control of the TCR allows a continuous variation from zero to 109 MVAr inductive. Taking into account the leakage reactance of the transformer (0.15 p.u.), the SVC equivalent susceptance seen from the primary side can be varied continuously from −1.04 p.u./100 MVA (fully inductive) to +3.23 p.u./100 MVAr (fully capacitive). Design a model and simulate the model in MATLAB/Simulink to show the transient response of the SVC with TCR, TSC and coupling transformer.
8. The transmission line parameters are given below:

   Length of transmission line = 180 km

   Line impedance: $Z_1 = Z_2 = 0.035744 + j0.507762\ \Omega/\text{km}$

   Capacitance: $C_1 = 0.00243\ \mu\text{F/km}$

   $Z_0 = 0.363152 + j1.326473\ \Omega/\text{km}$, $C_0 = 0.001725\ \mu\text{F/km}$

   Voltage level: 220 kV

   Source impedance:

   $Z_s = Z_r = 26.45\angle 80°$, $\delta = 200$

   The capacity of STATCOM is ±300 MVAr. The leakage reactance of the coupling transformer $X_T = 0.1$ p.u. Using simulation software, show the responses of the STATCOM when the following types of faults occur at 75 km away from its installed location: (i) single phase fault, (ii) three-phase short circuit fault, and (iii) double line to ground fault.

9.5 Conclusion

The operating characteristics of FACTS devices such as the SVCs/STATCOMs and their protection techniques were presented in this chapter. The SVC systems are available in a wide number of configurations; hence their means of protection do too, and SVC reactors can be custom designed for specific applications so does the reactors. For SVC applications, the control and protection system play an essential role in the overall performance of the power system network stability and reliability. From the protection standpoint, an extensive protection system is generally required for both the reactors and SVCs in order to optimise the equipment operational limits for maximum utilisation. It also includes an overview of the SVC protection methods, which covers the protection configuration for the reactor and SVC to substation busbars and transmission lines, impedance calculation for each system and their relevant control strategy. The chapter has discussed the protection of SVC, its components, e.g. TSC, TCR and harmonic filters and protection techniques for the FACTS devices like SVC and STATCOM. These branches are exposed to severe current and voltage transient during system disturbances. Insensitivity to harmonics and DC current are also essential. Thus, SVCs are normally ungrounded on the MV bus, and in this case the residual voltage protection is used to detect ground faults. If selective earth fault protection is required for the SVC, this can be accomplished by using either a grounding transformer or an automatic reclosure scheme.

SVC system are integrated with special protection scheme in their control system to detect abnormal operating conditions and to react rapidly to avoid damage and unnecessary tripping by the plant protection system. Those protection functions and their interaction with power system are important criteria for selection and application of each protection device, and they are covered in detail in this chapter. The objectives are: (1) provide fundamental concepts on protection relay characteristics in the presence of shunt-FACTS devices such as SVC/STATCOM, (2) investigate the impact of SVC and STATCOM on trip region for single-line-to-ground and double line to ground faults and (3) give students with first-hand tool on the study of power system protection.

REFERENCES

1. A. Abdolkhani, P. S. Bansaghiri and F. A. Rocky, Impact of STATCOM on distance relay operation for various types of faults, *International Journal of Advanced Research in Electrical, Electronics and Instrumentation Engineering*, 2(10), 4772–4779, 2013.
2. K. El-Arroudi et al., Operation of impedance protection relays with the STATCOM, *IEEE Transactions on Power Delivery*, 17, 381–387, 2002.
3. J. Francois and S. Datta, SVC protection and control basics (for energy porter SVC with external capacitor bank switching), *58th Annual Conference for Protective Relay Engineers*, College Station, TX, April 5–7, 2005.
4. R. Dubey, S. R. Samantaray and B. K. Panigrahi, Adaptive distance protection scheme for shunt-FACTS compensated line connecting wind farm, *IET Generation, Transmission & Distribution*, 10(1), 247–256, 2016.
5. A. Kazemi, S. Jamali and H. Shateri, Effects of STATCOM on distance relay tripping characteristic, in *IEEE/PES Transmission and Distribution Conference and Exhibition: Asia and Pacific*, 2005, pp. 1–6.
6. J. L. Blackburn, *Protective Relaying: Principles and Applications*, 2nd ed., New York: Marcel Dekker, 1998.
7. P. M. Anderson, *Power System Protection*, IEEE Press, Piscataway, NJ, 1999, pp. 673–708.
8. IEEE Std 1303–1994, Guide for Static var Compensator Field Tests, August 2011.
9. ANSI/IEEE C37.109–2006, IEEE Guide for Protection of Shunt Reactors, April 2007.
10. S. C. Chano et al., Static VAR compensator protection, *IEEE Transactions on Power Delivery*, 10(3), 1224–1233, 1995.
11. H. Mahdi, Technical specification and requirements of Static VAR compensator (SVC) protection consist of TCR, TSC and combined TCR/TSC, Electrical Engineering Department Gonabad University, Gonabad, Iran.
12. K. Wikström, Z. Gajic and B. Poulsen, The design of a modern protection system for a static VAR compensator, *CIGRE*, Jeju Island, Korea, October 19–24, 2009.
13. F. K. Basha and M. Thompson, Practical EHV reactor protection, *IEEE 69th Annual Conference for Protective Relay Engineers*, IEEE, College Station, TX, May 2014.
14. ANSI/IEEE C37.99 1990, Guide for the Protection of Shunt Capacitor Banks.

15. S. Samineni, C. Labuschagne and J. Pope, Principles of shunt capacitor bank: Application and protection, *63rd Annual Conference for Protective Relay Engineers*, Atlanta, GA, March 2010.
16. N. Hingorani and L. Gyugyi, *Understanding FACTS, Concepts and Technology of Flexible ac Transmission Systems*. New York: IEEE Press, 2000.
17. T. J. Miller, *Reactive Power Control in Electric Systems*. New York: Wiley, 1982.
18. A. Findley, M. Hoffman, D. Sullivan and J. Paramalingam, Lessons learned in static var compensator protection, *CPRE*, College Station, TX, November 2017.
19. P. Rao, M. L. Crow and Z. Yang, STATCOM control for power system voltage control applications, *IEEE Transactions on Power Delivery*, 15(4), 1311–1317, 2000.
20. J. Dixon, L. Morán, J. Rodríguez and R. Domke, Reactive power compensation technologies: State-of-the-art review, *Proceedings of the IEEE*, 93(12), 2144–2164, 2005.
21. M. Halonen, B. Thorvaldsson and K. Wikström, Protection of static var compensator, *CIGRE*, Jeju Island, Korea, October 19–24, 2009.
22. ANSI/IEEE C37.91–2000, Guide for Protective Relay Applications to Power Transformers, 2000.

Section III

Equipment Protection: Motor, Transformer, Generator, Substation Automation and Control; Overvoltage and Lightening Protection

10

Transformer Protection

Patrick T. Manditereza

CONTENTS

10.1 Introduction

A power transformer consists of insulated windings mounted on an iron core. The iron core is made up of laminations that have insulation between them that are then bolted together. The winding assembly is mounted inside a tank filled with oil. Usually a tap-changer in a separate tank is mounted to the transformer. Various types of faults may occur in the components that make up the transformer, such as winding short circuits, core faults, overheating, tank and accessory faults, or open circuits. Usually no relay protection is provided against open circuits and overheating, but thermal sensors may be added to control cooling fans and give an alarm to allow the network operator to reduce the load on the transformer, with the transformer being tripped only in a few exceptional cases. The main concern is to provide protection against short circuits in the transformers or their connections, and detection and removal of abnormal conditions before they develop into winding short circuits. The transformer is an important component in the power system, and any transformer fault should be removed as quickly as possible, especially for the higher rated transformers whose loss may lead to widespread loss of supply.

10.2 Origins of Transformer Faults

Transformer faults may originate within the transformer itself or may be due to abnormal external conditions.

10.2.1 Faults That Are Internal in Origin

Transformer faults may be due to [1]:

- Aging or sustained overloading that may lead to failure of the insulation of windings due to brittleness leading to inter-turn faults.
- Failure of the insulation of the laminations and core bolts that may lead to increased eddy currents causing excessive heating of the core. Hot spots may develop leading to damage of the winding insulation.
- Oil within the transformer tank has two purposes—to provide insulation and aid the removal of heat from the windings to the ambient. Poor quality oil or deterioration due to ingress of moisture may lead to low level partial discharges. The quality of the oil may also be compromised by decomposition because of overheating and the formation of sludge by oxidation as a result of bad electrical joints.
- Tap changer failure: The only moving parts in a transformer are those in the on-line tap changer mechanism. Defects within the tap-changer are a major cause of transformer faults.
- Tank failure: Loss of oil within the transformer tank reduces the winding insulation and may also lead to overheating due to loss of cooling.
- Cooling system failure: Transformer designs with forced cooling may also suffer from overheating due to defects in cooling system blocked ducts or failure of cooling fans.

10.2.2 Faults That Are External in Origin

External causes may also lead to transformer faults:

- Overload causes increased copper loss that may eventually lead to deterioration and failure of the winding insulation.
- Heavy through-faults also cause increased thermal and mechanical stresses that may damage transformer windings and insulation.
- Transient overvoltages may also cause fault. Transient surge voltages from lightning or switching with steep wave-fronts may cause insulation failure and inter-turn faults.
- Power frequency overvoltages are also a problem, especially with modern transformers. To minimise the amount of iron in the core, the modern transformer is designed to operate near the knee-point. Any increase in the power frequency voltage, in addition to the increased stress on insulation, will lead to an increase in working flux, in turn causing an increase in iron loss and magnetising current, a phenomenon referred to as overfluxing. (A drop in frequency may also cause overfluxing; in principle the V/f ratio should be maintained below a certain level in order to avoid overfluxing). This may result in flux being diverted from the laminated core into structural steel parts. Core bolts are subjected to large fluxes, which lead to temperature rise in

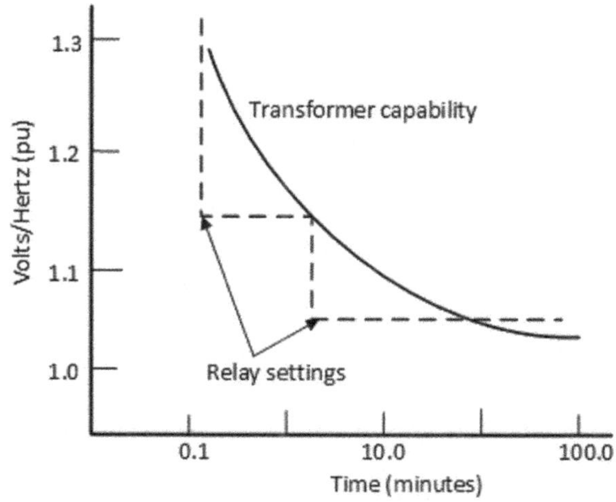

FIGURE 10.1 Transformer volts/hertz capability curve.

the bolts that can destroy their insulation. A transformer has a volts/hertz capability curve, as illustrated in Figure 10.1, and volts/hertz protection may be provided to protect the transformer against operation outside this capability. Many volts/hertz relays have two settings, a lower setting for alarm and a higher setting that may be used for tripping. The relay characteristic can be definite time or IDMT type. Overfluxing is characterised by a significant amount of fifth harmonic current and the relay may be designed to detect this harmonic.

10.3 Magnetising Inrush

Under steady-state conditions, the magnetizing flux in a transformer lags the induced voltage by 90°, as shown in Figure 10.2a. When the transformer is de-energised, a residual flux might remain. When the transformer is next energized, assuming zero residual flux, the resulting flux must start from zero rather than its negative maximum, as shown in Figure 10.2b and c. The flux will therefore rise to twice the normal peak value during the first half of the voltage wave. This level of flux drives the transformer into saturation and requires a very large exciting current, as shown in Figure 10.2c. If the transformer is energised with a residual flux, the saturation will even be greater, drawing magnetising currents several

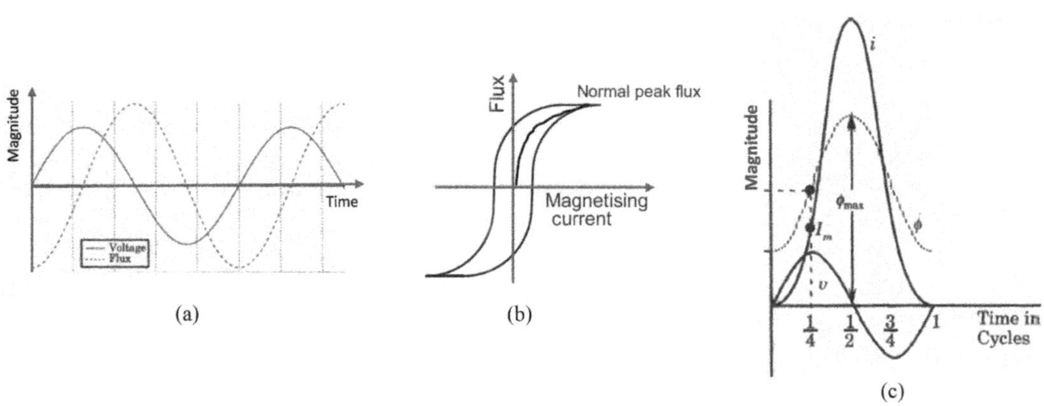

FIGURE 10.2 (a) Steady-state voltage and flux waveforms, (b) typical magnetising characteristic, (c) magnetising inrush current spike.

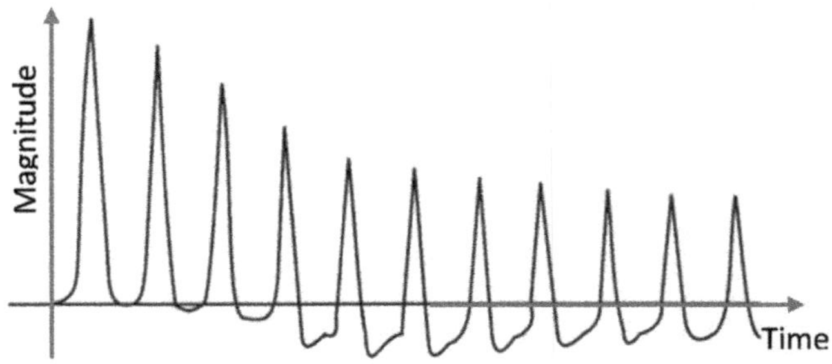

FIGURE 10.3 Magnetising inrush current waveform.

times the rated current of the transformer. The transformer protection must remain stable during the inrush transient condition.

The inrush current is nonsinusoidal and highly distorted, as illustrated in Figure 10.3. The waveform can be symmetrical or unsymmetrical about the horizontal axis, depending on the energising conditions. Fourier analysis shows that the typical inrush current contains both even and odd harmonics.

10.4 Overcurrent Protection

Faults may occur between windings (inter-winding faults) or between a winding and the core of the transformer (winding to core faults). These types of faults result in current flows that greatly exceed normal load currents. Protection is required against the overcurrents, and the type of protection provided depends on the application and importance of the transformer.

10.4.1 Fuses

Fuses are used on small distribution transformers, typically up to 1 MVA rating. The fuse should be rated to withstand the transformer short duration overloads and must withstand the magnetising inrush currents. The consequence is that fuse ratings are much higher than maximum transformer load current. This limits the sensitivity of the fuse such that fuses really provide only short circuit protection and not overload protection. As a result, the fault clearance times may be very long for low fault currents. The fuse operating characteristics are also not precise and may be affected by preloading and gradual damage from through-faults. However, fuses provide reasonable protection at low cost.

10.4.2 Overcurrent Relays

Use of overcurrent relays removes the excessive delays associated with fuses, especially for low fault currents. Overcurrent relays are used on larger transformers with associated circuit breakers, as illustrated in Figure 10.4a. When setting time overcurrent relays, it is necessary to specify pickup setting, time multiplier setting (TMS) and relay time-current characteristic [2]. Time-current curves (TCCs) are designated as standard inverse (SI), very inverse (VI), and extremely inverse (EI). The TCC characteristics are described by Equation (10.1):

$$t_{op} = \frac{A \times \text{TMS}}{\text{PSM}^B - 1} \tag{10.1}$$

where:

t_{op} is the operating time is seconds
TMS is the time multiplier setting
PMS is the plug setting multiple
A, B is the constants with values given in Table 10.1 for each curve type.

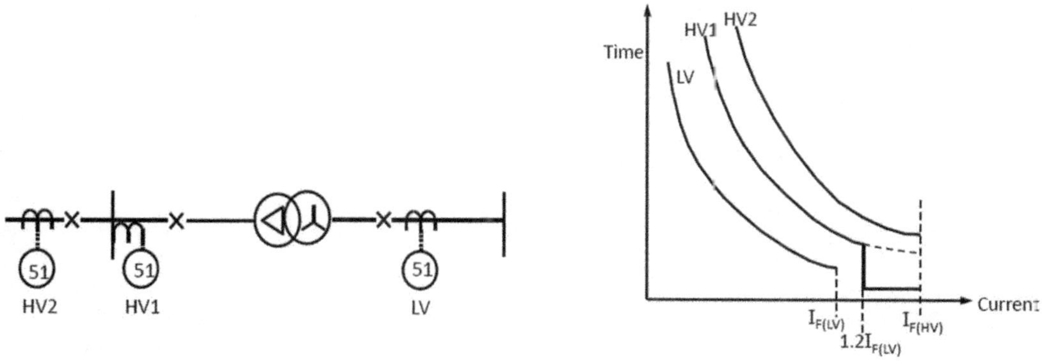

FIGURE 10.4 (a) Application of transformer overcurrent protection, (b) relay time-overcurrent curves.

TABLE 10.1
Operating Characteristic Constants

Curve Type	A	B
Standard	0.14	0.02
Very inverse	13.5	1
Extremely inverse	80	2

A detailed discussion of relay operating characteristics can be found in Glover et al. [3].

The relay pickup setting must be above the short time overloading rating of the transformer, and the relay characteristic must allow the inrush or cold load pickup current to flow without tripping. This, however, limits the sensitivity of the overcurrent relays. A high-set instantaneous relay is provided for rapid clearance of large fault currents, as illustrated in Figure 10.4b. The high-set element is usually set so that it does not reach beyond the LV busbar and into the response area of the downstream relay.

Example 10.1

Overcurrent relays with standard inverse (IDMT) characteristics are used for protection of the transformer (HV and LV sides), as shown in Figure 10.5. The current transformers on the HV and LV sides have ratios of 100/1A and 200/1A, respectively. The pickup (plug) settings on the HV and LV relays are 0.90 A and 0.70 A, respectively. The fault currents at the HV and LV bushings (referred to the HV side) are 708.56 A and 358.66 A, respectively.

1. Prove that the HV overcurrent relay will not trip at transformer full load (current).
   Transformer rated current = 65.61 A.
   The HV O/C minimum pickup is 90 A; HV O/C will not trip on full load.

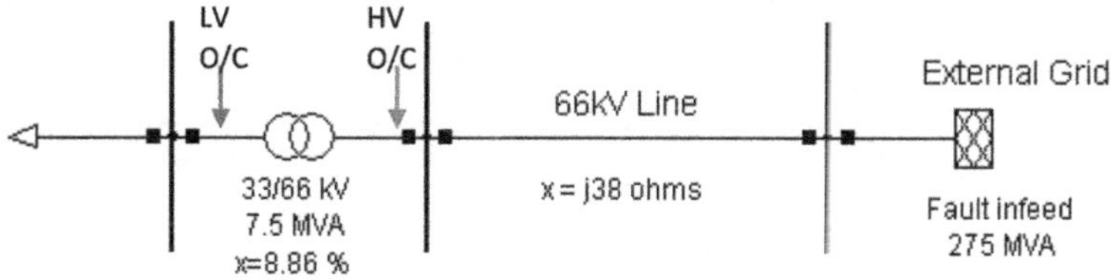

FIGURE 10.5 Radial feeder distribution system.

2. Determine the time multiplier setting (TMS) of the LV O/C relay given that the relay operates in 1.20 seconds for a 33 kV (LV) busbar fault.

   LV fault current = 717.32 A (referred to 33 kV).

   SI characteristic: $t_{op} = \dfrac{0.14 \times \text{TMS}}{\text{PSM}^{0.02} - 1}$

   $\text{PSM} = \dfrac{I_F}{I_{\text{p.u.}}} = 5.12.$

   Hence, TMS = 0.285
3. A high-set instantaneous element with a pickup setting of 4 A is incorporated in the HV overcurrent relay. Prove that this instantaneous element does not trip for a 33kV busbar fault but trips for a transformer HV (bushing) fault.

   33 kV busbar fault: $I_F = 358.66$ A (referred to 66kV)
   $I_F = 3.59$ A (< HV high-set element setting)
   Therefore, the high-set element will not trip.
   66kV bushing fault: $I_F = 708.56$ A
   $I_F = 7.09$ A (> HV high-set element setting)
   Therefore, the high-set element will trip.

Example 10.2

An overcurrent relay R with extremely inverse (EI) characteristic is used for protection of the transformer shown in Figure 10.6. The 400 V feeders are protected using 450 A fuses. The (three-phase) short-circuit fault level in kA at busbars 1 and 2 (referred to 11 kV) are 1.05 and 5.25 kA, respectively. Determine the appropriate time and current settings for relay R on the HV side of the transformer, assuming the following:

- EI relay characteristic given by $t = \dfrac{80 \times \text{TMS}}{\text{PSM}^2 - 1}$.
- Current transformer ratio is 200/1.
- The time-current characteristic of each fuse can be described as a curve passing through the following points: 8s at 6kA, 2.5s at 8kA, 0.4s at 12kA, 0.18s at 10kA, 0.07s at 16.27kA, 0.03s at 21kA, 0.01s at 29kA; 0.005s at 35kA.
- The minimum grading margin for relay-fuse coordination is given as ($\Delta t = 0.3t + 0.2$)s.
- The relay time multiplier setting is restricted to multiples of 0.01.
- The relay current setting is restricted to multiples of $0.1 \times I_{\text{nominal}}$.
- The relay setting is 1.3 × maximum load current.

Transformer rated current = 104.97 A.
Relay R setting (primary) = 1.3 × 104.97 = 106.46 A.
Relay B setting (secondary) = 106.46 × 1/200 = 0.68 A.
Select relay setting of 0.7 A.
$I_{F(\text{Bus\_1})} = 0.2$ p.u. = 1.050 kA. Referred to 400 V side, current equals 28.875 kA.

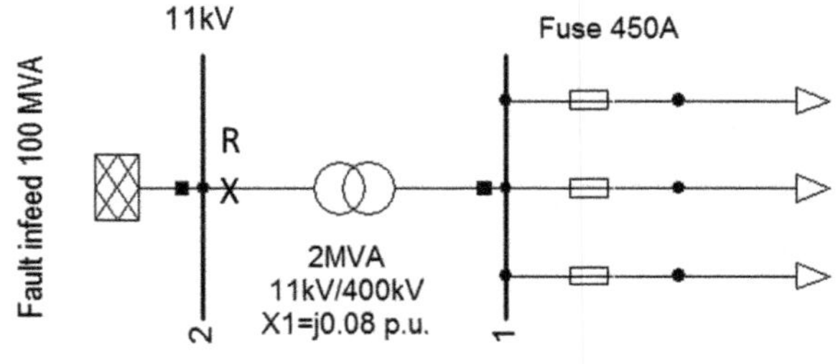

FIGURE 10.6 Simple radial distribution system.

Fuse operating time = 0.01 s.
Grading margin, $\Delta t = (0.3 \times 0.01) + 0.2 = 0.203$ s.
Required relay operating time = 0.01 + 0.203 = 0.210 s.

Relay characteristic EI: $t_{op} = \dfrac{80 \times \text{TMS}}{\text{PSM}^2 - 1}$.

PSM (at 1.05 kA) = 7.5.
Hence required TMS = 0.147.
Select TMS = 0.15.

10.5 Earth Fault Protection

An overcurrent relay sees all phase fault currents. However, some phase-to-earth faults may result in current flows that are smaller than full load current [4]. The sensitivity of the overcurrent relay may therefore be inadequate for earth faults. A dedicated earth fault relay may be added that sees earth faults only. The earth fault relay is connected in the transformer neutral circuit of Y/Y and Y/Δ transformers, with grounded Y-connections giving the neutral earth fault (NEF) protection, as illustrated in Figure 10.7a and b. The relay may also be connected to measure the residual current of the phase CTs (E/F), as shown in Figure 10.7c. With either connection, the relay sees earth faults only and may thus be set to pick up at currents well below full load current.

The conventional earth fault protection uses overcurrent elements. However, the simple earth fault scheme may fail to provide adequate protection for the entire transformer windings, especially with impedance-earthed neutral. In this instance, faults near the star-point produce very small currents (that are below relay setting). The restricted earth fault (REF) protection, discussed in Section 10.8, may be added to provide protection to the entire winding.

10.6 Differential Protection

Differential protection involves measurement of current at each end of an explicitly defined zone, as illustrated in Figure 10.8. Assuming the two sets of current transformers (CTs) have matching magnetisation characteristics, by Kirchoff's current law, no current should flow through the relay under normal or external fault conditions, as illustrated in Figure 10.8a and b. The relay operates only when there is an internal fault which causes a residual current to flow through the relay, as shown in Figure 10.8c.

10.6.1 Problems to Be Considered in the Implementation of Differential Protection

The operation of differential protection is simple in principle, but several issues need to be considered in the implementation of the differential scheme [5]. Spill currents may flow through the relay during normal conditions due to differences in magnetising characteristics of the current transformers. These differences are amplified at high external through-faults. Leakage or charging currents within

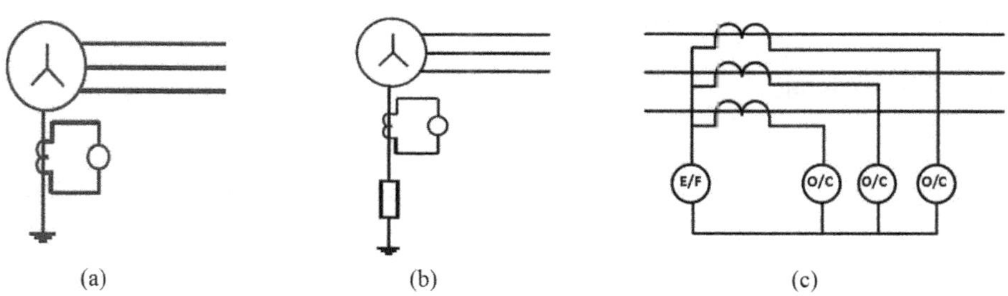

FIGURE 10.7 (a) Neutral earth fault (NEF) protection, (b) NEF protection with impedance-earthed star-point, (c) earth fault protection.

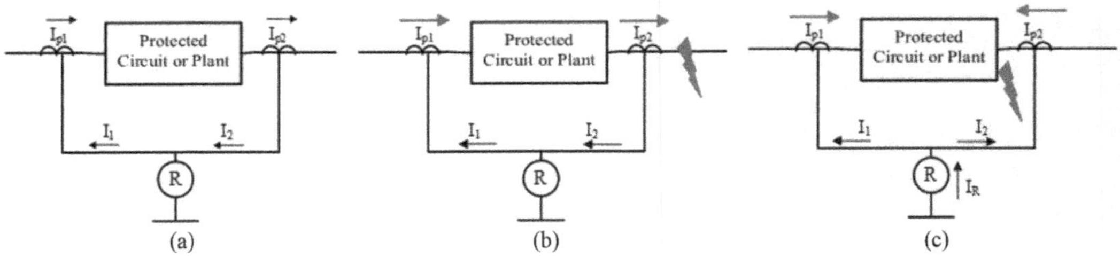

FIGURE 10.8 Differential protection current flows: (a) under normal conditions, (b) with external fault, (c) with internal fault.

the plant in the protected zone may also result in spill currents. Asymmetrical currents applied to a CT during large external faults may induce flux that is much greater than the peak flux, taking the CT into saturation. The other CT at the opposite end may still be in the linear range, resulting in current "spill."

The effect of magnetising inrush currents also needs to be considered. These inrush currents are seen only by one set of CTs causing large spill currents that may cause the relay to trip. Hence, stability needs to be provided against spill currents due to inrush and other causes, as discussed above.

Other issues to consider include the phase shift across the transformer windings. Phase correction will need to be implemented. It may not be possible to find CT ratios that make it possible to exactly match the pilot currents from either side of the transformer windings. In this case it will be necessary to correct for possible current magnitude imbalance. This is referred to as ratio correction. The ratio correction is complicated by the presence of a tap changer that adjusts the transformation ratio.

10.7 Differential Protection Types

Two methods are available to provide stability to prevent the differential relay from tripping when there is no internal fault. One method applies biasing, giving rise to what is called biased differential protection, while the other method involves inserting a high resistance in the relay circuit, giving rise to a scheme referred to as high impedance differential protection. Both methods allow the differential relay to retain sensitivity to internal faults.

10.7.1 Biased Differential Protection

A low threshold of spill current I_p may be allowed below which the relay should not operate. This may be expressed thus:

$$|I_1 - I_2| > |I_p|, \text{block} \tag{10.2}$$

$$|I_1 - I_2| > |I_p|, \text{trip} \tag{10.3}$$

where:

I_1, I_2 are the pilot currents
I_p is the spill current threshold

However, CT errors increase with magnitudes of I_1 and I_2. Hence, the threshold should be set to be dependent on I_1 and I_2. For example, can be made proportional to the average of I_1 and I_2. The operating principle becomes:

$$|I_1 - I_2| < k\left|\frac{I_1 + I_2}{2}\right|, \text{block} \tag{10.4}$$

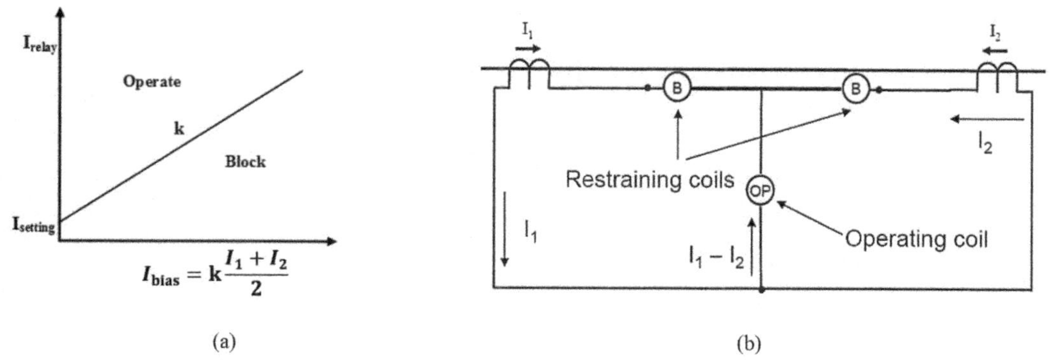

FIGURE 10.9 (a) Differential relay operating principle, (b) biasing coil arrangements.

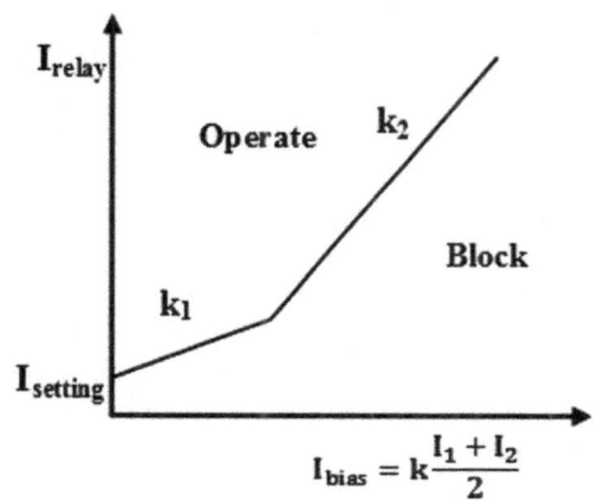

FIGURE 10.10 Characteristic of bias relay in differential protection.

$$\left|I_1 - I_2\right| > k\left|\frac{I_1 + I_2}{2}\right|, \text{trip} \tag{10.5}$$

$k\left|\frac{I_1+I_2}{2}\right|$ is called the biasing or restraining current or quantity.

This operating principle is represented graphically in Figure 10.9a. The relay operates when the operating current, I_{relay} $(= I_1 - I_2)$ exceeds a certain percentage of the restraining quantity, as determined by the value of k in the equation. A relay with such characteristic is also called percent differential relay. The restraining/biasing quantity is provided by restraining coils arranged as shown in Figure 10.9b. Each of I_1 and I_2 are passed through restraining coils. The restraining coils produce flux that opposes the effect of the operating coil.

Typically, the biased differential relay must be more sensitive (small bias) for low level unbalance such as that due to normal load and variation of tap change position [6]. The bias is required to be higher at high currents such as those due to through faults. The relay characteristic that achieves this operation is shown in Figure 10.10. The lower bias is provided by k_1 and k_2 provides stability at the higher currents.

Example 10.3

A single-phase two-winding, 125 MVA, 330/33 kV, Y/Y-connected transformer has differential relay protection. Select suitable bias (k) such that the relay blocks for up to 25% mismatch between the currents.

Condition for relay operation:

$$|I_2' - I_1'| > k\left|\frac{(I_1' + I_2')}{2}\right|$$

From which:

$$I_2' > \frac{2+k}{2-k} I_1'$$

25% mismatch implies operation for $I_2' > 1.25 I_1'$

i.e., $\frac{2+k}{2-k} = 1.25$

or $k = 0.22$

10.7.2 High Impedance Differential Protection

An alternative to biased differential protection is to provide stability using a technique referred to as high impedance differential protection [7]. An impedance, referred to as the stabilizing impedance is connected in series with relay, as shown in Figure 10.11. The objective is to minimise the spill current through the relay. For example, as can be seen in Figure 10.12, if one of the CTs saturates, the current from the unsaturated CT will flow mainly through the saturated CT rather than through the relay because of the relatively low impedance of the saturated CT compared to the high impedance in the relay circuit that includes the stabilizing impedance. This provides the required stability for external through-faults.

The distribution of potential around the pilot circuit of Figure 10.11 for a current through the protected zone is shown in Figure 10.13a. It can be seen that, with respect to the relay circuit, the output voltage of one CT (e1) is positive and that of the other CT (e2) is negative. It can be seen that the voltage across the relay element is almost zero. A similar voltage profile is obtained under external fault conditions but with higher magnitude of the voltages.

When one of the CTs saturates under heavy through-fault conditions, its voltage collapses to zero and the voltage profile changes to that shown in Figure 10.13b. A significant voltage now appears across the relay circuit, causing a spill current to flow through the relay. This represents the worst-case scenario, and a value should be chosen for the stabilizing resistance (R_{ST}) such that, under this condition (when one of the CTs saturates), the spill current is less than the pickup setting of the relay [8]. The value of R_{ST} is determined as follows.

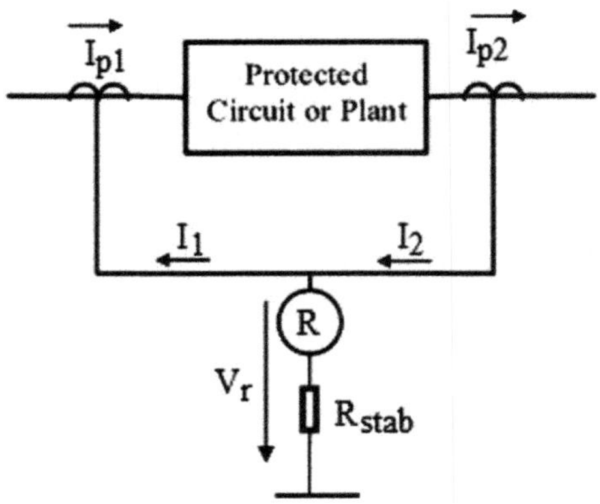

FIGURE 10.11 High impedance protection arrangement.

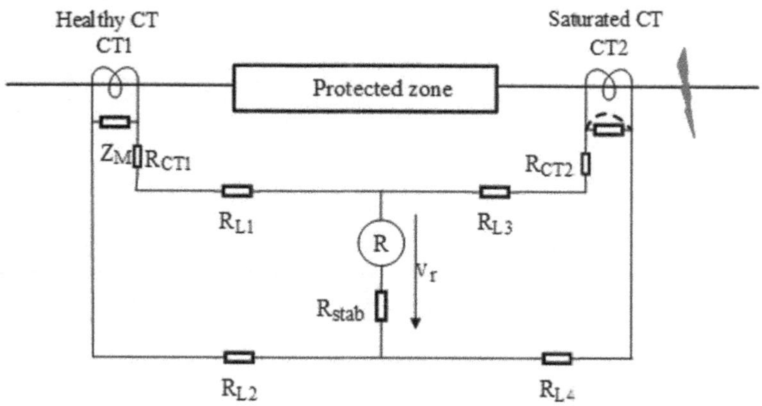

FIGURE 10.12 Circulating current flow path with saturated CT.

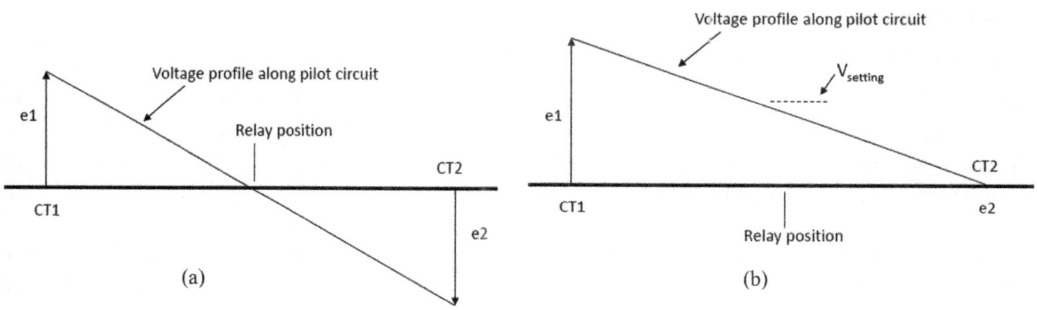

FIGURE 10.13 Distribution of potential around the pilot circuit: (a) under normal conditions, (b) with large external fault (one of the CTs saturated).

The voltage across the relay circuit (under the worst case scenario represented by Figure 10.13b:

$$V_R = I_F\,(R_{\text{CT}} + 2R_L) \tag{10.6}$$

where R_{CT} is the secondary leakage impedance of the CT and R_L is the resistance of the connecting leads from the CT to the relay location.

The stabilising resistor, R_{ST}, must limit the spill current to less than the relay pickup setting, I_S, for through-faults. Hence:

$$R_{\text{ST}} = (V_S \,/\, I_S) - R_R \tag{10.7}$$

where V_S is an arbitrarily chosen voltage setting ($\geq V_R$) and R_R is the relay burden.

When an internal fault occurs, the voltage profile is as shown in Figure 10.14. The CT output voltages are both positive, giving rise to a voltage profile which rises above the voltage setting V_S. Under this condition, current above the pickup setting I_S flows to operate the relay.

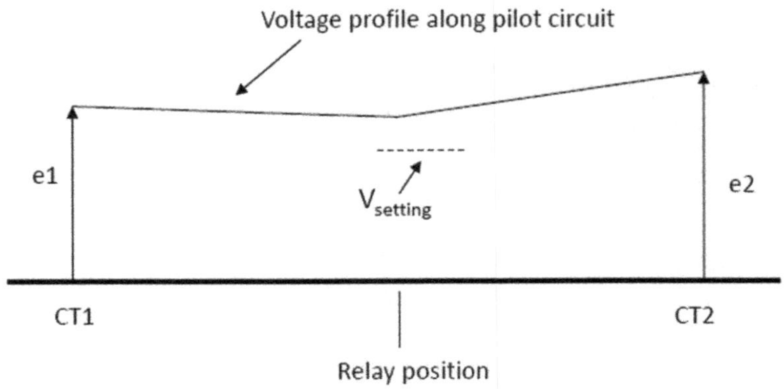

FIGURE 10.14 Distribution of potential around the pilot circuit with internal (in-zone) fault.

10.8 Restricted Earth Fault (REF) Protection

The sensitivity of the earth fault protection may be improved by applying restricted earth fault protection. In this arrangement, the relay is connected to measure the residual current of the phase and neutral CTs, as illustrated in Figure 10.15a. The protection is thus restricted to one winding of the transformer (between the CTs). The zone of protection is explicitly defined, hence the name restricted earth fault protection.

A high-impedance type or the biased low-impedance type relay may be used. The REF relay achieves instantaneous operation at low setting. This means a large percentage of winding is protected. The REF protection may be used on both windings of a Y/Y-connected transformer.

The schematic in Figure 10.15b shows that the simple REF connection can detect winding earth faults only. Interphase or inter-turn faults cannot be detected. However, if full access to the windings is available with CTs at both ends of the windings, the schematic in Figure 10.15c is obtained. This connection is able to detect winding to earth faults as well as interphase faults, but not inter-turn faults.

Depending on transformer connection and earthing, the conventional earth fault protection may be inherently restricted as for delta-connected winding, as illustrated in Figure 10.15a, or unearthed star winding. For both transformer connections, no zero sequence components can flow through the transformer to the other winding.

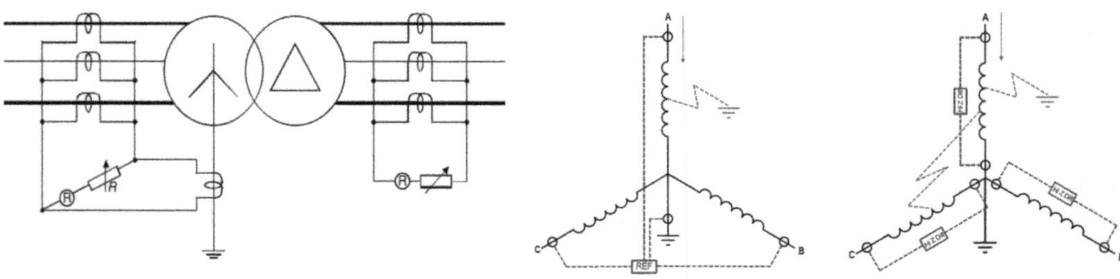

FIGURE 10.15 (a) Circuit diagram of restricted earth fault (REF) protection, (b) REF arrangement for detection of winding earth faults, (c) REF arrangements for detection of interphase faults.

Example 10.4

A high-impedance type restricted earth fault (REF) protection is to be applied to the star winding of a 20 MVA, 66/20 kV, Dyn11 transformer.

1. Select the ratios for the phase and neutral conductor current transformers. The available relay has a secondary rating of 1A.
   LV load current = 577.35 A
   Select CT ratio of 600/1A (for all four CTs)
2. Determine the value of the stabilizing resistance, given the following:
   - Maximum external phase-fault current of 3.5 k A.
   - The maximum earth fault current is limited to 1.0 kA by an earthing resistance.
   - The required earth fault operating current is 200A.
   - CT secondary resistance is 2.3 Ω.
   - The CTs have an exciting current of 1%.
   - Relay impedance is 1.5 Ω.
   - Connecting leads resistance is 0.05 Ω.

   Voltage across relay circuit, $V_r = I_f\,(R_{CT} + 2R_L)$.
   Max. phase-fault current, $I_f = 3.5$k A (primary) = 5.83 A (sec).
   Hence, the stability voltage, $V_s = V_r = 14.0$ V.
   Stabilizing resistance, $R_{st} = (V_S/I_S) - R_R$.
   Actual operating current and relay setting current related by the formula:
   $I_{op} = CT_{\text{ratio}} \times (I_s + nI_e)$
   $I_{op} = 200$ A and $n = 4$ (CTs)
   Hence $I_s = 0.293$ A.
   Hence, $R_{st} = 46.28\ \Omega$.

10.9 Transformer Differential Protection

The application of differential protection to the transformer is illustrated in Figure 10.16 [9]. Two sets of CTs are located on the HV and LV sides of the transformer. Several issues need to be considered in the application of differential protection to the transformer:

- Phase correction for possible phase shift across the transformer windings.
- Ratio correction for possible unbalance of signals from CTs on either side of the windings.
- Effect of inrush magnetizing currents.

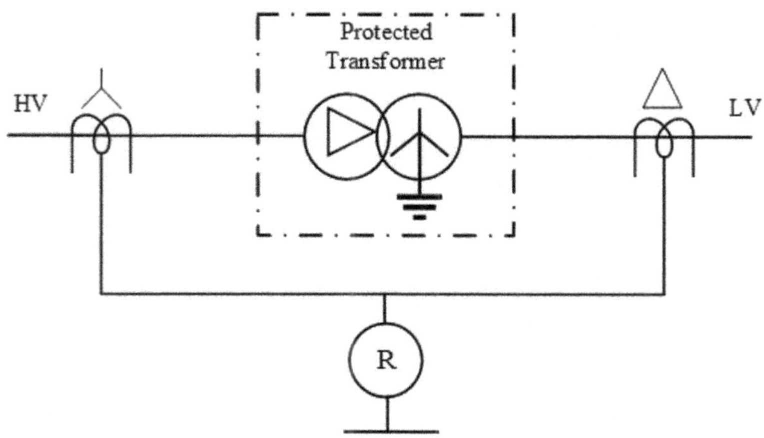

FIGURE 10.16 Application of differential relay to transformer protection.

10.9.1 Phase Correction

Phase correction is required to compensate for phase shifts across the transformer windings. Generally, if the transformer is star/delta, then the CTs must be connected in delta/star, as shown in Figure 10.17. Specifically, if the transformer has vector group Dy11 that introduces a +30° phase shift, then the CTs must be connected to give a –30° phase shift to compensate. Dyl and Yd1 provide the required –30° phase shift.

10.9.2 Ratio Correction

In most cases it is not possible to match the magnitude of the pilot currents through the ratios of the primary CTs. This is illustrated in Figure 10.18. The CT primary current magnitudes are labelled I_{LY} and $I_{L\Delta}$ for the star and delta connected sides of the transformer, respectively, as seen in the figure. The CT secondary current magnitudes are labelled $I_{S\Delta}$ and I_{SY} according to the CT connections—delta and star, respectively. Current balance on the pilot circuits requires the ratio of the currents $I_{S\Delta}$ to I_{SY} to be:

$$\frac{I_{S\Delta}}{I_{SY}} = \sqrt{3}\frac{N_{CY}}{N_{C\Delta}}\frac{I_{LY}}{IL\Delta} = 1.0 \tag{10.8}$$

That is, the CT ratios (N_{CY} and $N_{C\Delta}$, respectively) must be selected such that the ratio equals unity. However, this is not always possible and auxiliary or interposing CTs must be connected on the secondary circuits to provide ratio compensation between the pilot currents of the HV and LV sides of the transformer, as illustrated in Figure 10.19.

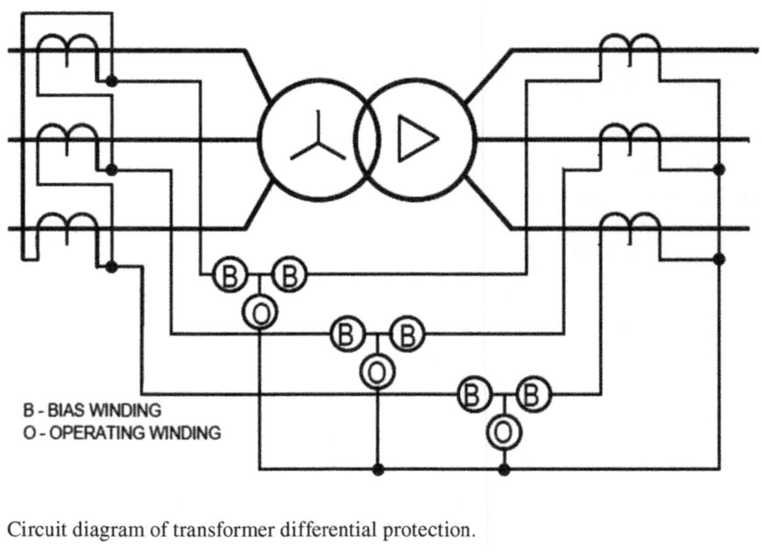

FIGURE 10.17 Circuit diagram of transformer differential protection.

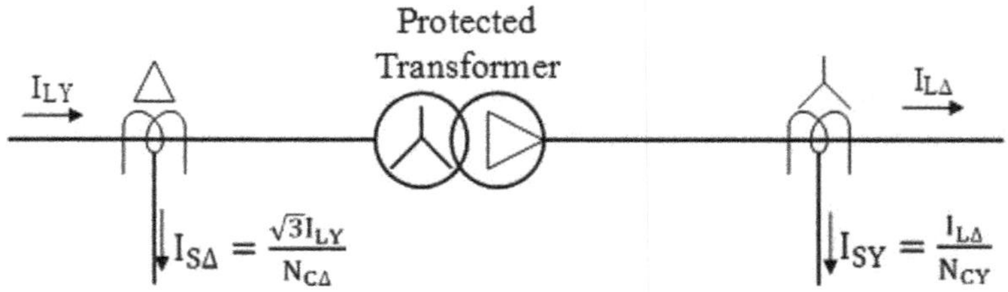

FIGURE 10.18 Magnitude of pilot currents.

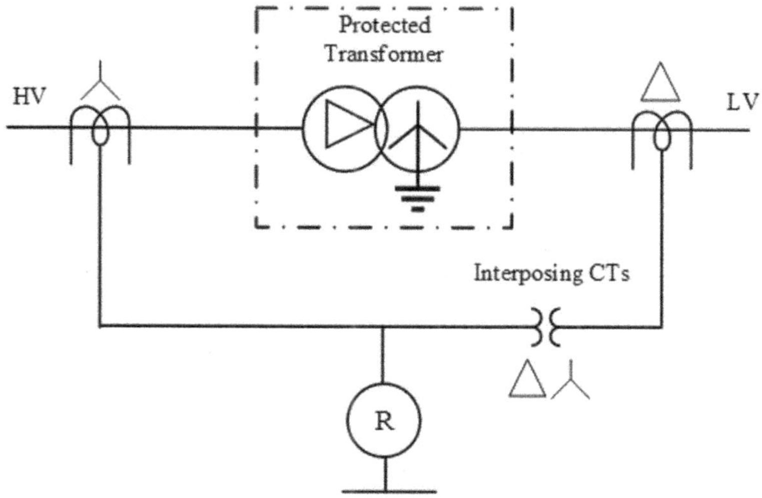

FIGURE 10.19 Application of interposing CTs to transformer differential protection.

Modern numerical relays perform the phase and ratio compensations in software and the primary CTs are connected in star, which is simple and also economical in terms of CT insulation requirement. This is illustrated in Figure 10.20.

10.9.3 Bias Settings

Typically, the bias setting for transformer protection is illustrated in Figure 10.21. The relay minimum pickup is provided by k_1. Stability against differential current due to maximum tap is provided by bias k_2 in the second section of the characteristic. Stability against heavy through faults is provided by the bias k_3. These biasing levels are adjustable according to the transformer operating environment.

10.9.4 Differential Protection Stabilisation During Inrush

The relaying pickup setting k_1 may be raised to a level above the inrush current, but then the relay becomes ineffective. Modern relays provide stability against current inrush using harmonic restraint [10]. It was noted in Section 10.3 that the inrush current is nonsinusoidal and rich in harmonics. The inrush current

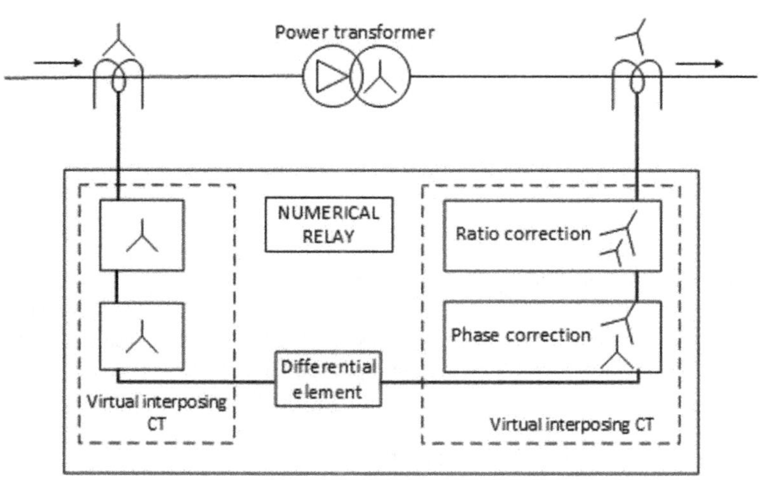

FIGURE 10.20 Implementation of differential protection with numerical relays.

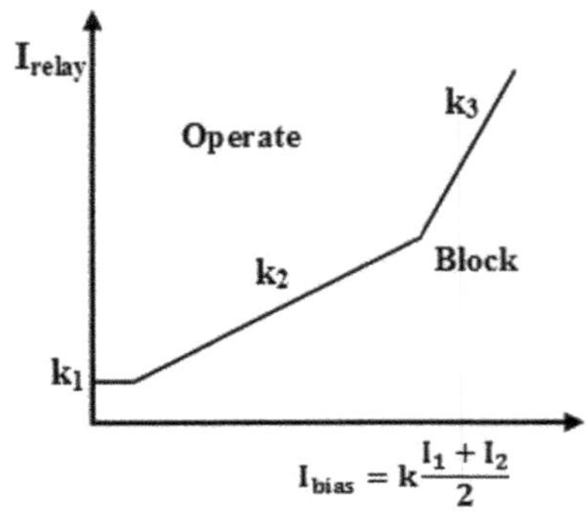

FIGURE 10.21 Characteristic of bias relay in transformer differential protection.

will therefore contain both even and odd harmonics. The differential current is filtered to extract the second harmonic component that is used to produce restraining quantity sufficient to prevent operation by inrush current. Normal fault currents do not contain the second harmonic and this provides effective differentiation between inrush and fault condition. The fifth harmonic is also used in some relays.

Example 10.5

Differential protection for a three-phase, 15 MVA, 102/33 kV, Δ/Y-connected transformer has to be designed. Multi-ratio current transformers with ratios of 600:5, 500:5, 400:5, 300:5, 200:5 and 100:5, are available. The differential relays have taps of 5.0, 5.5, 6.6, 7.3, 8, 9 and 10. Select the CT ratios and taps for the two sides of the transformer such that the percentage mismatch is less than 10%.

Line currents on the two sides:
102 kV side: $I = 65.61$ A:
Select CT ratio of 100/5 on primary side:
CTs Y-connected.
Pilot current = (65.61 × (5/100) = 3.28 A.
33 kV side: $I = 262.43$ A:
Select CT ratio of 400/5 on secondary side:
CTs Δ-connected.
Pilot current = (262.43 × (5/400) × sqrt(3) = 5.68 A.
Pilot currents not balanced. Use the 5.0 tap for the 3.28 A side, then the ratio is:

$$\frac{3.28}{5.68} = \frac{5}{x}$$

and:
$x = 8.66$
There is no tap of 8.66, but a tap of 9 is close. Use tap 9 for the 33 kV side.
% mismatch = (9−8.66)/8.66 = 3.93 %

Example 10.6

A wye-delta connected power transformer is protected, as shown in Figure 10.22, by a percent differential relay with 25% slope and 2 A minimum pickup. An external three-phase fault occurs, at the position shown, and fault currents flow through the transformer with the magnitudes as shown. Determine whether the differential relay will operate to trip.

Primary pilot current = 24000/(1200/5) = 100 A.
Secondary pilot current = sqrt(3) × 1156/(100/5) = 100.11 A.
Relay current, $I_2 - I_1 = 0.11$ A.
Relay will not operate.

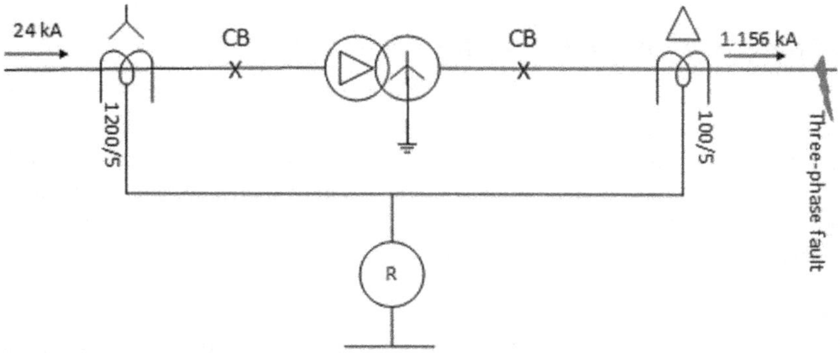

FIGURE 10.22 Application of differential relay to transformer protection.

Example 10.7

Repeat Example 10.6 but with a three-phase fault occurring between the high-voltage side breaker and the transformer. The fault infeed from the high-voltage side is 4 kA and that from the low-voltage side is 24 kA).

Primary pilot current = 2400/(1200/5) = 100 A
Secondary pilot current = sqrt(3) × 4000/(100/5) = 346.41 A.
Relay current, $I_2 - I_1 = 246.41$ A.
Relay minimum pickup = 2 A.
25% of restraining current = $0.25 \times (I_1 + I_2)/2 = 55.80$ A, that is, relay operating current = 55.80 + 2 = 57.80 A.
Relay will operate.

10.9.5 Differential Protection of Multi-Winding Transformer

Relays with more than two bias windings are available for the protection of three-winding transformers and other configurations with more than two circuits, as illustrated in Figure 10.23. Biasing is obtained by applying each CT current to its separate biasing winding. The separate bias windings are arranged so that their effects add to give the total bias.

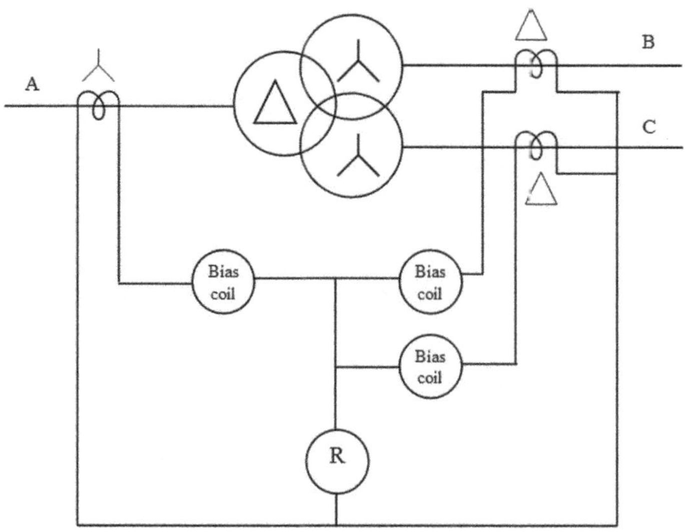

FIGURE 10.23 Application of differential protection to three-winding transformer.

10.10 Combined Differential and REF Protection

The differential relay pickup setting is recommended at 10% or 20% by most relay manufacturers [2]. Such a pickup level means that only a portion of the winding is protected against earth faults. Faults in the remainder of the winding will produce fault current below the minimum pickup. To offset this problem, the REF protection is added to compliment the differential scheme and provide protection to the last coil of the winding. This arrangement is illustrated in Figure 10.24. In this configuration, the differential and REF protection schemes share the same CTs.

10.11 Differential Protection Application with an Earthing Transformer

In some cases it may be desirable to earth the system fed from the winding of a transformer that is delta connected. This is done using an earthing transformer or reactor that is usually connected within the transformer protection zone, as illustrated in Figure 10.25. However, this arrangement may compromise the performance of the differential protection scheme.

During system earth fault, zero sequence currents which flow through the earthing transformer will flow through one set of the line CTs only. This may be seen by the relay as a fault and cause (unwanted) operation of the relay. In order to avoid this, the zero sequence currents of the earth fault need to be cancelled out and balance restored to the differential system. This is achieved by using three-phase CTs with a third tertiary winding that is delta-connected to perform the function of zero-phase-sequence-current-shunt. This winding shunts the zero-sequence component, thereby removing it from the differential relay circuit.

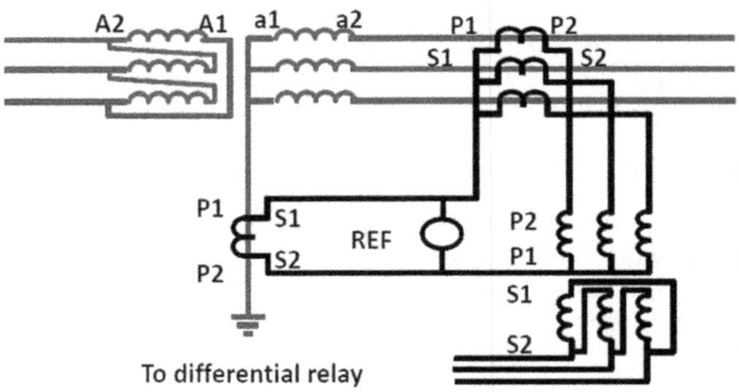

FIGURE 10.24 Circuit diagram of combined differential and REF transformer protection.

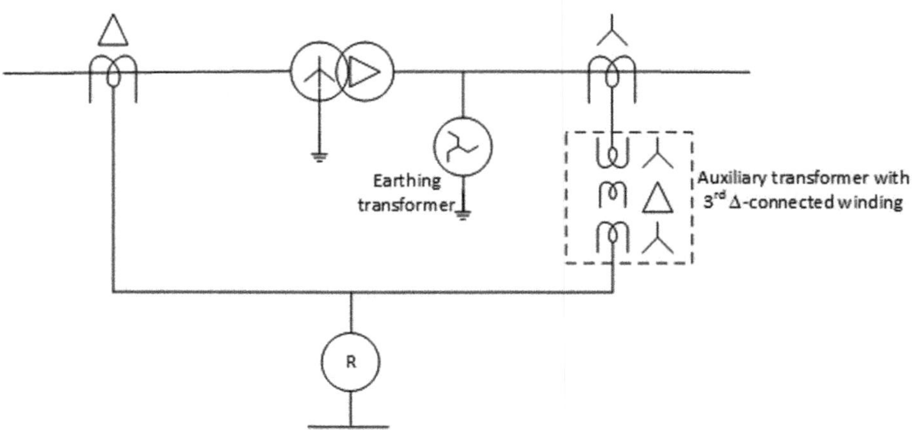

FIGURE 10.25 Circuit diagram of differential transformer protection with in-zone earthing transformer.

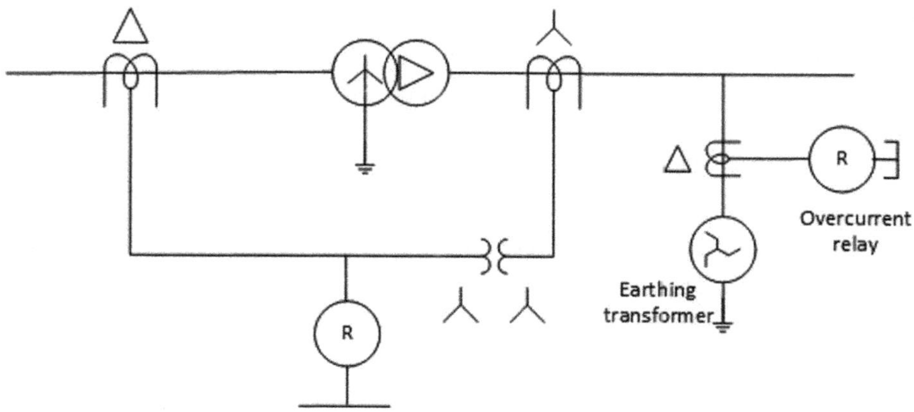

FIGURE 10.26 Circuit diagram of differential transformer protection with external (out-of-zone) earthing transformer

When the earthing transformer is not in the main transformer zone, it can be protected using overcurrent relays with the CTs connected as shown in Figure 10.26. Under normal conditions the earthing transformer passes zero sequence currents due to faults elsewhere in the system. For such external faults, the overcurrent relays must not operate. This requires that the CTs be connected in delta to circulate the transformed zero sequence currents without energising the relays. For a fault within the earthing transformer, the relays operate.

10.12 Buchholz Protection

Protective equipment reacts to faults with only a few exceptions, such as the Buchholz relay that can provide pre-fault alert [11]. The Buchholz relay consists of two elements contained in a housing connected to the pipe to the conservator tank, as illustrated in Figure 10.27a and b. The relay is designed to detect gas formation and large oil movements within the transformer tank.

One set of contacts operates for slow accumulation of gas from incipient faults to given an alarm. These incipient fault conditions include hot spots on the core due to lamination insulation failure, core bolt insulation failure, faulty joints, inter-turn faults and loss of oil due to leakage. These conditions cause decomposition of the oil with the resultant release of gas that rises and accumulates in the Buchholz relay chamber. When sufficient gas has accumulated, the upper float operates to give an alarm.

Severe winding faults, to earth or inter-phase, or sustained loss of oil, result in rapid generation of gas which increases tank pressure causing the displacement of oil towards the conservator tank. This violent movement operates the lower float of the Buchholz relay and trips the transformer.

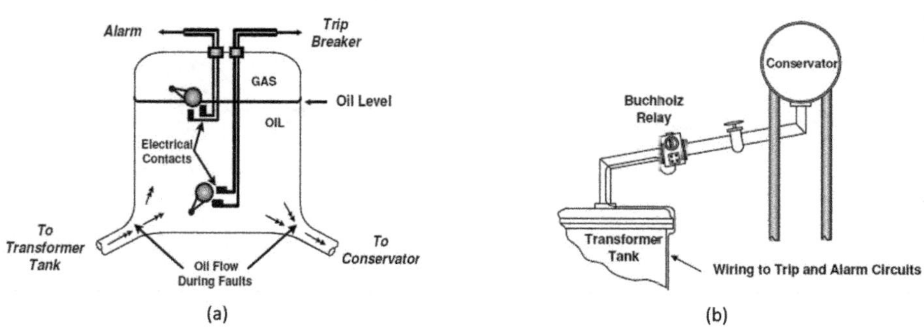

FIGURE 10.27 (a) Buchholz relay arrangement, (b) location of Buchholz relay.

10.13 Transformer Winding Temperature

A temperature gauge is used to sense the transformer winding temperature, as illustrated in Figure 10.28. The basic temperature sensing bulb, when placed in the oil, provides a measure of the oil temperature. In order to give a measure of the winding temperature, a heater supplied from a load sensing current transformer is installed around the sensing bulb to provide a local temperature rise above the general oil temperature. The effect of the heating coil, coupled with the heat of the oil on the bulb, allows the gauge to simulate the winding temperature. The thermal assembly has contacts that may be used to send an alarm or to control banks of fans. In some installations, extreme temperature rise may lead to automatic tripping of the transformer, though cooling and load reduction are preferred strategies to keep the transformer in service.

10.14 Pressure Release Valve

Major faults cause large amounts of gas to be generated within the transformer tank leading to high pressure increases. The high pressure might reach dangerous levels that may damage the transformer. In order to protect the transformer, the high pressures are relieved by a pressure relief value or explosion vent located at the top of the transformer tank. The operation of one type of pressure relief value is illustrated in Figure 10.29.

The pressure relief device consists of a spring which normally is uncompressed, and the spring force keeps the valve closed, as illustrated in Figure 10.29a. Pressure increase within the transformer

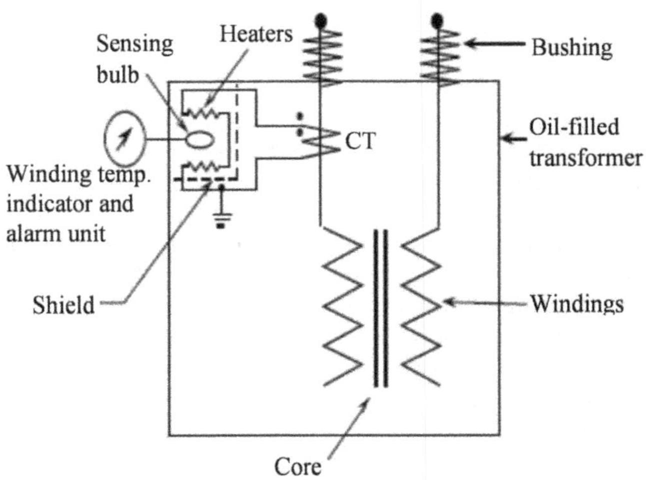

FIGURE 10.28 Transformer winding temperature monitoring arrangement.

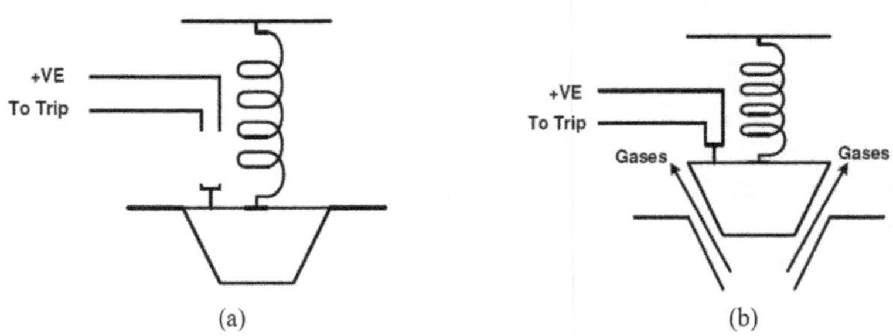

FIGURE 10.29 Pressure release valve: (a) normal state, (b) pressure release.

sufficient to compress the spring will cause the value to open, thereby provide an opening for pressure release, as illustrated in Figure 10.29b. The contact may be used to provide an alarm or trip the transformer.

10.15 Tutorial Problems

1. A transformer may be protected by a combination of differential and restricted earth fault protection relays. Explain why each of these functions is required.
2. An earthing transformer is used to provide the neutral on the secondary side of a Y/Δ-connected power transformer. Demonstrate, and illustrate with a diagram, the implementation of differential protection scheme that includes the earthing transformer within the differential protection zone.
3. Discuss two considerations that must be taken into account when designing transformer differential protection.
4. Transformers may suffer from winding short circuits, open circuits, or overheating. From a protection perspective, briefly explain the manner in which these problems are resolved.
5. A protection scheme for a 750 kVA, 33/11 kV, Δ/Y-connected transformer consists of three protection systems: an overcurrent relay, earth-fault relay and restricted earth fault relay. Explain why the three relay elements may need to be combined for protection of the transformer.
6. Some conditions provide currents that can contribute to a relay pickup when there is no fault present. Describe some of these currents and discuss what might be done to desensitize the relay to these types of currents.
7. An earthing transformer is used to provide the neutral on the secondary side of a Y/Δ-connected power transformer. Demonstrate, and illustrate with a diagram, the implementation of differential protection scheme that includes the earthing transformer within the differential protection zone.
8. Examine the problems of providing transformer protection using fuses. Can the fuses provide overload as well as fault protection?
9. A simple radial system is shown in Figure 10.30. The reactance of the transformer is 0.02 pu and the reactance of the feeder reactance 0.06 pu. The transformer is rated 100 kVA 11/3.3 kV. Time-overcurrent relays are used to provide protection at R_{ab} and R_{bc}. Assuming standard IDMT relay characteristics, determine the CT ratings, pickup values and the time multiplier settings for the two relays. All reactances are given to 100 kVA base.
10. A wye-delta connected power transformer is protected by a percent differential relay with 25% slope and 2A minimum pickup. A three-phase fault occurs on the primary (bushing) terminals

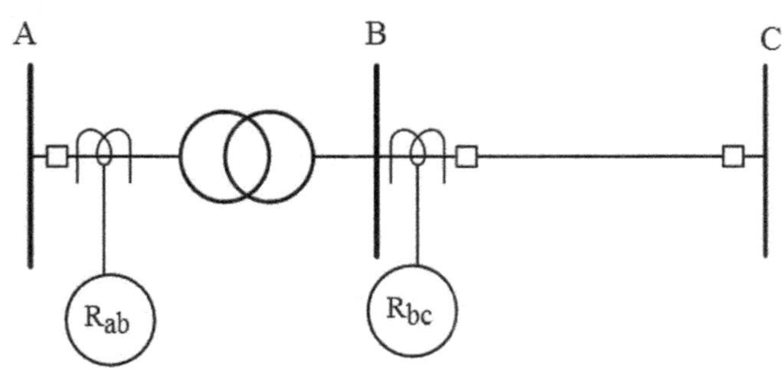

FIGURE 10.30 A simple radial system.

of the transformer. The resulting fault current infeed from the primary side is 4 kA and that from the secondary side is 24 kA. Assuming CT ratios of 100/5 A and 1200/5A on the primary and secondary sides, respectively, determine whether the differential relay will operate to trip the circuit breakers. Ensure proper connections for the CTs.

11. A 10 MVA 33/11 kV Dyn11-connected transformer with reactance of 10% is protected by a differential relay with 20% slope and 20% minimum pickup. The following primary CTs and interposing current transformers (ICTs) have been selected:

    HV primary CTs: Yy0-connected; ratio 250/1

    LV primary CTs: Yy0-connected; ratio 600/1

    HV side ICTs: Yy0-connected, ratio 1/1.43 (step-up)

    LV side ICTs: Yd1-connected, ratio 1/1.14 (step-up)

    A three-phase fault occurs on the HV bushings of the transformer resulting in fault infeed of 4 kA from the system and 10 kA from the transformer LV side. Determine whether the differential relay will operate to trip.

12. Figure 10.31 shows a 60 MVA, 102/33kV, Δ/Y-connected transformer whose star winding is protected by a restricted earth fault relay. An external stabilizing resistance of 33.55 ohms is connected in series with the relay of impedance 1.5 ohms. The relay is set to pick up at 20%. The current transformers are rated 1200/1 A, and the equivalent circuit parameters of the CT (referred to the secondary) are 2.3Ω (series impedance) and 1100Ω (shunt impedance), respectively. Determine whether the REF relay will pick up a 450A internal phase to earth fault.

13. A power transformer nameplate reads: 20 MVA, 33/11 kV, Dyn1. Biased differential protection is to be applied to this transformer. A numerical relay is available for the implementation. Suggest and provide the required solution, including the relay settings. Assume the following information:

    - Transformer has load tap changing of 10%.
    - CT error of 5%.
    - The relay has current setting range from 0.2 to 1.2A in steps of 0.01A.

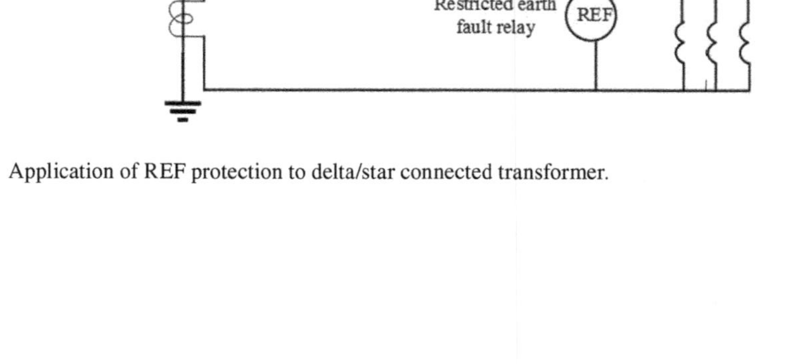

FIGURE 10.31 Application of REF protection to delta/star connected transformer.

14. The protection scheme for a 750 kVA, 33/11kV, Δ/Y-connected transformer consists of three protection systems: an overcurrent relay, earth-fault relay and restricted earth fault relay. Explain why these three protection systems may need to be combined for protection of the transformer.
15. A high-impedance type restricted earth fault (REF) protection is to be applied to the star winding of a 20 MVA, 66/20 kV, Dyn11 transformer.
    a. Select the ratios for the phase and neutral conductor current transformers. The available relay has a secondary rating of 1A.
    b. Determine the value of the stabilizing resistance, given the following:
        - Maximum external phase-fault current of 3.5 kA.
        - The maximum earth fault current is limited to 1.0 kA by an earthing resistance.
        - The required earth fault operating current is 200A.
        - CT secondary resistance is 2.3Ω.
        - The CTs have an exciting current of 1%.
        - Relay impedance is 1.5Ω.
        - Connecting leads resistance is 0.05Ω.
16. Differential protection for a three-phase, 15 MVA, 102/33 kV, Δ/Y-connected transformer has to be designed. Current transformers with ratios 1200:5 (taps of 1200, 1000, 900, 800 and 700) and 600:5 (taps of 600, 500, 400, 300, 200 and 100) are available. The differential relays have taps of 5:5, 5.5:5, 6.6:5, 7.3:5, 8:5, 9:5 and 10:5. Select the CT ratios and taps for the two sides of the transformer so that the percentage mismatch is less than 10%.
17. A single-phase transformer is rated 50 MVA, 110/33 kV. The tap changer with ±10% range is installed on the HV side. Percentage differential protection is used to protect the transformer.
    a. Determine the required rating of the CTs. Assume 20% transformer overload capability.
    b. Determine the appropriate taps that will minimize the required slope of the relay characteristic. Assume both relay windings are provided with taps for 3.0, 4.0, 4.8, 5.0, 5.2, 6.0 A. The maximum CT error is 5%.
    c. What pickup current setting for the relay would you recommend?

10.16 Conclusion

Transformers are susceptible to faults that can originate internally within the transformer or as a consequence of external causes. This chapter presented various schemes for the protection of the power transformer against these faults. The protection applied depends on the degree of importance of the transformer to the electrical system, which in turn depends on its size, cost, and application. The protection schemes discussed in the chapter range from the basic fuse used on the distribution transformer to the differential protection schemes used on the more important transformers. The chapter also provided detailed worked examples to enhance students' understanding of transformer protection and the calculation and selection of the relay settings.

REFERENCES

1. B. M. Weedy, B. J. Cory, N. Jenkins, J. B. Ekanayake, and G. Strbac, *Electric Power Systems*, 5th ed. Chichester, UK: John Wiley & Sons, 2012.
2. J. D. Glover, M. S. Sarma, and T. J. Overbye, *Power System Analysis and Design*, 4th ed. Toronto, ON: Thomson, 2008.
3. Alstom Grid, *Network Protection & Automation Guide*, 2nd ed. Stafford, UK: Alstom Grid, 2011.
4. L. G. Hewitson, M. Brown, and R. Balakrishnan, *Practical Power System Protection*. Oxford, UK: Newnes, 2004.
5. S. H. Horowitz and A. G. Phadke, *Power System Relaying*, 3rd ed. Chichester, UK: John Wiley & Sons, 2008.
6. B. A. Oza, N. C. Nair, R. P. Mehta, and V. H Makwana, *Power System Protection and Switchgear*. McGraw-Hill, India, 2010.
7. The Electricity Training Association, *Power System Protection*, 2nd ed. London, UK: The Institution of Electrical Engineers, 1995.
8. A. F. Sleva, *Protective Relay Principles*. Boca Raton, FL: CRC Press, 2009.
9. J. J. Grainger and W. D. Stevenson, *Power System Analysis*. New York: McGraw-Hill, 1994.
10. C. L. Wadwa, *Electric Power Systems*, 6th ed. New Delhi, India: New Age, 2009.
11. P. M. Anderson, *Power System Protection*. New York: Wiley-IEEE Press, 1998.

11

Generator Protection System

T. Adefarati and Ramesh Bansal

CONTENTS

11.1 Introduction

The main function of the protective devices in the generators is to minimise the number of unnecessary trips during the power system instabilities. A generator includes the stator, rotor, exciter, battery charger, control panel, field winding, fuel system, prime mover, lubrication system, voltage regulator, cooling system, exhaust system and other associated auxiliaries, as shown in Figure 11.1. Generators are designed by the original equipment manufacturers to operate at a high load factor for a long time, probably hours, weeks, months and years [1]. The operation of a generator for a long time allows it to be exposed to a number of faults due to abnormal working conditions. This requires various types of electrical and mechanical protection systems. The application of protective devices in a generator depends on some economic considerations, having considered the values of the machine together with its power output. Consequently, the optimal operation of a generator and its auxiliaries is controlled with the aid of monitoring devices that reduce the number of the incidences that are associated with the operating conditions. The protection system of a generator performs a significant function for the optimal operations of a power system to be achieved. It consists of relays, communication facilities and a scheme that sets up a planned strategy for operation of the protective devices.

In some situations, generators are designed by the original equipment manufacturers to export surplus power into the grid or to reduce the importation of power from the grid. The independent power providers want their generating units to be reliable, efficient and safe as well as protected from the faults that are associated with the operation of transmission and distribution lines. Owing to this, the power utilities focused on protection of their power systems because it mitigates damage to the

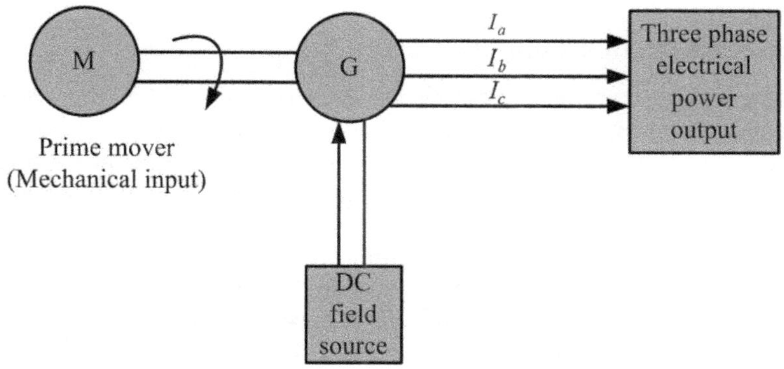

FIGURE 11.1 A simple block diagram for generator construction.

machines and enhances power generation security. Thus, a dependable power system needs an appropriate protection and control coordination between the generating units and load points. This depends on the prerequisites dictated by individual utility, the size of the generator, the size of interconnection voltage, the type of generator, the transformer connection, etc. In addition, several protection devices should be incorporated into the generator for specific applications and arrangements based on their economic justifications and to protect the personnel that operate the generating units from a number of incidences.

The presence of the protective devices minimises electrical faults that occur in the generators; owing to this, the generators must be incorporated with a number of protective devices that will initiate a disconnection of faults from the machine as soon as possible. In some cases, protection system can be utilised to completely shut down the machine in case of a fault that can cause catastrophic effects in the power system. To meet the power demand and specifications of the consumers at the load points without any interruption, the manufacturers of various protective devices must be able to offer a number of protection systems that can be modified at any time. These protection systems are designed and produced in accordance with the needs of independent power providers for various generating units based on the laid-down specifications. The protective devices of the generator can only be effective if the settings of the numerous relays and their coordination are appropriate. Otherwise, wrong settings and improper coordination of the protective devices will result in unnecessary power outages and loss of revenues owing to power interruption. A proper selection of protective devices guarantees a reliable operation of a generator and prevents damage of the major components that constitute the generating units. As a result of this, protection system must be carefully selected by the utilities during engineering design of a power system to avoid the economic losses associated with intermittent power outages.

Having considered the exorbitant cost of replacing any faulty components of a generator, it is mandatory that the generator's most important expenditure goes into the provision of the most comprehensive protection system. Protection systems can be used by the utilities to monitor the power system operating parameters such as current, terminal voltage, excitation voltage and stator and rotor windings. It can also be used to monitor the internal activities of the generator, which include partial discharge, water level, a cooling system that encompasses temperature and pressure, insulation breakdown, etc. The applications of the protective devices are discussed in subsequent subsections. This chapter describes in detail the function of protective devices and their numerous structures as well as their applications. This chapter is concerned mainly with the protection of generating units from abnormal operating conditions and minimum protection system required for a number of generating units. This chapter also covers protection system for small, medium and big generators such as reciprocating engines, gas turbines, hydro turbines, wind turbines, nuclear, steam turbines, etc.

11.2 Generator Protection Functions

The functions of generator protection system include measuring, ancillary functions, plausibility check, sequence of event recorder, disturbance recorder, self-diagnosis and supervision. The protective devices are selected by the utilities for a number of applications in the generator based on the conceptual understanding of the power quality, reliability, security, stability and economic aspect of power system. The protective devices are used in the generator to determine the level of its performance based on the manufacturer's specifications. This indicates that the performance of the generator depends on the number of protective devices utilised in the machine. The application of multiple protective devices improves the reliability of the system because of the higher likelihood of power outages from the generator owing to an unnecessary trip. The typical generator protection functions presented in Table 11.1 can be activated within the scope of the generator capacity for the purpose of protecting the machine. The functions listed in the table are so significant that they are always utilised in the protection scheme of many generators. The protection of the generator is a function of the size and cost of procurement. This shows that large

TABLE 11.1

Generator Protective Device Function Numbers

Protective Device Number	Protective Functions
15	Synchroniser
21	Distance protection
24	V/Hz (overexcitation protection)
25	Synchronism check
27	Undervoltage protection
27TN	Third harmonic neutral undervoltage
32	Reverse power anti-monitoring protection
38	Bearing over temperature
39	Bearing high vibration
40	Loss of field protection
46	Stator unbalanced current protection
49	Stator thermal protection
50B	Instantaneous overcurrent protection
51G	Time overcurrent ground overcurrent protection
51N	Time overcurrent neutral overcurrent protection
51V	Time overcurrent with voltage controlled or voltage constrained
58	Diode failure protection
59	Overvoltage protection for phase or ground faults.
59BG	Zero-sequence voltage protection; ground fault protection for an ungrounded bus
59GN	Voltage protection; primary ground fault protection for a generator
60	Voltage balance protection and voltage transformer supervision (fuse loss)
61	Time overcurrent protection; detection of turn-to-turn faults in generator windings
62B	Breaker failure protection
64F	Field (rotor) ground fault protection
78	Out-of-step protection
81	Over- and underfrequency protection
86	Hand-reset lockout auxiliary relay
87B	Differential protection
87GN	Generator ground differential
87T	Transformer differential protection
87U	Unit differential protection
94	Self-reset auxiliary tripping relay

TABLE 11.2

Different Types of Generators with Different Ratings

	Generator Rated Power (MVA)			
Protection	**<5**	**5–50**	**50–200**	**>200**
Impedance			x	x
Overexcitation		o	x	x
Undervoltage	y	y	y	y
Reverse power	x	x	x	x
Underexcitation	o	x	x	x
Unbalanced load		x	x	x
Stator overload	x	x	x	x
Rotor overload				x
Overcurrent time	x	x	o	o
Interturn fault			x	
Overvoltage	x	x	x	x
Stator earth fault 90%	x	x	x	x
Stator earth fault 100%			x	x
Rotor earth fault	o	x	x	x
Out-of-step				x
Above normal frequency	x	x	x	x
Below normal frequency				x
Differential	o	x	x	x
Dead machine		x	x	x
Stator earth fault	x	x	x	x
Rotor earth fault	x	x	x	x
Differential block (transformer)		x	x	x

and expensive generators and those being used for many critical applications must be highly protected from abnormal operating conditions and faults that are associated with their operations. Owing to this, some of the large generators use multiple relays or protective devices. The scope of protection function is reflected on the total cost and strategic importance of the generator such as availability, minimisation of damage, etc. The basic scope of protection functions based on the sizes of the generator is described in Table 11.2. The table shows the trend where the scope of protection scheme increases with the ratings of the generator. Moreover, protection relay integrates multiple functions required for the protection of generators, as presented in Figure 11.2. The multiple functions of the relays are utilised by gas turbines, steam turbines, hydraulic turbine and reciprocating engines that operate in parallel with the public network and/or stand-alone power system.

The scope of protection is given in Table 11.2, where *y* is only necessary for pump storage power station, i.e., motor/phase shift operation); *o* is the optional.

11.2.1 Types of Generator Protection

The application of protective devices is essential in power station operation due to the fact that protections against harmful abnormal conditions are provided. The protection system is a standout amongst the most fundamental factors in a power system that guarantees protected and dependable task of the generating units. The power system is mainly designed with the application of protection system to separate a defective segment of the power system from the entire system so that the healthy section can function at a reasonable accepted level without any harmful damage associated with the flowing of a fault current [2]. The continuous operation of a generator is prone to the electrical stresses enforced on the

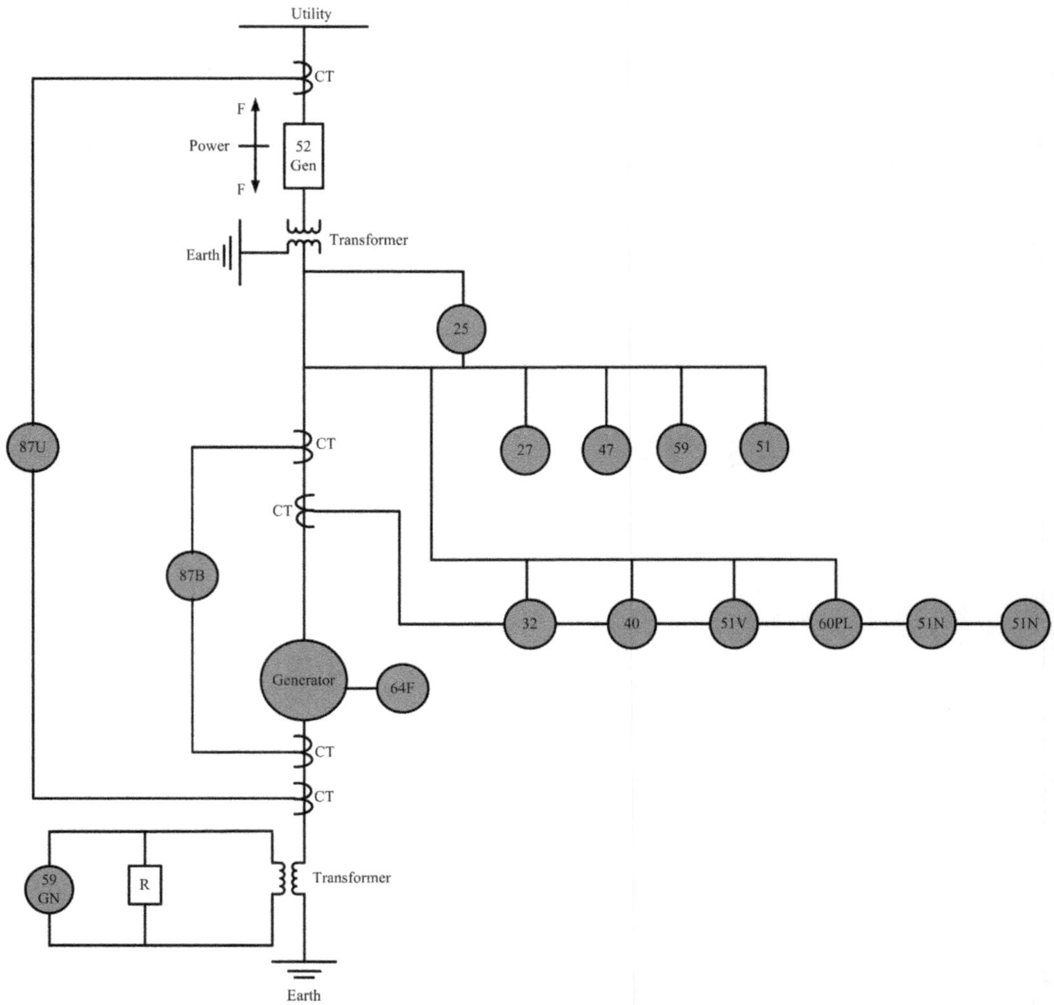

FIGURE 11.2 Multifunction generator protection device.

insulation of the rotor and stator, mechanical powers that act on the different sections of the machine and sudden increase of temperature. This means that a protection system must be integrated into the various generating units. Apart from this, some preventive measures must be put in place to prevent overloads and abnormal conditions of the generator with a goal that it can work optimally and securely. Regardless of the effective plan and design and preventive methods of protection, the hazards that are related to the operation of a generator can't be totally eliminated from the machine. Hence, the protective devices are utilised as part of the generator protection scheme that guarantees that faults will be removed quickly from the machine.

Generators are more prone to failure than any other equipment in the power system as a result of the numerous components such as stator winding, rotor winding, exciter, cooling system, etc. The optimal operation of the generators can be hindered owing to numerous electrical faults such as overload, overvoltage, motoring, phase fault, earth fault, unbalanced current, etc. The number of electric faults that occur in the generators are excessive; in view of this, the generator is protected with several electrical protective schemes at different phases of power system design. Moreover, mechanical protection constitutes the major part of a generating unit with the following faults: excessive vibrations, overspeed, underspeed, overheating, etc. The protection system of a generator is designed with the electrical and mechanical functions of its major components and other important factors in mind. The utilities should

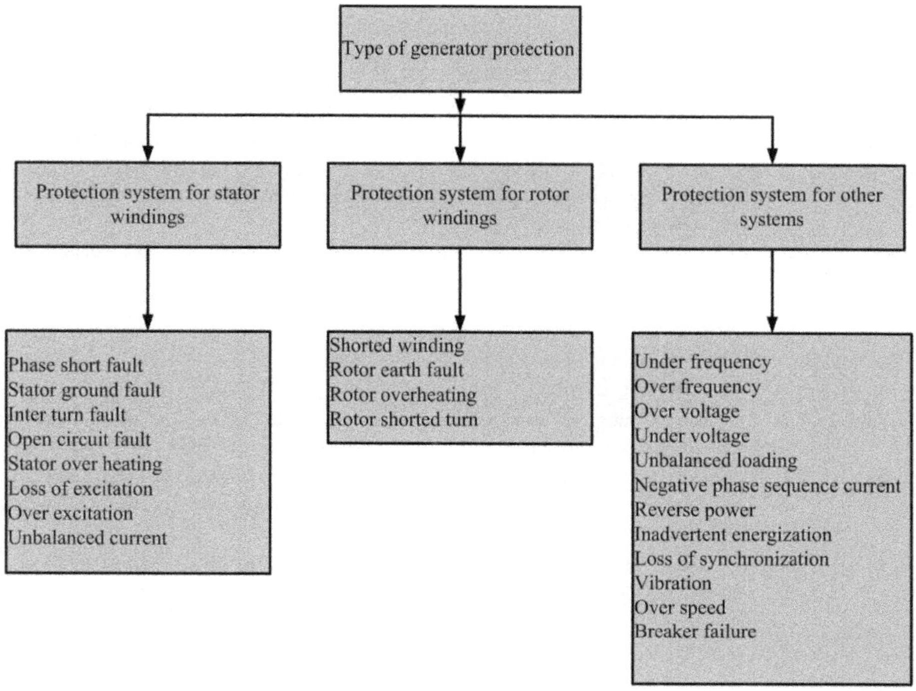

FIGURE 11.3 Protection systems for stator windings.

take into account the protection of generating units more than any other components of a power system by considering the abnormal working states of the generator. The generator protection systems are mainly designed by the original equipment manufacturers to protect the electrical and mechanical parts and other components of the machine. The protection schemes as applicable to the generator are briefly summarised in Figure 11.3.

11.3 Generator Stator Fault Protection

The majority of the faults reported in the generator are due to the insulation breakdown of the stator coils and other faults that are associated with the abnormal conditions of the generator. The stator faults encompass the fault currents in the stator of the machine and must be cleared as soon as possible to prevent unnecessary shutdown of the generator. The immense danger from the fault current is the likelihood of damage to the laminations of the core and stator windings and their auxiliaries because of the heat produced at the location of the fault. The stator would be disassembled and repaired in a situation where the damage to the stator is excessive. The process of dismantling and replacement of the damaged stator is lengthy and costly. This will escalate the operation and maintenance costs of the generator. The primary protection for the phase-to-earth fault and phase to phase is the application of differential protection in the generator. However, interturn fault protection is one of the most vital protection schemes for stator winding based on some benefits. The protection system is viewed as important due to prevention of insulation breakdown between the same phase windings. The interturn fault will eventually change to an earth fault in the long run if it is not well taken care of. It can be identified with the application of the stator differential and stator earth fault protection systems. With the increase in the power output of the generator together with its voltage, this type of protection is fundamental for all extensive generating units.

The generator stator faults can cause serious and expensive harm to the windings, insulation and the core of the machine. At the same time, the fault can produce severe mechanical vibration to the

TABLE 11.3

Type of Stator Winding Faults

Type of Faults	Descriptions
Phase-to-earth fault	This kind of fault is reduced with the application of the neutral grounding resistor (NGR).
Phase-to-phase fault	The chances that phase-to-phase fault would occur in a generator is limited because the insulation between the two phases is at least double, in terms of thickness, as the insulation between the winding and the iron core.
Ground fault	The fundamental driver of ground fault is attributed to insulation failure. Generators are typically grounded with the application of a resistor and an impedance to reduce the ground fault current and its consequences [5]. The ground fault causes extensive damage such as: • Short circuit out of the field winding • A severe unit vibrations • Overheating as a result of unbalanced currents
Phase-to-neutral fault	The phase to neutral fault can cause thermal damage to the stator's windings owing to insulation breakdown. The primary protection of phase faults can be provided by a differential relay.
Interturn fault	The interturn stator winding fault that occurs in a generator is caused by current surges with steep wave front. The fault can be identified by stator earth fault protection or stator differential protection.
Overheating	The overheating of the stator winding of the generator is attributed to overloading, failure of cooling systems and insulation breakdown of stator laminations. The fault can be detected with the integration of temperature detectors into the numerous points of the stator winding.

shaft and couplings. A sensitive and rapid response differential protection scheme is usually applied to the generators as a measure to react rapidly to a fault with a higher current. The differential relay is normally utilised as an essential scheme for protection of the generator stator windings against a phase fault. The function of the differential relay is discriminating and can be utilized with a very short tripping duration. The differential relays can be used in different capacities to detect phase-to-phase fault, double phase-to-earth faults and three-phase faults [3]. However, it is difficult for the differential relays to detect turn-to-turn faults in the same phase owing to the fact that there is no difference in the current that enters and leaves the phase winding [4]. Different types of stator windings faults are presented in Table 11.3.

11.3.1 Differential Protection

The internal fault inside the stator winding of the generating unit can be cleared with the application of differential protection scheme in conjunction with the differential relay of the machine [6]. For the optimal operation of the differential protection scheme to be obtained in the generator, instantaneous attracted armature relays are utilised due to their high-speed operations; in addition, they are not affected by the AC transient that emanates from the circuit of the generator [7]. A number of sensitive and high-speed differential protections are utilised by the utilities to protect their generators from numerous faults and heavy current that can damage the machine. Moreover, the protection scheme is very simple to construct and it operates with a substantial level of stability after clearing faults and external switching actions. This protection system may offer phase-to-earth and phase-to-phase fault protection for some generators.

Two sets of current transformers (CTs) are utilised for generator protection. The first set of CTs is tied to the line terminal of the stator's windings while the second set of CTs is tied to the neutral terminal of the generator per phase [6]. The attributes of the CTs installed per phase must match one another. Otherwise, it will cause a major mismatch in the characteristics of the CTs connected at both terminals of the generator. This will cause a high possibility of breakdown of differential relays during the fault that is external to the stator winding and during typical working states of the generator [7].

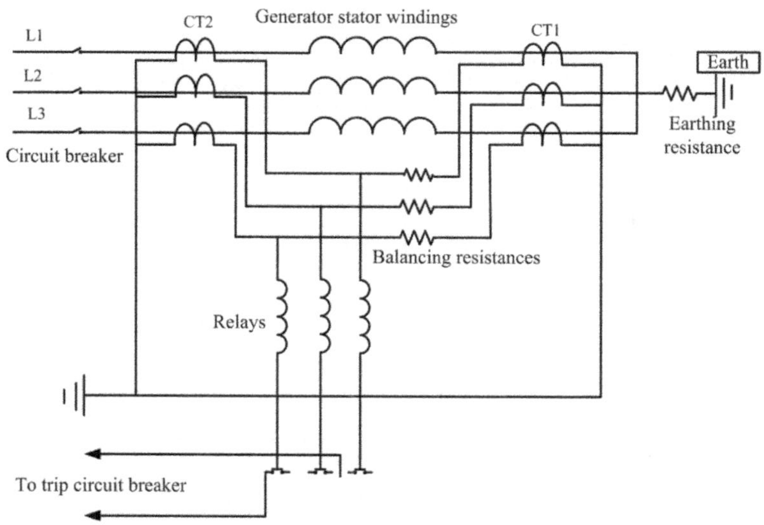

FIGURE 11.4 Differential protection for generators. (From Aujla, R. K., *Generator Stator Protection, Under/Over Voltage, Under/Over Frequency and Unbalanced Loading*, Department of Electrical and Computer Engineering, The University of Western Ontario, London, ON, 2008.)

The protection scheme allows the current at the two terminals to be compared. The currents at the two ends of the protection system are meant to be the same under normal operating conditions. Owing to the presence of faults in the system, there will be a slight deviation of the current recorded at the two ends of the protection system. The slight difference of current recorded at the two terminals of the protection system will flow via the operating coils of the relays due to mismatching of the CTs' characteristics. This makes the relays close their contacts and subsequently operate the circuit breakers or switches to trip. Hence, the circuit breaker operates to isolate the faulty segment from the entire power system, as shown in Figure 11.4.

The application of differential protection is restricted to certain areas owing to the fact that the ground fault is limited by utilising certain value of resistance in the neural earthing [2]. The occurrence of faults at the star point or neutral point makes a very small current flow in the operating coil of the relays. This current can further be reduced with the application of the neutral resistance. The presence of a small current in the operating coil of the relays is not adequate to operate the circuit breaker or switch. Owing to this, the protection system is simply designed to protect 80%–85% of the stator winding [8]. The application of the relays that operate with low settings will not provide a desired stability. The difficulty associated with the aforementioned scheme can be surmounted by utilising a modified differential protection system.

11.3.1.1 Modified Differential Protection

The modified differential protection can be used by the independent power providers to minimise the earth fault without affecting the stability of the system [6]. In this scheme, two types of relay are used in which the first type of relay is used for phase-to-phase fault and the second type of relay is utilised for the protection of ground fault. The operation of a modified differential protection allows the balancing resistance and two relays to be connected in star and the phase fault relays to be connected between the star point and the neutral pilot wire, as shown in Figure 11.5. The fault current that takes place in the system because of the phase-to-phase fault will be cancelled at the star terminal owing to the equal impedance. In addition to this, the star-connected circuit is balanced with reference to impedance. The modified differential protection provides up to 90% protection for the stator winding of the generator.

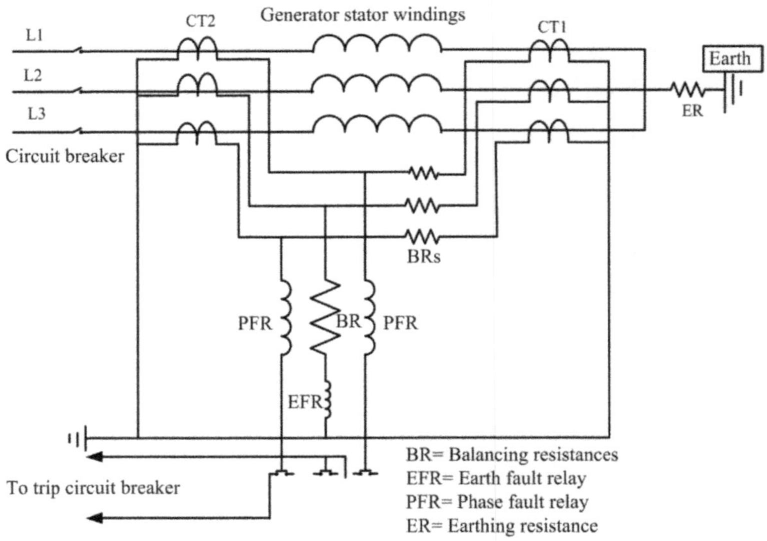

FIGURE 11.5 Modified differential protection for the generators. (From Aujla, R. K., *Generator Stator Protection, Under/Over Voltage, Under/Over Frequency and Unbalanced Loading*, Department of Electrical and Computer Engineering, The University of Western Ontario, London, ON, 2008.)

11.3.1.2 Biased Differential Protection

Biased differential protection has been reported to be the most effective scheme to prevent stator windings faults from the generator. The biased differential protection system consists of CTs, operating coils per phase, restraint coils and circuit breakers, as presented in Figure 11.6. The prerequisites of CTs for differential protection change in accordance with the relay used. With the integration of relays into the system, the specifications of CTs at the terminals of the stator windings of the generator are meant to

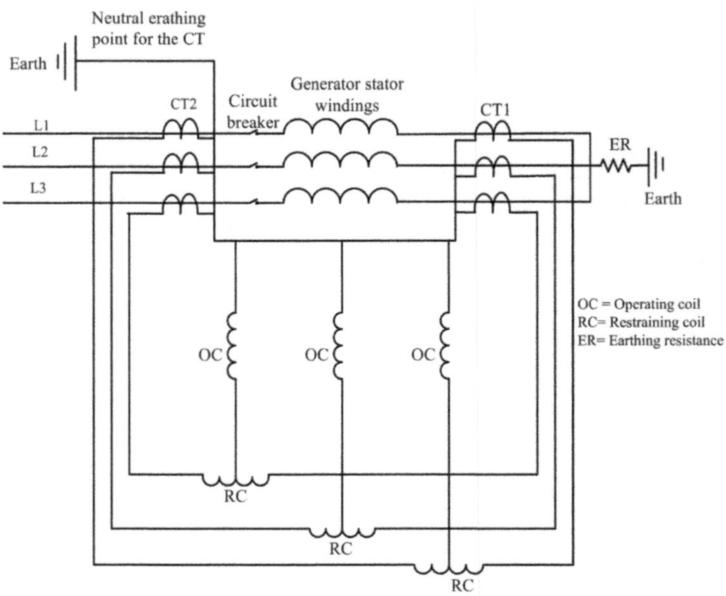

FIGURE 11.6 Biased protection of the stator winding. (From Aujla, R. K., *Generator Stator Protection, Under/Over Voltage, Under/Over Frequency and Unbalanced Loading*, Department of Electrical and Computer Engineering, The University of Western Ontario, London, ON, 2008.)

be the same. Change in the accuracy of the CTs tied to the busbar or stator winding of the generator will cause malfunctioning of the relays [6]. The challenge associated with the application of a modified differential protection scheme can be overcome by using biased circulating protection scheme. This will cause a sudden increment of the relay setting in accordance with the fault current. Any bias can be achieved by gradually changing the ratio of the relay restraining coil to the relay operating coil of the system. The current in the secondary of the CTs installed at the line will be the same as the current in the secondary of the CTs at the neutral terminal under typical working conditions. This will balance the flow of current in the restraining coils and there will be no current that flows in the operating coil [2]. The presence of phase-to-earth fault or phase-to-phase fault causes a significant deviation in the secondary current of the two CTs. The current that flows in the operating coil of the biased protection leads to instantaneous tripping of the switches or circuit breakers.

11.3.1.3 Self-Balanced Protection System

Self-balanced differential earth protection is employed for protecting the stator from the earth- and phase-to-phase faults. The aforementioned faults can damage the major components of the generator due to their severity. This results in a very expensive operation and maintenance costs and a very long downtime. The loss of substantial generating capacity in a power plant for a very long time would cause costly operation of a power system [6]. To detect an earth fault in the generator windings, phase and neutral, leads are connected to the circular aperture of the CTs positioned close to the generator output terminals [9]. The CTs are connected to the protective relays as shown in Figure 11.7. In typical operating conditions, the current that flows in the two leads of the cable is in the same direction and no magnetisation happens in the CTs [10]. The fault current only occurs once through the CTs. Once the earth fault takes place in any phase, magnetic flux will be induced. The emf that is induced in the relay circuit will cause the circuit breakers or switches to trip and isolate the entire system from the generating unit that has a variety of faults. Self-balancing protection is a very sensitive kind of ground fault protection for the generators.

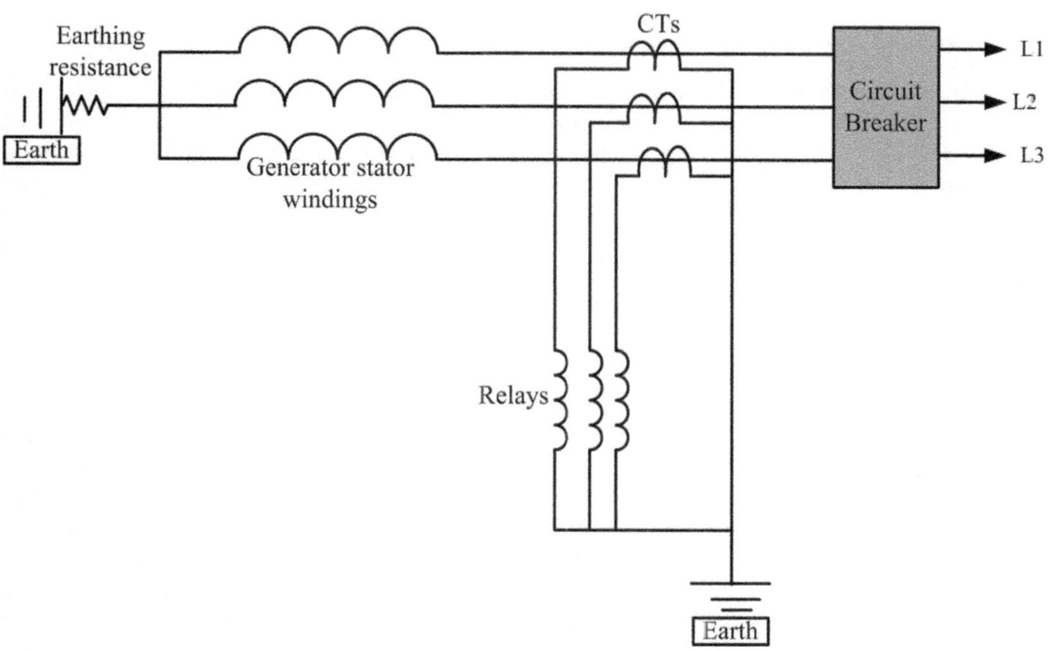

FIGURE 11.7 Self-biasing protection of the stator windings. (From Aujla, R. K., *Generator Stator Protection, Under/Over Voltage, Under/Over Frequency and Unbalanced Loading*, Department of Electrical and Computer Engineering, The University of Western Ontario, London, ON, 2008.)

11.3.2 Generator Grounding

A ground is a reference point of zero voltage potential that is typically an actual connection to the ground of the earth [5]. Owing to the importance of the earth system, it is a typical practice to ground all the various types and sizes of the generators by utilising external impedance. The reasons for the grounding are as follows:

1. To limit the thermal stress on the generators
2. To limit the mechanical stress on the generators
3. To limit or stop transient overvoltages during faults
4. To detect a ground fault
5. To limit fault damage in the generators
6. To reduce the earth fault current and related faults
7. To make available a source of grounding for the protection of the system
8. To improve the safety of the equipment
9. To avoid high operation and maintenance costs of the generator
10. To prevent stator damage after an internal ground fault
11. To detect ground faults within the machine

The exigency of grounding is very significant that an open ground condition can cause a serious safety problem to whoever operates the power generating units. Improper earthing of a generator can cause any person who touches any metallic parts of the machine to receive a severe electrical shock. In addition, a field earth fault will affect the optimal operation of a generator and causes a severe impact on the power system [5]. At the same time, it can cause extensive damage such as short circuit of the field winding, vibrations of the generator, rotor heating from unbalanced currents, insulation breakdown, severe damage to windings and generator core and arc damage at the points of the fault [11]. The grounding system of a generator is classified into stator and rotor earth protection.

11.3.2.1 Type of Grounding Schemes

The most common faults are the result of insulation failure that can cause mechanical damage to the core of the generator and other auxiliaries. Owing to this, earth fault protection is very crucial for the optimal operation of generators. Small, medium and large generators are purposely designed by the manufacturers not to withstand full fault current during a single phase-to-ground fault; that is why they are connected to any grounding scheme that is available in the power system. The common earthing techniques used in the real-time power system utilise the grounding resistor, neutral grounding transformer (NGT) and neutral grounding resistor (NGR), as presented in the following subsections [12].

11.3.2.2 Low Impedance Grounding

The advantage of using the low impedance grounding in the generators is to improve the sensitivity of the protection system. The damage associated with the presence of the earth fault in the generators can be quickly removed so that it will not cause more havoc in the power system. Owing to this, it is possible to use differential protection system with a directional ground overcurrent relaying to detect a ground fault in the power system. The scheme is essential to measure the sum of terminal current estimated at the neutral point. In the event that the two are identical, there will not be any internal ground fault. Conversely, any difference in the values of the current recorded at the neutral point and the sum of terminal current indicates the presence of an internal ground fault in the generator. Sometimes, the distribution transformer is discarded and a resistance with a high value is connected in the middle of the generator neutral and ground. The resistors are sized and selected on the basis of reducing the earth fault current to an acceptable level. The selection of the earthing resistor depends on the information of the system voltage and the type of protection scheme. The physical size, insulation level and cost of the

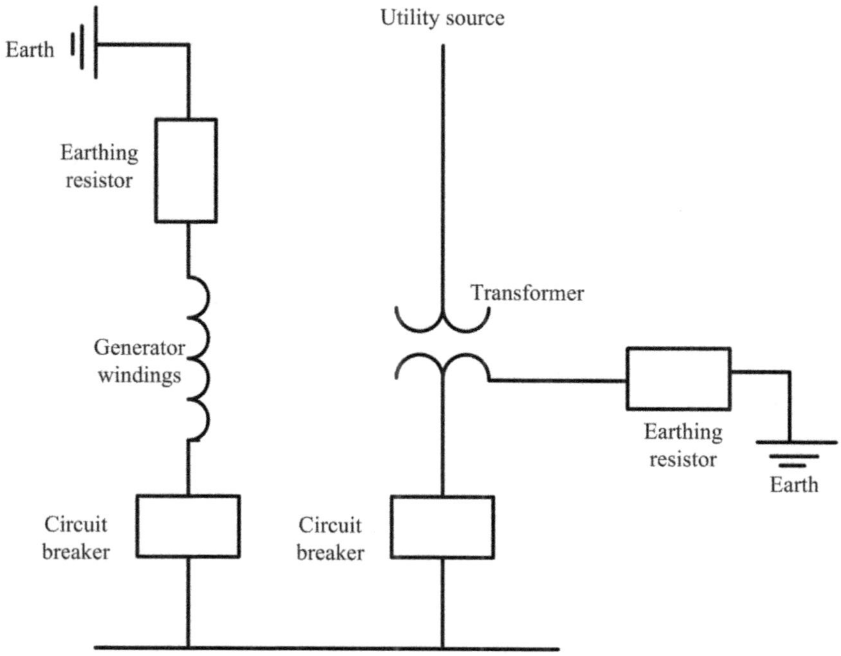

FIGURE 11.8 Simplified circuit.

resistors have prevented the usage of this method of grounding. The potential alternative to earth an MV generator is to utilise a low impedance grounding resistor technique. A low impedance grounding system is a connection of the neutral terminal of the generator to the ground via a low resistance impedance, as shown in Figure 11.8. The utilisation of a low impedance grounding technique lowers the magnitude of a fault current that flows during a single phase-to-ground fault, i.e., from 100A to 800A within a short period [13]. The application of low impedance grounding technique reduces the magnitude of the fault current and damage at the point of fault in the generator; it also provides a selective coordination of the protection system. In addition, a low impedance grounding scheme reduces the degree of electric shock associated with stray currents as well as thermal and mechanical stresses on the electrical components. It also controls transient overvoltages in the generator.

11.3.2.3 High Impedance Grounding

As the generation capacity of a power station is increasing based on the power demand, the value of the earth fault will also increase. As a result of this, the high impedance grounding is used by the utilities to reduce the magnitude of the fault current that flows in the windings of the generator. The fault current in the winding of the generator is limited to an acceptable level, probably 1–10 A primary with the application of the high impedance grounding technique. This minimises the degree of fault current in the generator and prevents the expensive costs associated with the replacement of faulty components in the system. The high impedance earthing technique also minimises the sensitivity of the differential protection in the stators of the generators. The high impedance grounding is the most common grounding technique for large generators. This technique allows a distribution transformer to be connected in the middle of the generator neutral and ground while a resistor is connected across the secondary for effective operation of the earthing system. This reduces the primary fault current of the generator to avalue in the range of about 3–25 A at the terminal of the machine. The high impedance grounding technique can be used for a variety of generators and industrial applications, as shown in Figure 11.9 [8]. The high impedance grounding method is used in a situation where the fault current is reduced to the accepted level, approximately 3 to 25 A primary. The application of the

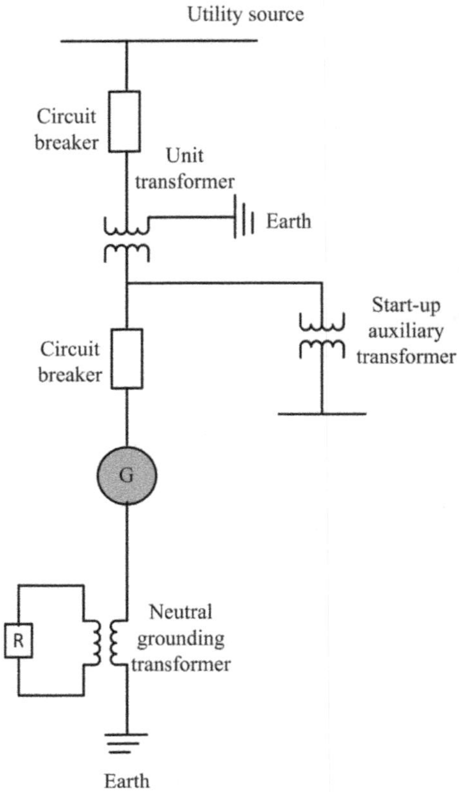

FIGURE 11.9 High impedance grounding.

high impedance grounding is to limit the fault current to a low value for an extended duration; it can also be used to prevent highly expensive repairs. The advantages of the high impedance grounding scheme are:

1. Reduction of transient overvoltages
2. Minimal fault current at the point of fault
3. Continuity of operation

11.3.2.4 Hybrid Impedance Grounding

The optimal solution to the earth fault problem in the generator can be obtained with the combination of the low impedance and high impedance grounded schemes. The hybrid impedance grounding system is designed to be low resistance for the external earth faults. Moreover, in a situation where there is an earth fault within the generator, the grounding scheme would revert to a high impedance grounding scheme as a way to minimise the damage associated with the earth fault. The generator is saved because it is never left in an ungrounded state and the stator windings are highly protected. This allows the low and high impedance schemes to be derived optimally by the utilities with a proper selective coordination to reduce damage at the point of fault. The ground fault current is limited based on the sum of the low and high impedance schemes adopted in the generator. The concept of hybrid impedance grounding technique as shown in Figure 11.10 employs a low resistance resistor and a high resistance resistor that are connected to the generator. The first grounding resistor has a low resistance, and it can only be switched off with the application a very fast isolator in the presence of the generator stator fault. This device is designed to isolate the first grounding resistor in three cycles. The second resistor is always in the circuit; it is designed to keep the generator earthed so that it will not isolate from the earth [13].

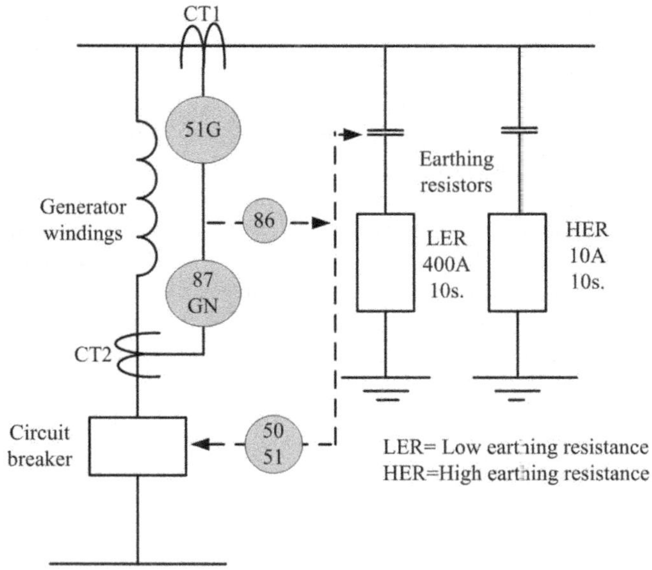

FIGURE 11.10 Hybrid impedance grounding.

11.3.3 Stator Earth Fault Protection

The fundamental causes of earth fault are initiated by the insulation breakdown owing to the thermal damage and mechanical stress on the insulation material or the anti-corona paint of the stator windings. The most common electrical faults in the generators are short circuit between the stator windings in the slots and short circuit between the stator core and stator windings [1]. To avert the danger associated with the stator earth fault, generators are generally earthed with an application of impedance as a measure to reduce the ground fault current [5]. The degree of stator earth fault protection relies on the technique adopted for grounding the generator and location of the fault. For effective operation of the generator and fault detection, a separate ground protection scheme is typically provided for different sections of the machine. In a situation where the neutral point of a stator is earthed with the application of a resistor, a CT is utilised at the neutral point to ground the connection. In addition, inverse time relay can be used in conjunction with the secondary terminal of the CT when the generating unit is directly connected to the supply busbar. The stator earth protection scheme is utilised in a situation where earthing impedance is applied to reduce the earth current in the generator to less than the magnitude that must not be exceeded owing to overcurrent. The type of stator earth protection scheme utilised by the utilities relies on the technique and connection of the generator to the busbar. In view of this, the neutral points of the generating units are typically grounded for protection of the stator windings and their related auxiliaries. The stator earthing points of the high voltage generators are usually earthed to reduce the amount of ground earth fault.

The common practice is to reduce the earth fault current to an acceptable value of 5–10 A by earthing the neutral point of the generation with the application of a resistor. In addition to this, an NGT can also be utilised to reduce the earth fault current to less than 1 A. The application of a resistor and NGT has kept the transient voltages within acceptable limits in the stator during the intermittent earth faults. This makes the earth faults trip the circuit breaker within a few seconds and causes an insignificant havoc on the laminations of the stator core. The grounding protection scheme prevents damage caused by transient overvoltages in a situation where there is an arcing earth fault in the stator. The protective plans that are mainly designed by the utilities to detect three-phase and phase-to-phase stator faults are not expected to give protection for phase-to-ground faults in the generator zone [10]. The earth resistor can also be utilised in a generator to limit the neutral voltage transmitted from the HV side of the transformer in a situation where there is a ground fault on the HV side to a maximum of 2%–3% of rated phase voltage of the generator [1].

The characteristics of the interturn fault in the generator make it difficult to detect in a power system. It will be easily changed to a ground fault. The danger of interturn can be prevented with the application of the stator earth-fault protection that will trip the generating units. The earth faults caused by mechanical damage may happen very close to the neutral point of a star-connected generator. The accessible phase-to-ground fault current turns out to be negligible irrespective of the grounding technique adopted. In a situation where the earth fault current is limited below the generator rated current, a substantial segment of the generator might be unprotected [10]. The low values of the earth fault current with the application of some devices may reduce the number of faults and havoc associated with such a fault; however, the detection of faults at a close proximity to the stator winding neutral point will become difficult. In view of this, there is a distinct trend towards providing earth-fault protection for the entire stator winding. The stator earth faults in close proximity to the star point or neutral point of the generator are difficult to detect owing to the fact that they do not produce a large current. The value of a fraction h of the winding that is not protected owing to closeness of the fault to the star or neutral point and the fault current can be expressed as follows [14]:

$$I_F = \frac{hV_a}{R_n} \tag{11.1}$$

$$= \frac{hV_L}{\sqrt{3}R_n}\,(A)$$

where R_n is the neutral resistance (Ω), V_a is phase voltage (V) and V_L is the rated line voltage (V).

The ground relay pickup current for the critical fault location can be computed as follows:

$$I_{PU} = k \times \text{CTR(A)} \tag{11.2}$$

where CTR is the neutral current transformer ration and k is the fraction of rated secondary current that gives pickup.

$$h = \frac{\sqrt{3}kI_{PU}R_n}{V_L} \tag{11.3}$$

The stator ground function 59GN is designed to detect an earth current fault in the stator winding of a generator connected to a delta-connected winding on the generator step-up transformer [4]. With typical settings, the stator ground function is capable of detecting faults within 2%–5% of the stator neutral. This function is susceptible to operate faults on the secondary of star-grounded voltage transformers connected at the generator terminals and for ground faults on the HV side of the generator step-up transformer due to zero-sequence voltage induced through the capacitive coupling between the windings of the generator step-up transformer [15]. The stator ground function is set with a time delay to coordinate the voltage transformer and transmission system ground fault protection as shown in Figure 11.11.

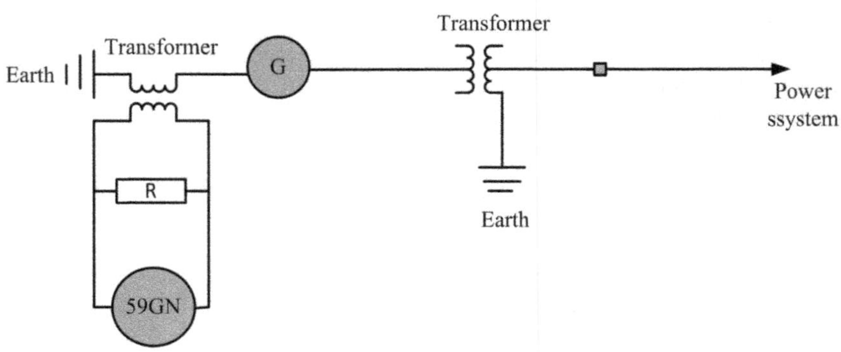

FIGURE 11.11 Stator ground protection.

11.3.4 Stator Interturn Fault Protection

The stator windings of generators are designed by the manufacturers with just a single turn per phase for each slot. The interturn faults for the medium and large generating units can only happen if there should be an occurrence of double earth faults or because of serious mechanical damage on the stator end winding of the generator [1]. The characteristics of differential protection scheme for stators don't give room for protection of the interturn faults in the same phase winding of the stator. The purpose for this is to allow the current produced as a result of the turn-to-turn fault to flow freely in the circuit between the associated turns. This doesn't produce any variance between the current that enters and leaves the windings at the two terminals where the CTs are installed because the stator windings of the generators are deliberately planned and designed with one turn for every phase per slot. The windings of some generators are intentionally designed with a single turn as a measure for provision of interturn fault protection scheme for the generator. However, interturn protection is highly applicable in the multi-turn generators such as hydroelectric generator, etc. Some of the stator windings utilise multiple coils to convey heavy current for their applications; for this reason, stator windings have two distinct paths.

In some cases, multi-turn windings are utilised for the machines that have capacities of about 50 MVA [6]. It is not easy to get dependable protection against short circuiting of one turn in a situation where the stator winding of the generator has a large number of turns for each phase. The interturn protection scheme is mainly designed with the CTs and relay, as presented in Figure 11.12. The primaries of

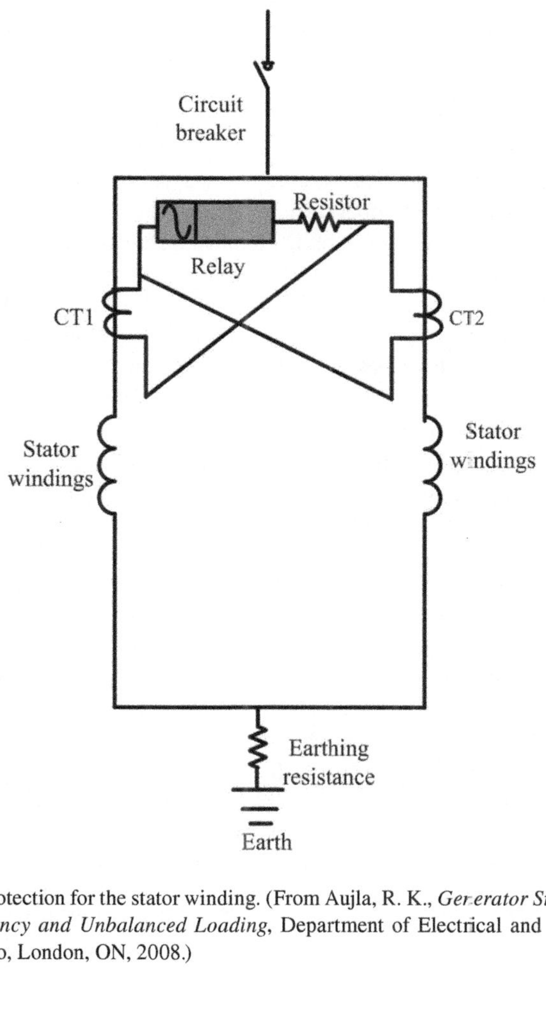

FIGURE 11.12 Interturn protection for the stator winding. (From Aujla, R. K., *Generator Stator Protection, Under/Over Voltage, Under/Over Frequency and Unbalanced Loading*, Department of Electrical and Computer Engineering, The University of Western Ontario, London, ON, 2008.)

the CTs are connected in the parallel paths. Also, the secondary of the CTs are interconnected as shown in Figure 11.12 for the purpose of preventing the generators from interturn faults. The system is designed so that the current that flows via the two parallel paths of the stator winding will be the same [6]. Since the system operates under normal operating conditions, no current flows in the relay operating coil [16]. The current that flows via the two parallel paths will be different under the interturn fault conditions. Hence, the distinction in the current that flows in the operating coil makes the circuit breaker trip and isolates the faulty generating unit from the power system.

11.3.5 Stator Overheating Protection

Overheating of stator is mainly caused by shorted laminations of the stator iron, overfluxing, overloads, iron core faults, insulation failures of the stator and cooling system failure [17]. The listed setbacks cause overheating of the stator winding and malfunctioning of the generator [2]. Overheating initiated by short-circuited laminations is much localised, and it has to be identified and removed before causing severe damage to the machine. The sudden increase in the generator's stator heat depends on the cooling system, seasonal changes, change in ambient temperature and other parameters. The reason for the stator winding overheating protection is to identify extremely high stator winding temperature before the event of generator damage. The stator overheating protection is designed to send a warning signal based on the temperature level of the generators. Once in a while two temperature settings are utilised: the lower setting is for the alert, and the higher setting is to trip the circuit breakers or electrical switches. Thermal protection is integrated into the stator winding to prevent the generator from the danger associated with overloading. The application of the stator overheating protection scheme in the generators allows incorporation of various temperature sensors to be utilised purposely to monitor the stator winding temperature. The common sensors usually utilised by the utilities for protection of their generating units against overheating are resistance temperature detectors (RTDs) and thermocouples (TCs). The sensors are utilised to continuously monitor the stator winding; in some cases, they can be connected to send alarm signals to appropriate protection schemes. Some generating units allow a current measuring device to be utilised in conjunction with a timing function to detect the temperature.

With a modern protection relay, temperature detection is frequently achieved by inserting thermocouples into the stator winding slots. The application of the thermocouples in the windings of the generators can be used to evaluate the temperature level of the stator windings, raise a warning signal, trip the circuit breakers or switches to prevent unnecessary damage to the generator. The thermocouples that sense the thermal level of the stator windings based on the sudden change in temperature are utilised to operate the temperature relay and to send a warning signal to the operators based on the operating conditions of the generators [18]. Thermocouples are only utilised to detect overheating initiated by overloading. They have no capacity to provide protection against overheating associated with the failure of the cooling system. To operate effectively and to prevent a short circuit of stator laminations, another measurement technique is incorporated to record the input and output of the cooling medium as a means of knowing any marked changes in the readings. Smaller generators are often provided with replica type temperature estimating devices that use stator current in a heat storage enclosure to estimate actual machine temperature. These devices are used to send a warning signal to the operators of the machines in case of possible serious problems. At unattended stations, the output of the temperature indicator may be used to sound an alarm and shut down the unit.

Generally, a few generators have various RTDs implanted in their stator windings, with at least two for each phase. The RTD is designed by the manufacturer to provide a constant current source to measure the voltage in the stator windings. Hence, the RTD inserted into the stator slot of the generator is utilised for temperature measurement. It is possible to have two settings: one is to send a warning signal, and the other is designed to trip the circuit breakers or switches. The RTDs are identified with the temperature level of the generator and its insulation level, as presented in Figure 11.13. In modern generators, RTDs are designed and routed to the control room through the supervisory control and

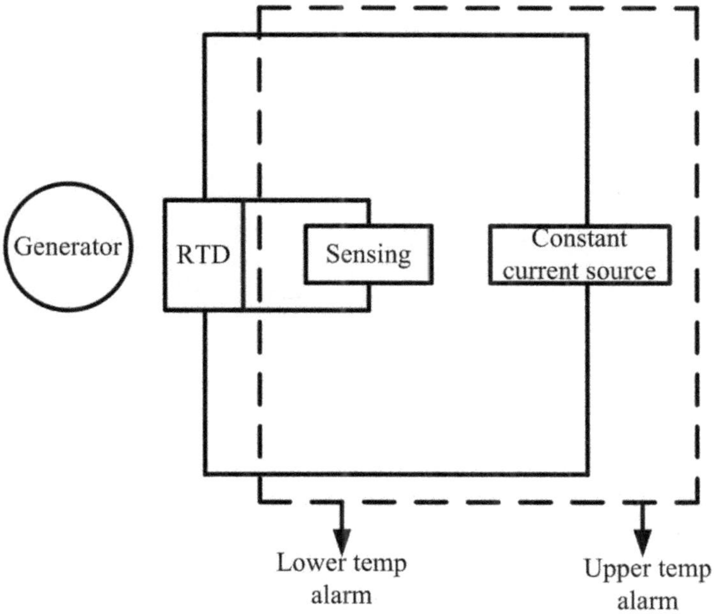

FIGURE 11.13 Resistance temperature detector system.

data acquisition (SCADA) and distributed control system (DCS). In manually operated power stations, RTDs are used to sensitise the operating team about the high and sudden overload conditions of the generator with the application of alarm and control systems. In remotely controlled power stations, the output of the RTDs is designed to remotely send a signal that trips the circuit breakers or other switches.

11.3.6 Phase Short Circuit Protection

The phase short circuit fault of a generator can result in insulation breakdown; thermal damage to the windings; and mechanical vibration of the bearing, shafts, couplings and other mechanical parts [2]. The phase short circuit fault protection systems are intended to protect the generating unit from the short circuit faults within or outside the windings of the generator. The phase short circuit protection is also designed to reduce short circuit faults that sandwich between the phases of the generator stator winding in a situation where there is a phase short circuit on the generator bus and there is an absence of a circuit breaker between the generator and the transformer. The potential option is to shut down and isolate the faulty generator from the entire power system as a prerequisite to limit the damage. It has been reported that phase short circuit is one of the uncommon faults in the generators. However, it is considered a matter of urgency to integrate a sensitive phase short circuit protective device into the generating units that are higher than 5–10 MVA. This will mitigate the damage associated with the aforementioned fault. The impacts of phase short circuit protection in a generator can be substantiated with the application of differential relays. The protection of the small generating units can be achieved via voltage/current relay. Thus, no differential relay is needed for the smallest sets of generating units [1]. Moreover, phase-phase faults, three-phase faults and double phase-to-ground faults can also be achieved with the utilisation of a differential relay [5]. A single phase-to-earth fault can be detected with a low impedance grounding of the generator [3]. The generator protective devices with function numbers 51 GN and 59 GN provide time overcurrent protection and voltage protection for the generator, as presented in Figure 11.14.

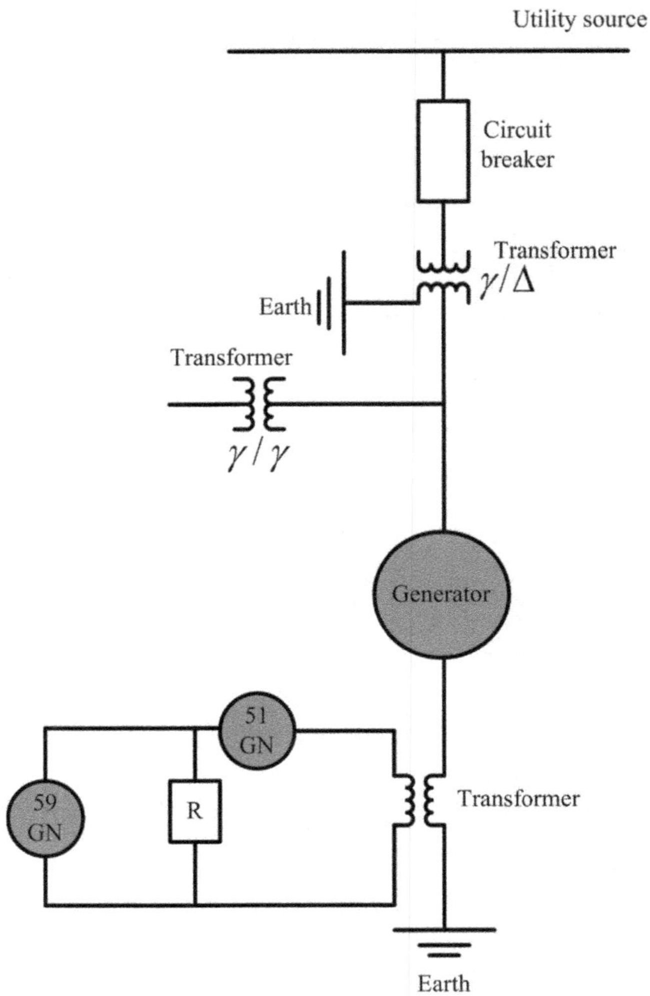

FIGURE 11.14 High-impedance grounding showing the main overvoltage protection and the backup overcurrent.

11.3.7 Excitation System

One of the most important components of the generator is the excitation system, which consists of automatic voltage regulator (AVR), exciter, protection unit, etc., as shown in Figure 11.15. The exciter is designed to provide a source to the field winding of the generator. The excitation level of the generator is regulated by the application of the AVR, which is very effective during steady state operations. The original equipment manufacturers make different options to power the AVR of various generators. Numerous excitation systems are available, and each one has its own unique features and benefits based on a number of applications. The selection of the best option by the independent power provider is based on some technical and economic considerations. One of the most important components utilized in the generators are the rotating and alternator rectifier exciter rectifiers. The AVR is connected to the stator winding to regulate the level of excitation of the generator and to sustain a constant voltage at the terminals of the generator with the fluctuating load conditions. The power source of the AVR is obtained from one of the following options: self-excited, excitation boost system, permanent magnet generator and auxiliary winding. The aforementioned methods are cost-effective approaches to provide the required input power to the AVR due to their simplicity. The input power source to the AVR

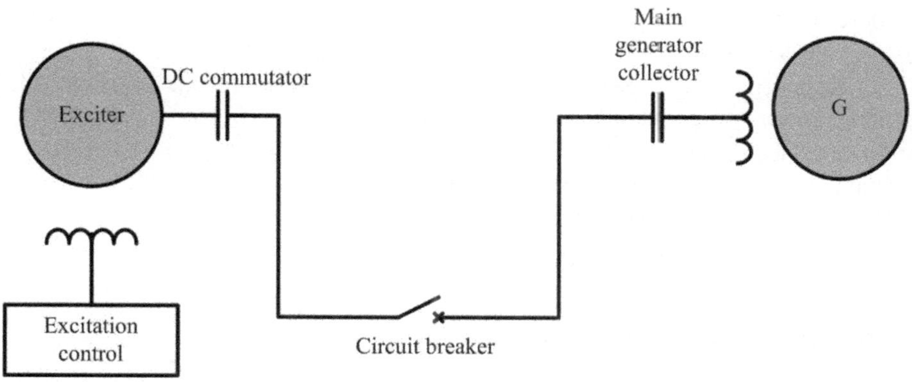

FIGURE 11.15 Excitation system.

is mainly designed to supply a DC output to the exciter stator, and the voltage level sensor is used to determine the level at which the AVR induces a DC voltage on the exciter rotor.

11.3.7.1 Causes of Loss of Excitation

A loss of excitation causes a severe operating condition for the generator, the operating team and the entire power system [4]. Unintentional disconnection of the source of excitation from the generator has various causes [19,20]. The reasons listed below can cause a loss of excitation in the generators:

1. An unplanned or intentional opening of the field breaker
2. A failure of the exciter
3. A failure of the AVR
4. A short circuit in the field winding
5. Zero field current owing to faults
6. Loss of power supply to the AVR owing to some faults

11.3.7.2 Characteristics of a Generator with Loss of Excitation

The features of generators that lost their excitation due to one fault or another are presented as follows [1,4]:

1. When a generator with sufficient active load loses its excitation due to some reasons, it exhibits the following characteristics: overspeeding, loss of synchronism, operation as an asynchronous machine; it obtains its excitation from the power system in the form of reactive power.
2. The parts of stator and rotor are subject to unnecessary overheating because the machine is allowed to keep running for a long time while at a high slip speed.
3. The generator terminal voltage changes in accordance with variation of the reactive current taken from the power system.
4. The loss of excitation causes excessive heating of the stator core during normal system conditions of the generator or while operating at full load.
5. There will be a high current induced in the rotor of the generator owing to a loss of synchronism. The high current may cause mechanical damage that results in overheating of the stator windings.

6. The level at which reactive power is being drawn from the power system may reduce the power system voltages and thereby affect the performance of the generators.
7. The increase in the reactive power flow in the power system causes voltage reduction. This will affect the power system stability.
8. A loss of excitation results in the loss of reactive power from a generator; this creates a substantial amount of reactive power to be obtained from the power system. It can cause severe machine damage due to large stator currents.
9. The power factor changes from lagging to leading.

11.3.7.3 Loss of Excitation Protection

A generator may lose its excitation owing to a variety of reasons, which were listed in Section 11.3.7.1. The moment a generator loses its excitation amid normal operation, it will overspeed by 3% to 5% of its rated speed and operate as an asynchronous machine [19,20]. The value of the speed increment is a function of the generator's load before losing its excitation. The generator will keep on supplying some power to the network and get its excitation from the power system as a reactive power. This method of operation is conceivable as long as the power system to which the generator is connected is sufficiently strong to provide an essential reactive power support [4]. This demonstrates that it will draw much of its reactive power from the power system for rotor excitation. During this condition, the stator currents will increase up to 100% of its nominal value since it has lost synchronism. Since the adequate reactive power support is not available due to loss of excitation, the best option is to shut down the generator. The severe impacts of an abrupt escalation of the line current can cause a great overheating of the stator windings, stator core and rotor core within a brief period of time. A fully loaded generating unit that operates under this condition will experience severe damage to its major component.

In view of this, a highly sensitive automatic disconnection device is needed to protect the stator windings from the heavy current and at the same protect the rotor from mechanical havoc caused by the induced slip frequency currents. The automatic disconnection device is utilized so that it will raise an alarm and the operators assigned to operate the power plant can restore the field excitation. Failure to achieve this would result in a total shutdown of the affected generating units. Therefore, protection against the loss of excitation in the generators is set to alarm and trip the circuit breakers as soon as possible. The most often utilised technique for loss of excitation protection is the application of distance relays to detect the change of impedance when viewed from the terminals of the generator [20]. Numerous things are needed to design a generator protection system against loss of excitation. This basically depends on the kind of generator arrangements used by the utilities [19,20]. The adopted technique is based on the fact that the active and reactive part of the impedance must be evaluated. The capability curve of 50 MVA generator with excitation capability at 0.8 lagging or 0.9 leading is presented in Figure 11.16.

11.3.7.4 Overexcitation Protection

Overexcitation occurs in the generator when the ratio of voltage to frequency is beyond the acceptable standard set by the manufacturers. This demonstrates that the proportion of the voltage to frequency utilised at the terminals of the generator surpasses 1.05 p.u. of the generator base. In this situation, the ratio of the voltage to frequency is more than a nominal value set by the manufacturer of a given generator [5]. This discussion shows that an excess voltage, or low frequency, or a combination of an excess voltage and low frequency can cause overexcitation of a generator. The excitation flux recorded in the core of the generator is directly proportional to the ratio of voltage to frequency at the terminals of the machine [1]. The level of excitation also depends on the change of temperature, eddy current loss and hysteresis loss. Whenever the ratios of voltage to frequency are exceeded, saturation of the magnetic core of the generator can happen. Consequently, the stray flux is induced in nonlaminated components that are not intended to convey flux. It can cause extreme interlaminar voltages between laminations at the ends of the core. The extremely high flux can likewise cause eddy currents in the generator laminations,

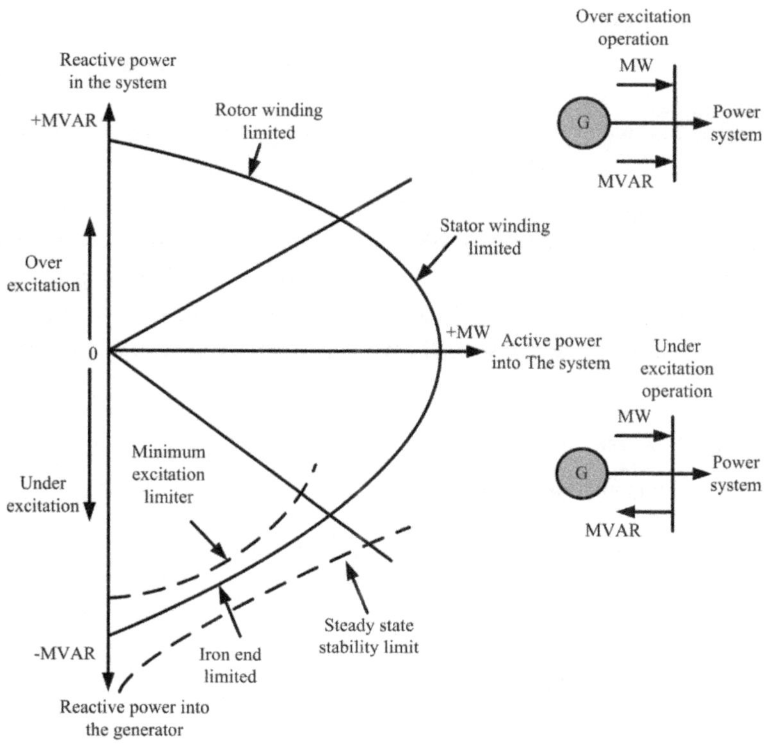

FIGURE 11.16 Generator capability curve.

which will result in a high voltage between laminations [4]. The core laminations are comparatively designed to withstand overexcitation without becoming extremely heated; however, nonlaminated metallic parts can encounter serious heat in a brief time.

Serious overheating can happen from saturation of the magnetic core of the generator; this will induce a stray flux in components not intended to convey flux. The field current that flows in the generator owing to the stray flux is excessive. This can result in severe overheating of the generator and insulation breakdown. The overexcitation protection is designed to measure the degree of voltage and frequency over an extensive scope of frequency. It can be used to determine the level of excitation of the generator based on the relationship between voltage and frequency. A volts/Hz relay that has an inverse time attribute is effectively utilised by the independent power providers to protect their generating units from the adverse effects of overexcitation. In addition, a V/Hz relay is also utilised to protect the generator against overfluxing, owing to the fact that it can measure the ratio between voltage and frequency effectively [1]. The protective device with function number 24 provides volt/hz protection for the generator, as presented in Figure 11.17.

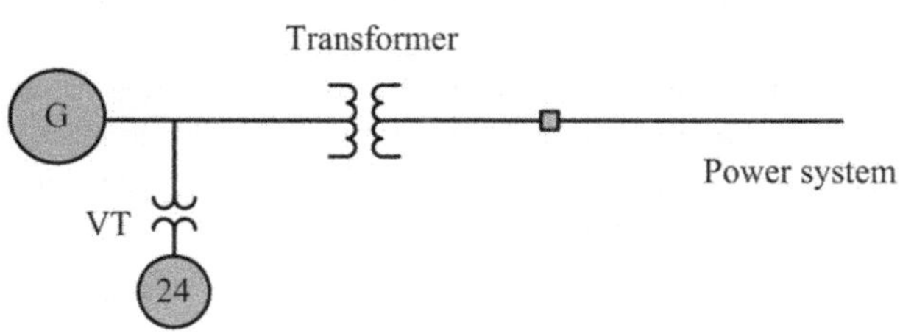

FIGURE 11.17 Generator system with overexcitation protection.

11.3.7.5 Overview of Overexcitation

An overview of overexcitation in the generator can be presented in the following list [1,8,12]:

1. The excitation flux in the core of a generator is directly proportional to the excitation voltage.
2. The excitation flux is inversely proportional to the frequency.
3. The danger of overexcitation is insignificant when the generator is tied to the power system.
4. The generator is exposed to a serious risk when it is disconnected from the power network.
5. Overexcitation happens in the generator during the startup and shutdown period.
6. Overexcitation occurs with higher voltage or at rated voltage and lower frequency.
7. The risk of overexcitation is high when the frequency of the generator is below the rated value.

11.3.7.6 Causes of Overexcitation

The causes of overexcitation in the generator are:

1. Running or shutting down of a generator with the AVR in service
2. Loss of the AVR voltage feedback signal

11.3.7.7 Effects of Overexcitation

The effects of overexcitation in the generator are the following:

1. Negative effects on the optimal operation of generators
2. Generator voltage regulator problems
3. Insulation breakdown
4. Severe overheating of the generator
5. Control component failure
6. Full load and partial load rejection
7. Stand-alone power operation during major disturbances

11.3.8 Stator Unbalanced Current Protection

Many factors can cause unbalanced currents at the terminals of the generators. The unbalanced currents that occur in the generators owing to some factors produce negative sequence currents that flow in the rotor winding, field windings and armature of the machine. The negative sequence currents have the capability of producing high temperatures within a short period, with serious unfavourable impacts on the particular regions of the rotor segments. The negative sequence current also creates a reactive field rotation that increases the synchronous speed of the generator in direct proportion with the rotor. This induces more frequency current in the rotor of the machine [19]. The current causes overheating in the rotor circuit, particularly in the alternator. The unbalancing current that happens because of a fault in the stator winding should be cleared as soon as possible with the application of differential protection incorporated into the generator. However, the unbalancing current that happens because of an unbalanced loading in the power system can continue for a long time based on the protection coordination of the network. The unbalanced fault currents can be cleared by using a negative phase sequence relay with the attributes to match the withstand curve of the generator.

11.3.8.1 Causes of Stator Unbalanced Current

The causes of stator unbalanced current in the generator can be listed as follows [16,19]:

1. Unbalance loads
2. Single pole opening of the circuit breakers
3. Asymmetrical transmission lines
4. Open phase or open circuit
5. Stator winding faults

11.4 Rotor Protection System

There are different types of rotor protection, and each one is designed to protect the rotor from a particular fault. The protection against unbalanced loading, shorted winding, ground winding, open field winding, overheating, loss of excitation and over-/underexcitation can be considered as a type of rotor protection since the effect of negative sequence currents is likely to damage the rotor.

11.4.1 Rotor Earth Fault Protection

The rotor winding of a generator is exposed to various abnormal conditions such as mechanical and thermal stresses from the presence of vibrations, bearing failure, extremely high current and failure of the cooling system. This will automatically cause an insulation breakdown between the field winding and the rotor core. The degree of insulation breakdown depends on the manufacturing failures, design failures, maintenance failures, material failures and the point where the stress becomes too much. As a result of this abnormality, the field winding is insulated from the earth. The first ground fault that happens in the field winding, exciter circuit and its related circuits has an insignificant fault current effect on the system. Owing to the negligible level of the earth fault, it does not generally endanger the operation of a generator [18]. However, the second earth fault will cause a heavy current to flow in the circuit, and a segment of the field winding becomes susceptible to an unwarranted short circuit. This will lead to serious damage owing to the production of unbalanced air gap fluxes that are detrimental to the operation of the machine. The violent vibration that results from unbalanced air gap fluxes can cause mechanical damage to happen in the bearing of the generator. This can damage the complicated parts of the machine. Thus, it is important that any insulation failure found during the operation of the generators should be repaired as quickly as time permits.

The unbalanced rotor winding can cause an uneven rotor current to flow in the machine; if adequate care is not taken, it will lead to mechanical damage as a result of vibrations. These conditions can cause severe and rapid damage to occur in the machine due to the fact that the field winding has been short circuited. The short circuit of the field winding of the machine has the ability to distort the flux across the air gap, and the flux can adopt a pattern like the one in Figure 11.18 [11]. This is because of the uneven ampere turns of flux in various sections of the field windings. It is mandatory to identify the earth fault that happens in the rotor field winding of a generator circuit and rectify such fault so that optimal operation of the machine will be obtained as quickly as possible. The ground check of the machine can be automatically carried out with the application of a sequencing timer and controller, or it can be manually checked by the personnel that operate the machine. The machine can also be protected from serious damage with the application of the vibration detector, which sends a warning signal to the operators and subsequently trips the affected generating units.

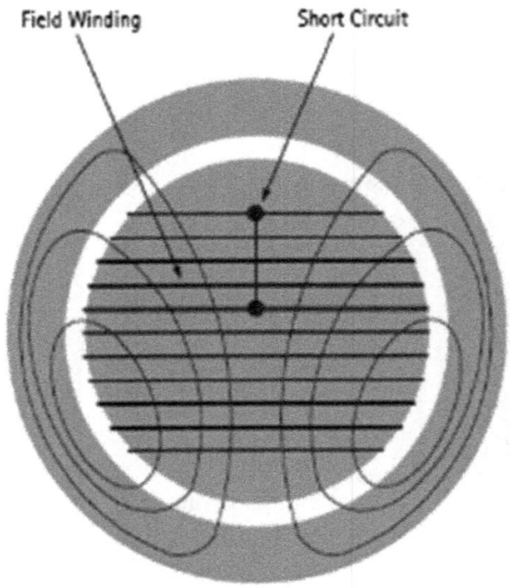

FIGURE 11.18 Field flux pattern with short field.

11.4.1.1 Effects of Rotor Earth Fault

The mechanism that causes the shorted winding is often due to grounding of the winding at two different places. The rotor ground fault may result in the following:

1. Uneven air gap fluxes
2. Prolonged rotor vibrations
3. Uneven overheating of the rotor
4. Changing of reactive power and output voltage
5. Major damage to the machine bearing
6. Serious damage to the rotor winding
7. Insulation breakdown

11.4.1.2 Type of Rotor Earth Fault Protection Schemes

Various techniques are available for the detection and tripping of generators because of the presence of field ground faults. These techniques can be utilised to detect a rotor earth fault in the generators. Based on the fundamental rule of machine protection, a relay is utilised via the earth fault path to protect the machine from the rotor earth fault. The three techniques of rotor earth fault protection for the generators are:

1. Potentiometer technique
2. AC injection technique
3. DC injection technique

11.4.1.3 Potentiometer Technique for Rotor Earth Fault Protection

The technique is appropriate for the generators that utilise brushes in their respective field windings. The scheme is illustrated in Figure 11.19, where one resistor is connected between the field winding and exciter. The system is designed so that the resistor is tapped at the centre and directly connected to the earth via a very sensitive voltage relay. The tapping point on the potentiometer varies with the application of a push button as shown in Figure 11.19. The relay setting is limited to approximately 5% of the exciter voltage.

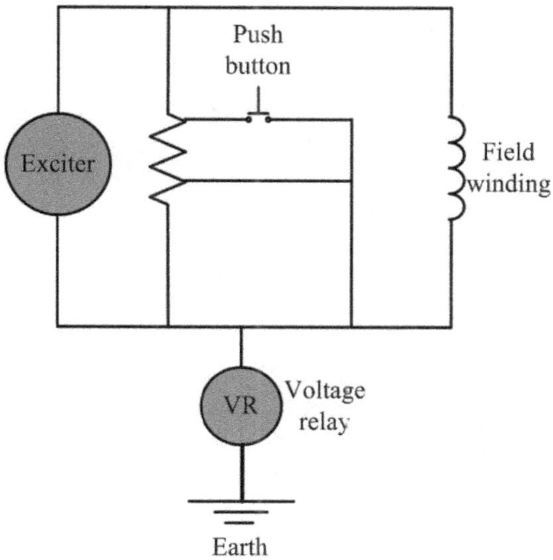

FIGURE 11.19 Earth fault protection by potentiometer technique.

The presence of an earth fault in the exciter circuit and field winding will automatically close the sensitive voltage relay via the earthed path. The earth fault in the field winding produces a voltage in the sensitive voltage relay because of the persistent change in the value of the resistance of the potentiometer.

11.4.1.4 Disadvantages of the Potentiometer Technique of Earth Fault Protection

The disadvantages of the potentiometer technique of earth fault protection in the generator are:

1. It can only be utilised to detect that the earth fault happened at any point apart from the focal point of the field winding.
2. Any point that is very close to the winding become grounded; the center tapped potentiometer would not detect it.
3. Because of the earth fault on the focal point, the field winding won't make any voltage appear in the relay.
4. The method is blinded to the earth faults at the focal point of the field winding.

The aforementioned setbacks can be reduced with the application of an additional tap on the resistor somewhere else besides the focal point of the resistor by means of a push button. To check that blinding to faults at the focal point does not happen, a manual switch is arranged to move the test point from the center to other points along the resistor. In the event that the push button is pressed, the center tap is moved. Owing to this, voltage appears in the relay even in the presence of the central arc fault that happens in the field winding [21].

11.4.1.5 AC Injection Technique of Rotor Earth Fault Protection

In this arrangement, the first terminal of the relay is connected to the field winding and exciter circuit, as shown in Figure 11.20. The second terminal is tied to the earth by means of an auxiliary transformer and a capacitor. The presence of the ground fault in the field winding completes the circuit owing to the fact that the relay circuit is closed through the earthed path. The system is designed so that the secondary voltage of the auxiliary transformer appears in the voltage relay. Hence, the relay will trip due to the presence of the earth fault in the system [21]. The system has no blind point, unlike the method discussed in Section 11.4.1.3.

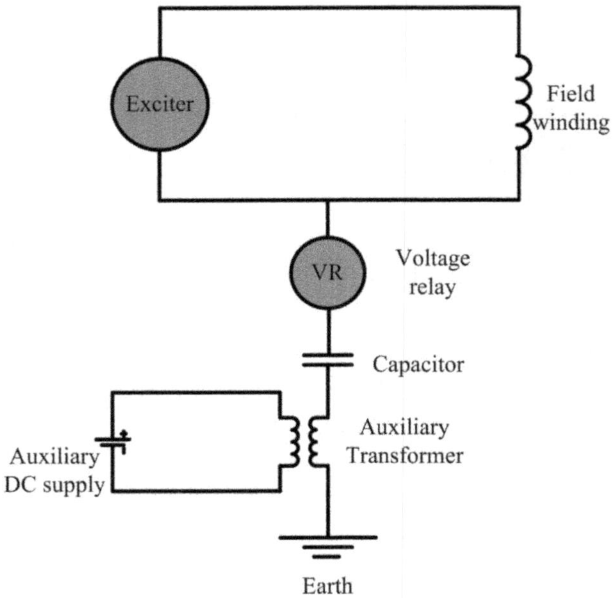

FIGURE 11.20 AC injection technique of rotor earth fault protection. (From Rotor earth fault protection of alternator or generator, https://www.electrical4u.com/rotor-earth-fault-protection-of-alternator-or-generator/, accessed April 2018.)

11.4.1.6 Disadvantages of an AC Injection Technique for Rotor Earth Fault Protection

The disadvantages of an AC injection technique for rotor earth fault protection in the generator are [21]:

1. There is a possibility of leakage current via the exciter and capacitors to the field winding.
2. Uneven magnetic field.
3. The current that passes through the major components of the system can cause corrosion of the bearing.
4. Mechanical stresses in the machine bearings.
5. Protection of rotor is not active in a situation where there is a failure of AC power supply to the circuit.

11.4.1.7 DC Injection Technique of Rotor Earth Fault Protection

The DC injection technique is the best among the three techniques owing to the fact the disadvantage of leakage current that is paramount in the AC injection method can be completely removed with application of additional components. The first terminal of the relay is tied to the positive terminal of the exciter, while the second terminal is tied to the negative terminal of an external DC source via an auxiliary transformer, as shown in Figure 11.21a. The external DC source is obtained by means of an auxiliary transformer with a bridge rectifier. For the effective operation of the bridge rectifier, its positive terminal is grounded [21].

It is observed from Figure 11.21b that, in a situation where there is a field earth fault, the positive potential of the external DC source will appear at the terminal of the relay that is tied to the positive terminal of the exciter. This allows the rectified voltage output of the bridge rectifier to appear in the voltage relay and operates it. A filter is effectively used to block AC currents that flow in the measuring circuit, as shown in Figure 11.21b [21].

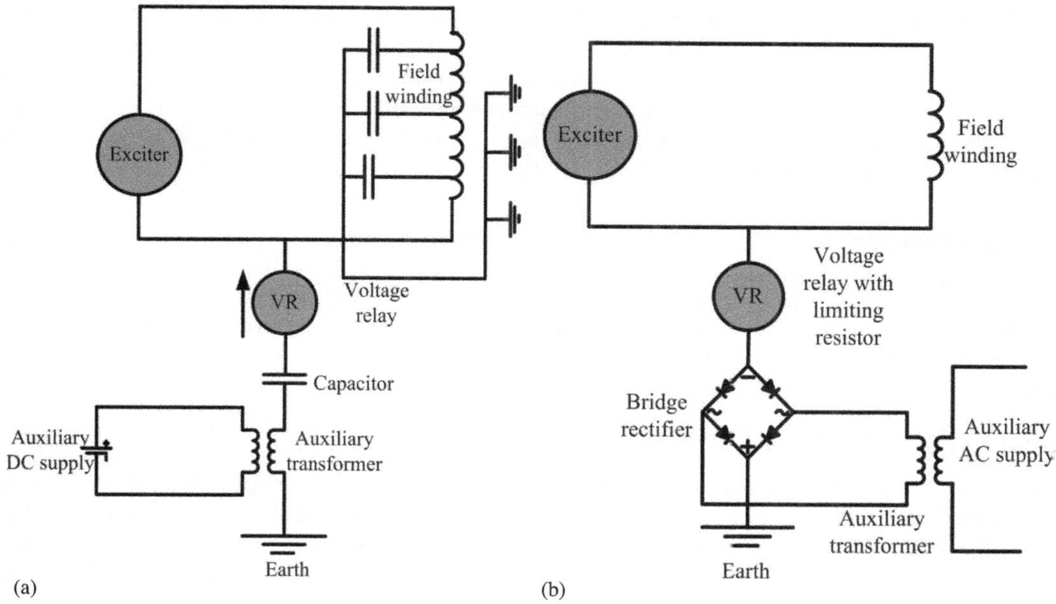

FIGURE 11.21 (a) DC injection technique of rotor earth fault protection. (From Rotor earth fault protection of alternator or generator, https://www.electrical4u.com/rotor-earth-fault-protection-of-alternator-or-generator/, accessed April 2018.) (b) The DC injection technique of rotor earth fault protection with a bridge rectifier. (From Rotor earth fault protection of alternator or generator, https://www.electrical4u.com/rotor-earth-fault-protection-of-alternator-or-generator/, accessed April 2018.)

11.4.2 Rotor Shorted Turn Protection

The shorted segment of the field winding of a generator results in an unsymmetrical rotor flux pattern. This initiates a potential vibration that can damage the rotor of the machine. The rotor shorted turn can be detected by utilising a test probe that is inserted into the air gap of the machine. The flux pattern of the machine that consists of positive and negative poles will be measured with the aid of a test probe that is placed in the air gap of the machine. Any substantial variation in flux pattern between the positive and negative poles demonstrates the presence of a shorted turn fault in the machine. Such an electrical fault can be eliminated by the application of automated waveform comparison techniques. These indicate that the waveform should be inspected remotely or manually at regular intervals. The short-circuited turns on the rotor have an adverse effect on the optimal operation of a generator. The detection of field faults due to various abnormalities is a serious challenge that has to be overcome within a short duration; otherwise, it can cause an extreme vibration. As a result of this, a vibration detection system is urgently needed for the mechanical protection of the generator.

11.4.3 Rotor Thermal Protection

The methods that are available to monitor temperatures of the rotor winding and cooling system allow the personnel that operate a generator to work effectively without exposing them to unnecessary risk. This can only be achieved by measuring the rotor winding's temperature with respect to field voltage and current. The temperature of a generator's winding can increase owing to numerous conditions. Among the conditions that lead to a sudden increase in the rotor winding's temperature are overload, unbalanced currents, winding failures and cooling system failure. The rotor thermal protection system has the capability to monitor and send a signal or alarm to the control room about a deteriorating level of the field winding of the machine because of overheating. The main technique for evaluating the sudden increase in the temperature of the winding is by estimating the field current and voltage at the collector. The comparison of the estimated resistance can be made with a known value at a known temperature. This method is limited to the generators that utilise collector rings. Another technique that is usually available is to observe the excitation current of the generator. The relays have been set to time the duration when the excitation will increase beyond the nominal value of the machine. The circuit breaker trips the machine the moment a certain setting is reached [19].

11.5 Protection for Other Systems

Other protection systems are meant to monitor overspeed, motoring, vibration, bearing overheating, undervoltage, overfrequency, cooling systems, etc.

11.5.1 Underfrequency Protection

The underfrequency of a generator sometimes occurs because it operates as an isolated power system. The major task of the utilities is to overcome the instability caused by a major loss of power generation from the faulty units and a serious overloading of the remaining generating units in the power station. The operation of the generators at low frequency can also happen when the machine together with its load is isolated from the entire network. This leads to underfrequency protection that can be assessed based on the characteristics of the generators and information about the network. A number of electric faults recorded in the power system may result from shortage of power generation when compared with the load demand at the load points. This demonstrates that inadequate power is being generated compared with the connected load. Since more power is being demanded at the load points, the generator is not capable of meeting the power demand at the reduced frequency. In this situation, underfrequency can cause an excessive load demand and overheating of the stator and rotor. The underfrequency operation of a generator can likewise cause cracking of the blade structure and other physical damage on alternators and turbines. The power output of the generator increases amid an overload and this reduces the frequency of the generator.

The main solution to overloading in a generator is to apply load shedding technique. Therefore, a load shedding facility needs to be incorporated into the system to avoid the effects of low frequency operations and to reduce step loading as the frequency drops. The load shedding system is designed to trip the main circuit breakers or feeders in the event of overload to reduce the system power demand. This will automatically coordinate the relationship between the load connected to the system and the power generation. The present practice is to overcome the effects of low frequency by shedding the non-essential loads in the system with the goal that the generator capability will be equivalent to or more than the connected load. This action creates the balance between the connected load and the power output of the generating units. The degree at which frequency of a generator changes depends on the time, degree of overload and variations of the generator in accordance with the frequency changes. Nevertheless, the prime mover of each generator must be designed so that it will be protected from the low frequency operation by tripping of the specific generating units. In some situations, a special relay is assigned to provide a warning signal or an alert and trips the circuit breakers or switches for protection of the generator.

11.5.2 Overfrequency Protection

The overfrequency operations of the generators can result from full or partial load rejection conditions or an imbalance between the power demand and generation. The load rejection which causes a mismatch between the available power generation and the load is initiated by a fault in the system and tripping of the generating units or failure of the distribution and transmission lines to operate effectively. The overfrequency of a generator depends on the governor droop characteristics and variation of the load. The governor droop characteristics allow numerous generators to keep running in parallel with the aim of sharing the loads among generating units based on their power ratings. For example, a governor with a 5% drop characteristic is intended to produce a 1.5% speed increment that can cause a 30% load

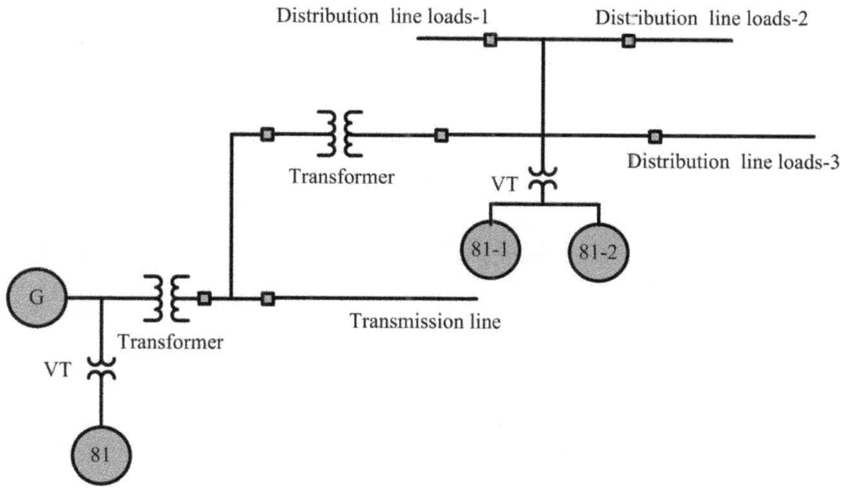

FIGURE 11.22 The coordination of the generator frequency and load shedding relays.

rejection [19]. In addition, full or partial load rejection can cause overspeed and overfrequency conditions. Overfrequency operation usually does not represent a severe overheating effect, except the nominal power of the machine is surpassed. The governor attached to the prime mover of the generator can be utilised to minimise the generator frequency and speed without tripping the circuit breaker or other protective devices of the generator. The control action of the governor can be utilised to restore the generator frequency and speed back to normal. The full or partial load rejection from the excess generation in the power plants can effectively be amended by reducing the power output of the generating units with the aid of the governor. The overfrequency protection scheme is fundamentally based on the protection of the prime mover as opposed to electrical protection of the generator. The protective device with function number 81 provides overfrequency protection for the generator, as presented in Figure 11.22.

$$\text{Droop}(\%) = \frac{S^i - S}{S^i} \tag{11.4}$$

where S^i is the no load speed (rpm) and S is the full load speed (rpm).

11.5.3 Overvoltage Protection

Overvoltage happens in the generator due to a sudden rise in the speed of the prime mover because of unexpected loss in the load demand [18]. It can also happen because of a defective voltage regulator and manual control mistakes. Overvoltage does not occur in some conventional generating units because the automatic generating control unit of the machines are exceptionally sensitive to variations in speed. The correct voltage is obtained at the terminal with the proper operation of the AVR during starting of the generating units. The accurate voltage is acquired before the synchronisation of the generating units. After synchronisation of the generating unit, the terminal voltage of the generator varies based on its AVR, the AVRs of other generating units in the power station and the voltage level of the power system. It is not imperative for a single unit to increase the terminal voltage of the entire power plant owing to the fact that it is connected to the system. The sudden change in the field excitation as a result of defective AVR will only increase the reactive power output. This will promptly trip the machine with the application of the impedance relay or the V/Hz relay. Moreover, a number of excitation limiters are installed to prevent the reactive output power and rotor field current from exceeding the design specifications. The protective device with function number 59 provides overvoltage protection for the generator, as presented in Figure 11.23a.

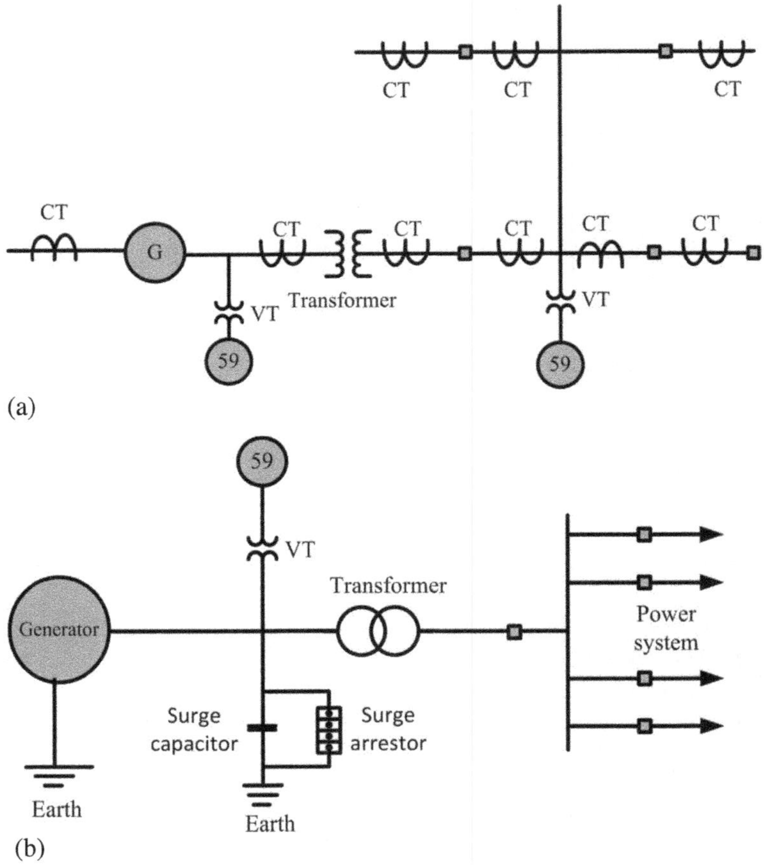

FIGURE 11.23 (a) Location of overvoltage relays requiring coordination, (b) overvoltage relay with surge arrestor and surge capacitor.

In a situation where the main feeder of the generator is tripped when the machine is running at full or partial load capacity [1], the sudden increment in terminal voltage of the generator above its nominal rating will be limited by a rapid operation of the AVR. However, due to the deficient operation of the AVR when the machine operates in an islanded mode, a severe overvoltage will occur. The terminal voltage will further increase with the overspeeding of the machine as a result of slow operation of the governor. A set of overvoltage relays should be incorporated into the power system as a measure to trip the generator and prevent it from the unwarranted effects associated with overvoltage caused by an abrupt loss of load and overspeeding. The protection for generator overvoltage is equipped with an overvoltage relay designed to trip the circuit breaker and the excitation circuit since tripping the circuit breaker is not enough to secure the machine against overvoltage. Moreover, overvoltage protection scheme is utilised in a power system for prevention of the insulation breakdown caused by a continuous overvoltage. The insulation system of the generator is designed by the manufacturers to continuously operate at 105% of the overvoltage. However, a continuous overvoltage condition that exceeds 105% should not to be permitted to happen owing to physical damage to the machine. Overvoltage can be caused by the following factors:

1. Malfunctioning of the AVR
2. An abrupt change of the load
3. Switching impulses
4. Loss of voltage transformer

5. Improper setting of AVR
6. Manual operation without the AVR in service
7. Sudden interruption of heavy load
8. Lightening impulses
9. Uneven breaker poles
10. Back flashover
11. Flashover in the power plant

11.5.3.1 Voltage Surge

Voltage surge is an abrupt change in the terminal voltage of a generator to a high magnitude within a very short time. The voltage surge is transient in nature and exists for a very short duration. The main causes of abnormal voltage surges in the power system are attributed to the presence of resonance, lightning and switching impulses, arcing ground, insulation failure or breakdown of the generator. The voltage surge that appears in the generator windings and other auxiliaries is very small in magnitude. The degree of voltage surges recorded in the generators hardly doubles or triples the nominal voltage of the system. The general damage associated with the voltage surges can be prevented with proper insulation of the rotor and stator windings and application of various protective devices. The voltage surges that happen in the generators are attributable to the lightning that is very high and severe. In the event that a proper protection system is not integrated into the power system, this will cause severe damage to the generators and other components of the power systems. Therefore, the surge protective devices that are used to prevent overvoltage from the generator are surge arrestors or lightning arrestors, as shown in Figure 11.23b. Surge arrestors are essential for protection of the power system against transient overvoltages. Lightning arrestors are utilised to send the high voltages caused by lightning strikes to the ground through conduction. Apart from the lightning arrestors, other techniques that are readily available for the protection of lightning are grounding screen and overhead ground wire. They operate rapidly because they provide a very low resistance route to the ground for lightning strikes. They are connected to the generators, substations, switchgear, distribution and transmission lines, etc.

11.5.4 Generator Undervoltage Protection

Generators are normally intended to work ceaselessly with at least 95% of their nominal voltage, when conveying rated power at the rated frequency [4]. The operation of a generator with a terminal voltage that is lower than 95% of its nominal voltage will bring about detrimental effects like importation of reactive power from the grid and reduction in stability limit. Owing to this, undervoltage relay is utilised to send a cautionary signal and not trip the generating unit. Consequently, the operators of the machine can take appropriate preventive measures to rectify the undervoltage condition. The following protection plans are utilised to provide supervision for other protection functions [22]:

1. Distance relay
2. Out-of-step relay
3. Inadvertent energising relay
4. Loss of field relay

11.5.4.1 Causes of Generator Undervoltage

When generators are designed to meet the power demand of consumers, undervoltage can happen due to the following reasons:

1. Overloading of the machine
2. Failure of the AVR
3. Transient reductions in voltage during the machine startup

4. Whenever a load that is applied to the generator exceeds the installed capacity
5. Failure of the AVR to respond to the load being applied
6. Failure of the governor to respond to the load being applied
7. Mechanical failure of some components

11.5.5 Protection of the Generator Due to Unbalanced Current

The generator unbalanced faults in conjunction with the system conditions can produce unbalanced three-phase currents and negative phase sequence currents in the generator. The negative sequence components of these currents increase the frequency currents in the rotor. This can cause overheating and damage to the machine in a very short time. The presence of an uneven three-phase current and uneven currents in the machine induces double frequency currents in the rotor core, field windings, retaining rings, etc [23]. The negative sequence protection that includes a time overcurrent is utilised to secure the machines before reaching their specific limits.

11.5.5.1 Effects of an Unbalance Current in the Generators

The effects of an unbalance current in the generators are [12]:

1. Rotor damage
2. Stator damage
3. Melting of the wedges in the air gap
4. Overheating of the rotor

The three-phase CTs are utilised for the protection of the system from the consequence of unbalanced current, as shown in Figure 11.24. The secondary terminals of the CTs are directly connected to the overcurrent relay via the negative filter. The major components that constitute the negative sequence filter are the resistors, capacitors and other valuable components. The negative sequence filter is designed and selected based on the specifications of the manufacturers so that negative sequence currents can flow through the relay, as presented in Figure 11.24. The operation of the selected relay is based on the fact that it can function at any specific value of the uneven current and the negative sequence component current [18]. The negative sequence filter and overcurrent relay are utilised in the circuit for protection of the machine against unbalanced current.

11.5.6 Negative Phase Sequence Current Protection

The negative phase sequence component of current is just like the positive sequence component. The only difference is the resulting reaction field that turns the other way to the DC field system. Consequently,

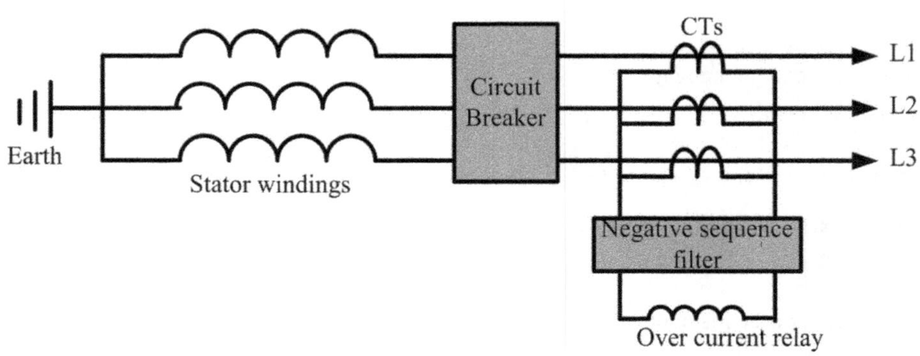

FIGURE 11.24 Protection against unbalance loading.

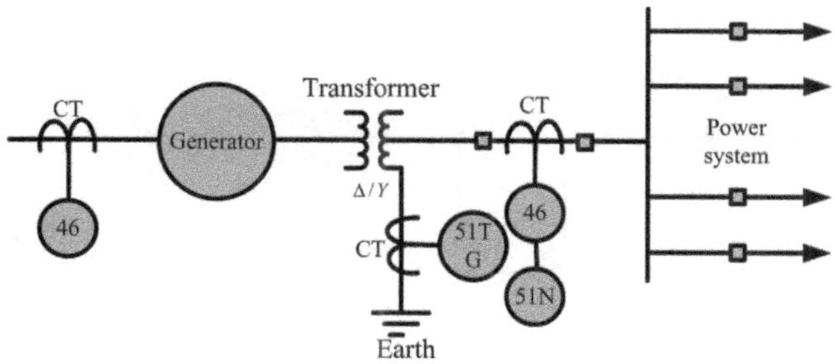

FIGURE 11.25 Negative phase protection coordination.

a flux is created that subsequently induces double frequency currents in the field system and other segments of the machine such as rotor body, etc. The subsequent eddy currents from the negative phase sequence current can cause extreme overheating of the rotor [4]. The negative sequence overcurrent protection regularly incorporates two settings that caution the operational activities and trip the system. The scheme also utilises an amount of negative sequence current produced by the unbalanced conditions of the system to which the generator is connected. The impact of a negative phase sequence is so serious that a single-phase load is equivalent to the typical three-phase rated current. This can rapidly increase the thermal heat of the rotor slot wedges. The negative phase sequence current protection is used for prevention of overheating that results from negative sequence currents [24]. The generator protective devices with function numbers 46, 51 TG and 51 N provide stator unbalanced, time overcurrent protection and backup for the generator earth fault protection, as presented in Figure 11.25.

11.5.6.1 Causes of Negative Phase Sequence Currents

The following conditions can cause negative phase sequence currents in a generator:

1. Open phases
2. System asymmetries
3. Unbalanced system faults
4. Unbalanced loads

11.5.7 Reverse Power Protection

The primary concern of the reverse power protection scheme is to protect the prime movers of the generators such as reciprocating engine, turbines, electric motors, etc., against mechanical and electrical damage during a motoring condition. The motoring operation of a generator is based on the fact that it operates while the power supply to the prime mover is disconnected when the generator is still on-line [4]. In this condition, the driving torque turns out to be not as much as the total losses in the generator and the prime mover. Owing to this, the generator begins to operate as an induction machine, absorbing the essential active power from the system instead of delivering active power into the network. This makes the generator act like an asynchronous machine and drives the prime mover. Hence, the reverse operation of the prime mover adversely affects its performance and results in overheating due to friction, windage loss owing to energy dissipation, distortion of turbine blades and physical damage to the damper windings. The consequences of reverse power in a power system has encouraged many utilities to incorporate reverse power protection relays into their generating units to protect the prime movers against reverse power actions. The protective device with function number 32 provides reverse power or anti-motoring protection for the generator, as presented in Figure 11.26. The application of a reverse power relay trips the circuit breaker when power flows from the network to the generating unit. This indicates that the reverse power relay will trip the generating unit when it is operating in the motoring mode. The reverse

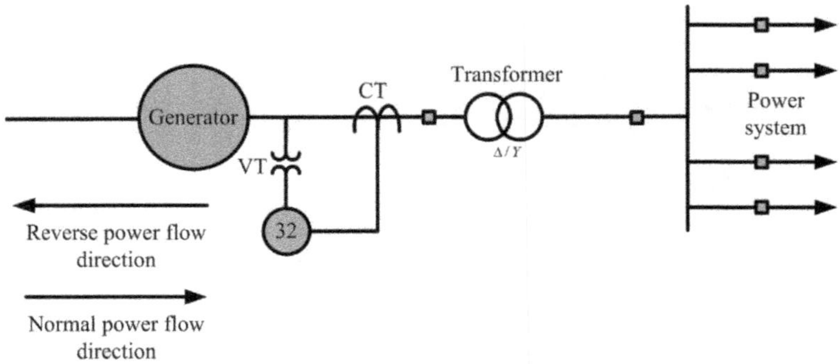

FIGURE 11.26 Reverse power flow detection.

power protection scheme is utilised by utilities for prevention of the generator against mechanical damage on shafts, gear boxes, turbine blades, etc., during the motoring operation of the machine [4]. The relay setting for reverse power protection can be estimated by the following equation:

$$\text{Setting} = \frac{(0.05 \times 5 \times 10^6)}{(\text{CT ratio} \times \text{VT ratio})} \tag{11.5}$$

11.5.8 Protection against Inadvertent Energisation

In the event that a generator is inadvertently energised while off-line, it operates as an asynchronous machine. The machine accelerates and a very high current is induced in the rotor components. This will probably damage the rotor, stator and other complex components of the generator and cause significant havoc such as overheating. At times, serious damage beyond repair has been reported in some generators due to inadvertent energisation. The low impedance of the machine at standstill causes currents of up to one to four per unit current of nominal rating to flow into the generating unit when the breaker operates [19]. The amount of the inrush current depends on equivalent system impedance. The terminal voltage of the machine can also increase from 20% to 70% of the rated value based on this condition. Obviously, these large currents can thermally damage the rotor owing to the excessive high currents and can create havoc in the windings as a result of insulation breakdown. The most important part of inadvertent energisation is damage to the bearing of the machine, which can be happen within a short time because of low oil pressure. Consequently, it is necessary that various protection schemes are assigned to protect the generating units at standstill against inadvertent energisation. An inadvertent energising protection system is presented in Figure 11.27. The generator protective devices with function numbers 27 and 50 provide

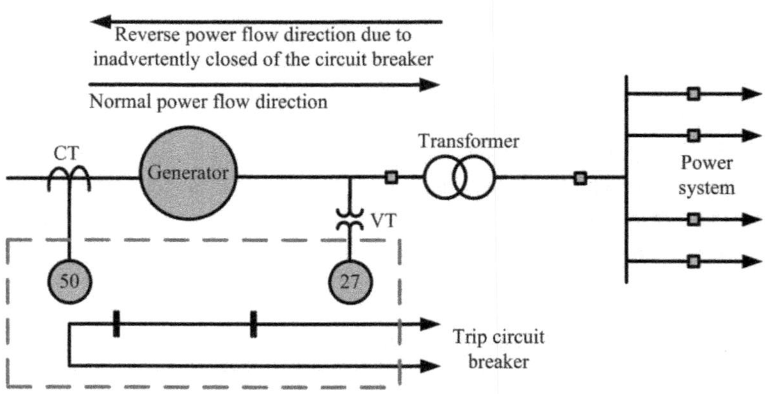

FIGURE 11.27 Inadvertent energising protection scheme.

undervoltage and instantaneous overcurrent protection for a current detector that is embedded in the circuit breaker, as presented in Figure 11.27.

11.5.8.1 Methods for Detecting Inadvertent Energisation

The typical plans utilised by utilities to detect inadvertent energising in the generators are:

1. Frequency administered overcurrent
2. Distance relay scheme
3. Directional overcurrent relays
4. Voltage administered overcurrent
5. System backup relays
6. Auxiliary contact scheme with overcurrent relays

11.5.8.2 Causes of Inadvertent Energising of the Generator

The causes of inadvertent energising of the generator are:

1. Operational errors
2. Circuit breaker head flashovers
3. Control circuit breakdowns

11.5.9 Circuit Breaker Failure Protection

Circuit breakers are designed to operate based on the fault current that is available in the power system. The failure of the circuit breaker to trip in the event of abnormal operating conditions or faults would result in loss of protection. Another abnormal operation of the circuit breaker is single phasing, which normally occurs if one or two poles of a three-phase circuit breaker operate [19]. This results in negative sequence currents operation of the power system. The problem can be overcome with the application of the breaker failure scheme in conjunction with the auxiliary switches, generator protective relays, timer, overcurrent relays, etc. In the event that the circuit breaker fails to clear the fault in a predetermined time, a timer can be utilised to initiate an action that trips the circuit breakers as a measure to isolate the affected generator from the network [20]. The protection system is designed to work against any flashover of two poles of the circuit breakers. This protects the circuit breakers from head flashover that occurs when arcing in the form of high voltage pass through the circuit breakers' contacts.

11.5.10 Loss of Synchronisation

There are various reasons that cause a generator to lose its synchronisation to the power system during operation. When a generator loses synchronism, it causes winding stresses, pulsating torques, turbine generator shaft damage, winding damage, winding support dislocation, rotor damage, coupling damage, etc. The aforementioned reasons have detrimental effects on the optimal operation of a generator. The adverse effect of loss of synchronisation can be removed from the generator with the application of a protection scheme. This can be achieved at the earliest opportunity, ideally during the first half-slip cycle [20]. The protection against the loss synchronism condition depends on the way that the impedance changes during an unstable condition. A dedicated relay can be utilised to protect the machine against loss of synchronism conditions. The utilisation of a loss synchronism protective device is to protect the generator in light of detailed stability studies and analysis. Stability studies are carried out to confirm the significance of the out-of-step relaying for these applications [4]. The conventional technique for out-of-step protection is an impedance relay that analyses the variation in impedance as viewed at the terminals of the power system component.

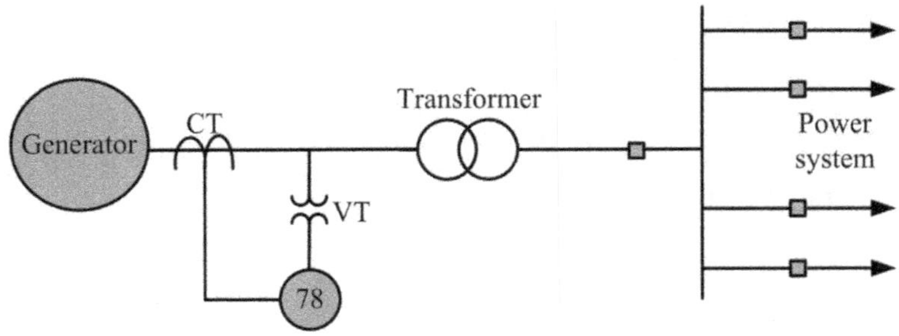

FIGURE 11.28 Generator out-of-step relay connection.

Protection devices like differential relaying and time-delay system backup can be used to detect the loss of synchronisation of a machine. The loss of excitation relay can also give some level of protection for loss of synchronism, but it cannot be relied on absolutely to detect loss of synchronisation in the generator. Hence, it is mandatory for the utilities to design the power system so that it would be highly protected in the event of loss of synchronisation where the electrical centre is positioned within the section of the high voltage ends of the transformers and generators [4]. In this situation, a different out-of-step relay is utilised to protect the generator. This type of protection is needed for the large generators that are tied to the HV sides of the power system. Moreover, this protection is also essential in a situation where the load centre is disconnected from the power system and the assigned relay fails to operate effectively owing to slow speed or failure to detect a loss of synchronisation fault. The protective device with function number 78 provides a loss of synchronisation protection for the generator, as presented in Figure 11.28.

11.5.11 Mechanical Faults

The importance of a continuous power supply has prompted independent power providers to protect their generating units from the conditions that could damage them. A number of faults happen in the mechanical section of a generator. The protection systems described in the previous sections provide protection to the generator but mechanical faults have not been covered. This section gives more details about mechanical faults associated with the operation of a generator. The mechanical faults of the generators owing to the prime movers and their auxiliary equipment can be protected to reduce detrimental effects on their operations.

11.5.11.1 Vibration Protection

Vibration protection has progressed from a significant startup purpose to an effective instrument during the generator's operation. A proper vibration protection improves the safety of the equipment and the reliability of the power system and reduces the operating costs over the lifetime of the generator. One of the major factors for optimal operation of a generation is vibration consideration. The operation of a generator is affected by vibration due to the rotating parts and other critical components. The vibration of the generator can damage the generator's rotor, armature winding, the core and frame. This indicates that more than 70% of the major segments of a generator are exposed to the risks associated with excessive vibration. Therefore, it is imperative to prevent the generator from unnecessary vibration. To achieve this objective, vibration detectors are incorporated into the bearings of the machine. The vibration detector is designed so that the voltage output from the coil is directly proportional to the amount of vibration of the machine. The voltage flows from the coil into the integrating circuits and after that into an interval indicating instrument. The three components of a vibration monitoring device that are commonly used in generators are transducers, monitors and machine diagnostic equipment.

11.5.11.2 Overspeed Protection

Based on the consequences of generator being operated at a high speed, it is suggested that all the generators driven by prime movers should be protected by an overspeed protection scheme. The overspeed component should be receptive to machine speed by electrical or mechanical identical connection. In the event that the overspeed element should not be badly affected by generator voltage, the overspeed component is configured as a major aspect of the prime mover together with the speed governor and the generator. It is designed to operate the governor so that it will shut down the prime mover and at the same time trip the circuit breaker of the generator. This measure prevents overspeed operation of the generator [5]. The overspeed component should be changed in order to function at approximately 3%–5% above the full load rejection speed. The speed governor is intended to keep an unsafe speed increment even with a 100% load rejection; however, an extra centrifugal overspeed trip protection system is utilised in the generator as a measure to start an emergency mechanical shutdown in a situation where overspeed is more than 10%.

11.6 Solved Examples

Question 1

A 6.6 kV, 5 MVA star-connected synchronous generator has a reactance of 1.45Ω/ph and negligible resistance. A differential protection system that operates when the out-of-balance current exceeds 25% of the full load current is utilised in the power system. The star point of the generator is connected to the ground by using a resistance of 7.5Ω. Estimate the portion of the winding that is not protected from the earth fault. Show that the effect of generator reactance is negligible in the system.

Solution

The effect of generator reactance is negligible since reactance of winding is directly proportional to N^2, where N is the winding turns; therefore:

$$x_G \, \alpha \, N^2$$

Let's assume that x% of the windings is unprotected. Therefore, the number of unprotected turns $= xN/100$

The reactance of unprotected turns will be proportional to the square of turns, i.e.:

$$\left(\frac{xN}{100}\right)^2$$

The reactance of the winding is 1.45Ω as stated in the question. The reactance of unprotected winding can be computed as:

$$\frac{1.45x^2N^2}{100^2}$$

The reactance is added vectorially to 7.5Ω resistance of x that is small in value; therefore, the effect of reactance is negligible in the system.

$$\text{Phase voltage} = \frac{V_L}{\sqrt{3}} = \frac{6600}{\sqrt{3}} = 3810\text{V}$$

The voltage of unprotected portion can be expressed as:

$$\frac{3810 \times x}{100}$$

Therefore, the fault current $= \frac{3810 \times x}{8 \times 100}$

Full load current $= \frac{5000}{\sqrt{3} \times 6.6} = 437.37\text{A}$

The percentage of balance current that is essential for the relay to operate effectively can be estimated as follows:

$$\{437.37 \times 0.25\} = 109.34$$

Therefore, $\frac{3810 \times x}{7.5 \times 100} = 109.34$

$$x = 21.52\%.$$

Question 2

A synchronous generator reactance at 10 kV is protected with the application of a balanced circulating current system so that its neutral point is connected to the earth/ground by using a resistance of 10Ω. The protective relay is designed to operate when there is an out-of-balance current of 1.75 A in the pilot wires. The pilot wires are connected to the secondary windings of 1000/5 ratio CT. Calculate:

1. The percentage of the winding that is not protected.
2. The minimum value of the grounding resistance required for protection of 75% of the winding.

Solution

The phase voltage of alternator is:

$$\frac{V_L}{\sqrt{3}} = \frac{10000}{\sqrt{3}} = 5773\text{V}$$

Let the x% of the winding be unprotected, and voltage of unprotected portion of the winding can be estimated as follows:

$$= \frac{5773x}{100}$$

The resistance of neutral is 10Ω. The effect of reactance of the generator winding is neglected.

The fault current $= \frac{5773x}{100} \times \frac{1}{10}\text{A}$

The current in the pilot wires that have a CT ratio of 1000 to 5 can be estimated as:

$$\frac{5773x}{100} \times \frac{1}{10} \times \frac{5}{1000}\text{A}$$

The pilot current ought to be equivalent to 1.75 A to operate effectively, i.e.:

$$\frac{5773x}{100} \times \frac{1}{10} \times \frac{5}{1000} = 1.75$$

Therefore, $x = 60.63\%$.

Protecting 75% of the winding indicates that 25% of the winding is unprotected. The voltage of unprotected problem can be computed as:

$$5773 \times 0.25 = 1443.25 \text{ V}$$

R is assumed to be the minimum value of earthing resistance required. Therefore, the fault current can be estimated as:

$$\frac{1443.25}{R}$$

The fault current in the pilot wire can be expressed as:

$$\frac{1443.25}{R} \times \frac{5}{1000} \text{A}$$

This should be equal to 1.75:

$$\frac{1443.25}{R} \times \frac{5}{1000} = 1.75$$

Therefore, the value of $R = 4.12\Omega$.

Question 3

A synchronous generator winding is protected through a differential relay. The relay is designed to operate with a minimum pickup current of 0.2 A and a percentage slope of 10%. A high resistance earth fault takes place a short distance away from the grounded neutral point of the generator winding, with CTs current on both terminals of the windings being 365 + *j*0.0 and 325 + *j*0.0 A. It is assumed that the CT ratio is 400 to 5. Determine if the relay will operate with these parameters or not.

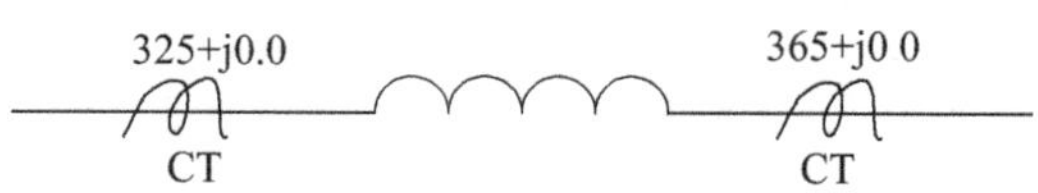

Solution

The differential current can be computed in Equation (11.24) as:

$$= 365 - 325$$

$$= 40\text{A}$$

The current in the operating coil can be expressed as follows:

$$\frac{40 \times 5}{400} = 0.5\text{A}$$

where the average current is

$$\frac{365 + 325}{2} = 345\,\text{A}$$

The average current through restraining coil $= \frac{345 \times 5}{400} = 4.31\text{A}$

Therefore, with 10% slope, the operating current can be estimated using Equation (11.29):

$$= \{0.1 \times \text{restraing current} + 0.2\}$$

$$= \{0.1 \times 4.31 + 0.2\}$$

$$= 0.631$$

Because the current via the operating coil is 0.5 A, this indicates that the relay will not operate effectively.

Question 4

A 14 kV generator has a ground relay pickup current of 1.0 A, a neutral resistance of 0.3 kΩ and a neutral CT rated as 200/5. Determine that percentage of winding that remains unprotected.

Solution

The fault current can be estimated as follows:

$$I_F = \frac{hV_a}{R_n}$$

$$= \frac{hV_L}{\sqrt{3}R_n} \ (A)$$

where R_n is the neutral resistance (Ω), and V_L is the rated line voltage (V). The ground relay pickup current for the critical fault location can be computed as follows:

$$I_{PU} = k \times \text{CTR (A)}$$

where CTR is the neutral current transformer ratio and k is the fraction of rated secondary current that gives pickup.

$$h = \frac{\sqrt{3}kI_{PU}R_n}{V_L}$$

The fraction k of primary rated current can be computed as:

$$I_{\text{pu}} = \frac{1}{5} \times \frac{200}{5} = 8\text{A}$$

$$k = \frac{1}{5} = 0.2$$

$$h = \frac{\sqrt{3} \times 0.2 \times 40 \times 300}{13800}$$

$$= 0.2969$$

The result indicates that about 30% of the winding is not protected based on the chosen values.

11.7 Tutorial Problems

1. What are the common types of generator faults?
2. Why is a protection system necessary in the generator?
3. Describe the type of protective devices that can be utilised in the generator for protection against the following faults:
   a. Overheating of the stator windings
   b. Inadvertent energisation
   c. Unbalanced current
   d. Negative phase sequence current
   e. Reverse power
   f. Undervoltage
   g. Overvoltage
   h. Underfrequency
   i. Overfrequency
   j. Overexcitation
   k. Underexcitation
   l. Stator earth fault
4. Explain why large alternators are grounded with large resistance.
5. Describe generator rotor earth protection.
6. What are the effects of the rotor earth fault?
7. List and explain the three type of rotor earth fault prediction techniques.
8. What are the disadvantages of the potentiometer technique of earth fault protection?
9. What are the disadvantages of an AC injection technique of rotor earth fault protection?
10. What are the causes of overexcitation in the generator?
11. What are the functions of the differential protection in the generator?
12. What type of protection does a resistance temperature detector provide?
13. What is the advantage of biased differential protection over simple differential protection?
14. Write a short description on the following:
    a. Differential protection
    b. Stator earth fault protection
    c. Stator interturn protection
    d. Stator overheating protection
    e. Phase short circuit protection
15. Explain rotor thermal protection.

11.8 Conclusion

A generator has to be protected not only from the electrical and mechanical faults but also from the adverse effects that arise from loss of synchronisation, loss of field winding, over- and undervoltage, over- and underfrequency, overheating, overspeed, loss of excitation, earth faults, etc. Owing to this, a protection system is designed under certain situations to monitor the abnormal operating conditions of the generators, like external and internal faults. The protective devices offer various functions in a power system to shut down the generating units during abnormal conditions. This prevents the damage associated with fault current that flows in the generator. This chapter described in details the functions of protective devices and their numerous structures and applications in the generator.

REFERENCES

1. Generator protection, http://docplayer.net/45677194-Abb-network-partner-ab.html, accessed April 2018.
2. C. Wadhwa, *Electrical Power Systems*. New Delhi, India: New Age International, 2006.
3. L. B. Sheng and S. Elangovan, A fault location method for parallel transmission lines, *International Journal of Electrical Power & Energy Systems*, 21(4), 253–259, 1999.
4. Power plant and transmission system protection coordination, https://www.nerc.com/docs/pc/spctf/Gen%20Prot%20Coord%20Rev1%20Final%2007-30-2010.pdf, accessed April 2018.
5. What are the common generator problems and its protection, docplayer.net/45677194-Abb-network-partner-ab.html, accessed April 2018.
6. R. K. Aujla, *Generator Stator Protection, Under/Over Voltage, Under/Over Frequency and Unbalanced Loading*, Department of Electrical and Computer Engineering, The University of Western Ontario, London, ON, 2008.
7. Differential protection of generator or alternator, https://www.electrical4u.com/differential-protection-of-generator-or-alternator/, accessed April 2018.
8. S. H. Horowitz and A. G. Phadke, *Power System Relaying*. Chichester, UK: John Wiley & Sons, 2008.
9. C. J. Mozina and D. C. Moody, Mill benefits from upgrading generator protective relaying, in *Proceedings of the IEEE Conference Record of the Annual Pulp and Paper Industry Technical Conference*, pp. 16–30, 2002.
10. N. Jenkins, Embedded generation, *Power Engineering Journal*, 9(3), 145–150, 1995.
11. J. Pyrhonen, T. Jokinen, and V. Hrabovcova, *Design of Rotating Electrical Machines*. Chichester, UK: John Wiley & Sons, 2013.
12. J. Simms and S. Pavavicharn, Generator stator protection, in *Proceedings of the IEEE 66th Annual Conference for Protective Relay Engineers*, pp. 505–528, 2013.
13. S. A. Panetta, Hybrid grounding, *IEEE Transactions on Industry Applications*, 51(6), 5058–5062, 2015.
14. A. V. C. Warrington, *Protective Relays: Their Theory and Practice, vol. 1*. New York: Springer Science & Business Media, 2012.
15. M. Baldwin, W. Elmore, and J. Bonk, Improve turbine-generator protection for increased plant reliability, *IEEE Transactions on Power Apparatus and Systems*, (3), 982–989, 1980.
16. T. Davies, *Protection of Industrial Power Systems*. Oxford, UK: Butterworth-Heinemann, 1996.
17. M. S. K. Elsamahy, Impacts of midpoint facts controllers on the coordination between generator phase backup protection and generator capability limits, PhD Thesis, Department of Electrical and Computer Engineering, University of Saskatchewan 2011.
18. Protection of the generator analysis, https://www.ukessays.com/essays/engineering/protection-of-the-generator-analysis-engineering-essay.php, accessed April 2018.
19. G. Klempner and I. Kerszenbaum, *Handbook of Large Turbo-Generator Operation and Maintenance*. Hoboken, NJ: John Wiley & Sons, 2011.
20. E. T. Gross and L. B. Levesconte, Backup protection for generators, *Transactions of the American Institute of Electrical Engineers. Part III: Power Apparatus and Systems,* 72(2), 585–592, 1953.
21. Rotor earth fault protection of alternator or generator, https://www.electrical4u.com/rotor-earth-fault-protection-of-alternator-or-generator/, accessed April 2018.
22. H. Sun, B. Zhang, Y. Lu, Z. Pan, and W. Wu, Modeling, simulating and online setting-checking for protective relay, in *Proceedings of the IEEE/PES Power Systems Conference and Exposition*, pp. 1–5, 2009.
23. J. H. Harlow, Practical cogeneration for the 1990s, in *Proceedings of the Industry Applications Society 38th Annual Petroleum and Chemical Industry Conference,* New York, pp. 109–117, 1991.
24. D. Graham, P. Brown, and R. Winchester, Generator protection with a new static negative sequence relay, *IEEE Transactions on Power Apparatus and Systems*, 94(4), 1208–1213, 1975.

12

Induction Motor Protection

N. T. Mbungu, Ramesh Bansal, Raj Naidoo, and D. H. Tungadio

CONTENTS

12.1 Introduction

The induction motor is one of the most popular machines; it is used for several applications in the industry due to its performance converting electrical energy to motion. The electrical driver requires a significant starting current and a specific consideration of the protection schemes [1–19]. The protection philosophy of an electrical motor is a function of the design structure of the machine and its ability to operate against the external and internal defaults [15,16]. When a fault occurs during a specified operating time, this philosophy aims to evaluate the impact of faults on the motor. In most cases, a power disturbance on the electrical system has a negative influence on the electrical part of an induction motor that could also be observed on some sensitive mechanical components of the entire process [15].

Smart grid technology has introduced a significant philosophy that has enhanced the performance of the electrical driver. An induction motor can operate efficiently when a fast fault detection is installed under a smart metering system, which consists of communicating with the protection scheme in real time. The objective of introducing an intelligent method on the protection schemes of the induction motor is to increase the motor life cycle and alleviate the motor damage caused by overload/thermal, start/stall, short-circuit, earth fault, drawing of negative phase sequence, rotor windings, and voltage disturbances.

This chapter discusses several types of motor protection philosophy. This consists of analysing the protection approaches of the induction motor that could further be implemented in the smart grid environment. The evaluations are based on first protection structures, which are applied in an asynchronous motor. These are overload/thermal protection, start/stall protection, short-circuit protection, earth fault protection, negative phase sequence protection, wound rotor protection, and under-/overvoltage protection, as well as loss-of-load protection. It is worthy to note that most motor protection schemes aim to inhibit overheating and avoid instability during the operating process of the motor, which can be caused by external and internal faults. Thus, a background in motor analysis and equivalent circuits will be provided. This consists of describing briefly the concept of different mathematical theories and models about how each component of the motor could be affected by excessive heating and loading problems, and how to protect the motor and increase the operating time of the system. It is also important to bear in mind that more than 80% of motor failures can be prevented by appropriate protection measures [16]. Figure 12.1 depicts the standard protection scheme of a direct, on-line induction motor protection.

12.2 Induction Motor Analysis

The section presents the principal induction motor analysis. The development of the system analysis is mostly based on the three-phase induction motor. The objective is to describe the fundamental models of the induction motor. This approach aims to assist the design of the induction motor protection philosophy of the selected disturbance phenomenon.

12.2.1 Base Voltage-Ampere

The base volt-ampere assists with determining the base impedance of an induction motor. The system analysis of the motor uses this strategy to describe the computation results of an induction motor. Due to the high range of values that some motors have, the per unit value is mostly used to analyse the motor behaviour. Note that the base volt-ampere derives from the values of current and voltage, which are measured from the motor nameplate at the full load. These values of voltage and current are considered as base voltage and base current. Equations (12.1) and (12.2) present the base three-phase apparent power or volt-ampere and the base of an impedance of a three-phase induction motor:

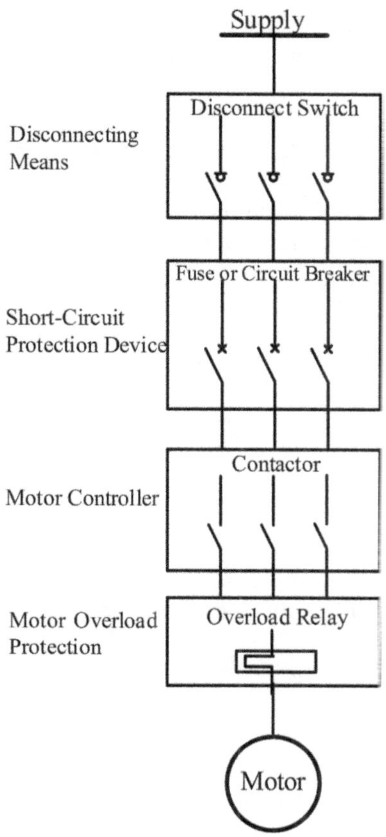

FIGURE 12.1 Induction motor protection structure.

$$S_{B3} = \sqrt{3} V_B I_B \tag{12.1}$$

$$Z_B = \frac{V_B^2}{S_{B3}} \tag{12.2}$$

where V_B is the base voltage or the rated line-to-line voltage and I_B is the base current.

12.2.2 Symmetrical Component Transformation

The symmetrical component coordinate of an induction motor is derived from the transformer of apparent power in phase coordinates. Suppose that the three-phase volt-ampere of Equation (12.2) is in per unit value; therefore, Equation (12.3) defines that three-phase apparent power in the function of three-phase active and reactive power as follows:

$$S_{3\phi} = P_{3\phi} + jQ_{3\phi} \tag{12.3}$$

Equation (12.3) could be rewritten as a function of voltage and current:

$$S_{3\phi} = V_{abc}^T I_{abc}^* \tag{12.4}$$

where V_{abc}, I^*_{abc} are the voltage and the conjugate value of current in the three-phase coordinate. By using the linear combination of matrices in Equation (12.4), and if this equation is compared with Equation (12.3), we can obtain the following:

$$S_{3\phi} = P_{3\phi} + jQ_{3\phi} = V_a I_a^* + V_b I_b^* + V_c I_c^* \tag{12.5}$$

Through the symmetrical component transformation, the three-phase voltage of Equation (12.4) is presented in the canonical form of symmetrical component as [15,20]:

$$V_{abc} = AV_{012} \tag{12.6}$$

where V_{012} is the symmetrical component of the voltage, and A is a phase rotation operator matrix, which given by:

$$A = \begin{bmatrix} 1 & 1 & 1 \\ 1 & a^2 & a \\ 1 & a & a^2 \end{bmatrix}$$

with $a = -\dfrac{1}{2} + j\dfrac{\sqrt{3}}{2}$

Through Equations (12.4) and (12.6), the voltage and current matrix in three-phase scheme can be rewritten as:

$$V_{abc}^T = (AV_{012})^T = V_{012}^T A^T \tag{12.7}$$

$$I_{abc}^* = A^* I_{012}^* \tag{12.8}$$

By substituting Equations (12.7) and (12.8) in Equation (12.4), the apparent power on the three-phase system can be described as:

$$S_{3\phi} = V_{abc}^T I_{abc}^* = V_{012}^T A A^* I_{012}^* = 3V_{012}^T I_{012}^* \tag{12.9}$$

The per unit value of apparent three-phase can be determined by dividing Equation (12.4) by Equation (12.1) as follows:

$$S_{3\phi \text{pu}} = \frac{S_{3\phi}}{S_{B3}} \tag{12.10}$$

Equation (12.10) can be simplified in the function of the symmetrical component of voltage and current given in Equation (12.9) as:

$$S_{3\phi \text{pu}} = \frac{3\left(V_{a0} I_{a0}^* + V_{a1} I_{a1}^* + V_{a2} I_{a2}^*\right)}{\sqrt{3} V_B I_B} \tag{12.11}$$

where V_{a0}, I_{a0}, V_{a1}, V_{a1}, V_{a2}, and I_{a2} are, respectively, voltage and current from zero, negative and positive sequences of phase a.

It is important to note that:

$$V_B = \sqrt{3} V_{B\phi} \tag{12.12}$$

By substituting Equation (12.12) into Equation (12.11) the per unit three-phase apparent power can be rewritten as:

$$S_{3\phi \text{pu}} = \frac{3\left(V_{a0} I_{a0}^* + V_{a1} I_{a1}^* + V_{a2} I_{a2}^*\right)}{3 V_{B\phi} I_B} \tag{12.13}$$

Considering the per unit values of the voltage and current in symmetrical component coordinates, the volt-ampere per unit of Equation (12.13) is:

$$S_{3\phi pu} = V_{a0pu} I_{a0pu}^{*} + V_{a1pu} I_{a1pu}^{*} + V_{a2pu} I_{a2pu}^{*} \tag{12.14}$$

It has been observed from Equation (12.14) that the representation of the apparent power of a three-phase induction motor does not allow the zero-sequence current to flow. This is caused by the different motor connections, which are usually wye with an ungrounded common point of the system configuration. Equation (12.14) is simplified by

$$S_{3\phi pu} = V_{a1pu} I_{a1pu}^{*} + V_{a2pu} I_{a2pu}^{*} \tag{12.15}$$

Equations (12.3) and (12.15) determine the per unit active power of three-phase induction motor in symmetrical component coordinates as:

$$P_{3\phi pu} = \Re (V_{a1pu} I_{a1pu}^{*} + V_{a2pu} I_{a2pu}^{*}) \tag{12.16}$$

where $\Re$ represents the real component of a complex number. This means that the protection of an induction motor is mostly relative to the real component.

12.2.3 Mechanical Torque

The most common cause of problems for the protective system of an induction motor with a load of the high moment of inertia is in the starting point. During the starting process, the whole system is affected by the inertia of driven load. Newton's law given in Equation (12.17) is considered the principal equation for acceleration of the motor, which is usually named the swing equation. This relation describes the dynamic model that can assist with analysis of the motor performance during the acceleration of shaft load:

$$J \frac{d\omega}{dt} = T_a \tag{12.17}$$

where J is the moment of inertia [Kg·m^2], ω is the shaft angular velocity [rad/s], t is the time [s], and T_a is the net accelerating torque [N·m].

Studying mechanical torque of an induction motor derives from the normalisation of the swing equation given in Equation (12.17), as well as the power system equation developed in Section 12.2.2. Therefore, relative to the base system developed in Equations (12.1) and (12.2), normalising the swing equation could be started by constraining the system frequency using the time as defined in Equation (12.17). Time is the inverse of the shaft angular velocity, and it can be written in the base model of the function of frequency as:

$$t_B = \frac{1}{2\pi f_B} \tag{12.18}$$

Note that, for a three-phase induction motor, the fundamental relationship between the system electrical power and the three-phase torque is defined as:

$$P_{3\phi} = T_{3\phi}\omega \tag{12.19}$$

Suppose that the three-phase power defined in Equation (12.19) is equal to the base apparent power defined in Equation (12.1); therefore the following relation in a base model can also describe Equation (12.19) as:

$$S_{B3} = T_{B3}\omega_B \tag{12.20}$$

From Equation (12.20), the three-phase base torque of the system can be written as follows:

$$T_{B3} = \frac{S_{B3}}{\omega_B} \tag{12.21}$$

Therefore, the swing equation given in Equation (12.17) can be written in base mode as:

$$J_B = T_{B3} t_B^2 = \frac{T_{B3}}{\omega_B^2} = \frac{S_{B3}}{\omega_B^3} \tag{12.22}$$

It becomes simple to normalise the swing equation. This is made possible by defining the normalised inertia constant, which is a function of the energy of the rotating mass when operating at rated angular velocity [15].

The per unit net acceleration and the constant of inertia are considered as normalised quantities. These variables are detailed in [15]. The accelerating torque is the function of the net torques that act to accelerate the shaft of the motor. Therefore, the per unit net accelerating torque can be described as:

$$T_{a\text{pu}} = T_{M\text{pu}} - T_{L\text{pu}} \tag{12.23}$$

where $T_{M\text{pu}}$ and $T_{L\text{pu}}$ are, respectively, the net developed mechanical torque of the motor and the net retarding load torque.

If the three-phase torque given in Equation (12.19) is taken in per unit, it can be equal to the per unit net developed mechanical torque. By substituting Equation (12.16) in that relation, we can write the mechanical torque in per unit as:

$$T_{M\text{pu}} = T_{3\phi\text{pu}} = \frac{P_{3\phi\text{pu}}}{\omega_{\text{pu}}} = \frac{1}{\omega_{\text{pu}}} \Re(V_{a1\text{pu}} I_{a1\text{pu}}^* + V_{a1\text{pu}} I_{a2\text{pu}}^*) \tag{12.24}$$

which is also equivalent to

$$T_{M\text{pu}} = T_{M1} + T_{M2} \tag{12.25}$$

with T_{M1} and T_{M2}, respectively, positive and negative mechanical torques.

Note that, during the balanced system voltage supply, the negative-sequence torque is roughly equal to zero. However, when the system voltage is unbalanced, the total torque that accelerates the shaft load is reduced. This reduction is caused by negative-sequence torque on the motor.

12.3 Equivalent Circuit of Induction Motor

Figure 12.2 represents the positive and negative-sequence equivalent circuit model of an induction motor. These networks are used to analyse the mechanical torque of each sequence. It is observed that the Thevenin equivalent resistance and reactance parameters for both positive and negative sequences are equal. These impedances are represented by R_{th} and X_{th}, and these parameters derive directly from the power supply of the motor, which depends, respectively, on the system input voltage V_{a1} and V_{a2} (positive and negative-sequence networks). As described in Figure 12.2, each network has its own motor terminal voltage, which is represented by V_{m1} and V_{m2}.

The stator and the mutual components resistance and reactance for both positive and negative-circuit are the same. These parameters are represented, respectively, by R_s, X_s, R_m, and X_m. However, in the rotor component, the resistance and reactance parameters are different for each network circuit.

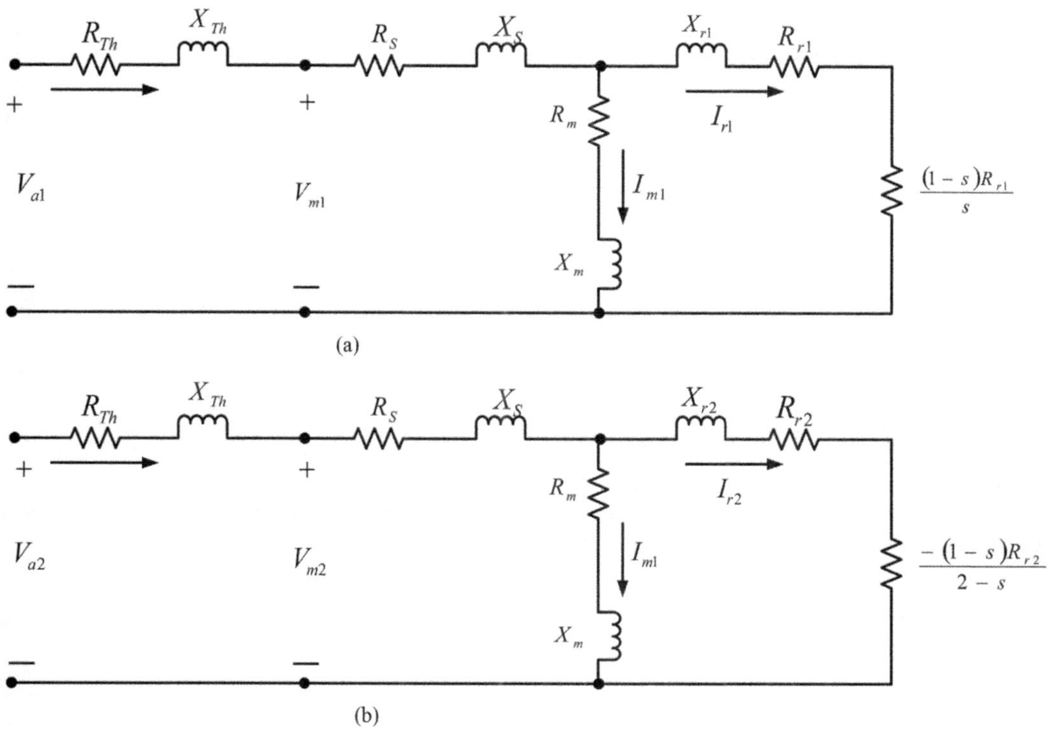

FIGURE 12.2 Induction motor Steinmetz's equivalent circuits model: (a) positive sequence and (b) negative sequence.

Using Kirchhoff's laws, the total rotor resistance for both positive and negative networks can be represented by the relations as:

$$\frac{R_{r1}}{s} = R_{r1} + \frac{(1-s)R_{r1}}{s} \tag{12.26}$$

$$\frac{R_{r2}}{2-s} = R_{r2} - \frac{(1-s)R_{r2}}{2-s} \tag{12.27}$$

where s represents the motor slip. Equations (12.26) and (12.27) show that the total rotor resistance is the function of the two parameters. The first parameter is the rotor power loss I^2R that is a function of R_{r1} for positive sequence and R_{r2} for the negative sequence. The second parameter represents the power transferred across the air gap that is available to drive the shaft load. The slip of positive network in Equation (12.26) is changed to negative network of Equation (12.27) from s to $2-s$.

12.3.1 Positive-Sequence Torque Analysis

The positive sequence of a three-phase induction motor as represented in Figure 12.2a can be summarised in Figure 12.3. The circuit consists of analysing the positive circuit impedance of the positive network rotor voltage.

Assuming all circuit parameters of Figure 12.3 are in per unit, we could write:

$$Z_{t1} = R_{t1} + jX_{t1} \tag{12.28}$$

where R_{t1} and X_{t1} are, respectively, the total positive sequence resistance and reactance of the induction motor. By using Equation (12.28) and Figure 12.3, the total positive sequence current can be determined. Therefore, the per unit core losses, stator copper losses, the positive-sequence air-gap power, and the

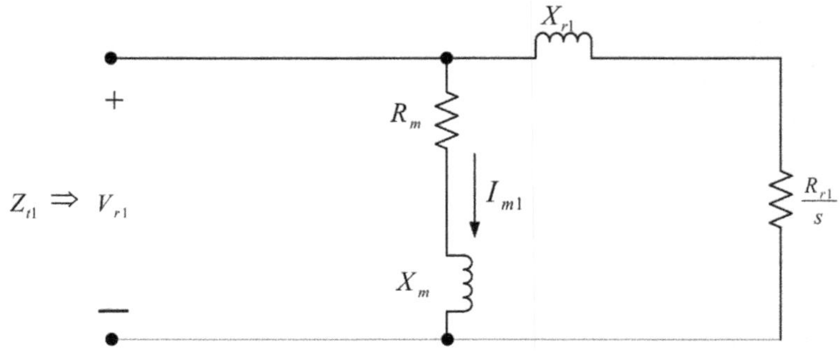

FIGURE 12.3 Positive circuit equivalent (impedance analysis).

rotor copper losses can be determined. It is important to note that per unit is used to determine these results. However, if the system is not in per unit value, the factor of 3 must be multiplied to these results. Equations (12.29) and (12.30) define the system losses on the stator and core as:

$$P_{sc1} = R_s I_{a1}^2 \tag{12.29}$$

$$P_{cl1} = R_m I_{m1}^2 \tag{12.30}$$

where I_{a1} is the total positive sequence input current, I_{m1} is considered as the positive sequence excitation current, P_{sc1} is the positive-sequence stator copper losses, and P_{cl1} is the positive-sequence core losses. Therefore, the air-gap power can be described as:

$$P_{ag1} = P_1 - P_{sc1} - P_{cl1} = \frac{R_{r1}}{s} I_{r1}^2 \tag{12.31}$$

Moreover, the rotor copper losses are defined to be the air-gap power divided by the slip, which is given by:

$$P_{rc1} = R_{r1} I_{r1}^2 \tag{12.32}$$

During the operation of an induction motor, there are system friction, windage and stay losses which are the function of the rotor angular velocity [15]. These losses are considered as the rotor core losses or rotational losses, and are represented in per unit by the relation below:

$$P_{\text{rot}} = P_{\text{con}}\omega = P_{\text{con}}(1-s) \tag{12.33}$$

where P_{con} is a constant that is estimated to be in the range of 0.05–0.08 per unit of the percentage of the motor rating. The power air-gap of the induction motor is also a function of the rotor core losses developed in Equation (12.32) and the positive sequence developed torque. Moreover, the positive sequence developed torque is described by:

$$P_{d1} = P_{ag1} - P_{rc1} = R_{r1} I_{r1}^2 \frac{1-s}{s} = P_{ag1}(1-s) \tag{12.34}$$

From Equations (12.29) and (12.34), we can determine the shaft torque output power as follows:

$$P_{s1} = P_{d1} - P_{rot} = (P_{ag1} - P_{\text{con}})(1-s) \tag{12.35}$$

The shaft mechanical torque is computed as a function of the developed power on the shaft. The computation gives the desired results that define the solution of positive-sequence torque term in the startup equation. Therefore, this torque can be rewritten as:

$$T_{m1} = \frac{P_{s1}}{\omega} = \frac{P_{ag1} - P_{con}}{\omega_s} = R_{r1} I_{r1}^2 \frac{1-s}{s} - P_{con} \tag{12.36}$$

where ω_s is the synchronous angular velocity in per unit, which is equal to 1. The shaft torque is mostly a function of the developed power because, in most cases, the rational power losses are not considered.

12.3.2 Negative-Sequence Torque Analysis

Figure 12.4 represents the negative-sequence network of an induction motor. The system analysis is the same as the computation methodology developed for the positive sequence. However, it is important to note the difference of the function of slip in Figure 12.4. Equation (12.37) describes the function of the total impedance of negative-sequence as:

$$Z_2 = Z_{th} + Z_s + Z_{t2} \tag{12.37}$$

Using the same procedure described in Section 12.3.1, the negative-sequence stator copper losses are determined as:

$$P_{sc2} = R_s I_{a2}^2 \tag{12.38}$$

Equation (12.39) describes the negative-sequence core losses as:

$$P_{cl2} = R_m I_{m2}^2 \tag{12.39}$$

From Equations (12.37) to (12.39), the negative-sequence air-gap power can be computed as:

$$P_{ag2} = P_2 - P_{sc2} - P_{cl2} = \frac{R_{r2}}{2-s} I_{r2}^2 \tag{12.40}$$

Moreover, the negative-sequence rotor copper losses are given by:

$$P_{rc2} = R_{r2} I_{r2}^2 \tag{12.41}$$

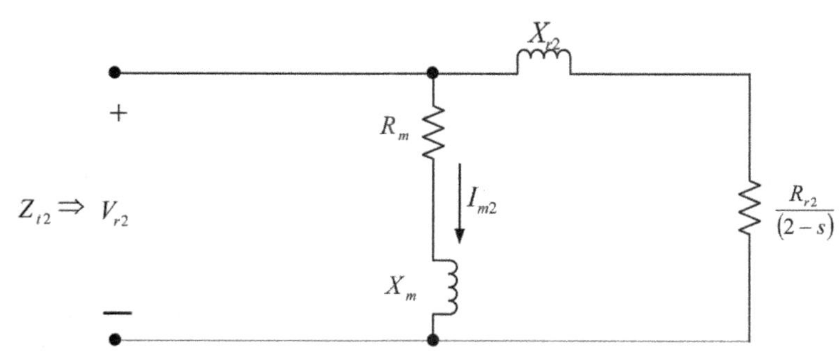

FIGURE 12.4 Negative circuit equivalent (impedance analysis).

The shaft torque for negative-sequence can be determined as described for positive-sequence network. The negative-sequence developed mechanical power can be determined as follows:

$$P_{d2} = -\frac{1-s}{2-s} R_{r1} I_{r2}^2 \tag{12.42}$$

The negative sequence mechanical torque is determined by:

$$T_{m2} = \frac{P_{d2}}{\omega} = -\frac{R_{r2}}{2-s} I_{r2}^2 \tag{12.43}$$

The negative-sequence mechanical torque is negative because the developed power is always negative.

12.4 Overload/Thermal Protection

The overload and thermal protection of a motor is designed by the principle of heat transfer of thermodynamics. The physical fundamentals of heat transfer, including a solid background in the induction motor thermal protections, have been evaluated in several textbooks [14–19]. An induction motor can contain all heating transfer phenomena either during startup or at steady state. This heat can be transferred via conduction, convection, and radiation.

12.4.1 Thermal Modelling for Drives

Figure 12.5 gives the simplified electric analog of the motor thermal system for rotor and stator. This is considered the third-order thermal mode for the drive. In this reduced model, the rotor temperature is completely decoupled from the stator winding and stator core temperature. The system of the simplified thermal model contains three components of the temperature, which are stator core temperature (θ_c), stator winding temperature (θ_s), and rotor temperature (θ_r). Studying the thermal protection of the induction motor consists of varying the temperatures of the system as a function of time [15,18]. Based on the reduced model of Figure 12.5, the thermal model of the motor can be written as follows:

$$\frac{d\theta_s}{dt} = a_s P_{sL} + b_{sc}\theta_c \tag{12.44}$$

$$\frac{d\theta_c}{dt} = a_c P_{cL} - b_c\theta_c + b_{cs}\theta_s \tag{12.45}$$

$$\frac{d\theta_r}{dt} = a_r P_{rL} - b_r\theta_r \tag{12.46}$$

where P_{sL}, P_{cL}, and P_{rL} are the power losses on the stator, core and rotor; q_s, q_m, and q_r are the heat flows on the stator, core, and the rotor, with $a_s = 1/C_s$, $b_{sc} = 1/R_{sc}C_c$, $a_c = 1/C_c$, $b_c = [(1/R_{cA}C_c) + (1/R_{sc}C_c)]$, $b = 1/R_{sc}C_c$, $a_r = 1/C_r$, and $b_r = 1/R_{sc}C_c$, where C_s, C_c, and C_r are the thermal capacitances on the stator, core and rotor; R_{sc}, R_{ca}, and R_{ra} are the thermal resistances on stator/core, core/air-gap, rotor/air-gap.

It is important to note that the analog of a thermal system for induction motors that is derived from the electrical system, as described in Figure 12.5, is related as follows: the generation of the heat flow is equivalent to the current of the system that introduces losses on the machine. As it is supposed to be the

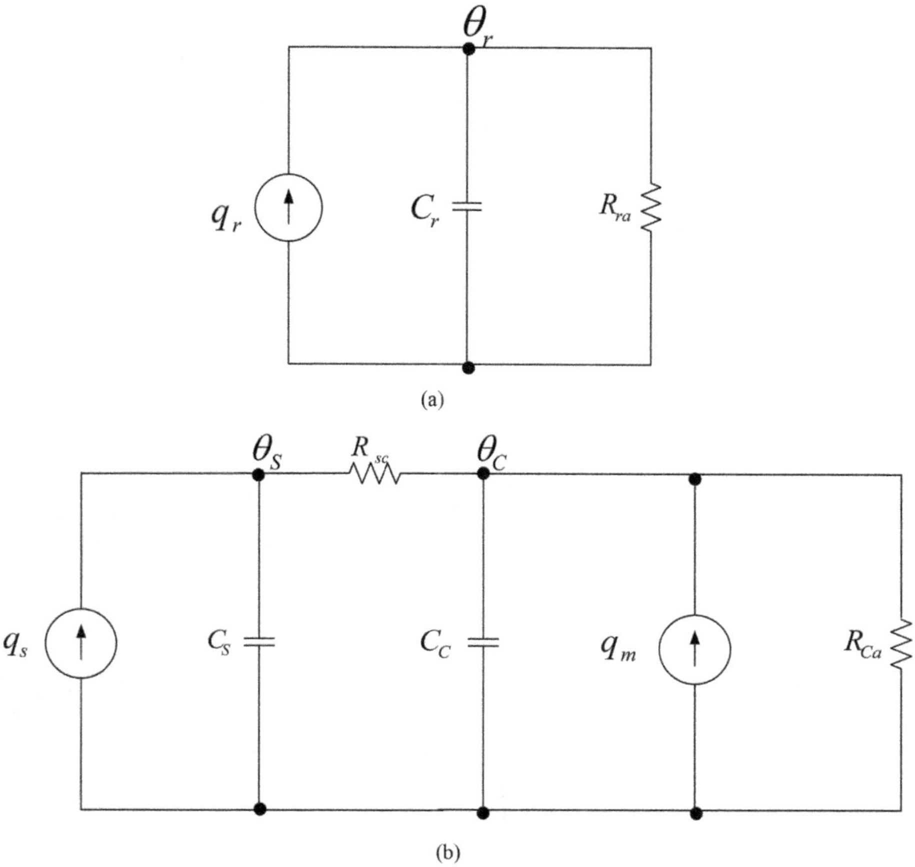

FIGURE 12.5 Simplified electric analog of the thermal system for an induction motor: (a) rotor and (b) stator.

electrical system of current generators, the system voltages are related to the temperature of the three nodes, as described in Figure 12.5. Therefore, we can write the equivalence between the power losses and the heat flow on the system as:

$$P_{1L} = I_s^2 R_s = q_s \tag{12.47}$$

$$P_{0L} = I_M^2 R_M = q_c \tag{12.48}$$

$$P_{2L} = I_r^2 R_r = q_r \tag{12.49}$$

It is worth noting that the temperatures of each component vary as a function of the heat flow, temperature and time.

12.4.2 Thermal System Parameters

Resolving the differential equations of the thermal system of an induction motor requires the determination of the system parameters, which are the thermal capacitance and resistances. Determining the

TABLE 12.1

Thermal Conductivity (k) and Specific Heat (c_p) of Common Materials

Material	Thermal Conductivity (W/Mk)	Specific Heat at 100 K (J/kg K)
Aluminium	204	481
Asbestos	0.166	
Carbon steel (0.5%)	54	
Carbon steel (1.5%)	36	
Copper		254
Copper at 20°C	386	
Gold		108
Iron		216

thermal capacitance of any component derives from the internal energy of the system ($dU = mc_p d\theta$), where the variation of the internal energy by time is called the heat flow. It is described as:

$$q = mc_p \frac{d\theta}{dt} \tag{12.50}$$

where m and c_{cp} are, respectively, the mass (kg) and the specific heat (J/kg·K). The specific heat of common materials at 100 K is given in Table 12.1. From Equation (12.50), and using the equivalence of the electrical system where the current is equal to the capacitance multiplied by the voltage; the thermal capacitance can be written as follows:

$$C = mc_p \tag{12.51}$$

Consider a cylinder that represents the end cross-sectional view of an induction motor. If we take the one-pole pitch of an induction motor, the basic heat conduction for a cylindrical geometry that is provided by the thick wall pipe equation can be written as:

$$q = \frac{2\pi kd}{\ln(r_o/r_i)}(\theta_i - \theta_o) \tag{12.52}$$

where k, d, r_i, r_o, θ_i, and θ_o are, respectively, the thermal conductivity, the length of pipe, the inside pipe radius, the outside pipe radius, the inside and the outside pipe temperatures. The thermal conductivity depends on the types of materials. Table 12.1 describes the thermal conductivity of common materials [15]. Equation (12.52) represents the heat flow of the only sector of the induction motor. Considering that each slot pitch of the motor is equivalent to one section, a constant n can be defined as the number of sectors or slots on the machine stator. By using the same analog of thermal variables to electrical variables, the thermal resistance of the nth sector can be written as:

$$R = n\frac{\ln(r_o/r_i)}{2\pi kd} \tag{12.53}$$

It is important to note that determining the thermal capacitance and the resistance of the core, stator and rotor must be computed following the relations given in Equations (12.51) and (12.53). Moreover, this determination must be made in the manner of each specific component.

Using the relations of variables that described the reduced model of the induction motor, we can write the thermal model given by Equations (12.44) through (12.46). The different coefficient parameters of this model are determined by computing thermal capacitance and resistance.

12.4.3 Thermal Energy Limit

The design approach of an induction motor is modelled with constraints on the thermal energy of the system, which are limited only on the power losses due to the motor windings and induced current in the iron parts. We know that the heat flow is equivalent to the power losses of the motor, and by considering the heat transfer which is affected by the thermal capacitance, Equation (12.50) for a three-phase electrical induction motor can be written as:

$$q = C\frac{d\theta}{dt} = 3I^2R \tag{12.54}$$

Resolving the temperature of Equation (12.54) requires the integration function of time varying between 0 and a given time *t*. This differential equation allows the determination of the thermal energy limit of the induction motor, which determines the lifetime of the machine. This model of thermal energy limit of an induction motor is widely investigated in [15].

The lifetime of the windings and the motor depends on the operating temperature of the machine. As shown in Figure 12.6 [19], the insulation resistance of the motor depends on the temperature. When temperature rises, the insulation resistance decreases. An increase of only 5% in the stator current leads to a temperature rise of approximately +10°, which halves the lifetime of the windings. Figure 12.7 describes the lifetime of the motor as the function of nominal current and temperature [19].

12.4.4 Thermal Relay Model

The rotor heating is the most critical component of the induction motor. Studying a thermal model rotor in the heat transfer modelling can resolve the heating problem of the motor. This approach aims to realise the thermal relay that could protect the motor during the surge starting current and high startup inertia

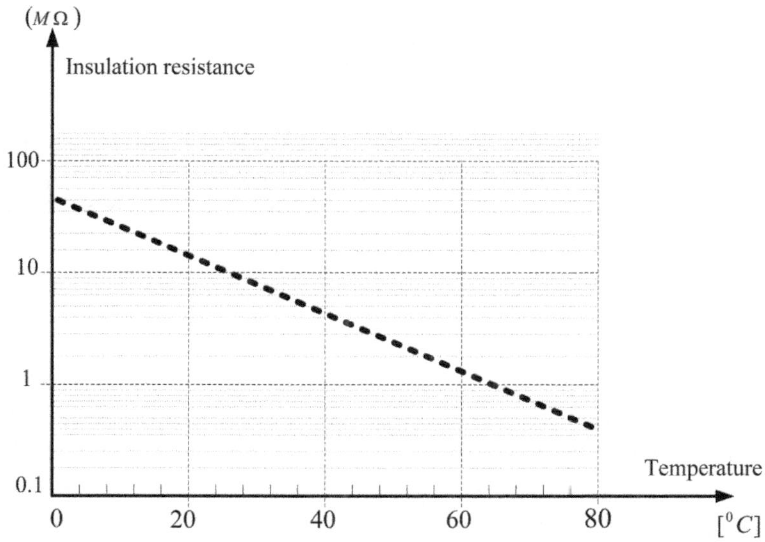

FIGURE 12.6 Insulation resistance temperature.

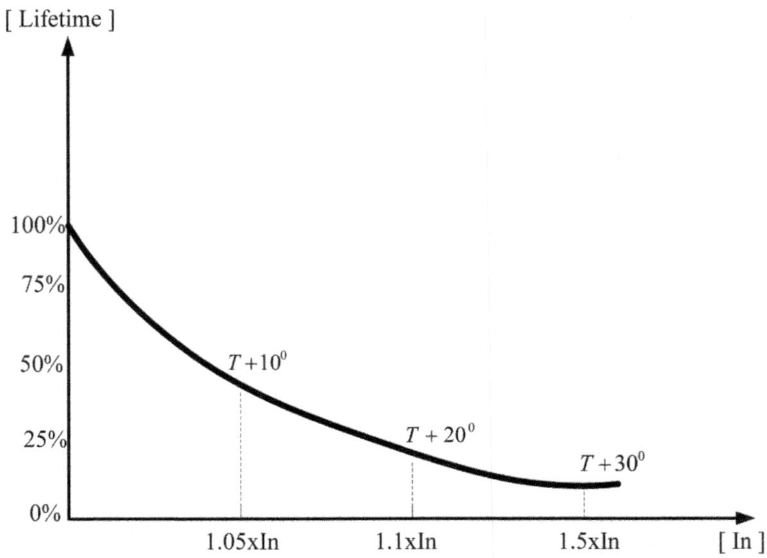

FIGURE 12.7 Lifetime of motor depending on operation.

loads [14–16]. Considering the heat flow model developed by Equation (12.54) in the input of the rotor component, this relation can be written, using Watts as the unit, as:

$$q_{in} = 3I_r^2 R_r = C_r \frac{d\theta_r}{dt} \tag{12.55}$$

where I_r, R_r, C_r, and θ_r are, respectively, rotor current (A), rotor electrical resistance (Ω), rotor thermal capacitance (J/K), and the rotor temperature (K).

By normalising Equation (12.55) using Equation (12.1), we can consider the following per unit equality between the electrical parameters and the thermal parameters as:

$$R_r I_r^2 = \frac{C_r}{\omega_B} \frac{d\theta_r}{dt} = H_c \frac{d\theta_r}{dt} \tag{12.56}$$

where ω_B is the base radian frequency (rad/s) and t is the time in seconds. In Equation (12.56), we have defined a new parameter H_c, which is called the thermal inertia constant (s) due to the time variable of the system in Equation (12.56), in seconds. It is necessary to notice that this parameter is analogous to the mechanical inertia constant in Equation (12.23). This inertia constant in a thermal manner is computed as:

$$H_c = \frac{C_r}{\omega_B} \tag{12.57}$$

We can compute the rotor temperature of Equation (12.56) as a function of the thermal inertia constant and the electrical parameters as follows:

$$\theta_r = \frac{1}{H_c} \int_0^t R_r I_r^2 dt \tag{12.58}$$

Assuming the rotor current reaches a constant value which is determined at locked-rotor, we can therefore find a limiting value of temperature. This is determined at a limiting time t, as defined in Equation (12.58). With a constant rotor current, the rotor temperature will be computed in Equation (12.59), as follows:

$$\theta_r = \frac{R_r I_r^2 t}{H_c} \tag{12.59}$$

When the time is considered at its limited value, and the constant rotor current is at locked-rotor, this will cause the rotor to reach its limiting temperature. Note that, when the induction motor allows the locked-rotor current to flow more than the limiting time, the rotor will overheat. Therefore, protecting the thermal circuit of an induction motor consists of controlling the rotor-locked current and the limiting time [15]. Based on Equation (12.59), we can construct the thermal rotor model that is used in a commercial protective relay for a large motor, and we can write the limiting rotor temperature as:

$$\theta_{\text{Lim}} = \frac{R_r I_{Lr}^2 t_{\text{Lim}}}{H_c} \tag{12.60}$$

By replacing the thermal inertia constant described in Equation (12.57) into Equation (12.60), the rotor thermal capacitance can be computed as:

$$C_r = \frac{R_r I_{Lr}^2 t_{\text{pu}}}{\theta_{\text{Lim}}} \tag{12.61}$$

where $t_{\text{pu}} = t_{\text{Lim}}\omega_B$, and it is the per unit time.

Assuming that we make an analog below:

$$I_{Lr}^2 t_{\text{pu}} = \theta_{\text{Lim}} \tag{12.62}$$

$$C_r = R_r \tag{12.63}$$

the increment of temperature is the function of the thermal resistance and the heat flow can be written as follows:

$$\Delta\theta = \text{R}q = \theta_r - \theta_A \tag{12.64}$$

where R is the thermal resistance. By replacing the temperature found by Equation (12.62) in Equation (12.64), we can find in the function of the reference temperature at a time t_r and the ambient temperature at a time t_A as:

$$\text{R}q = I^2(t_r - t_A) \tag{12.65}$$

When the motor is starting at a certain operating time t_o, with the operating temperature greater than the ambient temperature, the time difference of Equation (12.65) will decrease. Then Equation (12.65) can be rewritten as follows:

$$\text{R}q = I^2(t_r - t_0) \tag{12.66}$$

Based on the modelling of the motor protective relays that were developed from Equations (12.55) to (12.64) and Figure 12.5a, a new representation of the rotor thermal model can be depicted in Figure 12.8. This thermal model is considered the Zocholl rotor thermal model, which is mostly used in commercial motor protective relays.

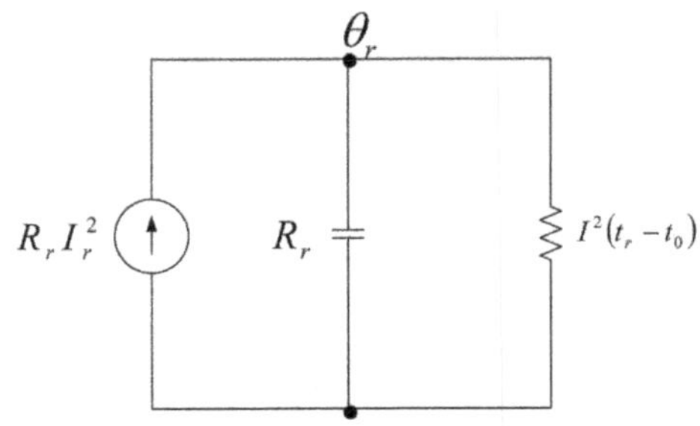

FIGURE 12.8 Simplified rotor thermal model for the thermal relay realisation.

The thermal model for a given temperature at any instant can be expressed as:

$$\theta = \theta_{\max}(1 - e^{-t/\tau}) \tag{12.67}$$

where τ is the heating time constant, and $\theta_{\max}$ is the final steady state temperature. As the temperature rise is proportional to the current, Equation (12.67) can be written in function of current as:

$$\theta = KI_{\max}(1 - e^{-t/\tau}) \tag{12.68}$$

where $I_{\max}$ is the current that flows continuously to produce the temperature $\theta_{\max}$ in the motor, and $K = R_{r2}/R_{r1}$, which is selected to be at rated speed. When this value is not provided, it can be set equal to 3. For a given overload current I, the permissible heating time for this current can be computed from Equation (12.68) as follows:

$$t = \tau \ln\left[\frac{1}{\{1 - (I_{\max}/I\}}\right] \tag{12.69}$$

Therefore, the influence rising heating from both positive and negative sequence currents can be taken into consideration as a current equation by:

$$I_{eq} = \sqrt{(I_1^2 + KI_2^2)} \tag{12.70}$$

The overload thermal model must consider the fact that the motor must cool down during the periods of light load and the initial state of the motor. Thus, the rate of cooling, τ_r, is defined as a constant cooling parameter of the motor. The final thermal model is given by:

$$t = \tau \ln\left[\frac{k^2 - A_1^2}{k^2 - 1}\right] \tag{12.71}$$

where τ is the heating time constant; A_1^2 is the initial state of the motor (cold or hot); and $k = I_{eq}/I_{th}$, with I_{th} is the thermal setting current.

12.5 Start/Stall Protections

During the starting period of an induction motor, there is often a significant value of current on the stator winding that is more than the full load rating current. The transient starting current can decrease the performance of the motor, which will lead to the augmentation of the rotor speed. The starting protection of the motor aims to ensure a certain constant limitation of the starting current. It is necessary to note

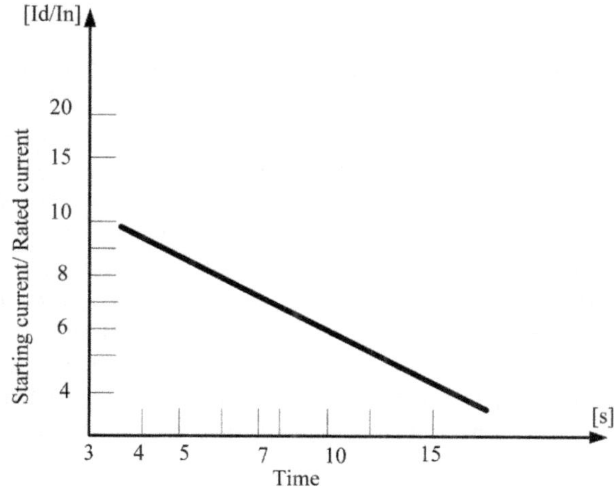

FIGURE 12.9 Starting time based on the ratio of starting current to rated current.

that the starting current of an induction motor depends on the type of design of the induction motor and the starting methods (direct-on-line, star-delta, etc.). For a direct-on-line method, the start current can vary between 4 and 8 times full-load current. A motor's starting phase is the duration required for it to reach its nominal rotating speed. Figure 12.9 describes the starting time as the function of the ratio of starting current and rated current [19].

The design setting of the motor overload and stall protective device depends on the starting current and starting time. The current that the motor will draw during the starting period and the time that it will take to accelerate its load to reach the rated speed is computed as [13]:

$$I_{st} = \frac{\text{kVA/hp} \times \text{motor hp} \times 1000}{1.73V} \tag{12.72}$$

$$t_{st} = \frac{WK^2 \text{r/min}}{308\,T(n)} \tag{12.73}$$

where I_{st} is the motor starting current (A), kVA/hp is the motor starting kilovolt-amperes per horsepower, motor hp is the motor nameplate horsepower rating, V is the motor rated voltage (V), t_{st} is the starting current, WK^2 is the combined moment of inertia of the motor and its load (lb/ft²), r/min is the motor-operating speed in revolutions per minute, and $T(n)$ is the motor torque (lb/ft).

During the starting or operating period of the motor, when the start current becomes larger than the locked-rotor current, the motor will fail to start or it will stall. This problem is considered one of the external hazard motor problems that is often caused by the excess load on the rotor [18]. Based on the current drawn on the system, it is quite difficult to determine the difference between a stall condition and a healthy start. Therefore, a different time concept of the current drawn must be discriminated between the starting and stalling conditions. These conditions are derived from the equations as follows:

$$t_{st} < t_{sw} \tag{12.74}$$

$$t_{st} > t_{sw} \tag{12.75}$$

with t_{st}, and t_{sw} are starting time and stall withstand time, respectively. In Equation (12.74), motor protection can be easily defined. However, in Equation (12.75), it is difficult to determine the motor protection due to the high inertia loads that motor drives. Based on these conditions, a locked rotor protection is introduced due to the excessive start time to decimate the protection philosophy.

12.5.1 Excessive Start Time and Locked Rotor Protection

There are several external hazards problems, such as loss of a supply phase, mechanical problems, low supply voltage, excessive load torque, etc. [15], that can cause a motor to fail to accelerate. These problems can cause overheating in the motor. Locked-rotor protection is one of the motor protection philosophies that can ensure a safe stall time of the induction motor. A given typical induction motor can safely hold locked rotor current for about 20 seconds. However, when an excessive start time occurs, a huge current will be drawn from the supply and that will cause the motor to generate an extremely high temperature. Due to the absence of the cooling system caused by not rotating, the motor winding can be damaged quickly. Protecting the motor against these problems requires a proper discrimination between Equations (12.74) and (12.75) because the excessive start time can be observed in both cases.

In the case of Equation (12.74), the protection is defined by the definite time overcurrent characteristic. This implies that the current setting is superior to full load current but inferior to the starting current. The protection time setting should be greater than the starting time but less than the acceptable safe starting time [15]. On the other hand, in Equation (12.75) the protection requires an important time delay that is greater than the maximum allowable time that the motor can safely carry the starting current. However, the defined characteristic cannot be used to determine the induction motor protection. The protection system requires a speed-sensing switch to detect the rotor movement that can indicate a safe start using a successful start detector to select the relay timer.

12.5.2 Starting Protection

The protection of the induction motor during starting is catered against excessive rotor heating due to unbalanced voltages. This excessive rotor heating can occur during accelerating and full speed operation. The starting protection of the motor also aims to disconnect the motor from the power supply and avoid damage on the machine, which will occur when the motor fails to turn at all or fails to reach full speed [13].

The starting protection of the induction motor is a function of motor start detection, the number of starts, start lockout and time. This protection function prevents motor temperatures from being exceeded and prevents damage to the motor. Any typical motor has a constraint on the number of starts. When a given motor reaches its permitted number of starts, the start protection should block the motor. A specific start protection must distinguish a cold start from a hot start due to the difference in temperatures between both starting types. This system protection must count them separately. It is necessary to notice that the number of cold starts is greater than the number of hot starts. Through a specific time delay, the number between consecutive starts of each type (cold or hot), the system protection can allow the motor to cool down between each start type. The motor protection assists the start time to inhibit only after the defined time by the motor specification.

12.5.3 Stall Protection

Stalling of any motor is caused by two different principal cases. First, in the starting process, the motor can stall due to mechanical failure, large load, low voltage, etc. Second, stalling can be observed during the running process of the motor due to low speed and variable load. The setting of the stall protection must therefore be close to the thermal overload and the limited locked rotor of the motor. This means that the device can limit the motor to operate before the thermal and locked rotors are exceeded [13].

During starting or the operation of the motor, if the current drawn becomes more than the starting current threshold that can be computed in Equation (12.73), this current is considered a stall current. The system protection must connect the motor if this stall current persists for a time greater than the setting of the stall timer. The stall protection can also ensure the operation and defeat the re-acceleration scheme of the motor during the surge current due to the restoration of the supply after transient voltage loss. It is important to note that the time delay setting is a function of the re-acceleration scheme and the characteristics of individual motors [15]. Through the study of the transient stability of the re-acceleration scheme proposed, the delay time of the relay can be determined.

12.6 Short-Circuit Protection

The short-circuit protection of a motor is set for two principal reasons: protecting the motor against the stator winding faults and protecting it from terminal flashovers. The most of motors are designed to have a significant amount of insulation between phase windings. This aims to avoid faults between two phases, which seldom occur. When the stator winding of the motor is connected to the grounded metal, during the short-circuit in the motor the fault would very quickly involve earth. Therefore, the short-circuit protection can instantaneously be ensured by earth fault protection. It is important to determine the time delay and a single definite overcurrent relay that can clear the short-circuit faults before damaging the motor. The time delay that prevents spurious operation due to current transformer spill currents is randomly set at 100 ms. The definite time overcurrent relay is set at approximately 125% of I_{st} [15]. Figure 12.10 presents the curves of the most magnetic circuit breakers for motor protection, which are current limiting devices and so contribute to coordination. The very short cutoff time can break the short-circuit current before it reaches its maximum amplitude. This strategy limits the thermal and electrodynamic effects and improves the protection of wiring and equipment [19].

The protection co-ordination during the short-circuit in the motor is made in a different strategy. The objective is to provide fault clearance by minimizing the damage due to that fault. When the motor is fed from a fused contactor, this system coordinates the protection that could involve the time delay. However, the circuit breakers can provide an effective protection on the motor. For the large motor that operates in high voltage supply fed via the circuit breaker, a differential unit protection can provide protection against the line-to-line faults and line-to-earth faults (power system resistance earthed). In this case, the fast-sensitive detection of the differential protection can play an important role to provide protection in the early stage of the faults that could minimize damage to the motor. Therefore, the sensitivity of the normal definite time overcurrent would be worthless, and the sensitive earth protection may not be provided to the motor.

12.7 Earth Fault Protection

Earth faults are considered to be electric faults in the stator windings. Earth faults can be caused by the breakdown in winding insulation [14–16]. The provision of the ground faults is important to avoid damage that can occur due to faults (line-to-neutral, line-to-ground and line-to-line) that could

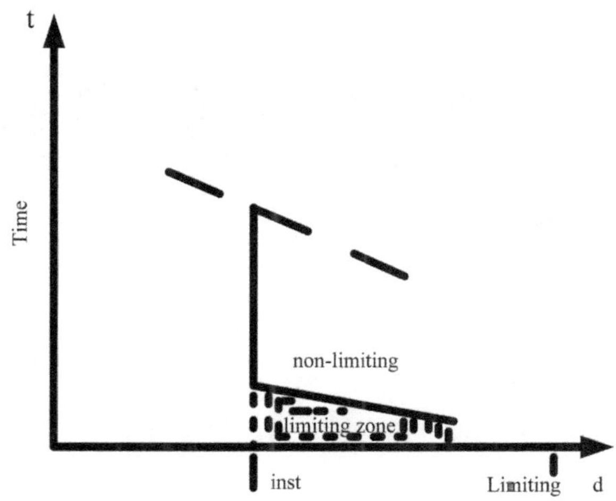

FIGURE 12.10 Curves of magnetic circuit breaker tripping.

affect the metal frame. Earth faults are easily detected using an instantaneous relay set at 20% of the motor full-load current transformers, which is connected to the neutral of the wye supply transformer [14,15]. There are several types and sensitivities of earth protections that are the function of the system earthing [14].

12.7.1 Solidly Earthed System

On a low voltage system, for reasons of personal safety, the electrical configuration mostly uses the solidly earthed system. It is a common type that depends on the required sensitivity.

When the acceptable sensitivity requires being more than 20% of the motor continuous rated current, the conventional configuration can use the residual current transformer connection for earth fault protection. The configuration setting needs to be imposed at a lower limit. The configuration of the residual current transformer connection for earth fault protection must ensure that the unequal current transformer saturation during the motor starting does not affect the relay operation. Therefore, a stabilising resistor in the series with the relay is used. The stabilising resistor aims to increase the setting system of the relay and provide an adequate delay in the protection system. This can also be avoided by using a definite time delay [14]. However, the stabilising resistor provides an instantaneous tripping characteristic of the protective scheme, and it is computed as follows:

$$R_{\text{stab}} = \frac{I_{st}}{I_0}(R_{ct} + k_1 R_l + R_r) \tag{12.76}$$

where R_{stab} is the stabilising resistor value (Ω); I_{st} is the starting current referred to the current transformer secondary (A); I_0 is the relay earth fault setting (A); R_d is the direct current resistance of the current transformer secondary (Ω); R_l is the current transformer single lead resistance (Ω); and k_1 is the current transformer factor, which is equal 1 for star point at current and 2 for star point at relay. Figure 12.11 shows the structure of motor protection with the stabilising resistor in ground relay [16].

The grading of the relay with the fused contactor is also considered one of the common coordination protective systems. In this structure, the power supply of the motor is connected to the fuse contactor. As the contactor cannot clear the faults, the fuse plays an important role in ensuring the safe operation of the motor. The protection coordination is made between the fuse and relay. The relay via the contactor

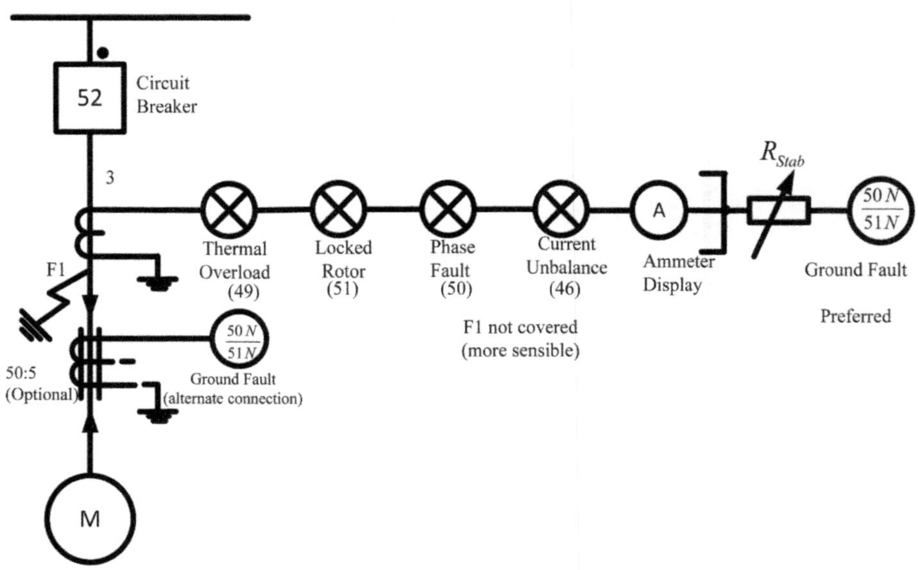

FIGURE 12.11 Induction motor earth fault protection structure.

cannot trip the system until the fuse has time to operate. Therefore, this configuration can operate efficiently by taking into consideration an intentional definite time delay in the relay.

The application of the core-balance current transformer is necessary when the system requires the more sensitive relay. This constitutes a ring current transformer where all four-wire systems (all phases and neutral) from the supply to the motor are passed. This transformer has a single core, which reduces the magnetising current requirement so that it can enable the low setting on the relay.

12.7.2 Resistance Earthed System

High-voltage configuration systems use this system. The objective is to constrain the damage that can occur during the earth faults by limiting the flowing of the earth current on the electrical system. There are two common methods of resistance protection system: low and high resistance earthing.

12.7.2.1 Low Resistance Earthing

The configuration limits the faults current between 200 and 400 A, and the resistance must be designed to handle this current. When the low resistance earthing system uses the residual current transformer connection, due to the possibility of current transformer saturation during the starting process, the system can accept a minimum sensitivity of 10% of CT rated primary current. On the other hand, for the core-balance current transformer, the sensitivity is constrained to be three times the steady-state charging current of the feeder. This system requires using a simple nondirectional earth fault relay with a setting no greater than 30% of the minimum expected earth fault current [14]. It is important to note that a considerable setting and time delay can be made for solidly earthed systems when the sensitivity does not meet the low resistance earthing requirement.

12.7.2.2 High Resistance Earthing

In this method, the protection scheme is used for high voltage and can limit the earth fault current to a few amps. High resistance earthing uses the sensitive direction earth fault relay when the system capacitive charging current is applied. High resistance earthing requires the utilisation of core balance voltage transformer instead of the current transformer with a relay characteristic angle setting of +45°. This will measure the residual voltage of the system. The protection engineer must ensure that the relay characteristic meets all operation and protection requirements. It is important to note that this system uses uncritical time delay that is fast enough to disconnect equipment in the event of the second fault immediately [14]. This disconnection removes the current limiting resistance that leads to the large fault current on the system. However, the first earth fault causes minimal damage to the system.

12.8 Negative Phase Sequence Protection

The negative sequence current flows on the three-phase induction motor when it is supplied by unbalanced voltage conditions. This fault can be caused by unbalanced loading, loss of a single-phase, and a single-phase short circuit. The negative phase sequence fault is considered one of the external hazard faults that can affect the starting and operation process of an induction motor. If the earth fault protection is of the sensitive variety, negative sequence faults can be detected by earthing protection.

As the negative sequence faults occur when the system voltage is unbalanced, the motor equivalent circuit for both the positive and negative networks as described in Figure 12.2 can be used to analyse the impact of unbalanced voltage on the system [14,15]. When the magnetising impedance is negligible, as described in Figure 12.2, we can compute the positive and negative impedance as follows:

$$Z_1 = \sqrt{\left[R_s + \frac{R_r}{s}\right]^2 + \left[X_s + X_r\right]^2} \tag{12.77}$$

$$Z_2 = \sqrt{\left[R_s + \frac{R_r}{2-s}\right]^2 + \left[X_s + X_r\right]^2} \tag{12.78}$$

During the unbalanced period, the rotor can stall; therefore the slip, $s = 1$, which means that the two impedances of the system are equal. However, when the motor is operating, the positive and negative impedance described in Equations (12.77) and (12.78) are not the same. Suppose the assumption of the slip is maintained and the motor is operating in the balanced condition; the ratios can be written as:

$$\frac{I_{st}}{I_{\text{run}}} \cong \frac{\sqrt{\left[R_s + \frac{R_r}{s}\right]^2 + \left[X_s + X_r\right]^2}}{\sqrt{\left[R_s + R_r\right]^2 + \left[X_s + X_r\right]^2}} \tag{12.79}$$

where I_{run} is the running current of the motor. This assumption introduces a difference between the resistance terms of both networks as described in Equations (12.77) and (12.78). Moreover, the resistance terms could be negligible due to their smaller magnitude compared to the reactance terms, and the total impedance of the system during the running period for a given slip is affected by the reactance terms. Therefore, we can express this assumption as follows:

$$\frac{I_{st}}{I_{\text{run}}} \cong \frac{X_1}{X_2} \tag{12.80}$$

Equation (12.80) for a typical induction motor can be considered to be in the range of 5–8.

By considering the total positive-sequence and negative-sequence input current, as described in Figures 12.3 and 12.4, and taking into consideration the ratio developed in Equation (12.80), the ratio of negative-to-positive current can be written as:

$$\frac{I_{a2}}{I_{a1}} = \left(\frac{V_{a2}}{V_{a1}}\right)\left(\frac{Z_1}{Z_2}\right) = \left(\frac{V_{a2}}{V_{a1}}\right)\left(\frac{I_{st}}{I_{\text{run}}}\right) \tag{12.81}$$

If the given motor is running with a ratio of stating and running current of 7, and the negative to positive sequence applied voltage is 0.04, i.e., 4% negative sequence, we can compute the ratio of negative-to-positive currents by using Equations (12.80) and (12.81) which will be 28%. Due the influence of summing positive and negative sequence current, the total torque of the induction motor is affected. We can write the mechanical torque as:

$$T_m = T_{m1} + T_{m2} = \frac{R_r}{\omega_s}\left[\frac{I_{r1}^2}{s} - \frac{I_{r2}^2}{2-s}\right] \tag{12.82}$$

From Equation (12.82) we can see that the negative-sequence current produces a little torque. Using Equation (12.81) it could also be observed that during the secondary unbalance, the negative sequence current is also small. However, the negative sequence current is at twice the supply frequency [15]. Through the influence of the skin effect on the rotor, the heating impact on the rotor at any given negative sequence current is larger than the same positive current. This can demonstrate that negative sequence current leads to rapid heating of the motor. Large motors, where the resistance is important, are susceptible to the impact of the negative sequence current. This is one of the important aspects of protecting the motor against negative sequence faults.

12.9 Protection of Rotor Windings

The hazardous external faults (unbalanced voltage) that cause the rotor current to increase about twice the supply frequency can also cause rapid heating of the rotor as well as serious damage to the motor. Therefore, rotor protection is a necessity to avoid some unexpected damage. It is important to note that these faults occur, an important frequency that involves a high a resistance compared to that derived from positive-sequence currents arises. It is well known that unbalanced voltage leads to negative-sequence faults; thus the heating impact that comes from one per unit positive-sequence current is less than that caused by one per unit negative-sequence current.

We can determine the total heating of the motor as a function of power loss, as described in Equations (12.47) through (12.49). By considering the assumption made in negative-sequence protection, where the magnetising impedance is negligible, the three-phase power losses may be written as:

$$P_{\text{loss}} = P_{\text{loss1}} + P_{\text{loss2}} \tag{12.83}$$

Considering the stator and rotor power of each network, as described in Figure 12.2, Equation (12.83) can also be written as follows:

$$P_{\text{loss}} = 3(I_{a1s}^2 R_{1s} + I_{a1r}^2 R_{1r}) + 3(I_{a2s}^2 R_{2s} + I_{a2r}^2 R_{2r}) \tag{12.84}$$

From the model developed in Figure 12.2, note that the stator resistance for negative- and positive-sequence current is the same; this can be simplified by writing the relation below:

$$R_{1s} = R_{2s} = R_s \tag{12.85}$$

On the other hand, the rotor resistances are totally different for both networks. It can be assumed that the rotor resistances for negative- and positive-sequence currents are linked in this relation as follows:

$$R_{2r} = gR_{1r} \tag{12.86}$$

where g is a positive parameter whose value must be more that one.

Therefore, the total power losses on the motor, when Equations (12.85) and (12.86) are introduced in Equation (12.84), becomes:

$$P_{\text{loss}} = 3R_s(I_{a1s}^2 + I_{a2s}^2) + 3R_{1r}(I_{a2s}^2 + gI_{a2r}^2) \tag{12.87}$$

It is not easy to determine the rotor resistance, and it is demonstrated in Equation (12.87) that there is a significant difference in rotor resistance for both networks. Therefore, studying the rotor heating system can be estimated by measuring the stator sequence currents. This approach is made for some rotor protection systems. In most cases, protection of the motor against the rotor heating consists of using the stator currents or temperatures to provide protection for the rotor. This protection scheme aims to avoid disconnection due to brief unbalances and the excessive rotor heating.

However, for wound rotor designs, this stator protection scheme described here may not be adequate [15]. Usually, any stator differential protection cannot detect the faults that occur in the rotor winding. An instantaneous stator current overcurrent relay device gives some degree of protection against the rotor winding faults. When this protection scheme is applied, if a slight time delay is about 30 ms, the instantaneous unit will be limited to three times the full load. Note that this approach considers the starting current constraint, which is set by resistance to a maximum of twice the full load.

12.10 Undervoltage/Overvoltage Protection, and Loss-of-Load Protection

12.10.1 Undervoltage/Overvoltage Protection

During the low voltage condition when the motor drives a shaft load, as for the normal voltage condition, the motor will slow down, which will draw more current from the supply. When the voltage dip occurs, that fault will also increase the slip of motor compared to driving the same load torque at normal voltage supply. When the low voltage supply becomes a large drop, the motor may stall. Therefore, protection of the motor during this transient event should be taken into consideration. Some protection schemes use a low voltage alarm of a specific dip value where the protection may trip the unit. For motors not in service, the protection systems require an undervoltage protection. When the power of a motor is less than 1500 hp, the undervoltage protection may be applied on at least one phase. However, for the motor with power more than 1500 hp, the undervoltage protection may be applied in all three phases [15].

The voltage and time delay settings are the main parameters of undervoltage protection that depend on the electrical system and motor. These parameters must be set carefully to avoid spurious trips due to all voltage drops during transient faults, starting of motors, etc. Therefore, the voltage must be able to handle the voltage drop below 80% during starting of a machine. While the motor can re-accelerate after a voltage dip that lasts from 0.5 to 2 seconds, the time delay must be set, based on motor system characteristics and taking into consideration this range of time. For many induction motor controllers, the undervoltage protection is considered a standard feature. Note that overcurrent and thermal relays can also provide some protection; however, undervoltage relays are preferable.

12.10.2 Overvoltage Protection of the Induction Motor

If a motor is connected by long-distance cables, its machine may often be exposed to overvoltage [21]. The overvoltage protection of an induction motor is the function of the nominal supply voltage of the power system and the voltage indicated on the motor nameplate. When a given motor is fed with a voltage 10% greater than the nominal operating voltage of the machine, this overvoltage will increase core losses associated with saturation of the motor. Therefore, the machine will be overheating. For an overvoltage of 10%, the motor will be overloaded by about 10%. Moreover, the same overvoltage percentage increases the core loss of 20%–30% greater than the nominal value, which leads to serious overheating of the machine. An overvoltage of 12% has a protection response of 8.3 ms [22]. Thus, protecting the motor against an overvoltage will consist of creating a protection philosophy that could be coordinated with the thermal and overload protection.

Table 12.2 describes the impact of unbalance in the voltage power supply of the same torque as a function of stator current, loss increase and heating on the machine [19].

12.10.3 Loss-of-Load Protection

The utility of loss-of-load protection can be found in several applications. This protection can protect a pump from becoming unprimed, stops the motor to avoid any mechanical transmission failure due to conveyor belts and/or other components and protects the synchronous motor against loss-of-supply condition.

TABLE 12.2

Effect of Voltage Unbalance on Motor Characteristics

Unbalanced Value (%)	Stator Current (A)	Loss Increase (%)	Heating (%)
0	I_n	0	100
2	$1.01I_n$	4	105
3.5	$1.04I_n$	12.5	114
5	$1.075I_n$	25	128

The protection scheme is implemented by a low forward power relay [14]. The structure of that protection must be interlocked with the motor starting device to prevent operation during tripping that will avoid the motor start. For a compressor, for example, if the starting is against a very low load to prevent maloperation of the motor, the loss-of protection can also apply to inhibit the starting duration. It is important to note that the setting of the relay depends on the type of application to be performed. For synchronous motor loss-of supply protection, after a pickup of the element, a time delay may be required to prevent operation during the transient event.

12.11 Motor Protection Solved and Unsolved Problems

12.11.1 Solved Problems

12.1 A 13 hp induction motor operates under a 230 V three-phase system. Calculate the approximate full-load current, no-load current and stating current (or locked-rotor current).

Solution

This is the approximate solution, where the full-load current is calculated by Wildi [23]:

$$I_{fl} = 600\frac{P_{hP}}{E} \tag{12.88}$$

Equation (12.88) is computed as follows:

$$I_{fl} = 600\frac{13}{230} \cong 33.913\text{ A} \tag{12.89}$$

Use the characteristics of a squirrel-cage induction motor given in Table 12.3 [23]. The 13 hp induction motor is classified as small, the no-load and locked rotor current can be determined as:

$$I_{nl} = 0.5I_{fl} = 16.9565\text{ A} \tag{12.90}$$

$$I_{Lr} = 6I_{fl} = 203.478\text{ A} \tag{12.91}$$

Note that, during the starting period, the current on the motor is equal to the locked rotor current. This current in the motor is approximatively between 5 and 6 times the rating current on the motor (as described in Table 12.3). The rise in the power losses, as described in Equation (12.29), is more than 25–36 times the normal condition. Therefore, longer starting time of the motor could limit or delay motor operation at locked condition for more than few seconds

TABLE 12.3

Typical Characteristics of Squirrel-Cage Induction Motors

Loading	Current (per unit)		Torque (per unit)		Slip (per unit)		Efficiency		Power Factor	
Motor size	Small	Big	Small	Big	Small	Big	Small	Big	Small	Big
Full-load	1	1	1	1	0.03	0.004	0.7 to 0.9	0.96 to 0.98	0.8 to 0.85	0.87 to 0.9
No-load	0.5	0.3	0	0	0	0	0	0	0.2	0.005
Locked-rotor	5 to 6	4 to 6	1.5 to 3	0.5 to 1	1	1	0	0	0.4	0.1

Small motor: under 11 kW or 15 hp; big motor: between 1,120 kW (1,500 hp) and 25,000 hp.

12.2 Use problem 12.1 to determine the short circuit protection for both conditions of the locked rotor as described in Table 12.3 when the protection relates to a current transformer ratio of 40/1.

Solution

The short circuit setting is given by:

$$I_{sc} = 1.25 \times I_{st} \tag{12.92}$$

The starting current is a function of the percentage of starting current times the full load current, as described in Equation (12.26). Therefore, the short-circuit element is set for each condition as:

$$I_{sc} = 1.25 \times 5 \times 33.913 = 211.965\text{A} \tag{12.93}$$

$$I_{sc} = 1.25 \times 6 \times 33.913 = 254.348 \text{ A} \tag{12.94}$$

By considering the relay nominal current, the setting value will be determined by the ratio of the short-circuit current and current transformation ratio. Using Equations (12.93) and (12.94), the setting value of the relay will be:

$$\frac{211.965}{40} = 5.299I_n \tag{12.95}$$

$$\frac{254.348}{40} = 6.3587I_n \tag{12.96}$$

The settings permit a minimal time delay of 100 ms when currents are less than or equal to 120% of setting, and 40 ms above this setting value.

12.3 When the earth protection uses a typical setting of 25% of motor rated current, determine the settings of earth fault relay, locked rotor/excessive start time and stall protections.

Solution

The setting current of earth fault relay will be as follows:

$$I_{\text{efr}} = 0.25\frac{33.913}{40} = 0.212I_n \tag{12.97}$$

The locked rotor and stall protection will use the same setting current, which has to be less than the starting current of the motor to ensure a start condition. Therefore, a setting of $2I_n = 80$ A can be used. For locked rotor/excessive starting time protection, the setting time delay should be 15 seconds which is less than starting and cold start time. The time delay of stall protection can be set less than the hot stall time but more than the starting time. This strategy of setting a time more than the start time aims to respect the margin to avoid a spurious trip in case the starting time is more than the anticipated time. Thus, a value of 6.5 seconds can be set as a delay time for stall protection.

12.4 Determine the voltage decay on the induction motor at open circuit time during the network disturbance when the supply voltage is 400 V at 50 Hz and 500 ms. The rotor reactance and resistance are 0.31 Ω and 0.26 Ω, respectively, with the magnetism reactance of 31 Ω.

Solution

Voltage decay is given as:

$$V(t) = \frac{1}{e^{t/t_{op}}} \tag{12.98}$$

where t and t_{op} are a given time and open circuit time constant, respectively. The open circuit time constant is determined by:

$$t_{op} = \frac{x_m + x_2}{2\pi f r_2} \tag{12.99}$$

where x_m, x_2, r_2, and f are magnetism reactance, rotor reactance and resistance, and system frequency, respectively. By subsisting all these variables, the open circuit time constant is calculated as:

$$t_{op} = \frac{31 + 0.31}{2 \times 3.14 \times 50 \times 0.26} = 0.3835\ \text{s} \tag{12.100}$$

At open circuit time, $t = t_{op}$. Thus the voltage decay is:

$$V = \frac{1}{e} 400 = 147.15\text{V} \tag{12.101}$$

At $t = 0.5$ s, the voltage decay will be:

$$V = \frac{1}{e^{(0.5/0.3835)}} 400 = 108.6\text{V} \tag{12.102}$$

12.5 Determine the voltage decay of problem 12.4 at 4 times the open circuit time constant.

Solution

At $t = 4t_{op}$

$$V(t) = \frac{1}{e^4} 400 = 7.33\ \text{V} \tag{12.103}$$

While the time is increasing, the voltage in the motor is decreasing to reach zero. This can be observed by analyzing Equations (12.101), (12.102), and (12.103).

12.6 Find the equivalent current on the motor during overheating when the real component of negative- and positive-sequences currents are 0.92 and 0.912, respectively, in per unit. Assume that the ratio of the negative-sequence rotor resistance and positive-sequence rotor resistance at rated speed is 3.

Solution

From Equation (12.70), the equivalent current on the motor is:

$$I_{eq} = \sqrt{(0.92^2 + 3(0.912)^2)} = 1.82\ \text{pu} \tag{12.104}$$

12.7 Find the percentage of the voltage unbalance and voltage unbalance factor of a three-phase motor, class EI2 when UAB, UBC, and UCA are 371, 374, and 395 V, respectively. It is assumed that the system phase angles are symmetric and the current flow in each phase are 18.3, 16, and 12 A.

Solution

Equations (12.105) and (12.106) determine the unbalanced voltage, which represents the percentage voltage unbalance and voltage unbalance factor.

$$V_{\text{unb}} = \frac{\max(|V_{ab} - V_{\text{avg}}|, |V_{bc} - V_{\text{avg}}|, |V_{cb} - V_{\text{avg}}|)}{V_{\text{avg}}} \times 100 \tag{12.105}$$

$$V_{\text{unf}} = \frac{V_1}{V_2}100 \quad (12.106)$$

where V_{avg} is the average of the system voltage, and can be found by:

$$V_{\text{avg}} = \frac{1}{3}(371+374+395) = 380 \quad (12.107)$$

By substituting Equation (12.107) in Equation (12.106), the percentage of voltage unbalance can be determined as:

$$V_{\text{unb}} = \max(0.0237, 0.0158, 0.0395) \times 100 = 3.95\% \quad (12.108)$$

From Equation (12.7), the symmetrical argument components of the voltage are:

$$\begin{cases} V_0 = 7.55 \\ V_1 = 7.55 \\ V_2 = 380 \end{cases} \quad (12.109)$$

Thus, the voltage unbalance factor will be computed, by using Equations (12.107) and (12.109), as:

$$V_{\text{unf}} = \frac{7.55}{380} \times 100 = 1.99 \approx 2\% \quad (12.110)$$

12.8 Determine the equivalent circuit current of problem 12.7 using the typical value of the negative- and positive-sequence rotor resistance ratio at rated speed.

Solution

From Equation (12.7), the symmetrical argument components of the current are:

$$\begin{cases} I_0 = 1.84 \\ I_1 = 1.84 \\ I_2 = 15.43 \end{cases} \quad (12.111)$$

From Equation (12.70) with $K = 3$, the equivalent circuit current can be computed as:

$$I_{eq} = \sqrt{\left(1.84^2 + 3 \times 15.43^2\right)} = 26.8 \text{ A} \quad (12.112)$$

12.11.2 Unsolved Problems

12.9 Determine the positive-sequence contribution to the input apparent power when the system is assumed wye connected.

12.10 Determine the overloaded and core loss percent when the overvoltage is the following:

a. 5%

b. 10%

c. 15%

d. 20%

e. 25%

12.11 Determine the cooling time constant of a motor when the ratio of the heating/cooling time constant is equal 2.5 for direct-online with the thermal time constant of 30 minutes, and autotransformer starting case when the thermal time constant is equal to 1.8 of direct-online.

12.12 Determine the thermal setting current of problem 12.11, when the motor is supplied by:

a. A balanced system with 250 argument value of current.

b. An unbalanced system when third phase has a current argument of 260 module with the angle of −5.

Discuss the results.

12.13 Determine the temperature of the motor in problem 12.11 when the full load current is equal to 220 A in 5 hours continuous operating after starting.

12.14 Use Table 12.2 to determine the stator current for a nominal current of 210 A when the unbalanced voltage is at

a. 4.5%

b. 6%

c. 1%

d. 2%

12.15 Use the information in problem 12.14 determine the percentage of loss increase.

12.16 Use the information in problem 12.14 determine the percentage of heating.

12.17 Discuss the results of problems 12.14 through 12.16 in the context of selecting the protection schemes.

12.18 Use the information in Table 12.3 to resolve the question of problem 12.1 when the voltage is 380 V and power rating of 50 hp.

12.12 Conclusion

This chapter has given an overview of the induction motor protections that can be implemented in the smart grid environment. Most motor protection schemes aim to avoid overheating during the operating time. Thus, intelligent design of the motor protection scheme means making sure that it can handle external and internal faults, prevent overheating of the system and thus increase the lifespan of an induction motor.

REFERENCES

1. Aswani, R., A. Awar, M. Deore, and R. Patel, "Induction motor fault detection, protection and speed control using Arduino," *IEEE International Conference on Innovations in Information, Embedded and Communication Systems* (*ICIIECS*), pp. 1–5, 2017.
2. Shivpuje, R. M., and S. D. Patil, "Microcontroller based fault detection and protection system for induction motor," *IEEE Conference on Intelligent Computing and Control Systems* (*ICICCS*), pp. 1187–1191, 2017.
3. Nath, P., J. Das, A. Rohman, and T. Das, "A fuzzy logic based overcurrent protection system for induction motor," *IEEE International Conference on Communication and Signal Processing* (*ICCSP*), pp. 755–759, 2016.
4. Garg, G., and A. Sinha, "An improved method for protection of three phase induction motor using microcontroller," *IEEE International Conference on Power, Control and Embedded Systems* (*ICPCES*), pp. 1–6, 2014.
5. El-Amary, N. H., F. A. Ezzat, Y. G. Mostafa, and W. A Ghoneim, "Thermal protection for successively starting three phase induction motors using particle swarm optimization technique," *IEEE 11th International Conference on Environment and Electrical Engineering* (*EEEIC*), pp. 788–793, 2012.

6. Devi, N. R., S. A. Gafoor, and P. R. Rao, "Wavelet ANN based stator internal faults protection scheme for 3-phase induction motor," *IEEE 5th IEEE Conference on Industrial Electronics and Applications (ICIEA)*, pp. 1457–1461, 2010.
7. Zhang, P., Y. Du, T. G. Habetler, and B. Lu, "A survey of condition monitoring and protection methods for medium-voltage induction motors," *IEEE Transactions on Industry Applications*, vol. 47, pp. 34–46, 2011.
8. Zhang, P., B. Lu, and T. G. Habetler, "An active stator temperature estimation technique for thermal protection of inverter-fed induction motors with considerations of impaired cooling detection," *IEEE Transactions on Industry Applications*, vol. 46, pp. 1873–1881, 2010.
9. Abdesh, M., S. K. Khan, and M. A. Rahman, "A new wavelet based diagnosis and protection of faults in induction motor drives," *IEEE Power Electronics Specialists Conference (PESC)*, pp. 1536–1541, 2008.
10. Du, Y., T. G. Habetler, and R. G. Harley, "Methods for thermal protection of medium voltage induction motors—A review," *IEEE International Conference on Condition Monitoring and Diagnosis (CMD)*, pp. 229–233, 2008.
11. Durán, M. J., J. L. Durán, F. Pérez, and J. Fernández, "Induction-motor sensorless vector control with online parameter estimation and overcurrent protection," *IEEE Transactions on Industrial Electronics*, vol. 53, pp. 154–161, 2006.
12. Cele, L. Z., A. M. Chol, and R. F. Chidzonga, "Problems on high failure incidence of induction motors and protection," *IEEE 7th AFRICON Conference in Africa*, vol. 2, pp. 1143–1147, 2004.
13. DeCastro, J. E., R. T. Beck, C. Cai, and L. Yu, "Stall protection of large induction motors," *IEEE Transactions on Industry Applications*, vol. 31, pp. 1159–1166, 1995.
14. Alstom, G., *Network Protection and Automation Guide*, Chapter 19, Stafford UK: ALSTOM GRID, May, 2011.
15. Anderson, P. M., "Motor protection," in *Power System Protection*, (Ed.) Anderson, P. M., Chapter 19. IEEE Press, 1999.
16. Hewitson, L., M. Brown, and B. Ramesh, "Motor protection relays," in *Practical Power Systems Protection*, (Eds.) Hewitson, L., M. Brown, and B. Ramesh, Chapter 17. Elsevier, 2004.
17. Dolby, P. M., "Protection of motors, reactors, boosters, capacitors," in *Power System Protection*, Edited by the Electricity Training Association, Vol. 1–4, Chapter 14. IET, London, UK, 1995.
18. Valenzuela, M. A., and P. Reyes, "Simple and reliable model for the thermal protection of variable-speed self-ventilated induction motor drives," *IEEE Transactions on Industry Applications*, vol. 46, no. 2, pp. 770–778, 2010.
19. Schneider Electric, "AC motors starting and protection systems," in *Automation Solution Guide: Practical Aspects of Industrial Automation Technology*, Chapter 4, Practical Aspects of Industrial Automation Technology, Institut Schneider Formation, 2007.
20. Naidoo, R., P. Pillay, J. Visser, R. C. Bansal, and N. T. Mbungu, "An adaptive method of symmetrical component estimation," *Electric Power Systems Research*, vol. 158, pp. 45–55, 2018.
21. Liu, Y., L. Wang, H. Gao, H. Zhang, and D. Xu, "Overvoltage mitigation of submersible motors with long cables of different lengths," in *EEE 17th International Conference on Electrical Machines and Systems (ICEMS)*, pp. 638–644, 2014.
22. Eldin, E. M. T., H. R. Emara, E. M. Aboul-Zahab, and S. S. Refaat, "Monitoring and diagnosis of external faults in three phase induction motors using artificial neural network," *IEEE Power Engineering Society General Meeting*, pp. 1–7, 2007.
23. Wildi, T., *Electrical Machines, Drives, and Power Systems*, 5th ed., Upper Saddle River, NJ: Prentice Hall, 2002.

13

Substation Automation and Control

Adeyemi Charles Adewole and Raynitchka Tzoneva

CONTENTS

13.1 Introduction

An electric power substation is a station within an electric power generation, transmission, and distribution system where transformation from one voltage level to another takes place. Typically, electric power is generated at a voltage level of 10–22 kV. At the transmission system, this generated voltage is transformed, or stepped-up, to voltage levels of 110, 220, 500, 765, or 1000 kV, and wheeled to the distribution system where they are stepped down to low voltages. Distribution substations are commonly fed by subtransmission lines or directly via transmission lines. The voltage level at the distribution system is usually in the medium voltage (MV) range, operates within the voltage range of about 6.6–66 kV, and may extend to the 132 kV voltage level in some instances. This serves as the last link to the customers, and the topology can be radial, ring, or mesh. The service to the secondary customers terminates at a voltage of about 400 V (three-phase circuits) or 230 V ± 10% (single-phase circuits) in Europe, Africa, most of Asia, most of South America and Australia, while 240 V (three-phase circuits) 120 V ± 5% (single-phase circuits) is used in North America.

Aside from voltage transformation, substations are used for the control and monitoring of the substation equipment, protective relaying, event/disturbance recording, metering, energy management, and asset management. The various equipment in a substation can be categorized according to their voltage level and function into primary and secondary equipment, respectively. Primary equipment is connected directly to the power system, while secondary equipment includes low-level voltage devices used for monitoring, metering, protection, and control. Examples of primary equipment include power transformers, current transformers (CTs), voltage transformers (VTs), merging units (MUs), circuit breakers (CBs), disconnectors and switches. Examples of secondary equipment are protective relays, intelligent electronic devices (IEDs), phasor measurement units (PMUs), revenue meters, power quality meters, and digital fault recorders (DFRs).

In modern power systems, the emergence of the smart grid concept has brought about the need for system-wide real-time observability, controllability, protection, and automation. Substation automation systems (SASs) are used for the monitoring, protection, control, measurement, and alarming of substation functions using modern IEDs. These IEDs are multi-functional and are activated through configuration to provide protection, automation, control, and monitoring functionalities. SASs are typically used in providing system reconfiguration after a fault or prior to maintenance. Also, SASs provide interfaces to the supervisory control and data acquisition (SCADA) systems and human machine interfaces (HMIs) for local and remote communication access. The rising adoption of SASs throughout the world serves as proof of the acceptance of such systems and the appreciation of their benefits by utilities [1–2].

This chapter covers substation automation systems using the IEC 61850 standard. The evolution towards the adoption of automation in substations, the structure of substation automation, and the possible substation logical interfaces are presented. This is followed by the data modelling approach of IEC 61850 functions and services. Communication network architectures and topologies are described. The system engineering of IEC 61850 devices including redundancy protocols like the parallel redundancy protocol (PRP) and the high-availability seamless redundancy (HSR) protocol are discussed. In addition, cyber security threats, vulnerabilities, and countermeasures in IEC 61850 system are described. Last, some IEC 61850 use cases, including sampled values (SV) and generic object oriented substation events (GOOSE) packet analyses and IEC 61850 communication-assisted protection scheme, are given.

13.1.1 Overview of Substation Automation Systems

SASs encompasses substation automation and control, and it can be defined as the act of managing, controlling, and protecting the substation segment of a power system by using the real-time information from protective relays, IEDs, PMUs, meters, programmable logic controllers (PLCs), and remote terminal units (RTUs). These components are harnessed together using communication, optimization, and intelligent systems to provide the following benefits [2]:

- Local and remote accesses
- Manual and automated data acquisition
- Improved decision making, system efficiency, and quality of service

- Increased grid performance, flexibility, and reliability
- Manual and automatic control
- Integrated and improved protection, control, disturbance and event recording
- Faster fault location and fault resolution times
- Remote switching, system reconfiguration, and restoration
- Application of optimization and intelligent-system techniques

Some factors that should be considered in the design and implementation of SASs include:

- Definition of the technical objectives of SASs.
- Cost assessment and feasibility of brownfield SASs installation in comparison with greenfield automation projects.
- The scale of control infrastructure and level of automation to achieve (central or distributed, system-wide or local, or a combinations of these), and its implications for the available communications system.
- The regulatory requirements that must be satisfied.
- The business case for SASs with respect to the improvement of the system's performance, quality of service, and technical and economic goals.

13.1.2 The Evolution Towards Substation Automation and Control

In the past, protection and control functions in substations were performed by a myriad of dedicated single-function electromechanical relays, discrete electronics, and push buttons. This required a lot of interwiring between the primary equipment at the substation yard and the secondary equipment in the substation local control house. Also, most of these devices have single functionality and are integrated using the Boolean algebra approach. With this, all the data from the protection and control devices at the protection, control, and metering (PCM) panels or marshalling kiosks are then sent from the substation local control house to the network control centre via RTUs.

The advances made with modern technology, computing, and communication have led to the development of communication-based protection and control using microprocessor-based multifunctional devices and various communication protocols. These multifunctional IEDs are programmable and can be used to perform protection and control tasks in the substation. The introduction of legacy microprocessor IEDs came with serial communication connections such as the RS-232 interface. This facilitated inter-wiring among the IEDs. With the first implementation of serial communication in substations, a new drive began for the interoperability of the various protection, control, and monitoring applications using a common protocol. Consequently, the IEC 60870-5-103 was developed as the informative interface for protective devices. This protocol is not only restricted to information but also to protection functions. Next was the introduction of more functionalities into a single device, thereby integrating protection, control or monitoring applications as commonly found these days.

The Modbus protocol was developed by Modicon (now Schneider) in 1979 for process control systems for transmitting control signals between PLCs using serial communication. Modbus is an application layer messaging protocol that is positioned at level 7 of the Open Systems Interconnection (OSI) model for client/server communication between devices connected to the same communication network. Its messages are of either query/response type or broadcast/no response type, which in either case can only be initiated by the client. The devices (PLC, HMI, control panel, driver, I/O device) can use Modbus to initiate a remote operation. Modbus can be implemented using Asynchronous Serial Transmission (ASCII or RTU) for serial connections (over RS-232, RS-422, RS-485, fiber, or radio) or TCP/IP over Ethernet (Modbus Plus: also referred to as Modbus+ or MB+). Modbus supports mainly four types of data: (1) read only 1-bit (discrete) inputs; (2) 2-byte input (read-only) registers, which can be provided by an I/O system; (3) 1-bit discrete (read-write) outputs/coils; and (4) 2-byte 3 holding registers, which are read/write enabled and can be altered by an application program [3].

The Distributed Network Protocol (DNP3) was initially developed as a proprietary protocol by Westronic Inc., (now GE Harris) in 1990 for client stations communications between RTUs and other IEDs. In 1993, it became a public domain protocol. It was originally designed for SCADA applications for the acquisition of information and sending of low- to medium-speed control commands between devices. DNP3 has been widely accepted for use in the electrical, water, oil and gas, and security sectors. The purpose of DNP3 is to transmit relatively small packets of data in a reliable manner, with the messages involved arriving in a deterministic sequence. It is a layered protocol with three layers (physical, data, and application layers) and one pseudo layer (pseudo transport layer). Application data may be any size, even zero (e.g., for a command signal). This is broken into multiple application protocol data units (APDUs), or fragments, with a size limit of 2048 bytes. Breaking longer messages into multiple packets helps optimize error control.

The APDU is broken down into multiple transport PDUs (TPDUs), with a size limit of 250 bytes to fit into the data link frame. Link layer header and the cyclic redundancy check (CRC) bytes are added to the TPDU, and the resultant link PDU (LPDU or frame) is sent to the physical layer, which transmits 8 bits of data plus an additional bit to indicate start/stop of each sequence. The physical layer could be RS-232 (for short distance point-to-point communications), RS-422 (a bidirectional extension of RS-232 for industrial environments), or RS-485 for multipoint communications. In recent times, the DNP3 protocol uses the Internet Protocol (IP) suite for the transport and network layers, and the Ethernet physical layer. This involves the encapsulation of the data frames from the DNP3 data link layer within the transport layer frames of the IP suite. DNP3 also supports peer-to-peer, multi-server, multi-master and hierarchical mode communications (using data concentrator) [4].

The IEC 61850 standard is an international nonproprietary standard for communication networks and systems for power utility automation. Its main objective is to specify the requirements and provide the framework in achieving interoperability between IEDs from different models or vendors. The IEC 61850 standard provides a single suite of protocols and services to address communications within and outside the substation through the integration of protection, control, and metering functions. It also provides the means for interlocking, inter-tripping, and other associated advantages of the Ethernet communication. The new functionalities feature object-oriented data modelling and a standardized configuration language. This is supported by the application of IEC 61850 logical nodes (LNs) and logical devices (LDs). This results in a significant improvement in the protection, automation, and control functions in the power system. The data models defined in IEC 61850 protocol can be mapped to various protocols, for example, to GOOSE (that allows for both analogue and digital peer-to-peer data exchange), mapping for sampled values (SVs), and mapping for manufacturing message specification (MMS), respectively [5].

13.2 Substation Automation and Control Using the IEC 61850 Standard

SASs are increasingly being implemented around the world using IEC 61850 compliant IEDs [1]. The IEC 61850 standard enables utilities to carry out protection, automation, and control functions by defining the data modelling and specifications for the exchange of information. The initial release of the standard [5] has been extended by various technical reports to cover intersubstation applications (IEC 61850-90-1), substation-to-control centre applications (IEC 61850-90-2), communication of IEEE C37.118 synchrophasor measurements (IEC 61850-90-5), and communication systems for the integration of distributed energy resources (IEC 61850-7-420).

Some of the benefits of the IEC 61850 standard include the following:

- A significant reduction in project cost as a result of the reduction in the use of copper cabling (hardwiring) between connected devices.
- Devices can exchange data using GOOSE messages over the station LAN without having to wire separate links for each IED. The use of the station LAN to exchange these signals reduces the cost of the infrastructure associated with wiring, trenching and ducting.
- Provision of better safety and isolation of live equipment and wiring.

- Ease of maintenance, and the reconfiguration of an IEC 61850-based network can be easily achieved with little impact on the existing equipment.
- Interoperability between multiple devices from the same or different vendors without the need for expensive protocol converters.
- Interchangeability of the devices in the substation without making changes to the other elements in the system.
- Lower commissioning costs because the manual configuration required by IEC 61850-compatible devices is less. Also, client applications do not need to be manually configured for each point they need to access, because this information can be directly obtained from the device or imported via a substation configuration language (SCL) file.
- The use of object names, which are standardised and are not dictated by the device vendor nor configured by the user.

The proceeding subsections present the structure and logical interfaces in IEC 61850-based systems, abstract modelling, and functional services.

13.2.1 Substation Automation Structure and Interfaces

The structure of substation automation systems can typically be discussed based on the functions carried out by the various types of equipment in the substation. Considerations should be given to the structure required at the data acquisition stage, and the structures required for protection, control, monitoring, supervision, and communication. Communication within the substation takes place over three logical levels, as shown in Figure 13.1: the process level, the bay level, and the station level.

At the process level, current and voltage measurements from instrument transformers in the switchyard are communicated to the protection, control, and metering (PCM) devices in the bay level. Typically, the current measurements are obtained from CTs with an output of 1 or 5 A, while the voltage measurements are from VTs with an output of 100–120 V. If conventional instrument transformers are used, the analogue measurements serve as inputs to MUs, where they are digitized and published as IEC 61850 SVs onto the substation local area network (LAN). Other types of information that can be obtained from the

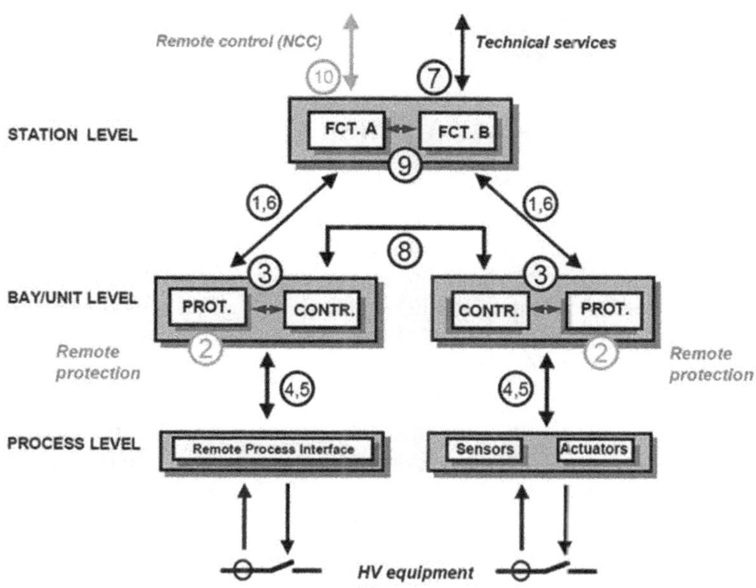

FIGURE 13.1 Levels in power system automation and logical interfaces. (From IEC 61850-1, Communication networks and systems in substations—Introduction and overview, *International Electrotechnical Commission (IEC)*, 2003.)

TABLE 13.1
Substation Logical Interfaces and Functions

Logical Interface Number	Function
IF1	Protection-data exchange between bay and station level
IF2	Protection-data exchange between bay level and remote protection
IF3	Data exchange within bay level
IF4	CT and VT instantaneous data exchange (especially samples) between process and bay level
IF5	Control-data exchange between process and bay level
IF6	Control-data exchange between bay and station level
IF7	Data exchange between substation (level) and a remote engineer's workplace
IF8	Direct data exchange between the bays especially for fast functions like interlocking
IF9	Data exchange within station level
IF10	Control-data exchange between substation (devices) and a remote control centre

process level are the positions of CBs, switches, and transformer tap-changers, and physical measurements such as temperature, pressure, and humidity. The bay devices subscribe to the current and voltage inputs from the process level via the logical interfaces IF4 and IF5 (Figure 13.1) to the bay level.

Bay level functions make use of the data from one substation bay and act mainly on the primary equipment of one bay. The functions commonly carried out at the bay level include protection, control, and metering. Communication takes place over the logical interface IF3 within the bay level, via the logical interfaces IF4 and IF5 to the process level, and via the logical interfaces IF1 and IF6 for communication with the station level. The interface IF8 is used for the horizontal communication between devices in different bays. Examples of bay devices include protection IEDs, DFRs, PMUs, bay controllers, PLCs, and meters.

Station level functions are related to the station-wide operation of the equipment within the substation. These can be process-related or interface-related station level functions. Process-related station level functions make use of the data and act on the primary equipment of more than one bay or of the entire substation.

Examples include protection and control functions. Communication is via the logical interface IF8. Interface-related station level functions include functions that interface the SAS to the substation HMI, to SCADA or to a remote engineering station. These functions communicate via the logical interfaces IF1 and IF6 with the bay level, and via the logical interfaces IF7 and IF10. A summary of the functions of the logical interfaces shown in Figure 13.1 is provided in Table 13.1.

13.2.2 Digital Substations Using the IEC 61850

Digital substation are substations whose process, bay, and station levels functions and operations are carried out using a communication network. This implies that the various devices within the substation are interconnected via a communication network. At the process level, nonconventional instrument transformers (NCITs) or MUs convert the analogue quantities to digital data streams that can be communicated via Ethernet communication. Two IEC standards play a very important role in the realization of a digital substation: (1) IEC 61850 and (2) IEC 61869 standards. Table 13.2 gives a breakdown of the various parts that make up the IEC 61850 standard.

In conventional systems, the analogue input module of the protection relay provides the interface between the onboard processor board(s), and the voltage and current quantities from the CTs and VTs located in the substation switchyard. These analogue signals are passed through anti-alias filters before being multiplexed onto an analogue-to-digital converter (ADC) chip.

The ADC provides a sampled data stream output that is transmitted to the protection module via the data bus. This is shown in Figure 13.2. Typically, a sampling rate within the range of 4–64 samples per cycle is usually used. Some algorithms also make use of frequency tracking to correct amplitude or phase error introduced by the transformers, analogue circuitry, or changes in the system frequency. In digital substations, the analogue signals from CTs/VTs are digitized in the MUs. An MU includes LNs

TABLE 13.2

Description of the Parts in the IEC 61850 Standard

Part Number	Description
Part 1	Introduction and overview
Part 2	Glossary
Part 3	General requirements
Part 4	System and project management
Part 5	Communication requirements for function and device models
Part 6	Configuration description language for communication in electrical substations related to IEDs
Part 7-1	Basic communication structure for substation and feeder equipment—Principles and models
Part 7-2	Basic communication structure for substation and feeder equipment—Abstract Communication Service Interface (ACSI)
Part 7-3	Basic communication structure for substation and feeder equipment—Common data classes
Part 7-4	Basic communication structure for substation and feeder equipment—Compatible logical node classes and data classes
Part 8-1	Specific communication service mapping (SCSM)—Mapping to MMS (ISO/IEC 9506-1 and ISO-9506-2) over ISO/IEC 8802-3
Part 9-1	Specific communication service mapping (SCSM)—Sampled values over serial unidirectional multidrop point to point link
Part 9-2	Specific communication service mapping (SCSM)—Sampled values over ISO/IEC 8802-3
Part 10	Conformance testing

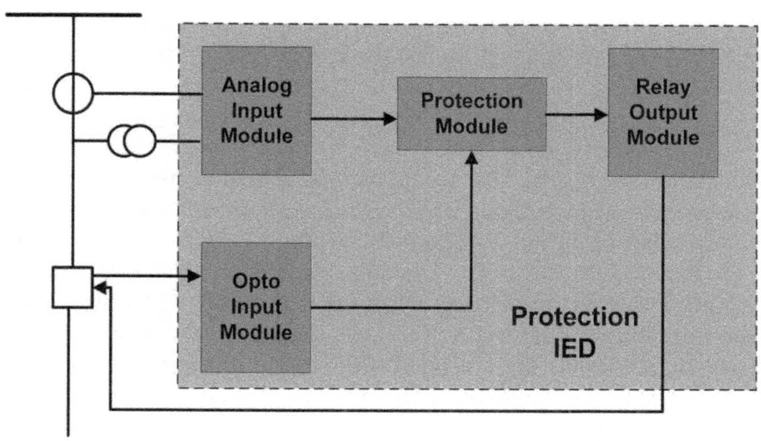

FIGURE 13.2 Conventional hardwired analogue inputs based IED.

TVTR (voltage transformer) and TCTR (current transformer), and serves as the interface unit that gathers multiple analogue information such as phase voltages and currents from the instrument transformers. All these analogue signals are converted to digital SVs by the MU.

This digitization process can also be achieved by using NCITs, which convert the power system analogue signals to digitized optic signals. The digital SVs are communicated to the bay level protection IEDs over the Ethernet communication network known as the process bus. The processes involved are shown in Figure 13.3.

The protection IED(s) receive the SVs from the process bus via its IEC 61850-9-2LE Ethernet module. The Ethernet module replaces the conventional analogue input module shown in Figure 13.2. It resamples the IEC 61850-9-2 samples published by the MU and transforms the signals to those used by the protection algorithm implemented in the protection LN. In addition to the IEC 61850-9-2 [7], the IEC 61869-9 standard [8] is equally important in digital substations. It defines the requirements for the use of NCITs that digitize the analogue current and voltage signals in the substation for the realization of digital substations.

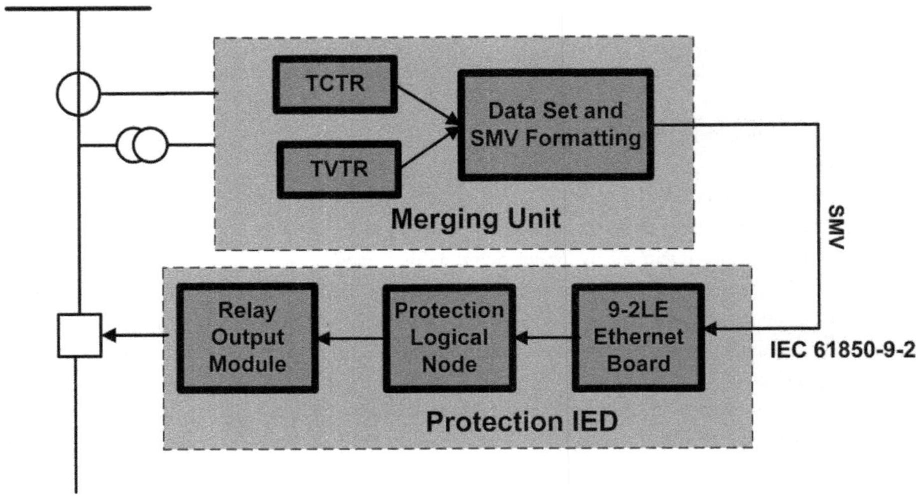

FIGURE 13.3 IEC 61850-9-2 sampled values-based IED.

13.2.3 IEC 61850 Data Modelling and Services

The IEC 61850 uses a data modelling approach which decomposes the application functions into the smallest entities possible using LNs that are contained in LDs. The decomposition of the functionalities of a physical IED into smaller entities enables the communication between services and multiple devices. The next subsection describes the data models and services in the IEC 61850 standard.

13.2.3.1 IEC 61850 Data Modelling

The logical nodes used in the IEC 61850 standard must be able to interpret and process the data subscribed to and the communication services used in order for there to be interoperability between the logical nodes. The associated data and services for an application can be modelled using three distinct levels, as defined in Part 7 of the IEC 61850 standard. Level 1 is the Abstract Communication Service Interface (ACSI) needed to exchange information between logical nodes [9], while levels 2 and 3 specify the application domain specific object model of the data classes. Level 2 of the modelling specifies the common data classes (CDCs), which are structured information consisting of one or more attributes [10]; Level 3 defines the compatible logical node classes and data classes [11].

The data model is a hierarchy of information given by the ACSI, which specifies the models and services used in accessing the elements of the domain specific object model. The data modelling approach in the IEC 61850 standard can be implemented using the conceptual structure given in Figure 13.4. This is made up of an IEC 61850-compliant physical device (PHD), LD, LN, data, and data attributes. The server shown in Figure 13.4 represents the external visible behaviour of an IEC 61850 device. It communicates with an IEC 61850 client and also sends information to peer devices. The LD enables the PHD to serve as a gateway. It contains the information produced and utilised by a group of domain-specific application functions defined using one or more LNs. LNs reside within one or more physical devices and serve as the container for various data with dedicated data attributes. The information in the data and data attributes are exchanged by the communication services according to defined rules. This approach makes it easier to have multiple LDs containing multiple instances of LNs, each of which contains a predefined set of data classes. The data provide the means to define specific information, e.g., the position (*Pos*) of a circuit breaker with quality information and timestamp. The data attributes can be the value to be controlled (*ctlVal*), information with origin, operate time (*operTim*), and control sequence number (*ctlNum*), or they can be the status value (*stVal*) information with the quality flag (*q*), time stamp (*t*) information. Some data related to the physical device are contained in a different LN referred to as logical node 0 (LLN0). The LLN0 contains information such as the physical device identification or nameplate,

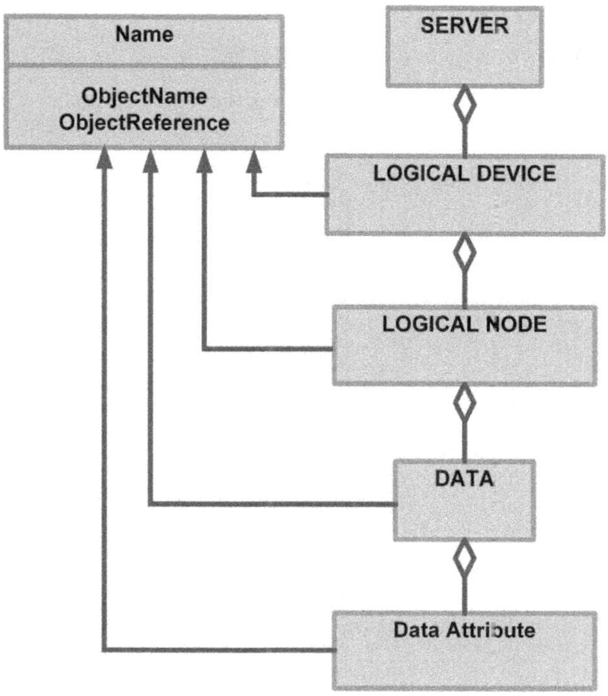

FIGURE 13.4 Basic conceptual class model of the IEC 61850 ACSI. (From IEC 61850-4, Communication networks and systems in substations—General requirements, *International Electrotechnical Commission (IEC)*, 2003.)

common mode settings, and common settings. LPHD represents the specific information for the physical device in which the LNs resides.

The Name class shown in Figure 13.4 is inherited by the LD, LN, data, and data attribute, and they should have an instance (unique) name known as the ObjectName. Also, they should have an ObjectReference (path name) which is a concatenation of all the object names from each container.

Table 13.3 gives a list of 13 LN groups and their descriptions as defined in the IEC 61850 standard. Ninety-two LNs were initially defined in the IEC 61850-7-4 standard; the number of LNs has increased to 233 LNs in the IEC 61850 Ed. 2 standard to account for distributed energy resources (DERs), amongst others. Out of the 92 LNs, 38 LNs (now 43 LNs in the IEC 61850 Ed. 2 standard) cover protection and protection-related applications. Some of the LNs introduced for DER applications include LNs for fuel systems (DFUL, DFLV), reciprocating engines (DCIP), fuel cells (DFCL, DSTK, DFPM), photovoltaics (DPVM, DPVA, DOVC, DTRC), combined heat and power (DCHC, DCHI, DCHX, DCHS), generators (DGEN, DRAT, DRAZ, DCST), battery systems (DBAT, DBTC), converters (YRCT, YINV), and physical measurements (MTMP, MFLW, MPRS, MVBR, MMET, MHET, MENV).

Tables 13.4 and 13.5 present some commonly used LNs for protection and protection-related functions. A description of the data model for IEC 61850-based protection, measurement, and switchgear control functions is presented in Tables 13.6 through 13.8. Note that the nomenclature for the LNs are standardized in the IEC 61850 standard.

The format is given as:

<Group designator><Mnemonic abbreviation of function><instance ID>

For example, <P> <TOC> <1> (PTOC1) represents the protection group with a time overcurrent protection. Figure 13.5 shows the various LNs and their interaction in a substation automation system divided into the process, bay, and station levels.

TABLE 13.3

IEC 61850 Logical Node Groups

Group Indicator	Logical Node Groups
A	Automatic control
C	Supervisory control
G	Generic function references
I	Interfacing and archiving
L	System logical nodes
M	Metering and measurement
P	Protection functions
R	Protection related functions
S	Sensors and monitoring
T	Instrument transformer
X	Switchgear
Y	Power transformer and related functions
Z	Further (power system) equipment

TABLE 13.4

Common Protection IEC 61850 Logical Nodes

Logical Node	IEC 61850	IEEE C37.2 (ANSI Codes)
Protection functions		
Distance protection	PDIS	21
Distance protection schemes	PSCH	
Volt per Hz protection	PVPH	24
Undervoltage protection	PTUV	27
Directional power/reverse power	PDOP/PDUP	37
Loss of field/underexcitation protection	PTUC/PDUP	40
Thermal overload protection	PTTR	49
Instantaneous overcurrent protection	PIOC	50
Time overcurrent protection	PTOC	51
Time overvoltage protection	PTOV	59
Directional overcurrent	PTOC	67
Frequency	PTOF/PTUF/PFRC	81
Differential	PDIF	87

TABLE 13.5

Common Protection-Related IEC 61850 Logical Nodes

Logical Node	IEC 61850	IEEE C37.2 (ANSI Codes)
Protection-related functions		
Synchronism check	RSYN	25
Breaker failure protection	RBRF	50BF/62BF
Directional element	RDIR	
Power swing detection or blocking	RPSB	68
Auto reclosing	RREC	79
Fault location	RFLO	
Digital fault recording	RDRE/RADR/RBDR	DDR/DFR
Circuit breaker	XCBR	32
Circuit isolating switch	XSWI	89
Substation metering	MMXU/MMTR	MET

TABLE 13.6

Example of an IEC 61850 Protection Dataset

	Logical Device	Logical Node	Data	Data Attribute
Object name	"SUBST_FDR1"	"PIOC1"	"Op"	"general"
Description	Station feeder 1	Instantaneous overcurrent protection 1	Operate	Three-phase trip

TABLE 13.7

Example of an IEC 61850 Measurement Dataset

	Logical Device	Logical Node	Data	Data Attribute
Object name	"SUBST_FDR1"	"MMXU1"	"TotW"	"mag"
Description	Station feeder 1	Measurement 1	Total real power	magnitude

TABLE 13.8

Example of an IEC 61850 Switchgear Control Dataset

	Logical Device	Logical Node	Data	Data Attribute
Object name	"SUBST_FDR1"	"XCBR1"	"Pos"	"ctlVal"
Description	Station feeder 1	Circuit breaker 1	Circuit breaker position	Control value (ON or OFF)

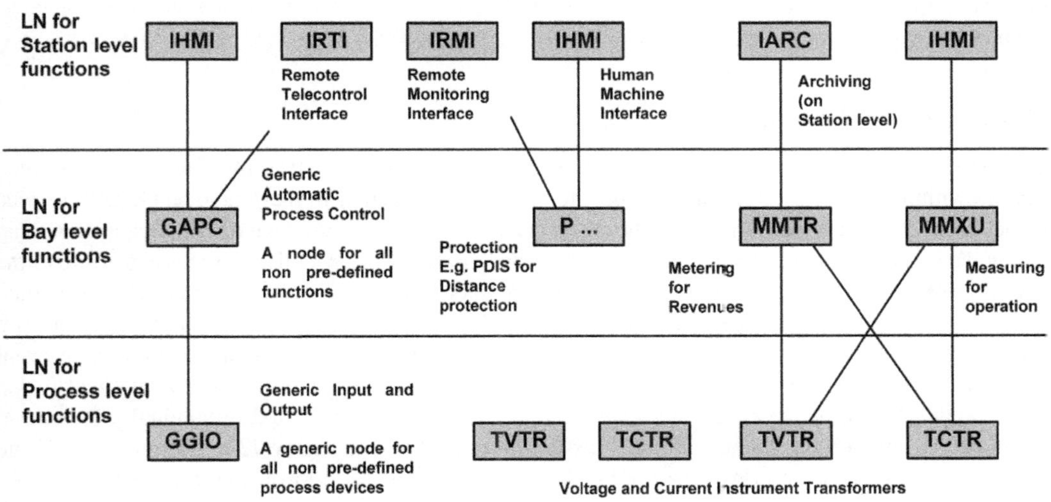

FIGURE 13.5 Typical LNs for a digital substation at the various levels of substation automation systems. (From IEC 61850-4, Communication networks and systems in substations—General requirements, *International Electrotechnical Commission* (*IEC*), 2003.)

In the communication of IEC 61850 services, multicast addressing with the following recommendations are used: The first three octets are assigned by IEEE with 01-0C-CD, while the fourth octet will be 01 for GOOSE, 02 for GSSE, and 04 for multicast sampled values. The last two octets is for individual addresses assigned by the range defined in Table 13.9.

13.2.3.2 Sampled Values (SVs)

Data acquisition using CTs and VTs takes place at the process level of the substation. The outputs of these CTs and VTs are sampled, converted to digital representation (SVs) by analogue MUs, and formatted for subsequent transmission via the process bus LAN. The analogue voltage and current signals can also be

TABLE 13.9

Recommended Multicast Addressing for IEC 61850 Services

	Recommended Address Range	
IEC 61850 Service	**Starting Address (in hex)**	**Ending Address (in hex)**
GOOSE	01-0C-CD-01-00-00	01-0C-CD-01-01-FF
GSSE	01-0C-CD-02-00-00	01-0C-CD-02-01-FF
Multicast sampled values	01-0C-CD-04-00-00	01-0C-CD-04-01-FF

directly digitized using NCITs. The sampling rate to use depends on the application. Eighty samples per cycle (equivalent to 4,000 samples for 50 Hz or 4800 samples for 60 Hz) is specified for protection application according to the IEC 61850-9-2LE, while 256 samples per cycle (equivalent to 12,800 samples for 50 Hz or 15,360 samples for 60 Hz) is used for measurement-related application [7]. These sampling rates have been standardized in the new IEC 61869-9 standard [8] irrespective of the system frequency to 4,800 samples per cycle (protection application) and 14,400 samples per cycle (measurement application), respectively.

The specifications for the peer-to-peer communication for SVs are defined in the IEC 61850-9-2 standard and the IEC 61850-9-2LE guidelines, respectively. MUs use the TVTR (voltage transformer) and TCTR (current transformer) LNs, respectively, and serve as the interface unit for gathering multiple analogue information such as phase voltages and currents from the instrument transformers. All these analogue signals are converted to digital SVs Ethernet packets by the MUs. The digital signals are communicated to the bay level protection IEDs over an Ethernet-based communication network known as the process bus.

13.2.3.3 GOOSE Messaging

GOOSE messages are time-critical event-based messages used in the transmission of data between devices within a multicast group using a peer-to-peer communication framework. The GOOSE service model of 61850-7-2 provides the possibility for a fast and reliable system-wide distribution of input and output data values. The SCSM specifies the syntax and encoding of GOOSE messages according to the IEC 61850-8-1, and uses a specific scheme of re-transmission to achieve the appropriate level of reliability.

When a GOOSE server generates a *SendGOOSEMessage* request, the current data set values are encoded in a GOOSE message and published as *T-DATA* on the multicast association. The event that causes the server to invoke a *SendGoose* service is a local application issue as defined in the IEC 61850-7-2. In order to ensure reliability, the same data is retransmitted by gradually increasing the sequence number *SqNum* and a retransmission time, as shown in Figure 13.6. Each message in the retransmission sequence carries a *timeAllowedToLive* parameter that informs the receiver of the maximum time to wait for the next retransmission. If a new message is not received within that time interval, the receiver shall assume that the association is lost. The time requirements for various message types are given in Table 13.10 [13].

13.2.3.4 Manufacturing Message Specification

The manufacturing message specification (MMS) is a communication service used in the communication of real-time data and control information between substation IEDs and services in the power system. MMS supports the services that provide the mapping to IEC 61850 abstract objects and services using a client-server communication framework [13]. The control model of ACSI is mapped to MMS read and write services. It is a mechanism through which the server can send data without an explicit request from the client. However, an open TCP socket connection already initiated by the client has to be in place. The MMS server can be the protection or control IEDs, while the MMS client represents the control centre applications. The client-server roles are interchangeable. That is, a device can be an MMS server at one point in time and can act like the MMS client at another time.

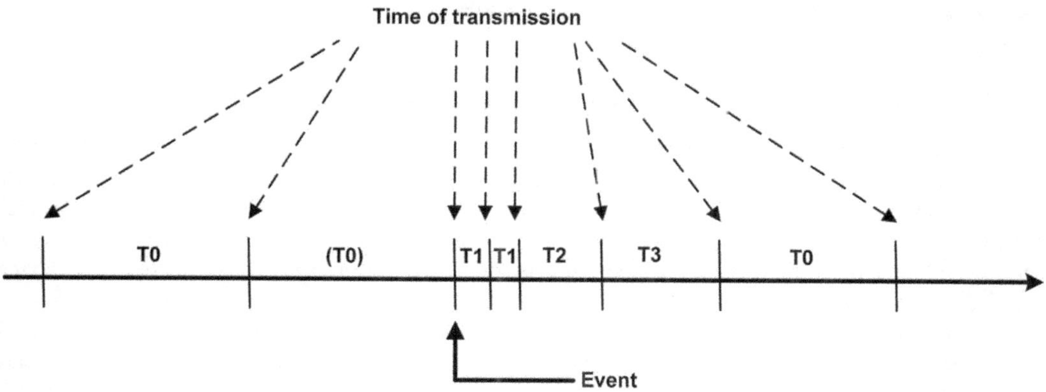

FIGURE 13.6 Transmission time for events: (T0) retransmission in stable conditions (no event in a long time), (T0) retransmission in stable conditions, (T1) shortest retransmission time after an event, (T1) shortest retransmission time after the event, (T2, T3) retransmission times until achieving the stable condition time. (From IEC 61850-8-1, Communication networks and systems in substations—Specific communication service mapping (SCSM)—Mappings to MMS (ISO 9506-1 and ISO 9506-2) and to ISO/IEC 8802-3, *International Electrotechnical Commission (IEC)*, 2004.)

TABLE 13.10

IEC 61850 Communication Networks Requirements for Functions and Device Models

Message Types	Definitions	Delay Requirements	Examples
Type 1	Messages requiring actions at the receiving IEDs	1a: 3 ms or 10 ms 1b: 20 ms or 100 ms	1a: Trips 1b: Commands, simple messages
Type 2	Messages requiring medium transmission speed	100 ms	Measurands
Type 3	Messages for slow-speed auto-control functions	500 ms	Parameters
Type 4	Continuous data streams from IEDs	3 ms or 10 ms	Instrument transformers and transducers output
Type 5	File transfer functions	1000 ms (not strict)	Large files
Type 6	Time synchronization messages	No requirement	6a: Station bus 6b: Process bus
Type 7	Command messages with access control	Equivalent to Types 1 or 3	Station HMI commands

The message flow for the MMS data exchange between the client and the server starts with a TCP three-way handshake between the MMS client and server. The connection oriented transport protocol (COTP) establishes a connection to the transport ISO protocol over TCP by means of the *Connection Request* message from MMS client and the *Connection Confirm* message from the MMS server. Afterwards, the MMS *INITIATE-REQUEST* is mapped onto the application association request (AARQ) of the association control service element (ACSE) layer. The MMS server replies with the *INITIATE-RESPONSE* message, mapping it onto the application association response (AARE) of the ACSE layer. Both messages are transported over the COTP data transport protocol data unit (TPDU). After the MMS client receives the INITIATE-RESPONSE from the MMS server, both MMS parties are associated and can begin the MMS data exchange, through *MMS Request* and *MMS Response* messages, respectively.

Some services offered by MMS include the *Read Service* that uses the V-Get function in order to transmit the current value of a real variable, the *Write Service* that uses the V-Put function to replace the current value of the real variable, and the *Define Named Variable Service* that creates a new named variable object that will be assigned real data. There are two transport profiles (T-profiles) that can be used by the client/server: (1) the TCP/IP profile, and (2) the ISO profiles. The TCP/IP profile uses ISO transport service on top of the TCP (RFC1006), ICMP (RFC792), and TCP (RFC793), while the ISO CO T-Profile uses connection oriented transport (ISO8072).

13.3 Communication Networks

13.3.1 Communication Network Architectures and Topologies

Communication network architecture is the logical and structural layout of the communication network consisting of transmission equipment, software, communication protocols and infrastructure (wired or wireless) transmission of data and connectivity between devices. The client/server architecture within the substation is a transaction whereby a request for information from a client to a server is followed by the delivery of the requested information by the server. An IEC 61850 client is a device or function that sends a message to an IEC 61850 server/device requesting the server to perform a specific task. For example, substation IEDs are servers when interfacing with bay or station level devices. Similarly, the devices at the bay level can act as clients when interfacing with other IEDs or as servers when communicating with station level applications. Master-slave communication is a polling scheme in which the master (device or function) requests for information from one or more IEDs (slaves). With this type of communication, only the master can initiate or issue unsolicited data or command. The peer-to-peer type of communication within a substation refers to the exchange of information between two or more substation devices, with each peer having equal rights and both acting as service providers or as a user of services.

Substation network topology refers to the way in which devices in a substation are connected together via the communication infrastructure. Five types of topologies are possible: (1) bus, (2) star, (3) ring, (4) mesh, and (5) hybrid topologies, respectively. In a bus topology, a common backbone link is used in connecting all the devices in the network. The network devices compete for access to the backbone for data transmissions. When a host gains access to the communication medium, it sends data messages which are then received by all the hosts connected to the same communication backbone. However, only the host to which the messages are addressed will respond to these messages, while the rest of the hosts will discard them. Bus networks work the best when a limited number of hosts are connected to the common bus; their efficiency is affected when a large number of stations require network access. Consequently, bus topologies are less popular today in the context of the increasing demand for network connectivity and large growth in data traffic.

In the star topology, a physical connection runs from each device on the network to a central communication network component, usually an Ethernet switch. The star topology is known for its robustness. If a particular link fails, only the devices using those links will be disconnected from the network, while all the other devices will not experience any disruptions in communication. The shortcomings of the star topology include the existence of a single point of failure and increased deployment costs. With the ring topology, each device is connected with two other devices in a daisy-chain arrangement to form a ring. In the basic ring network topology, the messages (data bits) travel in one direction only. Each host has a dual role serving as a host and as a communication relay. As a host, each host sends data messages to other hosts and receives messages addressed to it. Also, each host forwards messages addressed to other hosts to the next host on the ring. The main disadvantage of ring networks is their reliability. If a single link is broken, the communication between certain hosts is impeded. Dual ring redundancy solutions where communication is possible in both the clockwise and anticlockwise directions have been proposed to improve reliability. The increase in redundancy comes with higher deployment and maintenance costs. An example of the ring topology is the token ring protocol (IEEE 802.5) by IBM.

The hybrid topology is a combination of the bus and star topologies. In the hybrid topology, the hosts are connected to a network hub connected to other hubs in a tree-like structure. Routing messages in ring, bus, and star topologies is performed by broadcasting the messages to all hosts connected in the network. When tree topologies are used, messages originating at a host travel up the tree as far as necessary and then down the structure towards the destination host.

In a full mesh network topology, each host or network device is directly connected to any other device or host within that network. Although extremely robust, mesh topologies are very expensive, as they involve a high level of redundancy. Thus, they are hardly used for wired connectivity.

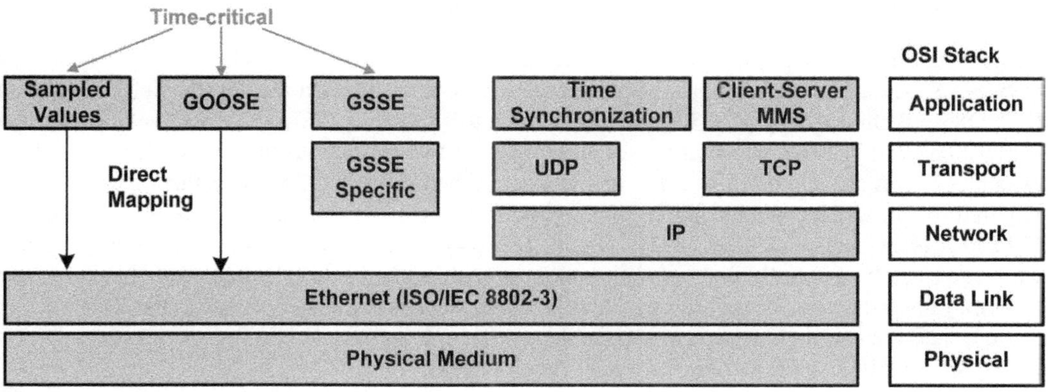

FIGURE 13.7 OSI layer and the IEC 61850 communication services.

However, mesh topologies are more popular for wireless networks. Also, full mesh topologies find application as backbone networks. Some of the advantages of mesh networks are their ability to withstand high data traffic since multiple independent paths can be formed, robustness, expansion and reconfiguration.

13.3.2 Communication Infrastructure for IEC 61850-Based Systems

Substation communication refers to the transfer of information from a source device to a destination device within a substation using a communication network. The OSI model divides the data communication process into seven distinct layers which define the data handling during the different stages of data transmission. These layers are the application, presentation, session, transport, network, data link, and physical layers. Each layer provides a service for the next layer above it. Figure 13.7 shows five of the layers of the OSI model and the IEC 61850 communication services. Five types of communication services are defined in the IEC 61850 standard: SVs, GOOSE, generic substation state event (GSSE), time synchronization, and MMS. The sampled values, GOOSE, and GSSE are mapped directly to the data link layer, while the time synchronization and MMS are mapped to the data link layer through the transport and network layers, respectively. At the physical layer, the specification allows connection to be made either electrically using an RJ45 connector (usually for testing), or by fibre-optic connection using LC or ST (BFOC2.5) type fibre connectors.

Standard signalling speeds for Ethernet communications are 10 Mbps, 100 Mbps, 1 Gbps and 10 Gbps. At present, MUs are implemented with 100 Mbps Ethernet communication cards. The data link layer (layer 2 of the OSI model), media access control (MAC) addresses take care of the physical addressing of devices on a network. These physical addresses are set in the manufacturing process using a 48 bit addressing scheme that allows a unique address for every piece of equipment manufactured.

13.4 Wide Area Data Exchange

With existing and new applications in the field of power system operation and protection, the requirement to exchange standardized information between devices in different substations is increasing. Also, communication between substations and control centres was recently covered in some of the technical reports of the IEC 61850 standard. The IEC 61850 standard provides the features to be used for such information exchanges. Substation-to-substation communication and substation-to-control centre communication are further discussed in the following subsections.

13.4.1 Substation-to-Substation Communication

IEC 61850 Ed. 1 standard was drafted and ratified for the information exchange between the SAS devices in the substation. This has been extended for communication outside the substation in the IEC 61850 Technical Report (TR) 90-1. Substation-to-substation (SS-to-SS) communication refers to functions within SASs which are distributed between two substations or between functions. Some examples of this are line protection and interlocking schemes. Figure 13.8 shows the logical interfaces required for SS-to-SS communication. The required interfaces for this function are logical interfaces IF2 and IF11, respectively. Interface IF2 is defined for protection-related functions between substations, while interface IF11 is dedicated to the communication of remote control devices. Note that these interfaces may not necessarily be according to the IEC 61850, but it is acceptable as long as the SAS functions on both sides are based on the IEC 61850 standard [14].

The IEC 61850-90-1 TR provides the following overview:

- Defines the use cases for substation-to-substation information exchange.
- Describes the communication requirements.
- Gives guidelines for the implementation of communication services and architecture.
- Defines the data required for interoperable applications.
- Describes the usage and enhancement of the IEC 61850 configuration language.

13.4.2 Substation-to-Control Centre Communication

The classical information exchange between substations and control centres involves the transmission of telecontrol information such as status information or measurements. This is usually achieved using classical protocols such as the IEC 60870-5-101 or 104, and DNP3. Recently, control centres are beginning to adopt object-oriented data modelling and standardized interfaces for data exchange based on the IEC 61968 [15] and IEC 61970 Common Information Model (CIM) [16]. For the interconnection of substations and

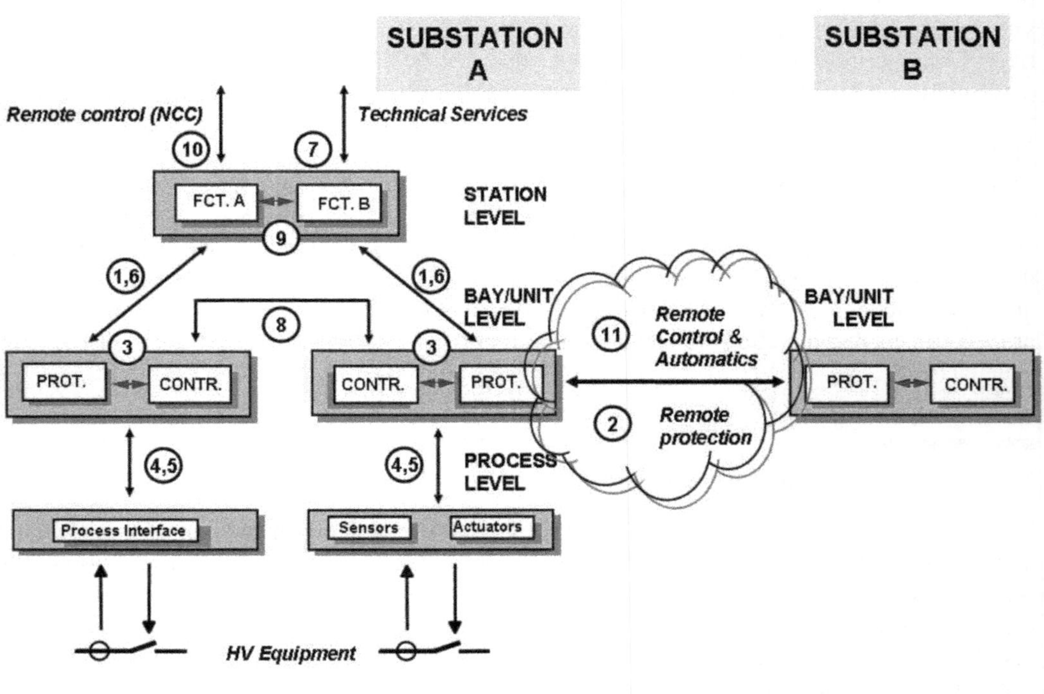

FIGURE 13.8 Logical interfaces between substations. (From IEC 61850, Communication networks and systems in substations, *International Electrotechnical Commission (IEC)*, 2003.)

control centres, the IEC 61850 standard was extended using the IEC 61850-90-2 for data exchange between SSs and control centres (CCs). The IEC 61850-90-2 TR provides a comprehensive overview of the requirements for IEC 61850 SS-CC information exchange or between substations and maintenance systems.

Control centre applications can access the substation devices (IEDs) using two methods: directly or via a gateway. The direct method grants full access to the control centre applications, while the gateway method uses proxies. The use of proxies restricts control centre applications because they only use the services that are provided to them by the gateway. Figure 13.8 shows that the logical interfaces of interest for SS-to-CC communication are the interface IF7 and IF10, respectively. Interface IF7 is defined for technical services, while interface IF10 is for CC remote control [17]. The engineering workflow for SS-CC communication requires two levels of communication: the substation LAN and the SS-CC WAN.

In particular, the IEC 61850-90-2 TR provides the following overview:

- Defines use cases and communication requirements that require an information exchange between substations and control or maintenance centres,
- Describes the usage of the configuration language of IEC 61850-6.
- Gives guidelines for the selection of communication services and architectures compatible with IEC 61850.
- Describes the engineering workflow.
- Introduces the use of a proxy/gateway concept.
- Describes the links regarding the specific communication service mapping.

13.5 System Engineering

13.5.1 System Engineering Tools

The IEC 61850 system engineering process describes the design, implementation, and commissioning of substation protection and automation systems, and their functional requirements. Typically, this involves the exchange of eXtensible Markup Language (XML)-based SCL files, which are used in the definition and configuration of the various IEC 61850 applications and devices. The representation of the IEDs and the SA functions using SCL files is one of the most prominent features of the IEC 61850 standard that enhances the communication capabilities between different IEDs towards a complete interoperable system. Part 6 of the IEC 61850 standard defines an XML-based SCL, which provides the formal description of the relationship between the substation automation system and the substation switchyard in an IEC 61850 project. Each device must provide an SCL file that describes its own configuration. The SCL contains four subsections: (1) the substation section describing the single line diagram and its LNs, (2) the communication section linking the IEDs, (3) the IED section describing the capabilities of the IEDs, and (4) the binding to LNs belonging to other IEDs, and the logical node type (LNType) section which defines the data objects within the logical nodes.

The IEC 61850 edition 2 defines six types of SCL files:

- IED capability description (ICD): describes the capabilities of the IED, and includes an IED section, data type templates, logical node type definitions, optional communication section and an optional substation section.
- System specification description (SSD): describes the specifications of the system, and includes a substation description section, data type templates, and logical node type definitions.
- Substation configuration description (SCD): describes the power system, and includes a description section for each IED, a communication configuration section and a substation description section. It presents the exchange of data from the system configurator to the IED configurators.
- Configured IED description (CID): used to communicate data and settings between an IED and its configuration tool. It includes a communication section that contains the current address of the IED.

- Instantiated IED description (IID): describes the exchange of data between IED configurator and the system configurator for a single preconfigured IED.
- System exchange description (SED): describes the exchange of data between the system configurators of different projects.

In the engineering of an IEC 61850-based project, the SCL files are used in exchanging the configuration data between the IED configuration tool and the system configuration tool from one or more vendors. The file type for this is the *.icd* file. Also, the exchange of data can be between the system specification tool and the system configuration tool. The file type of interest is the *.ssd* file. Furthermore, the exchange of data can be between the system specification tool and the IED configuration tools. The file type of interest for this is the *.scd* file. Another data exchange is that between the IED configuration tool and the IED. The extension for this file type is the *.cid* file. The IED configuration consists of the automation/protection configuration, and the IEC 61850 communication configuration. The automation/protection configuration is commonly done using the IEC configuration tool, including the programmable scheme logic (PSL). The communication configuration requires the *.icd* file, the definition of the network settings, the configuration of the GOOSE dataset, GOOSE transmit, GOOSE receive, and the configuration of the buffered/unbuffered reporting, respectively.

Two levels of communication are required for communication outside the substation: the substation LAN and the SS-CC WAN. This implies that the engineering process for each of the communication networks needs to be done separately. The *.icd* files are used by the system configurator as the IED template to instantiate project specific IEDs. The instantiation can also be done by importing the *.iid* file of the IEDs. All the data that are forwarded to the control centre are mapped as inputs to the proxy/gateway client IED. The SCD file generated by the system configurator tool is then imported by the IED configurator for the final IED instance configuration. In the IED configurator for the proxy/gateway server, all the restructuring and renaming using aliases may be done, and the LNs of the proxy/gateway server may be added.

A system integrator is required in the design, engineering, installation, testing and commissioning of IEC 61850-based projects. This is a multidisciplinary task requiring the knowledge of electrical engineering and communication networks. The former requires the design, configuration, and implementation of protection and automation schemes, while the latter involves the design, configuration, and implementation of the appropriate communication architecture, configuration of the IED *.icd* file, the configuration of *.cid/.scd* files, factory acceptance tests, and commissioning tests. In addition, a system integrator should also have broad technical knowledge of most vendor devices and engineering tools. The system configuration tool is an IED independent system level tool capable of importing and exporting configuration files defined by the IEC 61850 standard from several IEDs during system level engineering of an IEC 61850 project. The system configuration tool should be capable of generating a substation-related configuration file which may be sent to the IED configurator for system-related IED configuration. Figure 13.9 shows a system configuration tool (AcSELerator Architect) from Schweitzer Engineering Laboratories (SEL). Pane A contains the *.icd* files of various IED types, while Pane B contains the *.cid* files. The dataset tab in Pane C gives the datasets configured for the devices. Pane C also gives the transmit, receive, and reporting configuration, respectively. Figure 13.10 shows a MiCOM S1 Agile IEC 61850 IED Configurator Tool from Alstom Grid (now General Electric). Pane A gives the IED details, IED communication details, time synchronization, and details of the GOOSE messages. Pane B gives the details corresponding to the selected GOOSE parameter in Pane A.

13.5.2 Redundancy Schemes

Two types of communication network redundancies are available: the static and dynamic redundancies. The static redundancy provides seamless switchover but generally costs more, while the dynamic redundancy is cost-effective but requires switchover time. Some protocols such as the dual homing star protocol (DHSP) and the rapid spanning tree protocol (RSTP) both offer redundancy, but they are not effective

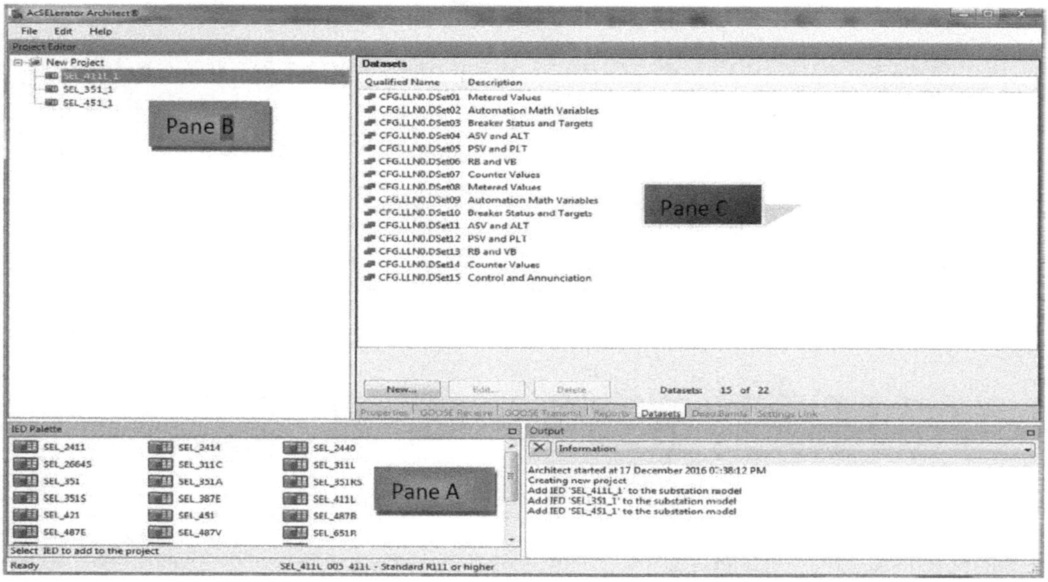

FIGURE 13.9 Example of IEC 61850 IED engineering tool from Schweitzer Engineering Laboratories (SEL).

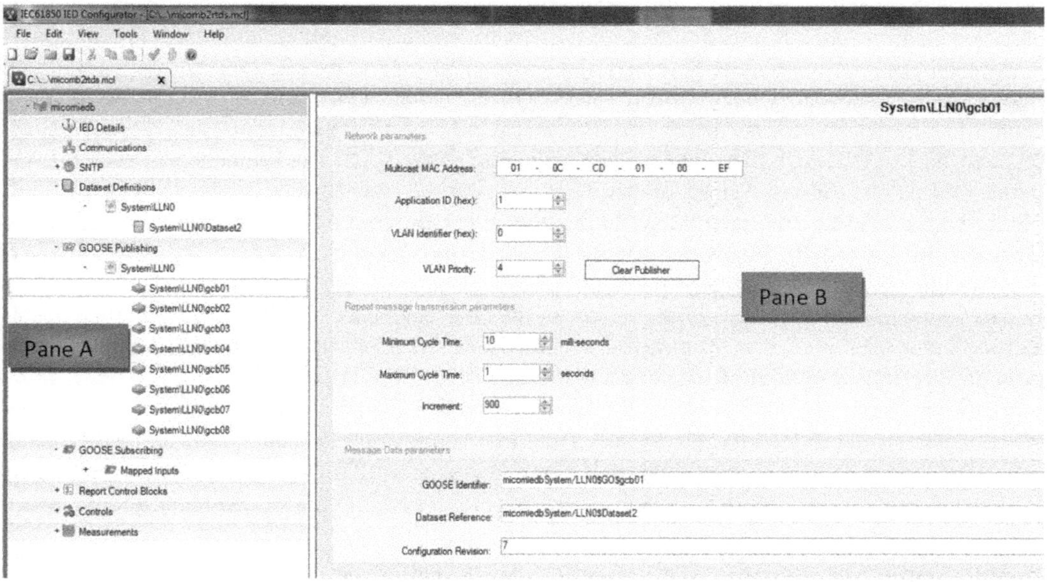

FIGURE 13.10 Example of IEC 61850 IED engineering tool from Alstom Grid.

for substation automation. To overcome these limitations, the IEC 62439-3 standard (industrial communication networks—high availability automation networks) was adopted for SASs. The advantage of the redundancy protocols defined in the IEC 62439-3 is that they do not experience reconfiguration timeouts for a single failure of one of their network segments.

13.5.2.1 Parallel Redundancy Protocol

The parallel redundancy protocol (PRP) is a redundancy protocol designed to provide Ethernet network availability for a single point of failure using static redundancy. The operating principle of the PRP is based on the use of dual independent parallel (redundant) networks (LAN A and LAN B) in which Ethernet

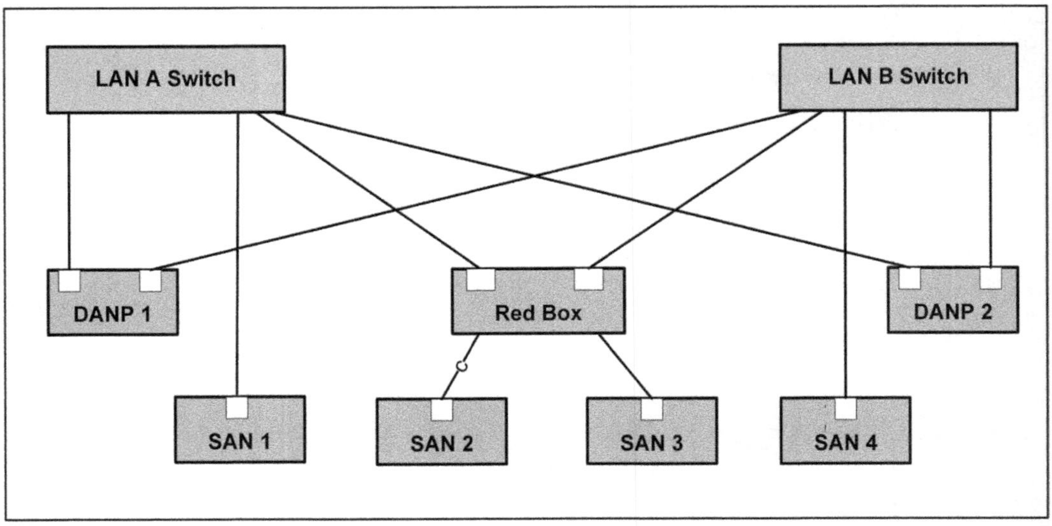

FIGURE 13.11 Example of a PRP network.

frames are duplicated and sent simultaneously over two separate networks. The IEC 62439-3 Clause 4 [18] defines the specifications for the PRP, and mostly finds application in star-topology networks in the substation environment.

PRP networks can have four node types: (1) double attached node with PRP (DANP), (2) single attached node (SAN), (3) redundancy box (RedBox), and (4) virtual doubly attached node (VDAN). As shown in Figure 13.11, the doubly attached nodes DANP 1 and DANP 2 have full node redundancy, while the singly attached nodes SAN 1 and SAN 4 do not have any redundancy. SANs can be connected to both LANs via the RedBox that converts a singly attached node to a doubly attached node, as shown in Figure 13.11.

Since both ports share the same MAC address, there is no impact on the functioning of the address resolution protocol (ARP). Thus, every data frame is seen by both ports. When a node sends a frame of data, the frame is duplicated on both ports and LAN segments, thereby providing a redundant path for the data frame in the event of the failure of one of the LAN segments. Note that during normal operating conditions, each port receives identical frames. However, duplicates are discarded by the receiving node. Typically, the RedBox appends a redundancy control trailer (RCT) on each Ethernet frame before transmission. The RCT consists of a sequence number (SeqNr), LAN identifier (Lanid), frame size (LSDUsize), and PRP suffix (PRPsuffix). The shortcoming of the current version of the PRP is that it was designed for layer 2 communication and is unable to support IP routing in routable-sampled values (R-SVs) and routable GOOSE (R-GOOSE) applications for SS-SS and SS-SC communication schemes.

13.5.2.2 High-Availability Seamless Redundancy Protocol

The high-availability seamless redundancy (HSR) protocol is based on the dual transmission of message frames over ring-topology networks in both directions. In the case of an error, the message frame will be transmitted without any delay. Also, no reconfiguration time is required for the network in the case of a single failure of the communication network. The HSR protocol is defined by the IEC 62439-3 Clause 5 [18]. The HSR protocol works on the premise that each device connected in a ring network is a doubly attached node running the HSR protocol. Such devices are referred to as the doubly attached node running HSR (DANH).

Similar to the PRP, SANs are connected via the RedBox. Figure 13.12 shows a simple HSR network, where a doubly attached node is sending a multicast frame. The frame is duplicated, and each duplicate frame is tagged with the destination MAC address and the sequence number. The frames differ only in their sequence number, which is used to identify one copy from another.

For convenience, the duplicate frames are labelled A frame and B frame, respectively. If the received frame is a duplicate, it will be discarded. Each frame is sent to the network via a separate port.

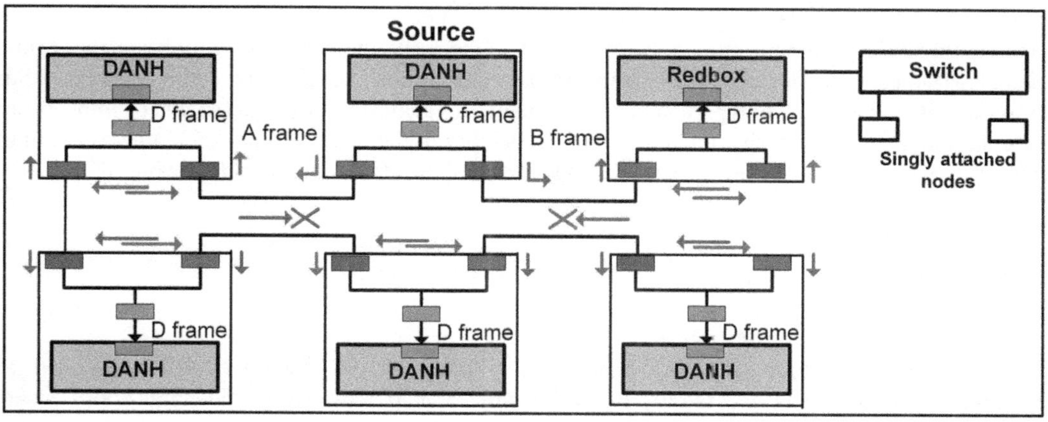

FIGURE 13.12 Example of an HSR network.

The destination DANH receives two identical frames from each port, removes the HSR tag of the first frame received and passes this to its upper layers. Also, in order to prevent the frames from circulating infinitely, the HSR node that initially placed the frame on the ring network will remove it on the completion of one cycle of frame transmission. A quadruple port device referred to as a QuadBox can be used to connect two HSR rings. The QuadBox behaves as an HSR node in each ring and is capable of filtering the traffic between the rings in addition to forwarding the traffic from one ring to the other.

13.6 Testing

Three types of tests are defined in this regard: (1) conformance testing, (2) performance testing, and (3) interoperability testing.

Conformance testing relates to the evaluation of an IED referred to as the device under test (DUT) in order to determine if the DUT conforms to the specifications of a particular standard. This is done by exchanging messages between a test system and the DUT. The test system sends a carefully selected array of test inputs to the DUT and records the responses of the DUT. IEC 61850 conformance testing is specified in Part 10 of the IEC 61850 standard. If the DUT is a server (such as a protective relay), then the test system emulates the networked combination of clients (like a substation data concentrator or historian) and other peer servers (like other protective relays on the LAN). To run a conformance test, it is important to first review the design information on the DUT as specified in the Protocol Implementation Conformance Statement (PICS). The PICS summarizes the communication capabilities of the system or device to be tested, as a subset of all that IEC 61850 offers. The Model Implementation Conformance Statement (MICS) gives details of the standard data object model elements supported by the system or device. The Protocol Implementation eXtra Information for Testing (PIXIT) is an optional document containing specific information on the communication capabilities that are beyond the scope of the IEC 61850 standard.

Performance testing refers to the specific requirements to be measured and evaluated in an IEC 61850-based system such that the claims specified in the DUT documentation can be compared with the measured results and with similar devices from other vendors. Interoperability testing is a test of the ability of a DUT to communicate seamlessly with another device from the same vendor or a different vendor.

13.7 Cyber Security

The security of the critical infrastructure in an IEC 61850-based system has become even more pronounced with the IEC 61850 Ed. 2 standard, which permits external communications such as SS-SS and SS-CC communications. Thus, communication is no longer isolated or confined within a substation.

Cyber attacks in the electric grid can take two forms: (1) threats to the physical security of the assets, and (2) threats to communication network security. Generally, these threats could be in the form of unauthorized access to devices, data or information; modification or theft of data or information; denial of service attack; and lack of repudiation. Countermeasures against these threats should be implemented in order to ensure the security, reliability, and availability of data and communication networks in the electric power grid. These countermeasures should be designed to maintain the following:

- Confidentiality
- Integrity
- Availability

Confidentiality refers to the ability to maintain the privacy of the contents of real-time and stored (archived) data, and limit access only to the intended user. If required, confidentiality also ensures that the sender and receiver of data are anonymous, and should remain unobservable by third parties. Integrity protects the authenticity of data or information within the system or over a communication network. A loss of information integrity results in an authorised modification or destruction of information. For example, the line currents of a normal operating condition could be "spoofed" or modified to a fault current greater than the pickup current setting of an overcurrent protective relay. Thereby, causing incorrect/nuisance tripping of the associated circuit breaker. Availability describes the continuous, uninterrupted, reliable, and timely access to a system, data or information by authorized users. A loss of availability in the form of denial-of-service attacks could result in the loss of grid situational awareness and reliability when system protection schemes fail as a result of lack of measurement inputs. The following subsections present the cyber security threats and vulnerabilities in SASs, and some detection and countermeasures against these threats.

13.7.1 Security Threats and Vulnerabilities in IEC 61850-Based Systems

The security threats and vulnerability in IEC 61850 systems can be physical attack, denial-of-service (DoS) attack, data spoofing, man-in-the-middle (MitM) attack, breach of packet confidentiality, and malicious code attack. Physical attack on an IEC 61850 infrastructure can be in the form of interruption or deliberate damage to equipment, communication network, and software applications. For example, the Ethernet cable connecting an IED to the communication network can be maliciously disconnected or slashed. Also, physical attack can be in the form of damage to the power supply to these assets. Some guidelines for mitigating physical intrusions can be found in IEEE 1402-2000 (R2008) [19]. A DoS attack on an IEC 61850-based system is an attack that compromises the availability of data to the services/applications that require such data by flooding a communication network with irrelevant traffic, thus causing communication network delays, packet losses, attenuation, and bandwidth limitations that will prevent the publishing and subscription of time-critical IEC 61850 messages. Data spoofing is the falsification of data by an attacker by masquerading as an authorized user of a service. In this way, the attacker sends false messages to the various components of the IEC 61850 system. For instance, an attacker could masquerade as a substation IED by streaming false data to control centre applications, thus causing the control centre protection/control applications to wrongly operate. This could lead to the loss in grid stability and reliability.

A MitM attack is an attack that could occur between the substation and control centre devices, especially during SS-to-SS or SS-to-CC communication. In such an attack, the attacker masquerades as legitimate. For example, the attacker could pretend it is the control centre application to the IED. Similarly, it could act as if it is the IED to the control centre application in order to gain access to the WAMS. With a MitM attack, an attacker maliciously integrates itself in the communication channel between two endpoints (hosts), with the attacker secretly impersonating both hosts and gaining access to the information being exchanged between the two hosts. It targets the confidentiality, integrity, and availability of information. MitM attacks can be divided into two types: (1) passive and (2) active. A passive MitM attack can be defined as one in which the attacker eavesdrops on the information transmitted between the hosts; an active MitM attack is the form of attack in which the attacker eavesdrops and also performs an action such as inserting, deleting, and/or modifying the information transmitted from one host to another. The confidentiality of the information contained in IEC 61850 messages can be breached through the sniffing of the R-SV or R-GOOSE packets. Typically, address resolution protocol (ARP) spoofing is used to sniff the traffic in the communication network. This type of attack would normally precede data spoofing and MitM attacks. Malicious code injection can be in the form of the injection of false measurements or command injection. False measurements would cause protection/control applications to operate incorrectly, e.g., trip when it is not expected to operate, or fail to operate when required. Also, commands in the form of malware introduced via the internet, removable USB drives, or via an engineering (maintenance) workstation can be inserted to erase or modify the IED settings, or steal IED login credentials.

13.7.2 Threat Detection and Countermeasures

To protect the critical assets in IEC 61850 systems, it is necessary to implement security measures that provide reliable and robust systems resilient to cyber attacks. It is best that the security measures are designed in the earliest stages of a project, and they should be considered through all phases of the process, from design to development to rollout. Prior to the deployment of an IEC 61850-based system, a set of cyber security requirements must be developed, new equipment must undergo vulnerability testing, and proper security controls must be designed to protect the IEC 61850 system from unauthorized access.

It is difficult for a single security measure to provide protection against all types of possible cyber threats. It is expected that multiple layers of security measures are implemented physically and on the communication network. For physical security, the communication network switches should be locked in racks/cabinets. Also, unused ports on the switches should be turned off. Switch management and port security can be further implemented. Access to a device via a port should be limited to MAC address and IP address, and only registered devices should be allowed to connect to the network. For network security, security mechanisms for ensuring the authentication, authorization and user access to network devices should be in place. It can be implemented by identifying and controlling who or what has access to which resources. Also, rules can be set by filtering incoming traffic against an access list. Furthermore, role based access control (RBAC) using the separation of duty approach can be implemented.

Some common DoS countermeasures include the use of large bandwidth connections to ensure that the network can handle the traffic, distributed or redundant infrastructure, filtering routers, disabling IP broadcasts, applying security patches, disabling unused ports, and performing intrusion detection. Code injection can be made difficult by randomizing the system address space, separating code and data, and monitoring the stack to detect buffer overflows. To counteract SQL injection attacks, inputs should be checked for characters that can be abused, and users should be mandated to use a static template with certain inputs that can be translated into queries. Also, databases should have strict access controls for allowing users to modify or manipulate data.

VLANs can be used to segment networks within a physical LAN such that devices connected to different physical switches can communicate with each other as long as they are in the same VLAN. Devices which are in different VLANs cannot communicate with each other and would require a VLAN-aware router to route packets. Aside from the use of VLANs, access control lists (ACLs) can be created in routers to prevent unauthorized access to the network. Also, firewalls can be applied to add security to an existing network. Firewalls can log, administer, and audit network access in order to create alerts during attacks and failures. They provide secure control functions like inspection, content inspection, access control, user control, protocol and services control, as well as data control for secure substation automation networking. Generally, firewalls can be used to create secure cells and secure zones inside a network and set restricted communication outside a cell and zone. Gateways may be applied to achieve cyber security against a variety of cyber attacks. A gateway collects metering, status, event, and fault report data from IEDs and RTUs and creates an interface between substation automation systems and external connections like a web browser or ERP systems. It can manage, filter and control data traffic and secure IEDs and other devices against external access. Normal gateways can be achieved by using virtual private networks (VPNs) and encryption.

13.8 IEC 61850 Use Cases

13.8.1 IEC 61850 GOOSE Messages Packet Analysis

The verification of the structure of IEC 61850 GOOSE messages can be carried out as part of the conformance testing procedure mentioned in Section 13.6 through the use of a network protocol analyser—Wireshark which sniffs the packets off a communication network to which IEC 61850-compliant IEDs are connected. Figure 13.13 shows a Wireshark capture obtained off a communication network to which actual IEDs publishing IEC 61850 GOOSE messages are connected. The Wireshark capture has been partitioned into Panes A–C for better analyses. Pane A gives the information of all the packets captured at the Ethernet interface in real time. A *goose* filter was applied to show only the captured IEC 61850 GOOSE packets. Pane A has five columns showing the number of entries, time of capture, the source, destination, type of protocol, and information on the captured packets, respectively. From Column 3 with header Source, three different IEDs can be observed to be publishing IEC 61850 GOOSE messages. These are the Areva P847 IED, SEL 451 IED, and the RTDS GTNET-GSE, respectively. Pane B presents the information corresponding to the selected packet in Pane A. This includes the information on the captured frame, the Ethernet destination and source addresses, and the details of the GOOSE messages. Pane C gives the hexadecimal values corresponding to the messages in the captured packets. For example, in Figure 13.13, the state number *stNum* in Pane B has a value of 34123 in decimal. The corresponding hexadecimal value is correctly shown in Pane C as 854b.

Figure 13.13 indicates that a frame message having 385 bytes was captured for the Areva P847 IED. Both the destination and source addresses are multicast addresses of 6 bytes, with the message type correctly indicated as 0 × 88b8 (GOOSE message type) as per the IEC 61850 standard. The application identifier (*APPID*) is 2 bytes in length and is used to select GOOSE messages from the frame and to distinguish the application association. The GOOSE Data Protocol Unit (*goosepdu*) contains the GOOSE control block reference (*gocbRef*), time allowed to live in ms (*timeAllowedtoLive*), dataset (*datset*), GOOSE ID (*goID*),

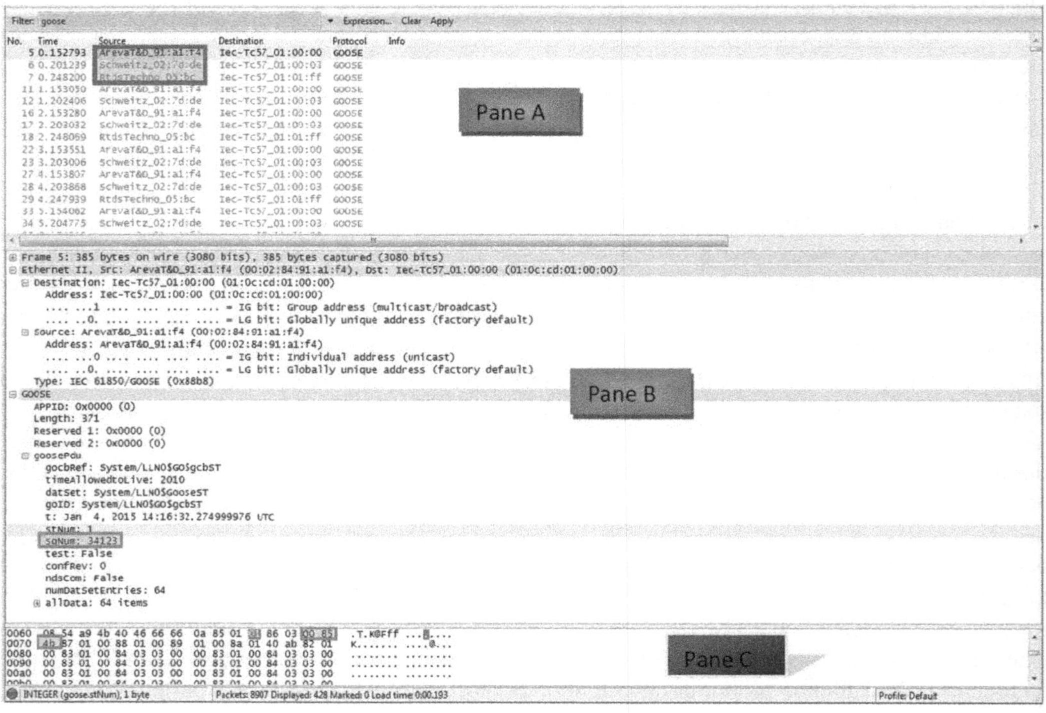

FIGURE 13.13 IEC 61850 GOOSE message analysis.

time of capture (t), status number (*stNum*), sequence number (*sqNum*), test bit, configuration revision (*confRev*), needs commissioning (ndsCom), and number of data set entries (*numDatSetEntries*). The last portion of the GOOSE message is the user-defined data content shown in Figure 13.13. The user-defined data attributes (*allData*) consist of the configured data items in the dataset.

Other parameters for GOOSE messages which are not shown in Figure 13.13 are the IEEE 802.1Q (for VLAN information), priority tagging (used to separate time critical and high priority bus traffic), TPID (Tag protocol identifier), and the TCI (Tag Control Information). The TPID is the Ethertype assigned for 802.1Q Ethernet encoded frames, while the TCI consists of the canonical frame indicator (CFI) and an optional VLAN identifier (VID). Further details can be found in the IEC 61850-8-1 [13].

13.8.2 IEC 61850 Sampled Values Packet Analysis

Similar to the verification of the structure of IEC 61850 GOOSE messages carried out in Section 13.8.1, sampled value messages can be analysed through Wireshark capture of the packets published by the merging units connected to a communication network. Figure 13.14 shows a Wireshark capture of a communication network with the *sv* filter applied. Pane A has five columns showing the number of entries, time of capture, the source, destination, type of protocol, and information on the captured packets, respectively. From Column 3, two merging units can be observed as highlighted. These are the Alstom MU (80:b3:2a:09:4c:c7) and the RTDS GTNET-SV (RtdsTechno_01:d4), respectively.

Pane B presents the information corresponding to the selected packet in Pane A. This gives the information on the captured frame, the Ethernet destination and source addresses, and the details of the SV messages. Pane C gives the hexadecimal values corresponding to the messages in the captured packets.

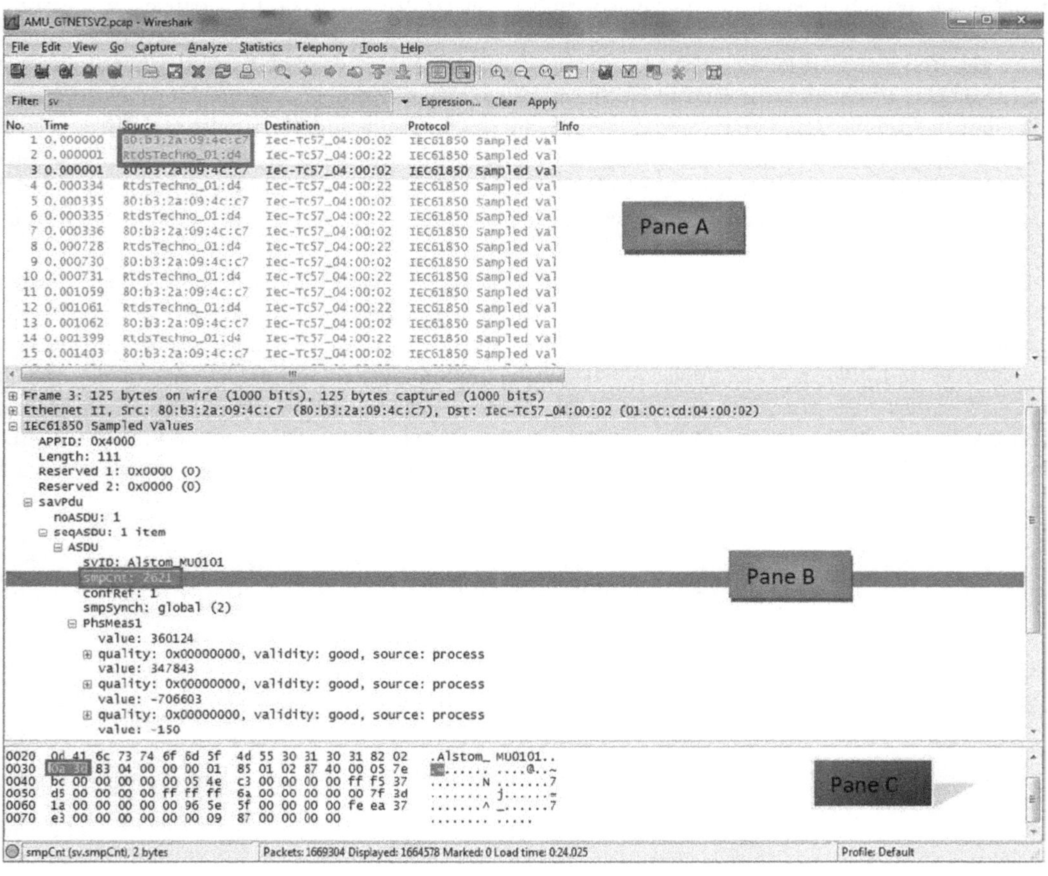

FIGURE 13.14 IEC 61850 sampled values message analysis.

In Figure 13.14, the sample count *smpCnt* (highlighted) in Pane B has a value of 2621 in decimal. The corresponding hexadecimal value is correctly shown in Pane C as 0a3d.

Frame number 3 of the captured packets bytes from the Alstom MU has a frame message of 125. Both the destination and source addresses are multicast addresses of 6 bytes, with the message type correctly indicated as 0 × 88ba (SV message type), per the IEC 61850 standard. The application identifier *APPID* is 2 bytes in length and is used to select SV messages from the frame and to distinguish the application association. The sampled protocol data unit (*savPdu*) contains the number of Application Service Data Unit (ASDU) *noASDU*, and the sequence of ASDU *seqASDU*.

The ASDU is contained in the *seqADSU* and it consists of the following: (1) sampled value identification (*svID*), (2) sample count (*smpCnt*), (3) configuration revision number (*confRef*), (4) sample synchronized by a clock signal (smpSynch), and (5) the SV data (PhsMeas1). A comparative analysis of the impact of IEC 61850-9-2 sampled values on the operating performance of protective IEDs was presented in [20]. Further details on sampled values can be found in the IEC 61850-9-2 standard, IEC 61850-9-2LE guidelines, and in the IEC 61869-9 standard [5,7,8].

13.8.3 IEC 61850 Communication-Assisted Protection Scheme

Communication schemes are used in speeding up the fault clearing times for remote faults. These faults would normally be cleared by the conventional step distance protection scheme after the lapse of an intentional time delay. During this time the fault conditions could lead to system instability or cascading blackout. Communication-assisted schemes use communication between the local and remote ends by transmitting relevant information from one end of the line to the other. Examples of some communication-assisted schemes are directional comparison blocking (DCB), direct under-reach transfer trip (DUTT), permissive under-reach transfer trip (PUTT), and permissive over-reach transfer trip (POTT).

An IEC 61850-based POTT distance protection scheme [21] is presented in this subsection as a use case. POTT schemes are used in facilitating accelerated tripping using the Zone 2 overreaching element from the local end to transmit a permissive trip signal to the remote end. This implies that if a local IED detects a local fault, it will send an inter-tripping permissive trip signal (using its Zone 2 element start signal) to the remote IED. The remote IED uses its Zone 2 element start signal in combination with the received permissive trip signal from the local IED to issue an instantaneous Zone 2 trip.

Thus, the remote end IED operates faster without having to wait for its configured intentional time delay before tripping. Figure 13.15 shows the LN, data, and data attributes for the POTT functionality as defined for a SEL-421 IED. Figure 13.16a shows the protection scheme logic for implementing a POTT scheme; Figure 13.16b shows its implementation using the IEC 61850 object modelling structure. When a Zone 2 fault is detected by the local IED, the status of the data attribute POTTPSCH1.Op.general changes from FALSE to TRUE, and a GOOSE message is published via Ethernet to the remote IED. The remote IED receives the POTTPSCH1.Op.general signal in its virtual input, and combines it with

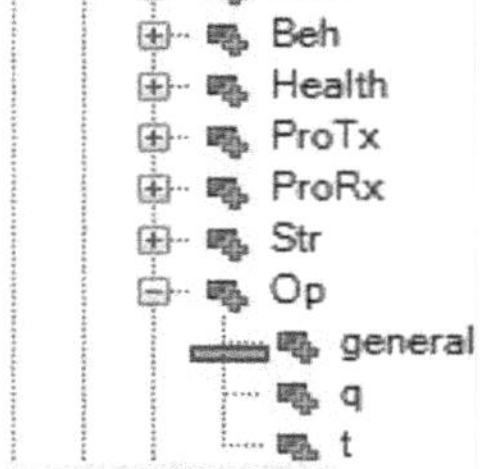

FIGURE 13.15 Communication scheme logical node POTTPSCH in the AcSELerator Architect engineering tool.

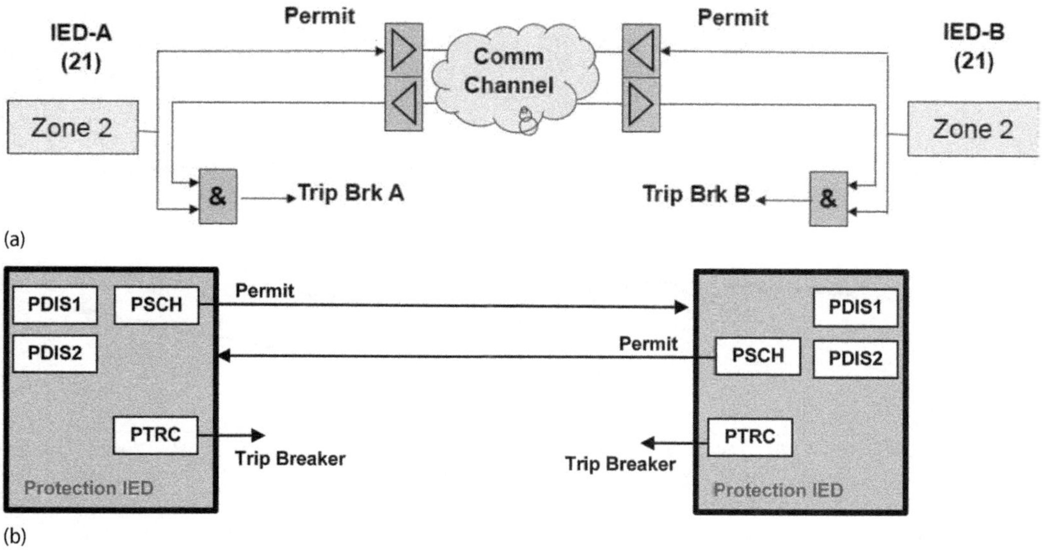

FIGURE 13.16 Architecture for (a) POTT communication scheme, (b) IEC 61850-based POTT implementation.

the Zone 2 start signal (Z1PPDIS1.Op.general). It then issues a trip signal to its associated circuit breaker (XCBR) via the protection LN PTRC. For the performance test carried out, the DUTs are fed with analogue signals from CTs and VTs using hardware-in-the-loop simulations conducted with the Real-Time Digital Simulator (RTDS®). This ensures that both IEDs are subjected to the same fault currents and voltages at the same instant.

Figure 13.17 shows the disturbance waveform of the analogue and binary signals extracted from one of the IEDs used. The local IED was configured to send a permissive trip signal to the remote end IED for local faults using IEC 61850 GOOSE messages. The results for the operate time obtained using the IEC 61850 GOOSE messages-based POTT scheme are shown in Figure 13.18.

The results obtained using IEC 61850 GOOSE messages were compared to those obtained using the SEL proprietary mirrored bits protocol, as shown in Figure 13.19. The IEC 61850 GOOSE messages proved to be as reliable as the mirrored bits protocol, and were faster in most cases. It should be noted that the IEC 61850 GOOSE messages went through multiple network switch hops, while the mirrored bits connection was via a direct serial-serial cable between the two DUTs.

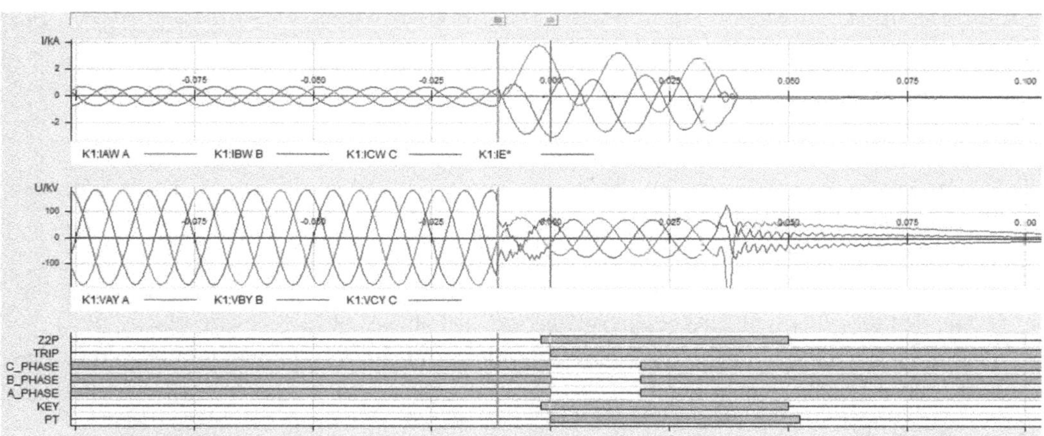

FIGURE 13.17 Disturbance record of a three-phase fault showing the analogue and binary quantities. (From T. Arnold et al., *IEEE Proceeding*, 1–7, 2005, © (2015) IEEE.)

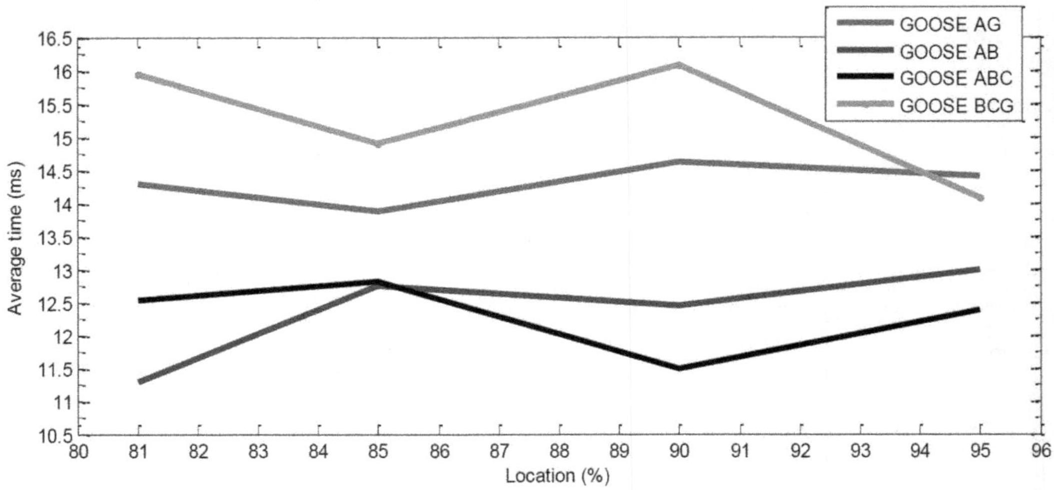

FIGURE 13.18 Case study 3: average trip time using IEC 61850 GOOSE messages for the POTT scheme. (From T. Arnold et al., *IEEE Proceeding*, 1–7, 2005, © (2015) IEEE.)

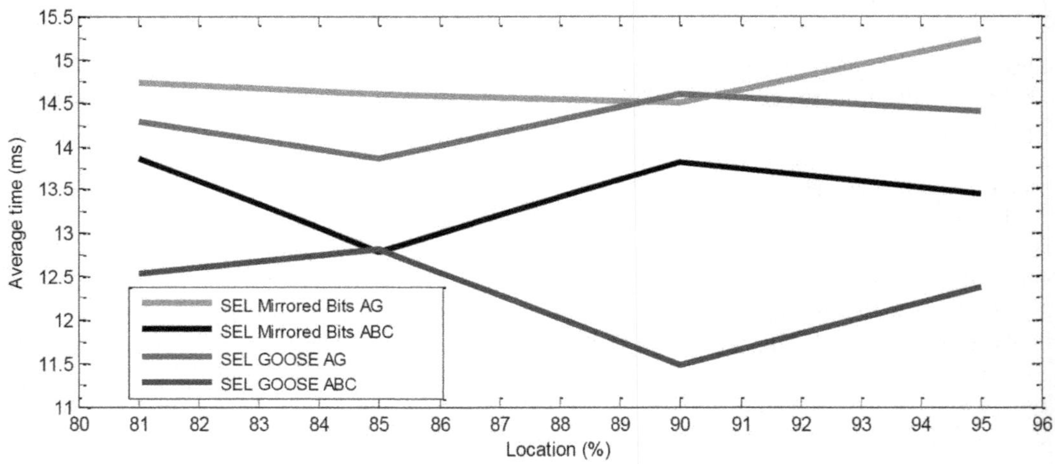

FIGURE 13.19 Comparison of the average trip time for IEC 61850-based POTT scheme and for mirror bit protocol-based POTT scheme. (From T. Arnold et al., *IEEE Proceeding*, 1–7, 2005, © (2015) IEEE.)

13.9 Tutorial Problems

1. List and discuss briefly any three proprietary protocols used in electric power substations for substation automation and control.
2. List five benefits of the IEC 61850 standard.
3. Communication within the substation generally takes place over three distinct logical levels. Mention these levels and the equipment that can be found in each of these levels.
4. With the aid of a diagram, compare and contrast the IEC 61850 SV-based protection IED with the conventional analogue input IED.
5. List and discuss the two redundancy protocols defined in the IEC 62439-3 standard.
6. List and discuss any five cyber security threats that can occur in IEC 61850-based systems and possible countermeasures for preventing them.

13.10 Conclusion

This chapter presented an introduction to substation automation and control in modern power systems. Also, the benefits and factors that should be considered in the design and implementation of SASs were given. The IEC 61850 standard, as a facilitator for achieving modern nonproprietary automation and control, was presented and discussed. The object-oriented data modelling approach used in the IEC 61850 standard and the various services are described. Furthermore, the application of the standardized configuration language in the actualisation of IEC 61850-based systems was elucidated.

Now that communication outside the substation is possible using the IEC 61850 Ed. 2 standard, it is important to ensure the confidentiality, integrity, and availability of the data through adequate cyber threat detection and countermeasures. In addition, two redundancy protocols adopted for use in power system automation (PRP and HSR) were presented. These redundancy protocols are reconfigurable with no switching delays. Three IEC 61850 use cases covering the packet analysis of GOOSE and SVs messages, and a communication-assisted protection scheme were presented.

ACKNOWLEDGMENT

The authors thank the International Electrotechnical Commission (IEC) and the Institute of Electrical and Electronics Engineers (IEEE) for permission to reproduce Information from its International Standards and publication. All such extracts are copyright of IEC, Geneva, Switzerland. All rights reserved. Further information on the IEC is available from www.iec.ch. IEC has no responsibility for the placement and context in which the extracts and contents are reproduced by the author, nor is IEC in any way responsible for the other content or accuracy therein.

REFERENCES

1. C. Brunner, Extending IEC 61850, *PAC World Magazine*, pp. 1–4, 2012.
2. J. A. Momoh, *Electric Power Distribution, Automation, Protection, and Control*, Boca Raton, FL: CRC Press, 2007.
3. Modbus Protocol, Modbus Organization. Retrieved August 2, 2013. http://modbus.org/docs/PI_MBUS_300.pdf
4. IEEE Standard for Electric Power Systems Communications-Distributed Network Protocol (DNP3), 2012.
5. IEC 61850, Communication networks and systems in substations, *International Electrotechnical Commission (IEC)*, 2003.
6. IEC 61850-1, Communication networks and systems in substations—Introduction and overview, *International Electrotechnical Commission (IEC)*, 2003.
7. IEC 61850-9-2, Implementation guideline for digital interface to instrument transformers using IEC 61850-9-2, UCA International Users Group. 2004.
8. IEC 61869-9, Instrument transformers–Part 9: Digital interface for instrument transformers, *International Electrotechnical Commission (IEC)*.
9. IEC 61850-7-2, Communication networks and systems in substations–Basic communication structure for substation and feeder equipment-Abstract communication service interface (ACSI), *International Electrotechnical Commission (IEC)*, 2003.
10. IEC 61850-7-3, Communication networks and systems in substations–Basic communication structure for substation and feeder equipment-Common data classes, *International Electrotechnical Commission (IEC)*, 2003.
11. IEC 61850-7-3, Communication networks and systems in substations—Basic communication structure for substation and feeder equipment—Compatible logical node classes and data classes, *International Electrotechnical Commission (IEC)*, 2003.

12. IEC 61850-4, Communication networks and systems in substations–General requirements. *International Electrotechnical Commission (IEC)*, 2003.
13. IEC 61850-8-1, Communication networks and systems in substations—Specific communication service mapping (SCSM)-Mappings to MMS (ISO 9506-1 and ISO 9506-2) and to ISO/IEC 8802-3, *International Electrotechnical Commission (IEC)*, 2004.
14. IEC 61850-90-1, Communication networks and systems for power utility automation—Use of IEC 61850 for the communication between substations, *International Electrotechnical Commission (IEC)*, 2013.
15. IEC 61968, Common information model (CIM)/Distribution Management, *International Electrotechnical Commission (IEC)*.
16. IEC 61970, Common information model (CIM)/Energy Management, *International Electrotechnical Commission (IEC)*.
17. IEC 61850-90-2, Communication networks and systems for power utility automation—Use of IEC 61850 for the communication between substations and control centres, *International Electrotechnical Commission (IEC)*, 2013.
18. IEC 62439-3 Industrial communication networks—High availability automation networks—Part 3: Parallel redundancy protocol (PRP) and high-availability seamless redundancy (HSR), *International Electrotechnical Commission (IEC)*.
19. IEEE Std 1402™, IEEE Guide for Electric Power Substation Physical and Electronic Security.
20. A.C. Adewole and R. Tzoneva, Impact of IEC 61850-9-2 standard-based process bus on the operating performance of protection IEDS: Comparative study, *19th IFAC World Congress (IFAC WC 2014)*, Cape Town, South Africa, pp. 2245–2252, August 24–29, 2014.
21. T. Arnold, A. C. Adewole, and R. Tzoneva, Performance testing and assessment of multi-vendor protection schemes using proprietary protocols and the IEC 61850 standard, *Industrial and Commercial Use of Electricity (ICUE) Conference 2015*, Cape Town, South Africa, pp. 284–290, August 17–19, 2015.

14

Overvoltage and Earthing Protection

N. T. Mbungu, J. J. Justo, and Ramesh Bansal

CONTENTS

14.1 Introduction

The electrical grid is currently vulnerable to internal and external faults. Smart grid application brings several improvements in the context of control, protection, and energy management for an electrical system control [1,2]. The overvoltage, earthing and lightning protection is considered one of the essential strategies of protecting the power grid against several faults. The intelligent system technology and development introduce the indispensable control philosophies on the power system, which can protect the entire electrical network before, during and after faults. This strategy aims to develop a faster fault localizer strategy that can clear the harmful event without affecting the system operation and performance through the communications, monitoring and control capabilities for a given type of power grid [3]. It is important to notice that the impact of faults due to overvoltage, earthing and lighting events differ in smart grid applications compared to conventional power grids. This impact is measured in the three following factors: the reduction of the number of cables, the shortening of wires and the presence of smart devices [4].

The overvoltage protection in the framework smart grid environment has recently been developed. In Schroeder et al. [3], overvoltage protection of data concentrators is extended under intelligent system application. This research work focused on protecting the link between the data concentrator and energy consumer for residential uses. The protection philosophy created the approach that can clear the overvoltage that comes from lightning strikes and current surges. In Wang et al. [5], a new grounding system is proposed to develop a flexible control strategy of neutral-to-ground voltage. This approach used the single-phase inverter to coordinate the relationship between injected current to the grounding system and neutral-to-ground voltage. It was observed that through the proposed active grounding system, the neutral-to-ground voltage was effectively constrained and the system overvoltage due to asymmetrically distributed parameters is avoided. In Lin et al. [6], the overvoltage testing system in smart grid application is presented for long and short distances. The overvoltage faults in an intelligent grid environment can affect not only the electrical grid but also the communication devices. The protection system must therefore be structured to protect the entire power grid as well as communication equipment. The study designed an overvoltage protection that can control the level of voltage flow on the smart grid devices.

All electrical apparatus and/or equipment require an earthing system, which creates safe operation of the electrical system. The earthing or grounding architecture consists of protecting personnel as well as the pieces of equipment against the undesired and unwanted electrical charge. This system mostly allows the negative sequences current to flow to the ground [7,8]. Based on the conventional earthing design methodology given by standard EN 50522, which is similar to that mentioned in British Annex NA, an application of a probabilistic approach that can evaluate the earthen system safely for a distribution transformer station is presented in Topolanek et al. [9]. This method aims to contribute to detect central probability aspect of the ground system such as human protection, respected risk scenarios, fault/contact coincidence, frequency and type of earth faults, and sensitivity analysis of all essential input variables. In Milioudis et al. [10], a more sophisticated method is developed under the smart grid architecture to support the safety and economic issues for both the public and the utility grid. The system design detects and clears the high impedance faults in rural overhead power distribution grid. An acute neutral point treatment and earth fault suppression are proposed in Siirto et al. [11]. The work aims to manage the earth fault in medium voltage networks that can avoid customer interruptions by resolving the challenges and issues of fault clearing process and safety.

The tropical regions around the earth's equator are most affected by the highest densities of a thunderstorm with lighting. In some sectors of the world, the storms usually appear in summer. While in some temperate climate zones, even in winter it can be observed that significant thunderstorms can affect the entire power grid by injecting severe overvoltages [12]. The protection of the electrical system against lightning flashover is first designed under bid data structures by using K-nearest neighbour and K-dimension tree in the smart grid environment [13]. The application aims to create real-time monitoring of lightning history so that the protection device can act in an intelligent controlled manner through a K-nearest neighbour search algorithm. In Liu et al. [14], the protection of high voltage (HV) in the smart grid for direct current (DC) is presented. The research aims at the ability of HV equipment to withstand overvoltage due to the lightning event. This chapter focuses on the conventional protection strategy that can be used in the smart grid application, which can protect the electrical grid during overvoltage, earthing and lighting events.

14.2 Overvoltage

The sources of overvoltage can be both external and internal in reference to the electrical network [12]. To provide an excellent protection philosophy of the electrical grid, it is necessary to address the different sources, causes, and effect of overvoltages. It is worthy to notice that an overvoltage in power system occurs when the amplitude or the r.m.s of the system voltages rises between 1.1 to 1.8 per unit (pu) value of the nominal rated voltage at power frequency periods at half cycles [15]. The overvoltage is also caused by system faults, which can be derived from the single line to ground fault that will raise the voltage r.m.s of other phases. In most the case, the overvoltage is observed in the sudden reduction in demands, lightning strikes, switching of transient loads, neutral displacement, and failure of control equipment such as voltage regulators. According to Schneider electrical in Ferracci [16], overvoltage is of three types: lightning, switching, and temporary. Overvoltage can occur between live conductors and neutral or phase-to-phase (in differential mode), and between phase and the exposed-conductive-part or earth (in standard mode). As a power disturbance problem, overvoltage is considered an occasional phenomenon for switching, ferroresonance, and lightning; this power quality problem is a frequent phenomenon for the converter and capacitor bank. However, according to IEEE 1159, overvoltage is classified into three types:

- Instantaneous overvoltage, which lasts from 0.5 to 30 cycles with an r.m.s of 1.1 to 1.8 pu.
- Momentary overvoltage, which lasts from 30 cycles to 3 sec with an r.m.s of 1.1 to 1.4 pu.
- Temporary overvoltage, which lasts from 3 sec to 1 min with an r.m.s of 1.1 to 1.2 pu.

14.2.1 External Overvoltage

The electrical network more often suffers from the external overvoltage caused by lightning, which is a phenomenon from the atmospheric disturbances in the vicinity of the power grid. A significant external overvoltage that negatively affects the electrical grid can come from solar eruptions, or solar super-storms. With this phenomenon, the injection of mass coronal in the direction of the earth can damage the entire the electrical system. The magnitude of the voltage becomes more and more important during this event, and from electrical line substances, it can observe the critical damage. However, in this chapter, only external overvoltage due to lightning will be considered, which occurs in the form of a surge and does not have a linear model or relation with the system voltage. The external overvoltages that can appear in the power grid are caused by the following atmospheric phenomena:

- Direct lightning strike.
- Electromagnetic induced overvoltages.
- Electrostatic induced overvoltages.
- Induced voltages due to changes in atmospheric conditions along a particular length of the transmission line.

14.2.2 Internal Overvoltage

Most of the overvoltage caused inside the electrical grid is the function of changing conditions of the operating system. In the worst scenario, some faults at given points of an electrical network can also introduce overvoltage into the electrical grid. Overvoltage due to changing conditions in the operation of the power system is classified into two groups: temporary overvoltage and switching overvoltage (transient over operation voltages of high frequency).

14.2.2.1 Temporary Overvoltages

If the heavy load is disconnected from the power grid during the steady-state condition of the operating electrical network, there will be a temporary increase in the voltage magnitude. This can be considered as an overvoltage event in specific points or buses of the electrical network, which will be regarded as a temporary fault on protection devices. The best way of protecting the electrical system and its equipment against the transient overvoltages consists of designing a control philosophy that can measure the impact of overvoltage into the system and coordinate relays and protection devices. This strategy aims to avoid an unwanted trip of the operating system and insulation failures [17,18].

14.2.2.2 Switching Overvoltages

The overvoltage due to switching is observed in steady state when a line or distributed generation such as diesel generator, wind power, PV power, and other energy resource is connected to the grid. The overvoltage is observed at interconnection level of distribution line due when an unloaded line is switched on. At the substation level, the overvoltage of a transient nature can result from the switching on of the primary side of transformer or reactors.

14.2.3 Effects of Overvoltages on Power Systems

A continuous overvoltage on the electrical network stresses the insulation and can cause damage to the electrical equipment or devices [15–17]. Pieces of equipment that can be negatively affected or damaged during overvoltages are cable, capacitor bank, circuit breaker, data processing load and numerical control, induction furnace, lighting, speed drive, and transformer [16]. The effects of overvoltages on power system are therefore variable, which is the function of the period of application, magnitude, repetitive, mode (common or differential), gradient and frequency. The consequences of overvoltages are classified as follows [16]:

- Degradation of equipment through operating that can be observed during repetitive rather than destructive overvoltages.
- Dielectric breakdown, which can essentially cause permanent damage to the electronic components.
- Disturbance in control system and low current communication circuits.
- Electrodynamic and thermal stress or fire that can be caused by:
  - Lightning: The overhead lines are most vulnerable to the lightning phenomenon. If the lightning strikes are close to the site of the installations that are supplied by underground systems, this, however, can also be affected by stress caused by high voltage due to lightning strikes.
  - Switching overvoltages: These are often observed on the electrical network, and they have a long duration with a substantial probability of occurrence compared to lightning. Switching overvoltages can also lead to severe degradation to the power grid, similar to that caused by lightning. The conventional switching systems that can be observed in the electrical networks are either by control or protection system for lines (transmission or distribution), cables, shunt/series capacitors, shunt reactors, transformers, and generators or motors [17].
- Long interruptions that are caused by the destruction of equipment such as loss of sales for the distribution company and loss of production for industrial companies.

14.2.4 Causes of Overvoltage

The top five overvoltages (that occur most often) in power system can be classified as follows:

- Arcing ground: This phenomenon can happen in a three-phase system when the presence of a sporadic arc in phase-to-ground fault occurs. It appears as short-live oscillations in the system that is caused by the change in voltage r.m.s and the current in the demand side. During the apparition of the arcing ground, a severe breakdown problem of the insulation can be observed. This may harm equipment connected to the power supply.
- External causes: It can be observed that the highest surges on power systems come from external events. Lightning is a rare phenomenon that leads toward high r.m.s of voltage surges. This can bring serious failure and damage to the electrical network.
- Insulation failure: In normal operation of electrical systems, the grounding of the conductor is the most common impact of insulation failure. This failure is observed when the insulation between phase and earth does not exist. All current from the live conductor flows downward, which creates a critical fault that in turn affects the insulation negatively.

- Power system surges: Poor system voltage regulation leads to the fluctuations of r.m.s voltage that can be over or under the per unit value. This voltage disturbance can cause severe damage to special sensitive loads such as electronic and computer-controlled equipment. A proper control and protection philosophy can ensure excellent operation of these pieces of equipment and avoid long-term breakdowns in system operation.
- Ferromagnetic resonance: At resonance condition of the power system, an overvoltage can occurs only if the value of inductive reactance becomes equal to the value of capacitive reactance. In the most case for the circuit that contains both capacitor and inductance with a saturable magnetic circuit. The resonance occurs particularly during the operation of the circuit when the devices must be closed and opened. When the pieces of equipment have either separate or nonsimultaneous operation, the ferromagnetic resonance can occur.

14.3 Insulation Coordination

Insulation coordination is the process of knowing the insulation levels of power system components. In other words, it is the process of determining the insulation strength of the equipment. The internal and external insulation of electrical equipment is exposed to continuous normal voltage and temporary abnormal voltage. The equipment insulation is designed so that it withstands the highest power frequency system voltage and the occasional temporary power frequency overvoltage due to occasional lightning surges. The highest degree of availability of the electrical grid is defined by its ability to coordinate the insulation. The insulation coordination ensures the excellent protection of any kind overvoltages (due to network or lightning) that occurs on the power grid. Through this method, a best protection philosophy in the context of technical and economic concerns to protect persons and equipment against overvoltages is ensured. This structure has an important value because it concerns high-voltage systems. For better control of insulation coordination, the following steps must be taken into consideration:

- Choose the correct overvoltage withstand level. This is a function of the various network components from among the insulating voltages satisfying the particular constraints.
- Know the level of the possible overvoltages that can occur on the power grid.
- Use of the corrected protective devices.

The equipment of the power system is assigned a rated insulation level, and the capability can be approved by applying different types of test. The requirement of insulation is determined by considering the factors discussed in the following sections.

14.3.1 Highest Power Frequency System Voltage

The AC power network has different nominal power frequency voltage levels, for example, 400 V, 3.3 kV, 6.6 kV, etc. When the system is lightly loaded, the power frequency voltage at the receiving end of the line rises. The equipment of the power system is designed and tested to withstand the highest power frequency system voltage (440 V, 3.6 kV, 7.2 kV, etc.) without internal or external insulation failure.

14.3.2 Temporary Power Frequency Overvoltages

The temporary overvoltage in the power system can be caused by the load throw-off, faults, resonance, etc. Inverse Definite Minimum Time (IDMT) relay protects against temporary power frequency overvoltage. The IDMT relay is connected to the secondary of the bus potential transformer and circuit breakers. The relay and circuit breaker take action within milliseconds and protect the system from temporary overvoltage. Figure 14.1 shows an example of an overvoltage for a temporary power frequency (TPF) insulation.

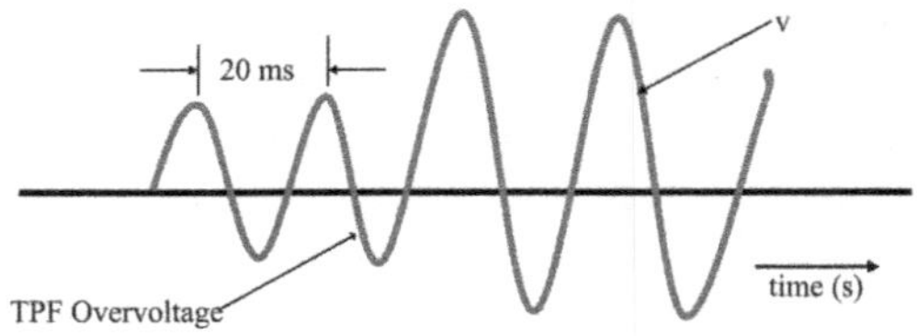

FIGURE 14.1 TPF insulation.

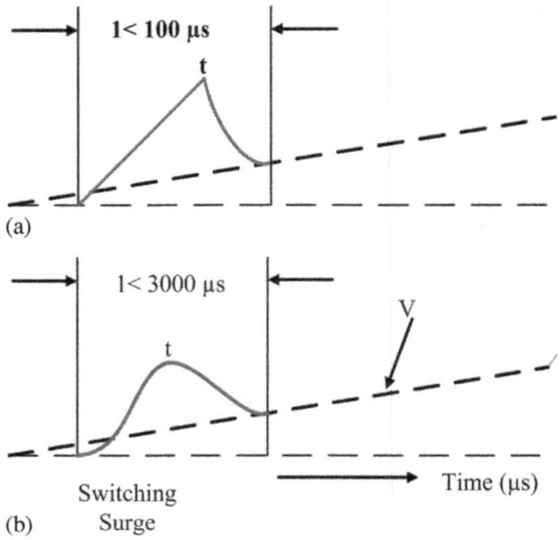

FIGURE 14.2 Typical surges due to (a) lightning and (b) switching.

14.3.3 Transient Overvoltage Surge

Lightning can cause transient overvoltage surges in the power system, switching, restrikes traveling waves, etc. The increase of voltage in the power system has the high peak, high rate of rising and last for a few tens/hundreds of microseconds and are therefore called the transients. This surge can cause spark overvoltage and flashover at sharp corners, between phase and earth, at the weakest point, the breakdown of gaseous/liquid/solid insulation, failure of a transformer and rotating electrical machines. Figure 14.2a and b describe the overvoltage transient event that can occur during lightning and switching surges.

Proper insulation coordination and surge arrester have minimized the failure rate due to lightning and switching. Several types of protective devices are installed on the network to intercept the lightning strikes and reduce the peak rate of rising of surge reaching the equipment.

14.4 Overvoltage Protection

Overvoltage protection devices are widely described in Fink [17]. In this brief subsection, we describe the different protection strategies that can be used during internal and external overvoltage events.

14.4.1 Protection Against Internal Overvoltages

Internal overvoltages are faults that occur inside the electrical network as described in Section 14.1.2. These faults are different from lighting or other external phenomena such as coronal mass ejection that comes from solar flares or solar superstorms. Coronal mass ejections can damage the entire electrical system. The modern world is vulnerable to solar eruptions, which are harmful to the electrical and communication systems and can create high overvoltage faults to the entire power grid. However, internal overvoltages can be caused by phase-to-phase, phase-to-neutral and phase-to-ground faults. Any given protection philosophy or device that limits the voltage on the power network can clear the internal overvoltage faults. Some of the overvoltage mitigation strategies are the following [19]:

- Controlled switching
- Line chokes
- Pre-insertion resistor
- Static automatic compensator
- Surge arrester
- Surge diverter

14.4.2 Protection Against External Overvoltages

Overvoltage may cause damage to insulators and substation equipment, so it is necessary to provide a scheme to protect the insulators and other apparatus from the harmful effects of overvoltage. Some devices are available to reduce the amplitude and front steepness of surges. The following protective equipment may be used to protect against the effects of overvoltage:

- Rod gap
- Surge diverter
- Overhead earth wire or shielding conductor

14.5 Grounding System

Earthing or grounding in a power system is an excellent technique for ensuring the stability of the system operation and safety of both people and equipment. An electrical system with a poor or missing earthing system is vulnerable to severe damage of the equipment and electrocution of people. According to the earth standards set by IEC, the term *earthing* means different things in different countries. However, the conductive mass of the earth at any potential point must be equal to zero [20]. The IEEE standard 80-2000 for grounding states that any equipment or electric circuit must be connected to the earth [21]. The grounding system therefore refers to equipment connected to the ground. This system can contain cable armour, cabinets, enclosures, frames, building steel, metallic raceways, and any nonconducting metals that are part of the electrical system.

The grounding system operates to maintain a zero potential in metal that can carry the current but are not part of the electrical network. This operation structure aims to eliminate faults on the system. When the system does not have grounding and a fault can cause electric shock if a person touches that component. The earthing system prevents the buildup of static charge. The grounding system also plays an important role in realizing the fault current from the electrical system. This system:

- Carries the electric currents into the earth. This system ensures the continuity of the service while respecting the operation and equipment constraint for both normal and fault conditions.
- Ensures the safety of the personnel working around the electrical system or in the vicinity of grounded facilities.

14.5.1 Safety Grounding System

The objectives for a safety grounding system are to [21]:

- Carry the electric currents into the earth and ensure the continuity of the service while repeating the operation and equipment constraint for both normal and fault conditions.
- Ensure the safety of the personnel working around the electrical system or in the vicinity of grounded facilities.

14.5.2 Safety Earthing System

The objective for a safety earthing system are to [20]:

- Fix a safety potential between phase earth in normal operation.
- Limit the voltage level between the non-current-carrying equipment into the electrical system and the earth.
- Implement protection devices to eliminate the risk of electric shocks for personnel.
- Limit the rising potential in low-voltage networks that comes from medium voltage faults.

14.5.3 Types of Grounding System

There are several types of grounding systems. They are classified according to the type of connection used.

14.5.3.1 Equipment Grounding

This system creates a zero potential for the non-current-carrying metal parts and then the earth and the non-current-carrying equipment. The strategy aims to interconnect all metals or equipment that are not part of the carrying current system and then connect that link to the earth.

14.5.3.2 System Grounding

The grounding for an electrical system has an earth conductor connected to non-current carrying metallic part. Several approaches for earthing system can be found in [17].

1. *Effects of lightning strikes on a grounding system:* When lightning strikes an object above ground, the current divides itself through the lightning protection system (LPS). This will depend on the impedance offered on its way to the LPS. Current is allowed to pass through each of the conductors present, and it is dissipated into the soil at each conductor intersection. The amount of current that can be handled by each of these conductors is crucial. Larger amounts current will be observed in electrodes with lower resistance to ground due to the energy passing through these conductors. The presence of both AC and DC current makes it possible to pass through structures at almost all frequencies ranging from 0 Hz to 10 MHz. The frequency corresponding to the lightning will not be distributed evenly, and this generated frequency is referred to as a spectrum. A proper model of the structure and its simulation using computer programming has to be in place to efficiently analyze the frequency spectrum profile.

   Another important phenomenon associated with lightning strike on a grounded system is the formation of electromagnetic interference due to the magnetic fields formed in the LPS and the grounding system. This results in the rise of the ground potential, which can subsequently affect the transfer voltages and currents of nearby pipelines, railways, communication lines, industries, etc. The damage can be reduced by considering and analysing the highest magnitude of frequency generated by the lighting through proper computer frequency spectrum analysis to calculate the electromagnetic and ground potential rise, which could have a devastating impact the surroundings. Different strategies for the electrical system against lightning strikes are described in Fink [17].

2. *Ground resistance:* It is important to continuously monitor the ground resistance using automated or timed measurement mechanisms in order to determine the current flowing through the system and the performance of the grounding system. The methods of testing include induced frequency test methods and automated data reporting. Induced frequency tests are limited to small electrodes because the tests use frequencies in the range of kilohertz and utility is connected to the grounding system. Long conductors act as inductive chokes and are not efficient in reflecting the 60 Hz resistance of the grounding system. Automated data reporting provides an added advantage of not disrupting the electrical system during measurement.

14.6 Solved Problems

1. Determine the substation critical striking distance of protected equipment using the rolling-sphere method at 10 kA.

Solution

The rolling-sphere method depends on the critical striking distance of the transmission line that is considered as the radius of the sphere. There are several methods to determine the striking distance. However, the transmission line as described in Fink [17] can also be used for the substation, and it is given by:

$$S_c = 10 I_c^{0.65} \tag{14.1}$$

where I_c is the critical stroke current in kA. Using Equation (14.1), the critical striking distance can be computed as:

$$S_c = 10 \times 10^{0.65} = 44.67 \text{ m} \tag{14.2}$$

Note that the equipment that is below the curved surface of the sphere of the radius given in Equation (14.2) is protected. However, the equipment that can not touch the sphere or penetrate it is unprotected.

2. Determine the allowable stroke current in a substation of 5 MVA, 66/11 kV with basic insulation level (BIL) of 350 kV for the three-phase secondary side. Note that the substation operates in a grounding system, and the surge impedance of the conductor through the surge is passing 280 Ω.

Solution

The allowable stroke current in kA is given by the relation below:

$$I_s = \frac{2.2(\text{BIL})}{Z_s} \tag{14.3}$$

where Z_s is the surge impedance in Ω. Thus Equation (14.3) will be computed as:

$$I_s = \frac{2.2(350)}{280} = 2.7 \text{ kA} \tag{14.4}$$

3. Determine the voltage wave during a lightning strike when its surge current is 3.8 kA and the line sure impedance has 80 Ω.

Solution

The voltage wave is computed using the following formula:

$$E_s = \frac{1}{2} I_s(Z_s) = \frac{1}{2} \times 3.8 \times 80 = 152 \text{ kV} \quad (14.5)$$

4. Determine the occurrence probability of peak current in any lightning strike when the specified crest current is:
   a. 55 kA
   b. 75 kA

Solution

The occurrence probabilities of peak current for any lightning strike is a function of the specified crest current, and it is determined by:

$$P(i) = \frac{1}{1 + \left(\frac{i}{31}\right)^{2.6}} \quad (14.6)$$

a. Use Equation (14.6) to determine probability of peak current for the specified crest current of 55 Ka:

$$P(i) = \frac{1}{1 + \left(\frac{55}{31}\right)^{2.6}} = 0.184 \quad (14.7)$$

b. Use Equation (14.6) to determine probability of peak current for the specified crest current of 75 Ka:

$$P(i) = \frac{1}{1 + \left(\frac{75}{31}\right)^{2.6}} = 0.0914 \quad (14.8)$$

5. Determined the flashover voltage of insulator string during a lightning strike of 3 μs when a substation contains the insulator string length of:
   a. 1 m
   b. 0.66 m
   c. 0.33 m

Solution

The flashover of insulator string is computed as follows

$$V_{fo} = (400 + 710t^{-0.75})l \quad (14.9)$$

where V_{fo}, t, and l are respectively flashover voltage in kV, time in μs, and insulator string length in m.

a. For the insulator string length of 1 m, Equation (14.9) can be solved as:

$$V_{fo} = (400 + 710 \times (3)^{-0.75})1 = 711.47 \text{ kV} \tag{14.10}$$

b. For the insulator string length of 0.66 m, Equation (14.9) can be solved as:

$$V_{fo} = (400 + 710 \times (3)^{-0.75})0.66 = 469.57 \text{ kV} \tag{14.11}$$

c. For the insulator string length of 0.33 m, Equation (14.9) can be solved as:

$$V_{fo} = (400 + 710 \times (3)^{-0.75})33 = 234.79 \text{ kV} \tag{14.12}$$

6. Determine the withstand voltage of the insulator string at 2 μs and 6 μs for 14.5a.

Solution

The withstand voltage of insulator string is given in Equation (14.13) for 2 μs and Equation (14.14) for 6 μs, as follows:

$$V_{(2i)} = 0.94(820l) \tag{14.13}$$

$$V_{(i6)} = 0.94(585l) \tag{14.14}$$

For 2 μs:

$$V_{(i2)} = 0.94(820) = 770.8 \text{ kV} \tag{14.15}$$

For 6 μs:

$$V_{(i6)} = 0.94(585) = 549.2 \text{ kV}$$

14.7 Tutorial Problems

1. Compare the methods used in Solved Problems 14.5 and 14.6.
2. From Solved Problem 14.2, determine the suitable lightning arrestor.
3. Determine the substation protected area at 55 Ka of Solved Problem 14.1.
4. What are the various sources of overvoltage in power systems?
5. Explain the various means of protection against overvoltage.
6. What are the differences between overvoltage, voltage dip and voltage sag as applied in power systems?

14.8 Conclusion

The chapter has presented several types of overvoltage and their corresponding sources of co-occurrence and earthing protection. Well-developed and well-implemented strategies can protect the electrical grid and equipment against overvoltage faults. Solved problems were included to give readers an opportunity to review the concepts of overvoltage and grounding.

REFERENCES

1. Naidoo, R., Pillay, P., Visser, J., Bansal, R. C., and Mbungu, N. T. (2018). An adaptive method of symmetrical component estimation. *Electric Power Systems Research*, 158, 45–55.
2. Mbungu, N. T., Bansal, R. C., Naidoo, R., Miranda, V., and Bipath, M. (2018). An optimal energy management system for a commercial building with renewable energy generation under real-time electricity prices. *Sustainable Cities and Society*, 41, 392–404.
3. Schroeder, J., Doherty, E., and Nager, M. (2011). Overvoltage protection of data concentrators used in smart grid applications. In *IEEE PES Innovative Smart Grid Technologies* (*ISGT*), pp. 1–4.
4. Gu, S., Li, D., Zeng, X., Su, J., He, Y., and Zhao, Z. (2017). Analysis of the characteristics and impact of lightning on smart substations. In *IEEE 5th International Symposium on Electromagnetic Compatibility* (*EMC*), Beijing, pp. 1–6.
5. Wang, W., Yan, L., Zeng, X., Fan, B., and Guerrero, J. M. (2017). Principle and design of a single-phase inverter-based grounding system for neutral-to-ground voltage compensation in distribution networks. *IEEE Transactions on Industrial Electronics*, 64(2), 1204–1213.
6. Lin, H., Wang, C., and Lei, W. (2016). The design of overvoltage testing system for communication equipment in smart grid. In *IEEE PES Asia-Pacific Power and Energy Engineering Conference* (*APPEEC*), pp. 1812–1816.
7. Lim, S. C., Gomes, C., and Ab Kadir, M. Z. A. (2013). Electrical earthing in troubled environment. *International Journal of Electrical Power & Energy Systems*, 47, 117–128.
8. Zhang, W., Ma, Z., Li, R., Zhang, Y., and Zhu, Z. (2017). Study on a sufficient method to make an smart test plan in grounding grid fault diagnosis. In *IEEE 29th Chinese Control and Decision Conference* (*CCDC*), pp. 6937–6941.
9. Topolanek, D., Vycital, V., Toman, P., and Carman, B. (2018). Application of the probabilistic approach for earthing system evaluation in distribution network. *International Journal of Electrical Power & Energy Systems*, 101, 268–279.
10. Milioudis, A. N., Andreou, G. T., and Labridis, D. P. (2012). Enhanced protection scheme for smart grids using power line communications techniques—Part I: Detection of high impedance fault occurrence. *IEEE Transactions on Smart Grid*, 3(4), 1621–1630.
11. Siirto, O., Loukkalahti, M., Hyvärinen, M., Heine, P., and Lehtonen, M. (2012). Neutral point treatment and earth fault suppression. In *IEEE Electric Power Quality and Supply Reliability Conference* (*PQ*), pp. 1–6.
12. Ribič, J. (2015). Impact of line length on the operation of overvoltage protection in LV networks. *Electric Power Systems Research*, 121, 216–226.
13. Gao, S., Che, R. F., and Meng, Q. M. (2017). Application of big data to lightning flashover warning in power grid. In *IEEE 2nd International Conference on Big Data Analysis* (*ICBDA*), pp. 464–467.
14. Liu, C. L., Shuai, Q., Qi, L., Cui, X., Fang, C., and Wei, X. (2014). Quantitative analysis of voltage distribution within ±1100 kv HVDC converter valve tower under various transient overvoltage conditions. In *International Conference on Lightning Protection* (*ICLP*), pp. 1558–1564.
15. Pawar, V. S., and Shembeka, S.M. (November 2013). Transient overvoltages in power system. *International Journal of Science, Spirituality, Business and Technology* (*IJSSBT*), vol. 2, no.1, 2277–7261.
16. Ferracci, P. H. (1990). Cahier technique no. 199 Power Quality, Schneider electrical.
17. Fink, D. G. (2007). *Standard Handbook for Electrical Engineers*. H. W. Beaty (Ed.). New York: McGraw-Hill.
18. Diesendorf, W. (2015). *Insulation Co-ordination in High-voltage Electric Power Systems*. Elsevier, London, UK.
19. Ferracci, P. (2001). Power Quality, Schneider Electric Cahier Technique 199.
20. Panetta, S. (2015). Shipboard electrical system grounding. In *IEEE Electric Ship Technologies Symposium* (*ESTS*), pp. 358–363.
21. Uzunlar, F. B., and Kalenderli, Ö. (2009). Three-dimensional grounding grid design. In *IEEE International Conference on Electrical and Electronics Engineering*, pp. I–139.

Section IV

Power Quality Issues, Reliability, Wide Area and System Protection; and Renewable DG Protection

15

Power Quality and Equipment Protection

Abhishek Chauhan, J. J. Justo, T. Adefarati, and Ramesh Bansal

CONTENTS

15.1 Power Quality

Power quality is defined in many ways according to the power utility and the power equipment manufacturer. For instance, a power utility can define power quality as the measurement of the reliability of the power network, but a power equipment manufacturer can define power quality as those set of characteristics of the power supply that assists the power equipment to work properly. Numerous publications define power quality in numerous ways. The increase in research in the field of power quality has been exponential between 1985 and 1996, and during that time, with about 2051 records, the term *voltage quality* has become known as *power quality* [1].

In this chapter various protection schemes are defined for the protection of power system components; hence in the context of this chapter, we define *power quality* as "the power system problem establishing deviations in the frequency, voltage and current that results in malfunctioning of customer load." Figure 15.1 shows the survey results from Dugan et al. [1], the survey stated the view point of customers and utility personals on power quality problems. Both the customers and utilities blame more than 60% of power quality issues on natural causes, for example, lightning, whereas customers and power utilities blame each other for causing power quality issues. It is important to understand that various abnormal events at the utility end are not observed in the utility's datasheet but have a serious impact on the end user equipment. One example is capacitor switching: this is one of the most common events at the utility side but it is responsible for transient overvoltage, which in turn has a negative impact on manufacturing machinery.

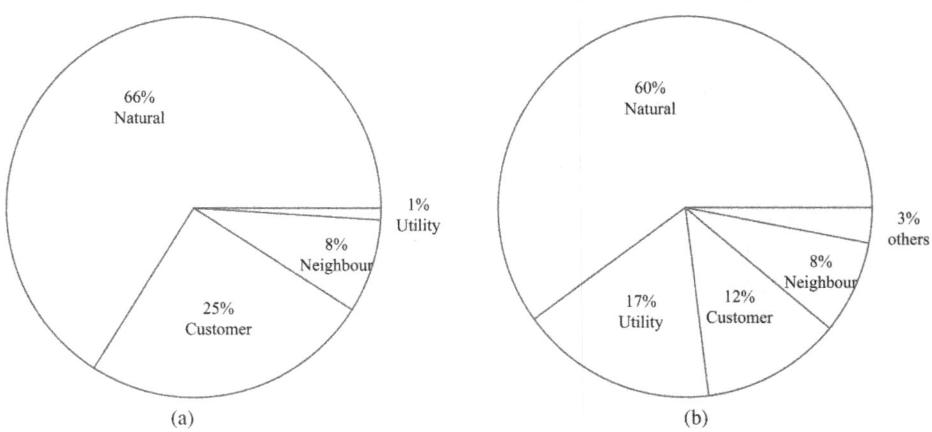

FIGURE 15.1 Causes of power quality problems: (a) view of utility and (b) view of customer. (From Dugan, R.C. et al., *Electrical Power System Quality*, 2nd ed., McGraw-Hill, New York, 2004.)

15.2 The Need for Power Quality Assessment

Power consumption is one of the most protuberant measures used for economic analysis, and hence countries are working hard to achieve most of it. For this reason, it is important to increase power generation, which in turn increases power transmission and distribution [1–3]. It is important to understand that the proper flow of power from the generator to the user, requires the stability of the power system. Power quality is a factor in that stability, and multiple reasons in support of need the power quality assessment are discussed next.

15.2.1 Role of Power Equipment

With advancements in power system networks, power equipment has become sensitive to voltage disturbances. When power electronic equipment used for various controlling applications is more sensitive to voltage variations, power quality assessment becomes necessary. Nonsinusoidal currents in rectifiers and inverters are also major concerns in power quality. The input current includes some high-frequency components, i.e., harmonics along with supply frequency components. These harmonic distortions produce harmonics in the supply voltage [1–3]. The power loads already produce harmonics, but adjustable speed drives, small consumer electronics equipment and power electronics converters are responsible for the exponential increase in harmonic voltage distortion and results in deterioration in the supply voltage. Advanced pulse width modulation (PWM) techniques also were out of question until semiconductor-based power control equipment was introduced. Although the semiconductor-based devices are more flexible and better controlled operations, but they generate more harmonics as well. Technologically, the power semiconductors-based peripherals started with the introduction of thyristors, followed by gate turn-off thyristors (GTOs), triacs, bipolar transistors (BJTs), metal oxide field effect transistors (MOSFETs) and insulated gate bipolar transistors (IGBTs). These semiconductor devices have high frequency switching and use PWM in power converters that have various applications in industries. Non-natural commutation of PWM results in high frequency current harmonics compared with previous versions of mercury-based converters. Change in design strategies also has a deep impact on the harmonics-generating nature of power peripherals [3]. Conventional equipment is underrated whereas, in the modern era, power equipment is deigned and pushed beyond its operating limits. For instance, some magnetic-material-based equipment are operated in the nonlinear region, which reduces the manufacturing cost and results in the extraction of more distorted current.

15.2.2 Standardization in Power Sector

Industrialisation comes with an increase in power generation that results in the establishment of more power utilities. These utilities view electricity as a product with certain characteristics which can be measured, enhanced and forecast. The open competition makes it more complicated because customers now can buy electrical energy from one source, transport it by another and pay for the actual connection to the local utility [2]. This makes it difficult for customers to identify who is responsible for the reliability of the power that is delivered. Most of the time it is observed that only the distribution utility is responsible for reliability, whereas the fault at the generation side can also have a serious impact at the consumer side. So this triggers the legalisation of the contract between the consumers and the power utilities where power quality characteristics are to be mentioned.

15.3 Evaluation of Power Quality

In order to mitigate power quality issues, evaluations are important. In the evaluation process the type of power quality issue is identified and, on the basis of data analysis, a range of solutions is identified. Possible solutions are targeted at all possible levels from the end user to the utility. Solutions that

1. Identify the Problem Category. (Voltage Unbalance, Sag, Harmonic Distortion or Transients)

2. Measurement and Data Acquisition.

3. Range of Solution. (Transmission Utility, Distribution Utility, End Customer System)

4. Evaluation of Technical Options.

5. Evaluation of Economics Associated with the Possible Solution.

FIGURE 15.2 Steps of power quality evaluation. (From Dugan, R.C. et al., *Electrical Power System Quality*, 2nd ed., McGraw-Hill, New York, 2004.)

are technically feasible but with high cost are discarded and more economic solutions are identified. Figure 15.2 shows the steps for the evaluation of power quality issues.

Note that both power utility and the end consumer must maintain power quality from their respective side. One way to control the harmonic current injections that are from nonlinear loads is by employing the delta-connected transformer using a 12 pulse inverter. The IEEE 519-2014 standard can be followed by both end users and electric utilities. The standard defines the limit of harmonic currents by controlling their reasonable harmonic goals. In this case, the harmonic currents can be controlled by the end user by monitoring their loads; power utilities can control their system impedances to keep the harmonic voltages at safer limits [1–3]. This ensures that the system is kept stable and the harmonics level in the power system remains at safer limits [1].

Interline dynamic voltage restorers (IDVRs) and flexible AC transmission systems (FACTS) are also employed for power quality improvement, especially to eliminate harmonic currents and voltage. The IDVR is responsible for refilling the energy to the DC link when the voltage injection in more than the one in the power lines. The voltage injection source is either a voltage source inverter (VSI) or a current source inverter (CSI). The IDVR is employed at the distribution side, and its performance clearly depends on the assumption that most of the sensitive loads are connected at the distribution side.

15.4 Frequency Variation as Power Quality Issue

Under normal conditions, the following relation is used:

$$\Delta f = f - f_r \tag{15.1}$$

where Δf is the allowing frequency deviation of slow frequency variation, the rated frequency is f_r (50 Hz or 60 Hz) and f is the real frequency value in Hz.

Relative frequency deviation is explained as a percentage, as follows:

$$\varepsilon_f(\%) = \frac{f - f_r}{f_r} \tag{15.2}$$

As per standard EN 50160/2006, the rated frequency of supply voltage is 50 Hz, and under normal conditions, the mean of fundamental frequency for a time of over 10s should stay within the limits as mentioned below.

1. System having synchronous connection with an interconnected system:
   50 Hz ± 1%, i.e., 49.5–50.5 Hz for 99.5% of the year.
   50 Hz + 4%/−6%, i.e., 47–52 Hz for 100% of the time.
2. Systems having no synchronous connection with and interconnected system:
   50 Hz ± 2%, i.e., 49–51 Hz for 95% of the week.
   50 Hz ± 15%, i.e., 42.5–57.5 Hz for 100% of the time.

15.4.1 Impact of Frequency on Motor Actuation

In industry, asynchronous and synchronous motors are employed for various applications, and the power-frequency characteristic $P = T.\ n$ depends on the load characteristics that are mechanically defined. $T = f(n)$, where T is the coupling torque to the motor shaft and n is the motor speed [2].

Curve 1 in Figure 15.3 explains the receivers that are independent of the frequency. Curve 2 showcases installations like conveyers, mine lifts, etc. These applications have uniform load and speed-independent coupling; hence, power consumption is proportional to the supply frequency. Curve 3 shows a parabolic characteristic for applications like ventilation pumps; here, a cubic power-frequency characteristic is achieved. It should be evident that the synchronous or asynchronous motors depend directly on the supply voltage frequency; hence, any change in supply frequency will definitely have a considerable impact on these industrial machines [2].

15.4.2 Capacitor Bank and Harmonic Filters

The variable Q is the reactive power generated by the capacitor bank, and it is directly dependent on the supply frequency:

$$Q = 2\pi fCV^2 \tag{15.3}$$

where C is the capacity of the capacitor bank and V is the voltage at the terminals.

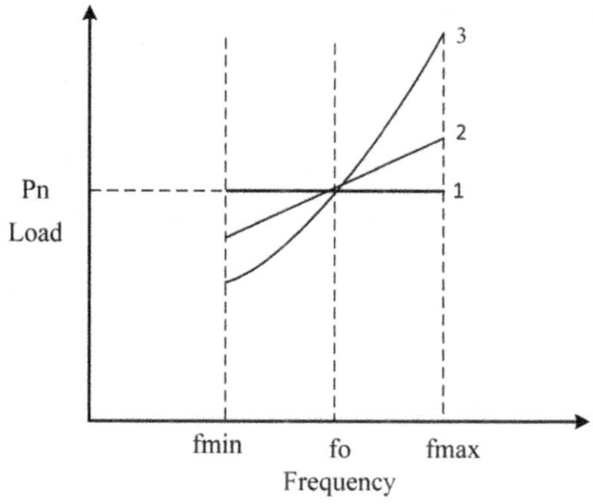

FIGURE 15.3 Power consumption of various types of receivers versus frequency variation. (From Baggini, A., *Handbook of Power Quality*, John Wiley & Sons, New York, 2008.)

Equation (15.3) shows that the reactive power is a frequency dependent parameter; hence, any variation in frequency will definitely create an appreciable change in the power factor at the supply busbar. Frequency variation has an impact on the harmonic filters when the capacitor is the part of the harmonic filters. Under normal operating conditions, L_h and C_h of a resonant circuit are designed for harmonic h; at this frequency, the circuit impedance actually leads to zero [2].

$$2\pi f_h L_h - \frac{1}{2\pi f_h C_h} = 0 \tag{15.4}$$

For high-order harmonics, the filter has considerable impedance that showcases a particular degree of distortion. Operation at any other frequency than the rated frequency will cause the filter to have an impedance other than zero, which shows a certain degree of discord (out of time tuning); the filter will fail to eliminate the harmonics under observation or will not filter them out.

15.4.3 Impact of Frequency Variation on Transformers

The magnetic flux produced in a magnetic circuit winding is calculated as follows:

$$\phi_\mu \cong \frac{\sqrt{2} \times V}{2\pi f w} \tag{15.5}$$

where w is the number of coils in the winding and f is the supply voltage frequency. Equation (15.5) reveals that a dip in the fundamental frequency leads to an increase in magnetic flux; hence, the magnetic induction slips into the nonlinear zone of magnetic characteristics. In other words, a transformer or a coil behaves like a nonlinear element under the frequency dip condition.

15.5 Voltage Unbalance as a Power Quality Issue

Voltage unbalance is of the most prevalent issues of relegation of power quality. It has a negative impact on industrial drives, especially on 3-Ø induction motors, which comprise 80% of industrial machines [4–20]. Uneven distribution of single-phase load, and improper transposition of transmission lines, adjustable speed drives and open delta transformers results in unbalance in voltage [4–11], [15–18]. Deterioration in operation of 3-Ø induction motors due to unbalanced supply has an impact on a company's manufacturing process and is also responsible for the cost of downtime in case of any machine failure [4].

15.5.1 Definitions of Unbalanced Voltage

In standards, several definitions of unbalanced voltage are available; few definitions are explained as follows.

15.5.1.1 National Equipment Manufacturer's Association (NEMA)

The line voltage unbalance rate (LVUR) is calculated as follows by NEMA [12]:

$$\begin{aligned} \text{LVUR}(\%) &= \frac{\text{Max voltage deviation from avg line voltage}}{\text{Average line voltage}}, \\ &= \frac{\text{Max}\left[\left|V_{ab} - V_{\text{avg}}\right|, \left|V_{bc} - V_{\text{avg}}\right|, \left|V_{ca} - V_{\text{avg}}\right|\right]}{V_{\text{avg}}} \times 100\% \end{aligned} \tag{15.6}$$

where $V_{avg} = \dfrac{V_{ab} + V_{bc} + V_{ca}}{3}$

This calculation is accepted in the industry because line voltage is easy to measure, but it is important to know that the impact of angle unbalance is ignored in this equation. It is evident from past research that peak copper loss, peak current and derating factor are subtle to angle unbalance. Hence, inclusion of the angle is required if performance of the induction machine is to be precisely monitored.

15.5.1.2 IEEE Definitions for Voltage Unbalance

The phase voltage unbalance rate (PVUR) is given by IEEE to explain the unbalanced voltage and is calculated as follows [13]:

$$\text{PVUR}(\%) = \frac{\text{Max voltage deviation from avg phase voltage}}{\text{Average phase voltage}},$$
$$= \frac{\text{Max}\left[\left|V_a - V_{\text{avg}}\right|, \left|V_b - V_{\text{avg}}\right|, \left|V_c - V_{\text{avg}}\right|\right]}{V_{\text{avg}}} \times 100\% \tag{15.7}$$

where $V_{\text{avg}} = \frac{V_a + V_b + V_c}{3}$

IEEE use phase voltages instead of line voltages, and here also the phase angles are ignored. PVUR can also be easily recorded in field.

15.5.1.3 International Electrotechnical Commission (IEC)

The calculation of unbalance voltage, i.e., voltage unbalanced factor (VUF), by the International Electrotechnical Commission is one of the finest calculations [14]. The value of VUF is calculated as:

$$VUF(\%) = \frac{|V_n|}{|V_p|} \times 100\% \tag{15.8}$$

where V_p and V_n are the positive-sequence and negative-sequence components, respectively, of voltage. VUF explains voltage unbalance more precisely because the three-phase has both the magnitude and the angle. The NEMA, IEEE and IEC calculations are not sufficient to paint a clear picture of VUF's impact on industrial machines.

15.5.1.4 Nonstandard Definitions

A new definition known as complex voltage unbalance factor (CVUF) is presented in Wang [15], where both the angle as well as magnitude is considered [18]. The CVUF is expressed as follows:

$$CVUF = \frac{V_n}{V_p} = K_v \angle \theta_v \tag{15.9}$$

where θ_v is the angle of unbalance. The information to calculate the phase angles cannot be found in the field; hence, it is a challenge to calculate CVUF in the field. In Jeong [16], the following relation is set up to estimate the phase angle from the line voltages:

$$\tan\theta_v = \frac{\sqrt{3}\left(V_{yz}^2 - V_{zx}^2\right)}{2V_{xy}^2 - V_{yz}^2 - V_{zx}^2} \tag{15.10}$$

15.5.2 Impact of Voltage Unbalance on Industrial Drives

Figure 15.4 shows the various negative impacts of voltage unbalance. Figure 15.5 shows that with an increase in voltage unbalance, i.e., *VUF*, there is decrease in efficiency of the 3-Ø induction motor [4–20]. As decrease in efficiency results in higher electric power usage for the same work, which has an appreciable impact on electric bills, as shown in Table 15.1. It also causes aging in the motor due to high winding temperatures under unbalanced voltage supply and leads to replacement of machines before their scheduled replacement time [4]. Premature shutdown of 3-Ø induction motors and spontaneous downtime cost more than expenditures accompanying the replacement and reinstallation of the induction machines [4–20].

There are two basic conditions of unbalance:. undervoltage (UV) and overvoltage (OV). These conditions are further classified into eight subcategories of unbalance: 3Ø -UV, 2Ø -UV, 1Ø -UV, 2Ø -A, 1Ø -A, 1Ø -OV, 2Ø -OV and 3Ø -OV [4].

If a 3-Ø induction motor operates under an unbalanced supply with a full load, the motor is forced to work at higher slip, which causes a rise in winding temperature, which in turn might result in premature failure of the motor. In order to protect the motor from any premature failure, the NEMA standard suggests derating the motor to condense its output horsepower (HP) load so that it can bear the additional heating forced by the unbalanced supply [5,10,11]. In addition, to protect the motor from any damage, it is not recommended to operate a 3-Ø induction motor above 5% of VUF [14].

The derating factor can be expressed as follows:

$$\text{Derating Factor} = \frac{\text{Mechanical Output Power Under Unbalanced Supply}}{\text{Mechanical Output Power Under Balanced Supply}} \tag{15.11}$$

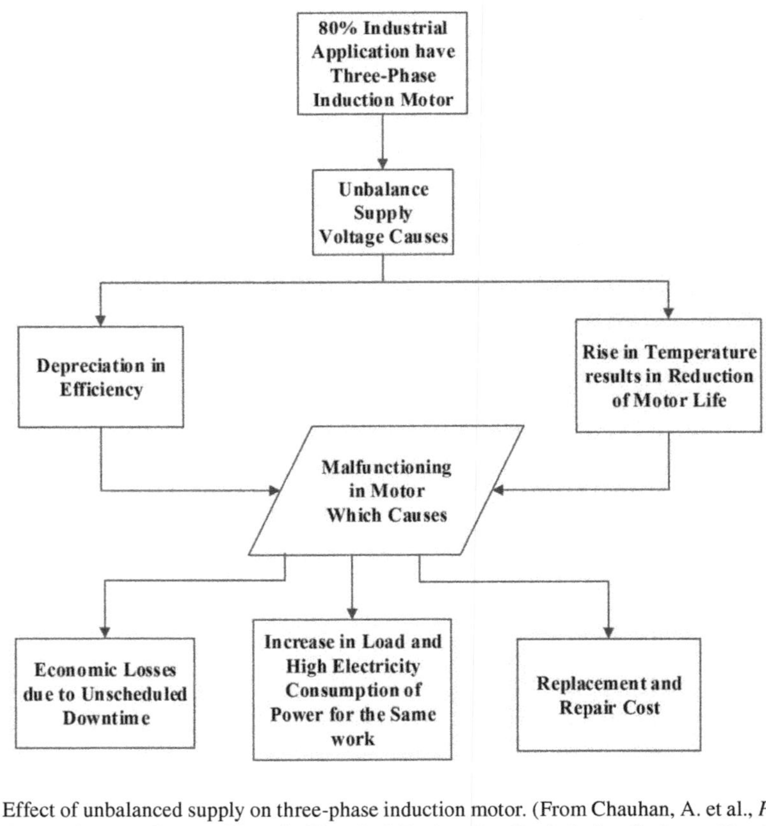

FIGURE 15.4 Effect of unbalanced supply on three-phase induction motor. (From Chauhan, A. et al., *Power Electronics and Renewable Energy Systems*, Lecture Notes in Electrical Engineering Book Series, Springer, New Delhi, India, 326, 47–55, 2014.)

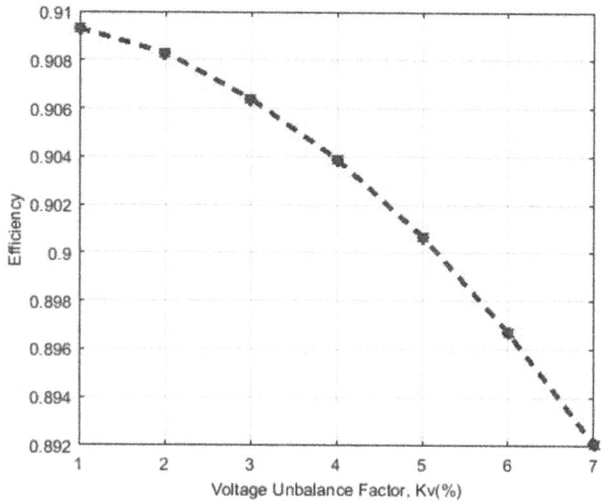

FIGURE 15.5 Efficiency of three-phase induction motor under unbalanced supply. (From Lee, C.Y., *IEEE Trans. Energy Convers.*, 14, 202–208, 1999.)

TABLE 15.1

Performance Analysis of 1–5 hp 3Ø Induction Motors when *VUF* is 4%

Voltage Unbalance Cases	Total Installed Capacity (kW)	Motor Efficiency (%)	Load Increase Rate (LdIR)	Extra Power Consumption per Year (kW/Yr)	Average Running Time per Year (Hour)	Extra Electricity Consumption per Year (kWh/Yr)	Extra Electricity Charge per Year ($M/Yr)
Balanced	534000	83.8	1	0	2500	0	0
3Ø -UV	534000	80.532	1.04058	21669.72	2500	54174300	3.142
2Ø -UV	534000	81.382	1029701	15860.33	2500	39650825	2.307
1Ø -UV	534000	81.506	1.023816	12717.74	2500	31794350	1.850
2Ø -A	534000	82.254	1.018786	10031.72	2500	25079300	1.459
1Ø -A	534000	83.041	1.009132	4876.48	2500	12191200	0.709
1Ø -OV	534000	83.225	1.0069	3684.65	2500	9211625	0.536
2Ø -OV	534000	83.402	1.004763	2543.44	2500	6358600	0.370
3Ø -OV	534000	83.584	1.002578	1376.65	2500	3441625	0.2

Source: Lee, C.Y., *IEEE Trans. Energy Convers.*, 14, 202–208, 1999.

Here, machine is loaded so that the current does not stray from the rated value [5,10]. The derating factor during different values of VUF is shown in Figure 15.6. During unbalance supply, the analysis is carried out by using the symmetrical component approach because unbalance supply results in positive-, negative- and zero-sequence components. As shown in Figure 15.7, a positive sequence component accounts for the required positive torque and also defines the condition of unbalance. In contrast, the negative-sequence component results in a negative torque and a negative-sequence current that in turn increases the winding temperature.

In Chauhan et al. [20], it is asserted that, in order to monitor the performance 3-Φ induction motor under unbalance supply, the degree of unbalance is not sufficient. In the condition of unbalance, undervoltage and overvoltage also play a vital role. Hence, the degree of unbalance along with the condition of unbalance should be considered. In Chauhan et al. [20], slip (*s*) can be considered to help identify the condition of unbalance and can be easily measured in the field, which is not possible in the case of a positive-sequence component. Figure 15.8 shows a change in efficiency using *s* under different degrees and conditions of unbalance.

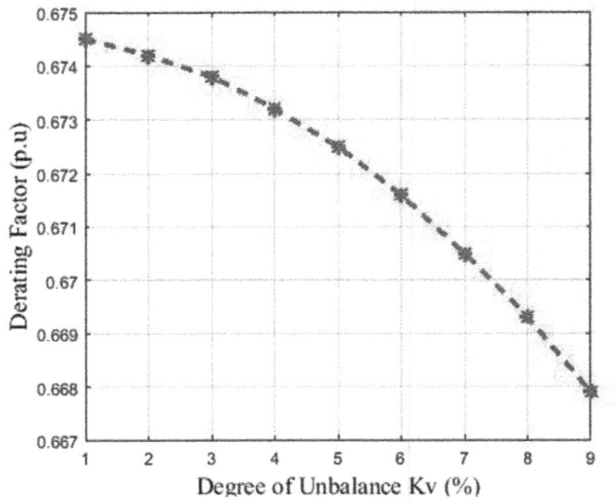

FIGURE 15.6 Derating of three-phase induction motor under different VUFs. (From Pillay, P. et al., *IEEE Trans. Energy Convers.*, 17, 485–491, 2002; Schmitz, N.L. and Berndt, M.M., *IEEE Trans. Power Appl. Syst.*, 680–686, 1963.)

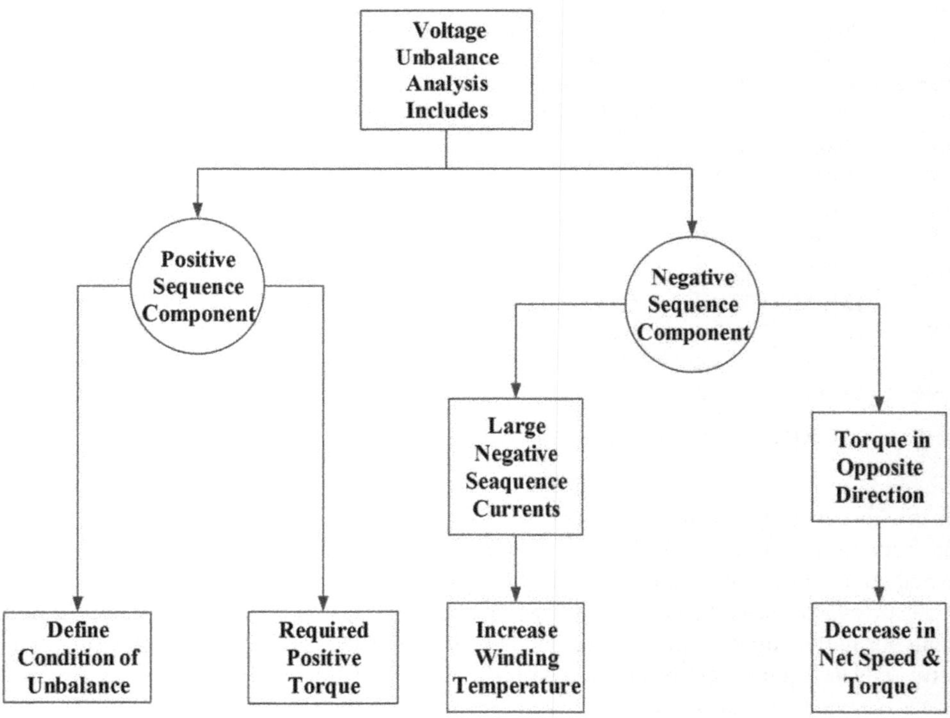

FIGURE 15.7 Analysis on the basis of symmetrical component theory. (From Lee, C.Y., *IEEE Trans. Energy Convers.*, 14, 202–208, 1999; Gnacinski, P., *IEEE Trans. Energy Convers.*, 23, 2008; Pillay, P. and Manyage, M., *IEEE Trans. Energy Convers.*, 21, 813–822, 2006.)

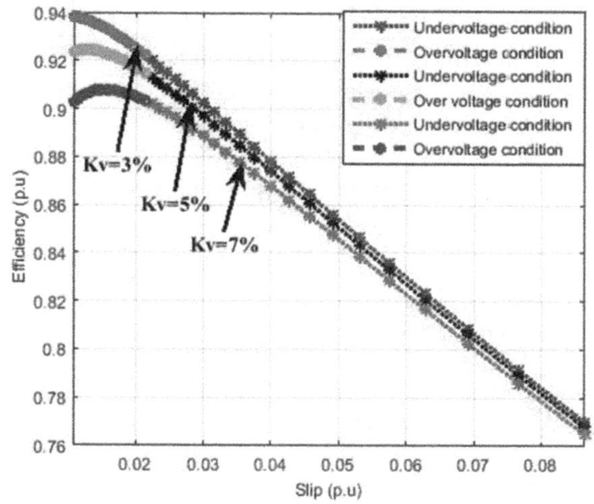

FIGURE 15.8 Variation of efficiency at different *VUF* and condition of unbalance. (From Chauhan, A. et al., *Power Electronics and Renewable Energy Systems*, Lecture Notes in Electrical Engineering Book Series, Springer, New Delhi, India, 326, 47–55, 2014.)

15.5.3 Mitigation of Unbalance Voltage

Several mitigation techniques are used to minimize voltage unbalance. Some are described in the following sections [21–25].

15.5.3.1 Uniform Distribution of Single-Phase Loads

Uneven distribution of single-phase loads through all three phases is the basic cause of supply unbalance. This can be rectified to an extent by uniform distribution of single-phase loads across all three phases [21]. Employment of automatic and manual feeder switching can also play an important role in relocating the load, which results in balance supply.

15.5.3.2 Static Voltage Ampere Reactive Compensator (SVC)

The SVC can be employed to minimize the VUF. In Campos et al. [23] a thyristor controlled shunt-connected SVC configuration connected with variable loads is explained. The main disadvantage of this method is the injection of harmonics in the AC network and the necessity of having large passive components that in turn increase the cost. The fixed capacitor thyristor controlled reactor (FCTCR) is also one of the SVCs employed in the distribution system to curtail the negative-sequence component. The addition of a microcomputer with a control circuit substitutes the traditional load switching and enables firm and active balancing.

15.5.3.3 Line Conditioners

In the previous section, it was stated that a negative-sequence component due to unbalance supply plays a vital role in the deterioration of the performance of industrial machines. According to Bhavaraju and Enjeti [21], injection of a correction voltage in any phase of three-phase system cancels the negative-sequence component.

15.5.3.4 Transposition of Transmission Lines

Transposition is known as the intermittent swapping of conductor positions of a transmission line. The problem of one-sided loads across three-phase systems can be mitigated by complete transposition. Multiple transpositions in long transmission lines are not possible because of the multiple substations that are connected to the transmission line. Moreover, urgent fault detection becomes a tedious task, so this mitigation technique is limited to short transmission lines only.

15.6 Harmonics

A harmonic is the frequency component that is the integer multiple of the fundamental frequency, where the fundamental frequency is 50 or 60 Hz. An order of a harmonic is expressed by n; for instance, if $n = 1$ (first harmonic = 50 Hz) and it is equal to the fundamental frequency itself, then if $n = 2$, the second harmonic = 100 Hz, and so on [2].

Harmonics are from nonlinear loads, and most of the loads today at the customer end are nonlinear. This results in n order harmonic additions on power networks, and the total harmonic distortion (THD) becomes so high that it makes the system unstable [2,26–28]. Table 15.2 lists various harmonic orders, frequencies, and the relative amplitude values.

In order to monitor the harmonic distortions added by various loads and power electronic peripherals, it is important to standardise a set of parameters. On the basis of these parameters, the acceptance of particular power peripheral or load is standardised. These parameters are listed in Table 15.3.

where V_n is the current harmonic amplitude n, V_1 is the amplitude of the fundamental component, V_N is the nominal voltage, V_{peak} is the peak value, $V_{average}$ is the average value and $V_{r.m.s.}$ is the r.m.s. value.

15.6.1 Sources of Harmonics

Harmonics sources are classified into three subcategories: (a) magnetic core equipment like motors, generators and transformers; (b) arc welding and arc furnaces; and (c) power electronics and electronics equipment.

15.6.1.1 Transformers

Transformers are power peripherals whose characteristic is highly nonlinear, which causes distortion in the saturation region, as shown in Figure 15.9. Transformers are designed so that the magnetising current does not exceed the nominal current (I_n) by more than 1%–2%. As shown in Figure 15.9, the nominal operating point lies below the knee point, i.e., in the linear region. Under normal conditions, there is

TABLE 15.2

Harmonic Orders and Relative Amplitude Values

Harmonic Order	Frequency (Hz)	Relative Amplitude Value
1	50	1
3	150	1/3
5	250	1/5
7	350	1/7
9	450	1/9
...	...	...
n	50n	1/n

Source: Baggini, A., *Handbook of Power Quality*, John Wiley & Sons, New York, 2008.

TABLE 15.3
Expressions for Standardization

Standard Expressions	
*n*th voltage harmonic ratio	$D_n = \frac{V_n}{V_1}, D_n = \frac{V_n}{V}$
Fundamental voltage harmonic ratio	$D_1 = \frac{V_1}{V}$
Peak factor: the ratio of peak and r.m.s. values for a periodic waveform. This selection parameter is ignored for some semiconductor devices according to their voltage and current peak values	$\frac{V_{\text{peak}}}{V_{\text{rms}}}$
Crest Factor: the r.m.s. to average value ratio. For sine wave, it is 1.11, whereas different value exhibits the waveform distortion	$\frac{V_{\text{rms}}}{V_{\text{average}}}$
Total distortion content	$\text{TDC} = \sqrt{Q^2 - Q_1^2}$
Harmonic r.m.s. value	$D_{hh} = \sqrt{\sum_{n\geq 2} D_n^{\,2}}$
Total distortion ratio	$\text{TDR} = \sqrt{\frac{V^2 - V_1^2}{V_1}}$
Total harmonic distortion factor: the ratio of r.m.s. harmonic component value to the fundamental component. The total harmonic current THD_I explains by using similar form	$\text{THD}_v = \frac{\sqrt{\sum_{n=2}^{n} V_n^2}}{V_1} \times 100\%$
Total distortion factor	$\text{TDF} = \sqrt{\sum_{n\geq 2}^{n\leq 50} \left(\frac{V_n}{V_N}\right)^2}$
Total demand distortion far a particular load current	$\text{TDD}_i = \sqrt{\sum_{n=2}^{40} \left(\frac{I_n}{I_1}\right)^2} \times 100\%$
Partial weighted harmonic distortion	$\text{PWHD} = \sqrt{\sum_{n\geq 14}^{n\leq 40} n\left(\frac{I_n}{I_1}\right)^2}$

Source: Baggini, A., *Handbook of Power Quality*, John Wiley & Sons, New York, 2008.

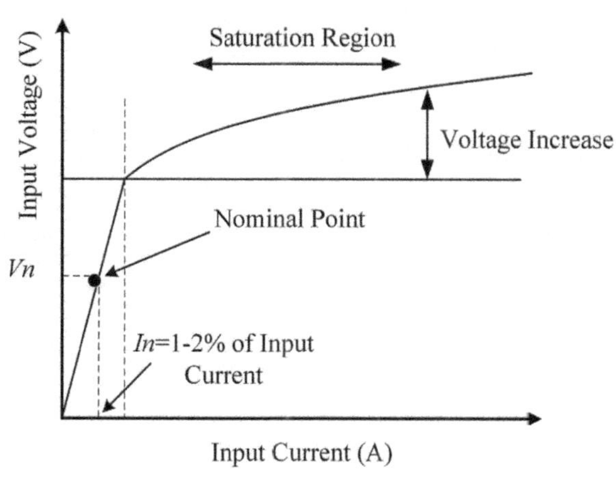

FIGURE 15.9 Transformer magnetization curve. (From Baggini, A., *Handbook of Power Quality*, John Wiley & Sons, New York, 2008.)

no significant source of harmonics. In the saturation region, however, even a slight increase in voltage above nominal voltage (V_n) results in a steep increase in magnetising current and a significant increase in the harmonics. Motors and generators are also a source of harmonics because of the large air gap. Their magnetizing characteristics are more linear when compared with transformers.

15.6.1.2 Arc Furnace

Because of technological changes, arc furnaces are operated today at lower power factors than they used to be. Thus, there is a need for stronger reactive power compensation, which is achieved by increasing the compensating capacitors rating, which in turn causes lower resonant frequency. The high harmonics amplitudes are significant in this spectrum range and result in the magnification of supply voltage harmonics [2].

15.6.1.3 Lighting Applications

Florescent lighting ballasts are popular because of their improved performance compared to magnetic ballasts. An electronic ballast circuit results in harmonics injected into the supply lines.

15.6.1.4 Power Electronics and Electronics Equipment

1. *Three-Phase Rectifier:* The UPS systems, adjustable speed controllers and converters are generally founded on the three-phase bridge, also known as a six-pulse bridge. This bridge adds harmonics on the order of $6n + 1$. In Baggini [2] the harmonics magnitude is shown to be significantly reduced by using 12-pulse converters.
2. *Switched Mode Power Supplies:* In most applications, various power peripherals are operated at different voltage levels. In order to provide variable DC to these peripherals, a single unit known as SMPS is used. The SMPS has single-phase rectifiers that use direct controlled rectification in spite of using different steps having transformers and rectifiers. Table 15.4 shows the magnitude of harmonic currents induced in single-phase applications; these values can change substantially under different operating conditions.

In direct control units, the reservoir capacitor is charged, and DC current is derived and fed to the loads as per their voltage and current requirements. These units draw pulsating current that contains third- and higher-order harmonics from the supply instead of continuous current. The effects of harmonics on the power system and power system parameters are explained in the next section.

TABLE 15.4
Magnitude of Harmonic Currents in Single-Phase Applications: Example

Harmonic Order	Welder (%)	PC (%)	Florescent Lamp (%)
1	100	100	100
3	29.6	75	12.3
5	8.8	47.3	13.8
7	2.0	22.9	3
9	2.3	9	1.1
11	2.3	3.3	0.7
13	1.1	3	0.5

Source: Baggini, A., *Handbook of Power Quality*, John Wiley & Sons, New York, 2008.

15.6.2 Effects of Harmonic Distortions

The heating loads are least affected by harmonic distortion, and electronic loads are designed for almost sinusoidal waveforms. Electric machines like motors lie between these two extremes. The effects are classified into short- and long-term effects. Short-term effects result in malfunctioning and failure of control, electronic and IT equipment. Long-term effects are thermal in nature and result in premature aging of equipment. It also depends on the values and harmonic order [2].

15.6.2.1 Phase and Neutral Overheating

Harmonics in the current create overloading issues in phase as well as in neutral conductors. Under current harmonics conditions, the heat starts deorbiting inside the cable and due to greater joule effect the line capacity is also reduced. This issue is more prominent in low-voltage systems because single-phase loads cause harmonics. Under normal conditions, the neutral current remains zero because the voltages are displaced by an angle of 120°, and the loads in all the three phases are equally loaded. When the loads are linear, a low magnitude neutral current apparently flows, although very low compared with line currents, because loads across the three phases are not equal [2]. Thus, installers keep the neutral wire at half size, but even when the balance nonlinear loads are supplied $3n$ harmonics add arithmetically in the neutral conductor results in high neutral currents up to 1.5 times of line currents.

15.6.2.2 Skin Effect

Skin effect is defined as the tendency of flowing alternating current on the outer surface of the conductor. This effect is prominent at high frequency supplies, and it is ignored at fundamental frequency. Above 350 Hz (seventh and above harmonics), this effect becomes significant and results in heating loss [2].

$$\delta = \sqrt{\frac{2\rho}{\omega\mu}} \tag{15.12}$$

where δ is the current penetration thickness, μ denotes the magnetic permeability, ρ is the resistivity and ω is the frequency.

15.6.2.3 Effect on Transformers

Eddy current losses equal about 10% of the total losses and increase with the square of the order of harmonic. If the transformer is fully loaded and feeds nonlinear loads, then the losses are twice the losses recorded while feeding an equivalent linear load. This additional eddy current loss results in a rise in temperature, which in turn decreases the working life of the equipment. These design specifications should be kept in mind when designing transformers.

15.6.2.4 Lighting Sources

Harmonics in the supply voltage shorten the life of incandescent bulbs. Mercury lamps and discharge light sources have parallel capacitors for power factor improvement and a series current limiting reactor that forms a resonant circuit. This close resonant circuit acts as a source of additional losses.

15.6.2.5 Circuit Breaker Operation

In circuit breakers, the switching ability is affected by the harmonic current while breaking small currents, and interruption of the short circuit current remains unaffected. The current derivative *di/dt* at zero interruption increases at high harmonics, and it hampers the process of zero current interruption. The name *residual current* itself suggests the working phenomenon of residual current circuit breakers (RCCBs). RCCBs operate on the basis of the residual current after adding the phase and neutral current. If the current is not found within the safe limits the circuit breaker gets tripped. The RCCB is an electromechanically operated CB and is unable to sum the high frequency currents correctly, which results in false tripping [2].

15.6.2.6 Electronic Equipment

Electronic equipment is very sensitive to harmonics and is also the cause. The following are two effects of harmonic distortion on electronic equipment:

1. *Zero-Crossing Noise:* In order to switch on the load, most of the electronic controllers identify the point of zero crossing in voltage waveform. This method reduces the transients and electromagnetic interference. Due to the presence of harmonics, the rate of variation in the voltage at crossing becomes faster, which is difficult to detect and in turn leads in faulty operation [2].
2. *Component Failure:* Due to harmonic distortion, the maximum voltage supply increases, and the equipment power supplies are noncontrolled rectifiers with a DC-side capacitor. Hence, with the change in input r.m.s. value, there is an equivalent change in output DC, and this will differ from its nominal value.

15.6.2.7 Relay and Contactors Protection

The sensitivity of a relay or contactor decreases with the order of the harmonic. In Datta and Nafsi [28] it is evident that some of the relays are not sensitive enough if the voltage distortion is above 20%. Above this point, operation is irregular in both normal and faulty situations. In low voltage power networks, the MCB operates on the basis of the peak value measurement. Because of harmonic distortion, there is an increase in the peak value, which results in tripping of the CB, even though the r.m.s. values are in the safer limits.

Underfrequency relays are also a victim of harmonics. The purpose of underfrequency relay is to detect the underfrequency event and trip the circuit accordingly. There are several cases where the frequency is measured on the basis of current zero crossings, but due to harmonics the frequency component undergoes more than two crossing in same interval. This event makes the frequency measurement inaccurate, and the relay does not trip, even in the actual condition of decrease system frequency

15.6.3 Techniques Used for Harmonics Mitigation

The effects of harmonics are discussed in Section 15.6, which shows some serious concerns about this problem, whether it is related to the relay or contactor systems, electronics equipment, transformer performance analysis, etc. In this section, various methods for mitigating the effects of harmonics are discussed. The end user cannot be responsible for all harmonic mitigation actions; the power utility must monitor and implement the necessary actions [2,26–28].

15.6.3.1 Custom Power Devices

Customer power devices (CPDs) use power electronic controllers in PDNs to provide reliable and good quality power for customers. Therefore, the CPD provides the necessary dynamic voltage restoration with hybrid active filters, which leads to improvement in power quality by improving the voltage profile and by mitigating the harmonics in the supply current. Figure 15.10 shows the configuration of the CPD based on a dynamic voltage restorer (DVR). It consists of an energy storage device such as a battery, an inverter, filter and interfacing transformer. Through constant monitoring of the supply current at the point of common coupling, a variation in the level of harmonic, i.e., the deviation from the sinusoidal may be detected by monitoring the zero crossing, which triggers the DVR controller to supply the necessary amount of compensating reactive current.

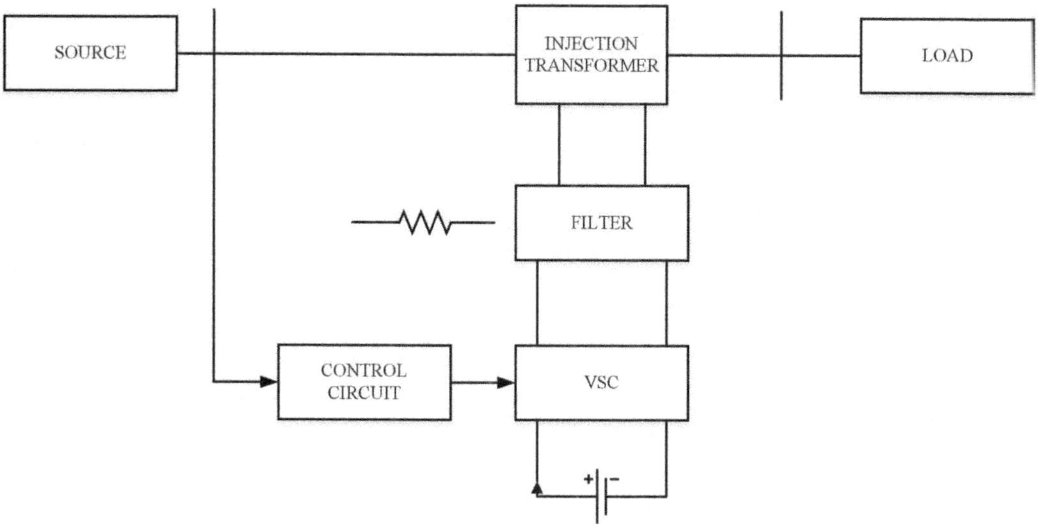

FIGURE 15.10 Configuration of CPD. (From Harmonics Mitigation, www.rroji.com, accessed July 1, 2018.)

15.6.3.2 Transient Voltage Surge Suppressors (TVSSs)

TVSSs clamp transient impulses (spikes) to a level that is safe for the electronic load and are a way to condition power. Employing a protection strategy for the entire facility safeguards the electrical system against most transients. Multi-stage protection entails using TVSSs at the service entrance, subpanel at the point of use. This co-ordination of devices provides the lowest possible let-through voltage to the equipment. TVSSs are used as an interface between the power source and sensitive loads so that the transient voltage is clamped by TVSSs before it reaches the load. TVSSs usually contain a component with a nonlinear resistance (a metal oxide varistor or a zener diode) that limits excessive line voltage and conducts any excess impulse energy to ground [29].

15.6.3.3 Isolation Transformer

Isolation transformers are used to isolate sensitive loads from transients and noise deriving from the mains. In delta-wye connection configurations, the isolation transformers keep harmonic currents generated by loads from getting upstream to the transformer. This particular part of isolation transformers is a grounded shield made of nonmagnetic foil located between the primary and the secondary. Any noise or transient that comes from the source is transmitted through the capacitance between the primary and the shield and on to the ground and does not reach the load. It provides a degree of isolation and filtering. These devices effectively reduce conducted electrical noise by physical separation of the primary and secondary through magnetic isolation. Isolation transformers reduce normal and common mode noises; however, they do not compensate for voltage fluctuations and power outages [29].

15.6.3.4 Filters

Several filters are designed to filter out the harmonics. The filter is designed for a given point in the supply and obtains the necessary frequency impedance characteristic of the supply system. The block diagram in Figure 15.11 shows the configuration of various filters that can be used. Low impedance can be achieved by using single frequency and double tune resonant filters. Damped filters provide low impedance in a broad frequency range; therefore, they are also known as broadband filters. Single frequency resonant filters along with broadband filters are the most commonly used filters.

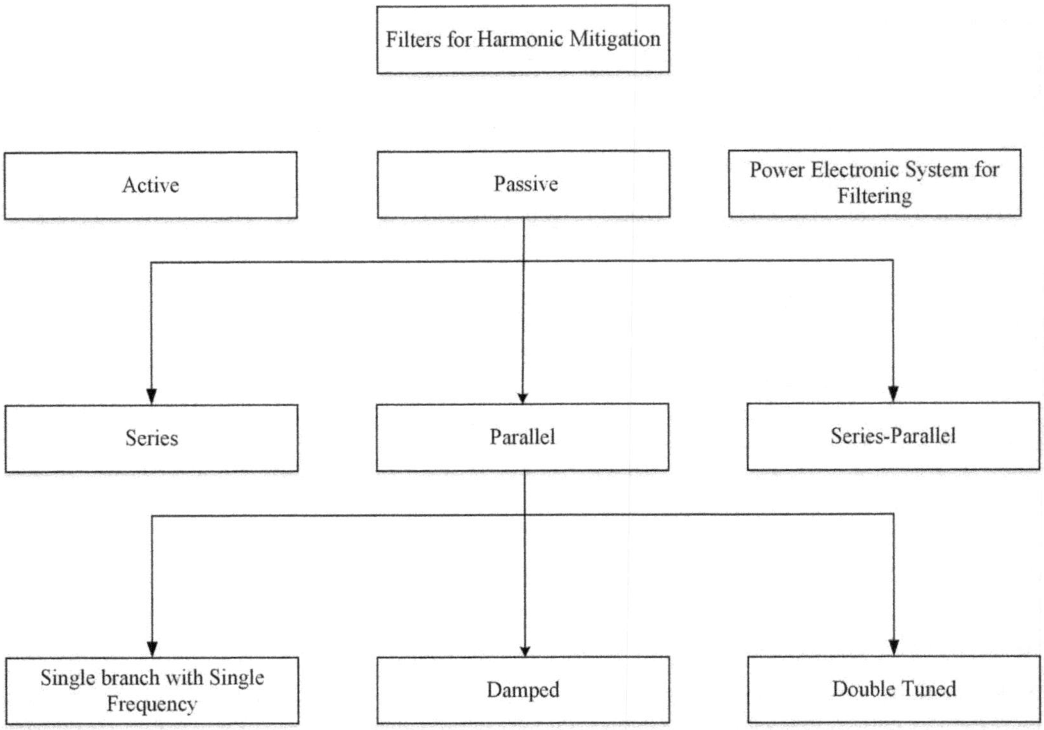

FIGURE 15.11 Classification of filter for harmonic mitigation. (From Baggini, A., *Handbook of Power Quality*, John Wiley & Sons, New York, 2008.)

15.7 Solved Examples

1. The 126.797 V/5.072 V and 113.541 V/6.81 V are the positive- and negative-sequence voltage sets, respectively, that are supplied to a three-HP, three-phase induction motor. Calculate the percentage of voltage unbalance factor. Suggest under which condition the motor should not be operated by the user.

   Solution

   Case 1: $V_{pl} = 126.797$ V, $V_{nl} = 4.885$ V

   $$VUF(\%) = \frac{|V_n|}{|V_p|} = \frac{4.885}{126.797} \times 100\% = 4\%$$

   Case 2: $V_{p2} = 113.541$ V, $V_{n2} = 6.81$ V

   $$VUF(\%) = \frac{|V_n|}{|V_p|} = \frac{6.81}{113.541} \times 100\% = 6\%$$

   From the calculation in Case 2, the VUF is above 5% and it is not recommend by NEMA, so the machine should not be operated under the unbalanced condition mentioned in Case 2.

2. A single-phase voltage source E is given by:

   $E = 141 \sin wt + 42.3 \sin 3wt + 28.8 \sin 5wt,$

The corresponding current is given by:

$I = 16.5 \sin(wt+54°) + 9 \sin(3wt-34°) + 6\sin(5wt-32)$

Find the power at the third and fifth harmonic and the total power delivered by the source.

Solution

Power at fundamental condition:

$$E_1I_1\cos\phi_1 = \frac{E_p}{\sqrt{2}} \times \frac{I_p}{\sqrt{2}} = \frac{141\times 16.5}{2} \times \cos 54 = 683\text{W}$$

Power at the third harmonic:

$$E_3I_3\cos\phi_3 = \frac{E_{3p}}{\sqrt{2}} \times \frac{I_{3p}}{\sqrt{2}} = \frac{42.3\times 9}{2} \times \cos 34 = 157.8\text{W}$$

Power at the fifth harmonic:

$$E_5I_5\cos\phi_5 = \frac{E_{5p}}{\sqrt{2}} \times \frac{I_{5p}}{\sqrt{2}} = \frac{28.8\times 6}{2} \times \cos 32 = 73.21\text{ W}$$

Total power = 683 + 157.8 + 73.21 = 914 W

3. A current drawn from a 50 Hz supply contains first, third, and fifth harmonics of 90 A, 12 A, and 10 A respectively. These are measure by an ammeter and cross through a coil of very small resistance; a voltmeter across the terminal shows a voltage of 7C V. What is the magnitude of current shown by the ammeter? What is the exact value of the inductance of the coil?

Solution

The r.m.s current is:

$$I = 0.707\sqrt{I_{1p}^2 + I_{3p}^2 + I_{5p}^2} = 0.707\sqrt{90^2 + 12^2 + 10^2} = 64.58\text{A}$$

$$E = 0.707\sqrt{E_{1p}^2 + E_{3p}^2 + E_{5p}^2}$$

$$I_{1p} = \frac{E_{1p}}{wL}, I_{2p} = \frac{E_{3p}}{3wL}, I_{5p} = \frac{E_{5p}}{5wL}$$

$$wLI_{1p} = E_{1p}, 3wLI_{3p} = E_{3p}, 5wLI_{3p} = E_{3p}$$

$$E = 0.707\sqrt{(wLI_{1p})^2 + (3wLI_{3p})^2 + (5wLI_{5p})^2} = 0.707wL\sqrt{i_{1p}{}^2 + 9I_{3p}{}^2 + 25I_{1p}{}^2}$$

$$70 = 0.707 \times 2\pi \times 50 \times L\sqrt{8100 + 9\times 144 + 25\times 100} = 0.0028\text{H}$$

15.8 Tutorial Problems

1. List various definitions of power quality in various standards.
2. How is the definition by IEC superior in explaining degree of unbalance?
3. What are the factors responsible for the voltage unbalance?

4. Explain the need for assessment of three-phase induction motors under voltage unbalance.
5. Explain the importance of slip for precise assessment of industrial machines operating under different sets of voltage unbalance.
6. Explain the differences between total harmonic distortion (THD), telephone form factor (TFF), distortion index (DIN), total demand distortion (TDD), power factor and displacement power factor.
7. Explain the various causes of poor power quality in industry and power systems. State the various means of mitigating these problems.
8. List various sources of harmonics, and explain the most common order of harmonics available in magnetic equipment.
9. How do harmonics affect the performance of relays and other protective devices?
10. Mention and explain the advantages of using UPS for power quality enhancement. Compare UPS with other approaches in terms of simplicity in implementation.
11. To drive a brushless DC machine, a duty cycle modulation drive with a voltage of 415 V and duty cycle of 0.75 is used. At a particular speed, the stator frequency fundamental component is 325 Hz. The switching frequency is 0 kHz. Calculate the amplitude of the two strongest harmonics in the region of 55 kHz.
12. High voltage side line-to-line voltages of a step-up transformer are 38, 33, and 100 kV for *R*, *Y* and *B* phases, respectively. Phase *R* leads phase *Y* by 100° and lags *B* by 170°. The transformer faces the angle and magnitude unbalance. Analyze the condition by using the symmetrical component approach.
13. Find the lowest possible frequency and voltage for proper operation of a four-pole, 50 Hz, 415 V, three-phase induction motor supplied by a variable frequency and voltage source. The motor has a rotor resistance of 0.1 Ω, and the stator resistance is neglected. The total inductance is 2×10^{-4} H.
14. The 115.877 V and 4.635 V are the positive- and negative-sequence voltage, respectively, supplied to a three-HP, three-phase induction motor. Calculate the percentage of voltage unbalance factor.
15. Due to a nonuniform air gap, a three-phase generator generated an e.m.f. of 230 V with 12% third harmonic and 15% fifth harmonic content. Calculate the r.m.s. value of the line voltage if it is Y-connected and Δ-connected.
16. A single-phase voltage source *E* is given by $E = 140 \sin wt + 11 \sin 3wt + 28.8 \sin 5wt$. The corresponding current is given by $I = 17 \sin (wt{+}55°) + 41.20 \sin (3wt{-}34°) + 6 \sin (5wt{-}32)$. Find the power at third and fifth harmonic.
17. An r.m.s current of 7 A, which has third harmonic content, passes through a 1 Ω resistance and a 12 mH coil. The r.m.s voltage across the coil is 32 V. Evaluate the magnitude of the fundamental frequency and the harmonic content of current if the fundamental frequency is $300/2\pi$Hz.

15.9 Conclusion

Power quality issues are controlled only when all power flow levels are monitored. The power quality issues at the consumer side can be mitigated using various techniques. The impact of voltage unbalance on industrial machines is one of the most emergent consequence of deprived power quality. Moreover, complete damping of harmonics from the power system is not possible, but by using harmonics filters, damping can be reduced to safer levels.

REFERENCES

1. R. C. Dugan, M. F. Mcgranaghan, S. Suntoso and H. W. Beaty, *Electrical Power System Quality*, 2nd ed., New York: McGraw-Hill, 2004.
2. A. Baggini, *Handbook of Power Quality*, New York: John Wiley & Sons, 2008.

3. M. H. J. Bollen, *Understanding Power Quality Problems*, Hoboken, NJ: John Wiley & Sons, 2000.
4. C. Y. Lee, Effects of unbalanced voltage on the operation performance of a three-phase induction motor, *IEEE Trans. Energy Convers.*, 14(2), 202–208, 1999.
5. P. Pillay, P. Hofmann, and M. Manyage, Derating of induction motors operating with a combination of unbalanced voltages and over or under voltages, *IEEE Trans. Energy Convers.*, 17(4), 485–491, 2002.
6. J. Faiz, H. Ebrahimpour, and P. Pillay, Influence of unbalanced voltage on the steady state performance of a three-phase squirrel cage induction motor, *IEEE Trans. Energy Convers.*, 19(4), 657–662, 2004.
7. J. Faiz and H. Ebrahimpour, Precise derating of three-phase induction motors with unbalanced voltages, *Proc. IEEE Ind. Appl.*, 485–491, 2005.
8. P. Gnacinski, Windings temperature and loss of life of an induction machine under voltage unbalance combined with over- or under-voltages, *IEEE Trans. Energy Convers.*, 23(2), 2008.
9. P. Pillay and M. Manyage, Loss of life in induction machines operating with unbalanced supplies, *IEEE Trans. Energy Convers.*, 21(4), 813–822, 2006.
10. N. L. Schmitz and M. M. Berndt, Derating poly-phase induction motors operated with unbalanced line voltages, *IEEE Trans. Power Appl. Syst.*, 680–686, 1963.
11. A. Chauhan and P. Thakur, *Power Quality Issues and their Impact on the Performance of Industrial Machines*, Hamburg, Germany: Anchor Academic Publishing, 2016.
12. Motors and Generators, ANSI/NEMA Standard MG1-1993.
13. IEEE Standard Test Procedure for Polyphase Induction Motors and Generators, IEEE Standard 112, 1991.
14. Testing and Measurement Techniques–Unbalance, Immunity Test, IEC Standard 61000-4-27, 2000.
15. Y. J. Wang, An analytical study on steady-state performance of an induction motor connected to unbalanced three-phase voltage, *Proc. IEEE Power Engineering Society Winter Meeting*, Singapore, 159–164, 2000.
16. S. G. Jeong, Representing line voltage unbalance, *37th IAS Annual Conference Meeting*, Pittsburgh, PA, 3, 1724–1732, 2002.
17. S. Nautiyal, A. Chauhan, P. Thakur and K. Govind, Estimating torque-speed characteristic of three-phase induction motor operating under unbalance supply, *Proc. IEEE Nirma University International Conference on Engineering*, Ahmedabad, India, 1–6, 2013.
18. A. Chauhan, P. Thakur and D. Raveendhara, Assessment of induction motor performance under supply voltage unbalance: A review, *Proc. IEEE Student Conference on Engineering and Systems*, MNNIT Allahabad, India, 1–6, 2013.
19. P. Giridhar Kini, R. C. Bansal and R. S. Aithal, Performance analysis of centrifugal pumps subjected to voltage variation and unbalance, *IEEE Trans. Ind. Electronics*, 55(2), 562–569, 2008.
20. A. Chauhan, P. Thakur and D. Raveendhra, Quantification of voltage unbalance conditions, *Power Electronics and Renewable Energy Systems*, Lecture Notes in Electrical Engineering Book Series, New Delhi, India: Springer, 326, 47–55, 2014.
21. V. B. Bhavaraju and P. N. Enjeti, An active line conditioner to balance voltages in a three-phase system, *IEEE Trans. Ind. Appl.*, 32(2), 287–292, 1996.
22. J. H. Chen et al., Using a static VAR compensator to balance a distribution system, *Proc. of IEEE Ind. Appl. Conf.*, 4, 2321–2326, 1996.
23. A. Campos, G. Joos, P. D. Ziogas and J. F. Lindsay, Analysis and design of a series voltage unbalance compensator based on a three-phase VSI operating with unbalanced switching functions, *IEEE Trans. Power Electron.*, 9(3), 269–274, 1994.
24. P. Lerley, Applying unbalance detection relays with motor loads, *IEEE Trans. Ind. Appl.*, 35(2), 689–693, 1999.
25. J. Driesen and T. V. Craenenbroeck, Voltage disturbances introduction to unbalance, *Copper Development Association IEE Endorsed Provider*, 5(3), 2002.
26. M. H. Rashid and A. I. Maswood, A novel method of harmonic assessment generated by three-phase AC-DC converters under unbalanced supply conditions, *IEEE Trans. Ind. App.*, 24(4), 590–597, 1988.
27. D. Li, Q. Chen, Z. Jia, and J. Ke, A novel active power filter with fundamental magnetic flux compensation, *IEEE Trans. Power Del.*, 19(2), 799–805, 2004.
28. S. K. Datta and A. Nafsi, Distribution relay performance under harmonics conditions. *PQA'92*, Atlanta, GA, 1992.
29. Harmonics Mitigation, www.rroji.com, accessed July 1, 2018.

16

Reliability Assessment of the Distribution System in the Presence of Protective Devices

T. Adefarati and Ramesh Bansal

CONTENTS

16.1 Introduction

The main function of the distribution power system is to deliver electric power from the supply points to various customers at the load points with a high degree of reliability. Therefore, a reliable distribution system must be designed by the distribution network operators so that it will have an adequate power supply to meet the power requirements of consumers at all times [1]. The number of power interruptions has a substantial effect on the quality of the power supply and the operations of the distribution system. For this reason, an optimal restoration technique that limits the number of customers affected by power outages should be the priority of the utilities [2]. The measures put in place by the distribution network operators with the application of protective devices to reduce the number of power outages in the distribution network have resulted in equipment safety and customer satisfaction. This has a considerable effect on the optimal operation of the distribution power system, with a minimum number of consumers affected by power interruptions. Consequently, the most important task of the distribution network operators is to operate their power systems with minimum interruption and the fastest power restoration procedure to return power to the areas as soon as possible.

Protective devices play a prominent role in the reliability of the distribution system by reducing the outage duration, annual failure rate and customers' interruption cost at each load point. The reliability of the distribution network can be enhanced with the application of protective devices such as circuit breakers, fuses, sectionalising switches, reclosers, etc., provided that the only faulty sections of the power system are isolated while other consumers get a reliable power supply by using the functioning sections of the power system. This reduces the number of affected customers and blackout areas so that power outages do not affect the entire distribution power system. The measures put in place by the power utilities to improve their reliability depend on the effectiveness of the power supply restoration procedure applied by the utilities. Owing to the economic impacts of power outages in the distribution system, the risk of electrical faults should be given more attention. The statistical data presented by many organisations show that many customers are affected by short-term outages and this results in substantial economic damage. This has prompted the power utilities to introduce quantitative analysis to measure the reliability of their power systems. Moreover, this will enable the stakeholders in the power sector to balance the utility reliability costs and customer interruption costs. The costs incurred by the distribution network operators to enhance the reliability of their power systems in the conceptual, design, planning and operation phases consist of replacement cost, capital cost and maintenance cost of the protective devices. It is anticipated that consumer interruption cost will continue to be reduced with the incorporation of protective devices to reinforce the existing distribution networks.

This chapter discusses ways to provide the optimal number and type of protective devices and to position protective devices in the distribution system to enhance the reliability of the system at the lowest cost. The proposed method can be used by the utilities to meet the level of reliability required by their consumers at reduced investment and outage cost. The reliability of the distribution system is a serious concern of the distribution network operators because it is usually assessed by utilising the key performing indicators (KPIs) set by the reliability regulatory commission of each country [3]. Violations of the KPIs by the attract stiff penalties that can become unbearable for the distribution network operators. For this reason, reliability indices are frequently utilised to measure the performance of the distribution network by focusing on the availability of uninterrupted power supply at the load points. The impacts of protective devices on a power system have been assessed, based on the interruption cost, total cost and the reliability index of the distribution system. The results obtained in this chapter demonstrate the effectiveness of using protective devices to minimise power interruption cost, which has been a challenge for many organisations. The reliability evaluation of the distribution system has become an important subject in the deregulated power sector owing to the increasing high demand for a reliable power supply with a reduced number of power outages. The power outages result in economic losses and socio-economic consequences based on the halt of major activities. Some policies such as renewable portfolio standard (RPS), fixed feed-in tariff and renewable obligation have been introduced in some countries as potential alternatives to reduce the consequence of power outages. This is important in making the power system more reliable, efficient and economically feasible.

This chapter also emphasises the cost benefits of the integration of protective devices for reducing the number of power outages and optimising the utilisation of the distribution networks effectively. The application of protective devices can be used in the power system to speed up the restoration process after a power outage. This considerably enhances the reliability of the distribution system by isolating a faulty section of the power system from the entire network. Hence, there is a provision for an alternative power supply through a functioning section of the power system to meet the power requirements of the consumers. However, the costs of installing and maintaining the protective devices should be the first priority of the distributed network operators when making any investment decision to improve the reliability of their networks. Proper planning must be put in place by the power utilities in order to make a significant investment decision based on the costs and benefits of power system automation with the integration of protective devices. The application of protective devices at optimal locations in the distribution systems can improve the reliability of the power system, but this depends on the economic analysis that must be carried out to obtain the best solution. The selection of the protective devices and their locations for improvement of the reliability of the distribution system depends on the following factors: type of customers connected, interruptions cost, installation cost and maintenance cost. To achieve the objectives of the power system with minimum power interruption, the proper economic analysis must be carried out on the application of the protective devices in the distribution system. The objective of using the economic analysis in the distribution system is to select the best choice that will achieve reliability and economic benefits.

The sudden increase in power consumption at minimum power outages has encouraged many power utilities to integrate protective devices into their power systems. This enhances the reliability of the system as well as the efficiency and quality of power supply. The tie switches, sectionalising switches, circuit breakers and fuses play important roles in the automation of the distribution systems. The optimal allocation of the aforementioned protective devices can be utilised to reduce the economic losses associated with the power outage as well as reduction of the duration, frequency and number of power outages. This chapter discusses protective devices that reduce the cost of customer service outage and the total cost of the system. The reliability model is formulated by considering the consumer outage cost and total cost of protective devices. The economic and reliability benefits of the distribution system are considered in the formulation of the cost model. The model is developed with the goal of incorporating cost-effective protective devices in the power system. The model is utilised in this work to analyse the impacts of the protective devices in a power system.

The proposed technique is implemented in the Roy Billinton Test System (RBTS) distribution system to verify its capability to achieve the aim of the research and to explore the cost benefit of protective devices in the distribution system. The KPIs obtained in the study are compared with and without the utilisation of the sectionalising switches and fuses. The application of protective devices in the distribution system can be utilised by the distribution network operators to reduce the impacts of power outages and total cost in the power system [4]. Benefits to customers of the utilities can also be increased by reducing the intermittent power outages with the application of protective devices. In the distribution power system, the presence of protective devices can substantially enhance the reliability of the system by minimising the duration and frequency of power outages at the load points. The application of protective devices can also be utilised to minimise the time required to restore power supply when a fault occurs in the power system. The proposed technique has been tested with the RBTS distribution network with satisfactory results, which clearly establishes the importance of incorporating the customer interruption cost in the design of a power system. The reliability worth analysis is implemented by using RBTS distribution network. The results obtained from different case studies show the technical and economic impacts of protective devices in a power system.

16.2 Distribution Power System

A power system is a network that includes several electrical components for delivery of electric power from the generating stations through the transmission and subtransmission lines to the load points [5]. The power system can be traditionally divided into a generation system, transmission system, subtransmission system and distribution system, as shown in Figure 16.1. The electrical power system is designed so that its operation depends on the power requirements and the system constraints. However, the function of a distribution system is to deliver electrical power at the consumer load points at the voltage that is commensurate with consumer requirements.

The distribution system can be classified into primary and secondary systems. The primary distribution system consists of substations, primary feeders, secondary feeders, service cables, sectionalising equipment, lateral and main feeders, protection equipment and secondary circuits. The primary distribution systems carry the medium voltage from the subtransmission lines to the distribution transformers, where it will be stepped down to 11 or 22 kV based on the voltage requirements of bulk purchase power customers. Some large customers are connected to the transformers that step down the distribution voltage to the voltage level required by their loads, as shown in Figure 16.1. The commercial, industrial and public institution customers that need bulk power requirements can be directly supplied from the primary distribution systems.

The secondary distribution system consists of the distribution transformers that step down the voltage to the level required by household appliances. A number of consumers can be fed through the secondary distribution lines at this voltage. The voltage required at the consumer load points is generally at 120/240 V for a single-phase system and 415 V for a three-phase system.

16.2.1 Network Configurations

The distribution systems are built as interconnected networks to provide uninterrupted power supply at all times. The performance of the distribution system is determined based on its configuration or how the networks are interconnected. These are generally operated in a radial, loop or mesh network. The configuration of the distribution system can be altered under normal operating conditions. This can be achieved by changing the topological structure of the distribution system by closing or opening the sectionalising switches to reduce the power outages while satisfying the system operating constraints.

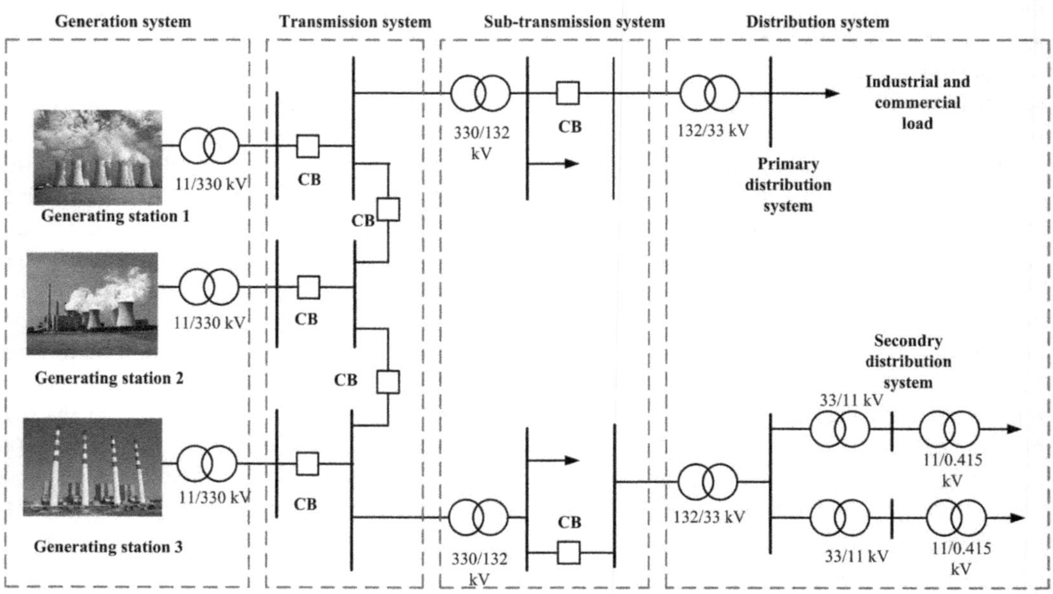

FIGURE 16.1 Schematic diagram of a power system.

16.2.2 Radial Distribution Power System

The radial distribution system is the simplest and cheapest configuration that can be utilised for power distribution in rural areas with a sparse population. It consists of a number of components such as transformers, lines, circuit breakers and busbars, as shown in Figure 16.2 [6]. A radial distribution system is a single power source; the failure of any component of the power system would ultimately interrupt power supply to the entire network because there is no alternative path to supply numerous consumers at the load points.

16.2.3 Meshed Distribution Power System

The distribution system can be configured by the power utilities so that load points in the system are connected by more than one line. A meshed distribution system is more reliable than a radial distribution system because it has multiple power sources, protective devices and many lines for power to flow in the system, as shown in Figure 16.3. However, it is highly expensive due to numerous

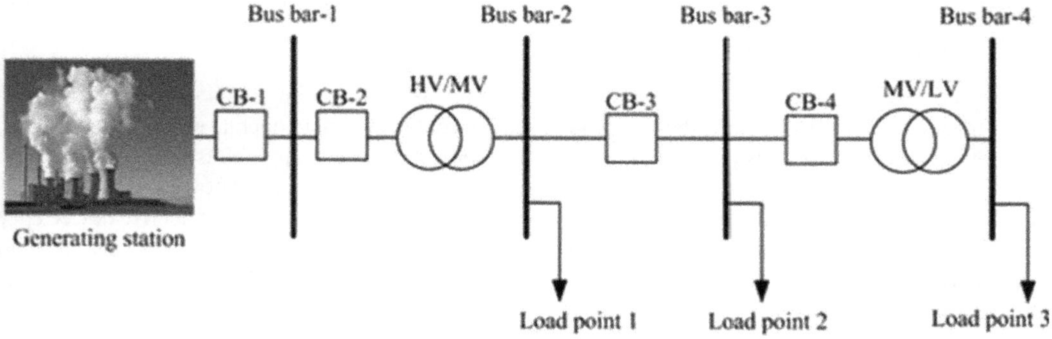

FIGURE 16.2 Radial distribution power system.

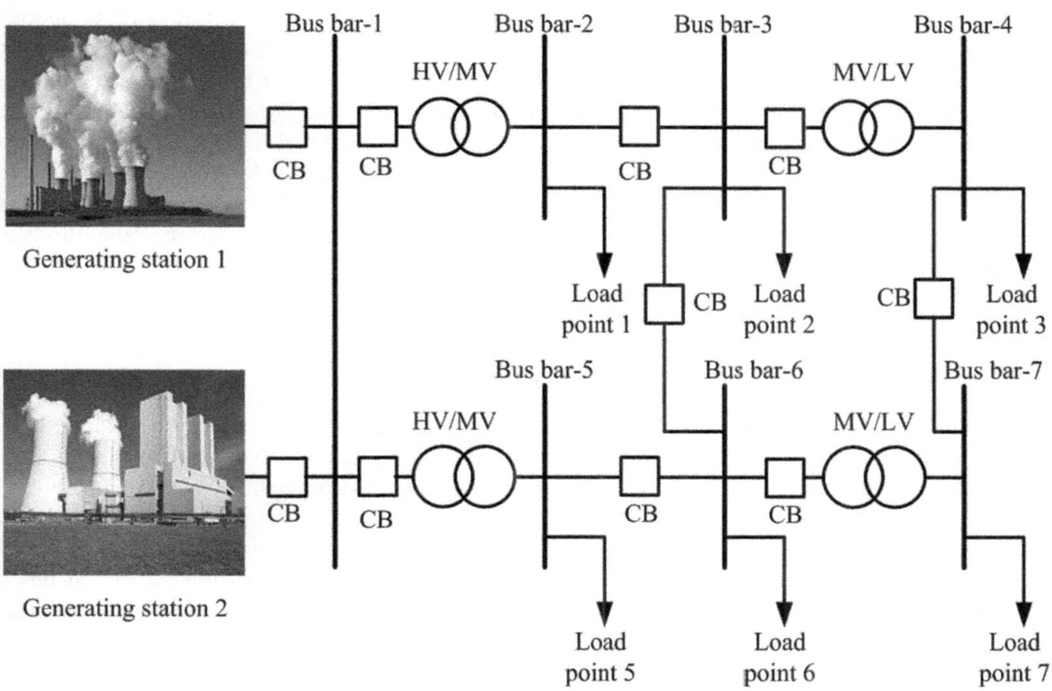

FIGURE 16.3 Distribution power system with the meshed configuration.

components such as protection devices, control devices, breakers, fuses, etc. A mesh configuration consists of two radial systems separated by a normally open switch; if any fault occurs in the distribution system, the sectionalising switches will be closed so that one section of the distribution system will be energised through the other networks. This reduces the number of customers exposed to a single component failure and the number of power outages per year [7]. The reliability of the distribution system can also be enhanced by increasing the investment cost of the operation and planning of the distribution system. The traditional planning of a power system allows upgrading of the existing distribution system or the addition of a number of electrical components to meet the load requirements due to annual load growth. Conversely, overinvestment in reinforcement of the distribution system will bring about unprofitable operation of the power system due to a high operating cost. The high operating cost will invariably be transferred to the end users in the form of a high tariff. As a result, the best option for reducing the number of power outages at the load points is to integrate protective devices in the distribution system, thus incurring economic and technical benefits.

16.3 Protective Devices

The objective of a protective device is to ensure a continuous power supply by isolating only the segments that are faulty while leaving the functioning sections of the system to operate effectively [8]. Thus, the protective devices are designed with a pragmatic approach to clear many faults in a power system. The protective devices are utilised manually or remotely by the utilities to control power supply interruptions when a fault occurs in the power system. In general, the economic impact of protective devices in a power system is to minimise the frequency and duration of power outages. For this reason, the technology employed in protection systems must be well established and sensitive to any electrical fault that can impair the reliability of a power supply at the load points.

16.3.1 Performance of Protective Devices

The performance of protective devices should be designed and practically coordinated to operate correctly and respond to faults in the coordinated power system. Protection systems are generally designed to operate correctly for protective functions in the protection zone. Some KPIs are utilised by the distribution network operators to compare the relative reliability of planned protection systems. These performance indicators can be used to quantify the reliability of the power systems and to make the best decisions at the minimum operating cost. Hence, for optimal operation of protective devices to be achieved in a power system, it must have the following performance and design criteria: dependability, speed, selectivity, sensitivity, cost, simplicity, security, reliability and accuracy.

16.4 Protective Devices in the Distribution Power System

Failures in the distribution system can cause an interruption of power supply at the load points. They can be categorised into momentary or sustained interruptions based on the duration of the outages. Consumers are more sensitive to outages because of the increase in sophisticated equipment and appliances that require a continuous power supply from the utilities. The impacts from momentary and permanent interruptions should be the priority of the power utilities. Power interruption can lead to a complete loss of service and cause a lot of economic damage to the consumers that depend on the utilities for power supply. For this reason, it is mandatory that the power utilities apply protective devices in their power systems to reduce the impacts of power outages to a minimum acceptable level. The main objective of utilising the protective devices in the power system is to reduce the number of people affected by the power outages. Different types of equipment are available for the protection of distribution systems. The particular type of protection used depends on the sensitivity of the load to power outage and equipment being protected. The protection system is a viable approach for enhancing the reliability of the distribution system.

The number of reclosers, circuit breakers, sectionalising switches, fuses, etc., that can be installed at the lateral and main feeders of the distribution system depends on the cost and investment implications. However, the application of protective devices in a power system also depends on the configuration of the networks and the acceptable standards laid down by the power reliability regulatory authorities. Sectionalising switches are always located on both sides of each line section to reduce the number, frequency and duration of power outages. The circuit breaker is generally situated toward the start of line segments of the feeders so that it will disconnect the faulty section from the entire power system. The economic analysis of the distribution systems can be assessed based on the type of consumers connected; load variation; reliability considerations; maintenance and installation costs; and type, number and location of the protective devices used in the power system.

Efficient planning of the distribution system involves the placement of protective devices in strategic places in the network for better operation and improvement of the quality of power supply. Protective devices in the distribution network improve the performance of a power system and guarantee a constant power supply at the consumers' load points. This considerably reduces power outages in terms of time required by the maintenance team to trace the fault and restore power to the affected areas. This results in meeting the objective of consumers' power requirements at highly reliable and low power outage cost at the load points. For this reason, a distribution power system is designed to identify the types, number and locations of the protective devices to reduce the total cost of the system. The optimisation of the outage cost is based on the capital cost of protective devices required to improve the reliability of the system, customer interruption and the cost of maintaining the protective devices.

The power utilities use protective devices in their networks to mitigate the socioeconomic effect of power outages at the load points. The protective devices considered in this section consist of circuit breakers, fuses, sectionalising switches, and automatic reclosing relays, as shown in Figure 16.4. The concept of power system reliability with the integration of protective device requires some terms to be defined. These terms have to do with the operation and protection of a power system. This section briefly explains various protection devices that can be used in the power system.

16.4.1 Application of Circuit Breakers in the Distribution System

Uninterrupted power supply at the load points has become a critical issue for the utilities and money spent on the reinforcement of their power systems is tantamount to improve power system performance. Due to this, a number of circuit breakers are installed at the substation for isolation of the faulty sections from the entire system, as shown in Figure 16.5. This has resulted in the reinforcement of the distribution system with the protective devices for better operation and improvement of system reliability. It also prevents the power system from the economic damage associated with the fault current. Circuit breakers are mainly classified on the basis of rated voltages, as presented in Table 16.1. More details about circuit breakers can be found in Chapter 3.

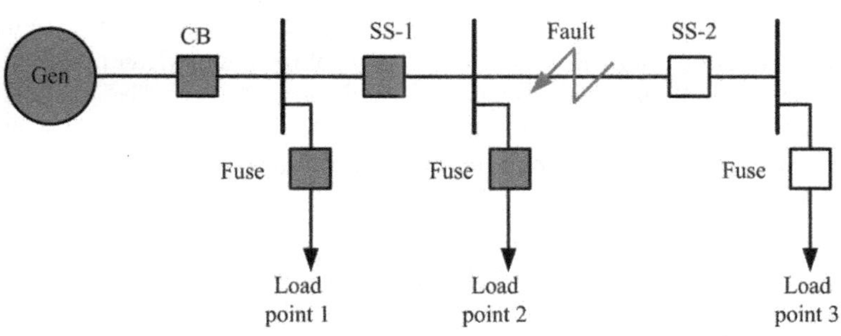

FIGURE 16.4 Coordination for radial feeders.

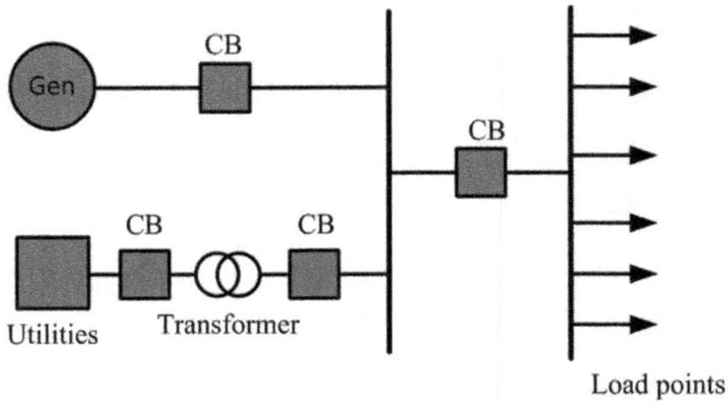

FIGURE 16.5 Single line of the distribution system with the application of the circuit breakers.

TABLE 16.1

Classification of circuit breakers

S/N	Type	Medium	Range
1	Air break CB	Air at atmospheric pressure	LV up to 1000 V
2	Tank type oil circuit breaker	Dielectric oil	Up to 33 kV
3	Minimum oil circuit breaker	Dielectric oil	36 kV, 1500 MVA 132 kV, 220 kV
4	Air blast CB	Compressed air (pressure 20 to 30 atmospheres)	132 kV, 220 kV 400 kV, 760 kV
5	SF_6 CB	SF_6 gas	132 k V, 220 kV 400 kV, 760 kV
6	Vacuum CB	Vacuum	11 kV, 33 kV
7	High voltage direct current CB	Vacuum or SF_6	± 500 kV DC

Source: Ali, A. S., *Smart Grids: Opportunities, Developments, and Trends*, London, UK: Springer, 2013.

16.4.2 Application of Fuses in the Distribution System

A fuse is utilised in the distribution power system to achieve a certain level of reliability and customer satisfaction with limited resources. Therefore, it is imperative to utilise fuses in conjunction with other protective devices for minimisation of total cost and maximising customer benefits. Fuses are installed at the lateral feeders of the distribution network for the protection of the loads. They have a breaking capacity but no reclosing capacity, like circuit breakers and reclosers. For this reason, they are always installed at the lateral of the distribution system. This chapter provides a brief highlight of how fuses can be utilised by the distributed network operators to improve the reliability, efficiency, economics and sustainability of electric services. A comparison of circuit breakers and fuses for high-voltage, medium-voltage and low-voltage applications is presented in Table 16.2.

The protection provided by a fuse in the lateral and main feeders of the distribution system is presented in Figure 16.6, in a situation where the primary fuses at the supply points are used for protection of the system. The fuses installed at the lateral feeders of the power system provide an overcurrent protection in conjunction with the sectionalising fuses at the main feeders. The sectionalising fuses are located at the main feeder to isolate the faulty lateral feeders from the entire power system. This prevents the entire power system from power interruption when a short circuit or fault current occurs between the lateral fuses and sectionalising fuses.

16.4.3 Application of Sectionalising Switches in the Distribution System

A sectionalising switch is a protective device that is designed to operate alongside with other protective devices such as circuit breakers, reclosers, etc., to isolate the faulty sections of a power system.

TABLE 16.2
Comparison of Fuses with Circuit Breakers

Fuses	Circuit Breakers
Not expensive	Expensive
Simpler	Complex and difficult to install
Fuses should be replaced once they are blown	Need to reset after operation
Fuses have a tendency to be quicker to interrupt the flow of fault current with the working time of 0.002 seconds.	Circuit breakers do not respond as quickly as fuses to electrical faults with the operating time of 0.02–0.05 seconds.
No maintenance cost	More expensive to repair and maintenance
Not safe owing to selection of a wrong fuse that has a current rating.	Safe because there is no doubt about picking the correct fuse rating.
Not sensitive to vibration of the buildings where fuses are being used.	More delicate to vibration and this causes the circuit breakers to trip for reasons unrelated to an electricity overload.
Fuses cannot be used in situations that involve a ground fault circuit interrupter.	Circuit breakers can be used in situations that involve a ground fault circuit interrupter.
Fuses work on the electrical and thermal properties of the conducting materials.	Circuit breakers work with the electromagnetic and switching principle.
Fuses can be used only once for electrical applications.	Circuit breakers can be used a number of times.
Fuses have only one pole and this has adverse effects on the operation of three-phase equipment.	Circuit breakers have multiple poles.
Fuses have a very low breaking capacity when compared with the circuit breakers.	A very high breaking capacity.
Fuses have the capability of detecting and interrupting electrical faults.	Circuit breakers are only designed to perform the interruption while the inbuilt relay system is utilised to detect faults.
Fuses have the ability only to protect the power system against overloads.	Circuit breakers have the capacity to protect a power system against overloads and short circuit faults.
The performance of fuses depends on the ambient temperature.	The performance of circuit breakers does not depend on the ambient temperature.
The configuration and design of fuses do not allow them to be used as switches.	The configuration and design of circuit breakers allow circuit breakers to be used as switches.
Fuses do not have auxiliary contact.	Circuit breakers have auxiliary contacts.

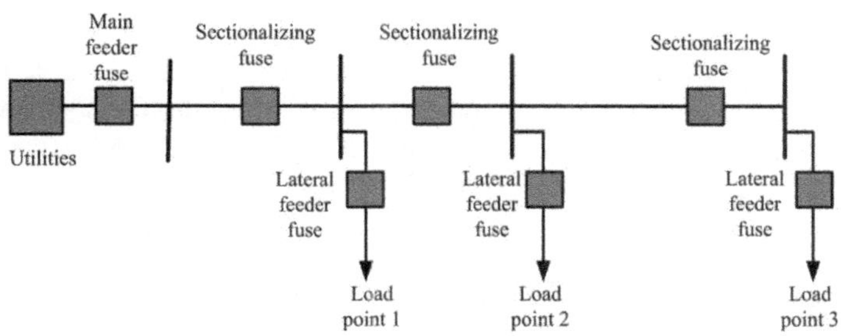

FIGURE 16.6 Operation of fuse in a radial distribution system.

A sectionalising switch is an automatic circuit opening device used for connecting or disconnecting sections or feeders of a power system. It is installed at the main feeder for isolation of the faulty segment of the distribution system once the upstream circuit breakers have interrupted the fault current in the system. The sectionalising switch is always installed downstream of other protective devices since it does not have the capacity to break fault current. It cannot interrupt a circuit, but it can be used to isolate and de-energise the circuit. As a result, it must be used in conjunction with a protective

FIGURE 16.7 Schematic diagram of sectionalising switch with a circuit breaker.

device that has breaking capacity. The operation of a sectionalising switch in conjunction with a circuit breaker at the upstream of the system is presented in Figure 16.7.

The following factors must be considered when selecting a sectionalising switch for the power system: system voltage, load current, short circuit level and the ability to work with other protective devices in the power system. The sectionalising switches have the following advantages:

1. Very simple in design and construction.
2. Relatively low initial cost.
3. Do not open accidentally under load and after a permanent fault.
4. Do not require replacement of fuse links when compared with the fuses.
5. Much lower outage time due to different mode of operation.
6. The possibility of error in the selection of the correct fuse link size and type is eliminated.
7. Operators can select different modes of operations based on the available situation.
8. Saves the time of operators for switching operations.

16.4.4 Application of Reclosers in the Distribution System

A recloser is a protective system equipped with a device that can close the circuit breakers upstream or downstream of the system after it has been opened owing to the fault current. It is used in the distribution system to detect and isolate momentary faults. A recloser is utilised with other protective devices to enhance the continuity of power supply at the load points by restoring power supply to the distribution system after a momentary fault. A recloser has the ability to detect a fault current in a power system by isolating the faulty section of the distribution system. It is designed by the manufacturer to break the circuit after the fault current continues for a programmed time. If the electrical fault that caused the operation of the recloser has not been cleared, the recloser will remain open until the fault is cleared. In order to prevent damage, each substation along the network is protected with a recloser that turns off power in the event of a short circuit. The recloser model is presented in Figure 16.8 for clearing of a fault in the distribution system.

16.4.5 Allocation of Protective Devices at Different Points in a Power System

The relationship between faults and interruptions is utilised by the distribution network operators based on the type of protective devices to clear the faults in different sections of the feeders. Therefore, the characteristics of a number of protective devices described below are applicable for protection of a power system. The reliability of the distribution system can be enhanced with proper allocation of the protective devices at appropriate points in the power system. The protective devices are selected according to the rules established by the stakeholders in the power sectors that understand the operation and planning of the network, for example:

1. Fuses must be installed at the beginning of the lateral feeders in conjunction with the circuit breakers and reclosers at the supply points or main feeders.
2. Fuses, sectionalising switches and reclosers are selected based on the permanent and temporary fault rates of each section of the distribution system.

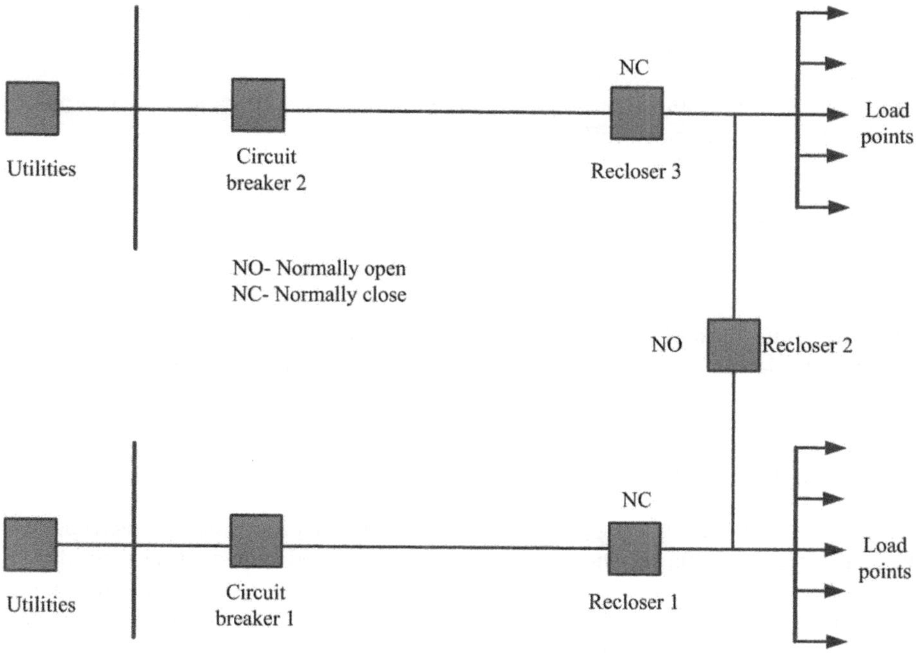

FIGURE 16.8 Schematic diagram of sectionalising switch with the circuit breakers.

3. Fuses and reclosers must be utilised at the end of a node in the electrical section that has the loads that require a continuous power supply at all times.
4. Sectionalising switches are utilised in a power system based on the nature and importance of loads at the respective feeders.
5. The allocation of sectionalising switches depends on the socioeconomic activities at each load point in the power system.
6. Sectionalising switches are utilised in a power system in accordance with the total amount of loads and the number of customers affected by power outages.
7. Reclosers must be used at the beginning of lateral feeders that have economic importance to the consumers.
8. Reclosers are installed at the beginning of main and lateral feeders that have important loads subjected to numerous permanent and temporary faults.
9. Reclosers must be installed at the lateral feeders in addition to the circuit breakers in the main feeders.

16.5 Power Interruptions

A power interruption can be defined as a loss of electric power supply to a particular area that consists of one or a group of customers. A power interruption is caused by faults at the power plants, damage to Transmission and Distribution (T&D) lines, substations, segment of the distribution system, a short circuit fault at T&D systems, overloading of electrical components, lightning, ice storms, wind storms, rain and flooding, car accidents, animals, scheduled outages, fallen trees, component failure, public damage, bush burning, digging too close to T&D lines, human error, etc. Power interruptions are critical at consumer load points where public safety is at risk. This will have a lot of economic consequences on consumers that depend only on the power source from the utilities. A power interruption causes economic losses for both the customers and the utilities. For this reason, institutions such as mines, airports, hospitals and hotels usually protect their power systems with a number of protective devices. These protective devices are designed to protect the power system whenever there is a power interruption from the utility sources. This will automatically reduce the socioeconomic effects that are associated with scheduled and unscheduled power outages. A power interruption is categorised into either momentary or sustained based on the duration and effect of the outage.

16.5.1 Momentary Power Interruption

Momentary power interruption is the short-term planned or accidental loss of energy to a number of customers triggered by the opening and closing operations of the protective devices. An electric power interruption with a duration of less than 5 minutes is categorised as a momentary power interruption based on IEEE Standard 1366. A momentary power interruption occurs due to a lightning strike or other factors and switching operations.

16.5.2 Sustained Power Interruption

Sustained power interruption is defined according to IEEE Standard 1366 as any interruption that lasts more than 5 minutes [9]. Sustained interruptions are usually caused by permanent faults due to storms, component failure, lightning, ice, trees striking lines, wildlife, etc. Such power interruption results in a total shutdown of the customers' facilities or equipment. The failure that occurs in a power system causes power outages that can be considered as permanent, a brownout and a blackout.

1. Permanent fault: A permanent fault is a massive loss of electrical power to a number of customers that results from a fault in the transmission and distribution lines. The power supply to the load points can only be restored once the fault has been cleared by the utilities.
2. Brownout: A brownout is a reduction in the voltage supply by the utilities to a particular area. It can be caused by the failure of generation, transmission and distribution facilities. This results in poor performance or poor operation of some equipment. In addition, a brownout that lasts for minutes or hours can cause the dimming of lighting when the voltage drops as a result of an effect of disruption of the grid. A number of regulatory policies can be enforced by the power system regulatory policies to reduce the load shedding during the peak or emergency period to prevent a blackout when power demand exceeds power supply.
3. Blackout: A blackout is a power interruption that occurs when a large number of customers are left without an electrical power supply for a determined duration of time. The duration of the power loss depends on the nature of the blackout and the configuration of the electrical network. Notable blackouts that occurred recently have been attributed to human error, equipment failure, animals, natural disasters, extreme weather conditions, bush burning, etc. Blackouts have become a national phenomenon in sub-Saharan and Middle Eastern countries due to persistent power outages that result in a severe socioeconomic impact on customers. The consequences of blackouts on the segment of the power networks are related to the economic damage, number of people that are affected and complexity of the power system.

In this context, the significance of power system reliability assessment is provided in this chapter to reduce the likelihood of power outage occurrence and the degree of their economic consequences. Recent power blackouts in the world are chosen for a concise review of the most essential related facts about the results of energy blackouts. The economic damage associated with blackouts in different parts of the world has made many countries pay more attention to a number of activities that enhance the reliability of their power systems. The reliability of the distribution system can be improved by channelling more resources into reinforcing the existing power system with protective devices. This will probably reduce the effects of a power outage on the populace. Recent power blackouts in the world are presented in Table 16.3.

TABLE 16.3

Recent Blackouts Around the World

Description	Date	Consequences
Moscow blackout	May 25, 2005	The blackout was due to a transformer failure that resulted in a power outage that affected 4 million people and cost about $70 million.
European blackout	November 4, 2006	The switching off of an electrical line over the River Ems to allow for the voyage of the *Norwegian Pearl* ship for safety reasons caused a power outage that affected 10 million people.
Chenzhou blackout	January 24, 2008	Affected 4.6 million people and more than 60 people died.
Brazil and Paraguay blackout	November 10–11 2009	A severe thunderstorm caused the failure of some transmission lines to operate effectively. This caused many hydro power plants to shut down and over 17,000 MW was lost.
Chile power blackout	March 14, 2010	The failure of a major transformer in the power system caused a power blackout that affected about 90% of the population of Chile.
North and Central Chile blackout	September 24, 2011	Affected 9 million people.
US blackout	October 29–30, 2012	A hurricane caused a power outage that left about 8 million people without electricity.
Southeast Queensland blackout	July 26–27, 2013	Severe weather that hit Southeast Queensland's electricity grid caused power interruption that affected 0.25 million people.
Egypt blackout	August 12, 2014	Technical failure that originated from routine maintenance caused a nationwide power outage that resulted in economic losses estimated to be over LE100 million.
Turkey blackout	March 31, 2015	The power interruption caused by technical problems affected 70 million.
South Australian blackout	September 28, 2016	A severe storm caused a power outage that affected 1.7 million people.
US blackout	March 1, 2017	A severe thunderstorm caused a power outage that affected 10 million people.

Source: Čepin, M., *Assessment of Power System Reliability: Methods and Applications*, London, UK: Springer Science & Business Media, 2011; Gungor, V. C., et al., *IEEE Trans. Ind. Electron.*, 57, 3557–3564, 2010.

16.6 Reliability

Reliability is the probability that a power system will perform its required function under express conditions for a particular time. The reliability assessment of a power system requires a range of methods, including detailed probability modelling of the entire power system by using a rule of thumb to reinforce the existing system with the integration of protective devices.

16.6.1 Importance of the Reliability Evaluation of a Power System

1. Power utilities must carry out a continue reliability assessment of their power systems to understand why and how failures occur in the system.
2. The goal of power system evaluation is to contribute to the enhancement of the reliability of a power system.
3. Power utilities must reduce the risk and economic consequences of the power outage and load shedding.
4. The performance and efficiency of a power system can be maximised by predicting the occurrence of failures in the power system.
5. The reliability assessment effectively reduces the occurrence of power failure in the power system.

6. The economic impacts of power outages can be estimated by using the reliability indices.
7. Reliable power plays a significant role in the economic and political well-being of a nation.
8. Reliability assessments boost economic activities of different sectors.

16.7 Reliability Indices

Various reliability indices can be utilised to quantify the reliability performance of the distribution system. These indices are applicable to the components, modules, load points and complete power system. Only the few of the most important reliability indices for the evaluation of the performance of a power system are presented in this section.

16.7.1 Failure Rate

The failure rate is the frequency at which a component in the power system fails to achieve its function. The failure rate of a component in a power system is associated with time, operating conditions, human error during design, installation, operation and maintenance, and environmental conditions such as humidity and ambient temperature. It is well known that vibration has a great influence on the failure rate of the electrical components. Moreover, the failure of the major electrical components in a power system are caused by either a single factor or multiple factors that are not predicted and time independent. The failure rate behaviour of an electrical component during its lifetime is presented in this section before the concepts of a power system reliability are discussed fully. The bathtub curve presented in Figure 16.9 is used in the reliability assessment of a power system. It describes the three distinct periods during the lifetime of a component: infant mortality, useful life time and wear-out.

1. Infant mortality: The infant mortality period of a component is characterised by an early high failure rate based on the bathtub curve presented in Figure 16.9. The initial failure rate is high due to poor design, the use of substandard materials to produce the components, material defects, noncompliance with the manufacturing specifications, errors in assembly, lack of adequate information on manufacturing process, wrong design, etc. In addition, early failure is also caused by the failure of the quality control and inspection units of the manufacturing companies to identify and correct any defect in their products before selling them. To reduce the early failure of electrical component, visual inspection and different kinds of electrical tests must be carried out. Early failures can also be minimised by using good design and maintenance practices. In the early life of an electrical component, the failure rate is high, but it will continue to diminish as defective components are found and disposed of, and early sources of potential failure, for example, installation and O&M errors are overcome. Hence, the early failure rate will decrease because the weak or defective components in the system have been removed.
2. Useful lifetime: The useful lifetime of a component is characterised by an essentially random failure with a constant failure rate that is low, as shown in Figure 16.9. The useful life

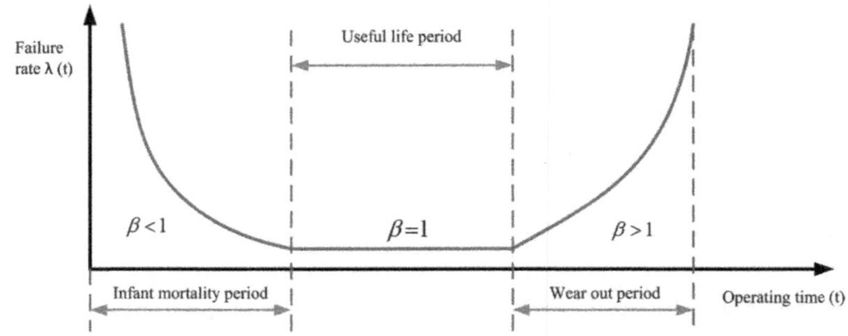

FIGURE 16.9 The bathtub curve.

of a component is dominated by random failures, mainly due to changes in temperature and voltage stress. During this period all the failure rates that are useful for estimation of the reliability are reordered. In a complex system that consists of many components which are randomly distributed, the system failure rate becomes constant as failed components of the system are replaced. The failure rate in the useful life of a component cannot be reduced or prevented by utilising good maintenance practice and prolonged burn-in periods. The power system components are designed to operate within a certain operating limit specified by the manufacturers. When these limits are exceeded, there is a chance of failure in the power system. Each component of the power system is prone to failure that cannot be anticipated. Nevertheless, the likelihood that failure will happen amid a given timeframe within the useful life can be determined by examining the component design. If the probability of random failure is observed to be too much in the component, some strategic measures must be introduced to change the design of the components and the operating conditions in which the components are being used.

3. Wear-out: During this period, the useful lifetime has ended. The wear-out time is characterised by an increasing failure rate owing to the component's aging, mechanical stress, deterioration due to worn-out parts, etc. In the later life of a component, the increase in failure rate is attributed to age and wear, which incur significant damage on the component as shown in the bathtub curve shown in Figure 16.9. The failure rate that is associated with wear-out can be reduced or prevented by persistent replacement of the faulty or failed components in the system. The most useful lifetime in reliability analysis is the application of Weibull parameters as a shape parameter (β). This parameter can be used to analyse the hazard function of a component as follows:
    a. $\beta < 1$: this indicates a reduction in hazard function.
    b. $\beta > 1$: this indicates an increase in hazard function.
    c. $\beta = 1$: this indicates constant hazard function.

16.7.2 Mean Time to Failure

Mean time to failure (MTTF) is the period of time that the component of a power system is projected to last in operation. MTTF can be used by the distribution network operators to evaluate the components of their power systems. The value of MTTF is the reciprocal of the effective failure rate of the system, and it is related to the mean time between failures (MTBF). The contrast between these terms is that while MTBF is utilised for components than that can be repaired and reused, MTTF is utilised for non-repairable components. As a metric, MTTF represents how long a component can reasonably be expected to perform in the field as a result of particular testing.

16.7.3 Mean Time to Repair

Mean time to repair (MTTR) is the time expected before a component in the power system will fail and need repair or replacement. The MTTR indicates the mean time required to replace a failed component in the power system. The value of the MTTR in a power system will be too high if it takes too much time to repair a faulty component. This will escalate the cost of installation as a result of downtime, until the time when the new component arrives and the conceivable window of time required to perform the replacement. One of the potential alternatives to reduce MTTR in a power system is to purchase some spare electrical components so that a replacement can be done quickly.

16.7.4 Mean Time between Failures (MTBF)

MTBF is the average amount of time that a component functions before failing. The MTBF includes the only operational time between failures and does not include repair times and other downtimes such as inspections and preventive maintenance. Assuming the component is repaired and begins functioning again, MTBF is intended to measure only the time a system is available and operating.

$$\text{MTBF} = \text{MTTF} + \text{MTTR} \tag{16.1}$$

16.7.5 Availability

Availability (A) is defined as the probability that a component of the power system will be available when required, or it can be defined as the aggregate of the total time that an electrical component is available for use in a power system. The availability of a component is a function of its failure rate (λ) and repair rate (μ):

$$A = \frac{\mu}{\lambda + \mu} \tag{16.2}$$

$$= \frac{\text{MTBF}}{\text{MTBF} + \text{MTTR}}$$

16.8 Concepts of Reliability

Power system reliability can be defined as the level to which the performance of the components of a power system results in the delivery of electricity to customers within accepted benchmarks and the quality anticipated. Power outages have adverse effects on the electric supply at the load points. The degree of the distribution system reliability may be measured by considering the following aspects:

1. The number of customers affected by the power outages
2. The number of loads connected at the load points
3. The duration of the power outages
4. The amount of power interrupted
5. The frequency of power outages
6. The magnitude of power outages

The reliability evaluation can be used to quantify the reliability attributes of power in line with the configuration and component reliability information. The model distinguishes zones that are characterised by good or poor reliability and furthermore recognises overloaded and undersized pieces of equipment that reduce the reliability performance of the system. The power utilities can develop a comprehensive reliability analysis model that can be utilised to spontaneously find potential problems and prescribe improvements in a power system. Some of the typical improvement recommendations that can be explored for reliability improvement projects include the following:

1. Load transfers among feeders
2. Upgrading of substations
3. Addition of new line reclosers, sectionalising switches, feeder tie points, automating feeders, undergrounding of circuits
4. Replacement of old components
5. Construction of new substations
6. Distribution system automation

16.9 Reliability Indices of the Distribution System

The reliability of a distribution power system is associated with a continuous power supply from the supply points to the load points. Power utilities and regulatory bodies analyse the performance of the power system by using some reliability indices. The reliability indices can be used to estimate the average length of power outages, the number of customers affected by power outages and the number of power outages [12]. Power system planners can also use the reliability indices to analyse the impacts of power interruptions on customers and the cost of the power outages [13]. The reliability performance of the distribution system depends on the component failures, restoration times and repair times of the system [14]. This section introduces the concept of the reliability indices such as basic load point's reliability indices, customer-oriented indices and energy-oriented indices. The reliability evaluation of a power system can be carried out with the concepts of the reliability indices and customers damaged function [13]. The substantial impacts of the protective devices in a power system can be assessed by utilising the reliability indices and other KPIs [14].

16.9.1 Basic Load Point Reliability Indices

Reliability evaluation of the distribution system determines the accessibility of power supply at the consumers' load points. The basic load point reliability indices such as the annual outage duration, average failure rate and average outage duration are used to estimate the reliability of the distribution system [13]. The basic load point reliability indices can be used to evaluate the energy-oriented and customer-oriented reliability indices as presented in Equations (16.3) through (16.5) [14]. These indices can be utilised to assess the reliability of various sections of the distribution system such as main feeders, lateral feeders, load points and the entire network [15]. The degree of service continuity from the supply points is measured with the application of the three reliability indices, which are briefly explained as follows:

1. Average failure rate: The average failure rate of a power system is the average number of failures that consumers experience at the load point during a giving period [16]. It can be expressed as follows, in Equation (16.3):

$$\lambda_s = \sum_{i=0}^{n} \lambda_i N_e \text{ (failure/year)} \tag{16.3}$$

   where λ_i is the average failure rate and N_e is the number of components whose faults will interrupt the power supply at the load point i.

2. Annual outage duration: Annual outage duration is the average total duration of failure power supply outage that consumers experience at the load point in a year [13]. It can be expressed as follows, in Equation (16.4) [17]:

$$U_s = \sum_{i=0}^{n} \lambda_i r_i \text{ (hour/year)} \tag{16.4}$$

   where r_i is the average outage duration and U_s is the annual outage time.

3. Average outage duration: Average outage duration is the average duration of outage at the load point i [18]. It can be expressed as follows, in Equation (16.5) [17]:

$$r_s = \frac{U_s}{\lambda_s} \text{ (hour)} \tag{16.5}$$

   where r_s is the average outage duration.

16.9.2 Customer-Oriented Indices

Customer-oriented indices are introduced in this section to study the impacts of power interruptions at the load points. These indices measure the level of dissatisfaction that individual customers experience due to a number of power outages. The majority of power outages that affect numerous customers at the load points are caused by the failure of electrical components in the distribution system [19]. The basic load point reliability indices discussed above can be used in conjunction with the customer-oriented indices to evaluate the reliability performance of the distribution system as well as the severity of power outages on the consumers. The customer-oriented indices can be estimated to obtain the overall performance of the distribution system. The following customer-oriented indices are defined in this section: Average Service Unavailable Index (ASUI), System Average Interruption Frequency Index (SAIFI), Average Service Available Index (ASAI), System Average Interruption Duration Index (SAIDI), Customer Average Interruption Frequency Index (CAIFI) and Customer Average Interruption Duration Index (CAIDI) [20]. The performance of the distribution system can be measured by using the duration and frequency of customers' outages at various levels of the power system. The customer-oriented indices are usually used to provide significant information about the performance of the distribution power system.

1. SAIFI: SAIFI is defined as the average number of times that a consumer experiences a power outage during a specific time. The SAIFI of a customer on a feeder or substation can also be defined as the average number of times that power to a consumer serviced by the feeder or substation is interrupted during a specific period [21]. It shows how regularly the average customer experiences power outage over a predefined time [22]. If the value of SAIFI obtained after carrying out the reliability evaluation of a power system does not fall within the acceptable standard, some strategic techniques such as proper maintenance, automation, integration of renewable energy resources and incorporation of protective devices into the system must be pursued to improve the SAIFI of the power system. SAIFI is expressed in Equation (16.6) as [17]:

$$\text{SAIFI} = \frac{\text{Total number of customer interruptions}}{\text{Total number of customers served}} \tag{16.6}$$

$$= \frac{\sum \lambda_i N_i}{\sum N_i} (\text{failure/customer} \cdot \text{year})$$

   where N_i is the number customers at load point i.

2. SAIDI: SAIDI is defined as the average interruption duration for customers served during a predefined period [21]. The reliability index allows the distribution network operators to estimate the duration that consumers would have been cut off from the power supply if all customers were out at one time [22]. It shows the aggregate duration of power outage for the average customer within a predefined time. SAIDI of a power system can be enhanced by reducing the number and duration of power outages experienced by the customers at the load points.

$$\text{SAIDI} = \frac{\text{Total number of interruption duration}}{\text{Total number of customers served}} \tag{16.7}$$

$$= \frac{\sum U_i N_i}{\sum N_i} (\text{hour/customer} \cdot \text{year})$$

   where U_i is the annual outage duration at load point i.

3. CAIFI: CAIFI measures the average frequency of sustained interruptions that customers experience at the load points. It is the average number of interruptions per customer interrupted per year [23,24]. The CAIFI of a power system can be improved by reducing the interval between

the beginning of a customer's interruption and when the service has been restored to the customer. It is expressed as follows, in Equation (16.8) [17]:

$$\text{CAIFI} = \frac{\text{Total number of customer interruptions}}{\text{Total number of customers interrupted}} \tag{16.8}$$

$$= \frac{\sum N_{ci}}{\sum N_{ca}}$$

where N_{ci} is the total customers interruption and N_{ca} is the total number of customers affected by power interruption.

4. CAIDI: CAIDI is the average duration of an interruption experienced by the customer [23,25]. It signifies the average time required by the utilities to restore power supply to the affected areas, once it has been lost. Consequently, it allows the distribution network operators to report the average duration of a customer outage for the customers affected. CAIDI of a power system can be enhanced by reducing the duration of power interruption and repair times.

$$\text{CAIDI} = \frac{\text{Total number of interruption duration}}{\text{Total number of customers served}} \tag{16.9}$$

$$= \frac{\sum U_i N_i}{\sum \lambda_i N_i} \text{ (hour/customer interruption)}$$

5. ASAI: ASAI is a measure of the average availability of the distribution systems that serves a number of customers [25]. It measures the part of time that a customer has power during the predefined reporting time [26]. It is the ratio of the total customer minutes or hours that service is available to the total customer minutes or hours demanded in a period. ASAI can be measured on an hourly, daily (24 hour), monthly (720 hours) and yearly (8760 hours) basis. A higher value of ASAI indicates a higher level of the reliability of the power system. Mathematically, it is normally expressed as follows:

$$\text{ASAI} = \frac{\text{Customer number hours service available}}{\text{Customer hours service demand}} \tag{16.10}$$

$$= \frac{\sum N_i \times 8760 - \sum U_i N_i}{\sum N_i \times 8760}$$

6. ASUI: One of the reliability indices that is commonly utilised by the utilities for performance evaluation of the distribution system is ASUI [25,27]. It is expressed mathematically as follows, in Equation (16.11):

$$\text{ASUI} = \frac{\text{Customer hours of unavailability service}}{\text{Customer hours demand}} \tag{16.11}$$

$$= \frac{\sum U_i N_i}{\sum N_i \times 8760}$$

$$= \frac{\text{SAIDI}}{8760}$$

$$= 1 - \text{ASAI}$$

16.9.3 Energy-Oriented Indices

The energy-oriented indices can be utilised by the distribution network operators to assess the reliability of their distribution systems and to offer profitable data for enhancements of the existing power system [28]. The Interruption Energy Assessment Rate (IEAR) indices, Expected Energy Not Supplied (EENS) and Expected Interruption Cost Index (ECOST) depict the general performance of the distribution system. The indices can be also used to forecast the future performance of the distribution system in view of the present performance of the system [15]. The sensitivity of customers with respect to failure of different components of a power system can also be measured with the application of energy-oriented indices. The reliability of a power system can be improved by using proper planning tools and good management policies. The impacts of component failures in the distribution system that result in the interruption of service at some load points can be investigated by using energy-oriented reliability indices. The energy-oriented indices can be used to measure the total energy lost due to power outages caused by the faulty components of a power system [29].

1. EENS Index: The EENS is the expected amount of energy that is not delivered at the consumer load points owing to an unexpected power outage or power interruption. The EENS can be utilised by the distribution network operators to carry out an economic and reliability evaluation of a power system. The EENS of the distribution power system can be expressed as follows, in Equation (16.12) [30]:

$$\text{EENS} = L_i U_i (\text{MWhr/yr}) \tag{16.12}$$

$$= \sum_{i=0}^{n} \lambda_i r_i L_i$$

$$= \lambda_x r_x L_x + (\lambda_x r_x + \lambda_y r_y) L_y + (\lambda_x r_x + \lambda_y r_y + \lambda_z r_z) L$$

$$= (L_x + L_y + L_z)\lambda_x r_x + (L_x + L_y)\lambda_y r_y + L_z \lambda_z r_z = PF_x \lambda_x r_x + PF_y \lambda_y r_y + PF_z \lambda_z r_z$$

$$= \sum_{i=0}^{n} \lambda_i r_i PF_i$$

where PF_x, PF_y and PF_z are the power that flows in the line sections x, y and z, L_i is the load point i, L_x, L_y and L_z are the load points x, y and z, respectively, as shown in Figure 16.10. Also, λ_x, λ_y and λ_z are the failure rates at load points x, y and z while r_x, r_x and r_z are the failure duration at load points x, y and z.

2. ECOST: A practical technique that can be used to estimate the reliability worth of a power system is based on the evaluation of the ECOST [31]. The performance of the distribution system can be estimated by using the ECOST, customer interruption costs and load point reliability indices. The ECOST can be utilised to assess the reliability of the distribution system

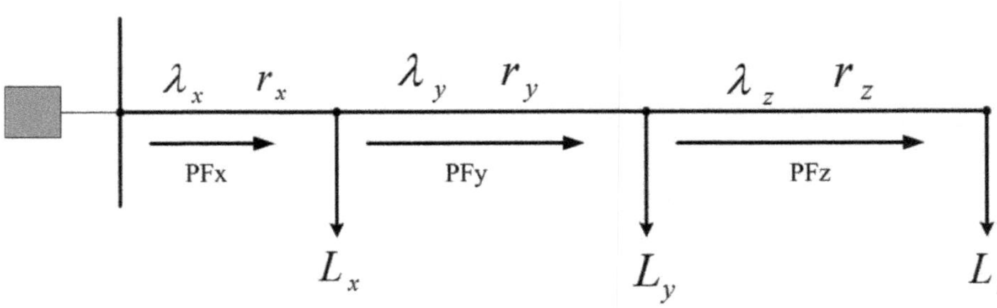

FIGURE 16.10 A distribution system with the section lines.

and to provide vital information that can be used for planning, upgrading and improvements to the system. The ECOST plays a significant role in the utilisation of the reliability concepts for the attainment of an acceptable level of a reliable power supply [13]. The ECOST depends on the characteristics of consumers and it can be expressed as follows:

$$\text{ECOST} = P_i \sum N_e f_{r,j} \lambda_i \text{ (\$/year.)} \tag{16.13}$$

$$= P_x N_e f_{x,j} \lambda_x + P_y N_e f_{y,j} \lambda_y + P_z N_e f_{z,j} \lambda_z$$

where $f_{r,j}$ is the cost of interruption; P_i is the average load at load point i; and P_x, P_y and P_z are average load at load points x, y and z as shown in Figure 16.10.

3. IAER: The IAER index can be used by the power utilities to estimate the expected annual economic damage borne by customers. The IAER can be expressed as follows, in Equation (16.14) [13]:

$$\text{IEAR}_i = \frac{\text{ECOST}_i}{\text{EENS}_i} \text{ (\$/kWhr)} \tag{16.14}$$

4. Average Energy Not Supplied Index (AENS_i): The severity of a power outage for consumers can be estimated by using AENS_i, which includes total energy not served and total number of customers served [6]. The AENS_i of the distribution power system can be expressed as follows [13]:

$$\text{AENS}_i = \frac{\text{Total energy not supplied}}{\text{Total number of customers served}} \tag{16.15}$$

$$= \frac{\sum \text{EENS}_i}{\sum N_i} \text{ (MWhr/Customer} \cdot \text{year)}$$

16.10 Objective Function

The main challenge in the power sector is figuring out how to design a system that can supply electric power to its numerous customers economically and proficiently. Reliability is an important benchmark in the conceptualisation, design and operation of a power system. It has also become a significant subject due to the abrupt increase in demand of a reliable power supply at minimal number of power outages. The objective function of this study is to enhance the reliability of the distribution system and also to minimise the total cost (TC) and reliability index (RI) while meeting the power requirement of consumers at the load points and the system constraints. The TC consists of ECOST, cost of protective devices (PC) and interruption cost at consumer' load points. The optimisation problem is formulated in this chapter to select the optimal number of protective devices and their locations to enhance the reliability of the system and to minimise the total cost of the power system. This can be achieved with the integration of protective devices such as circuit breakers, tie lines, sectionalising switches, etc., into a power system. The reliability indices and operational costs are the two important parameters when designing a power system. The minimisation of the objective function is implemented to achieve the optimal reliability and investment costs of a power system.

$$F = \min \sum_{i=1}^{n} (TC + RI) \tag{16.16}$$

The total cost can be expressed as:

$$TC = \sum_{f=1}^{N_f} \sum_{t=1}^{T} \sum_{q=1}^{N_q} \sum_{j=1}^{N_{LP}} \text{ECOST}_{qjtf} + \sum_{f=1}^{N_f} \sum_{t=1}^{T} \sum_{q=1}^{N_q} \sum_{j=1}^{N_{LP}} K_e \text{EENS}_{qjtf} + \sum_{p=1}^{N_p} PC_p \tag{16.17}$$

The cost of protective devices includes the annualised maintenance cost (AMC), annualised capital cost (ACC) and annualised installation cost (AIC) of the protective devices.

$$PC = \sum_{p=1}^{N_p} \mathrm{ACC}_p + \sum_{p=1}^{N_p} \mathrm{AIC}_p + \sum_{p=1}^{N_p} \mathrm{AMC}_p \tag{16.18}$$

The cost model of the protection system is developed by using the capital cost per unit. The annualised capital cost of the protective devices (ACC_p), which consists the cost of circuit breakers, fuses and sectionalising switches, is expressed as follows:

$$\mathrm{ACC}_p = C_{acap,i} \sum_{i=1}^{n} \left[FC \times X_{fc} + CB \times X_{cb} + SW \times X_{sw} \right] \tag{16.20}$$

where X_{fc}, X_{cb} and X_{sw} are the decision variables of fuses, circuit breakers and sectionalising switches; FC, CB and SW are fuses, circuit breakers and sectionalising switches, respectively.

The capital recovery factor (CRF) for the protective devices can be estimated as follows:

$$\mathrm{CRF}(i, P_{\mathrm{proj}}) = \frac{i.(1+i)^{P_{\mathrm{proj}}}}{(1+i)^{P_{\mathrm{proj}}} - 1} \tag{16.21}$$

where i is the annual interest rate (%) and P_{proj} is the lifetime of the circuit breakers, fuse and sectionalising switches.

The annualised capital cost of nth year for each protective device can be expressed as:

$$C_{\mathrm{acap},i} = C_{\mathrm{cap}} \times \mathrm{CRF}(i, P_{\mathrm{proj}}) \tag{16.22}$$

Annual interest rate

$$i = \frac{\left(i^{\mathrm{nom}} - f\right)}{(1+f)} \tag{16.23}$$

where f is the annual inflation rate, i^{nom} is the nominal interest rate, C_{cap} is the capital cost and $C_{\mathrm{acap},i}$ is annualised capital cost of nth year of the circuit breakers, fuse and sectionalising switches. The cost of the reliability is formulated by using decision variables that indicate the installation of circuit breakers, fuses and sectionalising switches on the section of a radial distribution system. The decision variables for installation of circuit breakers, fuses and sectionalising switches in the distribution system are defined as follows:

$$X_{fc} = \begin{cases} 1 \text{ Installation of fuse at the respective locations,} \\ 0 \qquad \text{otherwise} \end{cases} \tag{16.24}$$

$$X_{cb} = \begin{cases} 1 \text{ Installation of circuit breaker at the respective locations,} \\ 0 \qquad \text{otherwise} \end{cases} \tag{16.25}$$

$$X_{sw} = \begin{cases} 1 \text{ Installation of sectionalizing switch at the respective locations,} \\ 0 \qquad \text{otherwise} \end{cases} \tag{16.26}$$

The installation cost for each component of the protective device can be expressed as:

$$\mathrm{AIC}_p = C_{aic,i} \sum_{i=1}^{n} \left[FC \times X_{fc} + CB \times X_{cb} + SW \times X_{sw} \right] \tag{16.27}$$

$$C_{aic,i} = C_{ic} \times \text{CRF}(i, P_{proj}) \tag{16.28}$$

where C_{ic} is the installation cost and $C_{aic,i}$ is annualised installation cost of nth year of the circuit breakers, fuse and sectionalising switches. The annualised maintenance cost of the circuit breakers, fuses and sectionalising switches can be estimated by considering the yearly inflation rate. The annualised maintenance cost of nth year of the protective devices can be estimated by typically using 3% of the annualised capital cost.

$$\text{AMC} = Z\% \times C_{acap,i} \sum_{i=1}^{n} \left[FC \times X_{fc} + CB \times X_{cb} + SW \times X_{sw} \right] \tag{16.29}$$

The second component of the objective function is the reliability index, which is introduced to investigate the overall performance of the system with the integration of the protective devices.

$$RI = w_1 \frac{\text{SAIFI}}{\text{SAIFI}_T} + w_2 \frac{\text{SAIDI}}{\text{SAIDI}_T} + w_3 \frac{\text{EENS}}{\text{EENS}_T} + w_4 \frac{\text{ECOST}}{\text{ECOST}_T} \tag{16.30}$$

where RI is reliability index, w_1 is the weight of SAIFI, SAIFI_T is the base case value of SAIFI, w_2 is the weight of SAIDI, SAIDI_T is the base case value of SAIDI, w_3 is the weight of EENS, EENS_T is the base case value of EENS, w_4 is the weight of ECOST and ECOST_T is the base case value of ECOST.

$$\sum_{i=1}^{n} w_i = 1 \tag{16.31}$$

where n is the number of indices and w_i is the weight of i index. The weights of all the reliability indices present in Equation (16.30) are assumed to be 0.25. The values of the weighted factors depend on the importance of the reliability index to be considered.

16.10.1 Benefit to Cost Analysis

A benefit-cost ratio (BCR) is a financial indicator used in analysing the costs and benefits of a project. It summarises the overall value of money expended on a project and the expected balance of benefits made from the investment and costs incurred during the execution of the project. In other words, a benefit-cost ratio is the ratio of the benefits obtained from a project expressed in terms of the monetary value and relative to its costs. The BCR can be utilised by the utilities to determine the best investment decisions. It can be used to predict the feasibility of a project. This can be achieved by comparing the amount of monetary gain during a predefined time interval of a project and the amount it costs to execute the project. The project is a good investment if the value of BCR is high when the benefit is higher than the cost of the project. The BCR offers a comparison between the benefit and cost.

$$\text{Benefit} = \sum_{i=1}^{n} \left(\text{TOC}_{\text{base}}^{i} - \text{TOC}_{\text{new}}^{i} \right) \tag{16.32}$$

where $\text{TOC}_{\text{base}}^{i}$ and $\text{TOC}_{\text{new}}^{i}$ are the values of total outage cost before and after placement of protective devices in the power system.

$$\text{BCR} = \frac{\text{Benefit}}{TC} \tag{16.33}$$

According to the general rule of thumb, a high value of BCR shows that the benefit is higher than the cost of the project. This indicates a considerable benefit with respect to the costs. Hence, the investment return begins as soon as the BCR exceeds 1.0. The higher the BCR, the better the investment.

16.10.2 Test System

A radial 11 kV distribution network that is utilised in this study is a modified RBTS bus 2 system that consists of 22 load points, 20 step-down transformers, 4 circuit breakers at the supply points and numerous sectionalising switches and fuses at the respective points, as shown in Figure 16.11. The necessary reliability data for the study such as feeder types, feeder lengths, feeder section numbers, number of customers, type of customer, customer damage function, average load at each load point, maximum load at each load point, component failure and system reliability data are provided in [32,33]. The capital cost and cost of installation of the circuit breaker, sectionalising switch and fuse are \$538, \$461 and \$120, respectively, based on their specifications. The annual interest rate for the protective devices is assumed to be 6%. The cost of maintenance for the protective devices is assumed to be 3% of the capital

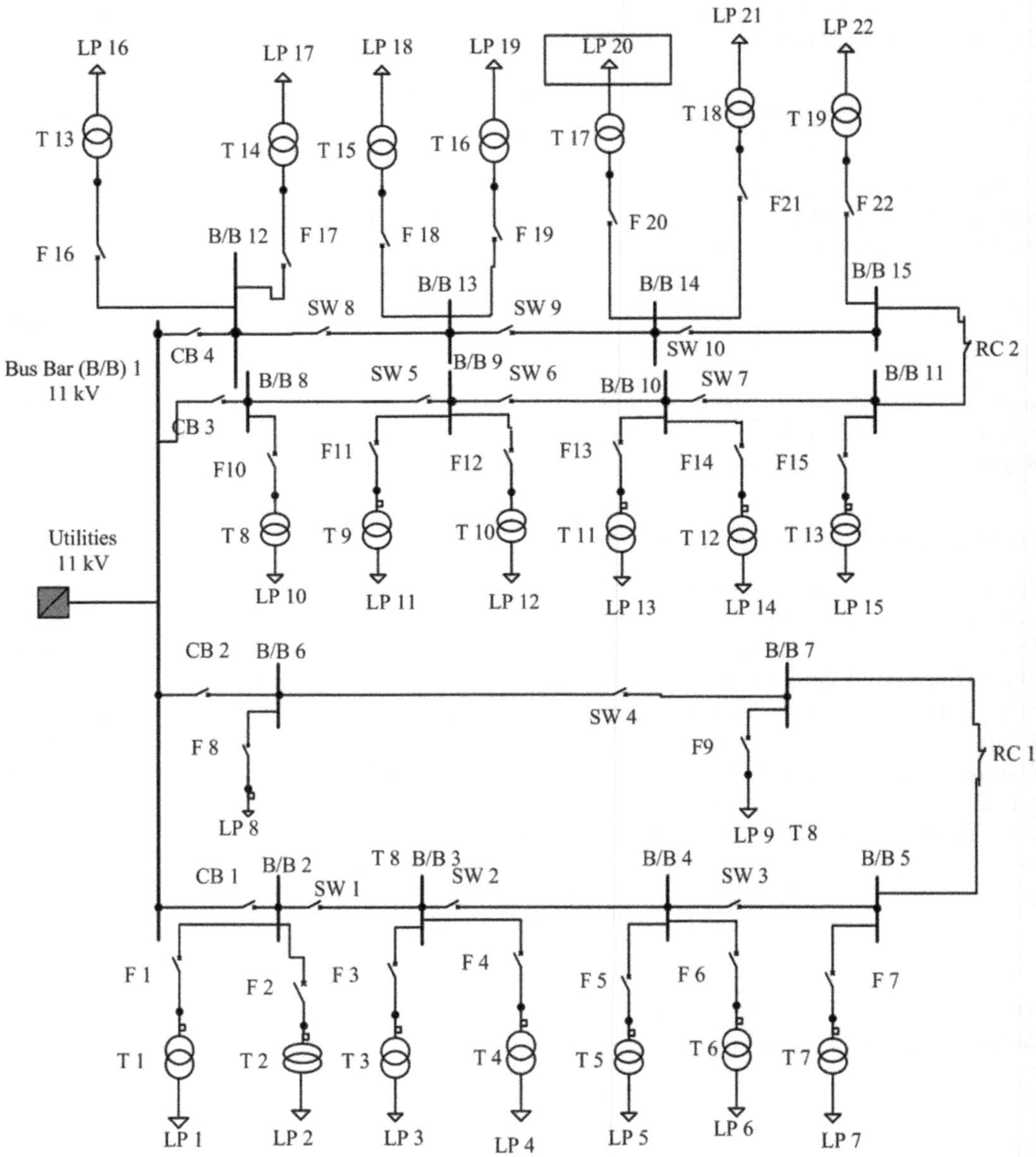

FIGURE 16.11 Modified RBTS bus distribution system. (From Sirisumrannukul, S., *Network Reconfiguration for Reliability Worth Enhancement in Distribution Systems by Simulated Annealing*, INTECH Open Access Publisher, London, UK, 2010; Allan, R. N., et al., *IEEE Trans Power Syst.*, 6, 813–820, 1991.)

TABLE 16.4

Specifications of the Protective Devices

Operating Parameters	Sectionalising Switches	Circuit Breakers	Fuses
Lifetime	25 years	15 years	5 years
Cost	$461 per piece	$538 per piece	$40 per piece
Other parameters	Type: High Voltage, Poles Number: 3, Ambient temperature: 25°C ± 40°C, Rated voltage (kV): 11, Maximum voltage (kV): 12, Rated current (A): 400/630, 1 min withstand voltage (kV): 2/48 42/48, Thermal stable current (virtual) (kA): 12.5/20, Dynamic stable current (peak) (kA): 31.5/50, Short circuit making current (kA): 31.5/50, Rated breaking current (A): 400/630 and Rated transfer current (A): 1000/1000	Type: Vacuum, Poles Number: 3, Rated current: 1250A, Ambient temperature: –25°C–40°C, Rated voltage: 11 kV, Highest working voltage: 12 kV, Rated short circuit breaking current: 31.5 kA, Rated short circuit making current: 80 kA and Rated Peak withstand current: 80 kA	Usage: High Voltage Breaking Capacity High Type: Drop out Rated voltage: 11 kV Rated current: 100/200A Function: Short circuit protection Weight: 230kg Temperature: –40°C to 40°C

cost. The life expectancy of the protective devices is mostly expected to be about 4–25 years, based on the appropriate maintenance and exposure to good operating conditions such as load operations, environmental conditions, overloads, short circuits, etc. The favourable operating conditions and proper maintenance in accordance with the specifications of the manufacturers enhance the optimal operations of most of the protective devices. Consequently, the conditions of service have a considerable effect on the time of useful service of the protective devices in a power system. The specification for each protective device in this study is presented in Table 16.4.

16.11 Results and Discussions

The impacts of protective devices in the distribution system are evaluated by using circuit breakers, sectionalising switches and fuses. The modified RBTS system with the aforementioned protective devices is illustrated by investigating their effects on the reliability of a power system.

16.11.1 Effects of Circuit Breakers

The first case study is illustrated by adding a number of the circuit breakers at the supply points of the distribution system shown in Figure 16.11. The application of the circuit breakers as the only protective devices in the system enhances reliability of the distribution system and reduces the economic loss associated with the risks of power outages. This will improve the rate at which the power is being delivered at the load points and the income made by the power utilities. The benefits of using only circuit breakers in a power system is limited due to some technical and economic reasons, which have been explained in the previous subsections. In case study 1, 4 circuit breakers are placed in the supply points of the distribution system shown in Figure 16.11 as a base case study. The value of TC recorded in this case study is $2,684,206/year, which includes the TOC and PC with the estimated values of $ 2,682,785.5/year and $1,420/year, respectively, as shown in Figures 16.12 through 16.14 and Table 16.5. The values of ECOST, EENS, SAIFI, SAIDI and RI in this case study are $1,315,480/year, 248.601 MWh/year, 1.2877 f/customer/year, 23.9151 hr/customer/year and 1, respectively, as shown in Figure 16.15 and Table 16.6.

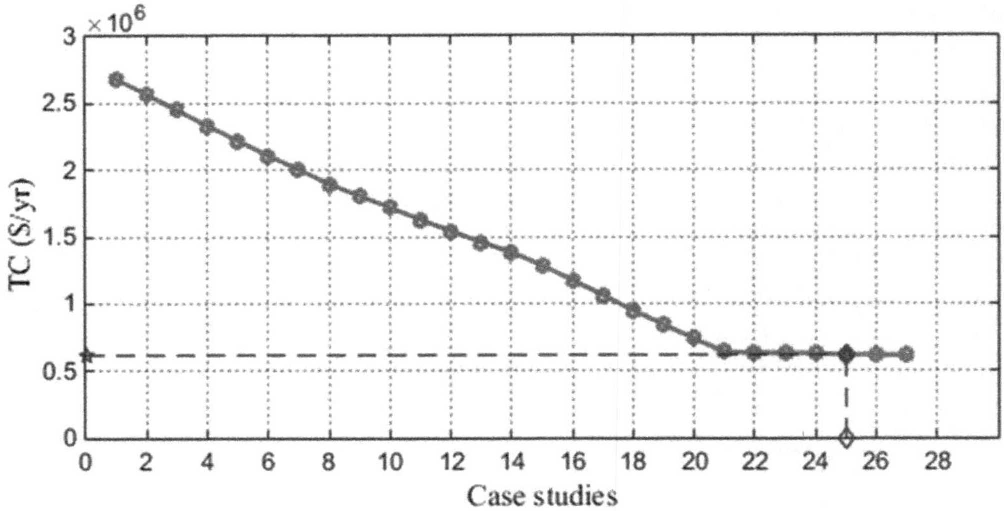

FIGURE 16.12 Number of protective devices and their impacts on the total cost.

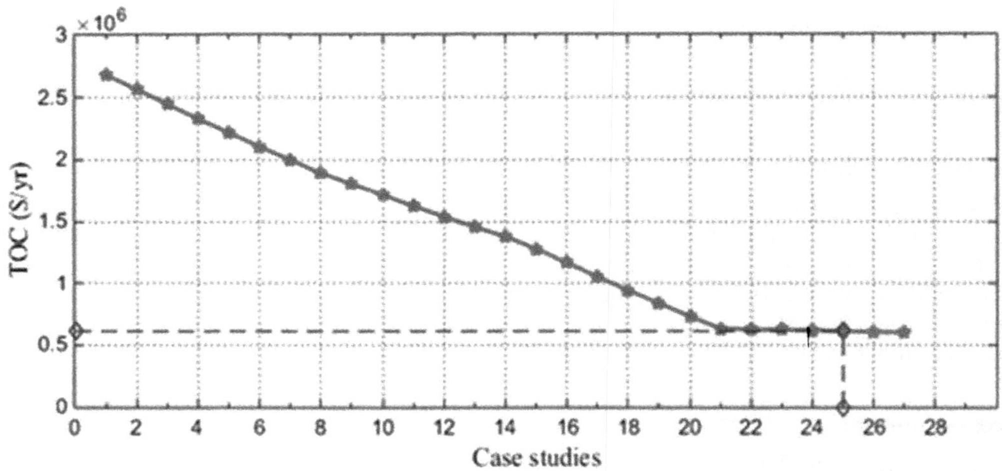

FIGURE 16.13 Number of protective devices and their impacts on the total outage cost.

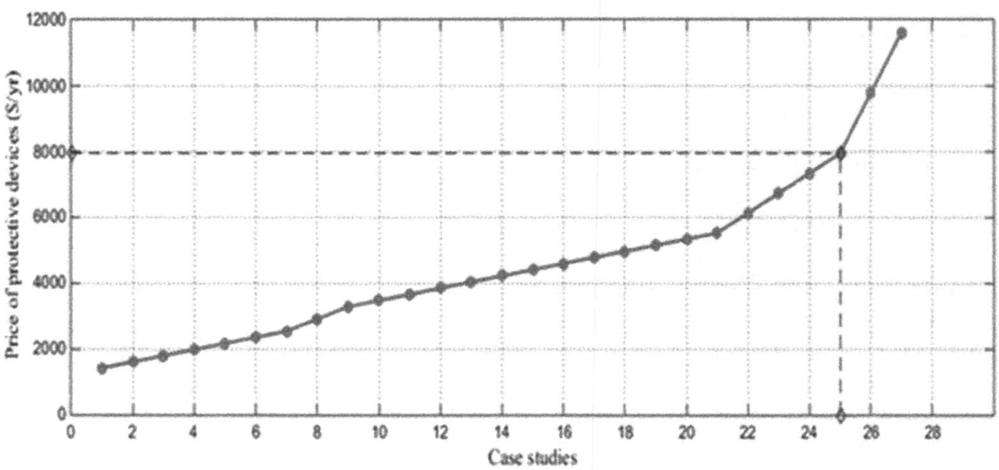

FIGURE 16.14 Prices of protective devices.

TABLE 16.5

Costs and Benefits of Protective Devices

Case Studies	Protective Device Location	TOC ($/year)	PC ($/year)	TC ($/year)	Benefit ($/year)	BCR	RI
1	CB1-CB4	2,682,785.5	1420	2,684,206			1
2	CB1-CB4 and F1	2,566,769	1607	2,568,376	116,016.5	0.04517	0.968565
3	CB1-CB4 and F1-F2	2,448,882	1794	2,450,676	233,903.5	0.09544	0.936023
4	CB1-CB4 and F1-F3	2,330,995	1981	2,332,976	351,790.5	0.1507	0.903501
5	CB1-CB4 and F1-F4	2,219,467	2168	2,221,635	463,318.5	0.20854	0.868048
6	CB1-CB4 and F1-F5	2,106,104.5	2355	2,108,459	576,681	0.27350	0.831168
7	CB1-CB4 and F1-F6	2,001,171.1	2542	2,003,713	681,614.4	0.34017	0.802883
8	CB1-CB4 and F1-F7	1,892,801.3	2916	1,895,717	789,984.2	0.41672	0.760249
9	CB1-CB4 and F1-F10	1,802,482.3	3289	1,805,772	880,303.2	0.48749	0.734558
10	CB1-CB4 and F1-F11	1,713,635.7	3476	1,717,112	969,149.8	0.56440	0.708128
11	CB1-CB4 and F1-F12	1,622,206.8	3663	1,625,870	1,060,579	0.65231	0.68081
12	CB1-CB4 and F1-F13	1,537,883.8	3850	1,541,734	1,144,902	0.74260	0.65012
13	CB1-CB4 and F1-F14	1,453,212.6	4037	1,457,250	1,229,573	0.84376	0.619115
14	CB1-CB4 and F1-F15	1,377,779.3	4224	1,382,003	1,305,006	0.94428	0.591129
15	CB1-CB4 and F1-F16	1,279,471.9	4411	1,283,883	1,403,314	1.09302	0.558305
16	CB1-CB4 and F1-F17	1,167,800.3	4598	1,172,398	1,514,985	1.29221	0.528099
17	CB1-CB4 and F1-F18	1,056,128.8	4785	1,060,914	1,626,657	1.53326	0.497912
18	CB1-CB4 and F1-F19	942,667.1	4972	947,639	1,740,118	1.83626	0.466679
19	CB1-CB4 and F1-F20	835,963.2	5159	841,122	1,846,822	2.19566	0.431759
20	CB1-CB4 and F1-F21	730,971.8	5345	736,317	1,951,814	2.65077	0.398193
21	CB1-CB4 and F1-F22	632,268.3	5532	637,801	2,050,517	3.21498	0.365047
22	CB1-CB4, F1-F22 and SW1	625,134.9	6141	631,276	2,057,651	3.25951	0.355514
23	CB1-CB4, F1-F22 and SW1 and SW5	623,095.8	6749	629,845	2,059,690	3.27015	0.355133
24	CB1-CB4, F1-F22 and SW1, SW 5 and SW 8	618,278.8	7358	625,637	2,064,507	3.29984	0.346153
25	CB1-CB4, F1-F22 and SW2, SW 4, SW6 and SW 9	610,160.4	7966	618,127	2,072,625	3.35307	0.335935
26	CB1-CB4, F1-F22,SW1- SW 6 and SW8 - SW 9	605,631.5	9792	615,424	2,077,154	3.37516	0.329168
27	CB1-CB4, F1-F22 and SW1- SW10	599,777.7	11,618	611,395	2,083,008	3.40697	0.328032

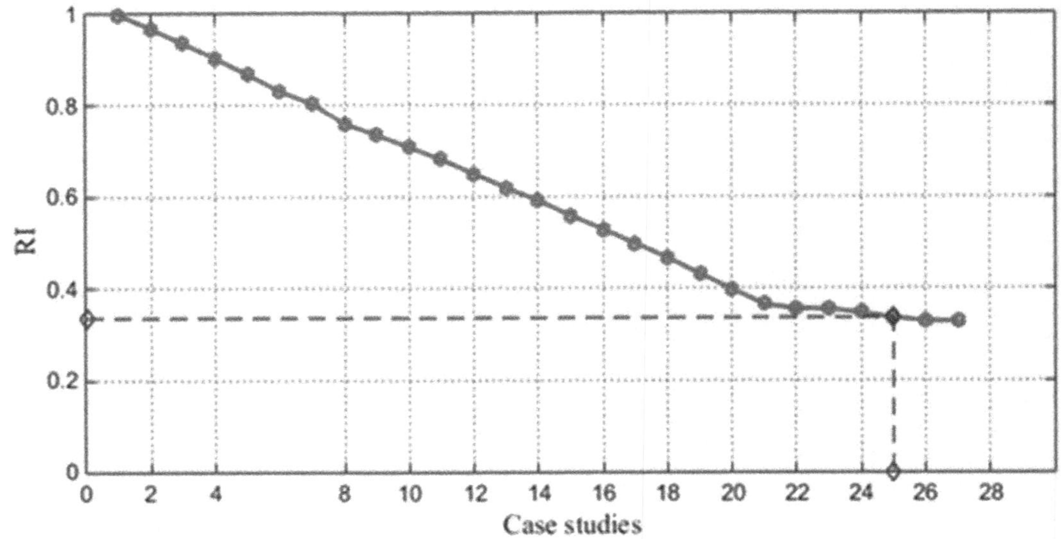

FIGURE 16.15 Number of protective devices and their impacts on the reliability of a power system.

TABLE 16.6

Impacts of the Protective Devices on the Reliability of a Power System

Case Studies	Number of Circuit Breakers	Number of Fuses	Number of Sectionalising Switches	EENS (MWhr/year)	ECOST ($/year)	SAIFI (failure/ customer/year)	SAIDI (hour/ customer/year)
1	4	0	0	248.601	1,315,480	1.2877	23.9151
2	4	1	0	238.744	1,253,677	1.2768	23.1822
3	4	2	0	228.684	1,191,120	1.2628	22.4343
4	4	3	0	218.624	1,128,563	1.2489	21.6864
5	4	4	0	208.864	1,070,715	1.2318	20.6031
6	4	5	0	198.903	1,012,138	1.2103	19.4976
7	4	6	0	188.629	963,711.6	1.1903	18,413
8	4	8	0	178.053	913,509.8	1.1691	17.3229
9	4	10	0	169.695	869,159.8	1.1587	16.6232
10	4	11	0	161.421	825,820.2	1.1461	15.9128
11	4	12	0	152.829	781,647.3	1.1325	15.1817
12	4	13	0	144.655	742,281.3	1.1127	14.1156
13	4	14	0	136.439	702,798.1	1.0919	13.0441
14	4	15	0	128.029	673,619.8	1.0756	12.0092
15	4	16	0	118.575	627,309.4	1.0565	10.9753
16	4	17	0	109.254	566,903.3	1.046	10.2756
17	4	18	0	99.933	506,497.3	1.0356	9.5759
18	4	19	0	90.421	445,351.6	1.0223	8.8618
19	4	20	0	81.285	388,895.7	1.0018	7.8073
20	4	21	0	72.332	333,145.8	0.9855	6.7739
21	4	22	0	62.83	286,703.3	0.9654	5.7348
22	4	22	1	61.8	285,234.9	0.9318	5.5727
23	4	22	2	61.631	284,125.3	0.9318	5.5727
24	4	22	3	60.922	283,207.8	0.8987	5.4133
25	4	22	4	59.987	280,231.9	0.8631	5.241
26	4	22	7	59.424	278,799.5	0.8389	5.233
27	4	22	10	58.700	276,927.9	0.8387	5.122

The results obtained in this case study indicate that the power system is not reliable due to the high values of the reliability indices and total cost. With these results, it is imperative to utilise another set of protective devices for minimisation of total cost and maximising customer benefits. The availability of an uninterrupted power supply at the load points has become a critical issue for the utilities because the money spent on the reinforcement of their power systems is to improve the performance of the power system. This indicates that it is mandatory to reinforce the distribution system with fuses and sectionalising switches for better operation and improvement of system reliability.

16.11.2 Effects of Fuses

The assessment of case studies 2–25 is based on the evaluation of the outage cost, investment cost and total cost as well as the economic impacts that the additional costs of the fuses have on the reliability performance of a power system. The impacts of incorporating protective devices into the distribution network are assessed by using various case studies. The number of fuses is increased from 1 to 22 to illustrate the effects of the aforementioned protective devices on the TC as presented in Figure 16.12. From Table 16.5, it can be observed that there is a considerable change in the value of TC with respect to the number of fuses utilised in the network. To increase the level of the reliability at the load points, a number of fuses are incorporated into the power system. This reduces the number of power outages at the load points as well as the cost associated with the power interruption, as shown in Figure 16.13. In fact, this also reduces the risk of using sophisticated equipment in the presence of power outages. A reliable power supply is difficult to achieve owing to the high cost of the protective devices. The application of the protective devices in a power system has become a serious challenge for the utilities due to the aforementioned setbacks. To maintain the reliability of a power system at a reasonable level, the optimal number of protective devices must be installed at different sections of the power system while meeting the cost constraints of the system. In case study 2, the impact of fuses on the reliability of the distribution system is investigated. The values of TC and TOC obtained in this case study have been reduced by 4.32% and 4.32%, while the cost of protective devices increased by 13%. In addition to this, the values of ECOST, EENS, SAIFI, SAIDI and RI have been reduced by 4.7%, 3.96%, 0.85%, 3.06% and 0.97%, respectively. This case study has a marginal benefit-cost ratio with a value of 0.0045171. It can be established from Figures 16.12 and 16.13 that the values of TOC and TC are reduced gradually with the integration of additional number of fuses at the lateral feeders of the system, while the values of PC are increased substantially, as presented in Figure 16.14.

In case study 15, the power system has 4 circuit breakers and 15 fuses at different sections. The results obtained in this case study by installing 4 circuit breakers and 15 fuses at the supply points and lateral feeders indicate that the values of TC and TOC have been reduced substantially — by 52.30% and 52.12% — while the cost of protective devices increased by over 201%. The reliability performance of the system has reduced the ECOST, EENS, SAIFI and SAIDI with the following values: 46.57%, 45.11%, 15.21% and 45.46%, respectively. The value of RI in this case study is estimated to be 0.558305. The benefits and RBC have been improved significantly to $1,403,314/year and 1.093023, as shown in Figures 16.16 and 16.17. The reliability assessment is carried out in case study 21 where 22 fuses are installed at the lateral feeders and 4 circuit breakers at the supplied points. The values of TC and TOC have been further reduced with the installation of an additional number of fuses and circuit breakers by 76.24% and 76.43%, while the cost of protective devices increased by over 289.5%. The analysis of the results presented in Table 16.6 shows that the values of ECOST, EENS, SAIFI and SAIDI and RI have improved by 78.21%, 74.73%, 25.03% and 76.02%, respectively. The value of RI in this case study has reduced to 0.365047, as presented in Figure 16.15. This indicates that the application of the fuses has improved the overall reliability of the entire network.

The BCR is increasing with an additional number of fuses; this can be established in case study 21 when the benefits and BCR are $2,050,517/year and 3.214981, as presented in Figures 16.16 and 16.17. In this case study, fuses are utilised because of their low capital cost compared with the total outage cost at the load points. This study has established the impacts of fuses on the reliability of the distribution system and the necessary capital investment and cost to ensure the sustainability of the power supply at the load points. This case study provides a brief highlight of how a number of fuses can be utilised by the distributed network operators to improve the reliability, efficiency, cost and sustainability of electric services.

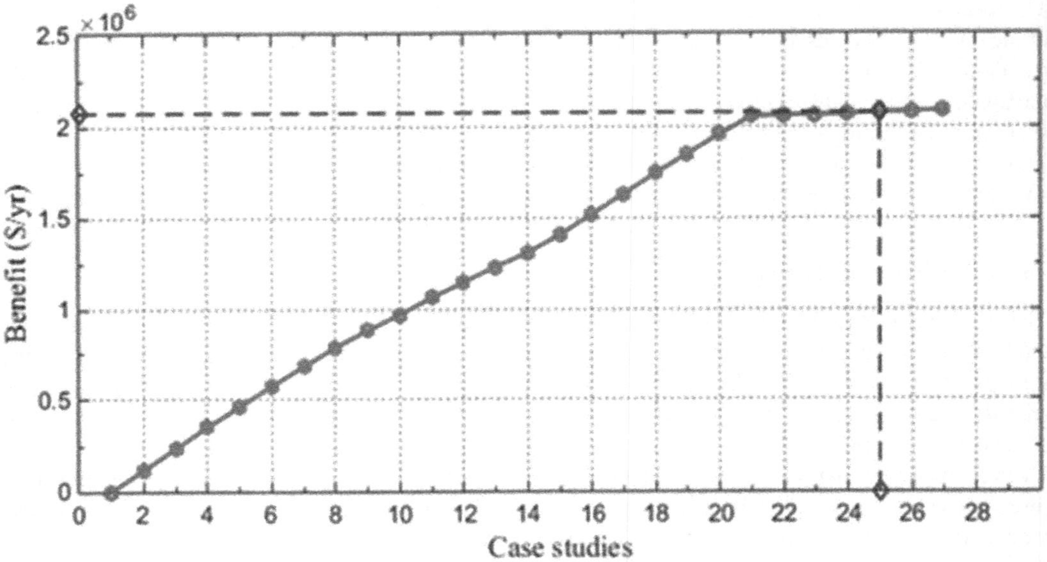

FIGURE 16.16 Number of protective devices and their benefits in the power system.

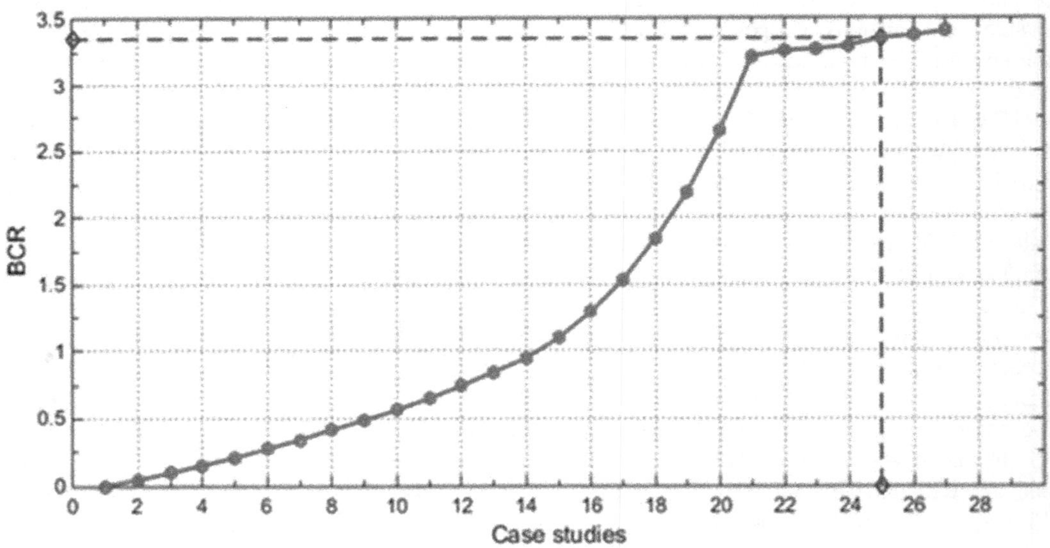

FIGURE 16.17 A number of protective devices and their benefit-cost ratio.

16.11.3 Effects of Sectionalising Switches

The sectionalising switches can be utilised by the distribution network operators to improve the reliability a radial distribution system. The sectionalising switches can provide a more remarkable reliability improvement by reducing the load points outage cost because they operate with automatic mode and manual mode. The benefits of sectionalising switches in a power system can be quantified in terms of reduction in the number of power outages. This is achieved by isolating the faulty section of the power system with the application of the sectionalising switches. The reliability effects of using sectionalising switches in a power system can be assessed by using ECOST and EENS indices and other reliability indices such as RI, SAIFI and SAIDI based on the type of customer, failure rate of the component, customer damage function, peak and average load, etc. All these factors enable the impacts of the sectionalising switches to be effectively investigated by the power utilities.

In this section, the application of the sectionalising switches is centred on minimising the total cost of the reliability, which consists of the protective devices capital investment cost and installation cost, O&M costs and power outage cost. In case study 25, the reliability impact of sectionalising switches in terms of the TC, TOC and PC is investigated based on the results presented in Figures 16.12 and 16.14. The values of TC and TOC in this case study have been reduced extensively — by 76.97% and 77.25% — compared with case study 1. These results have shown the major benefits of utilising sectionalising switches in the distribution system. The value of the PC increased by 461% with additional number four sectionalising switches connected to the existing number of fuses and circuit breakers in case study 21. Based on the results presented in Tables 16.5 and 16.6, the application of sectionalising switches in the modified RBTS reduces the objective function of the system by a considerable amount, despite the fact that the cost of the sectionalising switches is too high. With this, the values of ECOST, EENS, SAIFI and SAIDI have been reduced by 78.10%, 75.87%, 32.93% and 78.09%, respectively. In addition, the value of RI in this case study has been considerably reduced to 0.336. This case study has a significant impact on the benefits and BCR with the following values: $2,072,625/year and 3.350374, as presented in Figures 16.16 and 16.17. In addition, it has provided a considerable reduction in the number of power interruptions experienced at the load points and enhances the continuity of power supply. Comparing the results obtained in this case study with the results obtained in case study 1 shows that it is highly imperative to incorporate sectionalising switches into a power system as a prerequisite to reduce the restoration time, a vital way to decrease the power outage costs.

Figures 16.12 through 16.15 show that an additional number of sectionalising switches in the existing power system has reduced the values of TC, TOC and RI, respectively. This trend continues until the optimal number of sectionalising switches is reached. The optimal number of sectionalising switches has been positioned in Figures 16.12 and 16.17. Beyond this point, the benefit of each additional sectionalising switch is marginal; introducing more switches does not have much effect on the TC, TOC and RI, as shown in case studies 26 and 27. The optimal points for the sectionalising switches are located so that they will be able to cover their protection zones. It is difficult for the utilities to achieve a certain level of reliability and customer satisfaction with limited resources. Therefore, a good management policy is required for allocation of the budget for optimal operation of their power systems at minimal total cost and outage cost.

16.12 Tutorial Problems

1. Define the term *reliability*.
2. What is the importance of reliability in the power system?
3. What are the factors that affect the reliability of the power system?
4. How does reliability affect cost decisions for the utilities?
5. What practical solutions can be adopted by the utilities to improve the reliability of their power systems?
6. What are the reasons why utilities use the reliability indices to measure the performance of their networks?
7. Write a short paragraph about the high cost of unreliability.
8. What are the benefits of carrying out the reliability assessment of the power system?
9. Define the following reliability indices.
   a. Average Service Unavailable Index
   b. System Average Interruption Frequency Index
   c. Average Service Available Index
   d. System Average Interruption Duration Index

e. Customer Average Interruption Frequency Index
f. Customer Average Interruption Duration Index
g. Interruption Energy Assessment Rate
h. Expected Energy Not Supplied
i. Expected Interruption Cost Index
j. Average Energy Not Supplied Index

16.13 Conclusion

The availability of an uninterrupted power supply at the load points has become a critical issue for the utilities because the money spent on the reinforcement of their power systems is equivalent to improvement of power system performance. It is difficult for the utilities to achieve a certain level of reliability and customer satisfaction with limited resources. Therefore, a good management policy is required for the allocation of the budget for optimal operation of their power systems at minimal total cost and outage cost. This has resulted in reinforcement of the distribution system with protective devices for better operation and improvement of system reliability. It is imperative to utilise a number of protective devices for minimising cost and maximising customer benefits. This chapter provided a brief review of how protective devices can be utilised by the distributed network operators to improve the reliability, efficiency, cost and sustainability of electric services. This chapter described protective devices that reduce the cost of customer service outage and reduce the total cost of the system. The reliability model is formulated by considering the consumer outage cost and total cost of protective devices. The economic and reliability benefits of the distribution system are considered in formulation of the cost model. The model was developed to have an accurate and cost-effective way of using protective devices in the power system. The model is utilised in this work to analyse the impacts of protective devices in a power system.

REFERENCES

1. T. Adefarati and R. C. Bansal, "Reliability and economic assessment of a microgrid power system with the integration of renewable energy resources," *Applied Energy*, vol. 206, pp. 911–933, 2017.
2. T. Adefarati, R. C. Bansal, and J. J. Justo, "Reliability and economic evaluation of a microgrid power system," *Energy Procedia*, vol. 142, pp. 43–48, 2017.
3. R. Billinton and R. N. Allan, *Reliability Evaluation of Engineering Systems*. New York: Springer, 1992.
4. S. W. Min, J. K. Park, K. H. Kim, I. H. Cho, and H. J. Lee, "A fuzzy relation based fault section diagnosis method for power systems using operating sequences of protective devices," in *IEEE Power Engineering Society Summer Meeting*, vol. 2, pp. 933–938, 2001.
5. M. F. Islam, A. M. Oo, and S. H. Chowdhury, "The traditional power generation and transmission system: Some fundamentals to overcome challenges," in *Smart Grids*. Cham, Switzerland: Springer, 2013, pp. 1–21.
6. R. Allan, *Reliability Evaluation of Power Systems*. New York: Springer Science & Business Media, 2013.
7. R. Allan and M. Da Silva, "Evaluation of reliability indices and outage costs in distribution systems," *IEEE Transactions on Power Systems*, vol. 10, no. 1, pp. 413–419, 1995.
8. J. D. McDonald, B. Wojszczyk, B. Flynn, and I. Voloh, "Distribution systems, substations, and integration of distributed generation," in *Electrical Transmission Systems and Smart Grids*. New York: Springer, 2013, pp. 7–68.
9. A. S. Ali, *Smart Grids: Opportunities, Developments, and Trends*. London, UK: Springer, 2013.
10. M. Čepin, *Assessment of Power System Reliability: Methods and Applications*. London, UK: Springer Science & Business Media, 2011.
11. V. C. Gungor, B. Lu, and G. P. Hancke, "Opportunities and challenges of wireless sensor networks in smart grid," *IEEE Transactions on Industrial Electronics*, vol. 57, no. 10, pp. 3557–3564, 2010.
12. K. Narayanan, S. A. Siddiqui, and M. Fozdar, "Hybrid islanding detection method and priority-based load shedding for distribution networks in the presence of DG units," *IET Generation, Transmission & Distribution*, vol. 11, no. 3, pp. 586–595, 2017.

13. W. Li, *Reliability Assessment of Electric Power Systems using Monte Carlo Methods.* New York: Springer Science & Business Media, 2013.
14. T. Adefarati and R. C. Bansal, "Integration of renewable distributed generators into the distribution system: A review," *IET Renewable Power Generation*, vol. 10, no. 7, pp. 873–884, 2016.
15. "Distribution reliability assessment," https://etap.com/product/distribution-reliability-assessment, accessed June, 2018.
16. A. A. Alkuhayli, S. Raghavan, and B. H. Chowdhury, "Reliability evaluation of distribution systems containing renewable distributed generations," in *IEEE North American Power Symposium (NAPS)*, pp. 1–6, 2012.
17. T. Adefarati and R. C. Bansal, "Reliability assessment of distribution system with the integration of renewable distributed generation," *Applied Energy*, vol. 185, pp. 158–171, 2017.
18. S. Ray, A. Bhattacharya, and S. Bhattacharjee, "Differential search algorithm for reliability enhancement of radial distribution system," *Electric Power Components and Systems*, vol. 44, no. 1, pp. 29–42, 2016.
19. P. Commission, "Electricity network regulatory frameworks," *Inquiry Report*, vol. 1, p. 62, 2013.
20. H. Zhang, P. Bi, X. Zhong, and Q. Jin, "Development and application of reliability analysis system for medium and high voltage distribution networks," in *IEEE (DRPT) Third International Conference on Electric Utility Deregulation and Restructuring and Power Technologies*, pp. 937–940, 2008.
21. J. Gers, "*Distribution System Analysis and Automation*", *IET Power and Energy Series*, vol. 68. London, UK: Institution of Engineering and Technology, 2013.
22. "Evaluation of data submitted in APPA's 2013 distribution system reliability & operations survey," http://www.publicpower.org/files/PDFs/2011DSReliabilityAndOperations_Report_Final.pdf, accessed June, 2018.
23. "Enterprise web portal for outage and reliability reporting," http://www.intergraph.com/assets/plugins/sgicollaterals/downloads/EnterpriseWebPortal_WhitePaper.pdf, accessed June, 2018.
24. T. Adefarati, A. Babarinde, A. Oluwole, and K. Olusuyi, "Reliability evaluation of ayede 330KV/132KV substation," 2014.
25. A. Samui, S. Samantaray, and G. Panda, "Distribution system planning considering reliable feeder routing," *IET Generation, Transmission & Distribution*, vol. 6, no. 6, pp. 503–514, 2012.
26. "The electricity regulatory authority (standard of performance) rules," http://www.caymanera.com/attachments/article/11/Gazetted%20Electricity%20Regulatory%20Authority%20(Standard%20of%20Performance)%20Rules%202012.pdf, accessed June, 2018.
27. J. F. Fajardo, A. A. J. Perez, S. S. M. Alsina, and J. E. Ramirez-Marquez, *Simulation Methods for Reliability and Availability of Complex Systems.* London, UK: Springer Science & Business Media, 2010.
28. N. R. Godha, S. R. Deshmukh, and R. V. Dagade, "Application of Monte Carlo simulation for reliability cost/worth analysis of distribution system," in *Power and Energy Systems (ICPS), 2011 International Conference*, 2011, pp. 1–6, IEEE.
29. A. Chowdhury, S. K. Agarwal, and D. O. Koval, "Reliability modeling of distributed generation in conventional distribution systems planning and analysis," *IEEE Transactions on Industry Applications*, vol. 39, no. 5, pp. 1493–1498, 2003.
30. N. Chaiyabut and P. Damrongkulkamjorn, "Impact of customer scattering on distribution system reliability with distributed generation," in *IEEE Region 10 Conference on TENCON 2010*, pp. 568–573.
31. T. Adefarati and R. Bansal, "The impacts of PV-wind-diesel-electric storage hybrid system on the reliability of a power system," *Energy Procedia*, vol. 105, pp. 616–621, 2017.
32. S. Sirisumrannukul, *Network Reconfiguration for Reliability Worth Enhancement in Distribution Systems by Simulated Annealing.* London, UK: INTECH Open Access Publisher, 2010.
33. R. N. Allan, R. Billinton, I. Sjarief, L. Goel, and K. So, "A reliability test system for educational purposes-basic distribution system data and results," *IEEE Transactions on Power Systems*, vol. 6, no. 2, pp. 813–820, 1991.

17

Advances in Wide Area Monitoring, Protection and Control

Adeyemi Charles Adewole and Raynitchka Tzoneva

CONTENTS

17.1 Introduction

The operating state of a power system can be categorized into five types [1]: normal, alert, emergency, *in extremis*, and restorative state. An electric power system is said to be in the normal operating state when generation is adequate for supplying the load demand, and no equipment is being overloaded. In this operating state, the system is stable, all the power system quantities are within their technical nominal values, and the system can withstand an equipment outage (contingency), thus satisfying the N–1 design criterion. A system proceeds to the alert state if the security level drops below a certain limit or if a contingency occurs. In this state, a further increase in system loading or the occurrence of another contingency may threaten the secure operation of the power system, and remedial actions must be implemented to restore the system to its normal (acceptable) operating state. If the disturbance is sustained, then the system enters the emergency state. The emergency state is characterized by low voltages and equipment overload. Power systems can degenerate into the *in extremis* state from the emergency state if no corrective actions are implemented or if the implemented corrective action is ineffective, the result of which are cascading outages and possibly blackout. The next state after the system collapse is the restorative state, in which remedial protection/control actions are implemented to reconnect a lost load or system and to transfer the system to either the alert state or to the normal state [1].

Wide area monitoring, protection and control (WAMPAC) refers to the use of measurements obtained from geographically separated areas of the power system (usually collated at a central location) in providing actionable information that can be utilized for grid situational awareness/analytics, and for protection/control actions for counteracting the prevailing disturbance in the system. Most electric power utilities around the world use supervisory control and data acquisition (SCADA) systems to provide data telemetry and telecontrol functions. SCADA systems are typically made up of sensors, meters, dataloggers, remote terminal units (RTUs), master terminal unit (MTU), intelligent electronic devices (IEDs), and programmable logic controllers (PLCs). The RTUs are typically used to collect field data from sensors, meters, dataloggers, and IEDs, and transmit them to the MTU via a communication medium. The MTU at the control centre displays the acquired data on the human-machine interface (HMI) and also issues remote control signals to the field actuation devices. Some of the communication protocols used in SCADA systems include Profibus, Modbus, IEC 60870-5-101, IEC 60870-5-104, and Distributed Network Protocol 3.0 (DNP 3.0).

However, SCADA systems are limited by their slow data sampling rate and slow reporting rate, and because the measurements obtained from various locations are not time synchronized at the point of acquisition. The use of the measurements from the existing SCADA systems might fail for real-time monitoring, assessment, protection and control because some of the system dynamics that occur, especially during short-term instabilities, might not be captured by the SCADA system due to its slow reporting rate of about 1 measurement every 2–10 seconds. Also, SCADA measurements from the various segments of the power system, when collected at the control centre, might correspond to different time instants because these measurements are not time-synchronized or time-tagged at the point of acquisition.

One way of capturing fast events/system dynamics over a wide area is through the use of synchrophasor measurements from widely dispersed phasor measurement units (PMUs), time-synchronized to a reference time from the Global Positioning System (GPS). The advent of synchrophasors has brought about the use of synchronized measurements in WAMPAC systems, with the possibility of higher reporting rates of up to 240 frames per second (fps) for a 60 Hz system and 200 fps for a 50 Hz system. WAMPAC systems based on synchrophasor measurements provide faster sampling and reporting rates, and the measurements from all parts of the system are synchronized to a common time reference.

Synchrophasor technology was heralded by the work of Phadke et al. [2], in which algorithms utilizing the positive sequence components of voltage, computation of frequency and rate-of-change of frequency (ROCOF) were presented. After this was the era of the development of a PMU prototype at Virginia Tech. These prototype PMUs were afterward deployed in some American utilities.

The success of these field trials resulted in the first commercial production of PMUs through Virginia Tech's collaboration with Macrodyne in 1991. An increased awareness of the importance of synchrophasor measurements began after the 2003 blackout in the United States. The report of the joint US-Canada Task Force on this blackout emphasized the importance of real-time measurements for situational awareness and for maintaining grid reliability. This resulted in renewed support for the Eastern Interconnection Phasor Project (EIPP), which included utilities in the northeastern part of the United States who were working together to advance state-of-the-art in grid monitoring, reliability, and information sharing. In 2007, the North American SynchroPhasor Initiative (NASPI) was formed by a consortium that included the US Department of Energy (DOE), the North American Reliability Corporation (NERC), utilities, and other stakeholders. The working group within NASPI is made up of seven teams tasked with the development and deployment of synchrophasor technology [3]. Other synchrophasor initiatives in other parts of the world can be found in Aminifar et al. [4].

This chapter introduces readers to the advances in WAMPAC and describes synchrophasor technology, components in synchrophasor-based systems, synchrophasor standards, and message types. Also, the optimal placement of PMUs, real-time monitoring applications, and post-disturbance applications are presented. Also presented are the design and implementation of a synchrophasor-based system integrity protection scheme (SIPS).

The impact of adverse communication network conditions on synchrophasor measurements is discussed. Lastly, cyber security threats and vulnerabilities, relevant cyber security standards, and countermeasures are discussed.

17.2 Synchrophasor Technology

The processing steps performed by a PMU are defined in the IEEE Std. C37.118.1™-2011 reference model, as shown in Figure 17.1. In the reference model, analogue outputs from current transformers (CTs) and voltage transformers (VTs) serve as inputs to the anti-aliasing low pass (LP) filter, which is used in limiting the bandwidth of the analogue signals in order to satisfy the Nyquist criterion. The output of the first stage LP filter is then passed on to the analogue-to-digital converter (ADC). A fixed frequency sampling synchronized to an absolute time base is used. This is followed by the complex multiplication of the sampled input with the nominal frequency carrier. LP filters are then applied to the real and imaginary outputs of the complex demodulator. The outputs from the LP filters are the real and imaginary parts of the synchrophasor estimates.

A sinusoidal voltage or current signal $x(t)$, as shown in Figure 17.2, has a behaviour defined by:

$$x(t) = X_m(t)\cos(\omega t + \phi(t)) \tag{17.1}$$

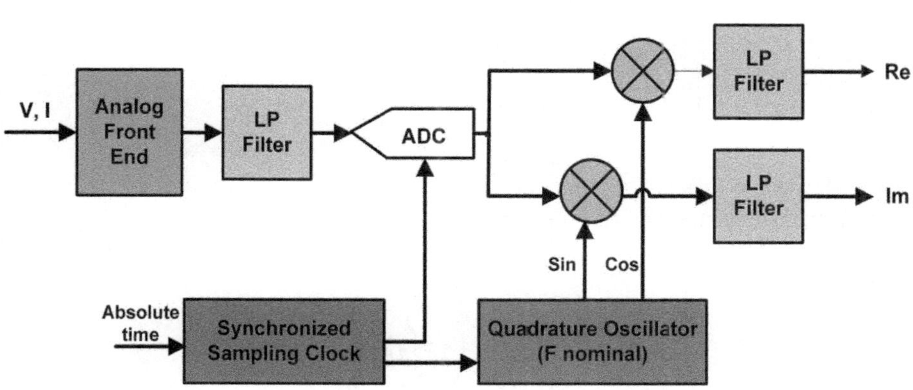

FIGURE 17.1 Signal processing steps of the PMU reference model. (From IEEE Standard C37.118.1™-2011, IEEE Standard for Synchrophasor Measurements for Power Systems, 2011.)

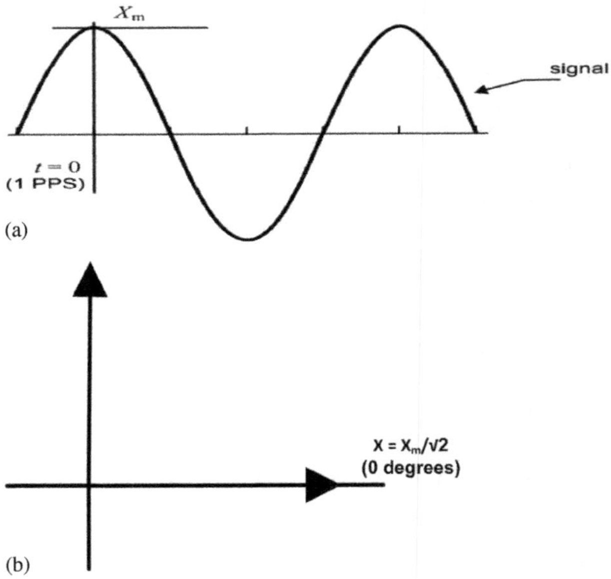

FIGURE 17.2 Voltage or current signal showing (a) sinusoidal representation; (b) phasor representation. (From IEEE Standard C37.118.1TM-2011, IEEE Standard for Synchrophasor Measurements for Power Systems, 2011.)

The phasor representation $X(t)$ of a signal $x(t)$ (Figure 17.2b) is given below:

$$X(t) = \left(\frac{X_m(t)}{\sqrt{2}}\right)e^{j\phi(t)} \tag{17.2}$$

$$= \left(\frac{X_m(t)}{\sqrt{2}}\right)(\cos\phi(t) + j\sin\phi(t)) \tag{17.3}$$

$$= X_r(t) + jX_i(t) \tag{17.4}$$

where X_m is the peak magnitude of the filtered synchronized measurement, ϕ is the phase angle relative to a cosine function at the nominal system frequency ω, and X_r and X_i are the real and the imaginary parts of the phasor measurement, respectively.

From Equation (17.4), a sinusoidal signal can be rewritten as:

$$x(t) = X_m(t)\cos(2\pi f_0 t + \phi(t)) \tag{17.5}$$

The deviation from the nominal frequency $g(t)$ is given as:

$$g(t) = f(t) - f_0 \tag{17.6}$$

where f is the actual frequency, and f_0 is the nominal angular system frequency.

The synchrophasor estimates over time are derived as follows (IEEE C37.118.1-2011; IEEE C37.118-2005):

$$x(t) = X_m(t)\cos\left(2\pi\int f(t)dt + \phi\right) \tag{17.7}$$

$$x(t) = X_m(t)\cos\left(2\pi\int \left(f_0 + g(t)\right)dt + \phi\right) \tag{17.8}$$

$$x(t) = X_m(t)\cos\left(2\pi f_0 t + 2\pi \int \left(g(t)dt + \phi\right)\right) \tag{17.9}$$

The phasor representation of Equations (17.1) through (17.9) is:

$$X(t) = \left(X_m(t)/\sqrt{2}\right)e^{j\left(2\pi \int g(t)dt + \phi\right)} \tag{17.10}$$

If $g(t) = \Delta f$ is equal to the deviation from the nominal system frequency and is constant, then

$$\int g(t)dt = \int \Delta f dt = \Delta f t \tag{17.11}$$

Thus, the synchrophasor measurement is given as:

$$X(t) = \left(X_m/\sqrt{2}\right)e^{j(2\pi\Delta f t + \phi)} \tag{17.12}$$

17.2.1 Components and Benefits of Synchrophasor-Based Systems

A typical WAMPAC system includes PMUs, IEDs, phasor data concentrators (PDCs), a time synchronization source (e.g., GPS satellite clock), data archiver/historian, and a communication network infrastructure. A comparison of SCADA-based measurements and synchrophasor measurements is presented in Table 17.1. From Table 17.1, it is clear that SCADA systems are limited by their slow data sampling rate, slow data communication rates, unsynchronized measurements, and time-consuming state estimation.

TABLE 17.1

Comparison of PMU-Based Systems and SCADA-Based Systems

S/N	PMU-Based Measurements	SCADA-Based Measurements
1	Measurements are synchronized and referenced to a common time base, usually the GPS.	Measurements are asynchronous and not acquired simultaneously.
2	Time synchronization is done at the measurement stage.	Time synchronization is done on arrival at the energy management system (EMS).
3	Data acquisition can be done with reporting rates of up to four times the nominal system frequency/second (240 fps or 200 fps).	Relatively slow. Scanning of devices done periodically (1 observation every 4 seconds) and takes 2–10 seconds.
4	Adequate for transient disturbances.	The low data rate of SCADA-based system could be too slow to capture transient disturbances.
5	Requires fewer measurements to provide the system state and does not require iterative state estimation. Angle measurements are obtained directly; no need for calculations.	Requires more system variable for SE. Measurements required are voltage magnitudes, real and reactive power, and system topology to calculate phase angles.
6	Having direct measurement of angle eliminates the errors due to the inaccuracies in network parameters.	The calculation of angles depends on system reactance, which in turn depends on system topology. Thus, SCADA-based SE is prone to inaccuracies.
7	Suitable for capturing dynamic changes.	Fails when the system state is changing quickly, especially during dynamic conditions.
8	Suitable for WAMPAC as a result of the time-stamped phasor measurements.	Only suitable for local monitoring and control.
9	Capable of oscillation detection.	Not suitable for oscillation detection.
10	Perfect for steady-state, dynamic, and transient conditions.	Steady-state only.

Some apparent benefits of synchrophasor technology include:

- Improved system visibility and situational awareness to aid decision making at control centres
- Real-time monitoring and assessment of system dynamics
- Improved validation of power system components using real-time measurements
- Improvement in state estimation (SE) through the combination of synchrophasor measurements with the traditional SE techniques
- Fast and easy collation of widely dispersed measurements during post-disturbance forensic analysis
- Easy detection of congestion in transmission lines by monitoring the phase angles of voltages along a transmission corridor.

17.2.2 Synchrophasor Standards

The first standard for synchrophasor measurements was developed after the success with the first PMU prototype. This was necessary in order to support the development of PMUs and their ancillary equipment (PDC, historian, HMI, and visualization tool). Also important was the need to establish a common platform for vendors to use in the design and manufacture of synchrophasor devices. The first synchrophasor standard for power systems was the IEEE Std. 1344-1995 [6]. The standard was intended to introduce synchrophasors and define the basic concepts of synchrophasor measurements. It specified the synchronizing input, phasor computation from sampled analogue measurements, and the phasor measurements data outputs. However, the IEEE Std. 1344-1995 did not specify the response time of PMUs, accuracy requirements, and the reference model for computing the phasors. Also, the IEEE Std. 1344-1995 did not include the command frame for dialogues between the synchrophasor PDCs (clients) and the PMUs (servers). Only the data frame, header frame, and configuration frame were defined. The second of century (SOC) timing was calculated as the UTC time in seconds from midnight of January 1, 1900.

The IEEE Std. 1344-1995 was reaffirmed in 2001 and replaced by the IEEE Std. C37.118-2005 [7]. The IEEE Std. C37.118-2005 provided some additional clarification for phasor and synchronized phasor definitions. Also, the standard introduced the calculation of total vector error (TVE) to verify the compliance of phasor estimates with theoretical equivalents for the same time instants. Furthermore, compliance tests and data reporting rates that are submultiples of the nominal system frequency were defined. The message formats in the IEEE Std. C37.118 was updated from the IEEE Std. 1344-1995 to improve the exchange of information with other systems in higher hierarchies (e.g., regional or national PDCs). This includes the addition of the sync, frame size, and station identification fields to the data, configuration, header, and command message types, respectively. The data frame also includes an additional field for analogue data. The SOC timing is calculated as the UTC time in seconds from midnight of January 1, 1970.

The IEEE Std. C37.118.1™-2011 for synchrophasor measurements replaced the IEEE Std. C37.118.1™-2005. The IEEE Std. C37.118.1™-2011 is divided into two parts: (1) IEEE C37.118.1™-2011 standard for synchrophasor measurements [5], and (2) IEEE C37.118.2™-2011 standard for synchrophasor data communication [8]. The IEEE Std. C37.118.1-2011 has clauses that cover the scope and the needs of the standard, the definition of the measurements, the measurement requirements, the quantification of the measurement, the testing of the measurements, and the definition of the accuracy limits. New definitions are provided for the phasors, frequency, ROCOF, frequency error (FE), ROCOF error (RFE), and performance requirement for dynamic conditions, and two performance classes (the M-Class and the P-Class) are introduced. The P-Class was defined for applications that require fast response (e.g., protection applications), while the M-Class was defined for applications that require higher accuracy (e.g., metering applications). TVE requirements and the compliance tests are expanded in this standard by adding temperature and dynamic performance tests, respectively. The IEEE Std. C37.118.2-2011 gives background information on synchrophasor measurements, synchrophasor measurement systems, communication protocol, and message formats. It defines the synchrophasor data transfer requirements, clarifications, and the modification of previous standards with respect to data transfer. Modifications to the IEEE Std. C37.118.1-2011 were published in [9]. This was intended to correct inconsistencies, provide further clarifications, and remove some of the limitations introduced by the IEEE Std. C37.118.1-2011.

The IEEE C37.118-2005 was primarily split into the IEEE C37.118.1-2011 and IEEE C37.118.2-2011 in order to be able to transport synchrophasor measurements using the IEC 61850 framework. The IEC 61850-90-5 [10] is an International ElectroTechnical Commission (IEC) Technical Report (TR) released in 2012 for the transportation of synchrophasor information according to the IEEE Std. C37.118. The IEC Technical Committee (TC) 57 WG10 chose the existing measurement (MMXU) logical node. In addition to the use of the existing IEC 61850MMXU, it was necessary to specify the performance class of the measurements as either protection class (P-Class) or measurement class (M-Class) as defined in IEEE C37.118.1. To accomplish this, the IEC 61850 Calculation Method (*ClcMth*) was extended to include P-Class and M-Class [10]. Also, the existing MMXU definition contained data objects that represented the voltage and current measurements, as well as frequency. However, IEC 61850 did not have provisions for ROCOF. Therefore, the IEC TR 61850-90-5 added the data object *HzRte* (Hertz Rate) to the MMXU logical node. Also, routable Generic Object-Oriented Substation Event (GOOSE) for routing event data, and routable sampled values (SVs) for routing periodic data was specified.

The IEEE Std. C37.242-2013 is a guide for PMU synchronization, calibration, testing, and installation [11]. The guide provides techniques detailing the role of time synchronization, various time synchronization sources, time synchronization accuracy, and availability. Also, the common causes of timing errors, the vulnerability of satellite-based timing system to radio-frequency interference (RFI)/disruption, performance testing and experimental tests for time synchronization sources were presented. The performance requirements for PDCs are covered in the IEEE Std. C37.244-2013 [12]. The IEEE Std. C37.244-2013 provides guidance on the functions and requirements of PDCs in terms of PDC latency, supported I/O streams, communication interface, transport protocol, time-alignment of PMU measurements, and error handling. Cyber security issues relating to availability/prevention of denial of service attacks, integrity and authenticity through the use of digital signatures/authentication codes, and confidentiality by using encryption/access control are discussed. Practical outlines for functionality testing of PDCs are provided in the standard. The tests covered include conformance testing, design testing, type testing, interoperability testing, commissioning testing, post-commissioning (field) testing, cyber security testing, and application-specific testing.

17.2.3 Synchrophasor Message Structure, Types, and Formats

Four message types are defined in the IEEE Std. C37.118: data, configuration, header, and command [7]. The data, configuration, and header messages are transmitted from a data source (PMU/PDC) to the data sink (PDC). While the command message type is sent from the data sink and received by the data source. Note that the PDC often serves as an IEEE C37.118 client. However, it can act as an IEEE C37.118 server when it outputs the concentrated time-aligned measurements, hitherto received from the PMUs, to other PDCs existing on a higher hierarchy. The data frame contains an identification header, message length, message source ID, status information, and the data itself. The data includes the phasors, frequency, ROCOF, analogue, and digital data types. The data from each PMU is organized as a block of data, as shown in Figure 17.3.

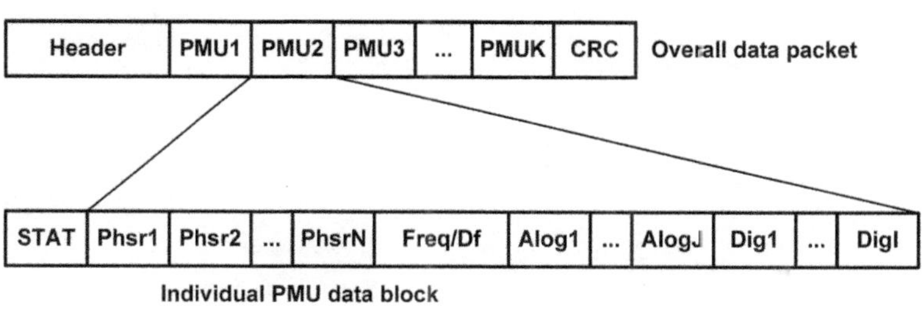

FIGURE 17.3 PMU data packet organization [IEEE C37.118-2005].

The header frame is a human-readable frame in ASCII format sent from the data source to the data sink/destination (PDC). It is user configured and contains the information relating to the PMU, type of algorithm, data sources, type of signal processing, and scaling. The command frame is sent from the data destination (PDC) to the data source (PMU/PDC) to start or stop the transmission of data, or to request the configuration data prior to the transmission of the data. The configuration frame is a binary file that describes the configured capability of a PMU. Figure 17.4 shows the typical communication that takes place between PMUs and PDCs. Figure 17.5 gives the packets captured using the Wireshark network protocol analyser. Figure 17.5 can be partitioned into three panes (Panes A–C) for better analysis. Pane A presents a list of all the packets captured at the Ethernet interface in real time. It has six columns

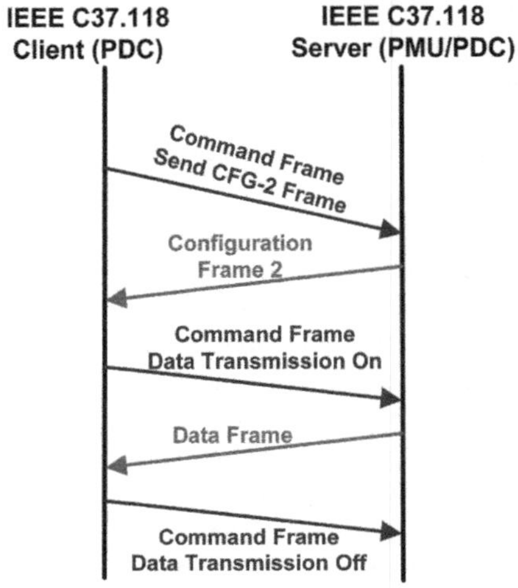

FIGURE 17.4 IEEE C37.118 synchrophasor client-server communication.

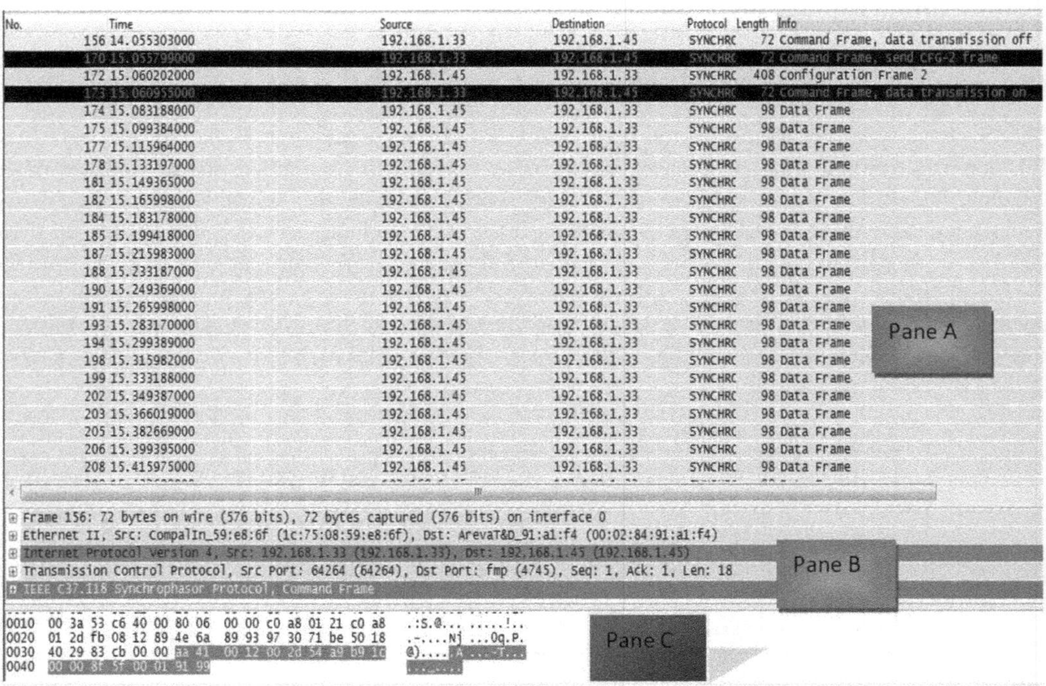

FIGURE 17.5 Synchrophasor messages for PMU-1 and PDC-1 captured using the Wireshark protocol analyser.

showing the time of capture, the source, destination, type of protocol, frame length, and the message type. Column 1 gives the time of capture of the packet. The IP address of the source device is given in column 2, while column 3 presents the IP address of the destination device. The type of protocol (SYNCHRC) is given in column 4, signifying the synchrophasor protocol. The frame length gives the size of the information contained in the message type as shown in column 5.

Column 6 of Pane A shows the information relating to the sequence of interaction between the PDC (data concentrator) and the PMU (data source). The PDC initiates a query (Command frame) requesting the CFG-2 Configuration message from the PMU.

The PMU responds to the Command frame query by issuing its CFG-2 configuration, and shortly afterwards commences the data transmission of synchrophasor measurements to the PDC. Pane B gives a layered classification of the selected packet in Pane A. The hexadecimal values corresponding to the messages in the captured packet are given in Pane C. The data frame packet captured using the network protocol analyser is shown in Figure 17.6.

17.2.4 Time Synchronization

Synchrophasor measurements must be synchronized to an accurate time source with better than 1 μs accuracy. The IEEE Std C37.118.1 specifies a maximum TVE of 1%. This corresponds to a phase error of 0.01 radian, or 0.57 degrees. A 0.01 radian phase error corresponds to a time error of ±26 μs for a 60 Hz system, and ±31 μs for a 50 Hz system. Time synchronization to PMUs or PDCs can be directly obtained from an accurate time reference such as the GPS. Inter-Range Instrumentation Group mod B (IRIG-B) is commonly used for local time dissemination. It may be provided in two formats: the level shift-1 kHz amplitude modulated signal (modulated), or the bi-phase Manchester modulated format (demodulated). With some clocks, the modulated time signal gives a guaranteed time accuracy of 100 μs, while the

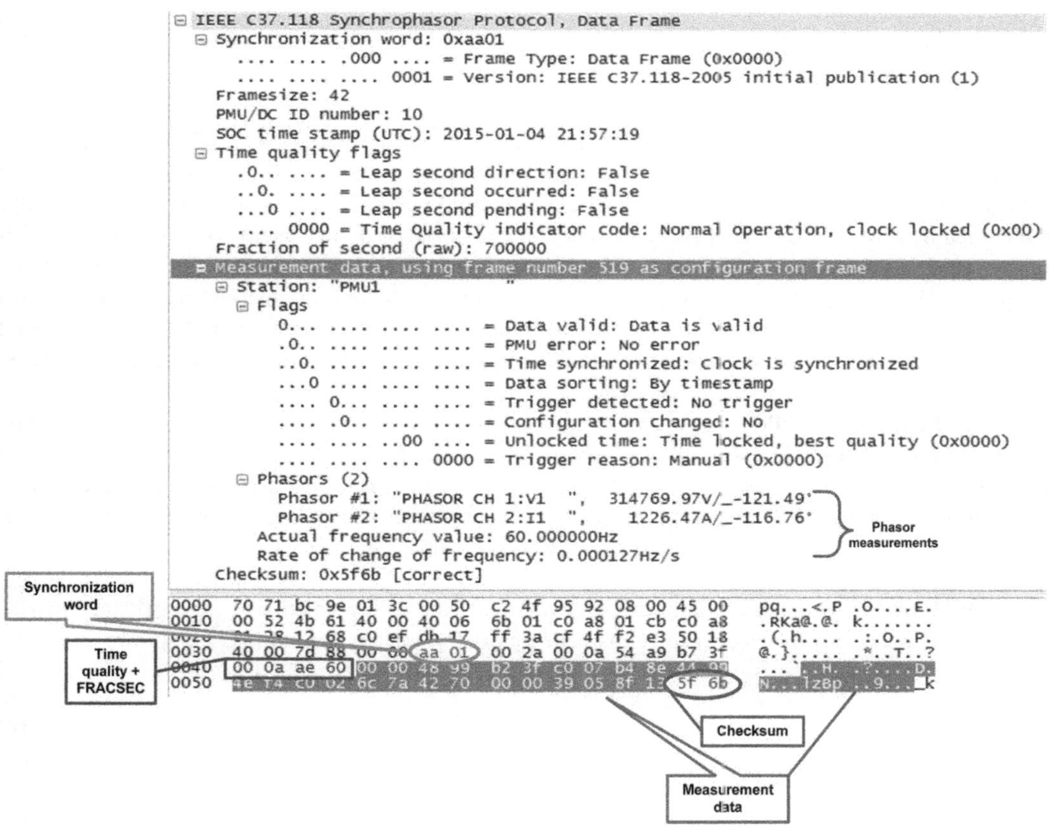

FIGURE 17.6 Data frame organization from the captured data frame (PMU-2).

demodulated time signal gives an accuracy of 1 μs. If the modulated time signal is used, it may need to be supplemented with a 1 pulse per second (PPS) pulse train to achieve the required accuracy.

A 1 PPS timing system has a pulse train of positive pulses, with the rising edge of the pulses coinciding with the seconds change in the clock. The pulse widths vary from 5 μs to 0.5 s, and the signal is usually a 5.0 V magnitude driving a 50 ohm load. The IEEE Std 1588-2008 Precision Time Protocol (PTP) [13] is a time synchronization method based on Ethernet communication architecture with a guaranteed accuracy of 100 ns. This profile was adopted for power system applications as specified in the IEEE C37.238-2011 Std. Aside from the GPS, other Global Navigation Satellite Systems (GNSS) time references can be obtained from the Russian-based GLONASS and the European-based Galileo systems. The particular GNSS time reference to use depends on the following: (1) accuracy, (2) coverage area, (3) duration of continuous availability, (4) reliability, (5) integrity, and (6) hold-over capability during loss of satellite synchronization.

17.3 System Planning and Functional Requirements

17.3.1 Optimal PMU Placement (OPP)

The success of any PMU-based application depends on the availability of data and the point of data acquisition. The OPP problem deals with the optimal location of PMUs in order to reduce the cost of PMU integration and the amount of PMU data that needs to be processed or analysed while maintaining complete system observability. Generally, a power system is observable if the available measurements are sufficient in the determination of the voltage magnitude and voltage phase angle at each of the load buses in the power system [14].

Power system observability can be classified into two categories:

- Numerical observability
- Topological observability

A numerically observable system is one in which the measurement matrix (H) is of full rank and connects all the system nodes together. Matrix H is an $m \times N$ matrix, where m is the number of voltage and current phasors, and N is the number of buses. The required computation for numerical observability is obtained by using the Gaussian elimination technique or the triangular factorization of the Jacobian matrix, gain matrix, or heuristic matrix. The null space of the gain or Jacobian matrices can also be computed [15]. Topological observability is the availability of one or more measurement tree(s) of full rank connecting all the system nodes together with either direct measurements from PMUs or calculated measurements. The calculated measurements can be obtained by applying Kirchhoff's and Ohm's laws, respectively.

Several studies on the OPP using mathematical programming, heuristic, and meta-heuristic methods have been covered in the literature. Methods for full network observability based on graph theory and tree search algorithm [16], Tabu Search (TS) algorithm [17], modified binary particle swarm optimization (PSO) [18], probabilistic approach method [19], and simulated annealing [20] have been proposed. Methods based on binary integer programming (BIP) have also been used [21–23].

1. OPP Problem Formulation: The basic formulation of the OPP problem using the BIP approach for any N-bus power system can be mathematically formulated as follows [21–23]:

$$S = \min_{x} \sum_{k=1}^{N} x_k \tag{17.13}$$

subject to:

$$Ax \geq b \tag{17.14}$$

$$x = \begin{bmatrix} x_1 & x_2 \ldots & x_N \end{bmatrix}^T \tag{17.15}$$

where $x_k = \{0,1\}$, $k = \overline{1,N}$, $A \in \Re^{NXN}$ is the connectivity matrix obtained from the binary transformation of the system's bus admittance matrix, N is the number of buses in the system, and x_k is the PMU placement variable and equals 1 if a PMU is sited at bus k and 0 if otherwise.

The connectivity matrix is given as $a_{ij} = 1$ if node i and j are linked or if $i = j$, $a_{ij} = 0$ if otherwise. $X \in \Re^N$ is the vector of the possible locations of the PMUs. The vector $b \in \Re^N$ is given as:

$$b = \begin{bmatrix} 1 & 1 \ldots 1 \end{bmatrix}^T \tag{17.16}$$

2. Example: For the IEEE 14-bus system shown in Figure 17.7 [24], the objective function for the OPP problem can be formulated as:

$$\min \left\{ x_1 + x_2 + \ldots + x_{14} \right\}$$

subject to the following observability constraints for buses 1–14, respectively, obtained from the system admittance matrix:

$$f_{n1} = x_1 + x_2 + x_5 \geq 1$$

$$f_{n2} = x_1 + x_2 + x_3 + x_4 + x_5 \geq 1$$

$$f_{n3} = x_2 + x_3 + x_4 \geq 1$$

$$f_{n4} = x_2 + x_3 + x_4 + x_5 + x_7 + x_9 \geq 1$$

$$f_{n5} = x_1 + x_2 + x_4 + x_5 + x_6 \geq 1$$

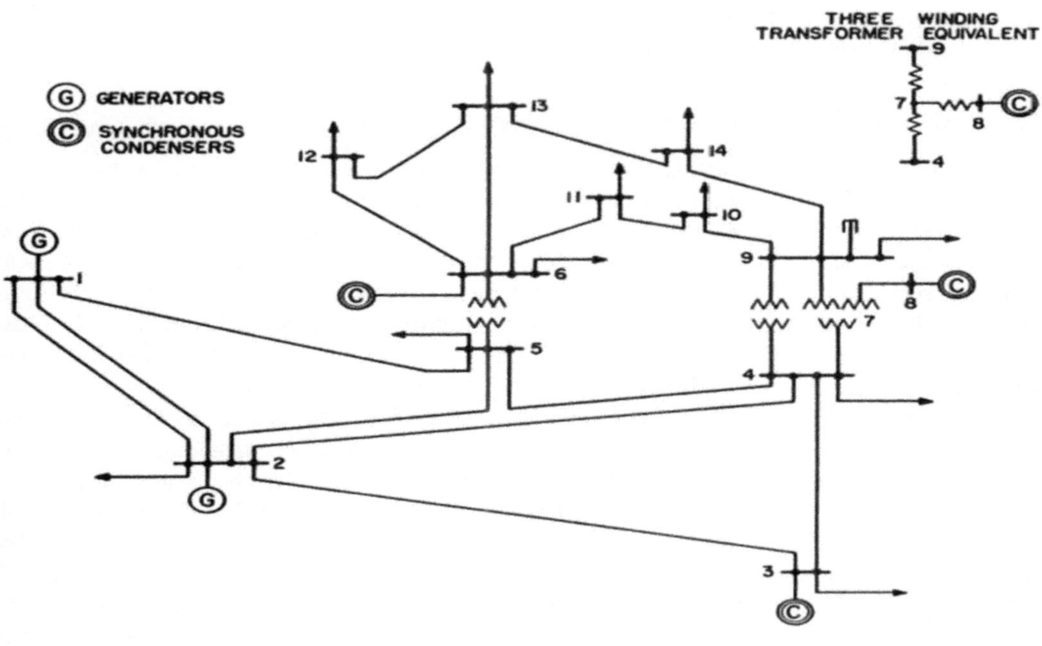

FIGURE 17.7 IEEE 14-bus system. (IEEE Power Systems Test Case Archive, http://www.ee.washington.edu/research/pstca/, accessed January 13, 2014.)

$$f_{n6} = x_5 + x_6 + x_{11} + x_{12} + x_{13} \geq 1$$

$$f_{n7} = x_4 + x_7 + x_8 + x_9 \geq 1$$

$$f_{n8} = x_7 + x_8 \geq 1$$

$$f_{n9} = x_4 + x_7 + x_9 + x_{10} + x_{14} \geq 1$$

$$f_{n10} = x_9 + x_{10} + x_{11} \geq 1$$

$$f_{n11} = x_6 + x_{10} + x_{11} \geq 1$$

$$f_{n12} = x_6 + x_{12} + x_{13} \geq 1$$

$$f_{n13} = x_6 + x_{12} + x_{13} + x_{14} \geq 1$$

$$f_{n14} = x_9 + x_{13} + x_{14} \geq 1$$

$$b = [1\,1\,1\,1\,1\,1\,1\,1\,1\,1\,1\,1\,1\,1]^T$$

The first observability constraint for Bus 1, f_{n1}, implies that at least one PMU must be located at either of buses 1, 2, or 5, in order to have complete observability. The second observability constraint for Bus 2, f_{n2}, implies that at least one PMU must be located at one of the buses listed by f_{n2}. The connectivity matrix A is obtained from the observability constraints given above.

The optimal solution obtained using BIP is given as:

$$\mathrm{x} = [0\,1\,0\,0\,0\,1\,1\,0\,1\,0\,0\,0]^T$$

Thus, PMUs would need to be installed at buses 2, 6, 7, 9 for complete topological observability. The above formulation in (17.13)–(17.16) can be further improved through the consideration of other constraints such as the critical buses in the system, the presence of zero injection buses, and PDC infrastructure.

17.3.2 Communication Architecture, Protocols, and Media

Two types of data concentration architecture can be implemented in synchrophasor-based systems: the central concentration and the local concentration architectures. In the central data concentration architecture, all the individual PMUs in all the substations send their synchrophasor measurements to the control centre. While in the local data concentration architecture, the PMUs within a substation send their synchrophasor measurements to a substation PDC. The substation PDCs then send a single output to the control centre. The advantages of the local concentration architecture include a reduction in the required bandwidth for synchrophasor communication from the substation to the control centre, and improved system security since a secured connection using an encryption method can easily be provided to the outgoing data at a common point.

The transmission of synchrophasor measurements from PMUs to PDCs can be done using serial communication protocols and Ethernet communication protocols. The serial communication protocol is via RS232 serial communication, while the Ethernet communication protocols can be the user datagram protocol (UDP) or transmission control protocol (TCP). The UDP is a connectionless protocol whereby the source broadcasts the data to the destination internet protocol (IP) address without handshaking and predetermining the transmission channel. Also, the source does not verify if the data were received at the destination or not. As a result, the UDP provides low overhead in the delivery of the data. However, an error-free data transmission is not guaranteed. The TCP protocol is a connection-oriented, full-duplex protocol whereby a handshaking process occurs between the source and the destination nodes, and the data are only transported upon the determination and acknowledgement of an end-to-end connection between the source and the destination. If some packets are not received, a request to re-transmit is sent by the destination. Typically, the TCP guarantees an error-free data transmission through checksum and packet re-ordering.

Although the TCP protocol is more secure, the added security introduces latency, which makes the UDP protocol faster. The TCP protocol is commonly used for intra-substation communication because it is a connection-oriented protocol, and it is suitable for the substation environment where the number of PMU measurements (and overhead) is less than what is obtainable at the control centre where a large number of measurements from several PMUs are concentrated. The UDP protocol is typically used for substation to control centre data transmission because of the need to reduce the overhead/bandwidth required. A higher communication bandwidth would be required if the TCP protocol were used, because of the extra overhead encapsulating the application layer data. Also, the TCP protocol introduces latency in the measurements because of the delivery and flow control mechanism implemented in it. This latency would be more pronounced when used for substation to control centre communication.

The communication medium used for synchrophasor data transfer can be utility-owned or leased infrastructure. Examples include optic fibre, power line carrier, leased line, satellite, microwave link, radio link, Synchronous Optical Network (SONET), or internet-based networks.

17.3.3 Bandwidth Planning

The bandwidth requirements for a synchrophasor-based communication network depend mainly on the number of PMUs, phasor format, PMU reporting rate, transmitted data, and the transport protocol used. An example is presented below for a power system with a total of 18 PMUs and reporting rates of 60 fps. The synchrophasor measurements published by each PMU include the positive sequence voltage and current, system frequency, ROCOF, two to four analogue measurements, and a 16-bit binary word. Table 17.2 shows the typical data frame size calculated for each PMU used.

The TCP/IP overhead per frame is given by:

$$\begin{aligned} \text{TCP/IP overhead} &= \text{TCP overhead} + \text{IP overhead} + \text{MAC overhead} \\ \text{TCP/IP overhead} &= (24 + 20 + 18)\text{ bytes} = 62\text{ bytes} \end{aligned} \tag{17.17}$$

TABLE 17.2

Data Frame Calculation per PMU

No	Field	Size (Bytes)
1	SYNC	2
2	FRAMESIZE	2
3	IDCODE	2
4	SOC	4
5	FRACSEC	4
6	STAT	2
7	PHASORS Positive-sequence voltage and current (floating-point format)	$8 \times$ PHNMR[a] $8 \times 2 = 16$
8	FREQ (floating-point format)	4
9	DFREQ (floating-point format)	4
10	ANALOG (floating-point format)	$4 \times$ ANNMR[b] $4 \times 4 = 16$
11	DIGITAL	$2 \times$ DGNMR[c] $2 \times 1 = 2$
	Repeat 6–11	For the number of PMUs in the data frame
12+	CHK	2
	Total	**60**

[a] Number of phasors.
[b] Number of analogue values.
[c] Number of digital status word.

The total PMU frame length for data transmission using TCP protocol is given as:

$$\text{TCP frame length} = \text{PMU data frame} + \text{TCP/IP overhead} \tag{17.18}$$

From Table 17.2, the data frame per PMU was calculated as 60 bytes. Therefore,

$$\text{TCP frame length} = (60 + 62)\ \text{bytes} = 122\ \text{bytes}$$

The bandwidth required per PMU based on the calculated frame size above is given as:

$$\text{Minimum bandwidth required} = \text{Framesize (bits)} \times \text{PMU reporting rate} \tag{17.19}$$

$$= 122 \times 8 \times 60\ \text{bps}$$

$$= 58.56\ \text{kbps}$$

For a 50 Mbps network, the maximum theoretical number of PMUs that can be used is given by:

$$\text{Number of PMUs} = \frac{\text{Available bandwidth}}{\text{Required bandwidth per PMU}} \tag{17.20}$$

$$\text{Number of PMUs} = \frac{50\ \text{Mbps}}{58.56\ \text{kbps}} = 853\ \text{PMUs}$$

The bandwidth required can be further reduced by using the fixed 16-bit (integer) format for the phasors, analogues, frequency, and ROCOF measurements, respectively.

For the PDC to control centre communication, UDP protocol was used. If the synchrophasor measurements from 18 PMUs used are published, the required bandwidth is given as follows:

$$\text{UDP frame length} = \text{PMU data frame} + \text{UDP/IP overhead} \tag{17.21}$$

$$\text{UDP/IP overhead} = \text{Source port} + \text{Destination Port} + \text{UDP length} + \text{UDP checksum} + \text{IP overhead} \tag{17.22}$$

$$\text{UDP/IP overhead} = (2 + 2 + 2 + 2 + 20)\ \text{bytes} = 28\ \text{bytes}$$

From Table 17.2, the data frame per PMU is calculated as 60 bytes. Therefore, the frame length using the UDP protocol is:

$$\text{UDP frame length} = (60 + 28)\ \text{bytes} = 88\ \text{bytes}$$

The bandwidth based on the calculated frame size above is given as:

$$\text{Minimum bandwidth required} = 88 \times 8 \times 60\ \text{bps} = 42.24\ \text{kbps}$$

17.4 Real-Time Wide Area Monitoring Systems

Real-time wide area monitoring systems are very important in the operation of a reliable and resilient grid. They provide unprecedented real-time information on the dynamic state and response of the power system to perturbations. Power system stability is the ability of a power system to operate in equilibrium with acceptable system variables under normal operating conditions and to maintain its equilibrium after being subjected to a disturbance. Real-time wide area monitoring systems consist of applications encompassing grid visualization, situational awareness, oscillation detection and monitoring, transient stability monitoring, voltage stability monitoring, frequency stability monitoring, and state estimation. All these applications are discussed in the following subsections.

17.4.1 Grid Visualization and Situational Awareness

Synchrophasor measurements can be applied to provide real-time system-wide visualization, early-warning alarms, and situational awareness through the collection and collation of phasor measurements from across an interconnected power system, measurement display, and alarms. Typically, these synchrophasor measurements are displayed at the control centre and are used to provide information to system operators. Events and alarms indicating the state of the power system, voltages, real and reactive powers, system frequency, phase angles, and phase angle difference can also be provided. Furthermore, diagnostic functions, applications for detecting system stress, reliability, archiving, and playback functionalities can be implemented. Some commercial off-the-shelf packages for grid visualization and situational awareness include Alstom PhasorPoint, SEL SynchroWave Central, OSIsost PI Systems, and EPG RTDMS.

A well-designed visualization and situational awareness tool should have the following features:

- Real-time measurements display
- Adjustable time-interval trending
- Functionalities for selecting the tags and variables to display
- User-defined or customable dashboards, data view, and plotting
- Events reporting and acknowledgement of alarms
- Event recording
- Continuous and triggered data archiving.

17.4.2 Real-Time Oscillation Detection, Monitoring and Assessment

Electromechanical oscillations are inherently present in power systems. They occur as a result of perturbations caused by load-generator dynamics, switching actions, power exchanges over weak tie-lines, and contingencies. Electromechanical oscillations become a problem when they are not properly damped or controlled. These oscillations can be: (1) inter-area mode oscillations, (2) local plant mode oscillations, (3) intraplant mode oscillations, (4) torsional (subsynchronous) mode oscillations, and (5) control-mode oscillations. The most common of these oscillations are inter-area and local mode oscillations. Inter-area oscillations refer to the phenomenon whereby a group of generators within an area swing against a group of generators in another area of the power system; local mode oscillations occur when a single generator swings against the rest of the generators in the system.

Long transmission distances and the existent of weak tie-lines between areas is a major cause of oscillations, and oscillation monitoring systems (OMSs) are required to detect poorly damped low-frequency oscillations. Power system OMSs are designed to provide system operators with actionable real-time information to detect and identify the oscillation modes present in a power system. Also, the frequency and damping of each of the modes, and the amplitude of the oscillations should be calculated. Several algorithms for estimating the oscillation mode properties from ringdown data have been proposed in the literature using the Prony algorithm [25,26], matrix pencil algorithm [27], and Eigen system realization algorithm [28]. Ambient data methods include mode-meter analysis methods like the Wiener-Hopf linear prediction [29], Yule-Walker [30], several variants of the least squares method [31], and the subspace methods [32]. Other methods are based on time-varying nonlinear analyses like the Hilbert-Huang algorithm [33], wavelet analysis [34], and energy tracking operators [35]. The mathematical principles behind most of these identification methods of electromechanical modes have been thoroughly described in the IEEE Power & Energy Society (PES) Task Force Report titled *Identification of Electromechanical Modes in Power Systems* [36], and are not repeated here.

These traditional methods for oscillation detection can easily be adapted for PMU-based applications using dynamic system-wide measurements and responses that can be captured by PMUs as presented in [35,37–39]. In Ning et al. [37], two modal estimation methods using the frequency domain decomposition and subspace methods, respectively, were explored. Modal eigenvectors were applied in Trudnowski [38]. This involves the use of the right eigenvector computed from synchrophasor measurements in the determination of the power system's mode shape. The Teager-Kaiser energy operator was used in

Kamwa et al. [35] in the prediction of oscillations in power systems. A multi-band filter was first applied to the raw PMU measurements, followed by time-frequency analysis using the energy separation algorithm. The Prony method was implemented in Nabavi et al. [39] using a set of decentralized algorithms

In some recent literature, spectral analytical methods were used to estimate the dominant frequency in system. These methods typically would require a preprocessing stage where the bad data in the PMU measurements are removed. Some of these spectral estimation methods are the Yule-Walker, Welch, Multi-taper methods. For mode damping estimation, the autoregressive Yule-Walker model, half-power point, random decrement technique, Prony's methods, and frequency domain decomposition have been proposed [40–42]. At present, synchrophasor-based monitoring tools from ABB (PSGuard), SEL, and RTDMS are operational in some utilities around the world. Figures 17.8 and 17.9 show snapshots from some of these tools.

An example of PMU-based oscillation mode estimation is described in Banga-Banga et al. [43]. A method based on wavelet transform (WT) and the half power bandwidth (HPB) method is applied to

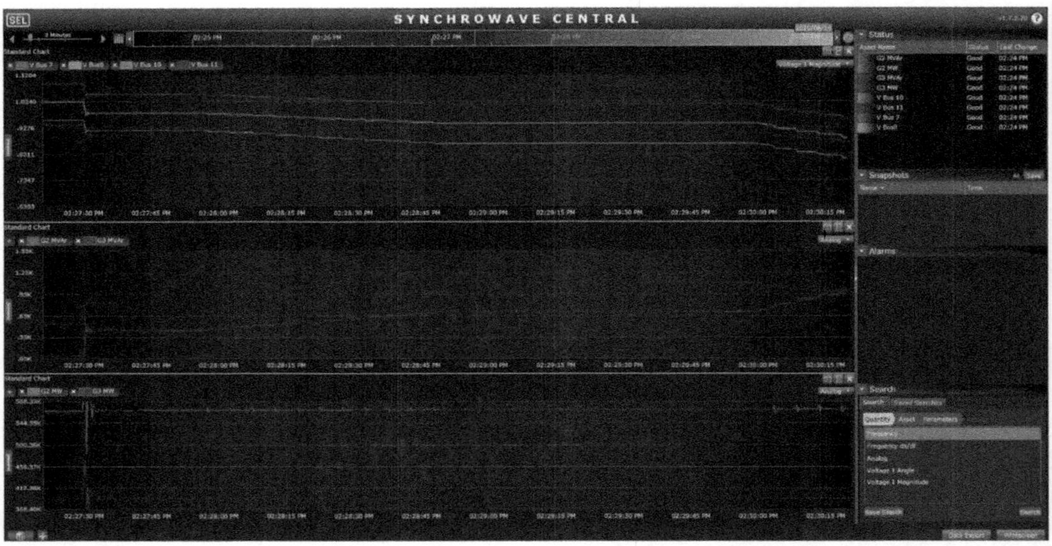

FIGURE 17.8 Screenshot of the SEL SynchroWave central software.

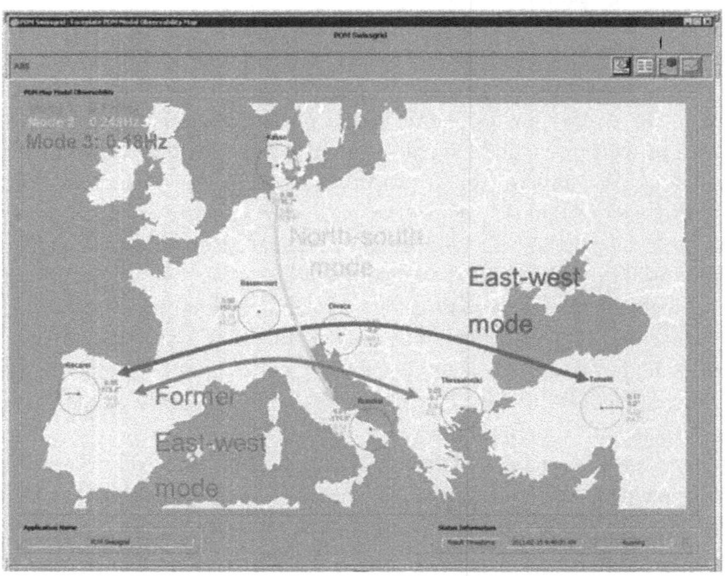

FIGURE 17.9 Screenshot of the ABB power damping monitoring (PDM) software.

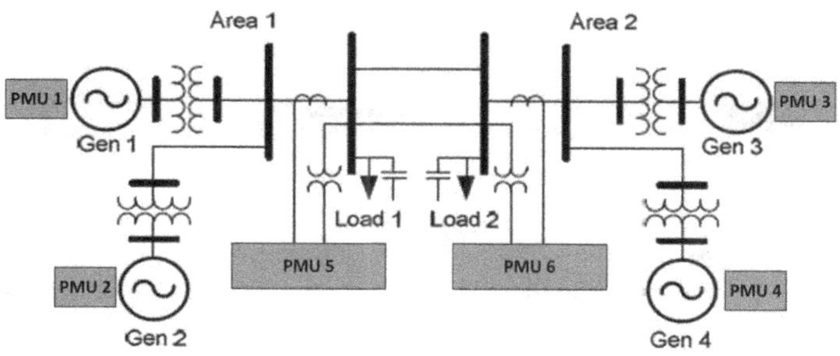

FIGURE 17.10 Two-area benchmark network.

the two-area benchmark model (shown in Figure 17.10) using the real-time digital simulator (RTDS) and PMUs. The two-area benchmark model consists of two similar areas, each with two synchronous machines of 900 MVA, respectively. These two areas are connected by a weak tie-line, causing the existence of inter-area and local mode oscillations. PMUs are placed at each of the synchronous generators, and at the buses between the two areas. The PMUs are configured to stream voltage and current phasors, frequency, ROCOF, and analogue measurements of real power at 10 fps. The oscillation frequency and damping ratios are obtained using a 3-minute ambient data time window of synchrophasor measurements.

Figures 17.11 and 17.12 shows the spectral analysis of ambient measurements. Table 17.3 gives the result for various discrete wavelet transform (DWT) mother wavelets and their comparison with results in Kundur [1].

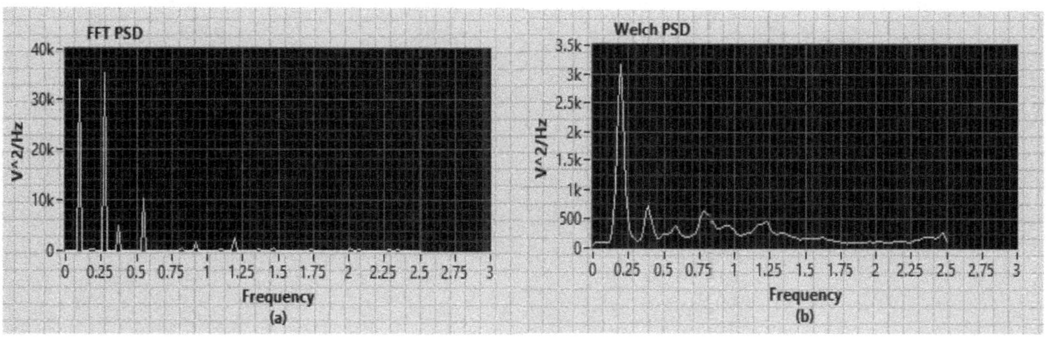

FIGURE 17.11 Signal spectra for (a) FFT at steady state and (b) Welch for transient.

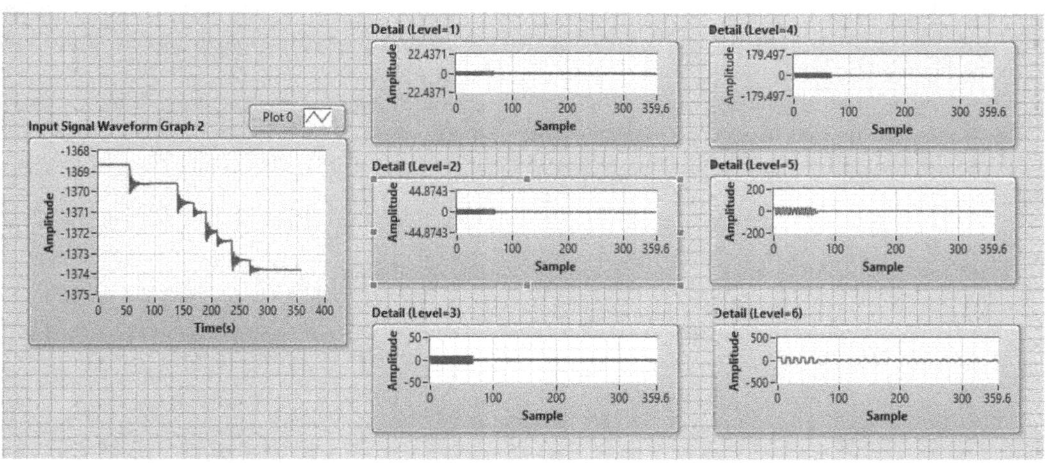

FIGURE 17.12 DWT for the active power transfer between area 1 and area 2.

TABLE 17.3

Wavelet Families in Frequency Estimation

Daubechies Family	Inter-Area Frequency	Biorthogonal Family	Inter-Area Frequency
db2	0.23611	bior1_3	0.379166
db3	0.5125	bior1_5	0.356945
db4	0.3875	**bior2_2**	**0.601389**
db5	0.359722	bior2_4	0.406944
db6	0.345834	bior2_6	0.370833
db7	0.331947	bior2_8	0.35469
db8	0.429167	**bior3_1**	**0.626316**
db9	0.359723	bior3_3	0.495832
db10	0.343058	bior3_5	0.4375501
db11	0.334723	bior3_7	0.390277
db12	0.4375	bior3_9	0.368058
db13	0.395883	bior4_(FBI)	0.390278
db14	0.351388	bior5_5	0.379167
		bior6_8	0.356943
Coiflets Family	**Inter-Area Frequency**	**Symmlets Family**	**Inter-Area Frequency**
coif1	0.543056	sym2	0.4277633
coif2	0.369145	sym3	0.515353
coif3	0.354561	sym4	0.376465
coif4	0.354565	sym5	0.361841
coif5	0.434909	sym6	0.354563
		sym7	0.347221
		sym8	0.347194
Harr family	**0.624968**		

The DWT is expressed mathematically by:

$$\mathrm{DWT}(m,n) = \frac{1}{\sqrt{2^m}} \sum_k f(k)\psi\left(\frac{n-k2^m}{2^m}\right) \tag{17.23}$$

where $f(k)$ is the discrete signal, $\psi(.)$ is the mother wavelet, m and n are the time scale parameters, k is both the discrete time and the number of coefficients in the DWT, 2^m is the variable for scaling, $k2^m$ is the variable for shifting, and $\frac{1}{\sqrt{2^m}}$ is the energy normalization component to ensure the same scale as the mother wavelet.

The damping ratio calculated using different system variables is presented in Table 17.4. The HPB method is applied in the calculation of the damping ratio, which is obtained as the distance between the two half-power points around the peak centre of the signal. This calculation is done using Equation (17.24).

TABLE 17.4

Damping Ratio Calculated Using Ambient Fata

Features	3 Minutes	10 Minutes	20 Minutes
MW exchanged between areas	0.17%	0.17%	0.16%
Speed of generator 1	0.30%	0.30%	0.16%
MW Generator 3	0.17%	0.17%	0.14%
Voltage angle difference	0.48%	0.48%	0.26%

$$\zeta = \frac{w_2 - w_1}{2w_n} \tag{17.24}$$

where w_2 and w_1 are the two 3db points below the inter-area mode frequency, and ζ is the damping ratio.

A comparison of the result obtained in Table 17.4 with that in Aminifar et al. [4] shows that the best result was obtained using the real power exchanged between the two areas shown in Figure 17.10.

17.4.3 Real-Time Transient Stability Monitoring and Assessment

Transient stability is the ability of a power system to remain in synchronism (in step) during large (severe) disturbances caused by faults. The post-disturbance stability of the power system depends on the initial operating state of the system as well as the severity of the disturbance. The period of interest for transient stability analysis is usually about 3 to 5 seconds after the disturbance [1].

Previously, transient stability assessments (TSAs) were commonly conducted for a set of predetermined credible contingencies using dynamic simulations. Recently, online TSAs are easily achieved using synchrophasor measurements. Some PMU-based methods in the existing literature considered the use of the partial energy function (PEF) [44], fuzzy-neural network [45], the energy function analysis method [46], self-adaptive decision tree [47], and Lyapunov exponents [48].

The classical approach for transient stability monitoring was first presented in 1954 [49] and was based on the equal area criterion (EAC). The EAC compares the generator capability to produce power with the transmission network capability to receive power using a power-angle plot. Figure 17.13 shows a simple one machine infinite bus (OMIB) that can be used in explaining the concept of the EAC. The OMIB is connected to the infinite bus by a parallel transmission line with the generator input and the reactance behind the source assumed to be constant. Figure 17.14 shows the pre-fault, fault, and post-fault power-angle curves of the generator output versus the displacement angle for a fault on line 2, which was subsequently cleared before the system became unstable.

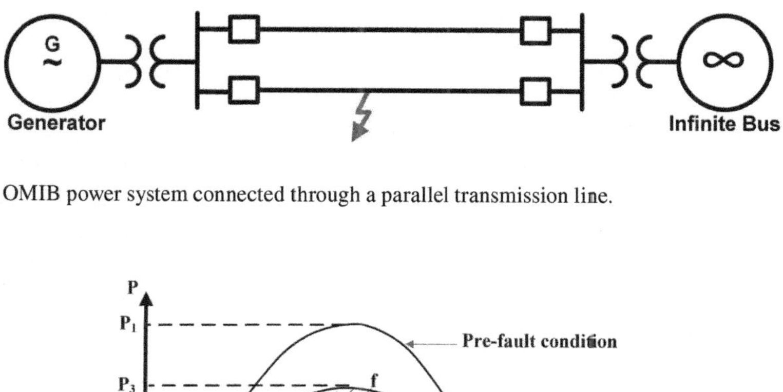

FIGURE 17.13 OMIB power system connected through a parallel transmission line.

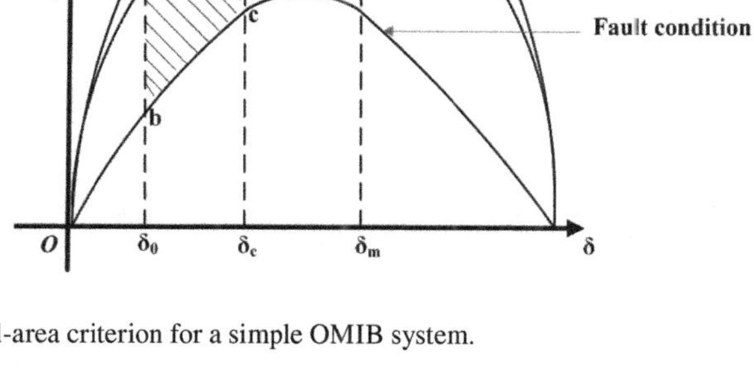

FIGURE 17.14 The equal-area criterion for a simple OMIB system.

For a system with one machine swinging with respect to (w.r.t.) an infinite bus, the swing equation of the finite machine may be written as:

$$M\frac{d^2\delta}{dt^2} = P_a = P_i - P_u \tag{17.25}$$

where M is the inertia constant of the finite machine, δ is the angular displacement of the machine w.r.t. the infinite bus, P_a is the accelerating power, P_i is the initial power (mechanical input), and P_u is the final power (electrical output).

Under normal operating condition, the generator mechanical input is equal to its electrical output. The initial displacement angle δ_0 is obtained as the point of intersection of the pre-fault curve with the horizontal input line (point *a*) in Figure 17.14. When a fault is applied at line 2, the operating point drops from *a* to *b* on the fault output curve. The operating condition moves along the curve from *b* to *c* due to the accelerating power causes. If point *c* is used as the critical clearing time to open the circuit breakers, the operating point would rise to point *e* on the post-fault output curve to *f*. The shaded area $A_2 = defg$ equals area $A_1 = abcd$. If the fault is cleared, a higher input (and initial output) would cause the operating point *f* to move along the curve to the right until it coincides with point *h*. The generator is stable if it can return to the power system the excess kinetic energy absorbed during the acceleration due to the fault before the displacement angle reaches the unstable equilibrium point (UEP).

A partial energy function (PEF) method derived from Kimbark's EAC was presented in Stanton [44] using the power-angle curve (Figure 17.14). The PEF is used to quantify the changes in the system as a result of the changes in the voltage phasors (magnitudes and phase angles). The mismatch between the generator capability to supply power and the transmission system's capability to receive power is given by the shaded portion, which represents the transient energy.

This transient energy function (TEF) calculates the energy associated with the rotor's swing for a range of angles for the power mismatch and may be calculated using Equation (17.26):

$$\text{Energy} = \int_{\delta_m}^{\delta_0} P\,d\delta \tag{17.26}$$

The shaded area in Figure 17.14 can be related to the rotational velocity of the generator. Thus, the computed area will give the transient energy of the system. If the rotational velocity of the generator is obtained using PMU measurements (via speed sensors), these measurements can be compared in real time to a critical predetermined threshold chosen through simulation studies. The above can be applied even to a multi-machine system The PEF computes the transient energy of an *i*th generator w.r.t. the excursion from a reference point. The transient energy V_i of a generator is obtained as the integral of the accelerating power P_i or the rotor displacement $d_{\delta i}$:

$$V_i = \int_{\delta_{im}}^{\delta_{i0}} P_i\,d_{\delta i} \tag{17.27}$$

For an *i*th generator to be stable, its fault energy V_{fault_i} must be less than the critical fault energy V_{critical_i}, as shown mathematically in Equation (17.28):

$$V_{\text{fault}_i} \le V_{\text{critical}_i} \tag{17.28}$$

When the transient energy of the generator is related to its kinetic energy, the relationship with velocity is established. Thus, Equation (17.28) may be rewritten as:

$$\omega_{\text{fault}_i} \le \omega_{\text{critical}_i} \tag{17.29}$$

where ω_{fault_i} is the ith generator speed relative to the inertia centre, $\omega_{\text{critical}_i}$ is its critical speed, and:

$$\omega_i = \bar{\omega}_i - \omega_{\text{coi}} \tag{17.30}$$

$$\omega_{\text{coi}} = \sum_{i=1}^{n} \frac{M_i \bar{\omega}_i}{M_T} \tag{17.31}$$

where ω_{coi} is the speed of the inertia centre, ω_i is the shaft speed of the ith generator relative to the inertia centre, $\bar{\omega}_i$ is the measured shaft speed of the ith generator, and M the generator inertia.

The two methods presented in Benmouyal et al. [50] for power swing detection can be adapted for transient stability assessment. The first method involves the calculation of the phase angle difference between the internal voltages of the two machines in a two-machine system. This phase angle difference can be computed in real-time using PMU measurements, and the equal area criterion can be applied to determine system stability at any given operating point. The second method uses the phasors of selected buses and the phase angle difference between any two phasors. This is premised on the fact that, during a disturbance, the phase angle difference changes. The system is stable if the phase angle difference settles to a new stable value. If it diverges, then the system is unstable.

17.4.4 Real-Time Voltage Stability Monitoring and Assessment

Voltage stability is the ability of the various segments of a power system to maintain an acceptable voltage profile during normal system operating conditions and after being subjected to a disturbance. Generally, voltage instability may be as a result of the inability of the combined generation and transmission infrastructure to meet the system demand at the load centres. This causes the power system to proceed to a state of voltage instability whereby a disturbance results in a progressive and uncontrollable voltage decline in a major part of the system. Voltage stability assessment involves the determination of the voltage stability level of a power system at the current operating point and for the anticipated system operating conditions. This involves the analysis of the system's proximity to voltage collapse, the driving force(s) behind the voltage instability/collapse, why it occurred, where the voltage critical (weak) areas are, and the effective countermeasures to implement in order to restore the system to an acceptable operating state.

Several methods for voltage stability assessment exist in the literature. Sensitivity-based methods were proposed in [1,51,52] using the reactive power reserve across widely dispersed reactive power sources in a power system. In [53], synchrophasor-based sensitivity analysis was proposed using algebraic equations fitted onto sampled states. Synchrophasor measurements were used to identify the point where the real power at the buses passed through the maximum loadability point. Reference [54] proposed a linearized voltage stability index based on the determinant of the load state matrix. Some of these methods assumed the availability of PMU measurements. The advantages of the proposed index include full linearity and less computation burden since there was no need for computing the network equivalents at each of the load buses. Similarly, Thevenin-based methods were proposed in [55–57]. Generally, the Thevenin method is based on the computation of the equivalent Thevenin voltage and Thevenin impedance as seen from the load bus of interest. The voltage collapse point is the maximum loadability point where the bus voltage is equal to the Thevenin voltage of the equivalent system, or the point where the apparent load impedance is equal to the Thevenin impedance.

In addition to the sensitivity-based and Thevenin-based methods, methods based on computational intelligence can also be found in the literature. The use of decision trees (DTs) in the detection of the onset of voltage instability was considered in [58–62]. Bagging and adaptive boosting algorithms were used in training the DTs in [63]. Reference [64] presented an artificial neural networks (ANN) method for estimating the system's margin to the maximum loadability point, while [65] suggested a support vector machine (SVM) for estimating the system's margin to its maximum loadability limit from a current operating point. The performance of the SVM was demonstrated for a normal loading condition and for different loading conditions.

A novel adaptive weighted-summation wide area monitoring algorithm capable of predicting the system state, identifying the voltage weak areas, and the system's margin to voltage collapse was presented in [66] for a group of generators in a large interconnected power system. This is given as:

$$vca\text{RVSA}_{rk} = \min_{r=1,N}\left\{\sum_{i=1}^{n_r} w_{ik}\text{RVSA}_{ik}\right\}, \quad i=\overline{1,n_r} \qquad r=\overline{1,N} \quad k=0,1,2,\ldots \tag{17.32}$$

where w_{ik} are the weights for the generators, RVSA_i are the computed generator-derived indices, i is the ith generator within a reactive power reserve basin (RPRB). *vcaRVSA* is the weighted summation of the RVSA in the RPRB. n_r is the total number of generators in the rth RPRB, and N is the total number of RPRBs. The RPRB is a group of coherent generators providing reactive power support (voltage control) to a group of load buses with a similar voltage collapse problem. This group of load buses are referred to as a voltage control area (VCA).

If generator reactive power reserve is used, the RVSA_{ik} index is replaced by the RVSA_{Qik} index. The RVSA_{Qik} index is computed as follows:

$$\text{RVSA}_{Q,ik} = \left(\frac{Q^c_{gmaxi} - Q_{gki}}{Q^c_{gmaxi}}\right) \times 100\% \quad k=0,1,2,\ldots \tag{17.33}$$

The weight of the ith generator is calculated on the basis of the real-time measurements from the PMUs as:

$$w_{ik} = \frac{Q^c_{gmaxi} - Q_{gki}}{\sum_{i=1}^{n_r}\left(Q^c_{gmaxi} - Q_{gki}\right)}, \quad \sum_{i=1}^{n_r} w_{ik} = 1 \qquad i=\overline{1,n_r} \quad k=0,1,2,\ldots \tag{17.34}$$

where Q^c_{gmaxi} is the maximum reactive power of the ith generator at the voltage collapse point, and Q_{gki} is the reactive power of the ith generator at the kth operating point. The operating state of the power system is predicted using the *vca*RVSA index given in Equation (17.32). Similarly, the system's percentage margin to voltage collapse and the identification of the voltage weak areas of the power system are predicted using the *vca*RVSA index.

17.4.5 Real-Time Frequency Stability Monitoring and Assessment

System frequency is a key indicator in the power system. It reflects the load-generator balance in the system and is a good indicator of the integrity of interconnections in large power systems. Frequency stability monitoring assesses the impact of generation real power-load demand mismatch on the system's frequency. In addition to the use of the system frequency, the ROCOF can also be applied as a monitoring index to indicate the operating state of the system because ROCOF is proportional to the real power mismatch and is dependent on the electric power system inertia.

A typical frequency stability monitoring application is the frequency monitoring network (FNET). FNET makes use of measurements from frequency disturbance recorders (FDRs). The FDR is a single-phase PMU that measures the voltage phase angle, amplitude, and frequency from a single-phase voltage source. FDRs are installed at the distribution level. This allows frequency monitoring at the distribution level, and can be used as a trigger to remedial action schemes such as underfrequency load shedding if certain thresholds are violated. The background information on FNET can be found in [67]. Wide-area frequency trending is another tool that is commonly implemented in utilities. Some platforms such as the BPA Stream Reader, SCE SMART®, and RTDMS implement frequency trending.

17.4.6 Other Applications of Synchrophasor Technology

17.4.6.1 State Estimation

State estimation (SE) is a critical application in most control centre energy management systems (EMSs). It is commonly used to estimate the bus voltage magnitude and phase angle based on measurements in order to provide information of the state of the power system with regards to the real and reactive powers. Most SEs (both static and dynamic) use the weighted least squares (WLS) approach in converting the nonlinear system of equations into a linear one through Taylor's series expansion. The procedure for this can be found in [68,69].

The traditional SE gives a measurement vector z for a set of nonlinear equations as follows:

$$z = h(x) + e \tag{17.35}$$

where x is a $(n \times 1)$ state vector consisting of voltage magnitude and phase angles. The measurement vector z $(m \times 1)$ consists of voltage magnitudes, active and reactive power, and power injections; $h(x)$ is a nonlinear vector measurement function; and e is the error vector of the uncorrelated measurement errors.

WLS algorithms are applied to minimize the measurement errors. The optimal solution is usually obtained after several iterations. With synchrophasor measurements from PMUs, direct measurements can be provided to the EMS in the control centre to give a direct and more accurate system state. With synchrophasor measurements, the phase angle measurements are obtained directly instead of through estimation. Also, the availability of PMU measurements allows the actual topology of the system to be known, thus giving improved observability of the system. Consequently, a new paradigm for linear state estimators (LSEs) can be implemented as presented in [70]. LSEs take advantage of the availability of voltage and current phasors to express the system states linearly without using iterative algorithms.

Hybrid SEs utilizing conventional SCADA measurements and PMU measurements were proposed in [71–73]. One advantage of the hybrid method is that it harnesses the benefits of synchrophasor measurements by including them in the SE process while keeping the existing SE model. In [71], the existing traditional SE was used after post-processing using synchrophasor measurements. A different methodology was presented in [72] where the voltage and current phasors are converted into equivalent branch flow measurements, which are then integrated to the traditional SE measurement set. Another alternative was the post-processing of the traditional SE results using synchrophasor-based LSE to improve the result obtained from the traditional SE. In [73], the synchrophasor measurements were used as the input to a pre-processor to carry out linear state estimation of the observable part of the system covered by PMUs. The results are then applied in the subsequent iterations of the traditional SE algorithm.

Three-phase state estimators utilizing both SCADA and phasor measurements have been proposed in [74]. However, such an estimator adds an extra term in the system of equations, which requires changes in the existing SE algorithms deployed at the control centre EMS. Other SE-related applications where synchrophasor measurements can be put to use include improving the network observability, aiding bad data processing when using the traditional SEs, and protective relaying.

17.4.6.2 Model Validation

The application of synchrophasor measurements for the benchmarking and validation of models of power system components have been reported in [75–78]. These include the determination of line parameters [75], load models [76], instrument transformer parameters [77], and machine/control system parameters [78]. For example, in order to determine the parameters of a transmission line, PMU measurements should be available from the sending and receiving ends of the line. With these, the line parameters can be calculated using the voltage and current phasors from both ends. A similar approach can be adopted for load model validation. In this case, the load model output would be compared with the measured output.

PMUs can be used to record dynamic disturbances, which can be applied to compare the actual response of the system with that obtained using power system simulation tools. The error calculated using the actual and the simulation measurements is minimized until an acceptable error margin is obtained. Model validation in power plants is very important because generator models and their controls need to be validated as a result of the effect of aging, saturation, operational wear and tear, etc. Thus, the parameters of the generator models and control tuning can be carried out as long as PMUs are located near or at the power plants.

17.4.6.3 Post-Disturbance Applications

1. Post-mortem analysis: The post-mortem analysis of blackouts such as the 2003 blackout in the northeastern United States was particularly difficult because many of the digital fault recorders (DFRs were triggered by local events. Correlating the records from such geographically separated DFRs for post-disturbance analysis of a wide area disturbance is particularly tedious and time intensive. However, post-mortem analysis has become easier due to the availability of real-time, wide area, time-synchronized PMU measurements from different parts of the system that can easily be concentrated, correlated, and compared. Disturbance recording using synchrophasor measurements can be done continuously or it can be triggered when certain conditions are true using logical functions. The continuous recording and communication of synchrophasor measurements from the substation to the control centre PDC is constrained by the available communication bandwidth and data storage facility. Thus, triggered recording/archiving of PMU measurements is preferable for disturbance analysis. For wide area disturbance recording, a cross-triggering technique whereby a local event at a PMU is used as a trigger to initiate recording at a PDC or at other PMUs across the system can be applied. This trigger can be published to the PMUs via IEEE C37.118 synchrophasor commands or by using the IEC 61850 GOOSE messages. Some of the tasks carried out during post-mortem analyses include the following:
    - Acquisition of disturbance records to identify the causes, origin, and time of a particular disturbance
    - Analysis of the sequence of events, types of fault, clearing and reclosing times.
    - Analysis of the pre-disturbance system operating conditions and system behaviour
    - Evaluation of the pre-fault, fault, and post-fault currents and voltages, and their deviation from calculated values
    - Analysis of protective relay pickup settings and actual protective relay response to a fault
    - Performance analysis of components of protective relaying and control systems, and their failure to operate as designed or incorrect operation
    - Identification of failed or malfunctioned protective and control system components
    - Evaluation of damages
    - Fault clearance, reconfiguration, and planning of appropriate sequence of restoration.
2. Fault location: Accurate fault location is an important aspect in the restoration process in the power system after a fault because the quicker the faulted section is identified or located, the shorter the restoration time. Synchrophasor-based fault location methods have been proposed in [79,80]. The general idea is based on leveraging wide area synchrophasor measurements to accurately pinpoint the faulted line in the system, even in active distribution systems, which are usually complex and difficult. A bus-impedance matrix was developed in [79] for the calculation of the fault location in a network using synchrophasor measurements from two remote buses in the network. A two-step technique was utilized in [80]. The first step involves the identification of candidate fault locations through an iterative process. In the second step, the actual location of the fault is identified through the comparison of the voltage phasors from two terminals with a predetermined threshold.
3. Controlled islanding: Controlled islanding refers to the deliberate creation and separation of a power system as a corrective maintenance action of last resort in response to cascading outages that could lead to a major system collapse or blackout. This would require a fast and reliable process designed to result in stable segments (islands) of the partitioned system.

The following three essential questions should be taken considered in the design of a controlled islanding scheme:

- Where to island?
- When to island?
- What post-islanding actions should be implemented?

The optimal locations to island can be identified using the slow-coherency method in [81], whereby the power system is represented as a graph. Afterwards, generators belonging to the same coherent groups are merged into a dummy node with subsequent splitting of the graphs into parts. The objective of the splitting algorithm is to produce islands with minimum active power mismatch, minimum power flow disruptions, or a combined minimization of both active and reactive power mismatches. Wide area synchrophasor measurements are essential in the real-time calculation of the generation-load balance required in maintaining system stability. The problem of when to island can be determined by detecting unstable inter-area modes and using this as a trigger to separate the system into islands. Post-islanding actions are intended to further enhance the reliability and security of the system and its islands through appropriate undervoltage and underfrequency load shedding, or through generation rejection. PMUs provide fast and reliable real-time measurements that can be used for monitoring and splitting the system in a timely manner before the system becomes unstable. Some PMU-based methods for controlled islanding can be found in [82,83].

4. System restoration: Real-time measurements from PMUs can assist operators in making informed decisions on the restorative process after a blackout. A post-blackout restorative process can be a sequence such as load restoration, control strategies, and when to restore. For example, the voltage magnitude, phase angle, and frequency information from PMUs can serve as indicators of the right time to close the circuit breaker. Another system restoration process is the generator black-start. This refers to the starting of the generating units in a system without using the power from the grid. Typically, this would involve energizing the system from a few generators at certain strategic locations in the grid. Afterwards, the islands around the generators can be reenergized, followed by the reconnection of the islands. In [84], a synchrophasor-assisted algorithm was proposed for aiding generator black-start restoration after a power outage. A generator's voltage must be synchronized to the grid voltage before it can be reconnected. It is important that the phase angle difference is monitored and within acceptable tolerances. Phase angle difference from PMUs at distribution substation buses can serve as an indicator to the distribution system dispatcher on when to tie the distribution feeders together in a loop system.

In the implementation of restoration schemes, synchrophasor measurements can be used to identify the system conditions in which the frequency, voltage, and voltage angles are out of synchronism with the rest of the system. Also, synchrophasor measurements can be used to quickly determine the sequence of events and the fault location, and to identify the possible causes of disturbances better than with SCADA measurements. Furthermore, with the accuracy provided by synchrophasor measurements, restoration can be easily done to return the system to an acceptable operating condition. In addition, PMUs can be used to provide operators with situational awareness of each of the respective restoration action. This would ensure that stability is maintained as lines and loads are reconnected.

17.5 Wide Area Protection and Control Schemes (System Integrity Protection Scheme)

Defence actions against power system instabilities can be generally classified under preventive and corrective actions, respectively. Preventive actions are used in pre-contingency situations to increase the system security margin. Corrective actions, on the other hand, are initiated in the post-contingency state in order to restore the system stability to its pre-contingency or an acceptable post-contingency margin. Corrective actions for system-wide disturbances are often implemented using remedial action schemes (RASs) or system integrity protection schemes (SIPSs) [85]. RASs or SIPSs belong to the class of automated remedial actions/special protection schemes (SPSs) designed to detect certain predetermined system conditions (abnormal system conditions) that have a high probability of jeopardising the secure or reliable operation of the power system. Some sort of pre-planned remedial action deemed as appropriate for the preservation of the integrity of the power system are afterwards initiated to mitigate the observed abnormal condition in a controlled manner [86,87].

17.5.1 Types of Protection and Control Schemes

SIPSs can be classified based into a number of groupings [85–87]:

- Classification according to the architecture: flat and hierarchical schemes
- Classification according to the location of the controller: centralized and distributed schemes
- Classification according to the input parameters used: event-based and response-based schemes
- Classification according to the decision-making process: rule-based and algorithm-based schemes
- Classification according to the control design: closed-loop and open-loop schemes.

For the flat architecture SIPS, the measurement and decision making are carried out at the same location. The hierarchical architecture for SIPS involves the use of a multi-layer measurement and decision-making process whereby information is transmitted to multiple locations within the power system, local SIPS, and a regional/national SIPS. Figure 17.15 illustrates the flat and hierarchical SIPS classification.

The centralized synchrophasor-based SIPS refers to the system protection schemes where measurements from substation PMUs/PDCs are transmitted to a central location via a communication infrastructure. The decision-making algorithm of the SIPS is implemented in a controller at the central location, and the signal for corrective action is issued from the control centre. For distributed SIPS, the measurement acquisition and decision/corrective actions are carried out by local controllers distributed across the power system. These distributed controllers may also operate independently without communicating with other similar controllers. SIPSs can also be classified based on their input type. For event-based SIPSs, outage events such as the status of major transmission lines, generators, transformers, and circuit breakers can be monitored and used to initiate pre-planned remedial actions. For response-based SIPSs, measurements obtained from the system's response to a dynamic system condition/disturbance are used in the detection and implementation of the predetermined remedial actions.

Rule-based SIPSs make use of a set of if-then rules. These rules use predetermined thresholds obtained from offline system studies, while algorithmic-based SIPSs are implemented using the analysis of the system model. For closed-loop SIPSs, the actions of the SIPSs are based on the response of the system to previous control action. This implies that there is a feedback path from the power system (plant) to the SIPS (controller). The controller continuously monitors the system's responses and takes the appropriate successive remedial actions based on this. On the other hand, open-loop SIPSs do not have a feedback path from the power system to the controller. Generally, event-based SIPSs belong to the class of open-loop control systems, while response-based SIPSs are closed-loop systems.

17.5.2 Functional Elements

The functional elements of SIPSs can be divided into three groups:

- System monitoring element
- Protection element
- Execution element.

The system monitoring element is used in the supervision and detection of the changes in the power system based on the phasors, analogues, and binary measurements received from PMUs. These changes can be in the system loading condition, topology, generator dispatch, and breaker or switch conditions. The protection element identifies system stability threats and arms the SIPS in order to mitigate such threats. The execution element receives the output signal from the SIPS protection element and performs the predetermined remedial action(s) to mitigate the instability. Figure 17.16 shows a typical block diagram of SIPS.

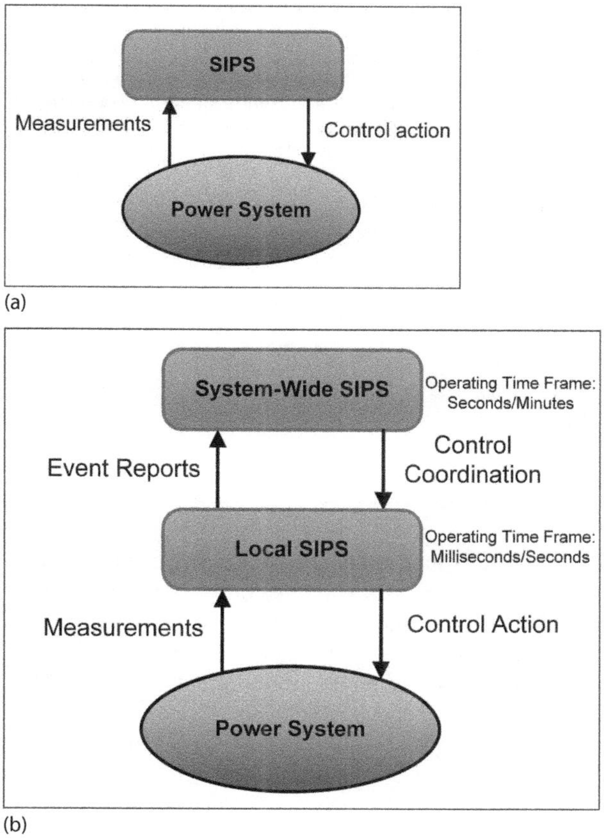

FIGURE 17.15 System integrity protection scheme: (a) flat architecture, and (b) hierarchical architecture.

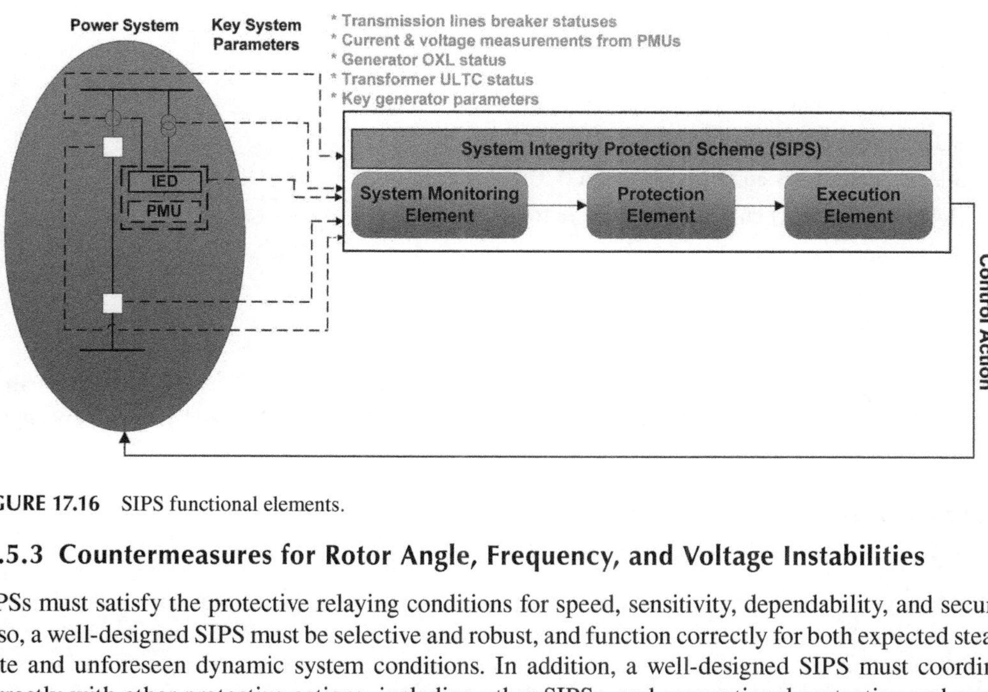

FIGURE 17.16 SIPS functional elements.

17.5.3 Countermeasures for Rotor Angle, Frequency, and Voltage Instabilities

SIPSs must satisfy the protective relaying conditions for speed, sensitivity, dependability, and security. Also, a well-designed SIPS must be selective and robust, and function correctly for both expected steady-state and unforeseen dynamic system conditions. In addition, a well-designed SIPS must coordinate correctly with other protective actions, including other SIPSs, and conventional protective and control

TABLE 17.5

Mitigating Actions for Various Power System Instability Problems

SIPS	Transient Instability	Small-Signal Instability	Frequency Instability	Voltage Instability
Generation rejection	×	×	×	×
Remote load shedding	×	×		
HVDC controls	×	×	×	
Braking resistor	×	×		
Underfrequency load shedding			×	
Turbine fast valving	×	×	×	
Automatic shunt switching	×	×		×
Undervoltage load shedding				×
Tap-changer blocking			×	×
AGC controls		×		×
Gas turbine startup			×	×

devices. The particular type of SIPS implemented in a power system depends on the probable threats expected. SIPSs can be said to encompass a combination of protective relays, IEDs, meters, control equipment, automation equipment, communication infrastructure, reactive power compensators, and flexible alternating current transmission system (FACTS) devices.

The various types of SIPSs for mitigating power system stability are [86,87]:

- Generation rejection
- Tie-line runback
- Transmission transfer trip
- Underfrequency load shedding
- Undervoltage load shedding
- Adaptive load mitigation
- Out-of-step tripping
- Shunt capacitor switching
- Tap-changer control
- Static VAR compensator (SVC) control
- Static synchronous compensator (STATCOM) control
- High-voltage direct current (HVDC) controls
- Series capacitor bypass
- Turbine fast valving
- Gas turbine startup.

Other SIPSs capable of mitigating wide area disturbance can be found in [86]. Table 17.5 summarizes the type of system disturbances for which some of the above-mentioned SIPSs can be effectively applied [86,87].

17.6 Cyber Security in Synchrophasor-Based Systems

The modern grid incorporates communication networks connecting substation devices and the control centres. Access to substation devices such as protective relays, IEDs, DFRs, PMUs, meters, AMI, RTUs, and substation computers requires network connection via local area networks (LANs), wide area networks (WANs), and the Internet, which makes it vulnerable to cyber attacks from hackers, terrorists, or disgruntled employees. Similarly, control centre infrastructure such as the SCADA system, EMS, databases, and other applications are also prone to attacks. These critical infrastructures should be protected against cyber attacks.

17.6.1 Cyber Threats and Vulnerabilities in Synchrophasor-Based Systems

Following the introduction presented in Chapter 13 on cyber security terminologies, threats, vulnerabilities, detection, and countermeasures, this chapter presents the threats specific to synchrophasor-based WAMPAC systems, cyber security standards, and an example of a typical cyber attack on a synchrophasor-based application. The security of the critical infrastructure within a synchrophasor-based WAMPAC system is particularly difficult because the assets within the WAMPAC system cuts across many entities spread across a wide geographical area. In some countries, the distribution system operators (DSOs) and the transmission system operators (TSOs) are required to share data amongst themselves, and with the system operator. A number of security threats and vulnerabilities can occur in the form of physical attack on the synchrophasor infrastructure when there is an interruption or deliberate damage to equipment, the communication network, and software applications. An example is the case whereby the network cable connecting a PMU to a PDC is disconnected or slashed. Also, a physical attack can be in the form of damaging the power supply to these assets. Some guidelines for mitigating physical intrusions can be found in IEEE 1402-2000 (R2008) [88]. Another type of attack on synchrophasor-based systems is the denial of service (DoS) attack, which compromises the availability of synchrophasor measurement to the services and applications that require such measurements. A common form of DoS attack is the attempt by an attacker to deny access to an authorized user to a particular service by flooding a communication network with irrelevant traffic, thus causing communication network delays, packet losses, attenuation, and bandwidth limitation that will prevent the publishing and subscription of time-critical synchrophasor messages.

In data spoofing attacks, an attacker could masquerade as a substation PMU by streaming false data to upstream PDCs so that control centre applications like SIPS operates incorrectly, thus leading to loss in grid stability and reliability. Another case is that of GPS spoofing, whereby an attacker sends wrong GPS signals capable of causing incorrect time synchronization of PMUs. In some PMUs, accurate time synchronization is required before the PMUs can stream synchrophasor measurements. GPS spoofing can prevent such PMUs from streaming measurements or could cause a PDC or synchrophasor-based application to fail.

A man-in-the-middle (MitM) attack is another threat that can occur between the substation PMUs and the substation or control centre PDCs. It can also occur between the substation PDCs and the control centre PDC. In such an attack, the attacker masquerades itself as legitimate. The confidentiality of information contained in synchrophasor measurements can be breached through the sniffing of the network packets between the synchrophasor client (PDC) and server (PMU/PDC). Typically, address resolution protocol (ARP) spoofing is used to sniff the traffic in the communication network. This type of attack would normally precede data spoofing and MitM attacks.

Malicious code injection can inject false measurements or false commands. False measurements would cause SIPSs to operate incorrectly, while command injection would inject false commands in the control centre applications. Packet injection is commonly used in MitM attacks and DoS attacks. An example of command injection is when SQL queries are inserted to add, delete, or modify a database where PMU measurements are being stored.

17.6.2 Cyber Security Standards

Foundational, regulatory, and normative cyber security standards have been drafted to provide guidance and specifications on the security of critical assets in the electric power grid. Substation communication security requirements are described in the IEC 62351 standard developed by working group WG15 of IEC TC57 [89]. The IEC 62351 standard defines security mechanisms to protect communication protocols and provide end-to-end security for substation systems based on the IEC 60870, IEC 61850, IEC 61970, and IEC 61968 standards, respectively. The objective of the IEC 62351 standard is to provide security features such as the authentication of data communication; authorization; prevention of eavesdropping, playbacks, and spoofing; and intrusion detection. Another important specification is the North American Electric Reliability Corporation Critical Infrastructure Protection (NERC CIP) standard [90]. NERC CIP is a cyber security framework for the identification and protection of critical cyber assets to increase reliability and protection from attacks. It consists of 9 standards and 45 requirements covering issues ranging from the security of electronic perimeters, procedures and requirements for technologies, asset databases, and access controls using different authentication methods and authorization policies in the identification of critical cyber assets and various access levels of authorization.

The IEEE 1686-2013 [91] is a security standard that defines the functions and features to be included in IEDs to provide the protection of these critical assets in accordance with the NERC CIP requirements. It also provides a table of mandatory requirements that vendors must use to indicate their level of compliance. IEC 61850 90-5 is a technical report for the transmission of synchrophasor data using the IEC 61850 framework. It provides specifications for asymmetric and symmetric key signature creation and tamper detection, and optionally specifies symmetrical key encryption to provide confidentiality. Another security standard is the National Institute of Standards and Technology (NIST) NISTIR 7628 [92]. It provides guidelines for implementing effective cyber security strategies within the smart grid. The standard proposes the framework for assessing risks in the smart grid, and identifies and applies appropriate security requirements to mitigate these risks. The IEEE C37.240 standard [93] provides the minimum requirements for achieving high levels of cyber security in substations. It was designed to provide a balance between technical feasibility and economic feasibility of the cyber security measures that can be implemented in substations so that legitimate operational activities are not impeded, especially during emergencies and restoration processes.

17.7 Example of a Cyber Security Attack

An example of a cyber security attack using a synchrophasor-based system is presented in [94]. A MitM interface served as a transparent proxy (invisible to the IEEE C37.118 synchrophasor clients and servers), which intercepted the synchrophasor communication between the IEEE C37.118 servers (substation PMUs/PDCs), then degraded them using adverse WAN conditions, before forwarding the synchrophasor measurements to the control centre IEEE C37.118 clients (superPDC) [95]. Thus, this type of MitM is an active one; it breaks the normally private client-server connection into two. The first connection is between the server and the MitM attacker, while the second connection is between the client and the MitM attacker (illustrated with dashed lines in Figure 17.17). All this happens without the server's

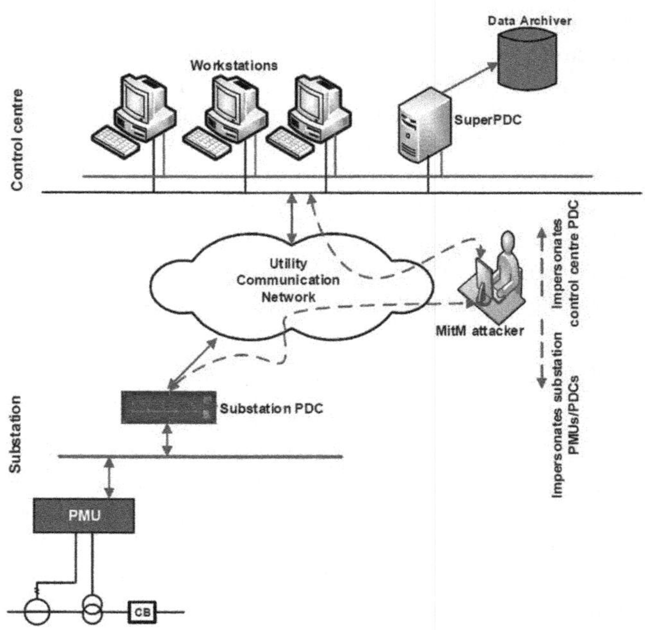

FIGURE 17.17 Man-in-the-middle attack for the developed co-simulation platform.

and the client's knowledge because the MitM impersonates and spoofs the communication between the PMUs/PDCs (IEEE C37.118 servers) and the control centre superPDC (IEEE C37.118 client).

The platform used for the MitM attack is the Kali Linux software package: an open source Debian-derived Linux distribution. The steps for the execution of the MitM attack are given below:

- Step 1: Initialise *Ettercap* in Linux and define the Ethernet interface of the cyber attacker machine to use for the MitM attack.
- Step 2: Scan the communication network and identify the IP addresses of the IEEE C371.118 servers and clients of interest.
- Step 3: Commence packet sniffing on the specified Ethernet interface.
- Step 4: Initiate ARP spoofing (ARP poisoning).
- Step 5: Activate IP forwarding to start forwarding the messages from the IEEE C371.118 servers to the clients via the MitM machine.
- Step 6: Emulate the adverse wide area communication network conditions using the *Netem* component in Linux.

Network devices use the ARP in resolving the network layer IP address to the link layer MAC address. With ARP poisoning, the MitM attacker sends a fake ARP message to the IEEE C37.118 servers and clients as a precursor to cyber attacks such as DoS, MitM, and session hijacking attacks, respectively. This enables it to associate its MAC address with the IP address of the IEEE C37.118 servers or clients. Thus, all the messages from the IEEE C37.118 servers to the IEEE C37.118 clients will be routed through the MitM attacker acting as a transparent proxy server for the duplex traffic between the IEEE C37.118 client (superPDC) and the IEEE C37.118 servers (substation PDCs). The emulated adverse communication network conditions are introduced into the communication network via the transparent proxy server (indicated by the dashed lines in Figure 17.17). The emulated WAN conditions considered is given in Table 17.6. These WAN conditions could result in the failure of the WAMPAC schemes during emergency conditions. Some of the results obtained for the investigations are presented and discussed below.

A co-simulation platform (shown in Figure 17.18), consisting of the integration of the RTDS and a communication network emulator, was used to investigate the impact of adverse network conditions on the performance of the WAMPAC algorithms proposed in [96]. The results obtained can be seen in Figures 17.19 through 17.21 for adverse network conditions such as latency, jitter, packet loss, random noise and their combinations.

An *N*–1 line contingency on the 10-bus multi-machine test system in [1], with and without the emulation of the network delay parameter, given in Table 17.6 was simulated. Figure 17.19 shows the plots of the synchrophasor voltage at bus-8 obtained without any network latency, and for the emulated network latencies of 250 ms to 750 ms with a jitter of 10%. Figure 17.20 shows that the network latencies ≥ 800 ms had adverse effects on the synchrophasor measurements. This is demonstrated by the increased losses in the synchrophasor measurements, as indicated by the highlighted dropped measurements.

Figure 17.21 shows the impact of the emulated packet losses and network corruption from noise using the same contingency given in Figure 17.20. From the results obtained, packet losses up to 2.5% did not have any adverse effect on the synchrophasor measurements published to the control centre using the emulated WAN conditions. However, packet losses greater than 2.5% had a greater impact, as indicated by the increase in the number of measurements dropped. Similarly, it was observed that corruption up to 1.0%

TABLE 17.6

Emulated WAN Conditions

S/N	Network Condition	Parameters
1	Latency	(100:50:1000) ms
2	Jitter	10% of latency
3	Packet loss	0.1%–10.0%
4	Noise	0.1%–10.0%

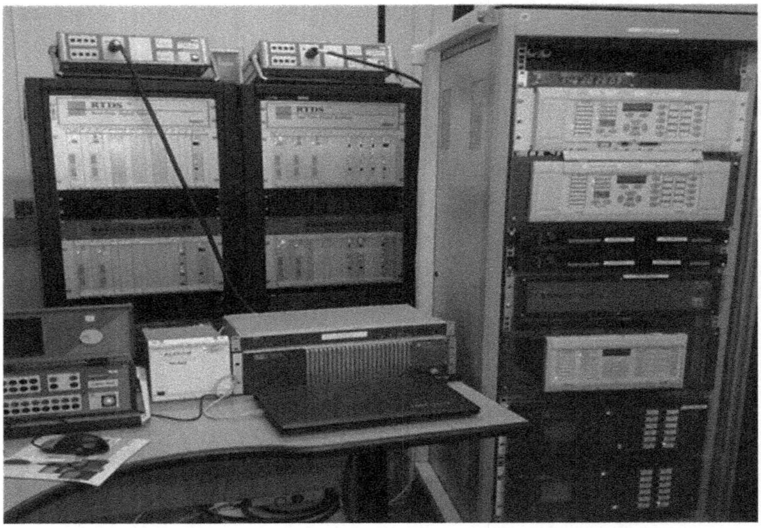

FIGURE 17.18 Real-time co-simulation platform.

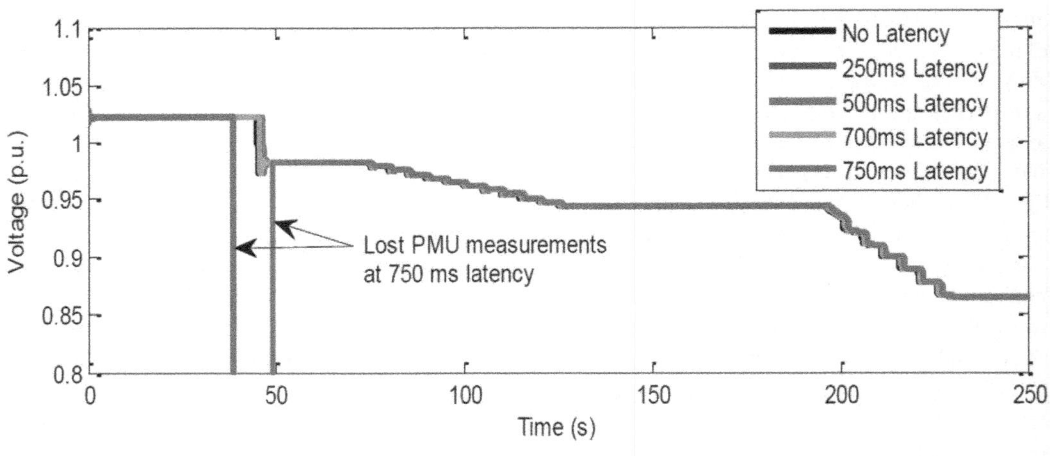

FIGURE 17.19 WAN emulation of network latency of 0–750 ms.

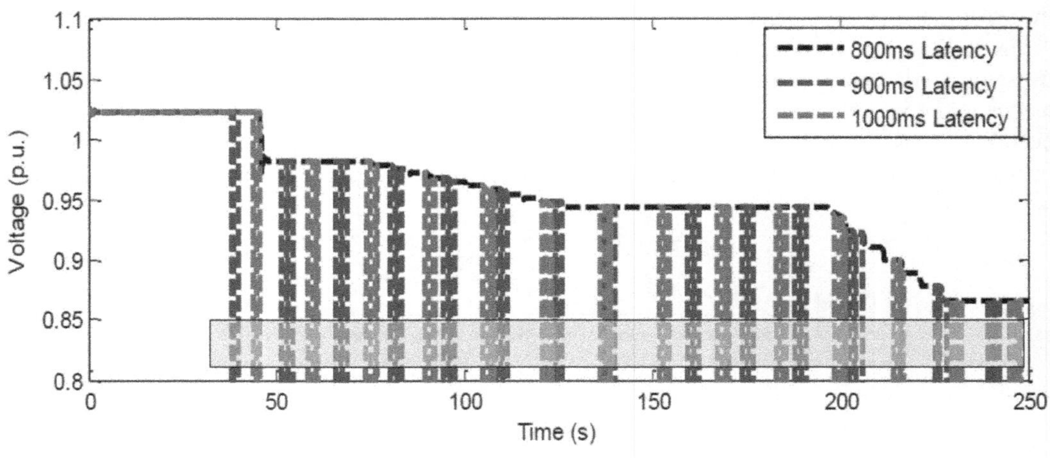

FIGURE 17.20 WAN emulation of network latency of 800–1000 ms.

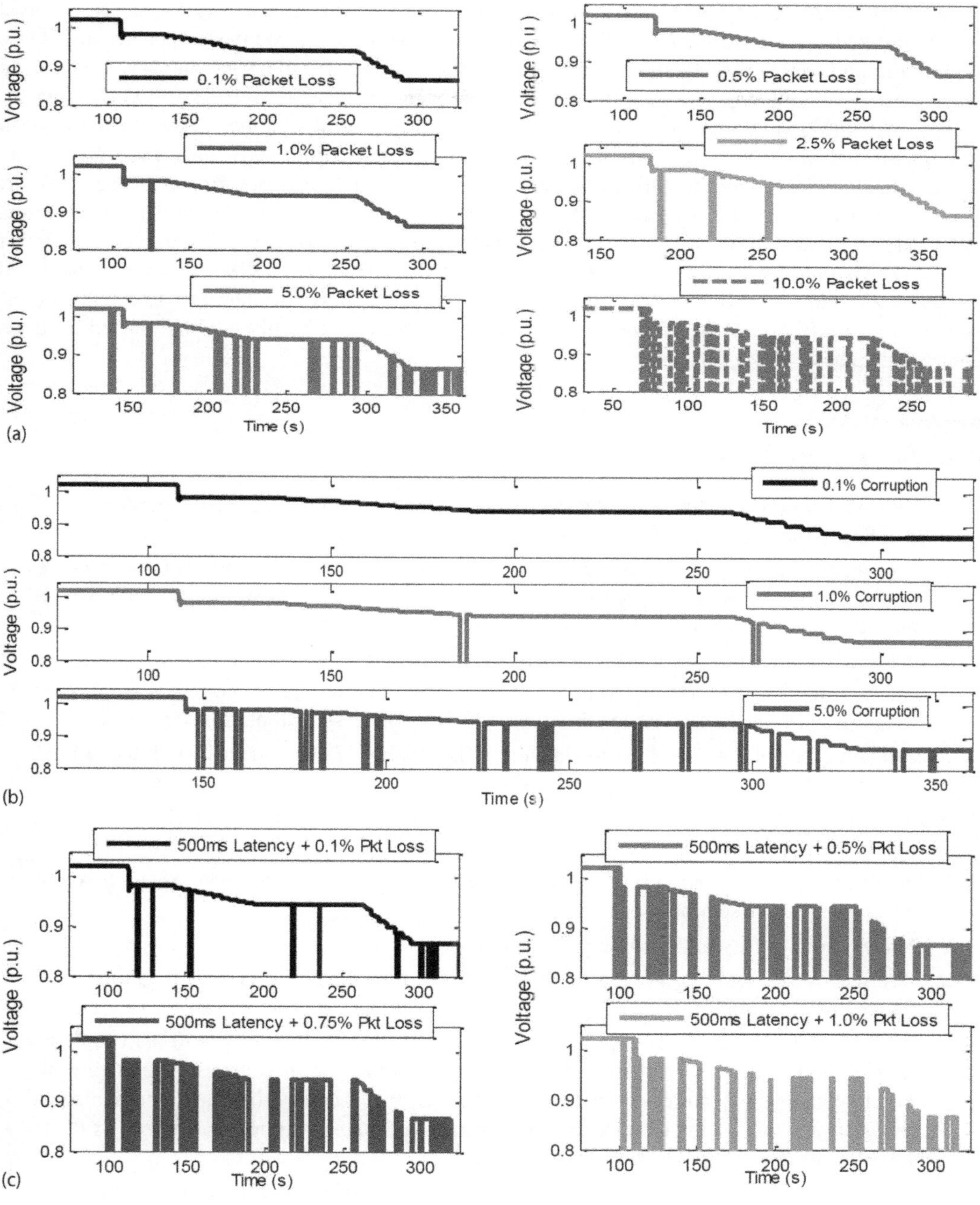

FIGURE 17.21 Emulated WAN conditions for (a) bus 8 voltages for various packet losses, (b) bus 8 voltages for corruption, and (c) bus 8 voltages for various combinations of adverse conditions.

was acceptable and did not have any adverse effect on the synchrophasors published from the substations to the control centre, as shown in Figure 17.21b. Figure 17.21c shows that, for a latency of 500 ms, the synchrophasor measurements could only tolerate a packet loss of about 0.1%. Beyond a packet loss of 0.1% and a latency of 500 ms, the lost packets increased and the network was severely degraded. This implies that the communication network implemented for this particular WAMPAC applications could only support a latency of 500 ms with ±10% jitter, 0.1% packet loss, and 1.0% random noise, respectively. Beyond this, the network degraded to an unacceptable level. Therefore, if a cyber attacker introduces adverse communication network conditions beyond that given above, it could result in the failure of the implemented SIPS.

Cyber attacks similar to the one presented above can be countered by using a detection algorithm based on the threshold of the maximum typical delay or adverse conditions in a communication network. Once the prevailing condition exceeds these quantified values, the detector will trigger an alarm and initiate logging, bad data or missing data algorithms, or other pre-planned countermeasures against cyber attack.

Also, the MitM cyber attack can be detected using access control whitelists. These whitelists can be implemented in the datalink layer using the source and destination MAC addresses, source and destination IP addresses in the network layer, and the source and destination ports in the transport layer, respectively. If access is requested by a host that is not in the corresponding whitelist, alarms and access blocking can be issued. Furthermore, public key encryption can be used to combat MitM attacks. This would involve the exchange of public key infrastructure (PKI) between the client and the server. In addition, MitM cyber attacks can be prevented in synchrophasor-based applications by retaining the IP and MAC addresses of legitimate devices, and by disenabling unused ports on Ethernet switches.

17.8 Tutorial Problems

1. With the aid of a diagram, describe the signal processing steps of the PMU reference model specified in the IEEE C37.118.1-2011 standard.
2. Draw the power-angle curve and discuss the condition required for the stability of two machines connected through series capacitive reactance. Explain how the power-angle curve can be implemented and used in real time for transient stability monitoring.
3. List five components and describe five benefits of synchrophasor-based systems.
4. State four existing protocols used in WAMS for substation-to-control centre communication. Compare and contrast such protocols with the IEEE C37.118 synchrophasor standard.
5. State the types of message formats defined in the IEEE C37.118 standard. With the aid of a diagram, describe the typical communication that takes place between a PMU and a PDC.
6. List the two classes of synchrophasor measurements in the IEEE C37.118.1 standard and explain their applications.
7. List five tasks carried out during post-mortem analysis of disturbances, and explain how synchrophasors aid these processes.
8. Define cyber security in the power system context.
9. Describe five critical synchrophasor infrastructures that should be protected against cyber attacks in electric power systems.
10. For the four-machine two-area power system network shown in Figure 17.10, an oscillation damping and frequency estimation algorithm required the phasor voltages, phasor currents, and three analogue quantities (apparent, real, and reactive powers) from the four machines in the system and the two interconnecting buses in each of the areas. Allow for a total of six PMUs and calculate the following:
    a. The bandwidth requirement for the six PMUs publishing three-phase currents and voltages, three analogue quantities, system frequency, and ROCOF (in polar and integer formats).
    b. The bandwidth requirement for the six PMUs publishing positive-sequence current and voltage, three analogue quantities, system frequency, and ROCOF (in rectangular and real formats).

17.9 Conclusion

This chapter presented the advances made in WAMPAC using synchrophasor measurements from PMUs. Synchrophasor components and benefits, standards, message structure and formats were presented and discussed. Also, system planning requirements with respect to the optimal placement of PMUs, communication architecture, and bandwidth requirements were presented. Various applications for real-time monitoring, grid visualization, oscillation detection and estimation, real-time transient stability monitoring, voltage stability monitoring and assessment, state estimation, model validation, and post-mortem analysis were further elucidated. System protection using synchrophasor-based SIPS was also considered. Lastly, cyber security threats and vulnerability in WAMPAC schemes, relevant cyber security standards, and an example of a cyber security MitM attack were presented and discussed.

REFERENCES

1. Kundur, P. 1994. *Power System Stability and Control*, New York: McGraw-Hill.
2. Phadke, A. G., Thorpe, J. S., and Adamiak, M. G. 1983. A new measurement technique for tracking voltage phasors, local system frequency, and rate of change of frequency. *IEEE Transactions on Power Apparatus and Systems*, vol. PAS-102, no. 5, pp. 1025–1038.
3. Overholt, P., Ortiz, D., and Silverstein, A. 2015. Synchrophasor technology and the DOE: Exciting opportunities lie ahead in development and deployment, *IEEE Power and Energy Magazine*, vol. 13, no. 5, pp. 14–17.
4. Aminifar, F., Fotuhi-Firuzabad, M., Safdarian, A., Davoudi, A., and Shahidehpour, M. 2015. Synchrophasor measurement technology in power systems: Panorama and state-of-the-art. *IEEE Access*, vol. 2, pp. 1607–1628.
5. IEEE Standard C37.118.1™-2011, IEEE Standard for Synchrophasor Measurements for Power Systems, 2011.
6. IEEE Standard 1344™-1995, IEEE Standard for Synchrophasors for Power Systems.
7. IEEE Standard C37.118™-2005, IEEE Standard for Synchrophasors for Power Systems, 2005.
8. IEEE Standard C37.118.2™-2011, IEEE Standard for Synchrophasor Measurements for Power Systems, 2011.
9. IEEE Standard for Synchrophasor Measurements for Power Systems—Amendment 1: C37.118.1a-2014 Modification of Selected Performance Requirements.
10. IEC 61850-90-5:2012. Communication networks and systems for power utility automation-Part 90-5: Use of IEC 61850 to transmit synchrophasor information according to IEEE C37.118.
11. IEEE Guide for Synchronization, Calibration, Testing, and Installation of Phasor Measurement Units (PMUs) for Power System Protection and Control: IEEE C37.242-2013.
12. IEEE Guide for Phasor Concentrator Requirements for Power System Protection, Control, and Monitoring: IEEE C37.244-2013.
13. IEEE Draft Standard for Use of IEEE 1588 Precision Time Protocol in Power System Applications: IEEE PC37.238/D21.1, 2016.
14. Krumpholz, G. R., Clements, K. A., and Davis, P. W. 1980. Power system observability: A practical algorithm using network topology. *IEEE Transactions on Power Apparatus and Systems*, vol. 99, no. 4, pp. 1534–1542.
15. Arefi, A., Hagfiham, M. R., and Fathi, S. H. 2011. Observability analysis of electric networks considering branch impedance. *International Journal of Electrical Power and Energy Systems*, vol. 33, pp. 954–960.
16. Nuqui, R. F., and Phadke, A. G. 2005. Phasor measurement unit placement techniques for complete and incomplete observability. *IEEE Transactions on Power Delivery*, vol. 20, no. 4, pp. 2381–2388.
17. Peng, J., Sun, Y., and Wang, H. F. 2006. Optimal PMU placement for full network observability using Tabu search algorithm. *International Journal of Electrical Power and Energy Systems*, vol. 28, pp. 223–231.
18. Hajian, M., Ranjbar, M., Amraee, T., and Mozafari, B. 2011. Optimal placement of PMUs to maintain network observability using a modified BPSO algorithm. *International Journal of Electrical Power and Energy Systems*, vol. 33, pp. 28–34.
19. Cepeda, J. C., Ramirez, D., and Colome, D. G. 2014. Real-time adaptive load shedding based on probabilistic overload. *Proceedings IEEE PES Latin America Transmission and Distribution Conference and Exposition*, September 10–13, 2014, pp. 1–6.
20. Mili, L., Baldwin, T., and Adapa, R. 1990. Phasor measurement placement for voltage stability analysis of power systems. *Proceeding of the 29th Conference on Decision and Control*, Hawaii, pp. 3033–3038.
21. Xu, B., and Abur, A. 2004. Observability analysis and measurement placement for systems with PMUs. *IEEE PES Power Systems Conference and Exposition*, vol. 2, October 10–13, pp. 943–946.

22. Gou, B. 2008. Optimal placement of PMUs by integer linear programming. *IEEE Transactions on Power Systems*, vol. 23, no. 3, pp. 1525–1526.
23. Adewole, A. C., and Tzoneva, R. 2014. Optimal PMU placement for voltage stability assessment in power system networks. *International Review of Electrical Engineering (IREE)*, vol. 9, no. 4, pp. 811–820.
24. IEEE Power Systems Test Case Archive, http://www.ee.washington.edu/research/pstca/, accessed January 13, 2014.
25. Hauer, J. F., Demeure, C. J., and Scharf, L. L. 1990. Initial results in Prony analysis of power system response signals. *IEEE Transactions on Power Systems*, vol. 5, no. 1, pp. 80–89.
26. Trudnowski, D. J., Smith, J. R., Short, T. A., and Pierre, D. A. 1991. An application of Prony methods in PSS design for multimachine systems. *IEEE Transactions on Power Systems*, vol. 6, pp. 118–126.
27. Hua, Y., and Sarkar, T. K. 1990. Matrix pencil method for estimating parameters of exponentially damped/undamped sinusoids in noise. *IEEE Transactions on Acoustic, Speech and Signal Processing*, vol. 38, no. 5, pp. 814–824.
28. Kamwa, I., Grondin, R., Dickinson, J., and Fortin, S. 1993. A minimal realization approach to reduced-order modeling and modal analysis for power systems response signals. *IEEE Transactions on Power Systems*, vol. 8, no. 3, pp. 1020–1029.
29. Pierre, J. W., Trudnowski, D. J., and Donnelly, M. K. 1997. Initial results in electromechanical mode identification from ambient data. *IEEE Transactions on Power Systems*, vol. 12, no. 3, pp. 1245–1251.
30. Peric, V. S., and Vanfretti, L. 2014. Power-system ambient-mode estimation considering spectral load properties. *IEEE Transaction on Power Systems*, vol. 29, no. 3, pp. 1133–1143.
31. Anderson, M. G., Zhou, N., Pierre, J. W., and Wies, R. W. 2005. Boot -strap-based confidence interval estimates for electromechanical modes from multiple output analysis of measured ambient data. *IEEE Transaction on Power Systems*, vol. 20, no. 2, pp. 943–950.
32. Zhou, N., Pierre, J., and Hauer, J. 2006. Initial results in power system identification from injected probing signals using a subspace method. *IEEE Transaction on Power Systems*, vol. 21, no. 3, pp. 1296–1302.
33. Messina, A. R., Vittal, V., Ruiz-Vega, D., et al. 2006. Interpretation and visualization of wide-area PMU measurements using Hilbert analysis. *IEEE Transactions on Power Systems*, vol. 21, no. 4, pp. 1761–1763.
34. Rueda, J. L., Juárez, C. A., and Erlich, I. 2011. Wavelet-based analysis of power system low-frequency electromechanical oscillations. *IEEE Transactions on Power Systems*, vol. 26, no. 1, pp. 260–268.
35. Kamwa, I., Pradhan, A.K., and Joos, G. 2011. Robust detection and analysis of power system oscillations using the Teager-Kaiser energy operator. *IEEE Transactions on Power Systems*, vol. 26, pp. 323–333.
36. IEEE PES Task Force, *Identification of Electromechanical Modes in Power Systems*: PES-TR14-2012.
37. Ning, J., Sarmadi, A. N., and Venkatasubramanian, V. 2015. Two-level ambient oscillation modal estimation from synchrophasor measurements. *IEEE Transactions on Power Systems*, vol. 30, no. 6, pp. 2913–2922.
38. Trudnowski, D. J. 2008. Estimating electromechanical mode shape from synchrophasor measurements. *IEEE Transactions on Power Systems*, vol. 23, no. 3, pp. 1188–1195.
39. Nabavi, S., Zhang, J., and Chakrabortty, A. 2015. Distributed optimization algorithms for widearea oscillation monitoring in power systems using interregional PMU-PDC architectures. *IEEE Transactions on Smart Grid*, vol. 6, no. 5, pp. 2529–2538.
40. Liu, G., and Venkatasubramanian, V. 2008. Oscillation monitoring from ambient PMU measurements by frequency domain decomposition. In *Proceedings of the IEEE Symposium on Circuits and Systems*, Seattle, WA, pp. 2821–2824.
41. Thambirajah, J., Thornhill, N. F., and Pal, B. C. 2011. A multivariate approach towards interarea oscillation damping estimation under ambient conditions via independent component analysis and random decrement. *IEEE Transactions on Power Systems*, vol. 26, no. 1, pp. 315–322.
42. Vanfretti, L., Peric, V., and Gjerde, J. O. 2013. Estimation of electromechanical oscillations in the nordic grid using ambient data analysis. Invited Paper, Panel Session on Identification of Electromechanical Modes in Power Systems, *IEEE PES General Meeting*, 2013.
43. Banga-Banga, T. P., Adewole, A. C., and Tzoneva, R. 2017. Oscillation mode estimation using spectrum analysis of synchronized phasor measurements. *Southern Africa Universities Power Engineering Conference (SAUPEC)*, Stellenbosch, South Africa, pp. 1–6.
44. Stanton, S. E. 1989. Transient stability monitoring for electric power systems using a partial energy function. *IEEE Transactions on Power Systems*, vol. 4, no. 4, pp. 1389–1396.

45. Liu, C., Su, M., Tsay, S., and Wang, Y. 1999. Application of a novel fuzzy neural network to real-time transient stability swings prediction based on synchronized phasor measurements. *IEEE Transactions on Power Systems*, vol. 14, no. 2, pp. 685–692.
46. Chow, J. H., Chakrabortty, A., Arcak, M., Bhargava, B., and Salazar, A. 2007. Synchronized phasor data based energy function analysis of dominant power transfer paths in large power systems. *IEEE Transactions on Power Systems*, vol. 22, no. 2, pp. 727–734.
47. Sun, S. K., Likhate, S., Vittal, V., Kolluri, V. S., and Mandal, S. 2007. An online dynamic security assessment scheme using phasor measurements and decision trees. *IEEE Transactions on Power Systems*, vol. 22, no. 4, pp. 1935–1943.
48. Yan, J., Liu, C., and Vaidya, U. 2011. PMU-based monitoring of rotor angle dynamics. *IEEE Transactions on Power Systems*, vol. 26, pp. 2125–2133.
49. Kimbark, W. K. 1995. Power system stability, Vol. 1, Elements of stability calculations. *IEEE Press Power Engineering Series*, New York: Wiley-IEEE Press.
50. Benmouyal, G., Schweitzer, E., and Guzman, A. 2004. Synchronized phasor measurement in protective relays for protection, control, and analysis of electric power system. *57th Annual Conference for Protective Relay Engineers*, pp. 419–450.
51. Ajjarapu, V. 2006. *Computational Techniques for Voltage Stability Assessment and Control*, New York: Springer.
52. Van Cutsem, T., and Vournas, C. 1998. *Voltage Stability of Electric Power Systems*, Boston, MA: Springer Science & Business Media.
53. Glavic, M., and Van Cutsem, T. 2009. Wide-area detection of voltage instability from synchronized phasor measurements. Part II: Simulation results. *IEEE Transactions on Power Systems*, vol. 24, no. 3, pp. 1417–1425.
54. Gu, W., and Wan, Q. 2010. Linearized voltage stability index for wide-area voltage monitoring and control. *Electric Power and Energy Systems*, vol. 32, pp. 333–336.
55. Vu, K., Begovic, M., Novosel, D., and Saha, M. M. 1999. Use of local measurements to estimate voltage-stability margin. *IEEE Transactions on Power Systems*, vol. 14, no. 3, pp. 1029–1035.
56. Julian, D. E., Schulz, R. P., Vu, K. T., Quaintance, W. H., Bhatt, N. B., and Novosel, D. 2000. Quantifying proximity to voltage collapse using the voltage instability predictor (VIP). *IEEE Power Engineering Society Summer Meeting*, vol. 2, pp. 931–936.
57. Warland, L., and Holen, A. T. 2002. Estimation of distance to voltage collapse: Testing and algorithm based on local measurements. *Proceedings of 14th Power System Computation Conference*, Sevilla, Spain, vol. 38, pp. 1–7.
58. Geurts, P., and Wehenkel, L. 2000. Temporal machine learning for switching control. *Proceedings of 4th European Conference of Principles and Practice of Knowledge Discovery in Databases (PKDD)*, Lyon, France, pp. 401–408.
59. Van Cutsem, T., Wehenkel, L., Pavella, M., Heilbronn, B., and Goubin, M. 1993. Decision tree approaches to voltage security assessment. *IEE Proceedings on Generation, Transmission & Distribution*, vol. 140, no. 3, pp. 189–198.
60. Nuqui, R. F., Phadke, A. G., Schulz, R. P., Richard, P., and Bhatt, N 2001. Fast on-line voltage security monitoring using synchronized phasor measurements and decision trees. *IEEE Power Engineering Society Winter Meeting*, vol. 3, pp. 1347–1352.
61. Diao, R., Sun, K., Vittal, V., O'Keefe, R. J., Richardson, M. R., Bhatt, N., Stradford, D., and Sarawgi, S. K. 2009. Decision tree-based online voltage security assessment using PMU measurements. *IEEE Transactions on Power Systems*, vol. 24, no. 2, pp. 832–839.
62. Adewole, A. C., and Tzoneva, R. 2014. Real-time deployment of a novel synchrophasor-based voltage stability assessment algorithm. *International Review of Electrical Engineering (IREE)*, vol. 9, no. 5, pp. 1021–1033.
63. Beiraghi, M., and Ranjbar, A. M. 2013. Online voltage security assessment based on wide–area measurements. *IEEE Transactions on Power Delivery*, vol. 28, no. 2, pp. 989–997.
64. Bahmanyar, A. R., and Karami, A. 2014. Power system voltage stability monitoring using artificial neural networks with a reduced set of inputs. *International Journal of Electric Power and Energy Systems*, vol. 58, pp. 246–256.
65. Suganyadevi, M. V., and Babulal, C. K. 2014. Fast assessment of voltage stability margin of a power system. *Journal of Electrical Systems*, vol. 10–13, pp. 305–316.

66. Adewole, A. C., and Tzoneva, R. 2016. Extended synchrophasor-based online voltage stability assessment using synchronous generator-derived indices. *International Transactions on Electrical Energy Systems*, vol. 27, no. 1, pp. 1–23.
67. Zhong, Z., Xu, C., Billian, B. J., Zhang, L., Tsai, S.-J.S., Conners, R. W., Centeno, V. A., Phadke, A. G., and Liu, Y. 2005. Power system frequency monitoring network (FNET) implementation. *IEEE Transactions on Power Systems*, vol. 20, no. 4, pp. 1914–1921.
68. Monticelli, A. 1999. Electric power system state estimation. *Proceedings of the IEEE*, vol. 88, no. 2, pp. 262–282.
69. Schweppe, F. C., and Rom, D. B. 1970. Power system static-state estimation, Part II: Approximate model. *IEEE Transactions on Power Apparatus and Systems*, vol. PAS-89, no. 1, pp. 125–130.
70. Phadke, A. G., Thotp, J. S., and Karimi, K. J. 1986. State estimation with phasor measurements. *IEEE Transactions on Power Systems*, vol. PWRS-1, no. 1, pp. 233–238.
71. Zhou, M., Centeno, V. A., Thorpe, J. S., and Phadke, A. G. 2006. An alternative for inclusive phasor measurements in state estimators. *IEEE Transactions on Power Systems*, vol. 21, no. 4, pp. 1930–1937.
72. Sodhi, R., Srivastava, S. C., and Singh, S. N. 2010. Phasor-assisted hybrid state estimator. *Electric Power Components and Systems*, vol. 38, pp. 533–544.
73. Kirincic, V., Skok, S., and Frankovic, D. A. 2016. Synchrophasor assisted hybrid state estimator. *Journal of Electrical Engineering*, vol. 67, no 2, pp. 103–110.
74. Haughton, D. A., and Heydt, G. T. 2013. A linear state estimation formulation for smart distribution systems. *IEEE Transactions on Power Systems*, vol. 28, no. 2, pp. 1187–1195.
75. Regulski, P., Valverde, G., Rebizant, W., and Terzija, V. 2009. Estimation of load model parameters for long-term voltage stability assessment studies. In *Proceedings of the International Conference on Advanced Power System Automation and Protection*, Seoul, South Korea.
76. Choi, B. K., Chiang, H. D., Li, Y., Li, H., Chen, Y. T., Huang, D. H., and Lauby, M. G. 2006. Measurement-based dynamic load models: Derivation, comparison, and validation. *IEEE Transactions on Power Systems*, vol. 21, no. 3, pp. 1276–1283.
77. Zhou, M., Centeno, V. A., Thorp, J. S., and Phadke, A. G. 2012. Calibrating instrument transformers with phasor measurements. *Electrical Power Components and Systems*, vol. 40, no. 14, pp. 1605–1620.
78. Kyriakides, E., Heydt, G. T., and Vittal, V. 2004. On-line estimation of synchronous generator parameters using a damper current observer and a graphic user interface, *IEEE Transactions on Energy Conversion*, vol. 19, no. 3, pp. 499–507.
79. Liao, Y. 2009. Fault location observability analysis and optimal meter placement based on voltage measurements. *Electric Power Systems Research*, vol. 79, pp. 1062–1068.
80. Ren, J., Venkata, S. S., and Sortomme, E. 2014. An accurate synchrophasor based fault location method for emerging distribution systems. *IEEE Transactions on Power Delivery*, vol. 29, no. 1, pp. 297–298.
81. Xu, G., and Vittal, V. 2010. Slow coherency based cutset determination algorithm for large power systems. *IEEE Transactions on Power Systems*, vol. 25, no. 2, pp. 877–884.
82. Sun, K., Hur, K., and Zhang, P. 2011. A new unified scheme for controlled power system separation using synchronized phasor measurements. *IEEE Transactions on Power Systems*, vol. 26, no. 3, pp. 1544–1554.
83. Franco, R., Sena, C., Taranto, G. N., and Giusto, A. 2013. Using synchrophasors for controlled islanding: A prospective application for the Uruguayan power system. *IEEE Transactions on Power Systems*, vol. 28, no. 2, pp. 2016–2024.
84. Koellner, K., Anderson, C., and Moxley, R. 2007. Generator black start validation using synchronized phasor measurement, *60th Annual Conference for Protective Relay Engineers*. pp. 498–504.
85. Moors, C., Lefebvre, D., and Van Cutsem, T. 2000. Design of load shedding schemes against voltage instability. *Proceedings. IEEE Power Engineering Society Winter Meeting*, vol. 2, pp. 1495–1500.
86. Horowitz, S., Novosel, D., Madani, V., and Adamiak, M. 2008. System-wide protection: Wide-area measurements for mproved power system protection. *IEEE Power & Energy Magazine, September/October*, pp. 34–42.
87. Madani, V., Novosel, D., Horowitz, S., Adamiak, M., Amantegui, J., Karlsson, D., Imai, S., and Apostolov, A. 2010. IEEE PSRC report on global industry experiences with system integrity protection schemes (SIPS). *IEEE Transactions on Power Delivery*, vol. 25, no. 4, pp. 2143–2155.
88. IEEE Std 1402™, IEEE Guide for Electric Power Substation Physical and Electronic Security.

89. IEC 62351, Power systems management and associated information exchange—Data and communication security.
90. NERC, Critical infrastructure protection (CIP), http://www.nerc.com/page.php?cid=2, accessed November 20, 2016.
91. IEEE Standard 1686™, IEEE Standard for Intelligent Electronic Devices Cybersecurity Capabilities 1.
92. NISTIR 7628 Guidelines for Smart Grid Cyber Security.
93. IEEE Standard Cybersecurity Requirements for Substation Automation, Protection, and Control Systems: IEEE C37.240-2014.
94. Adewole, A. C., and Tzoneva, R. 2017. Co-simulation platform for integrated real-time power system emulation and wide area communication emulation. *IET Generation, Transmission, & Distribution*, vol. 11, no. 12, pp. 3019–3029.
95. Adewole, A. C., Tzoneva, R., and Apostolov, A. 2016. Adaptive under-voltage load shedding scheme for large interconnected smart grids based on wide area synchrophasor measurements. *IET Generation, Transmission, & Distribution*, vol. 10, no. 8, pp. 1957–1968.

18

Protection of Renewable Distributed Generation System

Rishabh Dev Shukla, Ramesh K. Tripathi, Padmanabh Thakur, and Ramesh Bansal

CONTENTS

18.1 The Impact of Renewable Distributed Generation and Distributed Generation Penetration on Protection and Current Protection Practices

Distributed generations (DGs) can be described simply as small-scale electricity generations co-located with the loads. The distributed generation resources (DGRs) are co-generation (simultaneous production of heat and electricity) [1]. They can be connected on the customer side or to the distribution network. Initially, distribution systems were designed to operate near the load centre in stand-alone or isolated mode; that is, they are not connected to the utility grid or generation at the distribution level. In fact, the introduction of renewable generating sources to the distribution system is called renewable distributed generation (RDG). Generally, in RDG application, the customer consumes the total output generated from the RDG, and any excess power is delivered to the distribution network if the system is grid connected. If the customer demands more power than is available at RDG, additional power is supplied from the distribution network. It can cause a considerable impact on the network power flow, voltage conditions at various utility consumer equipment and fault ratings of switchgear. The harshness of the impact depends on the location and penetration level (or size) of the generating sources. RDG has turned out to be more common in power systems around the world as an alternative to constructing large generation units and building new transmission lines when highly reliable power is required. Due to technological advancement and changing economics and regulatory environment with electricity market liberalization, RDG has turned out to be more popular as the cost of small generation is decreasing day by day [2]. New policies related to climate change have engendered more attention in renewable energy application and the proficient use of cheap energy opportunities [3]. New generation systems should be planned and existing systems should be upgraded to allow RDG connections with non-violation of the allowable technical limits/standards towards phase imbalance, fault current limit, voltage variation etc. In order to decrease emissions of greenhouse gases from conventional plants, using distributed renewable sources for producing power seems to be a good choice [4]. DGs are positioned near the load centre, and properly positioned DGs in the distribution network also decrease in power losses. Only condensing distribution system/network losses add 1% to total greenhouse gas emissions [5]. The deployment in renewable energy technologies is leading RDGs into utilities frequently called hybrid systems. Hybrid systems consist of several generators connected in what is called a microgrid, as shown in Figure 18.1.

In a microgrid, the loads and microsources (wind, photovoltaic (PV), solar thermal, small hydro, biomass, etc.) are connected together and function as a single unit, with the possibility of producing both heat and power. The power electronic-based inverter is essentially part of majority of the micro-sources to make available the essential flexibility to assure function as a single aggregated unit [1,5]. For integrating renewable sources, a microgrid is more suitable, with the grid also being utilized in high reliability systems [6]. Large capacity RDG applications require comprehensive protection studies and impact assessments as they are considered similar to large generators associated with the transmission system of the utility. Smaller systems (5 MW or lesser) are incorporated into the utility protection system and are typically associated with the distribution and subtransmission systems. Adding RDG brings two aspects for protection necessities: those of the generation owners and the utility. The generator must be protected from abnormal situations and short circuits that

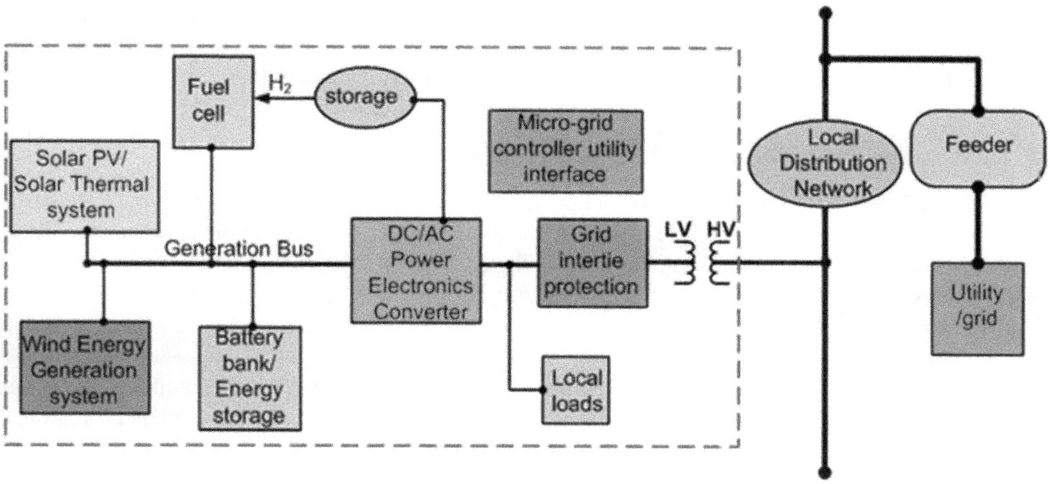

FIGURE 18.1 Typical configuration of renewable distributed generation connected to the utility.

might consequently damage it. Aberrant conditions can be enforced by the utility system on the RDG; these conditions include overvoltage, overexcitation, abnormal frequency, unbalanced currents, and shaft torque stress due to automatic reclosing of the utility breaker. The main concern of the utility is that RDG may cause considerable damage in the associated system and/or for their customers due to changes in the existing protection scheme or redundant fault current. The IEEE 1547 standard, about the interconnection of distributed resources with the power systems or grid, gives very limited authentic guidance and key requirements. Advanced standards are being developed in view of the necessity of integrating RDG with the distribution system.

18.1.1 Effect of RDG on the Power System/Grid/Distribution Level

The main effects of RDG on the power system or grid are: increased short circuit levels, change in load losses, change in voltage profile along the network, appearance of voltage transients, congestion may be noticeable in system, reliability, power quality can possibly be affected and the protections may not work appropriately. The close proximity to the consumer loads is one of the main advantages of RDG. RDG can play a key role in enhancing the reliability of the grid, reducing transmission losses, giving better voltage support and enhancing the power quality. In RDGs, the output is regulated in accordance with the variation in local load. Also, power output of the RDG can be regulated independently of the area under local loading. The implementation of this mode of control is dependent upon RDG operation, which follows price trends or the availability of natural recourses, like solar or wind power, and which might or might not correspond to the local load variations.

18.1.2 RDG/DG Penetration on Protection

The insertion of renewable generating sources advances the attributes of distribution system and has a significant impact on different technical parameters that are dependent upon its size and position in the network/system. The RDG can introduce considerable impact on the network power flow, voltage condition at different utility consumer equipment and fault ratings of the switchgear. The impact harshness depends on the site or position and penetration level or size of the generating sources [5]. The penetration level impact is realized by introducing the RDGs at the various weak nodes of the network [6]. By installing RDGs, the distribution system fault level will become higher due to the fault current contribution of RDG. The increased fault level may cause failure or malfunction between the circuit breaker (CB) and fuse, which will affect the reliability and safety of the distribution system [7]. The increased size or penetration level has a positive impact on the voltage profile and losses, i.e., reduced power losses, and enhances voltage profile of the system. This is due to the fact that the DGs are closer to the loads and avoid the huge power flow from substation to loads through several branches [5,8].

18.1.3 Current Protection Practices and Trends

The IEEE 1547 standard gives a common set of rules and regulations for interconnection of DRs/DGs with electric power systems. It supplies the requirements applicable to the operation, performance, safety considerations and protection, testing, and care of the connection [9]. This set of rules and regulations only provides partial real guideline and highlights only the important necessities. A new set of rules and regulations are being developed to cover further comprehensive necessities for the incorporation of DG with the distribution/power system. The present standard lacks the following: (1) no overvoltages or loss of utility relay synchronization, (2) detachment when no longer operating in parallel with the utility/grid, (3) the utility/grid remains un-excited when the utility/grid is de-energized, (4) unintentional islanding is not permitted, (5) relays are "utility graded," (6) no objectionable harmonics, (7) no loss of synchronization (flicker should be within the limit).

RDG protection is accepted from the point of common coupling (PCC) and the interconnection transformer. To protect the grid from the RDG on the grid-side during parallel operation, interconnection protection is used. It can be located on either side of the interconnection transformer, i.e., on the primary side, as illustrated in Figure 18.2, or on the secondary side, as shown in Figure 18.3. Usually, the utility establishes the requirements of interconnection protection, which can include the following: (1) disconnection of the RDG/generator on detection of an islanding condition (i.e., no longer functioning in parallel with the utility/grid); (2) utility protection from the burden originated by the RDG (i.e., fault current, transient overvoltages, etc.); (3) RDG protection from the burden originated by the utility (automatic reclosing).

Protecting the generator is introduced at the generator side of the PCC, and it protects the RDG/DG from aberrant operating conditions and internal faults. Normally, to protect the equipment, each RDG/DG would have its own protection placed at the terminals of the generator, as shown in Figure 18.4. In general, the main focus of this protection is to provide exposure of RDG aberrant operating conditions (reverse power, loss of field, unbalanced currents, and overexcitation) and internal short circuits.

In some cases, the utility may need RDG installations to accommodate particular areas. The utility may make use of detailed requirements available in the following areas [2–5]: (1) the interconnection of

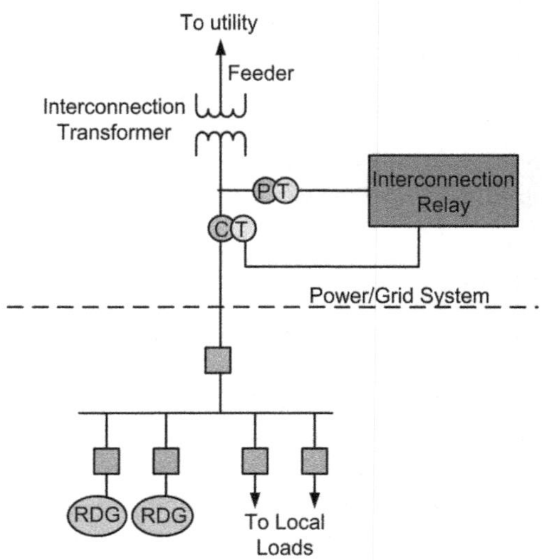

FIGURE 18.2 Protection interconnection at the secondary.

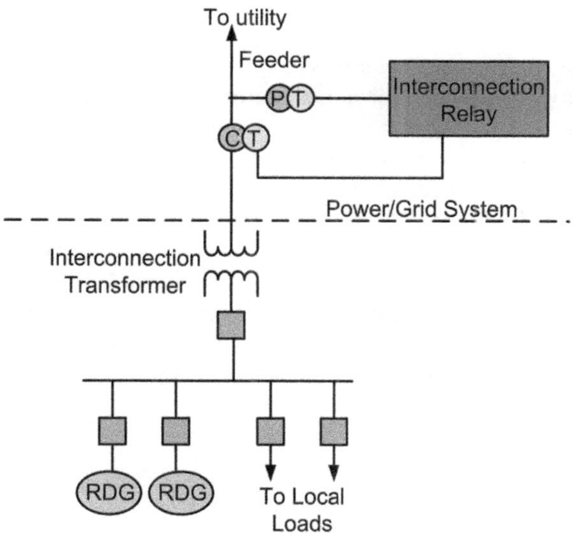

FIGURE 18.3 Protection interconnection at the primary.

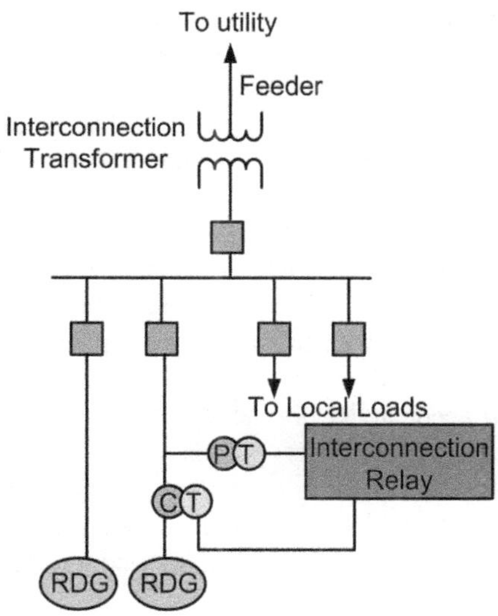

FIGURE 18.4 Generator side protection of RDG.

transformer winding configuration, (2) utility grade interconnection relays with common necessities, (3) requirements of current transformer (CT) and potential transformer (PT), (4) requirements of practical protection, and (5) operational speed.

18.2 Factors Affecting RDG/DG Protection

Various energy resources may be used; however, the interfacing method has been shown to have a major impact on the distribution system protection. A direct coupling-based scheme of rotary machines, such as squirrel cage induction generators or synchronous, is different from those that are connected through a power electronic converter. Also, the interconnection transformer type will have a large impact on the protection methods.

18.2.1 RDG/DG Types

Generators used in RDG/DG systems can be categorized into conventional and nonconventional generators. The generators based on combustion engines are called traditional generators and they are further divided into the following: microturbines (MTs), diesel engines, reciprocating type engines, and low speed turbines. Renewable machines such as PV devices, wind, small hydro, electrochemical machines such as fuel cells, and storage machines such as flywheels; batteries are categorized as nontraditional generators. The power electronic converter (inverter) for grid interfacing is a main part of the system used by the various nonconventional sources, for instance, PV systems. Synchronous generators are directly integrated to the utility grid. In addition, wind turbines, which are essentially based on induction generators, can be connected directly to the utility grid. In various applications, such as with synchronous machines, wind and micro turbines, utilization of power semiconductor devices for grid integration is beneficial with justification of the additional cost and complexity.

18.2.2 Voltage Regulation and Losses

The introduction of RDG in distribution systems has been shown to have an impact on voltage regulation. RDG/DG can support voltage in numerous cases. However, RDG can also cause an overvoltage or an undervoltage condition in the system. The origination of an overvoltage or undervoltage condition depends on different variables, including size and location of the RDG, load and line characteristics of the distribution system, and voltage regulation method [10]. The voltage level is lower with distance from the generator or transformer due to the unidirectional design of the traditional distribution network. These reductions are predictable and they are considered in the design of the network and network protection in order to ensure that the voltage is within tolerable limits in all normal conditions [4]. The current flows are altered or even reversed if an RDG/DG unit is connected, and the voltage will normally increase in an unpredictable manner. At low voltage levels, the prerequisites to meet standard voltage limits restricts the DG capacity that is connected to the system [11]. The voltage drop across the feeder is reduced due to reduction in the quantity of power that must be supplied from the substation because of the RDG installation. In this case, the RDG produces additional power than the local demand, and the differential power will flow reversed or upstream (i.e., towards the substation). This may result in an overvoltage and will conquer the voltage drop originated by the reactive power flow, if this reverse power flow is significant. Conversely, if adequate RDG is added near the substation, it can direct the load-tap-changing (LTC) transformer to activate at a lower secondary voltage, and this can cause an undervoltage towards the feeder end. There are various methods presented in the literature by which an overvoltage or undervoltage situation caused by RDG/DG can be compensated [12–14]. Few examples include introduction of voltage regulators in the line, adjustment in the sending end voltage, DG constraining, and network upgrade.

18.2.3 Power Quality Issues (Harmonics, Voltage Flicker, Unbalancing, etc.)

Various RDGs/DGs have different characteristics and hence they produce various power quality issues. Adding an RDG often leads to enhanced quality owing to an effect of increased network fault levels. A remarkable exception is that a single large RDG, such as wind turbine, on a weak network/system might cause power quality issues mostly during starting and stopping. Also increased use of power electronic devices and modern controls introduces power quality issues [15]. The RDG/DG connection to the network can affect the harmonics voltage distortion level, which depends on whether it is a power electronics converter or a rotating machine. Using power electronics interfaces gives advanced system support possibilities, but they will introduce harmonics currents into the system. Undue voltage harmonic levels may happen locally or elsewhere in the grid, depending on the system topology and the connection point impedance. Also, a rotating generator can introduce harmonics caused by the winding design and nonlinearities in the core. The significance of this condition depends on the layout of the grid and particularly on the details of the RDG/DG [4,11]. Various RDGs contribute to the network in single-phase only; this will create system unbalancing and it should be within the acceptable limit. Furthermore, RDG operation suffers from load unbalancing in different phases. This deteriorates the performance and has an adverse effect on the protection.

18.2.4 Short Circuit Levels

The appropriate synchronization of relays, fuses, re-closers and additional overcurrent equipment must be derived from the existing fault current. The fault current rating depends on the connection of the DR transformer. RDG infeeds, which are associated through the power electronic circuits, generate no significant concern on the network short circuit current and can be ignored on the side of protection [16]. The protection scheme should be intelligent enough to differentiate between faults that happen near the PCC and those that happen far from the PCC (i.e., on the utility side). Any utility side fault and fault within the RDG system require proper detection and the suitable protection schemes and/or devices with properly functioning disconnection equipment. If the RDR/DR (or combined RDRs/DRs) is capable enough compared to the utility substation source, it may produce a considerable impact on the utility substation fault current. It may produce failure to trip, co-ordination problems or sequential tripping [17]. The protection scheme may not identify the fault current if (1) it is subjected to phase-ground fault and a primary ground current source is not provided by RDG interconnection, (2) the single-phase RDG is coupled to a nonfaulty phase, (3) the self-quenching fault from the opening of the high-current utility source and the fault current falls to a low value sourced by the RDG.

The selection of the protection scheme depends on network cost and characteristics. Whenever a fault occurs in a system, islanding the distribution system with the same protection setting seems a better choice provided that the total load is less than generation for all time. Conversely, an adaptive protection scheme with two settings (i.e., one is applicable for grid connected and another for island conditions) and a situation detection technique that selects the setting can be used if there are not several plug and play generators. Sometimes the total load can be higher than generation. Adaptive protection with various settings can be used by means of communication to choose a suitable setting in case there are several plug and play generators and the fault power continuously changes.

18.2.5 Transformer Interface and Grounding

Single line to ground (SLG) fault is the most common in a power system. Different from the three-phase fault, the SLG fault current relies on the grounding. Commonly used grounding types in this power system are low impedance, isolated, and resonant groundings. In case of isolated grounding, large zero sequence capacitance of the lines is used for connection to ground. SLG fault currents are very small because the zero-sequence impedance (mostly capacitive in nature) is much larger than the short circuit positive-sequence impedance [18]. In the SLG case, the healthy phases see a rise in voltage. In case of resonant grounding, an inductor (referred to as Peterson-Coil) is used for neutral to ground connection; inductance is selected so that the capacitive component of the fault current is compensated to a small residual current. The largest fault current occurs in case of impedance grounding, and the voltage rise in the healthy phase is the least compared to other grounding for SLG fault.

The distribution system islanding may depart the RDG source without any grounding. In such a system, the SLG fault could result in customers, other than the faulty phase, supplied with much higher voltage that can increase up to 173%, at worst, or even more for an indefinite period [19]. This high voltage can produce significant damage in the utility and customer equipment [20]. Moreover, in case of unsymmetrical faults, negative-sequence currents are originated. The capability of machines/generators to survive with these fault currents for short durations depends on size. The smaller RDGs have less handling capability compared to larger machines/generators [21]. Thus, it is suggested that, for the island operation, the distribution system is grounded. Faults are cleared by using only overcurrent protection if effective grounding is provided and thus the requirement of other forms of protection against the earth faults (such as zero-sequence protection) can be avoided. Grounding of the ungrounded distribution

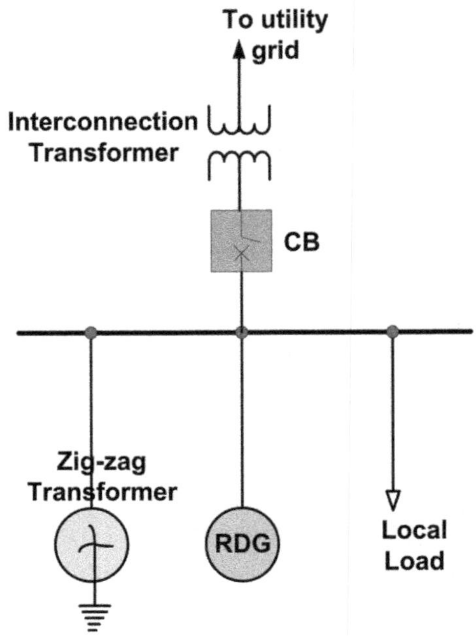

FIGURE 18.5 A zig-zag transformer grounding of an islanded distribution system.

system can be done through a grounding transformer which typically has a zig-zag or wye-delta windings configuration [22]. A simple, ungrounded, islanded distribution system grounded with a zig-zag transformer is illustrated in Figure 18.5.

18.2.6 Islanding and Unintentional Islanding

Islanding is the term to refer to the situation when the RDG is no longer functioning in parallel with the utility/power system or grid. It originates when part of the utility/power system, including generators and loads (with storage potential), becomes electrically isolated or disconnected from the main grid and enters a stand-alone or microgrid mode of operation. Islanding may occur due to the following conditions: (1) a fault that is only identified by the utility/power system, but not detected by the DG protection devices, and results in activating a disconnection of device; (2) incidental opening of the usual utility/grid supply by equipment malfunction; (3) switching of the utility for the distribution system and loads; (4) intentional disconnection for servicing; (5) human error; and (6) an act of the nature. The serious issue with island operation is that the DG works in its usual mode of operation (i.e., grid-connected operation) even when association with the utility/power system has been lost. In this situation the disconnection device is not activated leaving the microgrid connected to the utility/power system. This sustained function under inappropriate conditions is called unintended islanding. Normally, unintended islanding through distributed energy resources (DERs) outside utility control is undesirable for many reasons, including its capability to cause damage to consumer, system or DER equipment, and the probability for dangerous hazards.

18.2.7 Reliability and Intentional Islanding

As renewables are increasingly integrated into grid/power systems with the realization of smart grids, DG penetration is expected to increase sharply, and islanding can be taken as an opportunity to improve the power supply reliability. In addition, islanding can be economically advantageous [23,24]. By additional sale, islanding operation can increase the DG owners' revenue. As per custom prospective, it can reduce the frequency and interval of interruptions resulting from outages in the distribution and transmission systems [24]. Islanding operation of a distribution system with DG is a feasible choice provided the different issues with regard to islanded operation are appropriately addressed. Some important issues with islanding are state/situation detection (i.e., system is islanded or grid connected), load control, control of voltage and frequency and protection.

A control technique should be activated to handle both the operation and protection of the stand-alone or microgrid mode of operation after the identification of islanding (securely detached from utility/grid). Once islanding has been recognized, the process of auto-reclosing may happen to again connect the island to the main utility/power system. After the island has effectively reconnected to the grid, the control technique must turn back to grid-connected operating mode to manage operation and protection. The protection equipment operation requires different reactions during parallel operation with the utility and when functioning as a stand-alone or microgrid island mode because the fault current significantly changes between these two modes and impacts protection techniques that are short-circuit sensing based. The RDG/DG protection technique requires the following: (1) alternative setting groups of numerical relays for island operation, (2) an adaptive protection method suitable for both operating modes based communication practices, (3) time-delayed protection for undervoltage if traditional overcurrent protection will not function, (4) islanded system re-synchronizing, and (5) the expectation that the utility breakers/circuit reclosers will reconnect the island to the larger utility/power system when out of phase.

18.2.8 Load Shedding

The frequency of islanded distribution systems can be controlled by regulating the generator output power; i.e., if the frequency goes up, it can be regulated by reducing the output power of the generators, and vice versa. To increase revenue, however, the majority of RDG units function at maximum power point (MPP). PV generators employ MPP tracking, variable speed wind turbines to optimize power coefficient (Cp) in order to generate maximum power and joint power and heat plants are operated at maximum power. Thus, to maintain an islanding operation in underfrequency situations, with all RDGs operating at maximum power point, shedding some loads is the best way, and the technique is called underfrequency load shedding. It has been used in power systems for a long time. These techniques can mainly be categorized into static and adaptive or dynamic underfrequency load techniques. In static underfrequency load shedding, a fixed load at each frequency step is shed [25]. Conversely, in adaptive or dynamic load shedding, rate of change of frequency (RoCoF)–based load shed is used [26,27]. For islanded distributed systems, load shedding difficulty should be taken in a different path other than the large power system due to characteristic differences. Islanded distribution systems often have small generators with lower inertia and thus the frequency turns decay faster [28]. Therefore, it is very critical to shed loads quickly to stabilize the frequency in the islanded distribution systems. A dynamic underfrequency load shedding for an islanded distribution system is based on willingness of the customer to pay, RoCoF and load histories, as described in the literature [29]. When the islanded distribution system is effectively reconnected back to the grid or power system, the shed loads can be reconnected back to the system.

18.3 Protection with Islanding Operation (Anti-islanding)

Anti-islanding is defined as a feature of a grid tie converter/inverter that identifies the occurrence of a power outage and shuts itself down to stop electricity production. It is an essential requirement of the UL1741/IEEE1547 when the grid enters an islanded situation. The RDGs/DGs must identify islanding and instantly stop power feeding to the utility lines. Islanding is a condition in which a distribution system becomes electrically isolated from the rest of the power system and continues to be energized by the DG associated with it. At present, islanding is taken as a most important problem and all DGs need to shut down while a distribution system is islanded. According to the IEEE 1547-2003 standard, the DG is to be disconnected within 2 seconds of the detection of islanding [30]. Per the IEC 61727 standard, detection of the islanding and disconnection of DGs must be within 2 seconds [31]. The main reasons for avoiding islanding operation are: change in fault power level [32], reclosing when out of phase [33], voltage and frequency control [32], safety of line workers, need of proper grounding [19], etc. Numerous detection techniques for islanding have been developed [34,35].

18.3.1 Islanding Detection Method

Islanding detection is an area of much research and study. Generally, islanding detection techniques can be grouped into two broad categories: (1) remote or communication-based islanding technique and (2) local islanding techniques (active, passive, and hybrid). In the first category, the detection technique is located at the grid side, whereas in the second category, the detection technique is located at the inverter side. Local techniques can be further divided into: (1) passive ones, which are based on parameter measurement; (2) active ones, which generate disturbances at the inverter output; and (3) hybrid ones, which are the combination of both passive and active techniques. The following sections discuss some of the current islanding detection techniques [36,37].

18.3.2 Remote or Communication-Based Islanding Detection Technique

These techniques are mostly based on communication between DG units and the utilities. Generally, they use power line communication or Supervisory Control and Data Acquisition (SCADA) for the islanding detection. Conventionally, only power utility owned wires and channels subscribed from public telephone companies have been used. At present, radio transmitting (AM or FM) and optic fibres can be used. Communication-based techniques include: power line signalling (carrier communications), transfer trip, SCADA, and impedance insertion. In [38–41], power line, communication-based islanding, detection techniques are shown and the SCADA-based technique is presented in [42]. Although these techniques have improved reliability compared to others, they are quite complex and may be costly to implement, especially for small distribution systems [43]. In addition, transmitter problems may lead to failure in the entire system. Thus, other techniques, which include passive, active and hybrid techniques, are extensively used to detect islanding.

18.3.3 Passive Islanding Detection Technique

Passive islanding detection techniques require knowledge of system parameters like voltage, frequency, etc., which fluctuate greatly after islanding. Passive techniques basically check selected parameters such as voltage and frequency and/or their characteristics and cause the inverter to stop converting power when there is a change from normal specified conditions. Some passive islanding detection techniques presented in the literature are rate of change of frequency [44], rate of change of output power of DG [39], rate of change of frequency over power [45], harmonic distortion [46], voltage unbalance [47–49], data mining [50,51], wavelet and spectral analysis [52–54], and fuzzy logic based [55]. Even though passive methods are simple, the main problem with these techniques is islanding detection when the generation and load in an islanded distribution system closely match. The shortcomings of passive techniques can be improved by active techniques, which can detect islanding even in the case of a generator and demand a perfect match in an islanded distribution system.

18.3.4 Active Islanding Detection Technique

Active islanding detection techniques directly work together with the power system operation by introducing deliberate changes, disturbances, or perturbations. The main idea of active detection techniques is that a small disturbance or perturbation results in considerable change in the parameters of the system when a distribution system is islanded, at a time when the change is trivial and the distribution system is still connected to the grid. Some active islanding detection techniques presented in the literature are impedance measurement method [56], reactive power export error detection method [44], active frequency drift (AFD) [57], slip-mode frequency shift algorithm (SMS) [58], active frequency drift with positive feedback (AFDPF) [57], adaptive logic phase shift (ALPS) [59], automatic phase shift (APS) [60], and negative-sequence current injection [61]. These techniques introduce perturbations in the system at predefined intervals, even though it is not necessary in most operating conditions and it might affect the power quality.

18.3.5 Hybrid Islanding Detection Technique

Hybrid techniques are a combination of both passive and active techniques. They overcome the limitations with both passive and active techniques. Some of the hybrid islanding detection techniques present in the literature are voltage unbalance and frequency set point based method [62], average rate of voltage change and real power shift based method [63], and voltage/frequency and load switching based method [64]. The selection of the islanding detection technique depends mainly on DG technology. Every islanding detection technique has its own advantages and disadvantages. But a hybrid detection technique, with the combination of both passive and active techniques, has the advantages of both the techniques and seems to be the better solution for islanding detection.

18.3.6 Comparison of Islanding Detection Methods

The assessment of islanding detection techniques is completed through the nondetection zone (NDZ) identification, where there is a possibility of unintentional islanding. All techniques have some restrictions, which may include the need for coordination between the utility and the RDG/DG, high cost of implementation, susceptibility to false islanding detection (trouble tripping), probability of nondetection zone under some situations, and probability of reduction in utility power quality and voltage and frequency stability. The islanding detection techniques are divided into three categories: passive techniques, which depend on variations in DER terminal measurements to identify islanding; active techniques, in which the DER disturbs or perturbs its output so that it produces an observable change upon island creation; and communications-based techniques, in which information of the system status is transmitted to the DER and used to identify an island. Due to lack of impact on the system and easy implementation, passive techniques seem to be very attractive, but they generally have difficulty in excluding the NDZ with no false trips; i.e., they don't often attain both *selectivity* and *sensitivity* under the IEEE 1547-assigned trip times. Today, the active techniques can be used to eliminate this problem of NDZ to some extent as they have very small NDZs but only in the cases of low penetration. Active techniques are cost-effective, but they are inappropriate for high-penetration conditions. At present, the main problems with active detection techniques are: (1) they likely to conflict with the grid-support functions, which are becoming significantly important with the rise in DER penetration levels, and (2) they might degrade the transient response of the system at high-penetration conditions. The communications-based techniques are probably the best future contenders in prevention of unintentional islanding. Normally, it is well established that the NDZs of communication-based techniques can be very small, even though their island identification and cost efficacy are still under exploration. One key aspect of the communications-based techniques is the reliability of communications particularly. DER performance should be negotiated as small as possible if communications vanish and therefore requires the presence of a fall-back island identification technique in case of communications loss.

18.4 Protection of Microgrids

The microgrid concept has been projected as a way to integrate various renewable energy resources with distribution systems. The major elements of a microgrid are: distributed generators (DGs) such as wind turbines, solar photovoltaic arrays, rotating machines, fuel cells, etc.; distributed energy storage devices (DESDs) such as flywheels, batteries, supercapacitors, compressed-air systems, etc.; and critical/noncritical local loads. DESDs are primarily used as backup systems of energy to compensate for shortage of power inside the microgrid, mostly in the islanded operational mode when the generators might not be capable to provide the total demanded load power. Also, they are really helpful to smooth out the renewable energy resources intermittent power and/or to integrate considerable load changes [65].

Conventional rotating machine–based RDGs/DGs cannot quickly respond to the intermittencies of power owing to their inherent large time constants; thus, instability may happen in transient conditions. Therefore, the utilization of DESDs is essential.

Usually, the distribution systems/networks have been planned to function radially; i.e., the power flows from the higher voltage levels (utility grid side) down to the customers (local load side) connected in the radial feeders. This simple configuration has facilitated clear-cut protection schemes for typical distribution systems. Therefore, traditional distribution systems/networks are protected by the basic protective equipment such as overcurrent relays, fuses, and re-closers [66,19].

The introduction of the microgrid concept in the power system distracts the conventional protection techniques based on the radial network configuration theory showing high fault currents and unidirectional power flows. In the case of microgrids, the protection technique must able to ensure safer microgrid function in both the operating modes (namely, islanded and grid-connected modes). Thus, the issues related to protection with each operating mode should be addressed separately by the protection technique. Two main concerns [67] that need to be addressed in reference to its protection are (1) the time calculation when it should be islanded, such as in reply to abnormal situations that the utility might experience, and (2) the prerequisite of reliable and appropriately synchronized protection system so that it can reliably trip when a fault occurs.

Ultimately, the technical challenges in microgrid protection can be summarized as follows [68,69]: (1) making bidirectional power flow in both medium- and low-voltage systems; (2) two modes of operation: grid-connected mode, and stand-alone or islanded or isolated mode; (3) configuration changes in LV network because of switching of generators, storage systems and loads; (4) generation power intermittence due to the connection of several microsources in the microgrid; (5) an increase in penetration level of rotating machines, which may produce larger fault currents that exceed devices ratings; (6) inadequate short-circuit current level in the islanding mode of operation due to power-electronics coupled RDG; (7) decrease in the allowable timing of trips when faults take place in low- and medium-voltage systems to manage the microgrid stability; and (8) nuisance tripping as a result of faults on neighbouring feeders.

18.4.1 Microgrid Protection under Grid-Connected Mode

Fault currents are quite high because of the contribution of the utility grid in the grid-connected mode of a microgrid; therefore, the application of traditional overcurrent relays is possible. However, the traditional radial networks configuration is negotiated due to the presence of DERs, and protection synchronization might be affected or in some cases completely lost [70–72].

Introduction of a DER to a radial feeder anywhere downstream of the feeder re-closer may (1) vary the maximum and minimum feeder fault current, (2) reduce the fault current level of the upstream protective equipment compared to that of downstream equipment, and (3) cause a bidirectional power flow. Therefore, various issues may occur in the grid-connected operating mode depending on the type, size, and location of the DER. The important issues related to protection with operating mode are protection blinding, sympathetic or false tripping, malfunction or miscoordination between recloser and fuse, malfunction or miscoordination between fuse and fuse, and auto-reclosing failure [71,72–74]. Protection blinding is defined as a condition where a DER is tied anywhere between the feeding substation and the fault location. In this situation, the fault current calculated by the feeder relay located at the beginning of the feeder decreases due to the contribution of the DER compared to the condition that no DER is connected to the system. This may cause relay operational delay or even faults point unidentified. Sympathetic or false tripping is defined as a situation in which a fault occurs away from the feeder that encloses a DER, such as, a fault occurring at the neighbouring feeder. In this case, the DER share to the fault through its feeder and the upwards fault current starts flowing on that feeder. As a result, the nondirectional relay belonging to the healthy feeder may incorrectly identify a fault and cut off its feeder, which is problematic.

The installation of a DER to a feeder, whether downstream or upstream of the sideway connection points, varies the feeder maximum and minimum fault currents and therefore negotiates the synchronization between the protective equipment of the sideways. This condition is referred to as malfunction or miscoordination between a re-closer and a fuse and/or between a fuse and a fuse. Auto-reclosing failure may take place as the DER remains associated to the feeder, a situation called the dead time of the auto-reclosing mechanism. In this situation, the DER normally turns to quit from the synchronism relating to the grid, and thus the re-closure connects two asynchroncusly functioning systems after re-closing. This may cause severe damage to the generators in addition to the neighbouring networks. The DER manages the voltage in the network and the arc originates at the fault point all along the dead time of the re-closing sequences. This, in sequence, produces the fault to appear permanent and as a result, the failure of reclosing.

18.4.2 Microgrid Protection under Islanded Mode

As discussed in earlier sections, power electronic devices should be protected against overcurrent situations because they include solid-state switches, which have restricted current ratings. This causes a restricted fault current capability of electronically interfaced DERs. Thus, the number of fault currents are comparatively low in the islanded operating mode compared to those occurring in the grid-connected operating mode. By relying on the amount of power given by rotating-machine-based DERs compared to the power given by electronically interfaced DERs, the magnitude of fault current can fluctuate over a quite wide range in an islanded microgrid. As a result, the conventional overcurrent protection techniques are no longer sufficient for the islanded operating mode [75]. Additionally, since a fault might include considerable resistance, undervoltage protection functions may perhaps not succeed in identifying all types of faults. Thus, a fresh concern in the essentials of relaying is coming up with a protection technique that is capable of acting suitably for different faults inside an islanded microgrid.

18.4.3 Solutions for Microgrid Protection

In the literature, various solutions have been identified for microgrid protection. Broadly, the major protection techniques for microgrids can be categorized as (1) adaptive protection technique, (2) voltage-measurement-based protection techniques, (3) differential protection techniques, (4) distance protection techniques, (5) overcurrent protection and symmetrical components based techniques, and (6) protection technique improvement by using external devices/equipment. An overview of the major issues and features of these techniques are summarized in the following sections.

18.4.4 Adaptive Protection Technique

Adaptive protection techniques are primarily based on the utilization of adaptive relays, which can include their settings, logic functions or characteristics changed on-line in an appropriate manner, through externally control action or generated signals. According to the different solutions given in [75,76–83], the major issues regarding the possible implementation of an adaptive protection technique are (1) the prior knowledge of all possible configurations of the microgrid that is needed, (2) the prerequisite of doing extensive short circuit or power flows calculations when a configuration change is identified, (3) larger requirement of communication infrastructure, and (4) several protection equipment (fuses, etc.) in the existing power system need to be upgraded or updated.

18.4.5 Voltage-Measurement-Based Protection Techniques

Voltage-measurement-based protection techniques principally use voltage measurements to provide an appropriate protection system in microgrids. In general, the major issues that may arise with voltage-measurement-based protection techniques are as follows [84–88]: (1) slight deviations in voltage drop between the relays located at both ends of short lines cause protection function failures due to decrease in the voltage gradient; (2) comparatively large calculation complexity under the Park's transformation application; (3) identification problems under the high impedance faults; (4) practical application problems with some of the techniques, in addition with communication infrastructure, when large number of DGs are present; and (5) they are more reliant on the network/system architecture and the description of the relay protection region associated with each generator.

18.4.6 Differential Protection Techniques

Differential protection techniques are based on differential approach, i.e., some kind of comparison between measurements in various segments of the microgrid, and they are described in [89–92]. The main issues related to these techniques are as follows: (1) requirement of communication infrastructure that may not succeed at some point, leaving the microgrid without protection (therefore, few authors present various backup protections at levels); (2) requirement of coordinated measurements; (3) transients problems when switching the RDG/DG sources; (4) problems caused by unbalancing of systems or loads; and (5) comparatively high cost.

18.4.7 Distance Protection Techniques

Distance protection techniques use impedance or admittance measurements to identify the fault and trip appropriately [93,94]. The major issues related to these techniques are as follows: (1) restricted fault resistance that can be reliably identified; (2) inaccuracy in the measured admittance due to the fault resistance; (3) increased time of tripping as a result of the downstream source infeed; and (4) inaccuracy because of problems in extraction of fundamental caused by harmonics, time constant, and current transients and decaying DC magnitude.

18.4.8 Overcurrent Protection and Symmetrical Components Based Techniques

Overcurrent protection and symmetrical components based techniques attempt to improve the performance of conventional overcurrent protections, and they sometimes address measurement and calculations with symmetrical components [89,95,96]. Note that the major difficulty in these kinds of protection is typically associated with the requirement of an extensive communication system. In these cases when the communication system failure occurred, the entire overcurrent protection and coordination may be in peril.

18.4.9 Protection Techniques Improvement by Using External Devices/Equipment

When fault current levels are radically changed between the islanded (usually with inverter-coupled RDG) and the grid-connected operating modes, an appropriate protection scheme that executes accurately in both modes can be a genuine challenge. A different approach can dynamically modify the level of fault current when the microgrid changes its operating mode (i.e., from grid-connected to islanded, and vice versa) through certain externally installed devices such as fault current limiters (FCLs), flywheels, batteries, etc. These devices can either decrease or increase the level of the fault. The major issues related to using external devises and/or equipment in the microgrid are as follows [97–100]: (1) storage devices need a huge investment and must match the short circuit level of the main grid in order to ensure that faults are cleared in a timely manner; (2) the FCL-based schemes are applicable with a restricted number of RDGs, i.e., it is possible up to a certain number of DGs connected because it can be hard to calculate the FCL impedance value due to the DGs mutual influence; (3) sources having high short circuit capability (flywheels, etc.) need considerable investments, and their safe function relies on the accurate maintenance of the unit; and (4) the additional current-source-based techniques are highly reliant on the islanding detection technology and on the accurate operation of the current source.

18.4.10 Other Special Approaches and Techniques

Some other special approach techniques given are in the literature [101–108], but they are not used very frequently because of harder practical application, complexity, etc. They use the concepts of wavelet transformation (WT), artificial intelligence (AI) such as particle swarm optimization (PSO), artificial neural networks (ANNs), data mining, etc.

18.5 Protection of Wind Energy Conversion Systems [109–124]

On the basis of rotational speed, wind energy conversion systems (WECSs) can be generally categorized into two classes: fixed speed WECSs and variable speed WECSs. According to power converter rating with respect to the generator capacity, variable speed WECSs can be further classified into WECSs with a partial-scale power electronic converter and a full-scale power electronic converter.

18.5.1 Fixed Speed WECSs

Fixed speed WECSs have been used with several stages of the gearbox and a squirrel cage induction generator (SCIG) directly coupled to the grid through a transformer, as illustrated in Figure 18.6a. Due to the damping effect (large inertia of cage rotor), an SCIG can be used in fixed speed wind turbines. It operates only in a narrow range around the synchronous speed and always draws reactive power from the grid. An extended concept was proposed in the 1980s with a capacitor bank for reactive power compensation. Limiting of the high starting currents and smoother grid connection was achieved by incorporating a soft-starter, shown in Figure 18.6b. The soft-starter also efficiently damps out the torque peaks associated with the peak currents and thus reduces the loads on the gearbox. A pole-changeable SCIG has been used, which leads two rotation speeds.

The advantages of WECS in this configuration (i.e., SCIG) are the simple and cheap construction, and no need of synchronization. These advantages mean lower cost and better reliability. In addition, a WECS using an SCIG operates at constant speed and requires the stiff power grid for stable operation. In order to absorb high mechanical stress due to the wind gusts, which cause torque pulsations on the drive train, a more expensive mechanical construction is required.

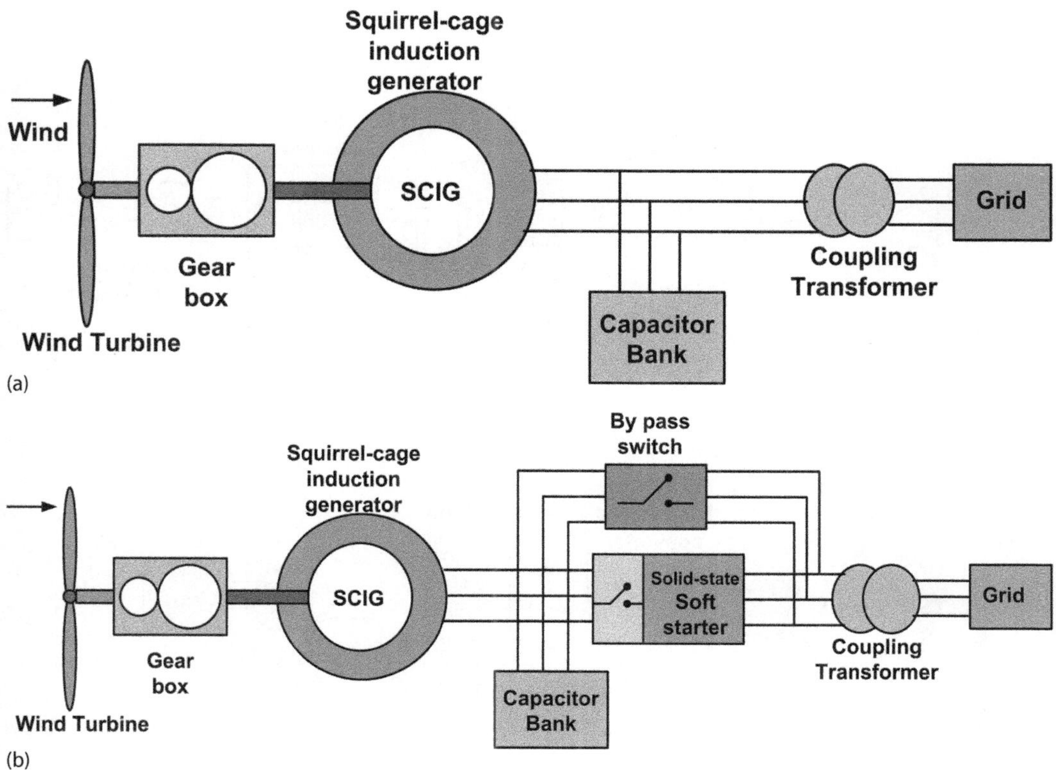

FIGURE 18.6 Schemes for fixed-speed WECS: (a) without soft starter and (b) with soft starter.

18.5.2 Variable Speed WECSs Using Full-Scale or Full-Rated Power Electronic Converter

Synchronous generators and SCIGs are employed in variable speed WECSs with full-scale power electronic converters. A full-rated PEC is connected between the generator and grid with the additional technical performance. In general, two back-to-back-connected voltage source converters (VSCs) with common capacitive DC link are used to achieve full control on the active and reactive power to and from the grid. With synchronous generators, diode rectifiers may be used on the generator side, as shown in Figure 18.7c, but in this case, the whole system is not fully controlled (i.e., the speed of the generator is not regulated with the wind speed to achieve the maximum power point tracking [MPPT]). As the generator is decoupled from the grid, a wide range of frequency and speed of the generator is possible for optimal operation. The grid-side converter can be used for controlling the active and reactive power independently and the improved dynamic response. Figure 18.7 shows five possible configurations of variable speed WECSs with full-scale power converters.

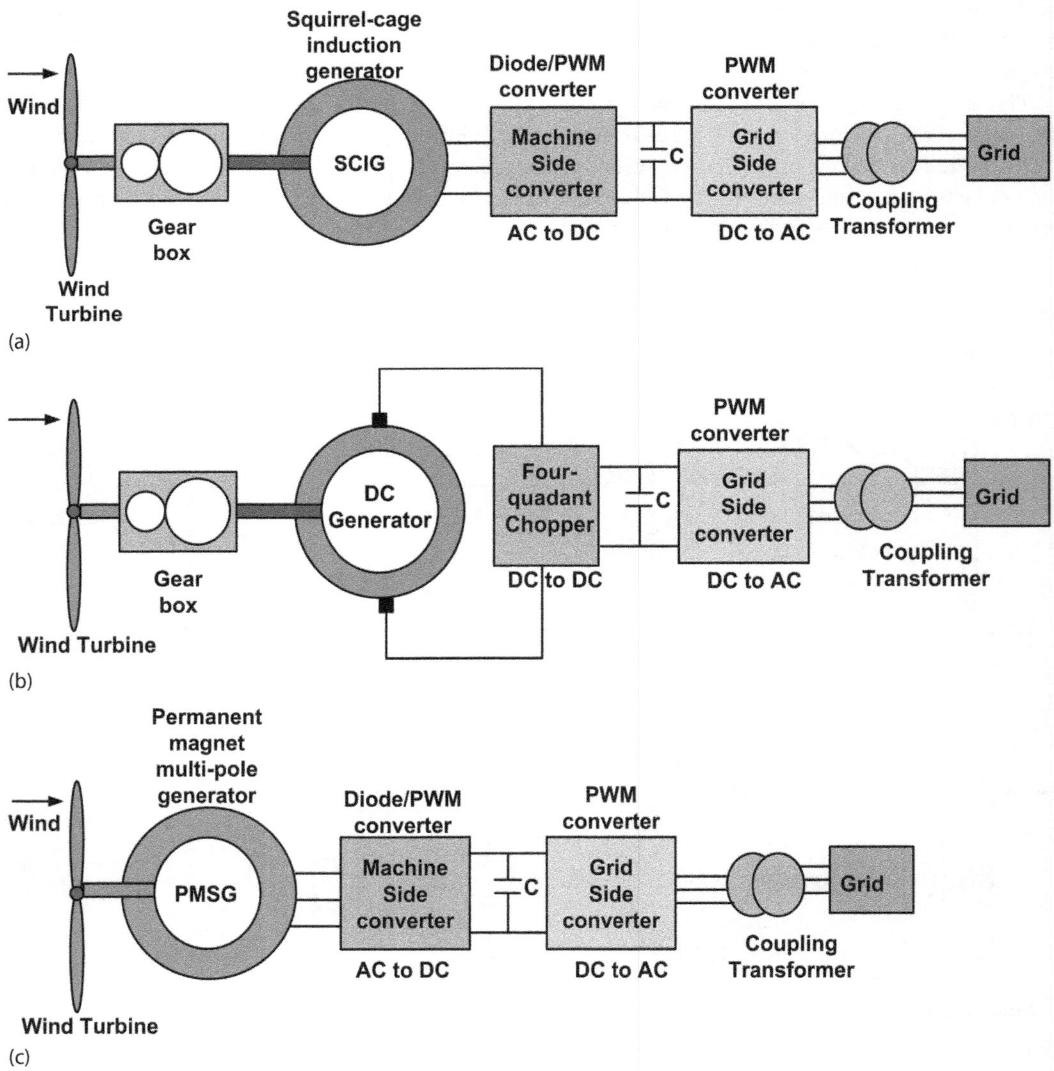

FIGURE 18.7 Schemes for variable-speed, full-scale, PEC-based WECSs: (a) using squirrel cage induction generator; (b) using DC generator; (c) using synchronous generator. *(Continued)*

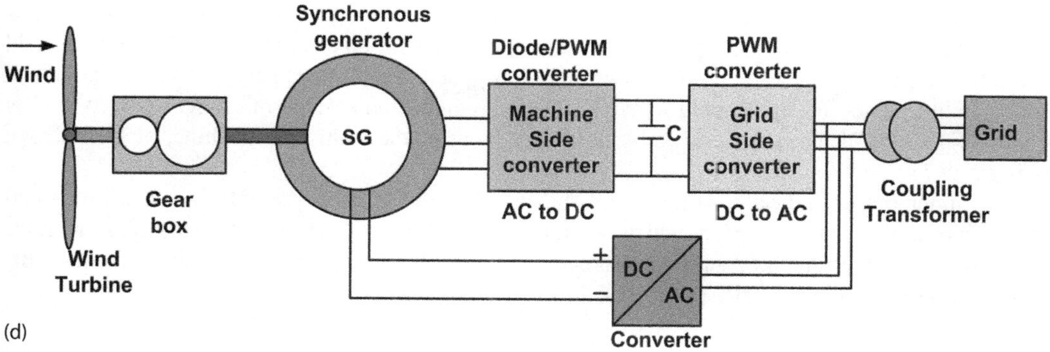

(d)

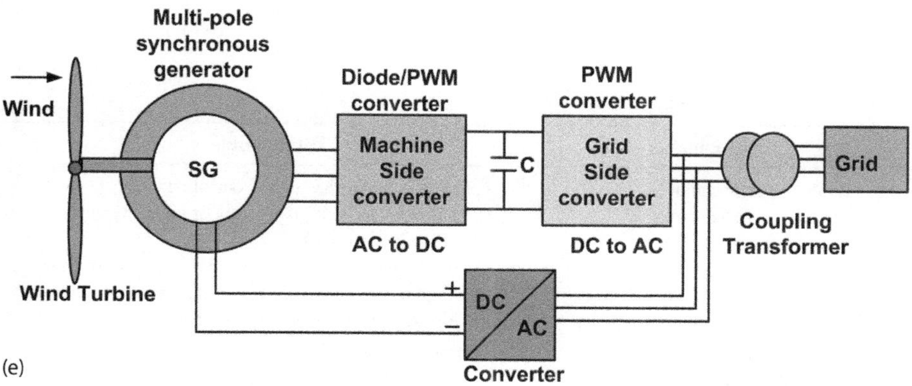

(e)

FIGURE 18.7 (Continued) Schemes for variable-speed, full-scale, PEC-based WECSs: (d) using multi-pole synchronous generator; and (e) using Permanent magnet multi-pole generator.

The variable speed configuration of a wind energy system by using a squirrel cage induction generator is shown in Figure 18.7a. The stator is connected to the grid by means of back-to-back three-level voltage source power electronic converter bridges. The power converter size depends on the stator power rating because it has to convert all the stator power. The gearbox is designed so that the maximum rotor speed corresponds to the rated speed of the generator. The merits of this variable speed configuration are its ability to make the best use of available wind power and to eliminate the need for a capacitor bank since it is able to draw its required reactive power from the grid. However, the drawbacks of this configuration are high cost and loss of the full-scale converter. The efficiency of the total system (gearbox, induction generator and converter) may be low. In the configuration shown in Figure 18.7b, a permanent magnet DC generator instead of a classic AC machine is used to provide the electromechanical energy conversion in the wind energy system. The DC machine has been selected due to its simpler electrical model. The drawbacks of this configuration concern the rating, economics and limited speed. In addition, DC machines have well known demerits in reference to power generation.

The configuration of a variable speed permanent magnet synchronous generator (PMSG) based WECS connected to the grid is shown in Figure 18.7c. Using of PMSGs as generators in wind energy technology is generally preferred in smaller WECS designs. The PMSGs have received much consideration in wind energy technology because of their self-excitation property, which allows operation at a high power factor and high efficiency. Additionally, the PMSGs have shown inherent economic and technical advantages: due to the absence of field losses, they have better thermal characteristics, higher reliability, solid field structure, high power to weight ratio, and better dynamic stability. Nevertheless, PMSGs also have some drawbacks, such as higher PM material cost, handling difficulties in manufacture, and demagnetization of the permanent magnet at high temperature.

Normally PMSGs are used for low-power applications; the more recent larger systems use a synchronous generator. The electrically excited synchronous generator (EESG) is typically designed with a rotor carrying DC excitation field windings system. The stator brings a three-phase winding quite similar to that of the induction machine. The rotor may either have salient poles or may be cylindrical. A field system based on salient poles is used most often in low speed synchronous generators and may be the most valuable version for application in WECS. A grid connection scheme of EESG based WECS is shown in Figure 18.7d. The gear box can be eliminated by using the multi-pole machine, which is shown in Figure 18.7e.

The common thread in the all configurations shown in Figure 18.7 is the use of full scale or fully rated power electronic converters (PECs), which degrades the efficiency of the overall system and increases the cost of the system. The brief summary shown in Table 18.1 gives a performance comparison of different generators used in variable speed WECSs.

TABLE 18.1

Performance Comparison of Different Generators Used in WECSs [112-121]

Generators	Advantages	Disadvantages
DC machine	• Simple electrical model • Low cost and effective	• Reactive power has to be controlled • If the wind velocity is too low or high, the wind turbine is turned off • The DC bus voltage is a sensitive variable for the design of the power converter and the DC capacitor. This voltage also has to be controlled
Permanent magnet synchronous generator	• Lower maintenance cost • Lower operating costs • Gearless • Rotor losses are less • High torque at low speeds • No need for separate excitation	• Higher initial cost • Not used for large-scale, grid-connected WECSs • Demagnetization problem in permanent magnets
Synchronous generator (with external excitation)	• Direct drive applicable • Does not require reactive power injection	• Higher maintenance costs in comparison to that of an IG • Need for external excitation • An addition rectifier is required • Overall high cost
Squirrel-cage induction generator (SCIG)	• Lower capital cost, rugged, and simple design • Used for large-scale, grid-connected WECSs • Excellent damping of torque pulsation	• Full rated power converter; thus, cost is increased • Increased losses through converter due to large converter size • SCIG always draws reactive power from the grid • Large in-rush current at starting
Doubly fed induction generator (DFIG)	• Reduced converter cost, converter rating is typically 25%–30% of total system power • Improved efficiency • Suitable for high-power applications • Generate or absorb reactive power	• Increased control complexity • Need for periodic slip ring maintenance • System cost is increased
Brushless doubly fed machine (BDFIG)	• High reliability and low-maintenance requirements • Best suited for offshore and difficult-to-reach places (such as isolated areas) • Brushless operation • Cost is likely to be less than that of an equivalent DFIG due to absence of slip-ring and the simpler structure of the rotor winding	• BDFIG is not stable over the whole operating speed range • Complex manufacturing
Switched reluctance machine	• More simple and robust • Variable speed generator	• Used only for small scale WECSs • High torque ripple and acoustic noise

18.5.3 Variable Speed WECSs Using Partial-Scale or Partial-Rated PEC

The DFIG concept belongs to a variable speed wind turbine with a wound rotor induction generator and power electronic converter (PEC) on the rotor side, as shown in Figure 18.8a. It is a four-quadrant system capable of supplying and consuming active and reactive power when connected to a grid. It has two modes of operation: subsynchronous and supersynchronous. These modes of operation depend on the angular speed of the rotor with respect to the synchronous speed. It is likely to control rotor current injection using fully controlled electronic converters to ensure effective operation in both sub- and supersynchronous modes. In the subsynchronous mode, the angular speed of rotor is less than synchronous speed. Real power is drawn from the stator/grid (or any other external source) and delivered to the rotor via power electronics converters, as shown in Figure 18.8b. In supersynchronous mode, the rotor angular speed is greater than the synchronous speed. Power is generated in the rotor (in addition to power generated in the stator) and is delivered to the grid/load via power converters, as shown in Figure 18.8b. DFIG based WECS uses the gear box to convert the low speed of the turbine shaft to the high speed at the rotor shaft, which must be acceptable for the DFIG.

The DFIG concept gives a wide operational speed range depending on the size of the power electronics converter. Typically, the speed ranges from +30% to –30% around the synchronous speed. The rating of the power electronic converter is only 25%–30% of the generator capacity, which makes this concept attractive and popular from a cost point of view. There are various kinds of topologies for the DFIG, such as the partial-scale power converter, conventional AC-AC converter, matrix converter or hybrid current-source/voltage-source power inverter. A brief topological summary is presented in Table 18.2.

The factors that influence the choice of power converters are the power supply circuit complexity, reliability, modularization, energy-saving condition, continuous run availability, technology stability etc.

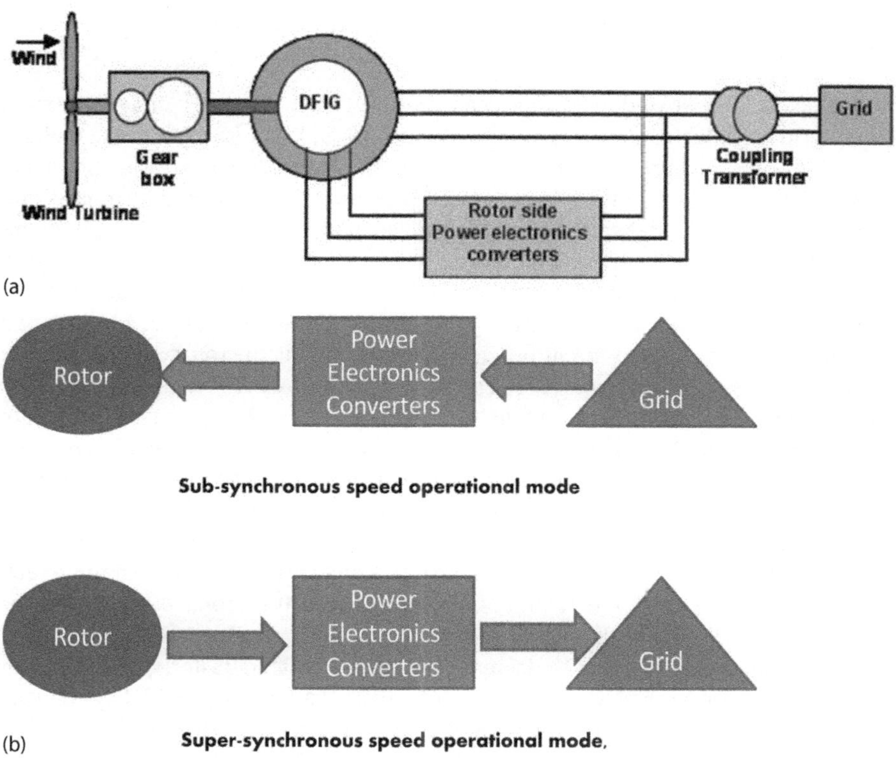

FIGURE 18.8 (a) Scheme of a variable speed concept with DFIG system and (b) power flow direction under the sub- and supersynchronous modes.

TABLE 18.2

Power Converter Topologies used in DFIG-Based WECSs

Power Converter Topology			
Rotor Side	**Grid Side**	**Advantages**	**Disadvantages**
Diode bridge	SCR inverter	• Easy controlling • Economical	• Provides power under supersynchronous mode only
SCR rectifier	SCR inverter	• Bidirectional power flow • More power output • Operate at both sub- and supersynchronous modes	• Firing and commutation problems • Harmonic distortion to the grid
Back-to-back hard-switching inverters		• Remove the firing and commutation problems • Very less harmonic distortion	• Need of bulky DC link capacitor, which is costly
Matrix converter		• No bulky energy storage or DC link is required • Single-stage AC/AC conversion	• Devices are more in number; thus, the switching of the many devices is complicated • Also the cost is increased due to thenumber of devices

The available research history shows that back-to-back PWM converters are a well-established technology used with DFIG. They have shown outstanding performance for high power applications in the MW range.

18.5.4 Protection Schemes of the Most Common Configuration (i.e., DFIG-Based) WECS [110]

In a grid connected DFIG based WECS, a deep drop in DFIG terminal voltage occurs when the power system/grid is subjected to any kind of fault (such LLG, LG, LL, three-phase faults etc.). The rotor side converter loses control over the rotor currents, so an additional hardware protection circuit is necessary. At present, the solutions given in the literature [110,124] and also used in the wind power industry, and they are shown in Figure 18.9a. The hardware circuitry can be implemented on the rotor side, the DC side or the stator side. The most frequently used protection technique is a rotor crowbar circuit that is short-circuiting the rotor winding when rotor currents of the DFIG or DC link voltage exceed their threshold value under the grid fault conditions. Therefore, a bypass path for the rotor overcurrent is given by short-circuiting the rotor winding via the rotor crowbar circuit, and the power electronics converter at the rotor side can be well protected. At present, the most commonly used protecting scheme is short-circuiting the rotor winding through the crowbar protection circuit when rotor currents of the doubly-fed generator or DC bus voltage exceed their rated value during the grid fault. Hence a path for the rotor overcurrent is provided so that the rotor-side converter can be well protected. The fact that the traditional crowbar circuits cannot be turned off immediately after the grid fault because of the thyristors means they do not meet the new grid codes. Therefore, new active crowbars, using active switches such as IGBT and GTO, are proposed to protect the system. The rotor-side converter with this kind of circuit can be still be connected to the rotor when a grid fault occurs. After the fault, the power system can be more flexible, taking less time to return to a normal operating mode. In order to decrease the rotor transients faster, the active crowbar circuit usually has a resistor on the DC side. In case of isolated/autonomous DFIG systems, rotor side crowbar (active) is sufficient. A block diagram of a control scheme for the protection of converter/inverter on the rotor side is shown in Figure 18.9b.

In order to reduce the rotor side transient faster, the active crowbar circuit (combination of a resistor and switch) is generally used on the DC side. A common technique for maintaining the voltage on DC

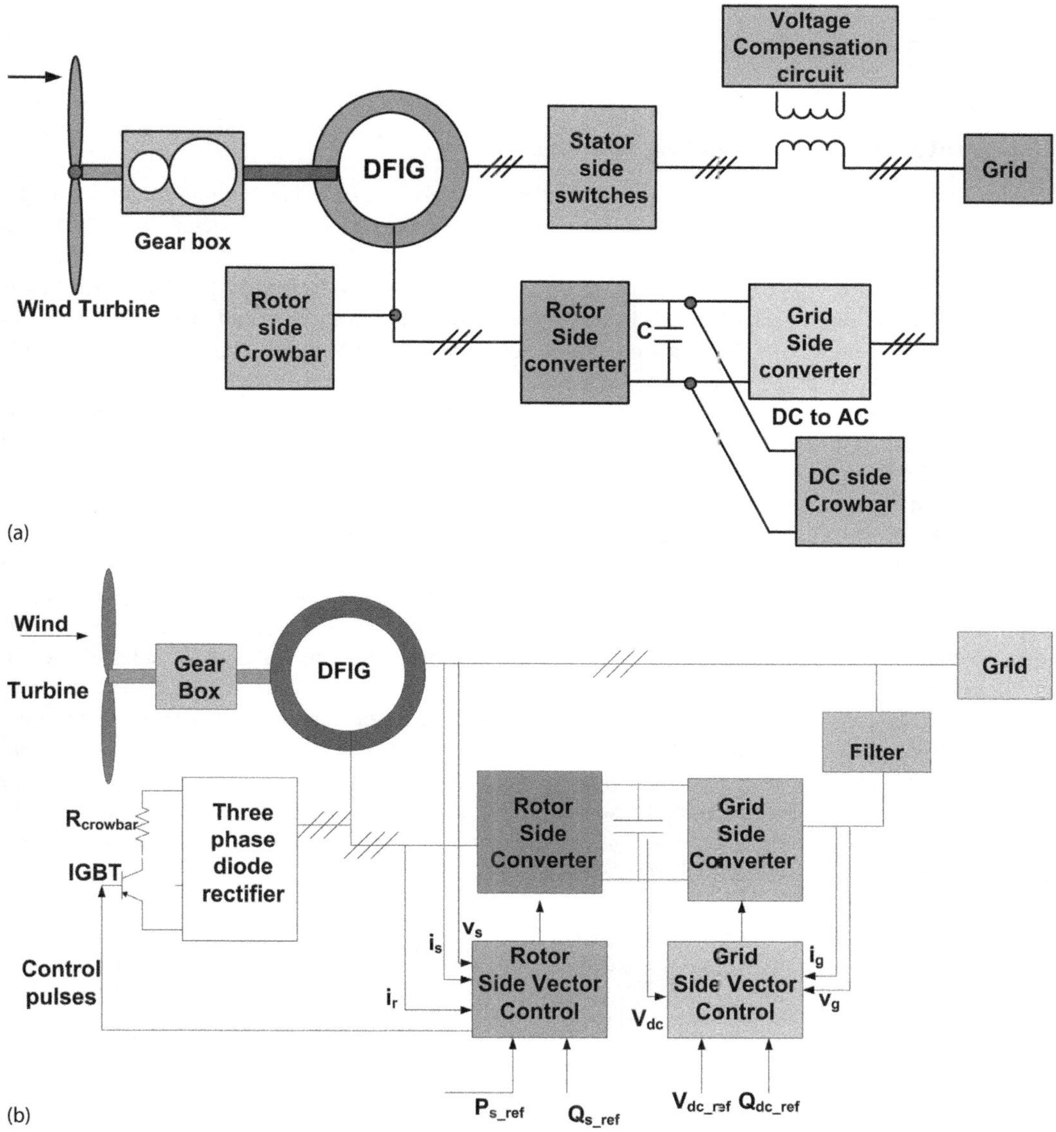

FIGURE 18.9 (a) Hardware protection circuits for DFIG based WECS and (b) active crowbar protection scheme for DFIG based wind energy system.

bus is to add a dissipation resistor on the DC side to consume the extra energy during the grid fault. In order to resolve this energy dissipation problem, an energy storage device (ESS) has been used instead of a DC-side crowbar. In accordance with the operating speed of the doubly-fed generator, the ESS could absorb/store extra energy during fault conditions and export this energy back to the grid after the clearance of the fault. The ESS can prevent the transient in switching operation and keep the system controllable at all times. However, the ESS cannot control the rotor current. Therefore large-capacity power electronics devices are necessary for the rotor-side converter in order to protect it from damage by rotor overcurrent. This will increase the cost of the system. An electronic switch should be inserted in the stator side besides the crowbar on the rotor terminals. It will disconnect the stator winding during the fault and thus prevent the reactive power absorption, but it completely interrupts the stator active power generation during the fault. This method requires an additional switch with the rating of the stator circuit. Compared to the traditional rotor-side crowbar protection circuit, it has advantages in the transient restriction. Protection enhancement is also achieved by inserting a series-connected voltage source converter at generator terminals that support the voltage during the fault.

18.6 Protection of Photovoltaic Systems

Photovoltaic (PV) cells are made of semiconductor materials that translate sunlight, or solar energy, into direct current electricity (DCE). The PV module is constructed by PV connected in series. Series connected PV modules constitute the PV string that generates higher power. Parallel connected PV strings, in a combiner box, are often used to generate higher current. Afterwards, this DC generation is converted into AC and fed into the grid. The PV system's installation and demand increase the requirement of effective protection. Like all electrical power systems, PV systems must include suitable overcurrent and overvoltage protection. As with all outdoor structures, PV installations are prone to lightning risk, which changes from region to region. Lightning strikes can influence photovoltaic generators and their installations, and also the electronic switches of the inverter. Therefore, it is essential to estimate the risk related to lightning strikes according to the Standard IEC (EN) 62305-2 to implement the right protective measures for the system [125]. Various considerations for estimating the risk of lightning and overvoltage are: risk increases with the increase in the size of the solar panel field; direct effect include lightning strikes on the modules; and indirect effects include overvoltages on the converter/inverter, modules, and other connections. Surge protection devices (SPDs) are mainly important for protecting sensitive electrical devices such as AC/DC inverters, PV modules and monitoring devices. Generally, technical procedures suggest protecting both DC and AC sides of the PV installation with SPDs.

In the past, the protection problem of the PV installations has been considered the prevention of direct lightning strikes through an external lightning protection system [126]. However, the protection problem is more complicated because if lightning strikes near the PV system, the installations could be injured due to overcurrents and overvoltages. These overvoltages may possibly reach significant values [127]; thus, overvoltage protection is important on both sides of the electronic devices (e.g., charge controllers and inverters), as shown in Figure 18.10 [128,129].

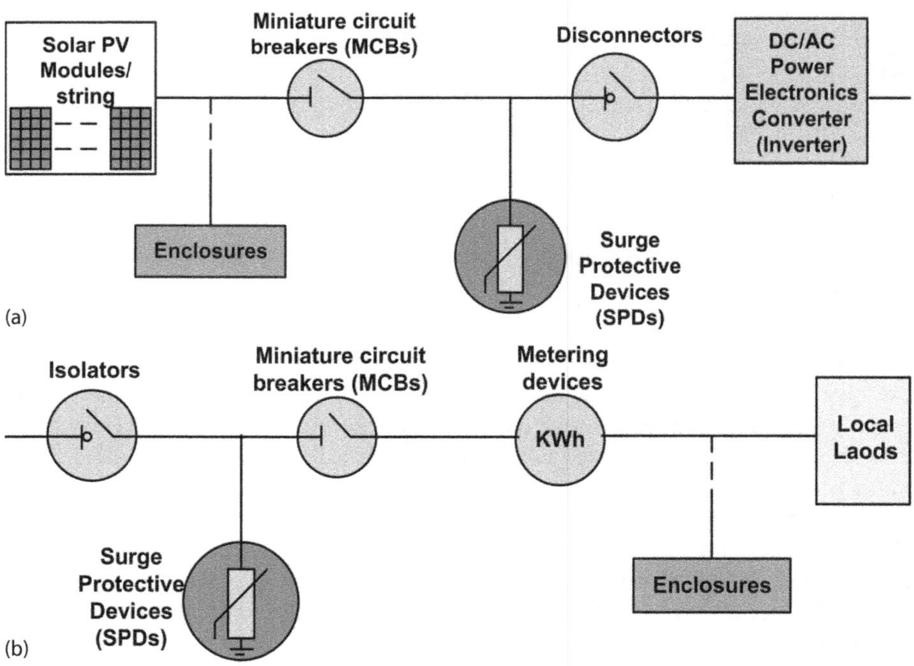

FIGURE 18.10 Protection of solar PV system: (a) DC side protection and (b) AC side protection.

18.7 Protection Considerations for Future Distribution Network/Systems

Today, distribution system utilities must face new vital challenges, such as electricity market deregulation, DR control and management, the probability of a turn into an active network from a passive one. Thus, the configuration of the distribution network/system will be transformed completely. The new distribution networks will require even more reliable and suitable communication links between customers and utilities, which will be important in providing controls and monitoring actions, metering signals, and dispatch signals. The recent dynamic role of the distribution system/network could tender new services to low and medium voltage customers and could permit an extensive utilization of renewable resources and dispersed generation. The newer role will be to transmit the energy coming from various producers with access to the electric market, communication and protection [130].

The bi-directional telecommunication system is one of the most important parts in making the intelligent grid/distribution system competent for real-time function. The requirements of communication between various operators and loads and RDGs/DGs of the distribution network are illustrated in Figure 18.11. The transmission system operator is accountable for power management between the transmission and distribution networks. The account of the power exchanges in the distribution network is performed by the distribution network operator (DNO), which enforces the constraints on the generated active and reactive power by RDGs. The trader, which is synchronized with the DNO, operates the transactions of the power and the services presented through dispersed generators, according to the rules and regulations given by the authority. The customers consume and/or produce the energy within the constraints of the distribution network operator and compliance with trader.

As discussed earlier, many distributed generators can enlarge the short circuit current levels and logically change the system impedance as seen by the fault location. Therefore, if the utility grid has had very high short circuit levels, it is essential to ensure that the CB breakoff capacity is not exceeded. The selectivity standard is heavily jeopardized because of the natural decline of fault current as the fault is moved away from the source. Considering the protection selectivity loss, the difficulty can be mitigated by: (1) varying the protection windows, which is achievable only if these changers do not create unnecessary stresses on the system equipment and on the fault clearing actions; (2) connecting the DG in a different grid position to vary its donation to the fault currents in the distribution lines; and (3) introducing appropriate series reactance by the DGs to minimize their donation to the fault currents.

The first option, i.e., varying the protection windows, appears to be the most sensible one, and a solution could be to use power line carrier (PLC) technology. PLC technology appears to be particularly suitable for these future medium-voltage distribution systems. The use of PLC in the distribution network (frequently known as distribution line carrier) is a comparatively new concept for the utilities.

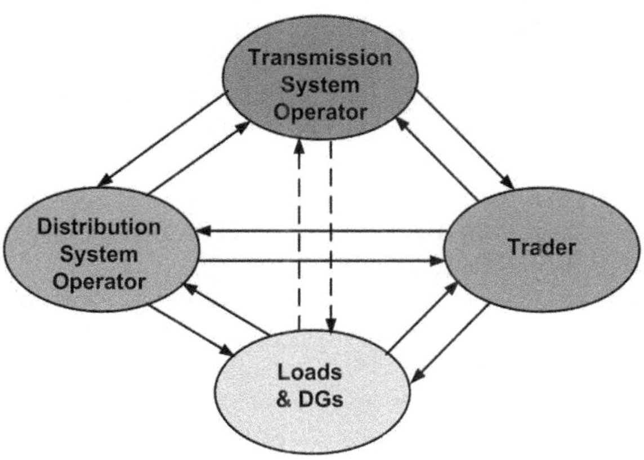

FIGURE 18.11 Requirements of communication between various operators and loads and RDGs/DGs of distribution network.

Now distribution companies are accomplishing the huge potential expressed by the advanced data communications such as in managing a large penetration of RDGs/DGs. The PLC can be effectively implemented as: (1) the coordinated voltage control, (2) a protection system to avoid DG islanding, and (3) the control of the coupled device at the PCC for a microgrid.

18.8 Tutorial Problems

1. Define the term *microgrid*. Draw a neat schematic for a typical configuration of RDG connected to the utility.
2. What are the different power quality issues to consider when protecting RDGs?
3. Differentiate between intentional and unintentional islanding of RDGs.
4. How is the islanding of an RDG detected? Discuss the classification of islanding detection schemes.
5. How are WECSs classified? Discuss briefly.
6. Sketch the working of a WECS with different AC generator.
7. Define the term *crowbar*. Describe an active crowbar protection scheme for a WECS.
8. What are the different hardware circuitries used for the protection of a WECS?
9. Draw and explain a schematic block diagram of a solar PV energy system with protection equipment included.
10. What is the role of the DNO in intelligent grid/distribution systems?

18.9 Conclusion

A detailed explanation of the protection issues concerning the RDG was given in this chapter. Different factors affecting the protection of the system were discussed. In addition, RDG islanding protection and anti-islanding methods were presented. Protection schemes commonly used in microgrids, wind energy conversion systems and photovoltaic systems were presented with the help of block diagram representations. Finally, protection aspects for future distribution systems were also presented and discussed in brief.

REFERENCES

1. L. J. Dulău, "Hybrid wind and solar power system," *2nd IFAC Workshop on Convergence of Information Technologies and Control Methods with Power Systems, Poster Session*, 2013.
2. G. Pepermans, et al., "Distributed generation: Definition, benefits and issues," *Energy Policy*, vol. 33, no. 6, 2005, pp. 787–798.
3. P. Rahmatallah, and J. Tooraj, "Distributed generation, storage, demand response and energy efficiency as alternatives to grid capacity enhancement," *Energy Policy*, vol. 67, 2014, pp. 222–231.
4. L. J. Dulău, M. Abrudean, and D. Bică, "Effects of distributed generation on electric power systems," *Procedia Technology*, Vol. 12, 2014, pp. 681–686.
5. K. Balamurugan, D. Srinivasan, and T. Reindl, "Impact of distributed generation on power distribution systems," *Energy Procedia*, vol. 25, 2012, pp. 93–100.
6. A. Canova, L. Giaccone, F. Spertino, and M. Tartaglia, "Electrical impact of photovoltaic plant in distributed network," *IEEE Transactions on Industry Applications*, vol. 45, 2009, pp. 341–347.
7. F. Wong, Y. S. Lim, P. Taylr, and S. Morris, "Optimal utilization of small-scale embedded generators in a developing country—A case study in Malaysia," *Renewable Energy*, vol. 36, 2011, pp. 2562–2572.
8. J. A. Martinez, and J. Martin-Arnedo, "Impact of distributed generation on distribution protection and power quality," *IEEE Power and Energy Society General Meeting*, Calgary, Canada, 2009, pp. 1–6.
9. "IEEE Standard for Interconnecting Distributed Resources with Electric Power Systems," IEEE STD 1547-2003, pp. 1–28.

10. Y. Baghzouz, "Voltage regulation and overcurrent protection issues in distribution feeders with distributed generation—A case study," *Proceedings of the 38th Annual Hawaii International Conference on System Sciences*, 2005, pp. 1–7.
11. "Distribute generation and renewable-integration and interconnection, quality and use of electricity-application guide," www.sier.ro, 2007.
12. J. H. Choi and J. C. Kim, "Advanced voltage regulation method at the power distribution systems interconnected with dispersed storage and generation systems," *IEEE Trans. Power Delivery*, vol. 15, no. 2, 2000, pp. 691–696.
13. C. L. Masters, "Voltage rise: The big issue when connecting embedded generation to long 11 kV overhead lines", *IEE Power Engineering Journal*, Vol. 16, No.1, 2002, pp. 5–12.
14. T. E. Kim and J. E. Kim, "Voltage regulation coordination of distributed generation in distribution systems," *Proceedings of the IEEE/PES Summering Meeting*, Vancouver, British Columbia, July 15–19, 2001.
15. S. N. Singh et al., "Distributed generation in Power Systems: An overview and Key Issues," Indian Engineering Congress, 2009.
16. J. Jager, T. Keil, L. Shang, and R. Krebs, "New protection co-ordination methods in the presence of distributed generation," *2004 Eighth IEE International Conference on Developments in Power System Protection*, 2004.
17. T. Seegers, K. Birt, R. Beazer et al., "Impact of distributed resources on distribution relay protection," *IEEE Power Engineering Society Working Group D3*, August 2004.
18. P. Hansen, J. Ostergaard, and J. S. Christiansen, "System grounding of wind farm medium voltage cable grids," *Nordic Wind Power Conference*, Roskilde, Denmark, 2007.
19. P. P. Barker and R. W. de Mello, "Determining the impact of distributed generation on power systems: Part 1—radial distributed systems", *IEEE Power Engineering Society Summer Meeting*, Seattle, WA, 2000, vol. 3, pp. 1645–1656.
20. "IEEE guide for the application of metal oxide surge arresters for alternating current systems," IEEE Standard C62.22-1997, 1998.
21. "Synchronous generation in distribution systems," http://www.essexenergy.ca/downloads/SynchronousGeneration.pdf.
22. M. Shen, L. Ingratta, and G. Roberts, "Grounding transformer application, modeling, and simulation," *IEEE Power Engineering Society Summer Meeting*, Pittsburgh, PA, July 2008.
23. F. Castro-Sayas and G. Clarke, "The cost and benefits of embedded generation islanding operation," *DTI Sustainable Energy Programmes*, Technical Report k/EL/00284/REP, 2002.
24. S. Abu-Sharkh et al., "Can microgrids make a major contribution to UK energy supply?" *Renewable and Sustainable Energy Reviews*, vol. 10, no. 2, 2006, pp. 78–127.
25. P. G. Harrison, "Considerations when planning a load-shedding programme," *Brown Bovcri Review*, no. 10, 1988, pp. 593–598.
26. V. V. Terzija, "Adaptive under-frequency load shedding based on the magnitude of the disturbance estimation," *IEEE Transactions on Power Systems*, vol. 21, no. 3, 2006, pp. 1260–1266.
27. D. Prasetijo, W. R. Lachs, and D. Sutanto, "A new load shedding scheme for limiting under frequency," *IEEE Transactions on Power Systems*, Vol. 9, No. 3, 1994, pp. 1371–1378.
28. P. Mahat, Z. Chen, and B. Bak-Jensen, "Control and operation of distributed generation in distribution systems," *Electric Power System Research*, Vol. 81, No. 2, 2011, pp. 495–502.
29. P. Mahat, Z. Chen, and B. Bak-Jensen, "Under-frequency load shedding for an islanded distribution system with distributed generators," *IEEE Transactions on Power Delivery*, vol. 25, no. 2, 2010, pp. 911–918.
30. IEEE Standard for Interconnecting Distributed Resources into Electric Power Systems, IEEE Standard 1547TM, June 2003.
31. "Photovoltaic (PV) systems—Characteristics of the utility interface," IEC 61727 Standard, December 2004.
32. S. P. Chowdhury, S. Chowdhury, P. A. Crossley, and C. T. Gaunt, "UK scenario of islanded operation of active distribution networks with renewable distributed generators," *Renewable Energy*, vol. 34, no. 12, 2009, pp. 2585–2591.
33. R. A. Walling and N. W. Miller, "Distributed generation islanding implications on power system dynamic performance," *IEEE Power Engineering Society Summer Meeting*, Chicago, IL, vol. 1, 2002, pp. 92–96.

34. R. S. Kunte and G. Wenzhong, "Comparison and review of islanding detection techniques for distributed energy resources," *40th North American Power Symposium*, Calgary, Canada, 2008, pp. 1–8.
35. P. Mahat, Z. Chen, and B. Bak-Jensen, "Review of islanding detection methods for distributed generation," *Third International Conference on Electric Utility Deregulation and Restructuring and Power Technologies*, Nanjing, China, vol. 1–6, 2008, pp. 2743–2748.
36. W. Xu, "An assessment of distributed generation detection methods and issues for Canada," University of Alberta–CETC, July 2004.
37. W. Bower and M. Ropp, "Evaluation of islanding detection methods for photovoltaic utility interactive power systems," International Energy Agency, March 2002.
38. M. Ropp, K. Aaker, J. Haigh, and N. Sabhah, "Using power line carrier communications to prevent Islanding," *28th IEEE Photovoltaic Specialist Conference*, Anchorage, AK, 2000, pp. 1675–1678.
39. W. Xu, G. Zhang, C. Li, W. Wang, G. Wang, and J. Kliber, "A power line signaling based technique for anti-islanding protection of distributed generators—Part I: Scheme and analysis," *IEEE Transaction on Power Delivery*, vol. 22, no. 3, 2007, pp. 1758–1766.
40. G. Wang, J. Kliber, G. Zhang, W. Xu, B. Howell, and T. Palladino, "A power line signaling based technique for anti-islanding protection of distributed generators—Part II: Field test results," *IEEE Transactions on Power Delivery*, vol. 22, no. 3, 2007, pp. 1767–1772.
41. G. Wang, J. Kliber, and W. Xu, "A scalable power-line-signaling-based scheme for islanding detection of distributed generators," *IEEE Transactions on Power Delivery*, vol. 24, no. 2, 2009, pp. 903–909.
42. M. A. Refern, O. Usta, and G. Fielding, "Protection against loss of utility grid supply for a dispersed storage and generation unit," *IEEE Transactions on Power Delivery*, vol. 8, no. 3, 1993, pp. 948–954.
43. S. Barsali, M. Ceraolo, P. Pelacchi, and D. Poli, "Control techniques of dispersed generators to improve the continuity of electricity supply," *IEEE Power Engineering Society Winter Meeting*, New York, vol. 2, 2002, pp. 789–794.
44. J. Warin and W. H. Allen, "Loss of mains protection," *ERA Conference on Circuit Protection for industrial and Commercial Installation*, London, UK, 1990, pp. 1–12.
45. F. Pai and S. Huang, "A detection algorithm for islanding-prevention of dispersed consumer-owned storage and generating units," *IEEE Transactions on Energy Conversion*, vol. 16, no. 4, 2001, pp. 346–351.
46. H. Kabayashi, K. Takigawa, and E. Hashimato, "Method for preventing islanding phenomenon on utility grid with a number of small scale PV systems," *2nd IEEE Photovoltaic Specialists Conference*, Las Vegas, NV, vol. 1, 1991, pp. 695–700.
47. S. I. Jang and K. H. Kim, "An islanding detection method for distributed generations using voltage unbalance and total harmonic distortion of current," *IEEE Transactions on Power Delivery*, vol. 19, no. 2, 2004, pp. 745–752.
48. S. Jang and K. Kim, "Development of a logical rule-based islanding detection method for distributed resources," *IEEE Power Engineering Society Winter Meeting*, New York, vol. 2, 2002, pp. 800–806.
49. S. I. Jang and K. H. Kim, "A new islanding detection algorithm for distributed generations interconnected with utility networks," *IEE International Conference Developments in Power System Protection*, Amsterdam, the Netherlands, vol. 2, 2004, pp. 571–574.
50. K. El-Arroudi, and G. Joos, "Data mining approach to threshold settings of Islanding relays in distributed generation," *IEEE Transactions on Power Systems*, vol. 22, no. 3, 2007, pp. 1112–1119.
51. K. El-Arroudi, G. Joos, I. Kamwa, and D. T. McGillis, "Intelligent-based approach to islanding detection in distributed generation," *IEEE Transactions on Power Delivery*, vol. 22, no. 2, 2007, pp. 828–835.
52. A. Pigazo, M. Liserre, R. A. Mastromauro, V. M. Moreno, and A. Dell'Aquila, "Wavelet-based islanding detection in grid-connected PV systems," *IEEE Transactions on Industrial Electronics*, vol. 56, no. 11, 2009, pp. 4445–4455.
53. N. W. A. Lidula, N. Perera, and A. D. Rajapakse, "Investigation of a fast islanding detection methodology using transient signals," *IEEE Power Engineering Society General Meeting*, Calgary, Canada, July 2009.
54. J. Yin, C. Diduch, and L. Chang, "Islanding detection using proportional power spectral density," *IEEE Transactions on Power Delivery*, vol. 23, no. 2, 2008, pp. 776–784.
55. S. R. Samantaray, K. El-Arroudi, G. Joos, and I. Kamwa, "A fuzzy rulebased approach for islanding detection in distributed generation," *IEEE Transactions on Power Delivery*, vol. 25, no. 3, 2010, pp. 1427–1433.

56. P. O'Kane and B. Fox, "Loss of mains detection for embedded generation by system impedance monitoring", *6th International Conference Developments in Power System Protection*, Nottingham, UK, 1997, pp. 95–98.
57. M. E. Rupp, M. Begonia, and A. Rohatgi, "Analysis and performance assessment of the active frequency drift method of islanding prevention," *IEEE Transactions on Energy Conversion*, vol. 14, no. 3, 1999, pp. 810–816.
58. G. A. Smith, P. A. Onions, and D. G. Infield, "Predicting islanding operation of grid connected PV inverters," *IEE Proceedings of Electric Power Application*, vol. 147, 2000, pp. 1–6.
59. J. Yin, L. Chang, and C. Diduch, "A new adaptive logic phase-shift algorithm for anti-islanding protections in inverter-based DG systems," *IEEE Power Electronics Specialists Conference*, 2005, Recife, Brazil, pp. 2482–2486.
60. G. Hung, C. Chang, and C. Chen, "Automatic phase shift method for islanding detection of grid connected photovoltaic inverter," *IEEE Transactions on Energy Conversion*, vol. 18, no. 1, 2003, pp. 169–173.
61. H. Karimi, A. Yazdani, and R. Iravani, "Negative-sequence current injection for fast islanding detection of a distributed resource unit," *IEEE Transactions on Power Electronics*, vol. 23, no. 1, 2008, pp. 298–307
62. V. Menon, and M. H. Nehrir, "A hybrid islanding detection technique using voltage unbalance and frequency set point," *IEEE Transactions on Power Systems*, vol. 22, no. 1, 2007, pp. 442–448.
63. P. Mahat, Z. Chen, and B. Bak-Jensen, "A hybrid islanding detection technique using average rate of voltage change and real power shift," *IEEE Transactions on Power Delivery*, vol. 24, no. 2, 2009, pp. 764–771.
64. W. Chang, "A hybrid islanding detection method for distributed synchronous generators," *International Power Electronics Conference (IPEC)*, Sapporo, Japan, June 2010, pp. 1326–1330.
65. S. Teleke, M. E. Baran, A. Q. Huang, S. Bhattacharya, and L. Anderson, "Control strategies for battery energy storage for wind farm dispatching," *IEEE Transactions on Energy Conversion*, vol. 24, no. 3, 2009, pp. 725–732.
66. *Cooper Power Systems, Electrical Distribution-System Protection*, 3rd ed. Cooper Industries, 1990
67. W. E. Feero et al., "White paper on protection issues of the microgrid concept," https://certs.lbl.gov/sites/all/files/certs-protection-issues-microgrid.pdf, accessed June 19, 2018.
68. R. J. Best et al., "Communication assisted protection selectivity for reconfigurable and islanded power networks," *Proceedings of the 44th International Universities Power Engineering Conference (UPEC)*, 2009, pp. 1–5.
69. A. Oudalov et al., "Novel protection systems for microgrids," 2009, http://www.microgrids.eu/documents/688.pdf, accessed June 24, 2018.
70. A. Girgis and S. M. Brahma, "Effect of distributed generation on protective device coordination in distribution system," *Proceedings of the Large Engineering Systems Conference on Power Engineering*, 2001, pp. 115–119.
71. N. Hadjsaid, J. F. Canard, and F. Dumas, "Disperse generation impact on distribution network," *IEEE Computer Application in Power*, vol. 12, no. 2, 1999, pp. 22–28.
72. K. Maki, S. Repo, and P. Jarventausta, "Methods for assessing the protection impacts of distributed generation in network planning activities," *IET 9th International Conference on Developments in Power System Protection*, 2008, pp. 484–489.
73. K. Kauhaniemi and L. Knmpnlained, "Impact of distributed generation on the protection of distribution networks," *Proceedings of the International Conference on Developments in Power System Protection*, vol. 1, 2004, pp. 315–318.
74. M. T. Doyle, "Reviewing the impacts of distributed generation on distribution system protection," *Proceedings of the IEEE Power Engineering Society Summer Meeting*, vol. 1, 2002, pp. 103–105.
75. H. Nikkhajoei and R. Lasseter, "Microgrid protection," *Proceedings of the IEEE Power Engineering Society General Meeting*, 2007, pp. 1–6.
76. R. M. Tumilty et al., "Approaches to network protection for inverter dominated electrical distribution systems," *3rd IET International Conference on Machines and Drives, Power Electronics*, 2006, pp. 622–626.
77. N. Schaefer et al., "Adaptive protection system for distribution networks with distributed energy resources," *10th IET International Conference on Developments in Power System Protection (DPSP 2010)*, 2010, pp. 1–5.
78. G. D. Rockefeller et al., "Adaptive transmission relaying concepts for improved performance," *IEEE Transactions on Power Delivery*, vol. 3, 1988, pp. 1446–1458.

79. A. Oudalov et al., "Adaptive network protection in microgrids," *International Journal of Distributed Energy Resources*, vol. 5, 2009, pp. 201–225.
80. Y. Han et al., "Study of adaptive fault current algorithm for microgrid dominated by inverter based distributed generators," *2nd IEEE International Symposium on Power Electronics for Distributed Generation Systems (PEDG)*, 2010, pp. 852–854.
81. D. Ke et al., "An adaptive protection method for the inverter dominated microgrid," *2011 International Conference on Electrical Machines and Systems (ICEMS)*, 2011, pp. 1–5.
82. T. S. Ustun et al., "A microgrid protection system with central protection unit and extensive communication," *10th International Conference on Environment and Electrical Engineering (EEEIC)* 2011, pp. 1–4.
83. M. Khederzadeh, "Adaptive setting of protective relays in microgrids in grid-connected and autonomous operation," *IET Conference Publications*, vol. 2012, pp. P14–P14.
84. H. Al-Nasseri et al., "Harmonics content-based protection scheme for micro-grids dominated by solid state converters," *12th International Middle-East Power System Conference MEPCON*, 2008, pp. 50–56.
85. H. Al-Nasseri, et al., "A new voltage-based relay scheme to protect micro-grids dominated by embedded generation using solid state converters," *19th International Conference on Electricity Distribution (CIRED 2007)*, Vienna, Austria, 2007, pp. 1–4.
86. C. Hou et al., "A study of voltage detection based fault judgment method in microgrid with inverter interfaced power source," *International Conference on Electrical Engineering*, Shenyang, China, 2009, pp. 1–5.
87. W. Xiao-ping et al., "Research on the relay protection system for a small laboratory-scale microgrid system," *6th IEEE Conference on Industrial Electronics and Applications (ICIEA)*, 2011, pp. 2712–2716.
88. T. Loix et al., "Protection of microgrids with a high penetration of inverter-coupled energy sources," *IEEE PES Joint Symposium Integration of Wide-Scale Renewable Resources into the Power Delivery System*, 2009, pp. 1–6.
89. H. Nikkhajoei et al., "Microgrid fault protection based on symmetrical and differential current components," Public Interest Energy Research California Energy Commission report, http://citeseerx.ist.psu.edu/viewdoc/download?doi=10.1.1.375.2266&rep=rep1&type=pdf, accessed June 20, 2018.
90. S. R. Samantaray et al., "Differential energy based microgrid protection against fault conditions," *IEEE PES Innovative Smart Grid Technologies (ISGT)*, 2012, pp. 1–7.
91. H. H. Zeineldin et al., "Distributed generation micro-grid operation: Control and protection," *Power Systems Conference: Advanced Metering, Protection, Control, Communication, and Distributed Resources*, 2006, pp. 105–111.
92. M. Dewadasa et al., "Protection of microgrids using differential relays," *21st Australasian Universities Power Engineering Conference (AUPEC)*, 2011, pp. 1–6.
93. M. Dewadasa et al., "Control and protection of a microgrid with converter interfaced micro sources," *International Conference on Power Systems*, 2009, pp. 1–6.
94. M. Dewadasa, "Protection for distributed generation interfaced networks," *Electrical Engineering, Faculty of Built Environment and Engineering, Queensland University of Technology*, Queensland, Australia, 2010.
95. M. R. Miveh et al., "Micro-grid protection by designing a communication-assisted digital relay," *American Journal of Scientific Research*, 2012, pp. 62–68.
96. M. A. Zamani et al., A protection strategy and microprocessor-based relay for low-voltage microgrids," *IEEE Transactions on Power Delivery*, vol. 26, 2011, pp. 1873–1883.
97. F. van Overbeeke, "Fault current source to ensure the fault level in inverter-dominated networks," *20th International Conference and Exhibition on Electricity Distribution—Part 1*, 2009, pp. 1–4.
98. N. Jayawarna et al., "Central storage unit response requirement in 'Good Citizen' Microgrid," *13th European Conference on Power Electronics and Applications*, 2009, pp. 1–10.
99. M. Khederzadeh, "Preservation of overcurrent relays coordination in microgrids by application of static series compensators," *11th International Conference on Developments in Power Systems Protection*, 2012, pp. 1–5.
100. R. M. Kamel et al., "Comparison the performances of three earthing systems for micro-grid protection during the grid connected mode," *Smart Grid and Renewable Energy*, vol. 2, 2011, pp. 206–215.
101. Y. Han et al., "Study on applying wavelet transform to the protection algorithm of microgrid dominated by inverter-interfaced DGs," *International Conference on Power System Technology (POWERCON)*, 2010.

102. S. Shi et al., "Protection of microgrid," *10th IET International Conference on Developments in Power System Protection-Managing the Change* (*DPSP*), 2010, pp. 1–4.
103. G. Joós et al., "Modeling protection for DGs and microgrids using data mining," *Nagoya Microgrid Symposium*, Nagoya, Japan, 2007.
104. H. H. Zeineldin et al., "Optimal coordination of overcurrent relays using a modified particle swarm optimization," *Electric Power Systems Research*, vol. 76, 2006, pp. 988–995.
105. M. Esreraig et al., "An observer-based protection system for microgrids," *IEEE Power and Energy Society General Meeting*, 2011, pp. 1–7.
106. H. Guang et al., "Brittleness entropy of complex system based protection of microgrid," *6th International Forum on Strategic Technology* (*IFOST*), 2011, pp. 433–436.
107. T. S. Ustun et al., "Implementation of Dijkstra's algorithm in a dynamic microgrid for relay hierarchy detection," *IEEE International Conference on Smart Grid Communications* (*Smart Grid Comm*), 2011, pp. 481–486.
108. J. Shi et al., "Fault diagnosis of micro-grid based on advanced Petri Net," *Applied Mechanics and Materials*, vol. 130–134, 2012, pp. 3015–3019.
109. R. D. Shukla, and R. K. Tripathi, "Maximum power extraction schemes and power control in wind energy conversion system," *International Journal of Scientific and Engineering Research*, vol. 3, no. 6, 2012.
110. R. D. Shukla and R. K. Tripathi, "Dynamic performance of DFIG based WECS under different Voltage Sag," *International Journal of Chemtech Research*, vol. 5, no. 2, 2013, pp. 980–992.
111. R. D. Shukla, A. Singh, and S. P. Singh, "Generators for variable speed wind energy conversion systems: A comparative study," *International Journal of Energy Systems Computers and Control*, vol. 3, no. 2, 2012, pp 103–117.
112. R. D. Shukla and R. K. Tripathi, "A novel voltage and frequency controller for standalone DFIG based wind energy conversion system," *Renewable and Sustainable Energy Reviews*, vol. 37, 2014, pp. 69–89.
113. R. D. Shukla, and R. K. Tripathi, "Speed-sensorless voltage and frequency control in autonomous DFIG based wind energy systems," *IEEE Australasian University Power Engineering Conference* (*AUPEC*), 2014, Perth, Australia.
114. R. D. Shukla, and R. K. Tripathi, "A stand-alone wind energy conversion system using wound rotor induction machine," *IEEE International Conference on Power, Energy and Control* (*ICPEC*), 2013, Tamilnadu, India.
115. R. D. Shukla and R. K. Tripathi, "Topologies for stand-alone DFIG based wind energy conversion system," *2nd IEEE International Conference on Power, Control and Embedded Systems* (*ICPCES*), 2012, Allahabad, India, pp. 1–6.
116. R. D. Shukla and R. K. Tripathi, "Low voltage ride through ability of DFIG based wind energy conversion systems—I," *IEEE Students' Conference on Engineering and Systems* (*SCES*) 2012, Allahabad, India, pp. 1–7.
117. R. D. Shukla and R. K. Tripathi, "Low voltage ride through ability of DFIG based wind energy conversion systems—II," *Proceedings of IEEE Students' Conference on Engineering and Systems* (*SCES*) *2012*, March 16–18, 2012, Allahabad, India, pp. 1–6.
118. R. D. Shukla, R. K. Tripathi, and S. Gupta, "Power electronics applications in wind energy conversion system: A review," *IEEE International Conference on Power, Control and Embedded Systems* (*ICPCES*), 2010, Allahabad, India, pp. 1–6.
119. Y. Ling and X. Cai, "Rotor current dynamics of doubly fed induction generators during grid voltage dip and rise," *International Journal of Electrical Power and Energy Systems*, vol. 44, no. 1, 2013, pp. 17–24.
120. M. Farshadnia and S. A. Taher, "Current-based direct power control of a DFIG under unbalanced grid voltage," *International Journal of Electrical Power and Energy Systems*, vol. 62, 2014, pp. 571–582.
121. E. G. Shehata and G. M. Salama, "Direct power control of DFIGs based wind energy generation systems under distorted grid voltage conditions," *International Journal of Electrical Power and Energy Systems*, vol. 53, 2013, pp. 956–966.
122. N. Mohan and T. K. A. Brekken, "Control of a doubly fed induction wind generator under unbalanced grid voltage conditions," *IEEE Transactions on Energy Conversion*, vol. 22, no. 1, 2007, pp. 129–135.
123. J. Morren and S. W. H. de Haan, "Ridethrough of wind turbines with doubly-fed induction generator during a voltage dip," *IEEE Transactions on Energy Conversion*, vol. 20, 2005, pp. 435–441.
124. J. M. Carrasco et al., "Power-electronic systems for the grid integration of renewable energy sources: A survey," *IEEE Transactions on Industrial Electronics*, vol. 53, 2006, pp. 1002–1016.

125. E. Pons and R. Tommasini, “Lightning protection of PV systems,” *2013 4th International Youth Conference on Energy* (*IYCE*), 2013, pp. 1–5.
126. D. C. Carmichael and G. T. Noel, “Development of low-cost modular designs for photovoltaic array fields,” *IEEE Transactions on Power Apparatus and Systems*, vol. 104, no. 5, 1985, pp. 1005–1011.
127. C. Zhang, J. Hu, Y. Tu, W. Sun, H. Jun Li and S. Wang, “Study of induced overvoltage on solar arrays,” *Proceedings of 7th Asia-Pacific International Conference on Lightning* (*APL*), 2011, pp. 852–857.
128. P. Vangala, M. Ropp, K. Haggerty, K. Lynn, and W. Wilson, “Field measurements of lightning-induced voltage transients in PV arrays,” *33rd IEEE Photovoltaic Specialists Conference*, 2008, pp. 1–4.
129. “Solar energy: Protecting and isolating PV systems.” ABB https://library.e.abb.com/public/40abe76927e9e01dc12576ba003c22bf/Solar%20Cat%20LR%20v.06.pdf, DOA, accessed May 1, 2018.
130. R. Benato and R. Caldon, “Application of PLC for the control and the protection of future distribution networks,” *IEEE International Symposium on Power Line Communications and Its Applications*, Pisa, Italy, 2007, pp. 499–504.

Index